Calculus For The Nonphysical Sciences

Calculus For The

HOLT, RINEHART AND WINSTON, INC.

Nonphysical Sciences

With Matrices, Probability and Statistics

Simeon M. Berman
NEW YORK UNIVERSITY

New York Chicago San Francisco Atlanta Dallas Montreal Toronto London Sydney

Library of Congress Cataloging in Publication Data
Berman, Simeon M.
Calculus for the nonphysical sciences.
1. Mathematics – 1961 – 2. Calculus.
I. Title.
QA37.2.B47 515′.15 73-18269
ISBN 0-03-088098-X

Printed in the United States of America
45678 071 98765432

This book was set in linofilm by Progressive Typographers
Printing and binding were done by Kingsport Press
The interior text was designed by Mel Haber
Eric G. Hieber Assoc. executed the art
Marion Palen supervised the production and editing
Renee Davis was the design supervisor
The text art and cover were based on designs by the author

To Jeremy, Daniel, Zachary

Preface

This book was written with the conviction that college mathematics will have to make itself more appealing to the students of the 1970s. The national interest in science and technology, so strong in the Fifties and Sixties, has been replaced by a growing pursuit of the humanistic professions. As teachers of mathematics, our main audiences used to be young people with an unquestioning commitment to the worth of mathematics and science. They followed us like soldiers, through the mountains and the swamps.

We are now teaching fewer mathematicians, physical scientists, and engineers. Our courses have become increasingly populated by students in the social, behavioral, commercial, and life sciences. Mathematics is only of secondary interest to them. Our work as promoters of mathematics is now much more challenging. These students want to be shown why mathematics is important and how they can use it. They have more options. If they do not like mathematics, they can take something else. But I believe that college mathematics *can* be made to appeal to students with a variety of interests.

Prerequisites and contents This text is designed for a course in mathematical methods for students whose specialty is not in the mathematical sciences. The prerequisite is three years of high school mathematics, not necessarily including trigonometry. The four main areas covered in the text are analytic geometry, matrix algebra, calculus of one and two variables, and probability-statistics. The material may be used for a one or two or even three semester course.

Features The point of the book is to stir the interest of the student by showing how mathematics is motivated by real situations and how it can be used. The three features of the book are

(i) *The pictorial approach.* Pictures and diagrams are extensively used to present mathematical concepts. There are nearly six hundred illustrations. Their purpose is to make the subject concrete and accessible to students who are not used to abstract reasoning with quantitative relations. Many of the illustrations are meant to give a pictorial representation of a mathematical idea. However, others are intended as caricatures of ideas in situations associated with the mathematics, and are meant to put more life in the text. As an amateur artist, the author drew the original illustrations as pencil sketches. These were redone for publication, professionally and very competently, by the Eric G. Hieber Associates, under contract from Holt, Rinehart and Winston, the publisher.

(ii) *Applications* Almost every theoretical concept is illustrated by applications. These are taken from areas of interest to the students: biology, business, chemistry, economics, environmental science, medicine, political science, psychology, and sociology. Classical physical applications of the calculus are included. These are shown to be analogous to applications in the relevant fields; for example, the physical concepts of velocity and acceleration are used in the description of the business cycle, the divorce trend, and population growth. Many numerical illustrations are given. These are drawn from data in United States Government reports of the Bureau of the Census, the Department of Health, Education and Welfare, and other agencies.

The students are viewed not just as specialists in their particular major subject, but as young men and women with their individual experiences and aspirations. Examples have been chosen not only from strictly academic fields but from many areas of human interest, from marriage to baseball. The range of applications included here is more than can be covered in a usual course. The applications should be selected by the instructor according to the interests of the students and his own. A note "See Example – " appears after most of the word exercises. This refers the reader to a similar example which has been explicitly done in the text; it should help the student do the exercise, and also enable the teacher to select the exercises for assignments.

(iii) *Intuition over rigor* I have found that students in this audience want to learn just enough theory – but no more – to convince them of the validity of mathematical methods. They are usually satisfied with informal indications of proofs, well below the rigor level of the professional mathematician. Therefore, rigorous mathematical proofs have here been replaced by intuitive or geometric discussions of major theoretical ideas. Some formal proofs are outlined with hints and guides as exercises for interested students.

I hope that the casual tone of the book will enable the student to read it with comfort. In accordance with the informal style, important concepts are usually stated within discussions and not formally displayed. As a help in locating them, important phrases and sentences appear in bold print, and other important but longer passages are accompanied by black bullets in the margin.

Structure The plan of the book is outlined in the diagram on page x. An arrow pointing from one topic to another indicates that the former is a prerequisite for the latter. Each chapter is divided into sections. Each of the latter is indexed by a double number: Section x.y is the y*th* section of Chapter x.

The text may be used to fit the needs of various courses. Note that the book may be started in any one of three ways, namely, with analytic geometry, matrix theory or probability. We describe three kinds of courses:

(i) A semester course along the track from elementary analytic geometry through functions and to differential and integral calculus of one variable.

(ii) A semester course in finite mathematics including topics in analytic geometry, matrix algebra, functions and probability.

(iii) A two or even three semester course covering all or most of the topics, and proceeding along the tracks in any order.

The subjects are presented in a conventional way, but with certain variations. The conics of analytic geometry are given with unusual applications. Certain topics in probability and statistics are introduced early, in the context of other areas. Markov chains are introduced with matrix theory. Least squares is presented in connection with the general idea of function. Population means, distributions and densities are major illustrations of the use of integral calculus. There is a big difference in the levels of the chapters on probability and statistics: the former is completely elementary, and the latter is at the highest level of the book. The most unusual chapter is Chapter 4, The Functions of the Calculus. Here various mathematical functions are shown to arise in real situations, and the student is taught to construct their equations in practical problems. In most calculus texts such construction is learned at the same time as mini-max theory. The novelty here is that the two topics have been separated for the purposes of learning, and each has been made more interesting and accessible.

Some of the exercises require a good deal of arithmetic computation. It is my own belief that such exercises make the problems more "real" for the student. If a calculator or computer is available, the computations can be satisfying. But even a little calculation with pencil can be good for the student. However, many instructors avoid computations, and such exercises may be omitted without loss of continuity.

Each section (with one exception) is followed by a set of exer-

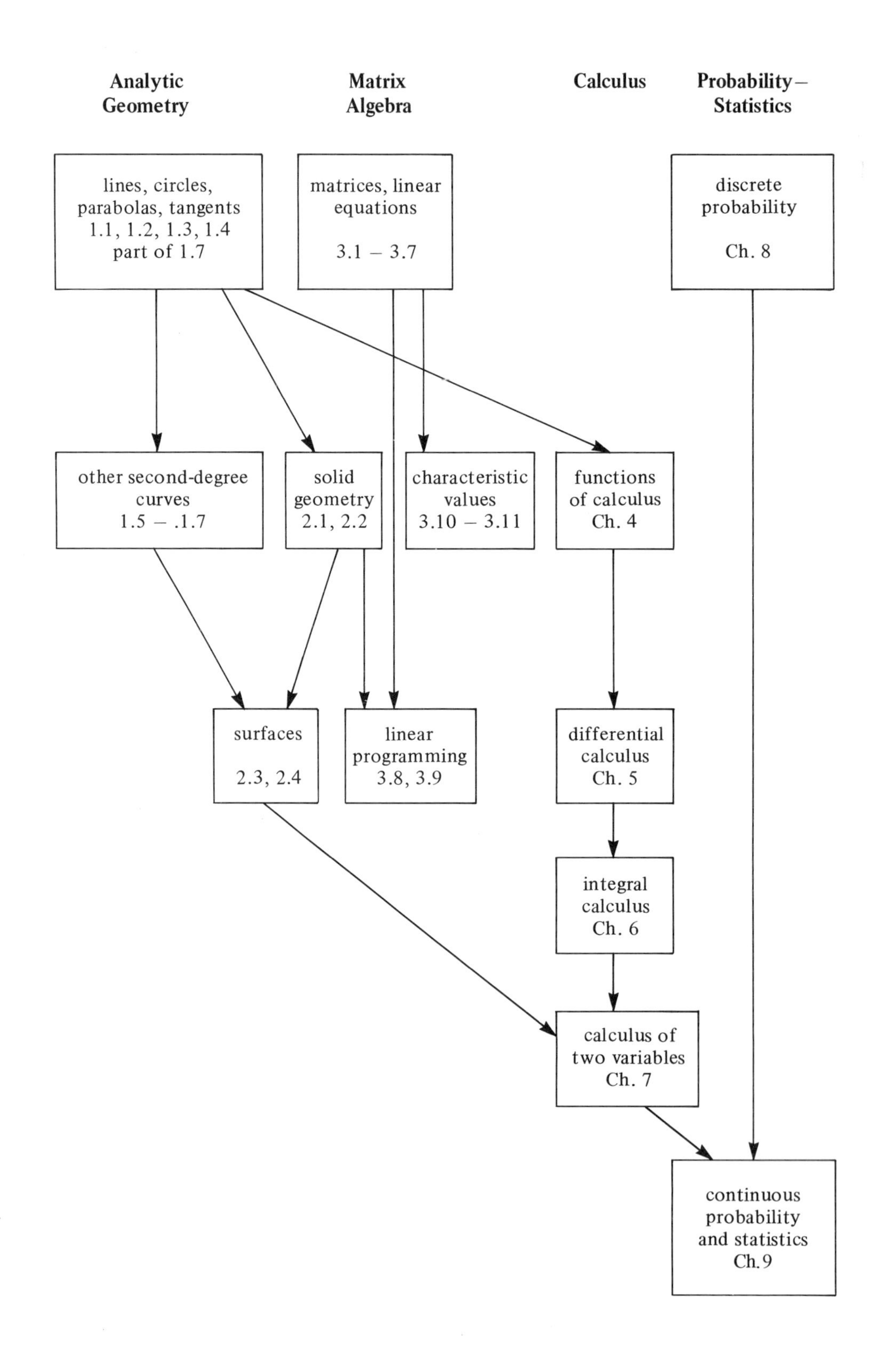

Analytic Geometry
Matrix Algebra
Calculus
Probability–Statistics
lines, circles, parabolas, tangents 1.1, 1.2, 1.3, 1.4 part of 1.7
matrices, linear equations 3.1 – 3.7
discrete probability Ch. 8
other second-degree curves 1.5 – .1.7
solid geometry 2.1, 2.2
characteristic values 3.10 – 3.11
functions of calculus Ch. 4
surfaces 2.3, 2.4
linear programming 3.8, 3.9
differential calculus Ch. 5
integral calculus Ch. 6
calculus of two variables Ch. 7
continuous probability and statistics Ch. 9

cises. The first several ones in each set are usually devoted to drill in techniques. These are followed by more specialized problems and applications. When proofs are assigned the general steps are given, and the student is asked to give the details of the justification. Any of these theoretical exercises may be omitted if desired. The answers to the odd-numbered exercises are given at the end of the book.

The manuscript for this book was completed while the author was on sabbatical leave from New York University in 1971–72.

February 1974 Simeon M. Berman

Contents

CHAPTER

1

Analytic Geometry

1.1 Rectangular Coordinates and the Slope of a Line

Plane geometry, as normally studied in high school, is called **synthetic geometry.** A synthetic subject or substance is one that is built from elementary parts. This kind of geometry is developed by starting with certain undefined elementary objects (points and lines), stating postulates, and then deducing the consequences. **Analytic geometry** is the study of the same subject with the aid of algebra. An analytic study is one where the **whole** is broken down into its parts and analyzed.

The central concept in analytic geometry is the **rectangular coordinate system.** It is also called the Cartesian coordinate system, after Rene Descartes (1596–1650), its inventor. The plane is divided into four **quadrants** by two perpendicular lines. These cross at a point called the **origin.** One line is horizontal and the other vertical, and they extend indefinitely in both directions. The horizontal line is the x-axis and the vertical line is the y-axis. The pair of lines is called a set of **coordinate axes.** The origin is labeled 0.

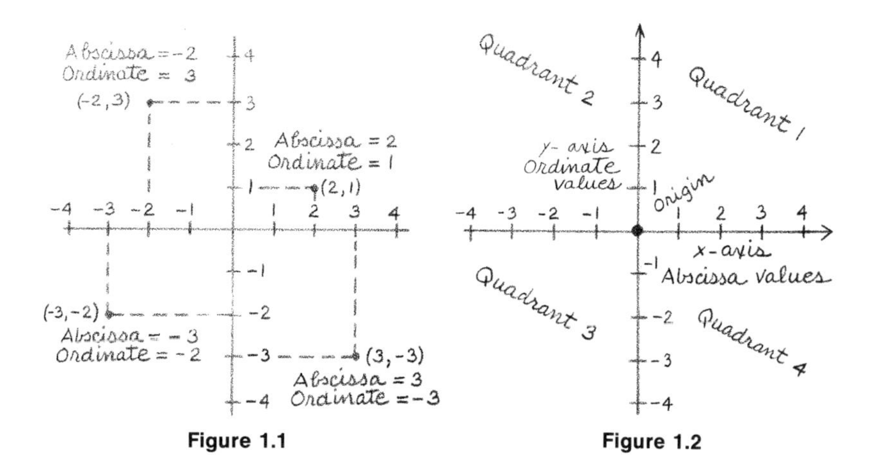

Figure 1.1 **Figure 1.2**

The axes are marked with a system of numbers. On the horizontal axis the algebraic values of the numbers increase as they move to the right. The spatial distance between two points on the axis is proportional to the algebraic difference between the corresponding numbers. The point at the origin is labeled 0. The numbers to the right of the origin are positive; those to the left are negative. On the vertical axis the numbers increase as they move up. The origin is marked 0. The numbers above the origin are positive; those below it are negative. The numbers on the x-axis are called **abscissa** values; those on the y-axis are called **ordinate** values (Figure 1.1).

Every point in the plane is labeled by a unique number pair consisting of an abscissa and an ordinate, or an x-value and a y-value. The abscissa of a point is that number on the x-axis where a vertical line through the point cuts the axis. The ordinate of a point is that number on the y-axis where a horizontal line cuts the axis through the point. The abscissa and ordinate are called the **coordinates** of the point. We let P stand for a given point. If its coordinates are x and y, then the point is denoted $P(x,y)$. The points $P(2,1)$, $P(-2,3)$, $P(-3,-2)$, and $P(3,-3)$ are plotted in Figure 1.2. Note that the abscissa of a point is positive when the point is to the right of the y-axis, and negative when the point is to the left. The ordinate is positive when the point is above the x-axis, negative when below.

The **distance *d*** between two points $P_1(x_1,y_1)$ and $P_2(x_2,y_2)$ is given by the formula

$$d = \sqrt{(x_2 - x_1)^2 + (y_2 - y_1)^2}\,, \tag{1}$$

the square root of the sum of the squares of the abscissa difference and ordinate difference. This is based on the Pythagorean Theorem, which states that the sum of the squares of the legs of a right triangle is equal to the square of the hypotenuse; and, conversely, if the sum of the squares of two sides of a triangle is equal to the square of the third side, then the triangle has a right angle. In Figure 1.3 we show that the distance from the point $P_1(x_1,y_1)$ to the point $P_2(x_2,y_2)$ is the hypotenuse of a right triangle whose legs are equal in length to the absolute differences of $x_2 - x_1$ and $y_2 - y_1$, respectively. (Recall that the absolute value of a number x is $-x$ when x is negative, and x when positive; for example, 8 and -8 have the same absolute value 8.) In Figures 1.2 and 1.3, the points are represented by the coordinates without the prefixed label P. When convenient, this will be done throughout.

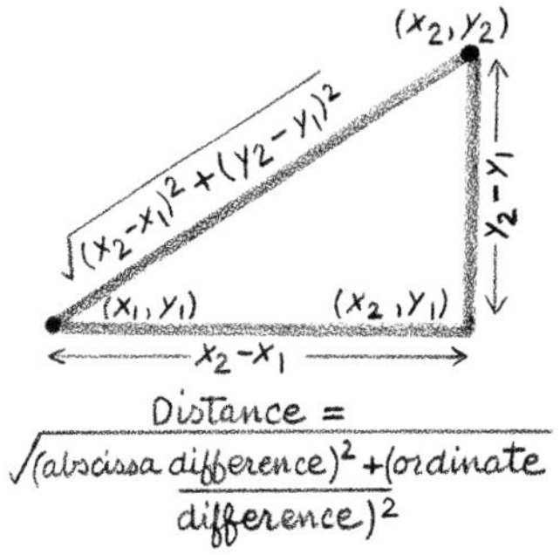

Figure 1.3

EXAMPLE 1

Find the distance between $P_1(-2,3)$ and $P_2(5,1)$. Apply formula (1) with $x_1 = -2$, $y_1 = 3$, $x_2 = 5$, and $y_2 = 1$:

$$d = \sqrt{(5-(-2))^2 + (1-3)^2} = \sqrt{49+4} = \sqrt{53}.$$

(Note the way the negative numbers are subtracted.) ▲

EXAMPLE 2

Show that the triangle with the vertices at $(-3,-2)$, $(4,-5)$, and $(5,7)$ is an isosceles triangle (has two equal sides). The triangle is drawn in Figure 1.4. By application of (1), the sides are found to be $\sqrt{58}$, $\sqrt{145}$, and $\sqrt{145}$; hence, it is isosceles. ▲

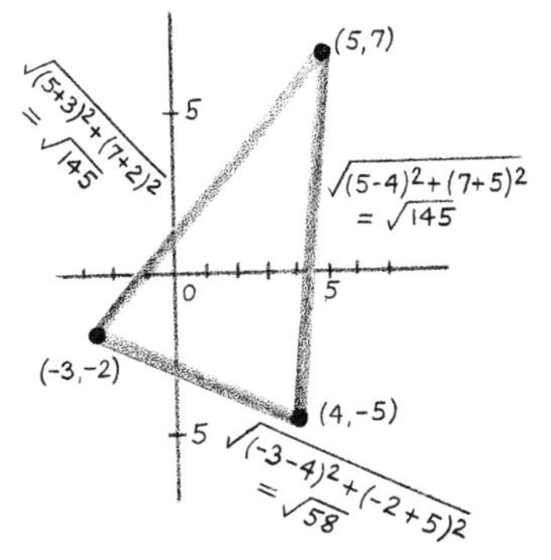

Figure 1.4

EXAMPLE 3

The shortest path between two points is along the line segment connecting them. If P is not on the segment from Q to R, then the sum of the distances from P to Q and from P to R is greater than the length of the segment. If P is on the segment, then the sum of the distances is **equal** to the length of the segment. We can use this criterion to determine whether or not three points lie on the same line. For example, the three points with coordinates (0,3), (6,0), and (4,1), respectively, have the mutual distances $2\sqrt{5}$, $\sqrt{5}$, and $3\sqrt{5}$. Since $\sqrt{5} + 2\sqrt{5} = 3\sqrt{5}$, the three points must be on one line (Figure 1.5). ▲

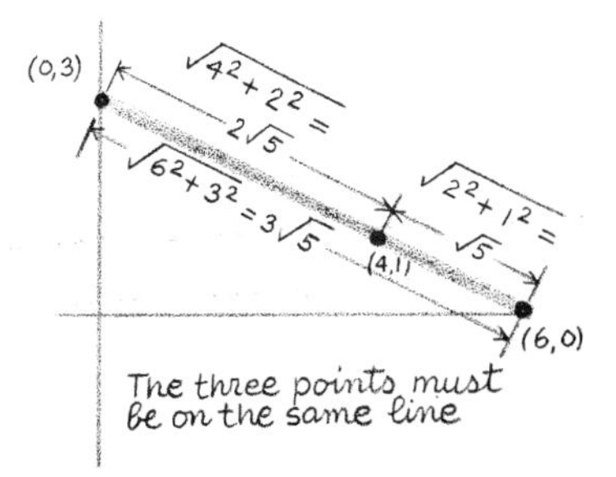

Figure 1.5

EXAMPLE 4

Show that the triangle whose vertices are (7,3), (10,−10), and (2,−5) is a right triangle, and find its area. By means of the distance formula we shall show that the square of one of the sides is equal to the sum of the squares of the other two. Then by the Pythagorean Theorem we shall conclude that the triangle must be a right triangle. The points are plotted in Figures 1.6(a) and 1.6(b), and the lengths of the sides of the triangle are computed. Finally, the area is obtained from the formula for any triangle, **area = $\frac{1}{2}$ base times height.** ▲

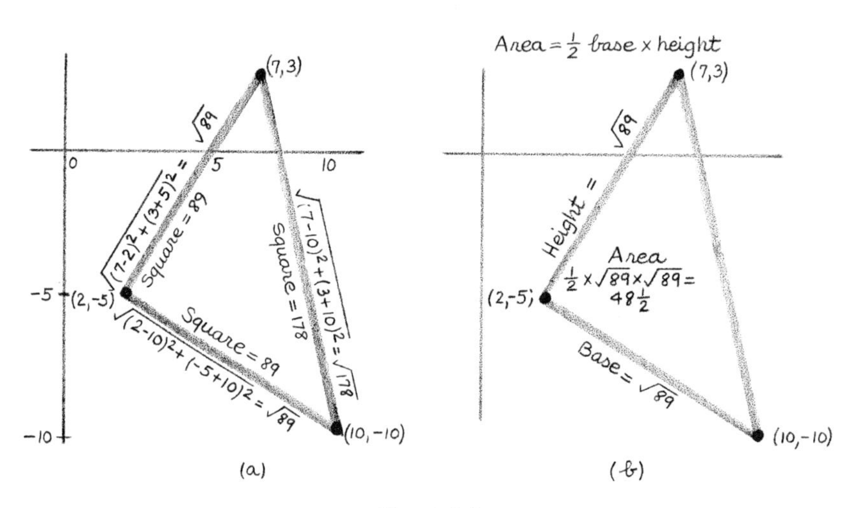

Figure 1.6

Consider a nonvertical line in the plane. For any pair of points $P_1(x_1,y_1)$ and $P_2(x_2,y_2)$ on the line, the ratio

$$\frac{y_2 - y_1}{x_2 - x_1}$$

is called the **slope** of the line. Each such line has a unique slope: it has the same value for **any** pair of points on it. The slope represents the ratio of the differences of ordinates to the differences of abscissas for any pair of points on that line. When the line rises from left to right, the slope is positive; when it falls from left to right the slope is negative; finally, a horizontal line has a slope equal to zero (Figure 1.7). Slopes are not assigned to vertical lines.

The reason that the ratio for the slope is the same for any pair of points on a line is based on similar triangles. If P_1 and P_2 are a pair of points on the line, and P_1' and P_2' another pair, then by similar triangles, the ratio of the ordinate to the abscissa difference is the same for both pairs (Figure 1.8).

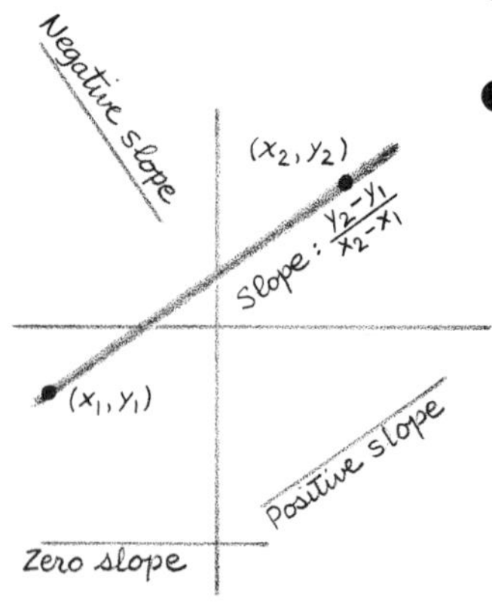

Figure 1.7

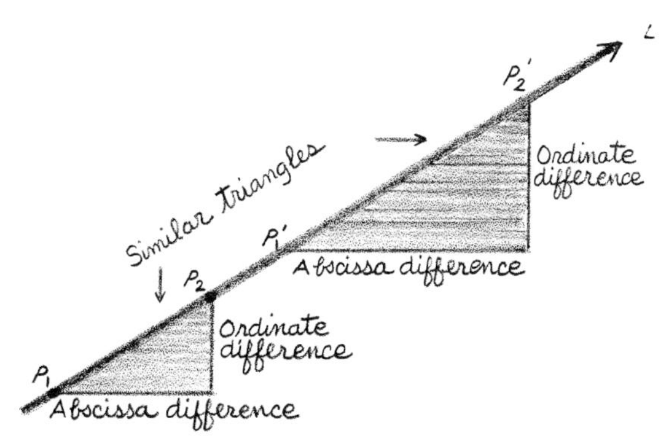

Figure 1.8

EXAMPLE 5

Find the slope of the line passing through $P_1(3,-2)$ and $P_2(-1,5)$. Here $x_1 = 3$, $y_1 = -2$, $x_2 = -1$, and $y_2 = 5$. The slope is

$$\frac{5-(-2)}{-1-3} = -\frac{7}{4}. \blacktriangle$$

Two nonvertical lines are parallel if and only if their slopes are equal. The reason is based on a simple construction of similar triangles [Figures 1.9(a) and 1.9(b)].

Two nonvertical lines are perpendicular if and only if the slope of one is the negative reciprocal of the slope of the other. A proof based on the Pythagorean Theorem is sketched in Exercise 13.

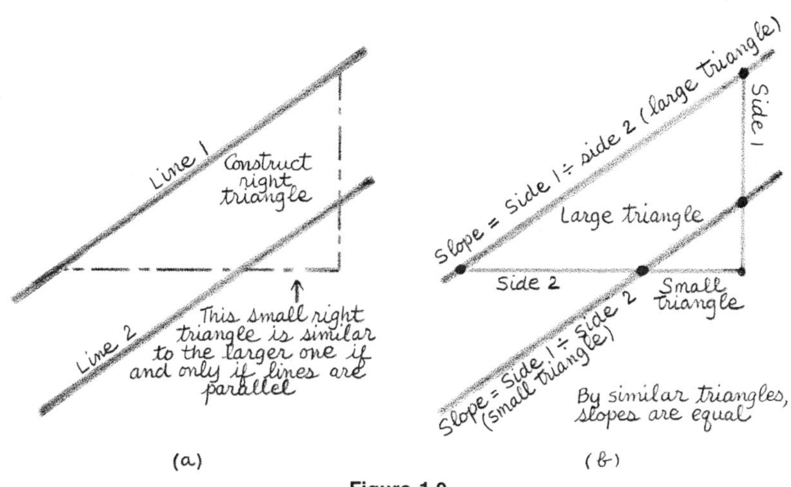

Figure 1.9

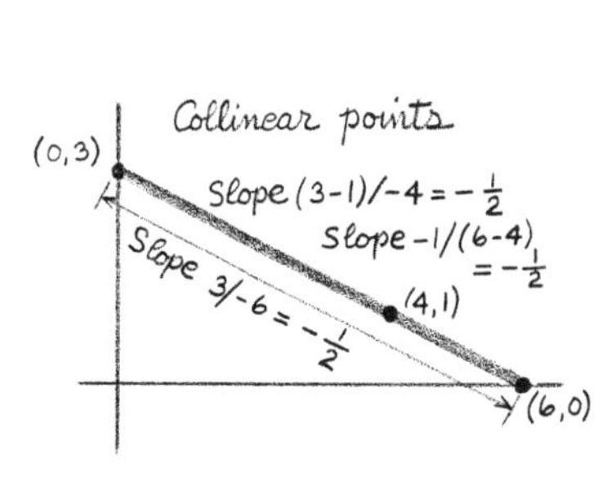

Figure 1.10

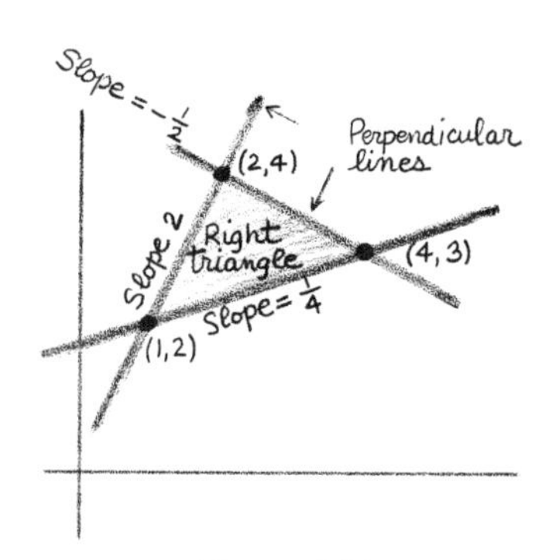

Figure 1.11

EXAMPLE 6

In Example 3 above we used the distance formula to determine whether or not three points lie on one line. Now we shall use slopes to determine the same. If any two lines joining any pair of the three points are parallel, then the points are on the same line. For example, the slopes of the lines joining pairs of points from the set (0,3), (6,0), and (4,1) are all equal to $\frac{1}{2}$ (Figure 1.10). Such points are called **collinear.** ▲

EXAMPLE 7

Show that the points (4,3), (2,4), and (1,2) are the vertices of a right triangle. It suffices to show that two of the connecting lines are perpendicular:

$$\text{slope from } (4,3) \text{ to } (2,4) = -\tfrac{1}{2},$$
$$\text{slope from } (4,3) \text{ to } (1,2) = \tfrac{1}{4},$$
$$\text{slope from } (2,4) \text{ to } (1,2) = 2.$$

The first and third lines are perpendicular; therefore, the points form a right triangle (Figure 1.11). ▲

1.1 EXERCISES

In Exercises 1 and 2 plot the pair of points and find the distance between them.

1 (a) (−3,1), (3,−1)
(b) (2,−6), (2,−2)
(c) (−1,−5), (2,−3)
(d) (0,3), (−4,1)
(e) (−4,5), (3,2)
(f) (−4,4), (10,2)

2 (a) (5,−3), (6,6)
(b) (2,−7), (−6,3)
(c) (0,3), (0,8)
(d) (5,2), (1,2)
(e) (4,−3), (1,0)
(f) (5,3), (−4,7)

In Exercises 3 and 4 show by means of the distance formula that the triangle with the given vertices is isosceles.

3 (a) $(-4,9)$, $(-3,2)$, $(1,4)$
(b) $(-2,3)$, $(-5,1)$, $(-4,6)$
(c) $(-3,-1)$, $(2,-2)$, $(1,6)$

4 (a) $(2,4)$, $(5,1)$, $(6,5)$
(b) $(6,7)$, $(-8,-1)$, $(-2,-7)$
(c) $(1,5)$, $(5,-1)$, $(9,6)$

5 By means of the Pythagorean Theorem show that the triangle with the given vertices is a right triangle.
(a) $(-2,2)$, $(8,-2)$, $(-4,-3)$
(b) $(-2,8)$, $(-6,1)$, $(0,4)$
(c) $(3,-2)$, $(-2,3)$, $(0,4)$

6 By means of the distance formula show that the given triples of points lie on a common line.
(a) $(0,2)$, $(3,-1)$, $(-2,4)$
(b) $(-1,3)$, $(-3,1)$, $(-5,-1)$
(c) $(1,2)$, $(-3,10)$, $(4,-4)$

In Exercises 7 and 8 find the slopes of the lines through the given pairs of points.

7 (a) $(4,1)$, $(7,-1)$
(b) $(4,6)$, $(-7,7)$
(c) $(5,-2)$, $(9,4)$
(d) $(-9,2)$, $(5,6)$
(e) $(7,-7)$, $(-6,5)$

8 (a) $(-2,9)$, $(5,-1)$
(b) $(-5,-9)$, $(3,6)$
(c) $(8,9)$, $(-6,3)$
(d) $(-8,3)$, $(3,7)$
(e) $(3,6)$, $(-2,-8)$

9 By using slopes show that these triples of points are vertices of right triangles.
(a) $(3,2)$, $(5,-4)$, $(1,-2)$
(b) $(3,4)$, $(-2,-1)$, $(4,1)$
(c) $(1,2)$, $(2,4)$, $(3,1)$
(d) $(6,5)$, $(1,3)$, $(5,-7)$
(e) $(-2,1)$, $(3,-1)$, $(7,9)$

10 By means of slopes show that these triples of points lie on a common line.
(a) $(-4,-6)$, $(0,-7)$, $(-8,-5)$
(b) $(-1,2)$, $(1,3)$, $(5,5)$
(c) $(1,4)$, $(0,1)$, $(-1,-2)$
(d) $(1,2)$, $(2,4)$, $(-1,-2)$
(e) $(4,1)$, $(5,-2)$, $(6,-5)$

11 Show that the following sets of points are vertices of parallelograms. (Recall that a parallelogram is a 4-sided figure with parallel opposite sides.)
(a) $(2,4)$, $(6,2)$, $(8,6)$, $(4,8)$
(b) $(-1,-5)$, $(2,1)$, $(1,5)$, $(-2,-1)$
(c) $(-1,-2)$, $(0,1)$, $(-3,2)$, $(-4,-1)$
(d) $(0,-1)$, $(-1,2)$, $(3,3)$, $(4,0)$

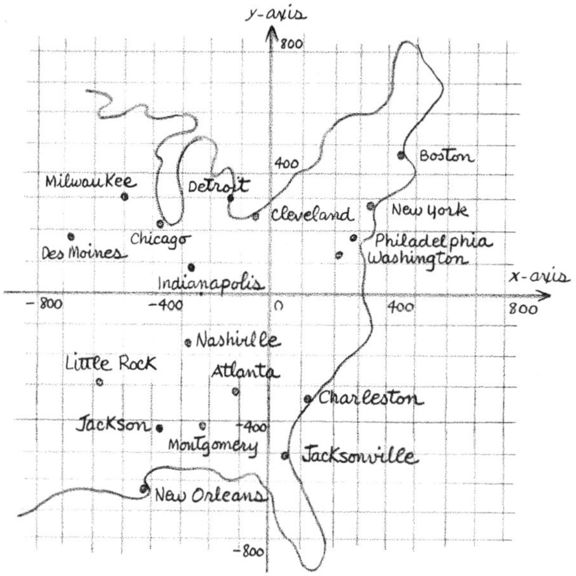

Figure 1.12

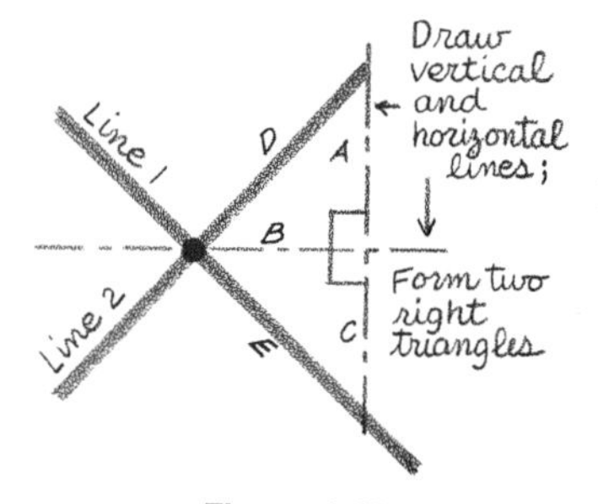

Figure 1.13

12 In Figure 1.12, the map of the eastern half of the United States, a set of coordinate axes is drawn, where the units are in miles. Estimate the coordinates corresponding to these cities: Boston, Chicago, Cleveland, Des Moines, Detroit, Jackson, Nashville, New Orleans, New York, and Philadelphia. Estimate these distances by means of the distance formula to the nearest 100 miles:

(a) from Indianapolis to Little Rock

(b) from Milwaukee to Charleston

(c) from Atlanta to Boston

(Use Table I for the square roots.)

13 For the proof of "Two nonvertical lines are perpendicular if and only if the slope of one is the negative reciprocal of the slope of the other," we refer to Figure 1.13. Fill in these steps of the proof:

(a) Prove that the slope of line 1 is equal to $-C/B$; and that the slope of line 2 is A/B.

(b) Lines 1 and 2 are perpendicular if and only if

$$(A + C)^2 = (A^2 + B^2) + (B^2 + C^2).$$

(c) The slopes of lines 1 and 2 are negative reciprocals of each other if and only if $AC = B^2$.

(d) From statements (b) and (c) deduce the assertion above.

1.2 The Straight Line and Its Equation

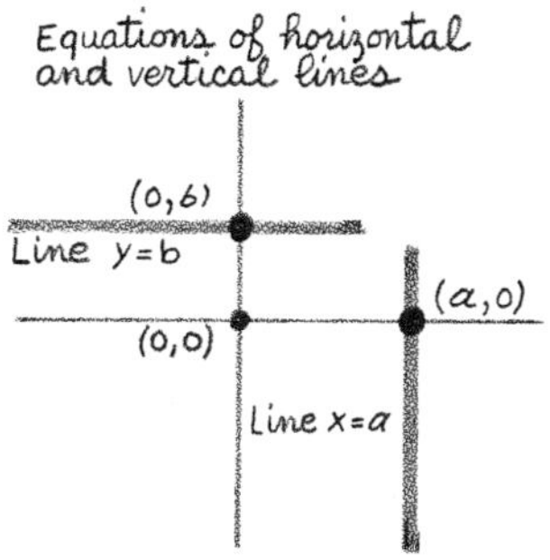

There are several ways to describe a particular straight line. In plane geometry a straight line is described by a pair of points that it joins. In analytic geometry we can specify a line by means of an algebraic equation. Let us start with the simplest cases. If a line is vertical in the plane, and if it passes through the x-axis at a point with abscissa a, then the abscissa of every point on that line is also equal to a. Conversely, any point in the plane with the abscissa a necessarily lies on this line. Therefore, a point with the coordinates (x,y) is on this line if and only if $x = a$. We define the equation of this line as $x = a$. For brevity, the line is called "the line $x = a$." For example, the line $x = 3$ is parallel to the y-axis and passes through the point on the x-axis 3 units to the right of the origin. Similarly, a horizontal line passing through a point on the y-axis with the ordinate b is called the line $y = b$ (Figure 1.14).

Equations of horizontal and vertical lines

Figure 1.14

There are several forms of the equation specifying a line parallel to neither axis. A line with slope m, passing through a point (x_1,y_1) has the equation

$$y - y_1 = m(x - x_1). \qquad (1)$$

This means that a point lies on this line if and only if its coordinates (x,y) satisfy Equation (1) (Figure 1.15). This is the **point-slope** form of the equation of the line. For example, the line through $(1,-1)$ and with slope 3 has the equation $y - (-1) = 3(x - 1)$, or $y + 1 = 3(x - 1)$.

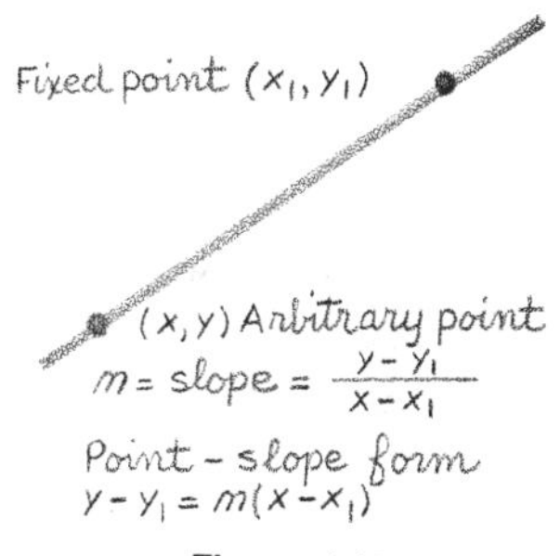

Figure 1.15

Equation (1) is based on the fact that if (x,y) is any point on the line distinct from (x_1,y_1), then its slope is, by definition, the ratio $(y - y_1)/(x - x_1)$, which is denoted m. On the other hand, if (x,y) is identical with (x_1,y_1), then (1) certainly holds, with both sides of the equation equal to 0.

We remark that when $m = 0$, then Equation (1) becomes $y = y_1$, and the line is horizontal.

The **slope-intercept** form is a special case of the point-slope form. If the point (x_1,y_1) is on the y-axis, then it is of the form $(0,b)$, where $x = 0$ and $y_1 = b$. If we put these in Equation (1), we obtain $y - b = mx$, or

$$y = mx + b. \qquad (2)$$

This is the equation of a line with slope m passing through the y-axis b units above the origin. If b is negative, it passes $-b$ units below the origin. For example, the line with slope $\frac{1}{2}$ and passing through the y-axis 7 units above the origin has the equation $y = \frac{1}{2}x + 7$. If it passes

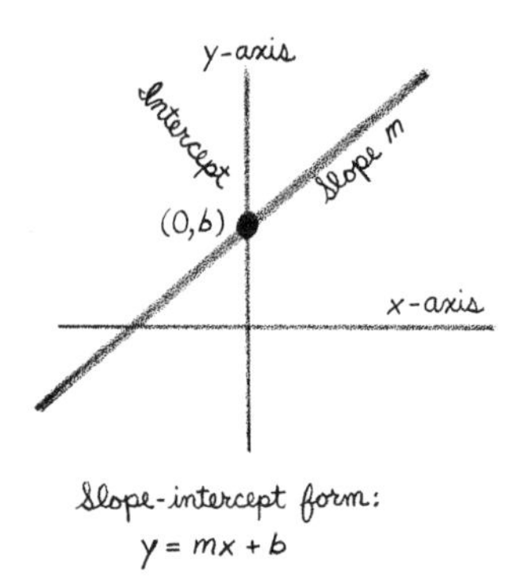

Figure 1.16

7 units below the origin, it has the equation $y = \frac{1}{2}x - 7$. The number b in Equation (2) is called the **y-intercept** of the line. Therefore (2) is called the slope-intercept form of the equation of the line (Figure 1.16).

Another useful form of the equation of the line is the **two-point** form. If a line passes through two given points (x_1,y_1) and (x_2,y_2), and if (x,y) is an arbitrary third point on the line, then the slope may be computed in three ways: It is equal to each of the ratios

$$\frac{y_2 - y_1}{x_2 - x_1}, \qquad \frac{y - y_1}{x - x_1}, \qquad \frac{y - y_2}{x - x_2}.$$

These ratios are necessarily equal. Therefore (x,y) satisfies both of the equations

$$\frac{y - y_1}{x - x_1} = \frac{y_2 - y_1}{x_2 - x_1} \qquad \text{and} \qquad \frac{y - y_2}{x - x_2} = \frac{y_2 - y_1}{x_2 - x_1}. \tag{3}$$

These are called the **two-point** forms. The two equations above are equivalent because they represent the same line (Figure 1.17). As an example, let us find the equation of the line through the points (1,3) and (4,−2). Here $x_1 = 1$, $y_1 = 3$, $x_2 = 4$, and $y_2 = -2$. Any third point (x,y) on the line satisfies both equations

$$\frac{y - 3}{x - 1} = -\frac{5}{3}, \qquad \frac{y + 2}{x - 4} = -\frac{5}{3}.$$

Either of these equations defines the line because (x,y) satisfies one if and only if it satisfies the other.

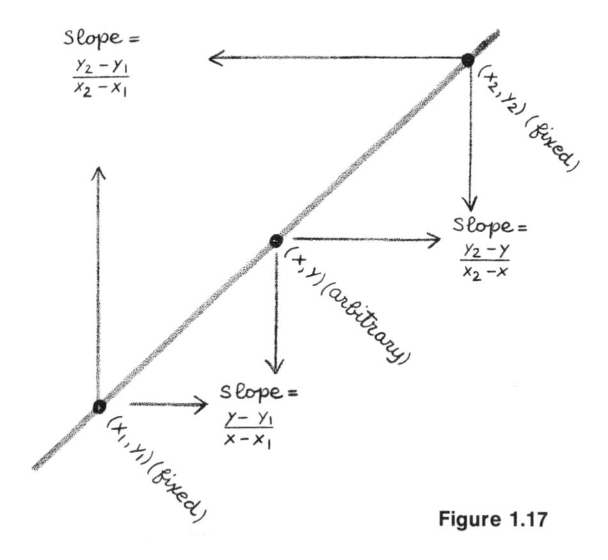

Figure 1.17

The various forms of the equation of the line may be changed to the **general linear equation**

$$Ax + By + C = 0, \tag{4}$$

where A, B, and C are constants. For example, the point-slope equation $y - 3 = -(\frac{5}{3})(x - 1)$ is equivalent to $5x + 3y - 14 = 0$; here $A = 5$, $B = 3$, and $C = -14$. Conversely, the form (4) can be easily changed to any of the other forms; for example, if $Ax + C$ is subtracted from both sides, and both sides are divided by B, then the equation becomes

$$y = -\left(\frac{A}{B}\right)x - \left(\frac{C}{B}\right).$$

This is the slope-intercept form (2) where $m = -A/B$ and $b = -C/B$.

Given the equation of the line, we can draw its graph by finding a pair of points satisfying the equation and passing a straight line through them. For example, suppose the equation is in the general linear form $3x + 2y + 6 = 0$. Put $x = 0$ and solve for $y: y = -3$. Thus, $(0,-3)$ is a point on the line. Next we put $y = 0$ and solve the equation for $x: x = -2$. Therefore, $(-2,0)$ is another point on the line. The graph of the line is drawn through these points (Figure 1.18).

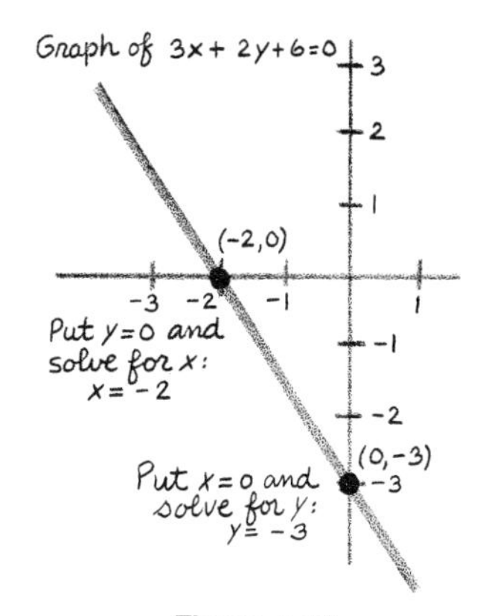

Figure 1.18

The point-slope form can be used to derive the equation of a line through a point, parallel or perpendicular to another given line. If a line passes through (x_0,y_0) parallel to a line with slope m, then it has the same slope m; therefore, it has the point-slope equation $y - y_0 = m(x - x_0)$ (Figure 1.19). If it passes through (x_0,y_0) perpendicular to a line with slope m, then it has the slope $-1/m$; therefore, its equation is $y - y_0 = (-1/m)(x - x_0)$ (Figure 1.20). For example, consider the line passing through $(1,-2)$ parallel to the line $(y - 4) = -2(x + 1)$. The latter has the slope -2, and so the former also does; hence, its equation is $y + 2 = -2(x - 1)$. If a line passes through $(1,-2)$ perpendicular to $y - 4 = -2(x + 1)$, then it has the slope $\frac{1}{2}$; hence, it has the equation $y + 2 = \frac{1}{2}(x - 1)$.

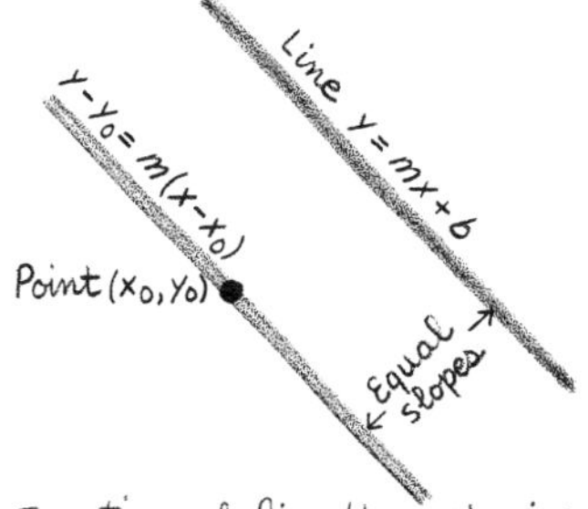

Equation of line through given point parallel to given line:
$y - y_0 = m(x - x_0)$

Figure 1.19

If two straight lines are not parallel, then, as we know from plane geometry, they intersect at exactly one point. We can determine the point of intersection by solving the equations of the lines simultaneously (Figure 1.21).

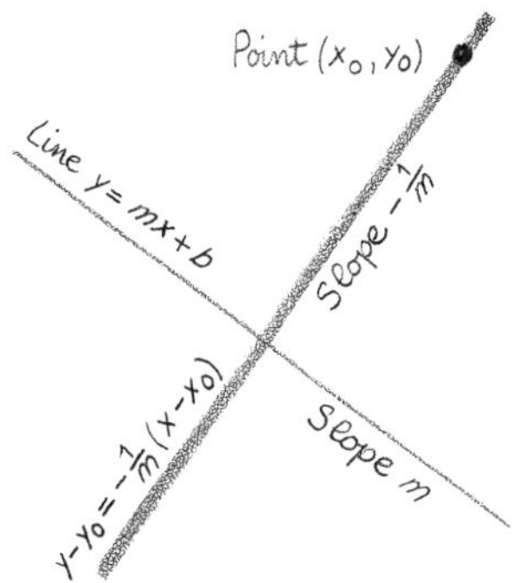

Equation of line through given point perpendicular to given line: $y - y_0 = -\frac{1}{m}(x - x_0)$

Figure 1.20

EXAMPLE 1

Find the point of intersection of the lines with the equations

$$2x + y - 2 = 0,$$
$$3x + 2y - 5 = 0.$$

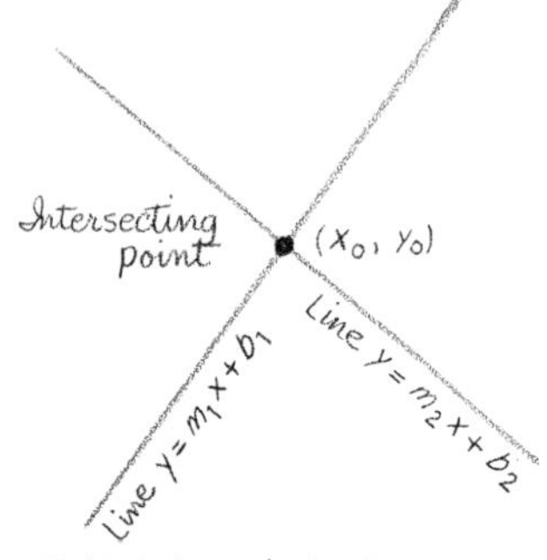

To find the point where two nonparallel lines cross solve the simultaneous equations

$$y = m_1x + b_1$$
$$y = m_2x + b_2$$

Figure 1.21

We solve this system by the method of elimination. Multiply the first equation by 2 and then subtract the second one from it:

$$\begin{array}{r} 4x + 2y - 4 = 0 \\ 3x + 2y - 5 = 0 \\ \hline x \qquad + 1 = 0, \end{array} \qquad \text{or} \qquad x = -1.$$

By substitution in either of the original equations, we get $y = 4$. So the two lines intersect at $(-1,4)$. ▲

EXAMPLE 2

When two lines are parallel but not identical the system of the two simultaneous equations has no solution. It is called an **inconsistent** system. To show this let us write the equations in slope-intercept form:

$$y = mx + b_1,$$
$$y = mx + b_2.$$

Here m is the common slope, and b_1 is not equal to b_2. Subtracting the second equation from the first, we obtain $b_1 - b_2 = 0$. This contradicts the assumption that the b's are not equal. It follows that there is no point (x,y) which satisfies both equations. ▲

Straight lines and their graphs are often useful in describing relations between two quantities. Such relations will be analyzed in more detail when we study **functions.** However, we now give an example for the purpose of illustration.

EXAMPLE 3

A college student works at a summer job during the vacation months. His net salary, after individual taxpayers exemption and allowable deductions, ranges from \$500 to \$1000. The amount of federal income tax that he must pay varies with the amount earned. According to the rates of 1970, a single taxpayer in the \$500–\$1000 range pays \$70 plus 15 percent of the excess over \$500. For example, if he earns \$600, his tax is $\$70 + 0.15(\$100) = \$85$ because the excess is \$100. In general, let x be the amount earned and y, the tax liability. Then the equation relating the two is

$$y = 70 + 0.15(x - 500), \qquad \text{or} \qquad y - 70 = 0.15(x - 500),$$

for values of x from 500 to 1000. This is the equation of the line passing through the point (500, 70) with slope 15/100 (Figure

1.22). To draw the graph, we first find a second point on the line. Put $x = 800$ and solve the equation of the line for $y: y = 115$. Then we pass a line through the two points (500, 70) and (800, 115). The line should not extend below $x = 500$ nor above $x = 1000$ because the tax rates change at those points. ▲

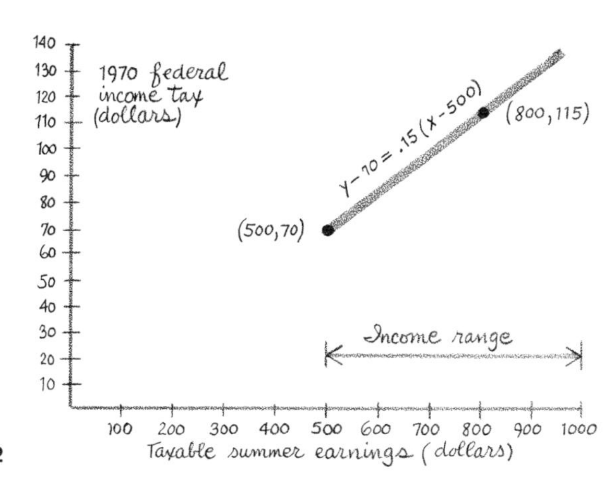

Figure 1.22

1.2 EXERCISES

In Exercises 1 and 2 write the equation of the line passing through the given point with the given slope.

1 (a) (2,−4), $m = 5$
(b) (−1,3), $m = 0$
(c) (1,3), $m = 2$
(d) (3,0), $m = -\frac{1}{4}$
(e) (0,0), $m = \frac{2}{3}$

2 (a) (1,2), $m = 3$
(b) (−1,−3), $m = -2$
(c) (4,5), $m = 3$
(d) (−4,2), $m = -\frac{1}{2}$
(e) (−1,2), $m = -1$

In Exercises 3 and 4 write the equations of the lines through the given pairs of points. Put the equations in the general linear form.

3 (a) (3,2) and (−5,2)
(b) (5,2) and (5,−1)
(c) (0,0) and (17,4)
(d) (4,0) and (0,−3)
(e) (−4,1) and (1,7)

4 (a) (4,−3) and (2,−1)
(b) (1,2) and (−3,5)
(c) (2,0) and (−1,−5)
(d) (3,1) and (−2,4)
(e) (−7,−4) and (2,−7)

5 Find the slope and y-intercept of each of these lines:
(a) $4x + y - 4 = 0$
(b) $y + 3x - 5 = 0$
(c) $2x - y + 3 = 0$
(d) $2y - x + 1 = 0$
(e) $2x - 5y + 7 = 0$

6 Find the point of intersection of the lines with the given equations.
(a) $4x - 7y - 5 = 0$ (b) $2x + 3y = 0$
$2x + 5y - 3 = 0$ $6x - 7y + 8 = 0$

7 Find the points of intersection of these pairs of lines:
(a) the line through $(-1,3)$ with slope -2
the line through $(4,5)$ with slope $-\frac{3}{2}$
(b) the line through $(1,3)$ with slope $\frac{2}{3}$
the line through $(3,0)$ with slope $-\frac{1}{4}$
(c) the line through $(0,0)$ and $(2,1)$
the line through $(3,2)$ and $(4,4)$
(d) the line through $(-\frac{1}{2},1)$ and $(5,-2)$
the line through $(2,-3)$ and $(1,0)$

8 Find the equation of the line through $(2,-3)$ and parallel to the line through $(4,1)$ and $(-2,2)$.

9 Find the equation of the line through $(-2,3)$ and perpendicular to the line $2x - 3y + 6 = 0$.

10 Find the equation of the line with y-intercept 4 and parallel to $2x - 3y - 7 = 0$.

11 At what point does the line through $(6,1)$ and $(1,4)$ intersect the line with the equation $2x - 3y + 29 = 0$?

12 A line is drawn from $(-4,-2)$ perpendicular to the line joining $(-3,7)$ and $(2,11)$. Where do they intersect? What is the distance from $(-4,-2)$ to the point of intersection?

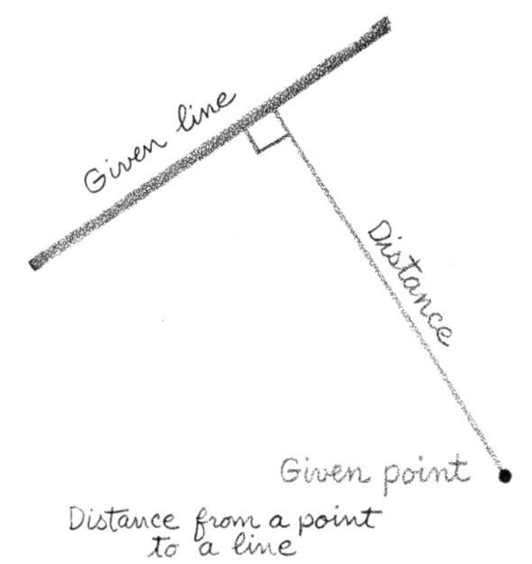

Figure 1.23

● **13** The distance from a point to a line is defined as the length of the perpendicular dropped from the point to the line (Figure 1.23). Using the 3-step method outlined in Figure 1.24 calculate the distance from the point $(-3,0)$ to the line $y = -x + 5$.

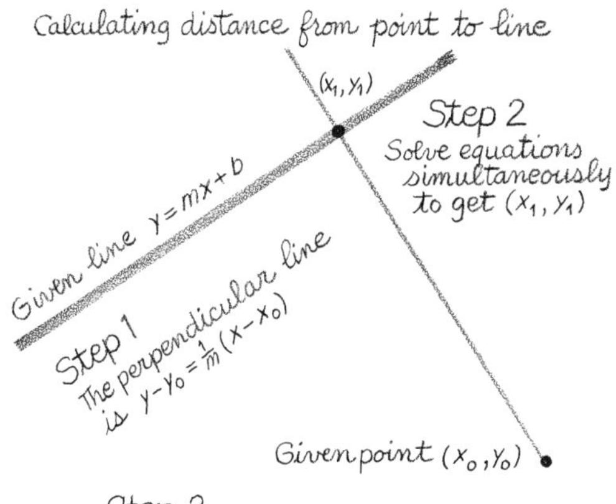

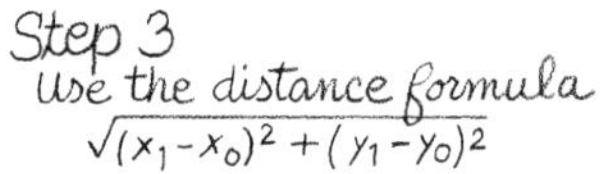

Figure 1.24

14 Find the distance from the origin to the line $2x + y + 1 = 0$.

15 On the 1970 federal tax schedule a single taxpayer earning a net taxable income of at least \$100,000 must pay \$55,490 plus 70 percent of the excess over \$100,000. Find the equation of the line relating the income and the tax liability.

1.3 The Circle

In plane geometry a circle is defined as the set of points at a fixed distance r from a given point C. The radius of the circle is r and the center is C. Now we give two forms of an algebraic equation which characterizes the coordinates of the points on a circle.

A circle with its center at the point (h,k) and of radius r has the equation

$$(x - h)^2 + (y - k)^2 = r^2. \qquad (1)$$

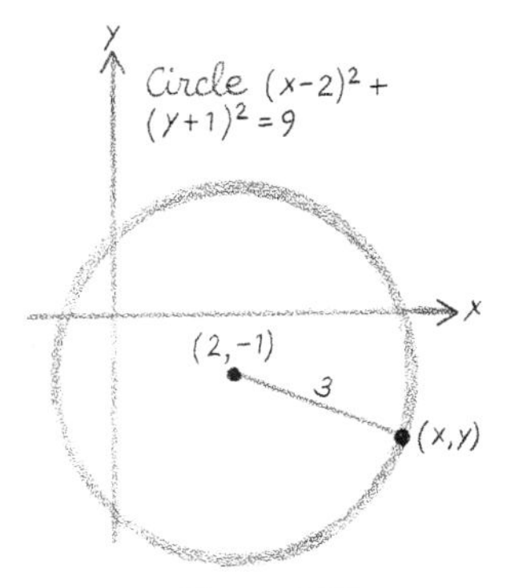

Figure 1.25

A point (x,y) lies on the circle if and only if it satisfies Equation (1). The reason is that the distance from the center (h,k) to the point (x,y) is $\sqrt{(x-h)^2 + (y-k)^2}$, and (1) states that this distance is equal to r. (More exactly, it states that the square of the distance is equal to r^2.) Equation (1) is called the **center-radius** form of the equation of the circle. For example, the circle with center at $(2,-1)$ and of radius 3 has the equation $(x - 2)^2 + (y + 1)^2 = 9$ (Figure 1.25).

When the center of the circle is at the origin, the numbers h and k are both equal to 0 and the equation takes the special form

$$x^2 + y^2 = r^2. \qquad (2)$$

For example, the circle with the center at the origin and of radius 3 has the equation $x^2 + y^2 = 9$.

EXAMPLE 1

Let us obtain the equation of the circle whose center is $(5,-2)$ and which passes through the point $(-1,5)$. The radius must be equal to the distance between these two points; namely, $\sqrt{(5 + 1)^2 + (-2 - 5)^2} = \sqrt{36 + 49} = \sqrt{85}$. Since the center is $(5,-2)$ and the radius is $\sqrt{85}$, it follows from (1) that the equation of this circle is

$$(x - 5)^2 + (y + 2)^2 = 85.$$

The general procedure for determining the equation of a circle with a given center and passing through a given point is outlined in Figure 1.26. ▲

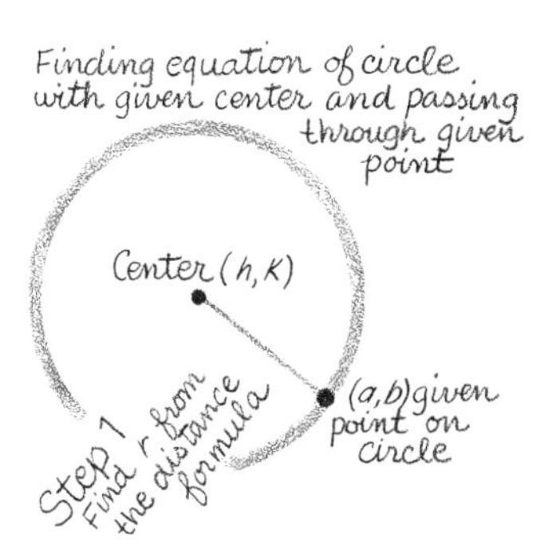

Figure 1.26

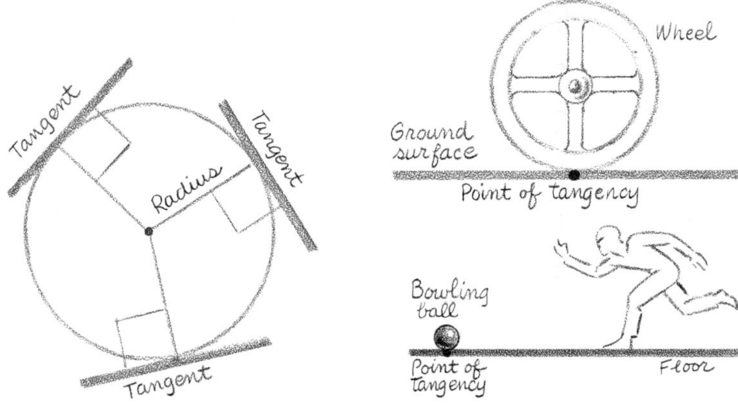

Figure 1.27 **Figure 1.28**

Recall from plane geometry that a line is **tangent** to a circle if they intersect at exactly one point. Such a line is called a tangent line, or simply a tangent. (This is from the Latin word **tangere:** to touch.) It is proved in plane geometry that the tangent is perpendicular to the radius at the point of tangency (Figure 1.27). Practical examples of tangent lines appear in Figure 1.28.

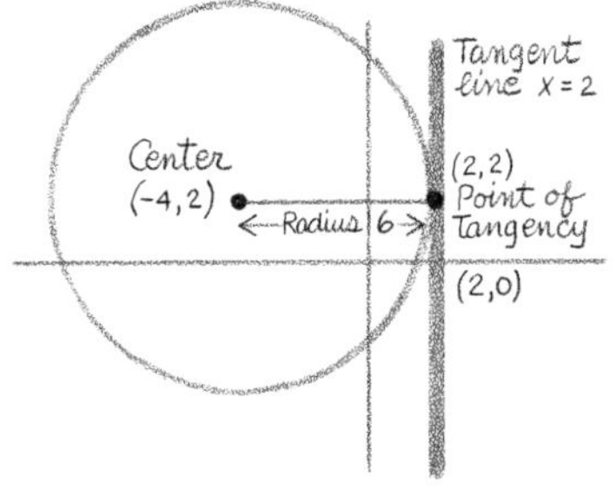

Figure 1.29

EXAMPLE 2

Find the equation of the circle with center at $(-4,2)$ and which has the tangent line $x = 2$. Since the center is given we have only to find the radius. The radius is equal to the distance from the point $(-4,2)$ to the line $x = 2$ (compare Section 1.2, Exercise 13). The point of tangency has to be at $(2,2)$; indeed, the tangent line is vertical so that the radius to the point of tangency is horizontal (Figure 1.29). It follows that the radius is of length 6. Therefore the equation of the circle is $(x + 4)^2 + (y - 2)^2 = 36$. ▲

EXAMPLE 3

A circle with center $(4,-1)$ passes through the point $(-1,3)$. Let us find the equation of the line tangent to the circle at the point $(-1,3)$. The line from the center to this point has the slope $-\frac{4}{5}$. The tangent line is perpendicular to it and so has the slope $\frac{5}{4}$. By the point-slope form [Equation (1) of Section 1.2] the equation of the tangent line is found to be $y - 3 = (\frac{5}{4})(x + 1)$ (Figure 1.30). ▲

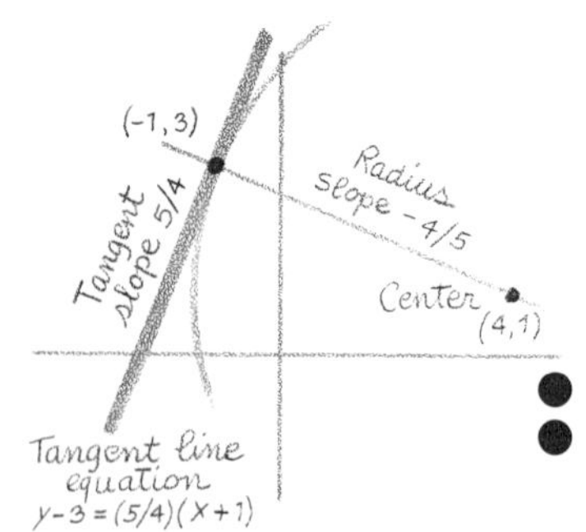

Figure 1.30

We now give a second form of the equation of the circle, the **general form.** If we expand the squares in (1) and rearrange the terms, we find that the equation may be rewritten as

$$x^2 + y^2 - 2hx - 2ky + (h^2 + k^2 - r^2) = 0.$$

Put

$$D = -2h, \qquad E = -2k, \qquad \text{and} \qquad F = h^2 + k^2 - r^2;$$

then the equation above is of the form

$$x^2 + y^2 + Dx + Ey + F = 0. \tag{3}$$

This is the general form of the circle. For example, the circle with center at $(2, -1)$ and radius 3 has the center-radius form $(x-2)^2 + (y+1)^2 = 9$, which can be changed to the general form

$$x^2 + y^2 - 4x + 2y - 4 = 0.$$

Every center-radius equation of the form (1) can be transformed into the general form (3). A natural question is: Is the converse true; that is, can every equation of the general form (3) be converted to a center-radius equation representing a circle? The answer is that every equation of the form (3) **can** be converted into the form (1); however, the number r^2 in (1) may turn out to be positive, negative, or zero. If positive, then the equation represents a circle of radius r; if zero, it is a circle of radius zero, namely, a point; finally, if it is negative, then there are no real numbers x and y satisfying Equation (1) so that it does not represent any circle.

The conversion of (3) to (1) is done by the method of **completing the square.** By this method a polynomial of the second degree in x is written as a perfect square plus a constant; for example, the second-degree polynomial $x^2 + 4x + 1$ may be expressed in the equivalent form $(x+4)^2 - 3$. The method is based on the simple formula

$$x^2 + Bx = (x + \tfrac{1}{2}B)^2 - (\tfrac{1}{2}B)^2. \tag{4}$$

It states that the square is completed by adding half the coefficient of x to x, squaring the sum, and then subtracting the square of half that coefficient. For example, $x^2 + 3x$ can be made a complete square by adding $\frac{3}{2}$ to x, squaring the sum, and then subtracting $\frac{9}{4}$:

$$x^2 + 3x = (x + \tfrac{3}{2})^2 - \tfrac{9}{4}.$$

EXAMPLE 4

A circle has the equation

$$x^2 + y^2 - 8x + 10y - 12 = 0.$$

Complete the square on x:

$$x^2 - 8x = (x-4)^2 - 16.$$

Complete the square on y:

$$y^2 + 10y = (y+5)^2 - 25.$$

The equation for the circle now becomes

$$(x-4)^2+(y+5)^2=53.$$

The circle has its center at (4,−5) and the radius is $\sqrt{53}$. ▲

EXAMPLE 5

We are given the equation

$$x^2+y^2-6x+8y+29=0.$$

Does it represent a circle? Complete the squares on x and y:

$$x^2-6x=(x-3)^2-9,$$
$$y^2+8y=(y+4)^2-16.$$

The given equation becomes

$$(x-3)^2+(y+4)^2=-4.$$

There is no pair of real numbers x and y satisfying this equation because the left-hand side is never negative because it is a sum of squares while the right-hand side is the negative number −4. It follows that there is no circle with this equation. ▲

1.3 EXERCISES

In Exercises 1 and 2 write the center-radius equation of the circle with the given center C *and radius* r*:*

1 (a) $C=(0,0)$, $r=2$
(b) $C=(2,-5)$, $r=3$
(c) $C=(-2,0)$, $r=2$
(d) $C=(-2,-4)$, $r=4$
(e) $C=(4,-3)$, $r=1$

2 (a) $C=(2,-5)$, $r=7$
(b) $C=(-3,-1)$, $r=5$
(c) $C=(0,4)$, $r=4$
(d) $C=(0,-4)$, $r=4$
(e) $C=(-4,6)$, $r=\frac{3}{2}$

In Exercises 3 and 4 complete the square in each of the equations and convert it to the form (1). State whether it represents a circle. If so, find the center and radius.

3 (a) $x^2+y^2+2x-4y-4=0$
(b) $x^2+y^2-4x+6y+9=0$
(c) $x^2+y^2-6x-4y-3=0$
(d) $x^2+y^2+8x+4y+11=0$
(e) $x^2+y^2-\frac{4}{3}x+\frac{2}{3}y+2=0$

4 (a) $x^2+y^2-x+3y+3=0$
(b) $x^2+y^2-6x-2y-6=0$

(c) $x^2 + y^2 + 5x = 0$
(d) $x^2 + y^2 - 8x - 7y = 0$
(e) $x^2 + y^2 + 5x - 6y + \frac{9}{4} = 0$

In Exercises 5 and 6 find the equation of the circle with the center C *and passing through the point* P.

5 (a) $C = (0,3)$, $P = (0,0)$ (c) $C = (2,3)$, $P = (3,-1)$
(b) $C = (0,0)$, $P = (0,1)$ (d) $C = (2,1)$, $P = 4,-2)$

6 (a) $C = (4,-1)$, $P = (-1,3)$ (c) $C = (0,-1)$, $P = (0,1)$
(b) $C = (3,-4)$, $P = (0,1)$ (d) $C = (8,5)$, $P = (5,-2)$

In Exercises 7 and 8 find the radius of the circle with the given center C *and tangent line* L.

7 (a) $C = (2,-3)$, $L\colon x = 0$
(b) $C = (-2,-1)$, $L\colon y = 0$
(c) $C = (1,4)$, $L\colon x + y = 2$
(d) $C = (-2,3)$, $L\colon 20x - 21y - 42 = 0$

8 (a) $C = (3,2)$, $L\colon y = 3$
(b) $C = (-1,2)$, $L\colon x + 2y = 4$
(c) $C = (0,-2)$, $L\colon 3x - y = 10$
(d) $C = (-4,2)$, $L\colon 3x + 4y - 16 = 0$

In Exercises 9 and 10 find the equation of the line tangent at the point P *to the circle with center* C.

9 (a) $C = (3,0)$, $P = (1,4)$ (c) $C = (3,-1)$, $P = (2,2)$
(b) $C = (0,-2)$, $P = (-3,1)$ (d) $C = (-2,-4)$, $P = (3,4)$

10 (a) $C = (-2,5)$, $P = (1,1)$ (c) $C = (1,1)$, $P = (-4,-2)$
(b) $C = (5,-4)$, $P = (0,0)$ (d) $C = (4,-3)$, $P = (0,1)$

11 A circle passes through the two points $(2,3)$ and $(-1,1)$, and has its center on the line $x - 3y - 11 = 0$.
(a) Show that if (h, k) represents the center of the circle, then it satisfies the equation

$$(h-2)^2 + (k-3)^2 = (h+1)^2 + (k-1)^2.$$

(b) Simplify the equation in (a) by squaring and cancelling to obtain a linear equation in the unknowns h and k.
(c) Show that in addition to the latter equation, h and k also satisfy $h - 3k - 11 = 0$.
(d) Solve the two simultaneous equations for h and k.
(e) Using (d) derive the radius of the circle and then its equation.

12 Using the method of Exercise 11 find the equation of the circle which passes through the points $(1,-4)$ and $(5,2)$, and has its center on the line $x - 2y + 9 = 0$.

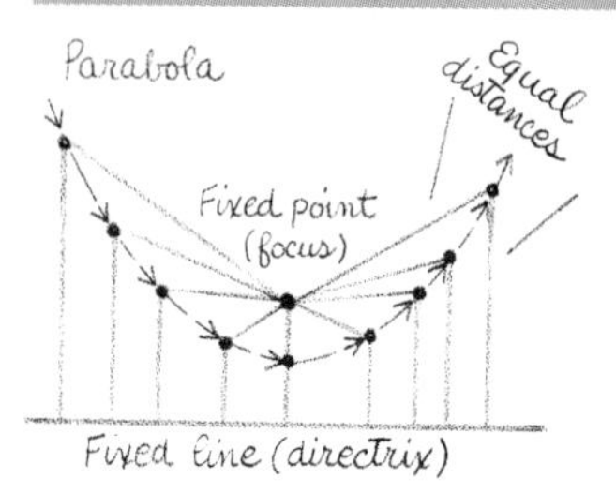

Figure 1.31

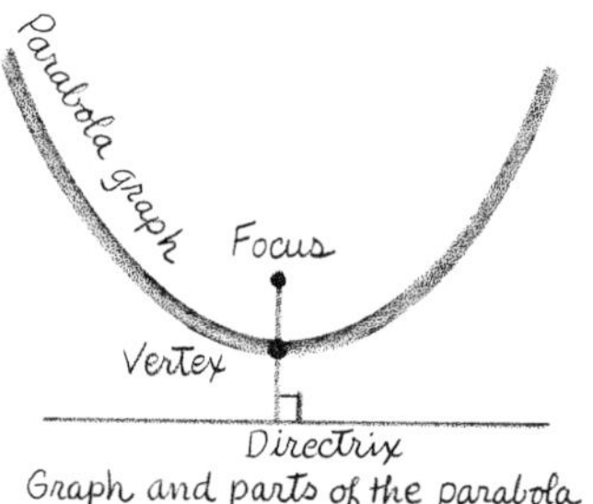

Figure 1.32

1.4 The Parabola

Suppose a point moves along a path in the plane in the following manner: Its distance from a fixed point is always equal to its distance from a fixed line. Such a path forms a geometric curve called a **parabola.** In other words, a parabola is the set of points equidistant from a fixed point and a fixed line. [Recall that the distance from a point to a line is the length of the perpendicular dropped from the point to the line (Section 1.2, Exercise 13).] The fixed point is called the **focus** of the parabola, and the line the **directrix** (Figure 1.31). The midpoint of the perpendicular from the focus to the directrix is called the vertex (Figure 1.32).

Although it is very easy to draw a straight line or circle with ruler and compass, the sketching of the parabola is more complicated. One method is outlined in Figures 1.33(a)–1.33(e). Here the directrix is horizontal and the focus is below it. The procedure yields a rough sketch of the parabola, an approximation by straight line segments. By choosing the points on the directrix sufficiently close to each other, we can get an approximation which is very close to the parabola. The justification of this sketching procedure is based on elemen-

Figure 1.33

Drawing a parabola with given focus and directrix

Directrix

Vertex

Step 1 Focus

Mark equally spaced points on directrix and join these to focus with lines

(a)

Step 2

Draw line through vertex, parallel to directrix, forming midpoints of the segments

(b)

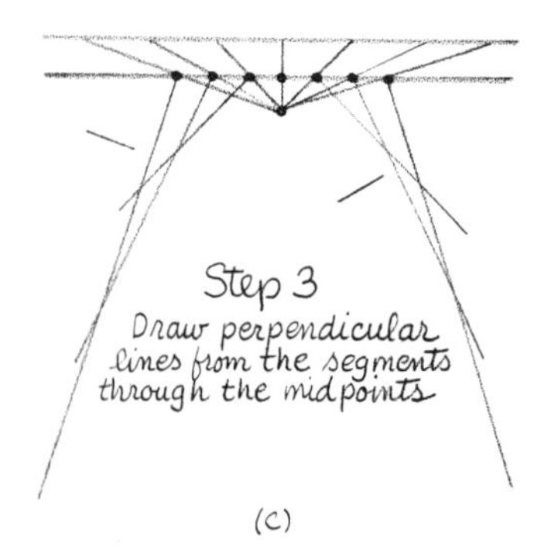

(c)

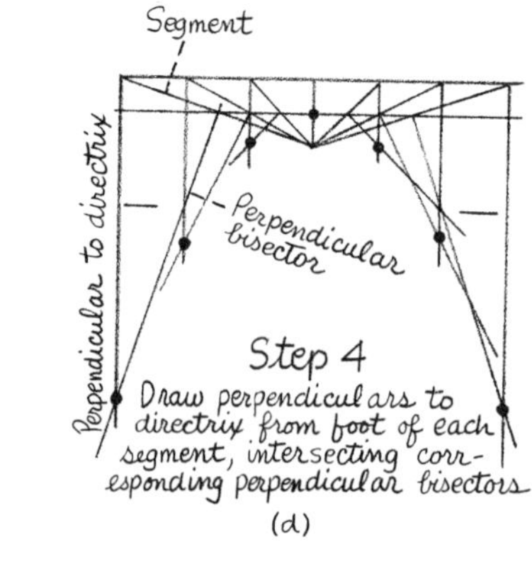

(d)

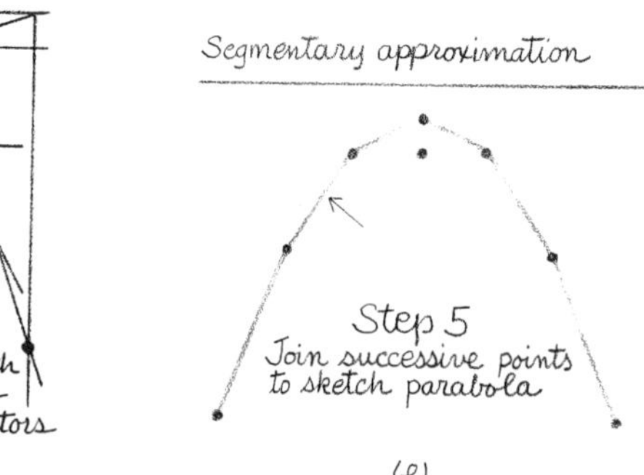

(e)

tary geometry and is outlined in Figure 1.34: Since the two right triangles are obviously congruent, their corresponding sides are equal, and so the distance from the point to the focus is equal to that of the directrix.

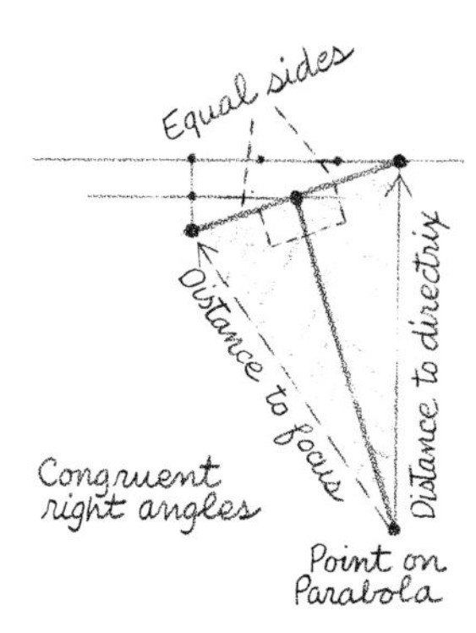

Figure 1.34

Parabolas occur in many situations in the real world. A ball thrown up at an angle to the ground rises and then falls along a path which has approximately the shape of a parabola (Figure 1.35). If equal weights are placed in a line, equally spaced, and then suspended from a thin cable, the cable will assume the approximate shape of a parabola (Figure 1.36). This principle is used in the design of a suspension bridge: The roadbed is of homogeneous weight and so the cable is designed to hang in the shape of a parabola (Figure 1.37).

A closely related property of the parabola is used in the design of an arch. Since the parabolic shape of the cable is natural for supporting a uniform load from above, it follows by symmetry that a parabolic arch will give uniform support to a load from below (Figure 1.38). This principle is used in the construction of arch-supported bridges (Figure 1.39), as well as in the human foot (Figure 1.40). The nose of a large airplane is shaped like a parabola so that the air pressure on the front of the plane will be equally distributed (Figure 1.41). The parabolic shape of the end of the eggshell gives it remarkable structural strength (Figure 1.42). A most fascinating arch appearing in a natural setting is the flight formation of a group of migrating birds. It is thought that the birds actually support each other flying against the horizontal resistance of the air. A bird drifting out of the formation immediately feels a drop of support and gets quickly back into place (Figure 1.43).

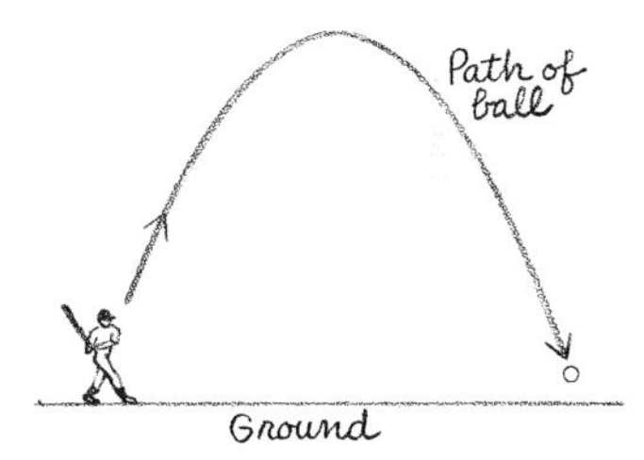

Figure 1.35

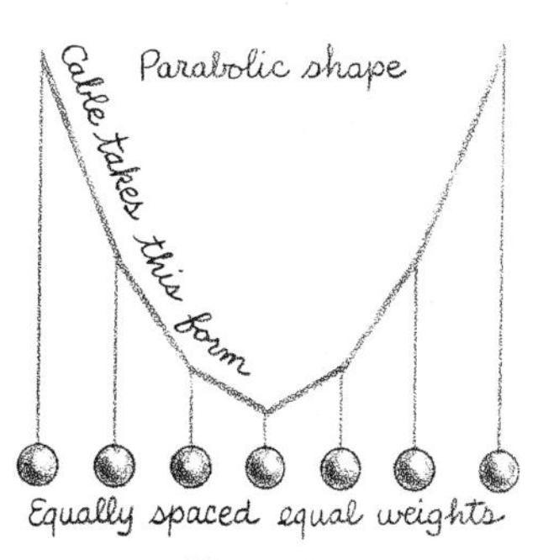

Figure 1.36

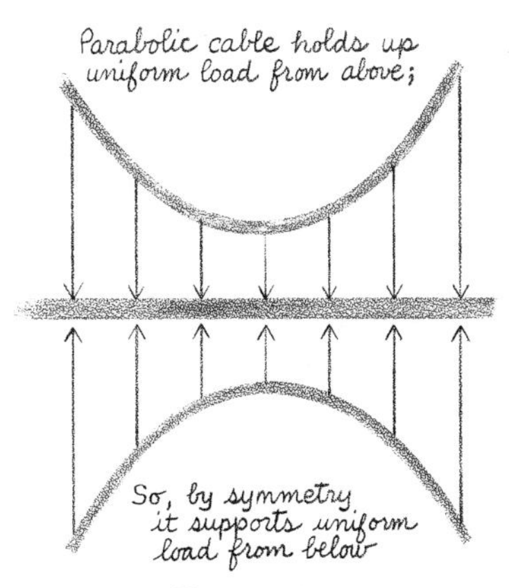

Figure 1.38

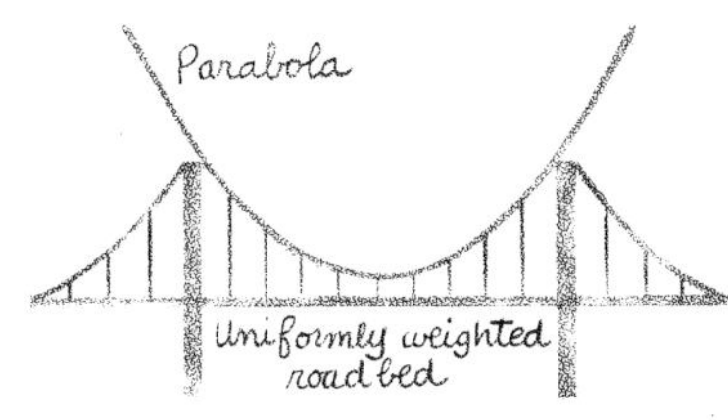

Figure 1.37

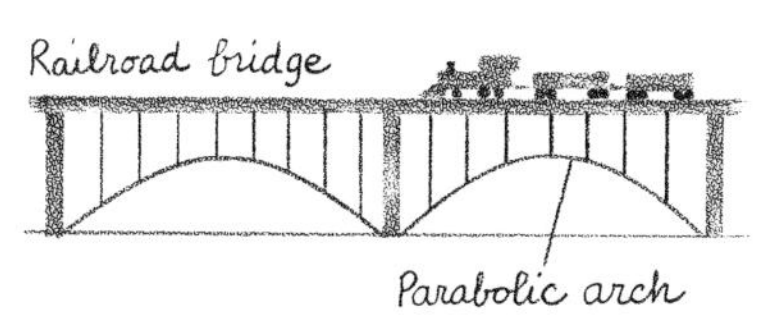

Figure 1.39

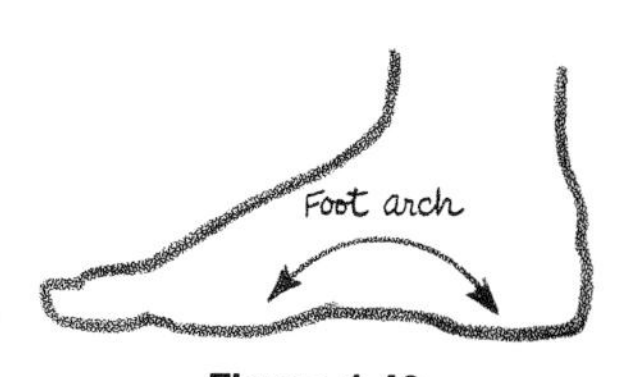

Figure 1.40

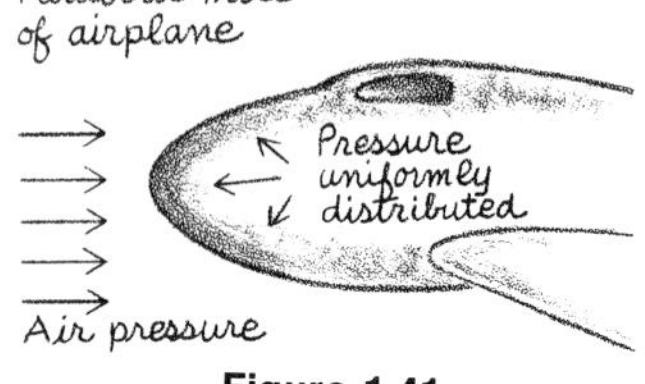

Figure 1.41

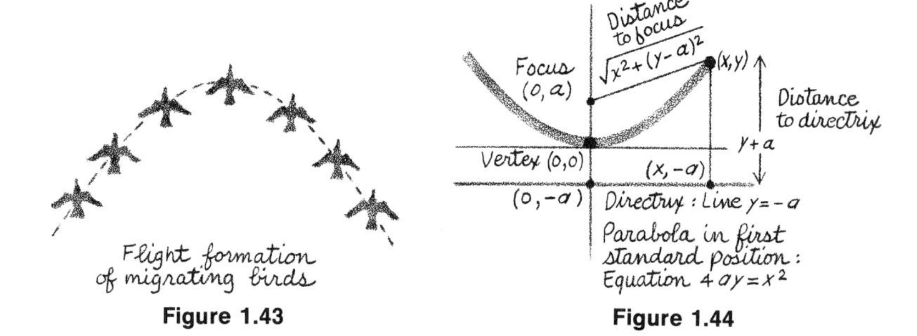

Figure 1.43

Figure 1.44

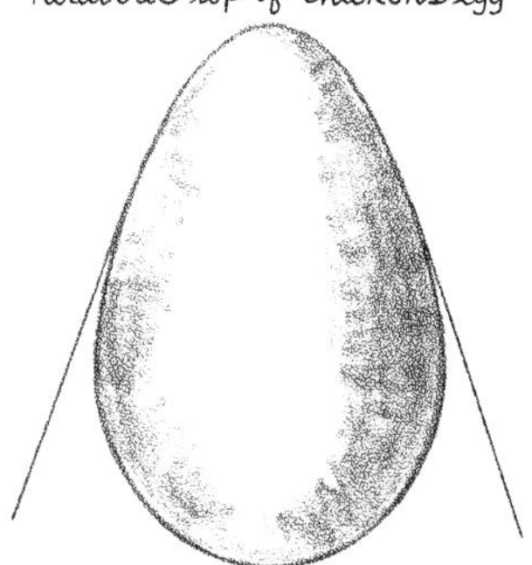

Figure 1.42

We now consider the algebraic description of the parabola. Its equation has several forms, depending on the position in the plane. We shall consider four types of positions: The directrix is horizontal, and the focus either above it or below it (two positions); and the directrix is vertical and the focus is to the left or to the right of it (two positions).

The first **standard** position for the parabola is with the focus on the y-axis, a units above the origin, where $a > 0$, and the directrix horizontal, running a units below the origin. In other words, the focus is at the point $(0,a)$ and the directrix is along the line $y = -a$. The equation of the parabola in this position is

$$4ay = x^2. \tag{1}$$

The derivation is based on the definition of the parabola as the set of points equidistant from $(0,a)$ and the line $y = -a$ (Figure 1.44). The distance from a point (x,y) on the parabola to the focus is $\sqrt{x^2 + (y-a)^2}$. The distance to the directrix is $y + a$. Therefore, the coordinates of the point on the parabola satisfy the equation

$$x^2 + (y-a)^2 = (y+a)^2.$$

When simplified, this equation is seen to be equivalent to (1).

The second standard position is obtained from the first by reversing the positions of the focus and the directrix: The focus is put below the x-axis at $(0,-a)$, and the directrix along the line $y = a$ (Figure 1.45). By similar reasoning the equation is found to be

$$-4ay = x^2. \tag{2}$$

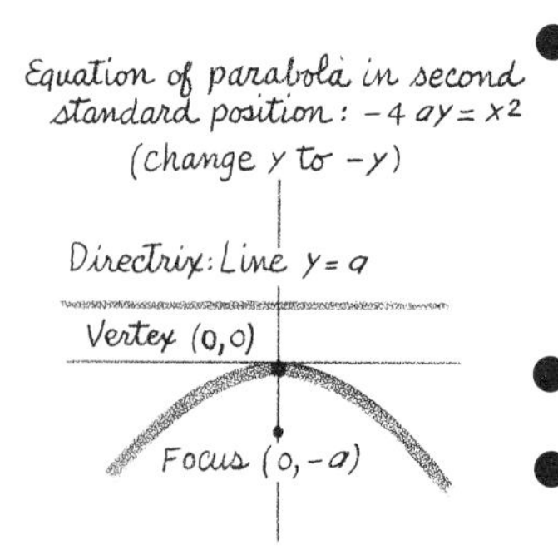

Figure 1.45

The third standard position is with the directrix vertical and the focus to the right of it. The directrix is the line $x = -a$, and the focus is the point $(a,0)$. The equation is $y^2 = 4ax$. The fourth standard position has the focus at $(-a,0)$ and the directrix the vertical line $x =$. Here the equation is $y^2 = -4ax$ (Figure 1.46).

In each of the above standard positions the vertex of the parabola is at the origin.

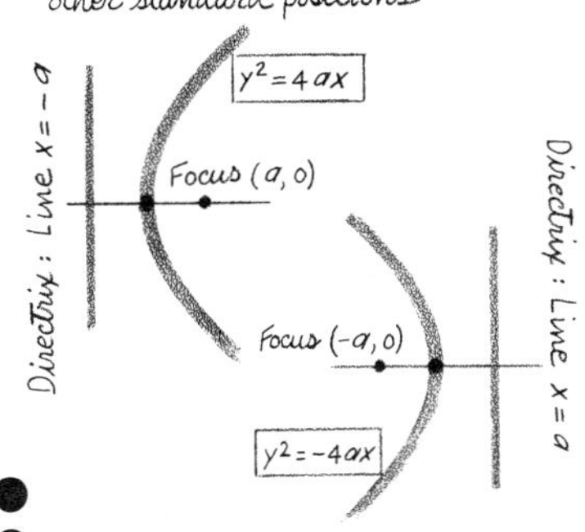

Figure 1.46

EXAMPLE 1

Find the focus and directrix of the parabola with the equation $-8y = x^2$. This parabola is in the second standard position. We find a by setting $-4a$ equal to the coefficient of y: $-8 = -4a$, or $a = 2$. The focus is at the point $(0, -2)$, and the directrix is the horizontal line $y = 2$. ▲

Next we consider parabolas with vertical or horizontal directrices, but where the vertex is not necessarily at the origin. If the directrix is horizontal and the vertex is at a point (h, k), a units above the directrix, then the equation of the parabola is

$$4a(y - k) = (x - h)^2. \tag{3}$$

This is obtained from that of the first standard position [Equation (1)] by substituting $x - h$ and $y - k$ for x and y, respectively (Figure 1.47). The equations of the other positions are similarly obtained from the standard ones (Figure 1.48).

EXAMPLE 2

The parabola $-8(y + 3) = (x - 2)^2$ has been obtained from the standard parabola $-8y = x^2$ by a change of vertex. The latter is now at $(2, -3)$. Its distance from the directrix is still 2 (compare Example 1). This is the parabola with a horizontal directrix 2 units above the vertex, along the line $y = -1$. It follows that the focus is 2 units below the vertex, at $(2, -5)$ (Figure 1.49). ▲

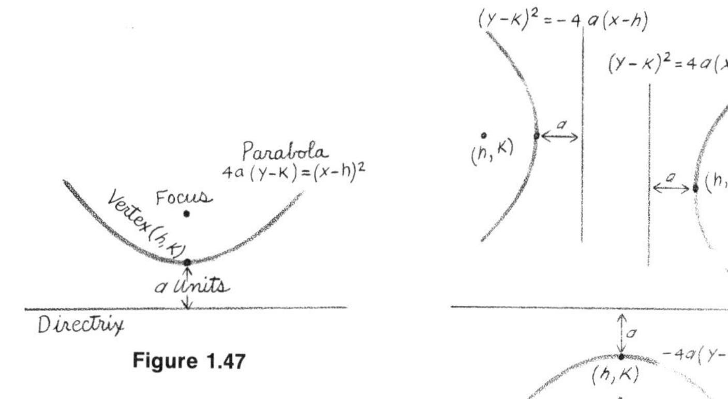

Figure 1.47

Figure 1.48

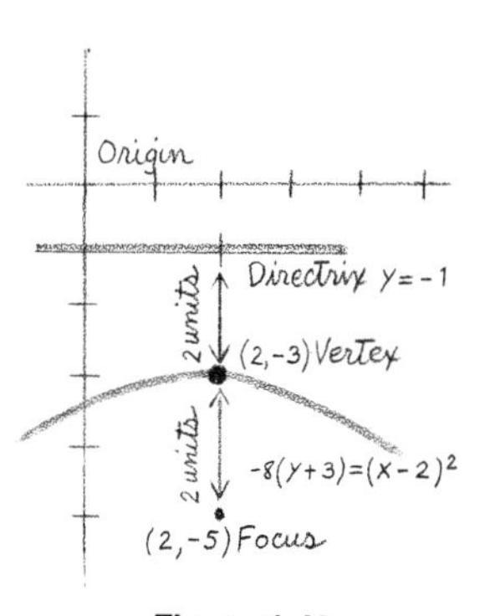

Figure 1.49

When the square in Equation (3) is expanded, it becomes

$$4a(y - k) = x^2 - 2hx + h^2.$$

This equation contains x, x^2, and y terms. Now we shall show how any second-degree equation of one of the forms

$$x^2 + Dx + Ey + F = 0 \qquad (y^2 \text{ missing}) \tag{4}$$

or

$$y^2 + Dx + Ey + F = 0 \qquad (x^2 \text{ missing}) \tag{5}$$

can be transformed into the equation of a parabola. This is done by completing the square (see Section 1.3).

EXAMPLE 3

Find the focus and directrix of the parabola with the equation $-2x^2 + 3x - y - 5 = 0$. We first put this in the form (4) by clearing the coefficient of x^2 by dividing by -2:

$$x^2 - \tfrac{3}{2}x + \tfrac{1}{2}y + \tfrac{5}{2} = 0.$$

Complete the square on x:

$$x^2 - \tfrac{3}{2}x = (x - \tfrac{3}{4})^2 - \tfrac{9}{16}.$$

The equation for the parabola becomes

$$(x - \tfrac{3}{4})^2 = -\tfrac{1}{2}y - \tfrac{5}{2} + \tfrac{9}{16},$$

or

$$\left(x - \frac{3}{4}\right)^2 = -\frac{1}{2}\left(y + \frac{31}{8}\right). \tag{6}$$

The vertex of this parabola is at the point $(\frac{3}{4}, -\frac{31}{8})$. The number a is obtained from the relation $4a = \frac{1}{2}$, or $a = \frac{1}{8}$. From (6) we see that the directrix is horizontal and that the focus is below it. Since the focus is $\frac{1}{4}$ unit below the directrix ($a = \frac{1}{8}$), it follows that the focus has the coordinates $(\frac{3}{4}, -\frac{33}{8})$. ▲

A parabola in a given position is determined by the vertex and one other point on it. Given the vertex (h,k) we can find the other unknown a in the equation of the parabola by substituting the coordinates of the other given point for x and y.

EXAMPLE 4

A parabolic arch has a span of 12 feet and is 9 feet high at its vertex. How high is the arch at a distance of 4 feet from the center of the span? We choose a convenient coordinate system, taking the x-axis along the base of the span with the origin at the center (Figure 1.50). The equation of the parabola is of the general form

$$-4a(y-k)=(x-h)^2.$$

Let us determine the exact values of a, h, and k. Since the vertex is at (0,9), the equation becomes

$$-4a(y-9)=x^2.$$

The curve passes through the point (6,0); hence, this pair of coordinates satisfies the last equation. We substitute these values for x and y, and then solve for the unknown value of a: $-4a(-9)=36$, or $a=1$. So the equation of the parabola is $-4(y-9)=x^2$.

To find the height of the arch 4 feet from the center, substitute $x=4$ and solve for y: $-4(y-9)=16$, or $y=5$. ▲

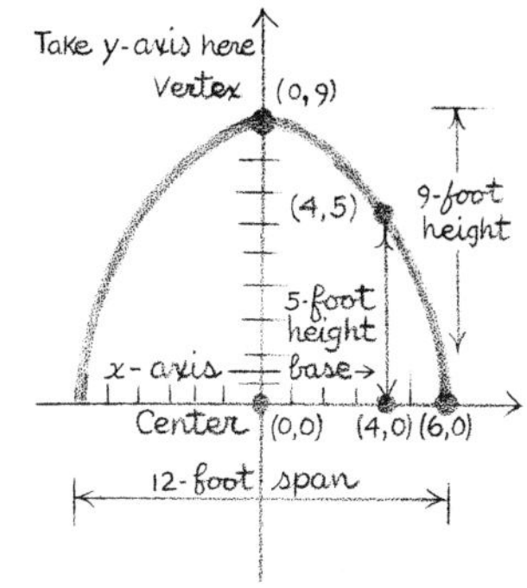

Figure 1.50

EXAMPLE 5

Sketch the graph of the parabola whose vertex is at (3,2) and focus at (5,2). The distance from the focus to the vertex is 2. Since the focus is 2 units to the right of the vertex, the directrix is a vertical line 2 units to the left of the vertex, namely, the line $x=1$. The equation takes the form

$$(y-2)^2=8(x-3).$$

The graph is symmetric about the line $y=2$, and opens to the right. In order to draw a rough sketch, we pick several values of y not too far from the value 2 at the vertex. We substitute these in the parabola equation and solve for x. For example, when $y=1$, we find $(1-2)^2=8(x-3)$, or $x=3\frac{1}{8}$. When $y=3$, we also find $x=3\frac{1}{8}$. When $y=0$ or 4, we have $x=3\frac{1}{2}$; and, when $y=-1$ or 5, we have $x=4\frac{1}{8}$. A sketch of the graph can now be drawn by plotting the vertex and the 6 other points (Figure 1.51). ▲

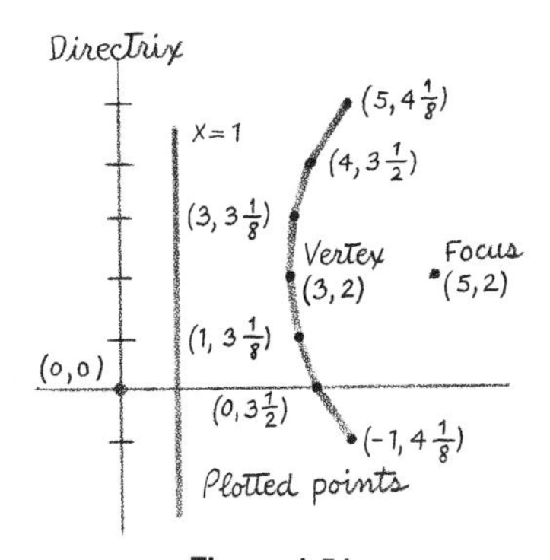

Figure 1.51

1.4 EXERCISES

1 Using the method outlined in Figure 1.33 draw an approximate parabola with a horizontal directrix with the focus 1 inch above the directrix. Place 7 equally spaced points on the directrix, with 1 inch between the successive points.

2 Repeat the construction in Exercise 1 with a total of 13 points on the directrix, $\frac{1}{2}$ inch apart. (Simply add points midway between those in Exercise 1.) Compare the graph to the previous one.

In Exercises 3 and 4 locate the focus and directrix of the parabola with the given equation.

3 (a) $20y = x^2$
(b) $-16y = x^2$
(c) $y^2 = -4x$
(d) $4(y-9) = (x+8)^2$
(e) $-4(y-8) = 7x^2$
(f) $(y-4)^2 = 16(x-6)$

4 (a) $y^2 = -8x$
(b) $20(y-5) = (x+11)^2$
(c) $8(y+2) = (x-2)^2$
(d) $(y+1)^2 = 20(x-7)$
(e) $(y-7)^2 = -28(x-6)$
(f) $2(y+5) = -5(x-2)^2$

In Exercises 5 and 6 find the equations of the parabolas with the given information.

5 (a) focus (3,0), directrix $x = -3$
(b) focus (0,−5), directrix $y = 5$
(c) vertex (−3,−1), focus (−3,1)
(d) vertex (5,1), directrix $y = 0$
(e) focus (1,1), directrix $y = 3$

6 (a) focus (−2,0), directrix $x = 2$
(b) focus (1,1), directrix $x = 4$
(c) vertex (0,2), directrix $x = 0$
(d) vertex (2,2), focus (1,2)
(e) vertex (−1,1), focus (−3,1)

In Exercises 7 and 8 find the focus and directrix of the parabolas defined by these second-degree equations.

7 (a) $x^2 - 4y + 20 = 0$
(b) $4y^2 - x - 4y = 0$
(c) $x^2 + 3x + 2y - 1 = 0$
(d) $4x^2 + 12x - 8y + 17 = 0$

8 (a) $y^2 + 3x - 6 = 0$
(b) $3x^2 + 2x - 2y - 7 = 0$
(c) $y^2 + 3x + 4y - 8 = 0$
(d) $y^2 + x + y + 1 = 0$

9 The two towers of a small suspension bridge are 60 feet high (above the roadbed). They are 500 feet apart. The lowest point on the cable is 10 feet above the road. The cable hangs in the shape of a parabola. What is the location of the focus of the parabola? How high above the road is the point on the cable 40 feet from the center of the bridge? (See Example 4.)

10 The ends of a parabolic arch are 100 feet apart and the arch is 30 feet high. Find the height of the arch 10 feet from one end of the arch. (See Example 4.)

11 A baseball is hit up into the air at an angle to the ground. Suppose that it travels along a parabolic path. If the maximum height of the ball is 80 feet and the range of the ball is 150 feet (the horizontal distance traveled) how high is the ball after it has gone 30 feet horizontally? (See Example 4.)

12 A ball is thrown up at an angle to the ground and moves along a parabolic path. It reaches a height of 120 feet and then starts to fall. As it falls 8 feet from its highest point it moves 4 feet horizontally. Find the total distance along the ground from the point the ball was thrown to the point where it falls. (See Example 4.)

1.5 The Ellipse

An ellipse is the curve seen by the eye when the circle is viewed in a plane at an angle to the line of vision (Figure 1.52).

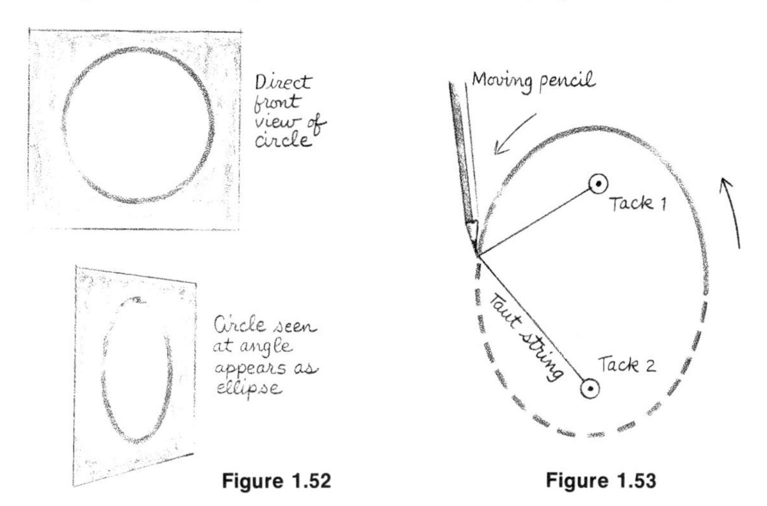

Figure 1.52

Figure 1.53

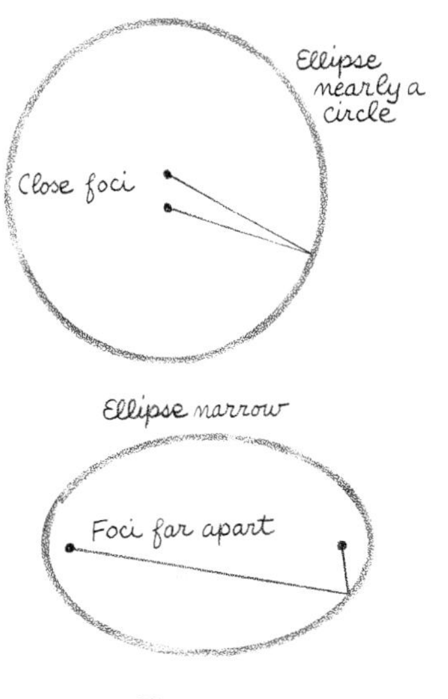

Figure 1.54

First we shall give the classical definition of the ellipse. Later we shall show by means of analytic geometry why circles at angles look like ellipses. Let there be two distinct fixed points in the plane. If a third point moves along a path in the plane so that the sum of its distances to each of the latter points remains constant, then the moving point traces an ellipse. This is our definition of the ellipse. The two fixed points are called the **foci** of the ellipse. (This is the plural of **focus.**) The construction of the ellipse is illustrated in Figure 1.53. Place a sheet of paper over a wood or cardboard surface and put two tacks through the paper. Tie the ends of a string to them. Pull the string taut with the tip of a sharp pencil or pen. Hold the latter vertical to the paper and move it around the tacks. It will trace an ellipse. The reason is that the sum of the distances from the point of the pencil to each of the tacks is equal to the fixed length of the string.

If the foci are relatively far apart (the distance between them is large relative to the length of the string), then the ellipse is long and narrow. If, on the other hand, the foci are relatively close to each other, then the ellipse is shaped nearly like a circle (Figure 1.54).

Instead of using tacks it may be more convenient to use 3 pencils and a closed string, as in Figure 1.55. Place the first 2 pencils at the foci, and pulling the string taut, trace the curve with the third.

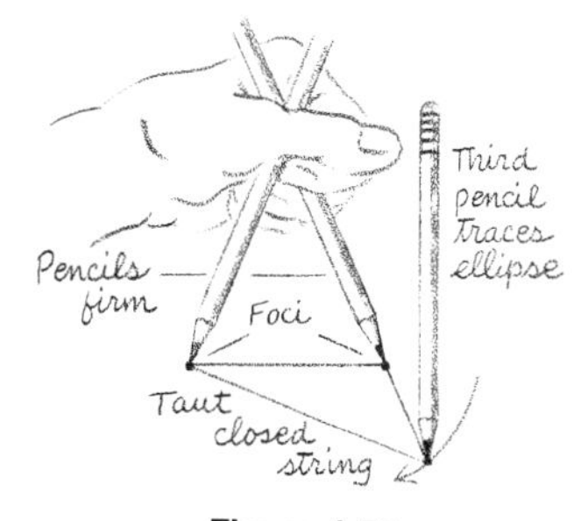

Figure 1.55

It is clear from the definition and the construction of the ellipse that the perimeter of the triangle formed by the foci and the moving point never changes as the point goes around the ellipse (Figure 1.56).

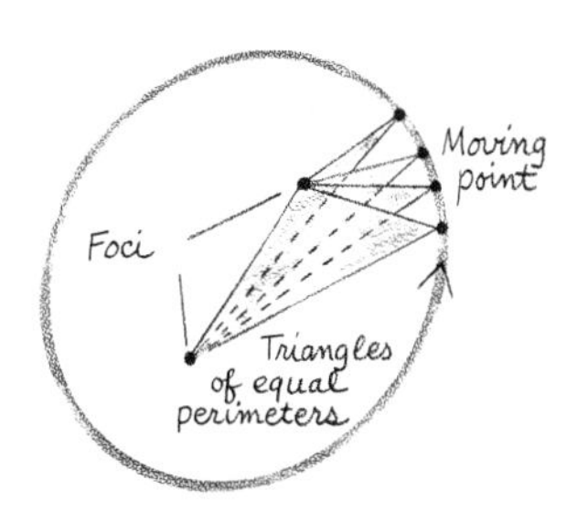

Figure 1.56

The ellipse arises in the physical world as the path of a planet or satellite in orbit around a larger body. The path of the earth around the sun is an ellipse with the sun at one focus. (In this particular case the foci are relatively close to each other so that the ellipse is nearly circular.) Similarly, when an artificial satellite is placed in orbit around the earth, it goes on an elliptic path with the center of the earth as one focus (Figure 1.57). This explains why the satellite has a high point (apogee) and a low point (perigee) in its orbit around the earth.

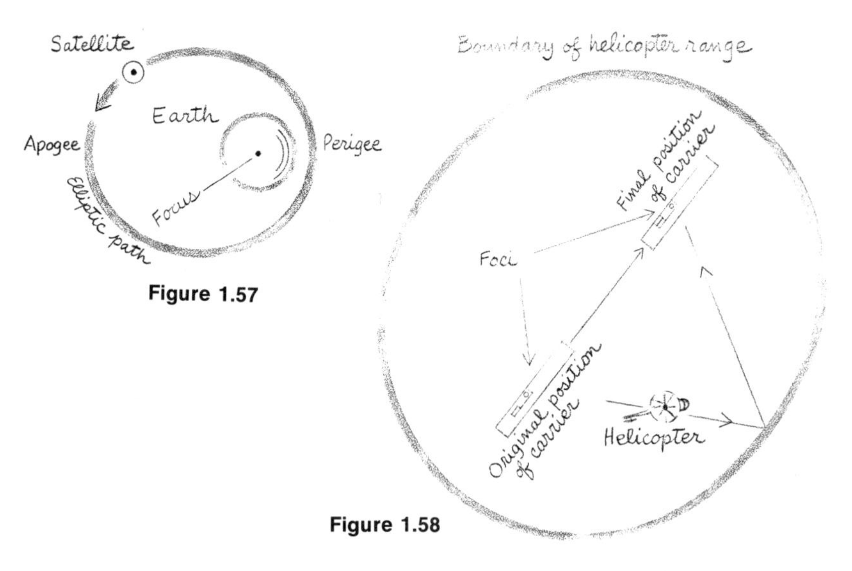

Figure 1.57

Figure 1.58

The ellipse also occurs as the boundary of the flight range of a plane or helicopter based on a moving carrier. This means that if a helicopter leaves a moving carrier and then returns to it after it has moved to a new position, then the path of the helicopter is bounded by an ellipse. The foci of this ellipse are the original and final positions of the carrier. The reason is that the helicopter has a fixed flying range, and the distance from the carrier and back cannot exceed it (Figure 1.58).

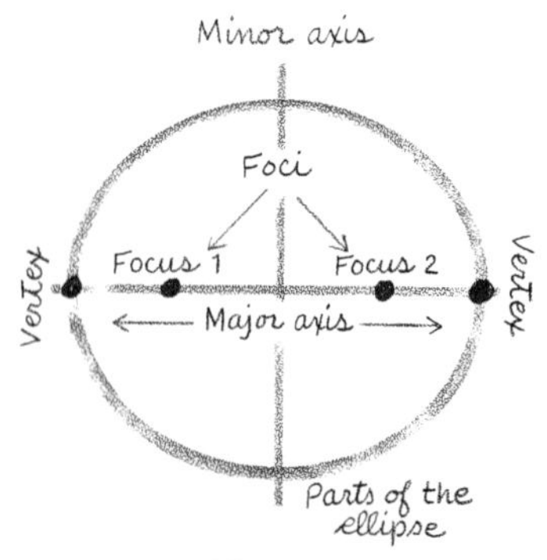

Figure 1.59

The line segment from one tip of the ellipse to the other, passing through the foci, is called the **major axis** of the ellipse. The segment perpendicular to it through the midpoint, running between the other two tips, is the **minor axis.** The points on the ellipse touching the major axis are the **vertices** (Figure 1.59). The **center** is at the intersection of the axes.

The ellipse has two standard positions: The first is with the center at the origin and horizontal major axis; and the second is also with the center at the origin but with a vertical major axis. In the first position the foci are at $(-c,0)$ and $(c,0)$, and the vertices at $(-a,0)$ and $(a,0)$, where a is larger than c. The distance from $(-c,0)$ to $(a,0)$ is $c+a$ and the distance back to the other focus is $a-c$; therefore, the sum of the distances from the foci to any point on the ellipse is the sum $(c+a)+(a-c)=2a$. Let $(0,b)$ and $(0,-b)$ be the points where the ellipse touches the minor axis. Then the distances from either focus to $(0,b)$ are equal; hence, since their sum is $2a$, each must be equal to a. It now follows from the Pythagorean Theorem that $a^2=b^2+c^2$ (Figure 1.60).

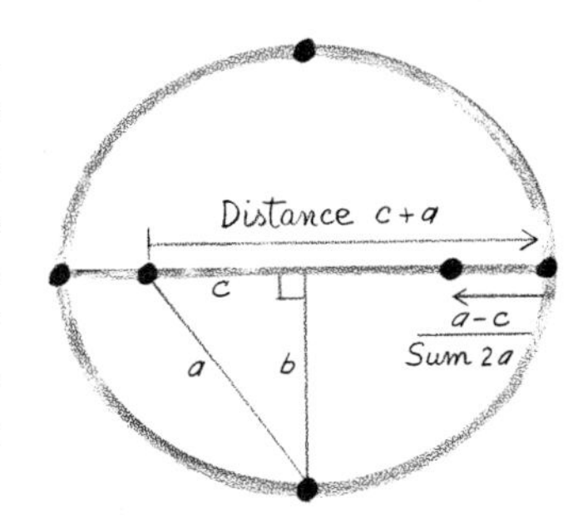

Figure 1.60

Now we derive the equation of the ellipse in the first standard position. The distance from the focus $(-c,0)$ to a point (x,y) on the ellipse is equal to $\sqrt{(x+c)^2+y^2}$; similarly the distance from the other focus is $\sqrt{(x-c)^2+y^2}$. Since the sum of the two distances is equal to $2a$, we get the equation

$$\sqrt{(x+c)^2+y^2}+\sqrt{(x-c)^2+y^2}=2a$$

or

$$\sqrt{(x+c)^2+y^2}=2a-\sqrt{(x-c)^2+y^2}\,.$$

Square both sides of the last equation:

$$(x+c)^2+y^2=4a^2-4a\sqrt{(x-c)^2+y^2}+(x-c)^2+y^2.$$

Expand the squares and cancel common terms on both sides of the last equation:

$$cx-a^2=-a\sqrt{(x-c)^2+y^2}\,.$$

Square again and simplify:

$$(a^2-c^2)x^2+a^2y^2=a^2(a^2-c^2).$$

Divide both sides by $a^2(a^2-c^2)$:

$$\frac{x^2}{a^2}+\frac{y^2}{a^2-c^2}=1.$$

Since $b^2=a^2-c^2$, the last equation becomes

$$\frac{x^2}{a^2}+\frac{y^2}{b^2}=1, \tag{1}$$

which is the equation of the ellipse in the first standard position (Figure 1.61).

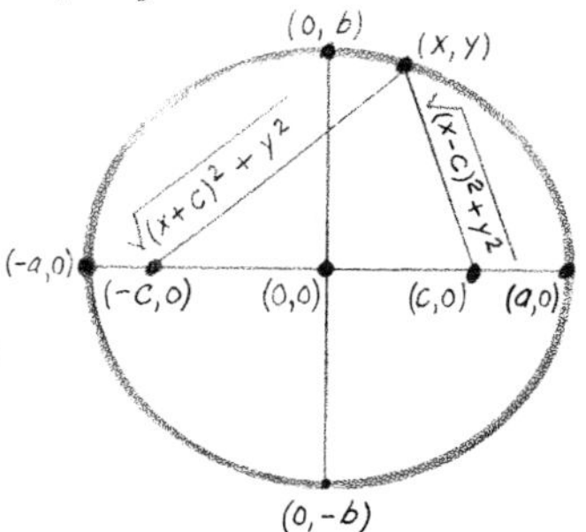

Figure 1.61

The second standard position for the ellipse is with the major axis on the y-axis. The foci are at $(0,-c)$ and $(0,c)$, and the vertices at $(0,-a)$ and $(0,a)$ (Figure 1.62). The equation is the same as (1) except that a and b are interchanged:

$$\frac{x^2}{b^2}+\frac{y^2}{a^2}=1. \tag{2}$$

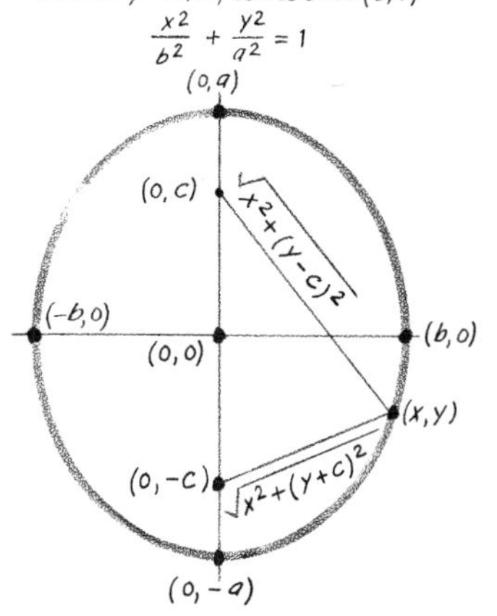

Figure 1.62

In Equations (1) and (2) the number a^2 is larger than b^2. Therefore, the magnitude of the numbers appearing under x^2 and y^2 indicates whether the ellipse is in the first or second standard position. If the number under x^2 is the larger, the major axis is horizontal. If the number under y^2 is the larger, the major axis is vertical. When they are equal, the ellipse is really a circle.

Now we explain why the tilted circle looks like the ellipse. Suppose we multiply Equation (2) by b^2 to get

$$x^2 + \left(\frac{by}{a}\right)^2 = b^2.$$

This is almost the equation of a circle of radius b except that y has been replaced by by/a. It is a circle with a magnified scale along the y-axis. When a circle is tilted along a vertical line through the origin, as in Figure 1.52, the x-axis is squeezed in the viewer's eyes. This is analogous to magnifying the other axis, the y-axis.

EXAMPLE 1

Find the equation of the ellipse with center at the origin, major axis of length 10, and foci $(-3,0)$ and $(3,0)$. The major axis is horizontal so that the ellipse is in the first standard position. The distance from vertex to vertex is $2a$; hence, $2a = 10$, so that $a = 5$. Half the length of the minor axis is b. This is computed from the relation $b = \sqrt{a^2 - c^2}$. Here $c = 3$, so that $b = \sqrt{25 - 9} = 4$. It follows that the equation of this ellipse is

$$\frac{x^2}{25} + \frac{y^2}{16} = 1. \blacktriangle$$

When the center of the ellipse is not at the origin but at a point (h,k), Equations (1) and (2) are modified in this way: x and y are replaced by $x - h$ and $y - k$, respectively. When the major axis is horizontal, the equation is

$$\frac{(x-h)^2}{a^2} + \frac{(y-k)^2}{b^2} = 1. \tag{3}$$

When the major axis is vertical, the equation is

$$\frac{(x-h)^2}{b^2} + \frac{(y-k)^2}{a^2} = 1. \tag{4}$$

EXAMPLE 2

Sketch the graph of the ellipse with the equation

$$\frac{(x-1)^2}{4} + \frac{(y+1)^2}{9} = 1. \tag{5}$$

First we identify a, b, and c. Since a^2 is the definition larger than b^2, it follows that $a^2 = 9$ and $b^2 = 4$, so $a = 3$ and $b = 2$. Since $c^2 = a^2 - b^2$, we have $c^2 = 5$, so that $c = \sqrt{5}$. Equation (5) is of the form (4) with $h = 1$, $k = -1$, $a = 3$, and $b = 2$. The center is at $(1,-1)$. The major axis is vertical, along the line $x = 1$, and is of length $2a = 6$. The minor axis is along the line $y = -1$ and of length $2b = 4$. This information is sufficient for the sketch (Figure 1.63). ▲

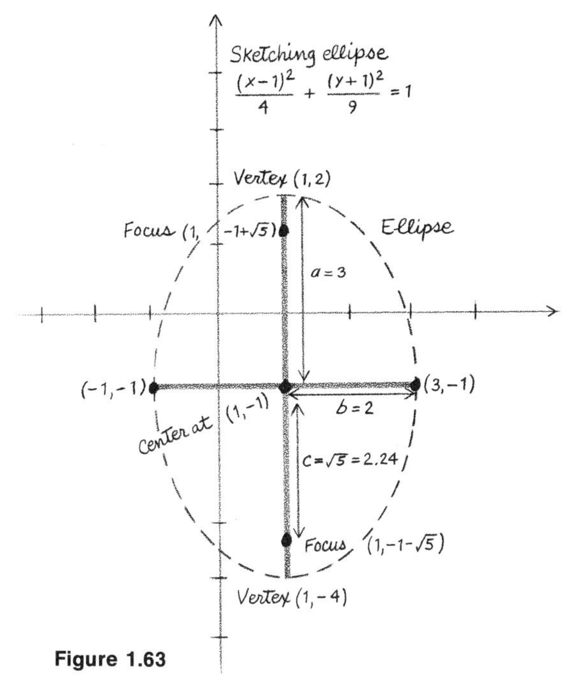

Figure 1.63

In the following example, we show how an ellipse is determined by means of secondary information about it.

EXAMPLE 3

Find the equation of the ellipse with its center at the origin, one focus at (5,0), and passing through the point (3,4). It follows that the major axis is horizontal and that the other focus is at the point (−5,0). The equation must of the form (1). As noted above, the sum of the distances from the point (3,4) to the foci is equal to $2a$; therefore,

$$2a = \sqrt{(3-5)^2 + 4^2} + \sqrt{(3+5)^2 + 4^2} = \sqrt{20} + \sqrt{80}\,.$$

Squaring both sides, we get

$$4a^2 = 20 + 2\sqrt{1600} + 80 \qquad \text{or} \qquad a^2 = 45.$$

Substitute $x = 3$, $y = 4$, and $a^2 = 45$ in Equation (1) and solve for the unknown b:

$$\frac{9}{45} + \frac{16}{b^2} = 1 \qquad \text{or} \qquad b^2 = 20.$$

It follows that the equation of the ellipse is

$$\frac{x^2}{45} + \frac{y^2}{20} = 1$$

(Figure 1.64). ▲

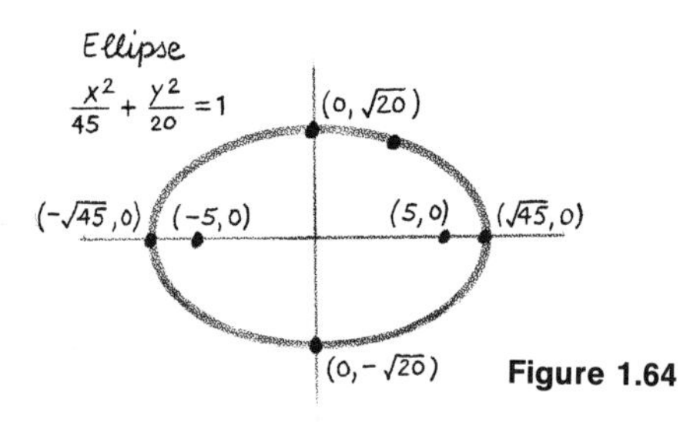

Figure 1.64

The **general form** of the equation of an ellipse with horizontal or vertical axes is

$$Ax^2 + By^2 + Dx + Ey + F = 0, \tag{6}$$

where A and B have the same sign. This is analogous to the general forms for the circle and parabola. It can be transformed into the form (3) or (4) by the method of completing the square. As in the case of the circle, such an equation does not always represent a real ellipse because there may not necessarily be real numbers that satisfy the equation.

EXAMPLE 4

Consider the ellipse with the equation

$$9x^2 + 4y^2 - 18x - 16y - 11 = 0.$$

Factor 9 and complete the square on x:

$$\begin{aligned} 9x^2 - 18x = 9(x^2 - 2x) &= 9((x-1)^2 - 1) \\ &= 9(x-1)^2 - 9. \end{aligned}$$

Complete the square on y:

$$\begin{aligned} 4y^2 - 16y = 4(y^2 - 4y) &= 4((y-2)^2 - 4) \\ &= 4(y-2)^2 - 16. \end{aligned}$$

Substitute in the original equation and simplify:

$$9(x-1)^2+4(y-2)^2=36 \qquad \text{or} \qquad \frac{(x-1)^2}{4}+\frac{(y-2)^2}{9}=1.$$

This is the ellipse with center at (1,–2), vertical major axis of length 6, and horizontal minor axis of length 4. ▲

1.5 EXERCISES

In Exercises 1 and 2 sketch the graph of the ellipse with the given equation.

1 (a) $\frac{x^2}{121}+\frac{y^2}{49}=1$

(b) $\frac{x^2}{4}+\frac{y^2}{9}=1$

(c) $\frac{x^2}{36}+\frac{(y-4)^2}{49}=1$

(d) $\frac{(x-3)^2}{25}+\frac{(y+4)^2}{16}=1$

(e) $(x-1)^2+4(y-2)^2=1$

2 (a) $\frac{(x+5)^2}{64}+\frac{(y-6)^2}{100}=1$

(b) $\frac{(x+3)^2}{3}+\frac{(y-7)^2}{4}=1$

(c) $\frac{4(x-5)^2}{9}+\frac{9(y+4)^2}{16}=1$

(d) $\frac{(x-1)^2}{169}+\frac{(y+2)^2}{144}=1$

(e) $9(x+.1)^2+4(y-2)^2=1$

In Exercises 3 and 4 derive the equation of the ellipse with the given property.

3 (a) center (0,0), focus (2,0), major axis 8
(b) center (0,2), focus (0,4), minor axis 4
(c) center (0,0), vertical major axis 8, minor axis 3
(d) center (1,–1), focus (3,–1), vertex (4,–1)
(e) foci (2,3) and (2,–1), major axis 6

4 (a) center (0,0), focus (0,–1), major axis 6
(b) center (1,1), focus (–1,1), major axis 5
(c) center (2,–1), vertical major axis 7, minor axis 1
(d) center (5,4), focus (7,4), vertex (2,4)
(e) foci (4,4) and (4,0), minor axis 2

In Exercises 5 and 6 find the equation of the ellipse passing through the given point P.

5 (a) foci (1,0) and (–1,0), $P=(1,1)$
(b) focus (1,–1), center (4,–1), $P=(8,0)$
(c) focus (0,1), center (0,0), $P=(3,-1)$

6 (a) foci (1,2) and (1,6), $P = (2,4)$
(b) focus (4,1), center (2,1), $P = (0,0)$
(c) focus (0,4), center (0,1), $P = (-2,0)$

In Exercises 7 and 8 find the center, major axis, and minor axis of the ellipse defined by the given equation.

7 (a) $9x^2 + 16y^2 - 144 = 0$
(b) $x^2 + 4y^2 - 2x - 3 = 0$
(c) $9x^2 + 4y^2 - 36x + 8y + 39 = 0$
(d) $25x^2 + 50x + 9y^2 - 72y + 169 = 0$

8 (a) $x^2 + 2y^2 - 12x + 16y + 66 = 0$
(b) $4x^2 + 7y^2 - 8x - 28y + 4 = 0$
(c) $25x^2 + 9y^2 - 50x + 36y - 164 = 0$
(d) $5x^2 + 2y^2 + 10x + 4y - 3 = 0$

9 A certain bird can fly a maximum of 20 miles in 1 hour without rest. He is released from a boat drifting in a straight line at 5 miles per hour. Suppose the boat starts $2\frac{1}{2}$ miles west of a given point on the line in the river. We take this point to be the origin. In order that the bird be able to get back to the boat at the end of 1 hour, within what ellipse should he remain? (See Figure 1.58.)

10 A young man is capable of swimming 400 feet. He plans to swim in a straight line between two points which are 300 feet apart. How far off the course is he permitted to stray and still reach his destination? [*Hint:* Find half the minor axis of the ellipse within which he must stay.]

11 In this exercise we outline the proof of the geometrical fact that the point on the ellipse nearest to one focus is the nearest vertex. Take the ellipse in the first standard position. We shall show that the distance from the focus $(c,0)$ to an arbitrary point (x,y) on the ellipse is at least equal to the distance $a - c$ to the vertex $(a,0)$ (Figure 1.65).

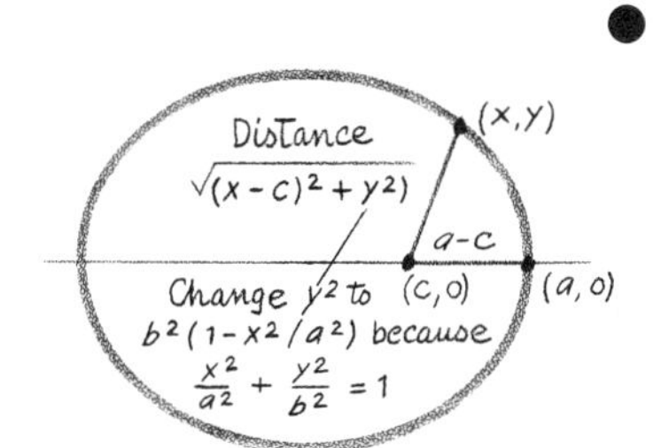

Figure 1.65

(a) Prove that the square of the distance from $(c,0)$ to (x,y) is $(x-c)^2 + b^2(1 - x^2/a^2)$. [*Hint:* Use the distance formula and then substitute for y^2 from Equation (1).]
(b) The square of the distance in (a) **minus** the square of the distance from $(c,0)$ to $(a,0)$ is equal to $(x-a)\left[\frac{c^2}{a^2}(x+a) - 2c\right]$. [*Hint:* Substitute $b^2 = a^2 - c^2$ in (a), subtract $(a-c)^2$, expand the squares, cancel terms, and factor.]
(c) Show why the product in (b) is positive if $x < a$.
(d) Why does the statement (c) complete the proof?

12 As the earth goes around the sun in its elliptic orbit, the longest distance between the bodies is 94.5×10^6 miles, and the shortest 91.5×10^6. Find the length of the major axis of the ellipse. (See Figure 1.57 and Exercise 11.)

13 An artificial satellite is placed in orbit around the earth. The apogee (highest point) of the orbit is 800 miles above the surface of the earth and the perigee (lowest point) 500 miles. Find the distance between the foci of the elliptic orbit. Use the fact that the radius of the earth is approximately 4000 miles. (See Figure 1.57 and Exercise 11.)

14 Here is a geometric method of constructing the ellipse with compass and ruler. We want to draw an ellipse with major axis $2a$ and minor axis $2b$. Draw concentric circles about the origin with radii a and b, respectively. Then draw a ray from the origin intersecting the circles at points A and B, respectively. Pass a vertical line through A and a horizontal line through B. Their point of intersection is on the ellipse. To prove this let (x,y) represent the point of intersection and justify these statements:

(a) A has the coordinates $(x, \sqrt{a^2 - x^2})$.

(b) B has the coordinates $(\sqrt{b^2 - y^2}, y)$.

(c) The ratio of ordinate to abscissa is the same for A and B, that is, $\sqrt{a^2 - x^2}/x = y/\sqrt{b^2 - y^2}$.

(d) If (x,y) satisfies the equation in (c), then it also satisfies Equation (1).

The construction is sketched in Figure 1.66.

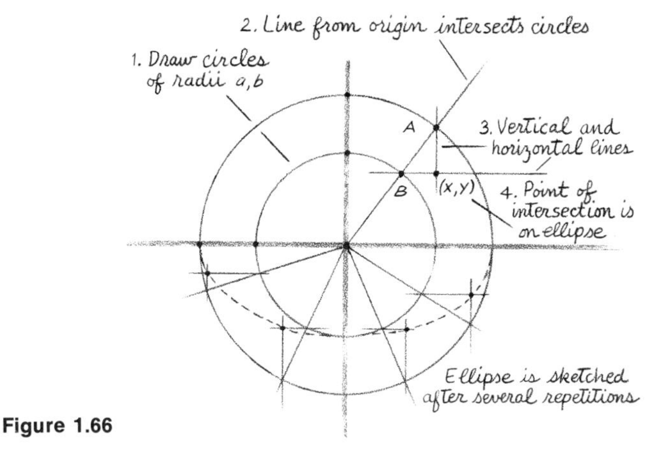

Figure 1.66

1.6 The Hyperbola

Let there be two fixed points in the plane which we shall call **foci.** If a third point moves in the plane in such a manner that the absolute differences between its distances from the two foci remains constant, then its path traces a curve called the **hyperbola.** This differs from the definition of the ellipse in that the absolute difference, not the sum, of the distances remains constant. The curve consists of two **branches,** one containing points closer to one focus, and the other containing points closer to the other focus. The **center** of the hyperbola is midway between the foci. The **vertices** of the hyperbola are the points

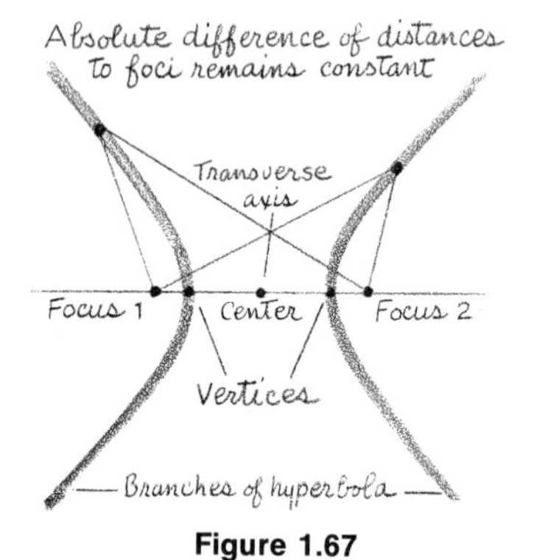

Figure 1.67

on it which intersect the line joining the foci. This line segment is called the **transverse axis** (Figure 1.67).

Here is the geometric basis for the construction of the hyperbola. Put two foci in the plane. Around these draw concentric circles of various sizes. A hyperbola consists of those points which are on the intersections of pairs of circles around the respective foci, and whose radii differ in absolute value by a constant amount (Figure 1.68). For example, let the point P_1 be on a circle of radius 7 about one focus, and on a circle of radius 4 about the other. Let P_2 be on a circle of radius 10 about one focus, and on a circle of radius 13 about the other. Then P_1 and P_2 are on the same hyperbola because the absolute differences between the radii of the pairs of circles are equal to the same number 3.

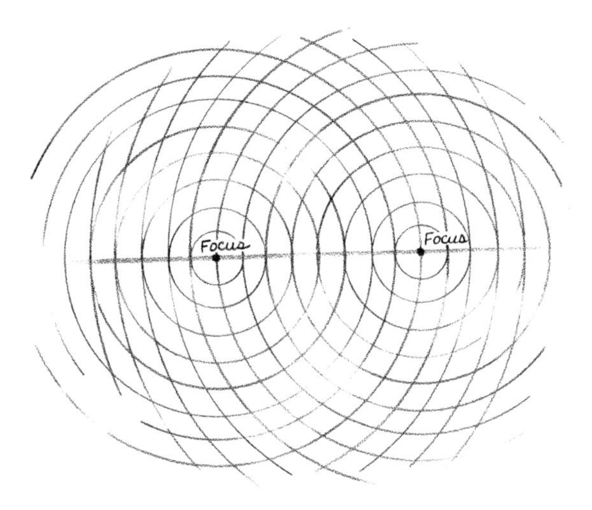

Figure 1.68

EXAMPLE 1

One of the most controversial studies of recent American history was the official investigation of the assassination of the American president on November 22, 1963 before hundreds of witnesses. The main point of the study was to determine whether the shooting was done from one site or more; for the latter implied that more than one person acted in the killing, and that a conspiracy existed. One of the surprising aspects of the testimony was that witnesses had varying recollections of the number of shots they heard, their timing, and the directions from which they came. Our purpose in discussing this event is not to present a new theory or dispute an old one. The author defers to the more informed in this matter. All that we attempt to do here is to use the mathematical theory of the hyperbola to show how single shots fired simultaneously from two different positions are heard in various orders and at different intervals by listeners in various positions. Furthermore, we shall show how the location of the origins of the shots can be determined from what was heard at different positions. One cannot infer from the theory that more than one gunner actually took part in this assassination. However, if one assumes or knows that it was so, then the theory would enable him to locate the positions of the gunmen.

As the reader probably knows from high school science, sound waves travel from a noise source like the ripples from the point a stone hits the water in a still pond. The waves travel through the air at a rate of approximately 1100 feet per second. Suppose that two guns are shot simultaneously from different positions; for example, suppose that they are 2200 feet apart. The sound waves move out from the sources at the same rate of 1100 feet per second. The waves look like the concentric circles in Figure 1.68; there the waves are separated by 275 feet, so that they represent the waves at intervals of $\frac{1}{4}$ second.

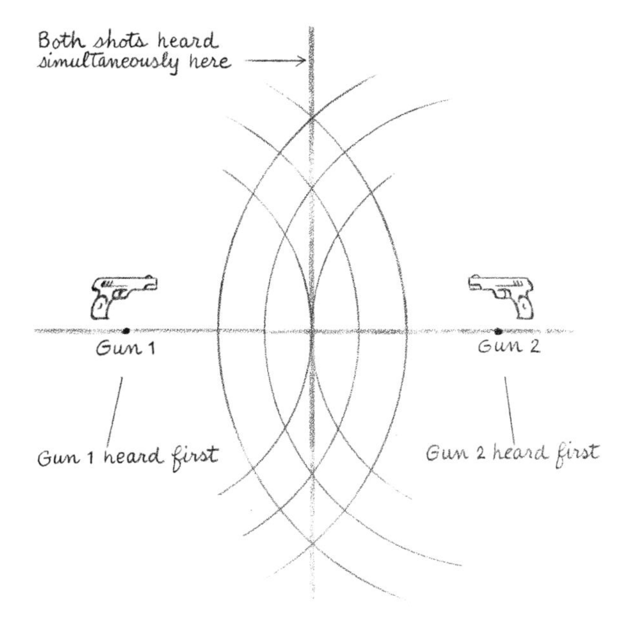

Figure 1.69

A person standing at an equal distance from the two guns will receive the two shot waves at the same time and so will not be able to distinguish them. He will hear just a single shot, but the noise will be twice as loud. The set of such points, those equidistant from the two sources, is the perpendicular bisector of the line segment joining the two. A listener not on this perpendicular bisector will first hear that gun to which he is closer, and then hear the second gun (Figure 1.69).

Now let us consider the set of all points in the plane where the shots are heard $\frac{1}{2}$ second apart. Since the sound travels at 1100 feet per second, the distances from such points to either gun must differ in absolute value by 550 feet; in other words, either the distance to gun 1 is 550 feet **more** than that to gun 2, or **less** than it. Suppose that a certain listener hears gun 1 first, $\frac{1}{2}$ second before gun 2. Such a listener must be at a point at the intersection of two circles about the respective guns, and where the radius of the circle about the closer gun is 550 feet less than the radius about the far gun. The set of such points is the branch of the hyperbola in the left-hand portion of Figure 1.70. If a listener is at a point where gun 2 is heard $\frac{1}{2}$ second before gun 1, then the point is on the opposite branch of the hyperbola. If the branches of the hyperbola are known, then the foci can be located. It follows that if witnesses to the shooting at various places could correctly report the time intervals between the shots, then the foci of the hyperbola (the location of the gunmen) could be accurately determined. With this example as our motivation, we commence our study of the hyperbola. ▲

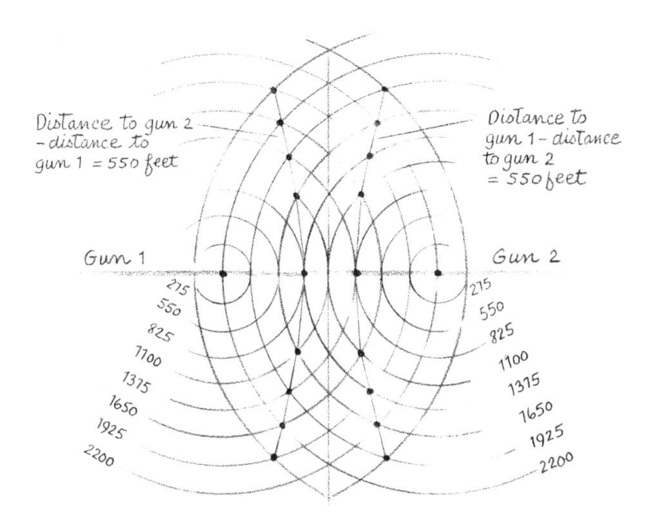

Figure 1.70

The hyperbola is in the first standard position when its center is at the origin, and the transverse axis is along the x-axis. Let the foci of the hyperbola be at the points $(c,0)$ and $(-c,0)$, and the vertices at

$(a,0)$ and $(-a,0)$; here a and c are positive. Put $b^2 = c^2 - a^2$; then a point (x,y) is on the hyperbola if and only if it satisfies the equation

$$\frac{x^2}{a^2} - \frac{y^2}{b^2} = 1. \tag{1}$$

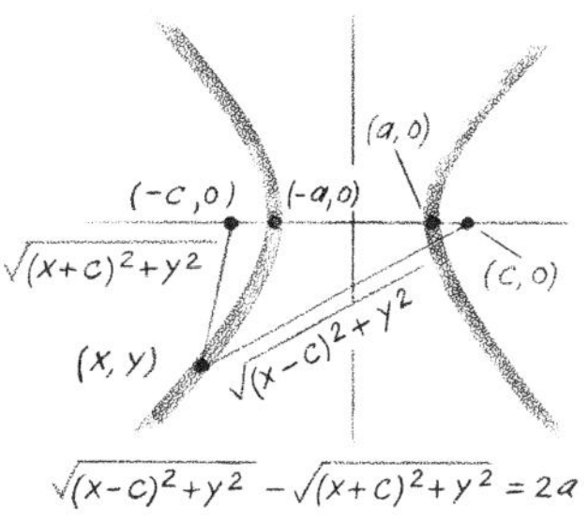

Figure 1.71

This is similar to the equation for the ellipse; only the sign between the terms is different, − instead of + (Figure 1.71).

The derivation of this equation is similar to the derivation of the equation of the ellipse. The absolute differences between the distances from a point (x,y) to the foci is the absolute value of

$$\sqrt{(x-c)^2 + y^2} - \sqrt{(x+c)^2 + y^2}.$$

We set this equal to $2a$, which is the distance between the vertices, or the **difference of the focal radii.** Then we square both sides of the resulting equation, separate the radicals that arise, square both sides again, and obtain (1). The details of the calculation are left as an exercise (Exercise 11).

The **conjugate axis** is the segment from $(0,b)$ to $(0,-b)$. The **asymptotes** are the two straight lines passing through the origin with slopes b/a and $-b/a$, respectively. Their equations are $y = (b/a)x$ and $y = -(b/a)x$, respectively. The branches of the hyperbola get closer and closer to the asymptotes, but never reach them, as the point on the curve moves away from the center (Figure 1.72).

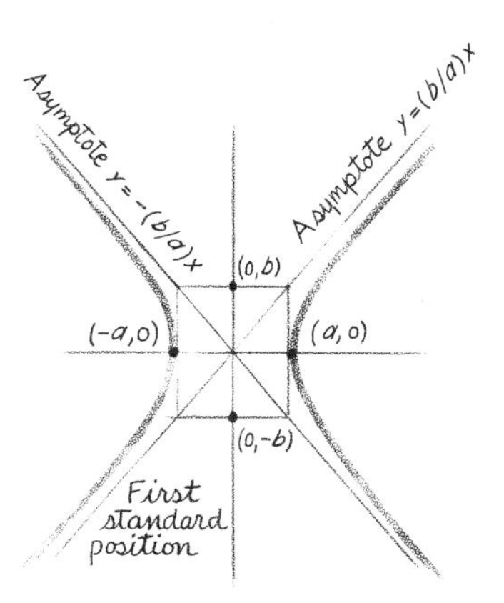

Figure 1.72

The second standard position for the hyperbola is with the center again at the origin, but with a vertical transverse axis. The foci are at $(0,c)$ and $(0,-c)$ and the vertices are at $(0,a)$ and $(0,-a)$. The asymptotes have the equations $y = (a/b)x$ and $y = -(a/b)x$, respectively (Figure 1.73). The second standard position is similar to the first except that the x- and y-axes have been interchanged. Thus the equation for the second standard position is obtained from that for the first by interchanging x and y:

$$\frac{y^2}{a^2} - \frac{x^2}{b^2} = 1. \tag{2}$$

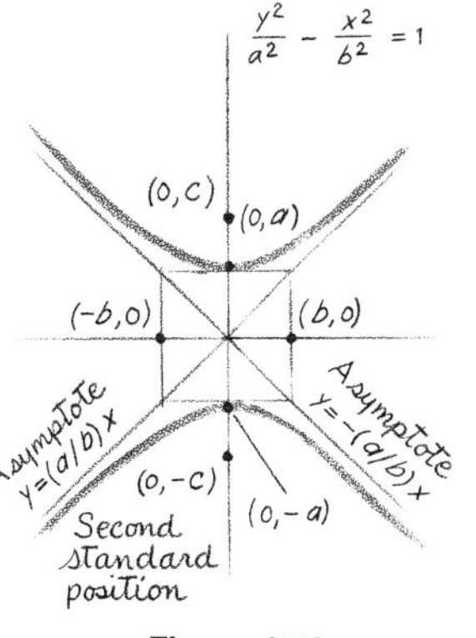

Figure 1.73

If the center of the hyperbola is at a point (h,k) which is not necessarily at the origin, and if the transverse axis is horizontal, then the equation of the hyperbola is transformed from (1) to

$$\frac{(x-h)^2}{a^2} - \frac{(y-k)^2}{b^2} = 1. \tag{3}$$

When the transverse axis is vertical, Equation (2) becomes

$$\frac{(y-k)^2}{a^2} - \frac{(x-h)^2}{b^2} = 1. \tag{4}$$

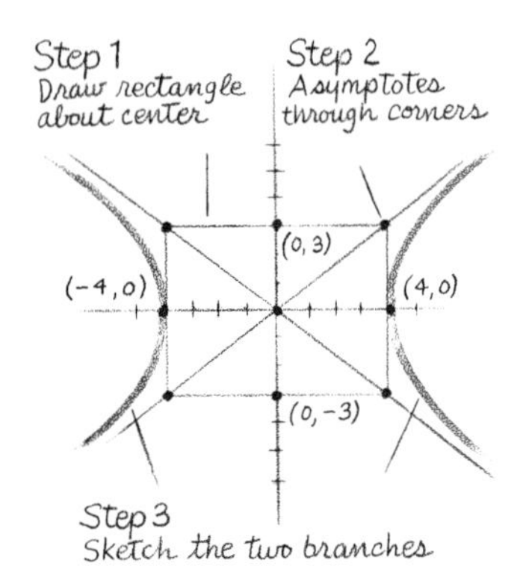

Figure 1.74

EXAMPLE 2

Sketch the graph of the hyperbola with center at the origin, foci at (5,0) and (−5,0), and vertices (4,0) and (−4,0). Here $c = 5$ and $a = 4$. The number b is obtained from the relation $b^2 = c^2 - a^2 = 9$, or $b = 3$. The transverse axis is the segment from (−4,0) to (4,0), and the conjugate axis is the segment from (0,−3) to (0,3). To sketch the hyperbola, first draw the **fundamental rectangle** about the origin, determined by the quantities $a = 4$ and $b = 3$. The asymptotes are drawn through the opposite corners of this rectangle. Finally, the two branches are filled in (Figure 1.74). ▲

EXAMPLE 3

Determine the equation of the hyperbola having foci at (1,5) and (1,−1), and conjugate axis 5. Since the center is midway between the foci, it is at the point (1,2). Since the formula for the distance between the foci is $2c$, and here in particular it is equal to 6, it follows that $2c = 6$, or $c = 3$. Since the conjugate axis is 5, the quantity b is equal to $\frac{1}{2}(5) = \frac{5}{2}$. Then a is computed:

$$a^2 = c^2 - b^2 = 9 - \frac{25}{4} = \frac{11}{4}.$$

The equation of this hyperbola is of the general form (4); hence, by substituting the values of a, b, h, and k, we find that the equation of this particular hyperbola is

$$\frac{4(y-2)^2}{11} - \frac{4(x-1)^2}{25} = 1. \blacktriangle$$

EXAMPLE 4

Here is a simple example of the determination of the positions of two guns shot simultaneously from two different positions. The guns are located at the unknown but symmetrically placed points $(c,0)$ and $(-c,0)$, where c is a positive number. At a point (−300,100) the shots are heard 5/11 second apart. What are the exact positions of the gun; in other words, what is the value of c? Here the ordinates are given in units of feet.

Since sound travels at 1100 feet per second, the distances of the point (−300,100) to the guns differ by 1100(5/11) = 500 feet. This point is on the hyperbola with guns at the foci; therefore, for this hyperbola, we have a difference of focal radii $2a = 500$, so that $a = 250$. The general equation of this hyperbola is (1) with a

given above. We can find c by first determining b. Substitute $x = -300$, $y = 100$, and $a = 250$ in Equation (1), and solve for b^2:

$$b^2 = \frac{y^2a^2}{(x^2 - a^2)} = \frac{(10)^2(250)^2}{(300^2 - 250^2)}$$
$$= \frac{2500}{11}.$$

It follows that $c^2 = a^2 + b^2 = 62{,}500 + 2500/11 = 62{,}727$, so that c is approximately 250.5 feet. ▲

This example is admittedly artificial because it is not usually known that the gunners are symmetrically placed about the origin on a given axis. A more likely situation involves two guns in completely unknown positions. If many witnesses heard the shots, then the positions can be determined from their reports of the time intervals between the sounds of the shots. Each witness is at a point on some hyperbola with foci at the guns (Figure 1.75). It can be shown (but we shall not prove it) that the reports of at most 5 witnesses usually suffice to determine the two foci common to all the hyperbolas. The reason is that the most general form of the equation of the hyperbola in any position in the plane contains 5 unknown quantities (2 center coordinates, 2 axis lengths, and an "orientation" angle); these can be obtained by solving a system of 5 equations in 5 unknowns.

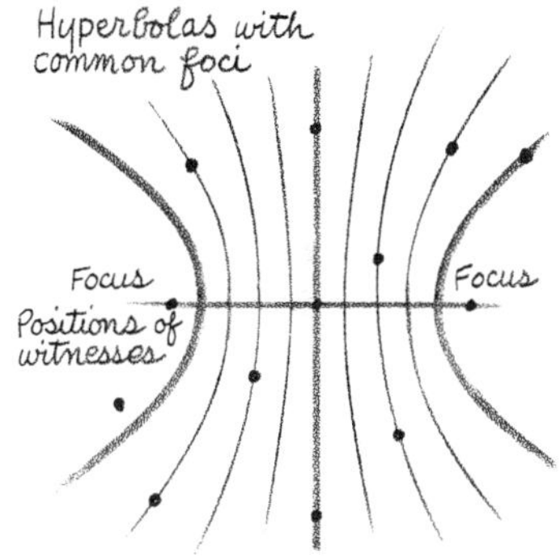

Figure 1.75

EXAMPLE 5

The hyperbola can be used to locate the position of a single gun from the reports of the exact times when it was heard in three different positions. (For example, this may be used to locate the position of an enemy battery in combat. More peaceful uses of the hyperbola will be given below.) Suppose the shot is first heard at a point P, then at a point Q s seconds later, and finally at a point R t seconds after it was heard at Q. The gun must be $1100s$ feet closer to P than Q; therefore, it lies on the branch of some hyperbola with foci at P and Q and with the difference of focal radii equal to $1100s$. It must be on the branch closer to P. By the same reasoning the gun must also be on the branch of a hyperbola with foci at Q and R and with the difference of focal radii equal to $1100t$, on the branch closer to Q. Therefore, the gun must be at some point where the branches of the two hyperbolas intersect. Sometimes there will be two such points, sometimes one (Figure 1.76). In the former case the determination of the exact position will have to be done by more detective work, by the trial-and-error method, or by means of given information.

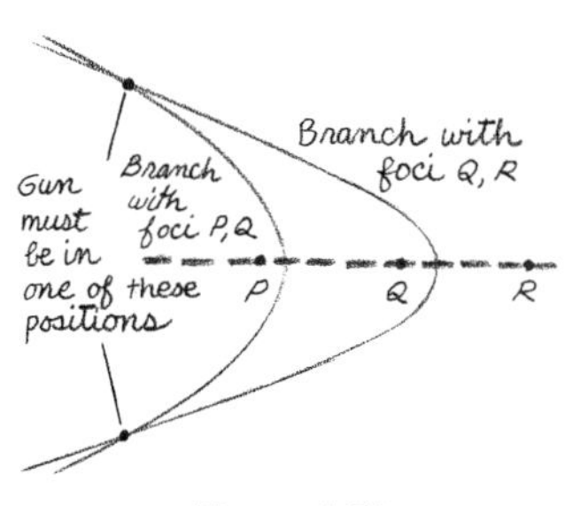

Figure 1.76

The determination of the points of intersection of the branches of the hyperbola is done by solving the pair of equations representing the pair of hyperbolas. This is harder than it sounds because the equations are both of degree 2. However, in many cases we can get an approximate solution by graphing the hyperbola or by using the asymptotes as an approximation to the actual curve. We shall show how this can be done in the next example. ▲

EXAMPLE 6

The hyperbola has many nonmilitary uses. It can be used to trace leaks of gas in the air or other pollutants of the atmosphere or water. Suppose that a certain kind of gas is known to diffuse from its source outward in a circle at a rate of 12 miles per hour when the wind is calm. On a windless night the smell of the gas is reported at various points and at various times along a road in the open countryside: first at a point P, then at a point Q 2 miles down the road 2 minutes later, then at a point R 1 mile further down the road $\frac{3}{2}$ minutes later. Find the points where the gas could be coming from.

Let us first find the lengths of the transverse and conjugate axes of the hyperbola having P and Q as foci. The distance between the foci is 2; hence $2c = 2$, so that $c = 1$. The gas travels $\frac{12}{60}$, or $\frac{1}{5}$ mile in 1 minute; therefore, the point P is $(\frac{1}{5})2 = \frac{2}{5}$ mile closer to the source than Q; hence, the difference of focal radii is $2a = \frac{2}{5}$, so that $a = \frac{1}{5}$. The length of the conjugate axis is

$$b = \sqrt{c^2 - a^2} = \sqrt{1^2 - \left(\frac{1}{25}\right)^2} = \frac{\sqrt{24}}{5} = \left(\frac{2}{5}\right)\sqrt{6}.$$

Place the center of this hyperbola at the origin; then its equation is

$$25x^2 - \frac{25y^2}{24} = 1. \tag{5}$$

Its graph appears in Figure 1.77. The quantity b is approximately 0.980; this can be calculated with the assistance of the square root table, Table I of the Appendix.

The same is done for the hyperbola having Q and R as foci. The distance between the foci is 1; hence $2c = 1$, so $c = \frac{1}{2}$. The difference of focal radii is $2a = (\frac{1}{5})(\frac{3}{2}) = \frac{3}{10}$, so that $a = \frac{3}{20}$. The length of the conjugate axis is

$$b = \sqrt{c^2 - a^2} = \tfrac{1}{2}\sqrt{0.91}.$$

The center of the hyperbola is now at $(\frac{3}{2},0)$, and the equation is

$$\frac{400}{9}\left(x-\frac{3}{2}\right)^2-\frac{4y^2}{0.91}=1. \tag{6}$$

The graph also appears in Figure 1.77. The quantity b is approximately 0.477.

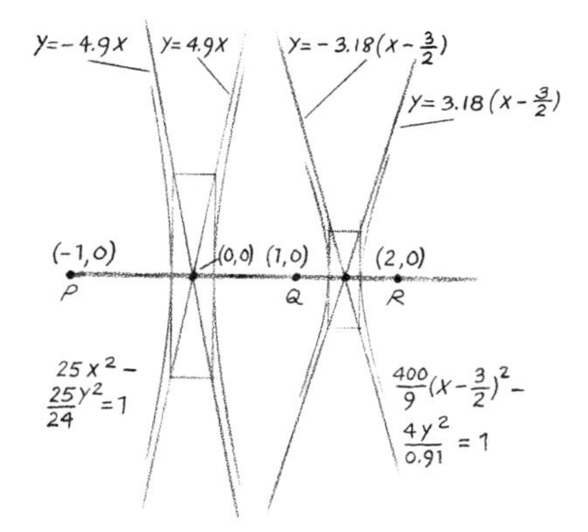

Figure 1.77

The asymptotes of the hyperbola (5) are the lines

$$y=-4.9x \qquad \text{and} \qquad y=4.9x.$$

The asymptotes of (6) are

$$y=-3.18(x-\tfrac{3}{2}) \qquad \text{and} \qquad y=3.18(x-\tfrac{3}{2}).$$

The points where the branches of the two hyperbolas intersect are obtained by solving (5) and (6). Since this kind of computation is, in general, difficult, we use an approximate method. Since the asymptotes of the hyperbola are very close to the curve itself when the part of the curve is far from the center, we may, as an approximation, substitute the equation of the asymptote for that of the curve. These are very simple linear equations so that it is very easy to find the solution. When we substitute the asymptotes for the left branches of the hyperbolas, we obtain the pair of linear equations

$$y=-4.9x \quad \text{and} \quad y=-3.18(x-\tfrac{3}{2}),$$

whose solution is $x=-2.77$ and $y=13.59$. When we solve the pair of linear equations with the positive slopes, the solution is, by symmetry, $x=-2.77$ and $y=-13.59$. It follows that the source of the gas is approximately 14 miles above or below the road and approximately 3 miles west of P (according to the way the graph has been drawn). Only further (nonmathematical) investigation can establish which of the two points it is. ▲

In this last example, we show how the equation of the hyperbola can be obtained from the data furnished by the coordinates of points which lie on it.

EXAMPLE 7

Find the equation of the hyperbola with center at the origin, transverse axis on the x-axis, and passing through (6,4) and (−3,1). The general equation is (1); we have to find a^2 and b^2. Since we are given two points that satisfy this equation, we substitute their coordinates for x and y, and solve the resulting two equations simultaneously for a^2 and b^2:

$$\frac{36}{a^2} - \frac{16}{b^2} = 1,$$

$$\frac{9}{a^2} - \frac{1}{b^2} = 1.$$

We change this to a simple linear system by putting $c = 1/a^2$ and $d = 1/b^2$:

$$36c - 16d = 1,$$

$$9c - d = 1.$$

The solution is $c = 5/36$ and $d = \frac{1}{4}$, so that $a^2 = 36/5$ and $b^2 = 4$. The equation of the hyperbola is now found to be

$$\frac{5x^2}{36} - \frac{y^2}{4} = 1. \blacktriangle$$

1.6 EXERCISES

In Exercises 1 and 2 sketch the graph of the hyperbola with the given equation.

1 (a) $x^2 - y^2 = 16$
(b) $x^2/4 - (y-1)^2/9 = 1$
(c) $4(y-2)^2 - (x-1)^2 = 1$
(d) $(y+2)^2/100 - (x+6)^2/64 = 1$
(e) $4(x-5)^2/9 - 9(y+4)^2/16 = 1$

2 (a) $x^2/16 - y^2/9 = 1$
(b) $(x+4)^2/16 - (y-3)^2/25 = 1$
(c) $(y-5)^2/144 - (x-1)^2/169 = 1$
(d) $(y+2)^2/16 - (x-1)^2/9 = 1$
(e) $(x+5)^2/4 - (y-3)^2/25 = 1$

In Exercises 3 and 4 derive the equation of the hyperbola with the given property.

3 (a) center (0,0), focus (3,0), transverse axis 4
(b) center (0,0), horizontal transverse axis 2, conjugate axis 1

(c) vertices $(0,4)$ and $(0,-2)$, focus $(0,-4)$
(d) center $(4,5)$, focus $(6,5)$, transverse axis 3
(e) foci $(-2,-2)$ and $(-2,4)$, transverse axis 4
(f) vertical transverse axis 4, center $(1,-4)$, passing through $(5,1)$
(g) center $(0,0)$, horizontal transverse axis, passing through $(5,5)$ and $(0,-1)$
(h) asymptotes $y = \pm\sqrt{3}\,x$, passing through $(4,6)$

4 (a) center $(0,0)$, focus $(0,-2)$, transverse axis 3
(b) vertex $(1,0)$, horizontal transverse axis, asymptotes $y = \pm\frac{2}{3}x$
(c) center $(1,1)$, vertical transverse axis 4, conjugate axis 6
(d) foci $(2,-1)$ and $(2,3)$, passing through $(1,-1)$
(e) horizontal transverse axis 1, center $(1,2)$, passing through $(0,0)$
(f) center $(0,0)$, horizontal transverse axis, passing through $(-3,\frac{1}{2})$ and $(4,-1)$
(g) center $(-4,1)$, vertex $(2,1)$, conjugate axis 4
(h) center $(0,0)$, vertex $(3,0)$, one asymptote $2x + 3y = 0$

5 Two guns are fired simultaneously from the points $(c,0)$ and $(-c,0)$, where c is measured in feet. At a point $(50,40)$, the shots are heard 4/55 second apart. How far apart are the guns? (See Example 4.)

6 What if, in Exercise 5, the guns are heard 3/55 second apart at the point $(-50,30)$?

7 Three points P, Q, and R are located along a straight seafront, to the west of the water. Q is 2200 feet north of P, and R is 2200 feet north of Q. The foghorn of a boat is first heard at the point P, then at the point Q 1 second later, and finally at R $\frac{3}{2}$ seconds after that. Find the approximate position of the boat. (See Example 6.) [*Hint:* Take units of 1100 ft. on each axis.]

8 A certain gas is known to diffuse at a rate of 100 feet per minute in a calm atmosphere. Three points P, Q, and R are separated by distances of 200 feet along a straight line. The gas is detected first at P, then at Q $\sqrt{2}$ seconds later, and finally at R $\sqrt{3}$ seconds after that. Find the two points from which the gas might be coming. Let P be the origin, with P, Q, and R along the x-axis. (See Example 6.)

9 An industrial waste is pumped into a large lake from one factory situated in a row of factories along the waterfront. A conservationist wants to determine which of the factories is doing the pumping. On the opposite side of the lake, which is 2 miles wide, there is a long straight shore, parallel to the row of factories. All the factories begin production at the same time every morning. The waste is known to diffuse through the water at the rate of 1 mile per hour. The waste is first detected at a point P on the opposite shore, and then at a second point Q 2 miles downshore 15 minutes later. What is the exact location of the factory which is the source of the waste? [*Hint:* Find the point where a hyperbola intersects a straight line. Let the opposite shore be the x-axis, and the factory shoreline $y = 2$.]

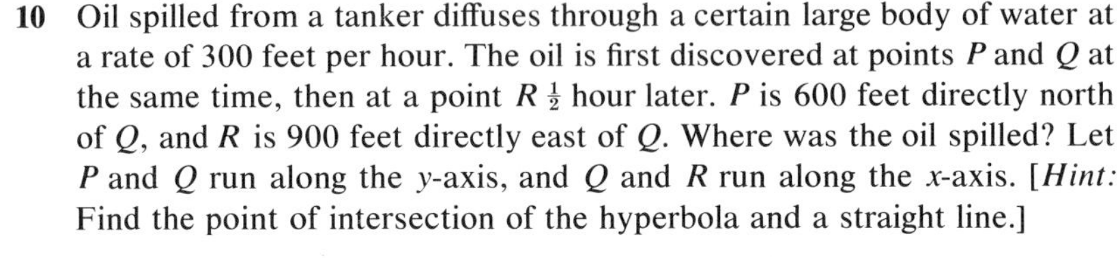

10 Oil spilled from a tanker diffuses through a certain large body of water at a rate of 300 feet per hour. The oil is first discovered at points P and Q at the same time, then at a point R $\frac{1}{2}$ hour later. P is 600 feet directly north of Q, and R is 900 feet directly east of Q. Where was the oil spilled? Let P and Q run along the y-axis, and Q and R run along the x-axis. [*Hint:* Find the point of intersection of the hyperbola and a straight line.]

11 Complete the details of the verification of Equation (1).

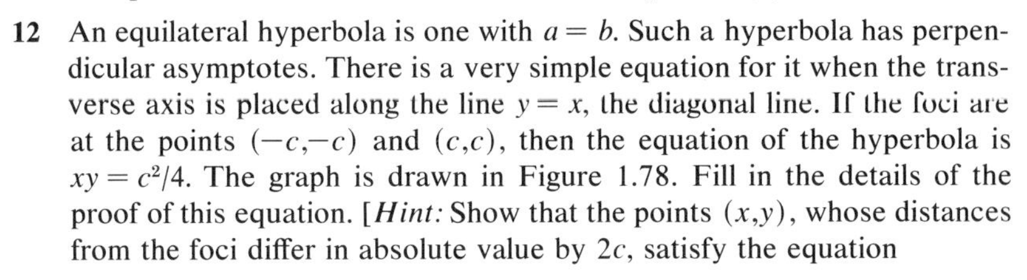

12 An equilateral hyperbola is one with $a = b$. Such a hyperbola has perpendicular asymptotes. There is a very simple equation for it when the transverse axis is placed along the line $y = x$, the diagonal line. If the foci are at the points $(-c,-c)$ and (c,c), then the equation of the hyperbola is $xy = c^2/4$. The graph is drawn in Figure 1.78. Fill in the details of the proof of this equation. [*Hint:* Show that the points (x,y), whose distances from the foci differ in absolute value by $2c$, satisfy the equation

$$\sqrt{(x+c)^2+(y+c)^2} = 2c + \sqrt{(x-c)^2+(y-c)^2}.$$

Then proceed as in Exercise 11.]

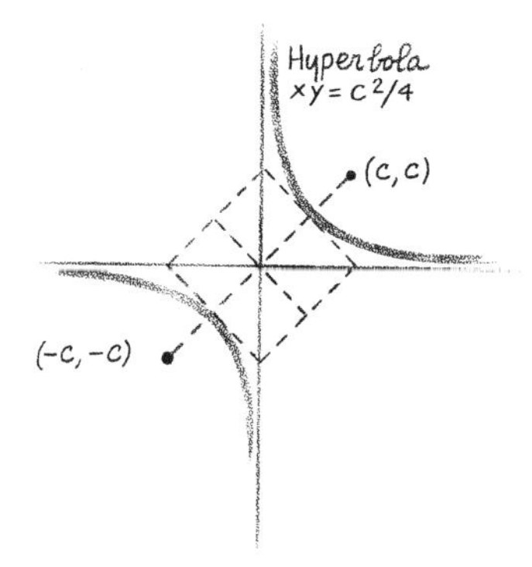

Figure 1.78

1.7 Tangent Lines and Their Slopes

In Section 1.3 we discussed the line tangent to the circle. The slope of the tangent was found to be the negative reciprocal of the radius line drawn to the point of tangency. In this section we give a more general method of defining the tangent line and its slope. It is equivalent to the previous definition in the case of the tangent to the circle, but it is more general because it can be used to define the tangent for many other curves. It is the fundamental definition in the theory of differential calculus, which is to be studied later.

Let us construct the tangent to the circle by our new method. The tangent is to pass through the point P on the circle. Take another point on the circle near P and draw a line through the two of them. This is called a **secant** line. (A secant line is a cutting line. The Latin root word **secare** means to cut.) Then take another point closer to P and draw another secant line through these. Continue in this way with a sequence of points getting closer and closer to P. It is geometrically apparent, as in Figure 1.79, that the corresponding sequence of secant lines approaches a limiting position which coincides with the one we had earlier defined to be the tangent to the circle at the point P.

Using this construction, let us derive the equation of the tangent. (This can also be found directly as the equation of the line perpendicular to the radius; however, the latter is applicable only in the case of the circle, not other curves.) For simplicity, consider a circle centered at the origin and of radius r; its equation is

$$x^2 + y^2 = r^2. \tag{1}$$

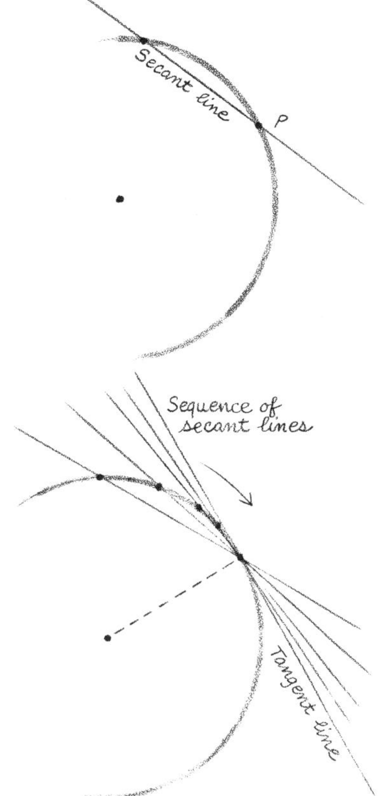

Figure 1.79

We shall find the equation of the tangent line through $P(x,y)$, where y is not equal to 0. Let $P^*(x^*,y^*)$ be another point on the circle. The slope of the secant line is, by definition,

$$\frac{y^* - y}{x^* - x}. \tag{2}$$

We estimate the value of this slope as P^* gets closer and closer to P. At first there appears to be no way to do this because both the numerator and the denominator of (2) become smaller and smaller. We solve this by using the special relation (1) between the coordinates of all points on the circle. Write the numerator in (2) as

$$\frac{y^{*2} - y^2}{y^* + y}.$$

Both (x,y) and (x^*,y^*) satisfy (1):

$$x^2 + y^2 = r^2, \qquad x^{*2} + y^{*2} = r^2;$$

hence,

$$x^2 - x^{*2} = y^{*2} - y^2.$$

Therefore,

$$y^* - y = \frac{x^{*2} - x^2}{y^* + y} = \frac{(x - x^*)(x^* + x)}{y^* + y}$$

Divide both sides of this equation by $x^* - x$ to obtain

$$\frac{y^* - y}{x^* - x} = -\frac{x^* + x}{y^* + y}.$$

As x^* and y^* move closer to x and y, respectively, the numerator on the right approaches $2x$, and the denominator $2y$; hence, the quotient above on the right approaches $-x/y$. This is the **slope** of the tangent line (Figure 1.80). The equation of the line itself can then be obtained from the point-slope equation of the line (Section 1.2). The slope is undefined when $y = 0$ because the tangent line is then vertical. Note that this is consistent with the fact that the slope of the tangent line is the negative reciprocal of the slope of the radius line; the slope of the line through the origin and the point (x,y) on the circle is y/x, and the negative reciprocal is $-x/y$.

If the circle has its center at (h,k), then its equation is $(x - h)^2 + (y - k)^2 = r^2$, and the formula for the slope of the tangent line at (x,y) is

$$\frac{-(x - h)}{y - k}. \tag{3}$$

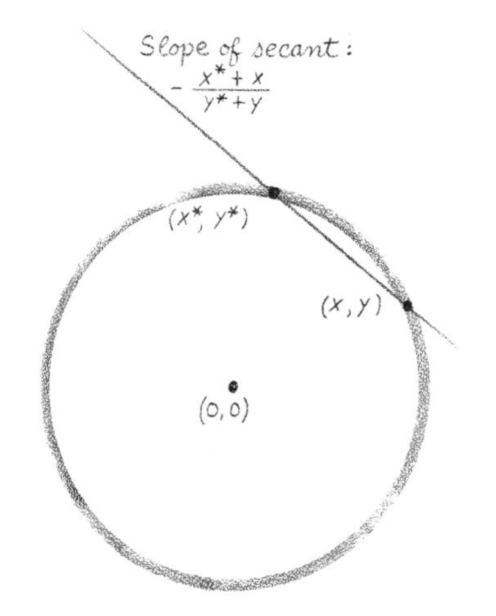

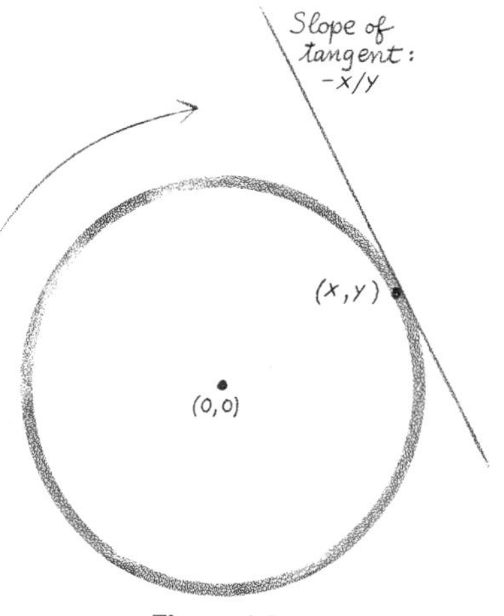

Figure 1.80

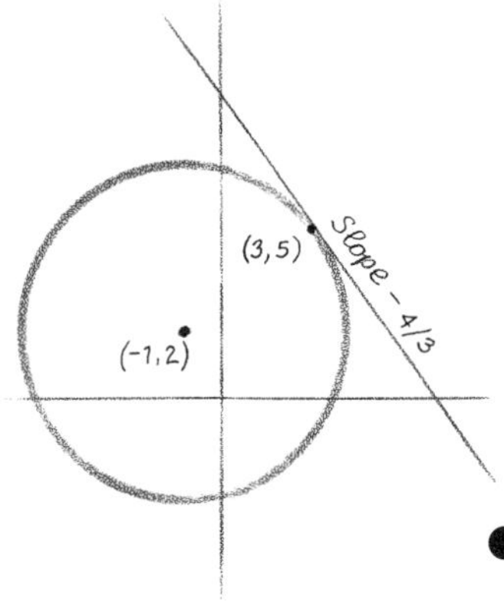

Figure 1.81

EXAMPLE 1

A circle is centered at $(-1,2)$ and has the radius 5. From formula (3) we find that the slope of the tangent line drawn to the circle at the point $(3,5)$ is

$$\frac{-(3+1)}{5-2}=\frac{-4}{3}.$$

By the point-slope formula, the equation of the tangent line is $y-5=-(4/3)(x-3)$ (Figure 1.81). ▲

We extend the definition of the tangent line to all the curves we have studied: circle, ellipse, parabola, and hyperbola. Let P be a given point on the curve; let us construct the tangent line through this point. Take another point on the curve near P and draw a secant line through P and this point. Then repeat this for a sequence of points on the curve getting closer and closer to P. The secants apparently approach a "limiting" position. This is the tangent line for the curve. The example of the parabola is typical; it is illustrated in Figure 1.82. The same limiting line is reached on the approach from either side of P.

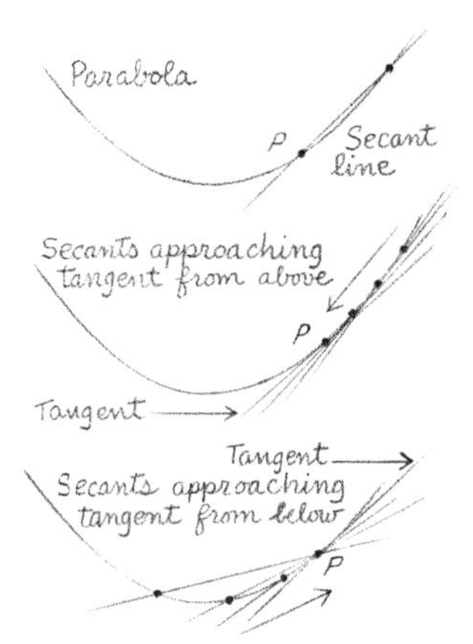

Figure 1.82

Now we derive the slopes of the lines tangent to the various curves.

Consider the parabola in the first standard position. Its equation is of the form $4ay=x^2$. Let $P(x,y)$ and $P^*(x^*,y^*)$ be two points on the parabola; then these satisfy the equations

$$4ay=x^2,$$
$$4ay^*=x^{*2}.$$

Subtract corresponding sides of the equations and factor the difference of squares:

$$4a(y-y^*)=(x+x^*)(x-x^*),$$

or

$$\frac{y-y^*}{x-x^*}=\frac{x+x^*}{4a}.$$

The left-hand side is the slope of the secant line through P and P^*. The right-hand side approaches $2x/4a$, or $x/2a$, as x^* approaches x. It follows that the slope of the line tangent to the parabola at the point (x,y) is

$$\frac{x}{2a}. \tag{4}$$

When the vertex of the parabola is at the point (h,k), the formula for the slope becomes

$$\frac{x-h}{2a}. \tag{5}$$

When the parabola is in the second standard position (with equation $-4ay = x^2$), formula (4) for the slope is modified by a sign change; and when the vertex is at (h,k), formula (5) is similarly modified. In the third and fourth standard positions (with equations $y^2 = 4ax$ and $y^2 = -4ax$, respectively) the roles of x and y are reversed, and formula (4) is changed to $y/2a$ and $-y/2a$, respectively.

EXAMPLE 2

Consider the parabola with the equation $(x+2)^2 = 4(y-4)$. It passes through the point $(1,6\frac{1}{4})$. The parabola is in the first standard position except that the vertex is at $(-2,4)$. We apply formula (5) with $x = 1$, $h = -2$, and $a = 1$; the slope of the tangent line at this point is $\frac{3}{2}$. This tangent line and several others are shown in Figure 1.83. ▲

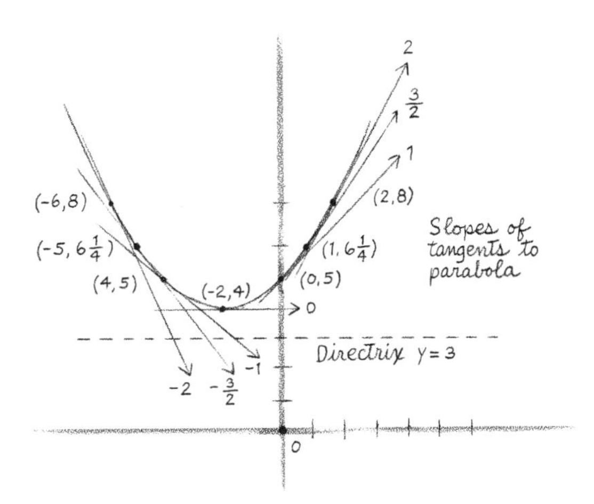

Figure 1.83

The derivation of the formula for the slope of the line tangent to the ellipse is similar to that for the circle. Let $P(x,y)$ and $P^*(x^*,y^*)$ be two points on an ellipse in the first standard position; then

$$\frac{x^2}{a^2} + \frac{y^2}{b^2} = 1,$$

$$\frac{x^{*2}}{a^2} + \frac{y^{*2}}{b^2} = 1.$$

Subtract the second equation from the first, factor the differences of squares, and divide each side of the resulting equation by $x - x^*$:

$$\frac{x + x^*}{a^2} + \frac{(y - y^*)(y + y^*)}{(x - x^*)b^2} = 0.$$

Solve for the slope of the secant line:

$$\frac{y - y^*}{x - x^*} = -\frac{b^2(x + x^*)}{a^2(y + y^*)}.$$

As P^* approaches P, the ratio on the right-hand side tends to

$$\frac{-b^2x}{a^2y}, \tag{6}$$

which is the slope of the tangent line.

When the ellipse is in the second standard position, the roles of a and b are reversed, and the slope is $-a^2x/b^2y$. When the vertex is at (h,k), the factors x and y are replaced by $x - h$ and $y - k$, respectively.

EXAMPLE 3

Find the slope of the tangent line to the ellipse

$$\frac{(x + 1)^2}{16} + \frac{(y - 2)^2}{9} = 1$$

at the point $(1, 2 + \frac{3}{2}\sqrt{3})$. This ellipse has a horizontal major axis, and its center is at $(-1,2)$. Here $a^2 = 16$, $b^2 = 9$, $h = -1$, and $k = 2$; therefore, the slope of the tangent line at the given point is

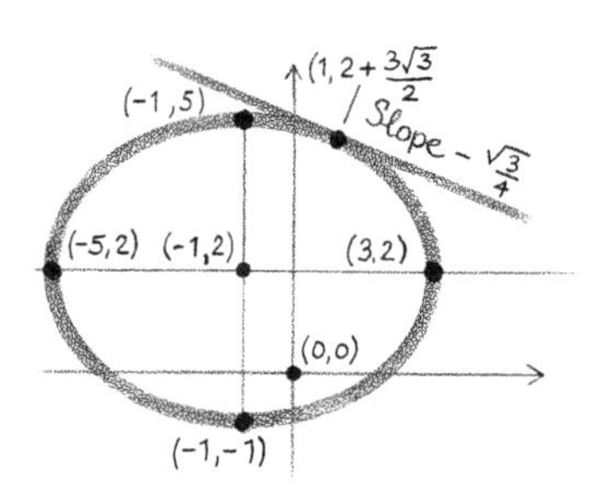

Figure 1.84

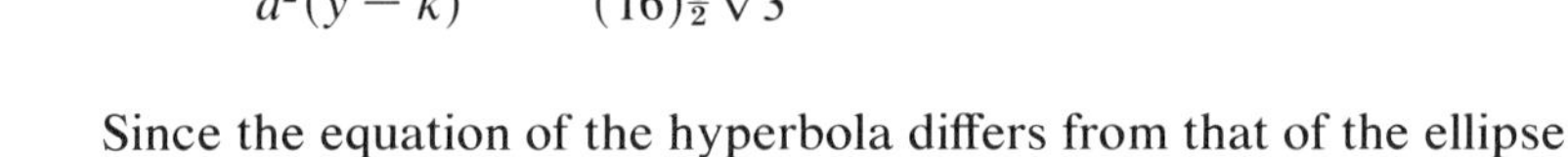

$$-\frac{b^2(x - h)}{a^2(y - k)} = -\frac{(9)(2)}{(16)\frac{3}{2}\sqrt{3}} = -\tfrac{1}{4}\sqrt{3} \qquad \text{(Figure 1.84).} \ \blacktriangle$$

Since the equation of the hyperbola differs from that of the ellipse only in the sign separating the two terms on the left side of the equation (for example, $x^2/a^2 - y^2/b^2 = 1$) the slope of the line tangent to the hyperbola is obtained from the formula for the slope of the line tangent to the ellipse by simply changing the sign; for example, the slope of the line tangent to the hyperbola in the first standard position at the point (x,y) is b^2x/a^2y.

EXAMPLE 4

Find the slope of the line tangent to the hyperbola $x^2/25 - y^2/36 = 1$ at each of the four points $(\pm 35/6, \pm\sqrt{13})$. The formula for the slope at the point (x,y) is b^2x/a^2y because the hyperbola is in the first standard position; here $a^2 = 25$, $b^2 = 36$, $x = \pm 35/6$, and $y = \pm\sqrt{13}$. The slopes are $\pm 42\sqrt{13}/65$ (Figure 1.85). ▲

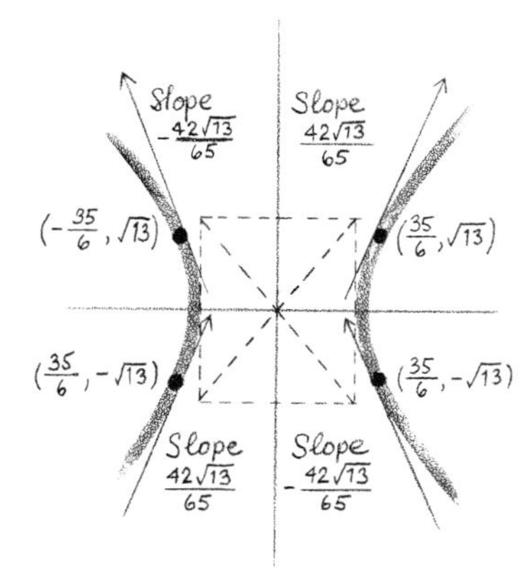

Figure 1.85

Here are several more examples concerning tangent lines.

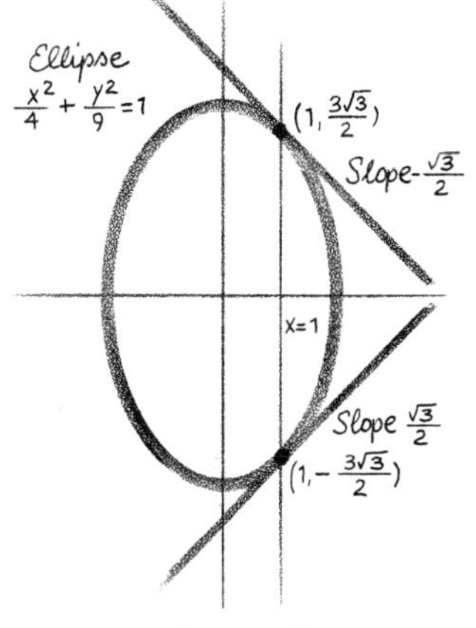

Figure 1.86

EXAMPLE 5

Find the equation of each tangent to the ellipse $x^2/4 + y^2/9 = 1$ at the points where it is cut by the line $x = 1$. First we determine the ordinates of these points. Put $x = 1$ in the equation of the ellipse and solve for y:

$$\frac{1}{4} + \frac{y^2}{9} = 1 \qquad \text{or} \qquad y = \pm\tfrac{3}{2}\sqrt{3}.$$

The two points are $(1, \pm\frac{3}{2}\sqrt{3})$. According to our formula the slopes of the tangent lines are $\pm\frac{1}{2}\sqrt{3}$. By the point-slope formula, the equations of the lines are

$$y = \pm\tfrac{3}{2}\sqrt{3} - \tfrac{1}{2}\sqrt{3}(x - 1)$$

(Figure 1.86). ▲

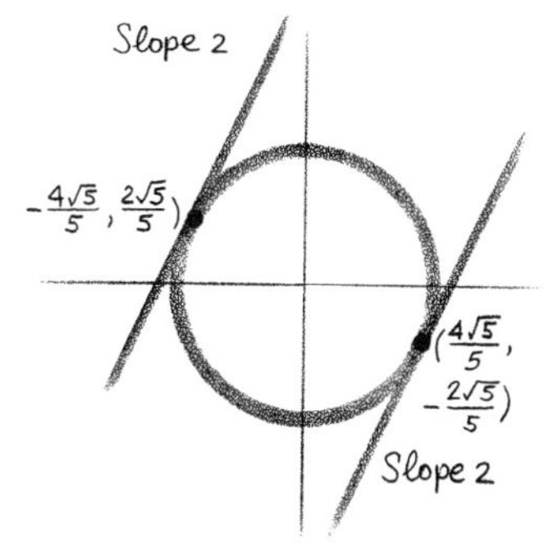

Figure 1.87

EXAMPLE 6

At what points on the circle $x^2 + y^2 = 4$ do the tangents have slope 2? If (x,y) is such a point, then by the formula for the slope of the line tangent to the circle, x and y must satisfy $-x/y = 2$, or $x = -2y$. Since the point also satisfies the equation of the circle, we substitute $x = -2y$ in this equation and solve for y: $(-2y)^2 + y^2 = 4$, or $y = \pm\frac{2}{5}\sqrt{5}$. Substitute in the equation $x = -2y$ to get the abscissas of the two points: $(-\frac{4}{5}\sqrt{5}, \frac{2}{5}\sqrt{5})$ and $(\frac{4}{5}\sqrt{5}, -\frac{2}{5}\sqrt{5})$ are the points (Figure 1.87). ▲

EXAMPLE 7

Find the point on the parabola $4y = x^2$ at which the tangent line has the slope -1. It was shown above that the slope of the line tangent to the parabola at a point (x,y) is $x/2a$. Here $a = 1$, so that the abscissa of the point where the slope is -1 satisfies $(x/2) = -1$, or $x = -2$. Then $y = 1$ at this point, so that the point of tangency is $(-2,1)$ (Figure 1.88). ▲

Figure 1.88

We conclude this section by describing the role of the tangent line in certain reflection properties of curves. We observe in nature that light travels from one point to another along the path that takes it the least time; for example, it travels through a uniform medium (air, water) along a straight line. (Of course, the path may change if it goes from one medium to another; this will be analyzed later in the study of calculus.) One consequence of this is the Law of Reflection: When a ray of light hits a reflecting surface such as a mirror at an angle, the angle of **incidence,** then it is reflected from the surface at the same angle, the angle of **reflection** (Figure 1.89). The reason is sketched graphically in Figure 1.90: The path from the source to the target is equal in length to the path from the source to the image of the target in the mirror. Since the latter path is minimized by a straight line path, the former is minimized by the reverse image of that path. It can be shown by elementary geometry that the angle of incidence is then equal to the angle of reflection (Exercise 13).

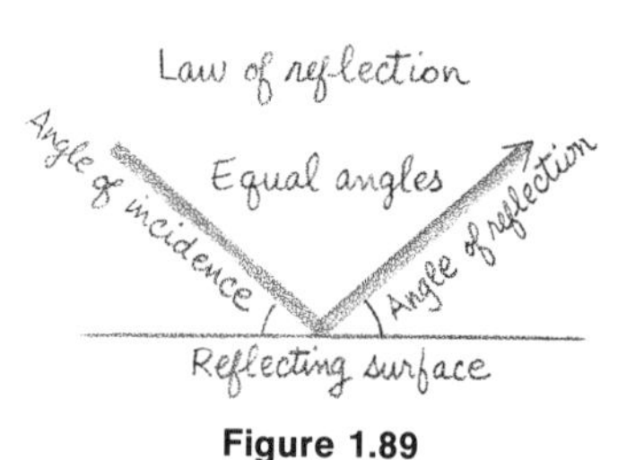

Figure 1.89

The mirror is an example of a **straight** surface. How is light reflected from a curved surface? Near a given point on a curve, the line tangent to the curve is nearly the same as the curve itself. Therefore, light is reflected from a point on the curve as if it were reflected from the tangent line at that point; the angle of incidence is equal to the angle of reflection (Figure 1.91).

If a mirror is shaped like the sector of a circle and a light source is placed at the center, then all the rays are reflected back to the source (Figure 1.92).

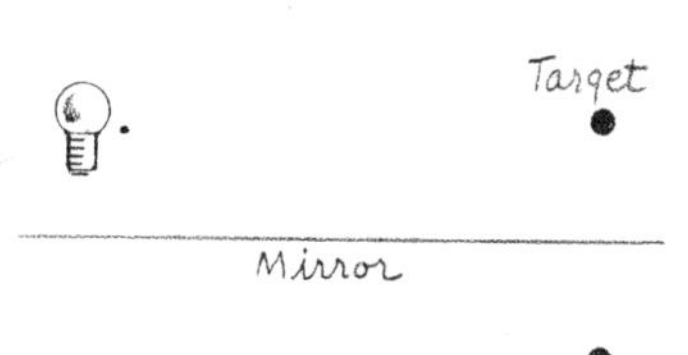

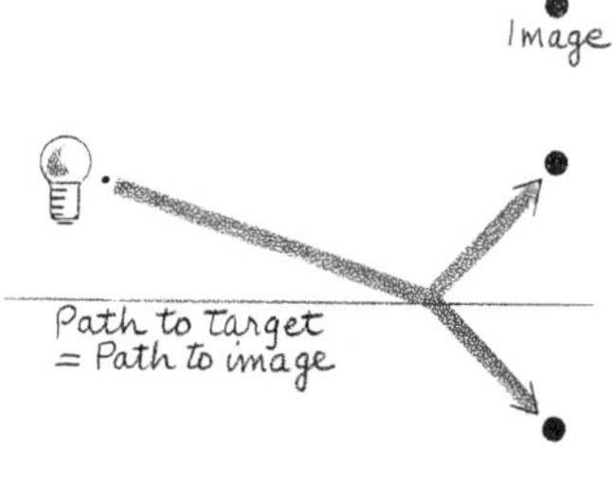

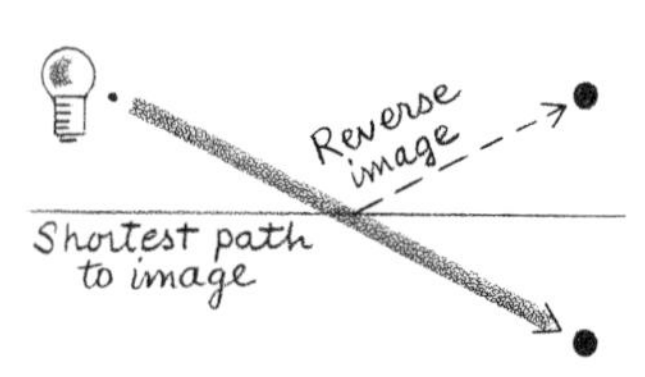

Figure 1.90

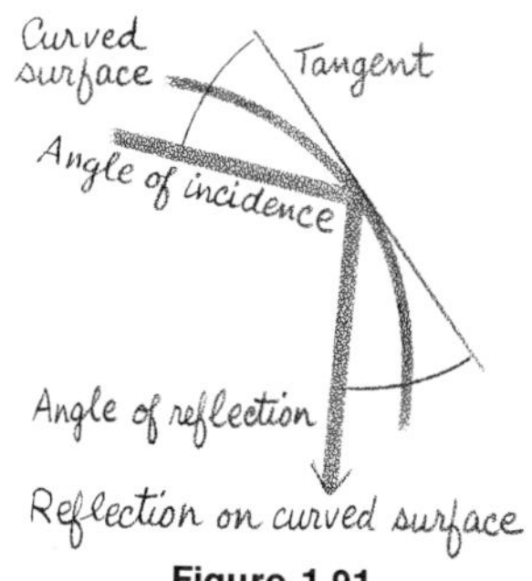

Figure 1.91

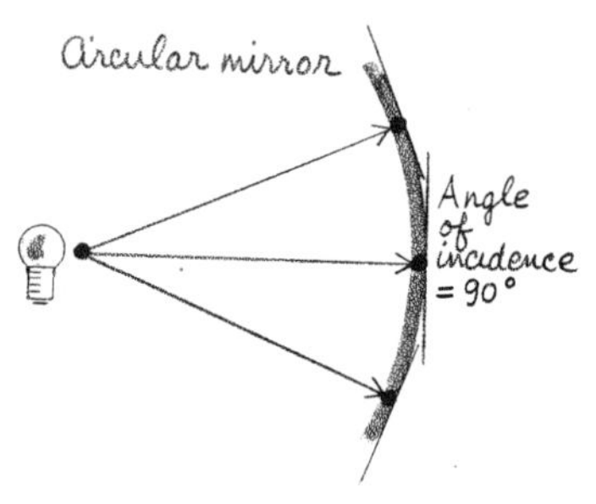

Figure 1.92

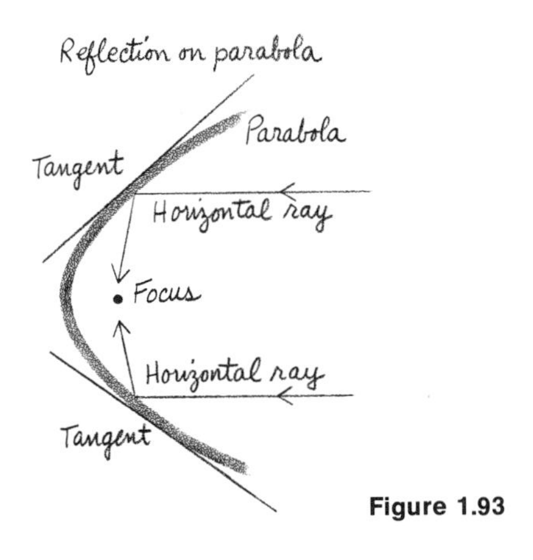

Figure 1.93

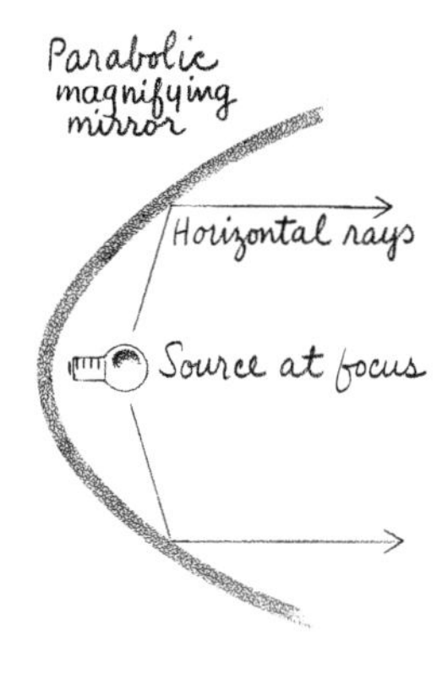

Figure 1.94

A parabolic mirror reflects all horizontal rays into the focus; this is the principle used in the design of large reflector telescopes (Figure 1.93). On the other hand, when the source is placed at the focus, then all rays are reflected from the inside surface of the parabola along parallel lines. This is used to magnify the light from a source, as in a car headlight (Figure 1.94).

An elliptic mirror reflects all rays coming from one focus into the other focus (Figure 1.95). A "whispering gallery" is a room whose walls and ceiling are shaped like a half-ellipse. When a person stands at one focus and whispers, his voice is reflected off the walls and sent to the other focus. A person standing there will hear a magnified version of the whisper.

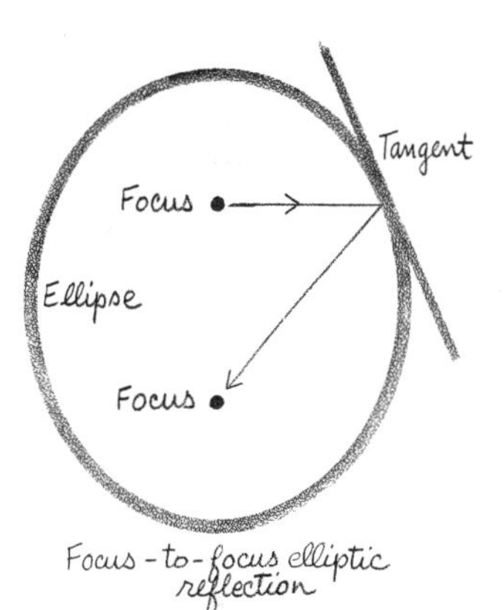

Figure 1.95

The reflection property of the hyperbola is different from the others. If a straight mirror is placed at any point on the curve perpendicular to the tangent line, then a ray from one focus is reflected into the other (Figure 1.96).

We shall give the proofs of some of these facts about reflecting surfaces as advanced exercises. The reader can see that the mathematical assertions above are consistent with what appears in the accompanying diagrams.

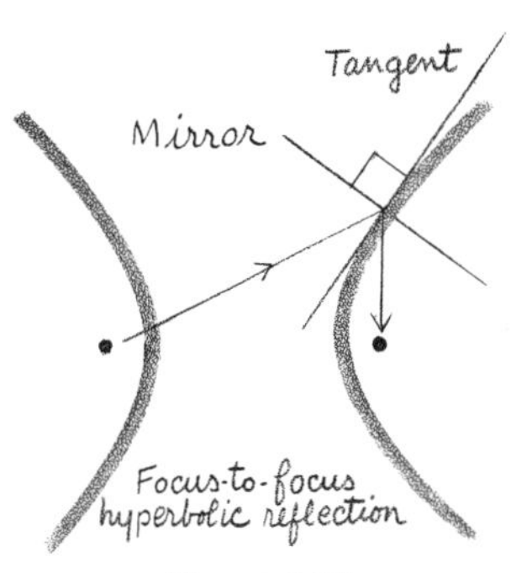

Figure 1.96

1.7 EXERCISES

In Exercises 1 to 4 find the slope of the line tangent to the curve at the point.

1 (a) $x^2 + y^2 = 9$ at $(-1, 2\sqrt{2})$
(b) $(x-1)^2 + (y+2)^2 = 16$ at $(4, -2+\sqrt{7})$
(c) $(x-2)^2 + y^2 = 25$ at $(-1, 4)$
(d) $8y = x^2$ at $x = 2$
(e) $-16(y-4) = x^2$ at $x = -3$

2 (a) $x^2 + (y+3)^2 = 36$ at $(0, 3)$
(b) $(x+3)^2 + (y-4)^2 = 10$ at $(-4, 7)$

(c) $y^2 = x - 1$ at $y = -2$
(d) $y^2 = -\frac{1}{2}(x - 2)$ at $y = 1$
(e) $-24(y - 2) = (x + 1)^2$ at $x = 3$

3 (a) $x^2/52 + y^2/13 = 1$ at $(-6,2)$
(b) $(x - 1)^2/45 + (y - 2)^2/20 = 1$ at $(4,6)$
(c) $(x - 2)^2/25 + (y - 1)^2/100 = 1$ at $(-2,7)$
(d) $x^2 - y^2 = 1$ at $(-\sqrt{2},1)$
(e) $x^2/6 - y^2/2 = 1$ at $(9,-5)$

4 (a) $(x - 1)^2 + 4(y + 2)^2 = 1$ at $(\frac{3}{2},\frac{1}{4}\sqrt{3} - 2)$
(b) $(x - 1)^2/39 + (y + 1)^2/64 = 1$ at $(0,(\sqrt{38} - 8)/8)$
(c) $(x - 1)^2/9 - (y + 2)^2/4 = 1$ at $(6,-14/3)$
(d) $y^2/3 - x^2/2 = 1$ at $(-\sqrt{26/3},4)$
(e) $5y^2/36 - (x - 1)^2/4 = 1$ at $(2,-3)$

5 Find the slopes of the lines tangent to the circle with center $(1,-1)$ and radius 5 at the points where it is cut by the line $x = -2$.

6 An ellipse has center $(2,0)$, horizontal major axis of length 8, minor axis of length 6. Find the slopes of the tangents to the points where the ellipse is cut by the lines $y = \pm 1$.

7 An ellipse has foci at $(0,0)$ and $(0,4)$, and minor axis of length 2. Find the slopes of the tangents (if not vertical) at the points where the ellipse is cut by the line $y = 2x$. [*Hint:* Substitute $y = 2x$ into the equation of the ellipse, solve for x, put this value back into the ellipse equation, and solve for y.]

8 Find the slopes of the tangent lines to the hyperbola $(x - 1)^2 - y^2/4 = 1$ at the points where it is cut by the line $x = -2$.

9 At what points on the circle $x^2 + y^2 = 9$ do the tangents have the slopes $\pm\sqrt{2}$?

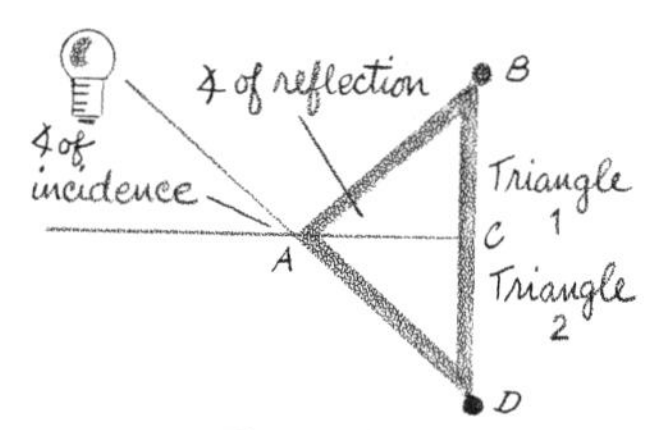

Figure 1.97

10 Find the points on the ellipse $(x - 1)^2 + y^2/4 = 1$ where the tangents have the slope 3.

11 Find the point on the parabola $16y = (x + 1)^2$ where the tangent has the slope $\frac{1}{3}$.

12 Find the points on the hyperbola $(x - 3)^2/4 - (y + 2)^2/3 = 1$ where the tangent line has slope 4 and -4.

13 We refer to Figure 1.97. Show that the angle of incidence is equal to the angle of reflection when the path from the source to the image is along a straight line. [*Hint:* Prove that the triangles are congruent, and cite the result about equality of alternate angles formed by two straight lines.]

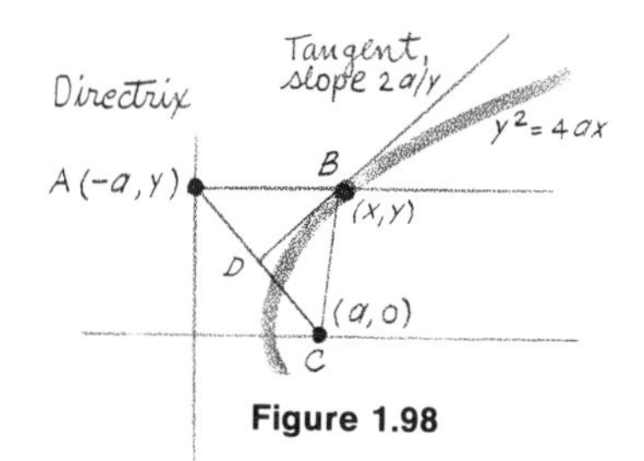

Figure 1.98

14 We outline the proof of the reflecting property of the parabola; the reader is asked to fill in the details. The parabola $y^2 = 4ax$ appears in Figure 1.98.

(a) Show that the tangent line to the parabola at B is perpendicular to the line from A to C.
(b) Show that triangles ABD and BDC are congruent.
(c) Conclude from (b) that the angles of incidence and reflection are equal.

CHAPTER

2 Solid Analytic Geometry

2.1 Rectangular Coordinates in Space, Distance, and Direction

Solid analytic geometry is analytic geometry in three dimensions. In this chapter we give a short discussion of its elements, with few applications. This is to provide the technical material needed for two subjects which themselves have important applications: Sections 2.1 and 2.2 are for linear programming (Section 3.9), and, in addition, Sections 2.3 and 2.4 are for functions of two variables (Chapter 7). Solid geometry is put here as a logical sequel to Chapter 1; however, the study of the various sections may be postponed or cancelled to suit the needs of various courses.

The position of a point in the plane is determined by a pair of coordinates with respect to two perpendicular axes. In **space,** three dimensions, the point is described by a set of three coordinates with respect to three mutually perpendicular axes. The axes intersect at a common point 0, the **origin.** A number scale is defined on each axis, and 0 represents zero on each. The axes are called the x-, y-, and z-axis, respectively. These determine three mutually perpendicular planes, the xy-, xz-, and yz-plane, respectively. The axes are drawn so that the yz-plane is parallel to the surface of the paper, and with the x-axis pointing outward from the paper, directed at the reader. However, the x-axis is drawn at an angle so that it will be visible to the eye (Figure 2.1). The three mutually perpendicular planes cut the space into eight **octants.** The first octant is the one in which the coordinates are all positive (Figure 2.2).

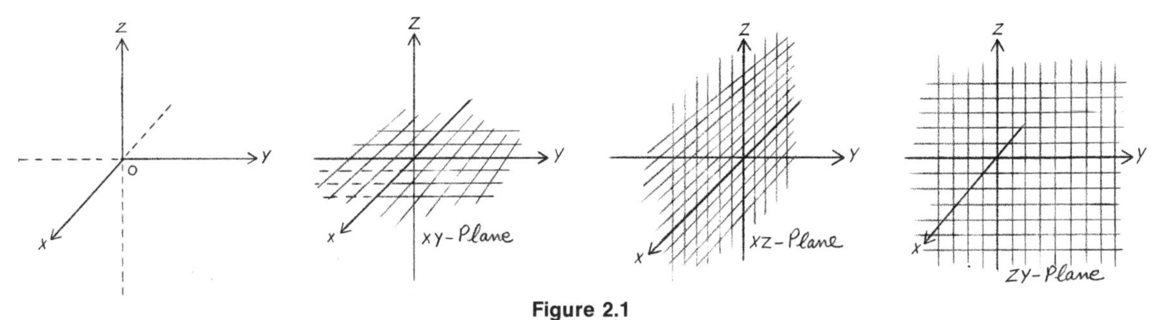

Figure 2.1

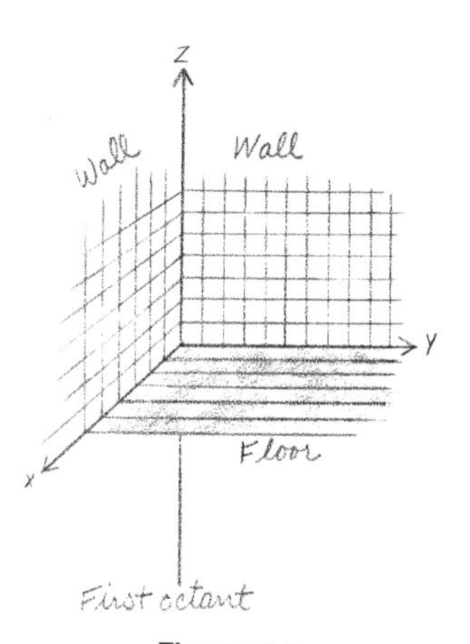

Figure 2.2

We denote by $P(x,y,z)$ or more simply by (x,y,z) the point in space whose coordinates are x, y, and z. The point (0,0,0) is the origin. A point of the form $(x,0,0)$ is on the x-axis; similarly, points of the forms $(0,y,0)$ and $(0,0,z)$ are on the y- and z-axes, respectively. For example, (1,0,0) is on the x-axis, 1 unit from the origin. If one of the three coordinates is equal to zero, then the point is in the plane spanned by the other two corresponding axes. For example, (2,1,0) is in the xy plane and (0,2,1) is in the yz-plane. The point (x,y,z) with $z \neq 0$ is z units above the point $(x,y,0)$ if z is positive, below if z is negative (Figure 2.3).

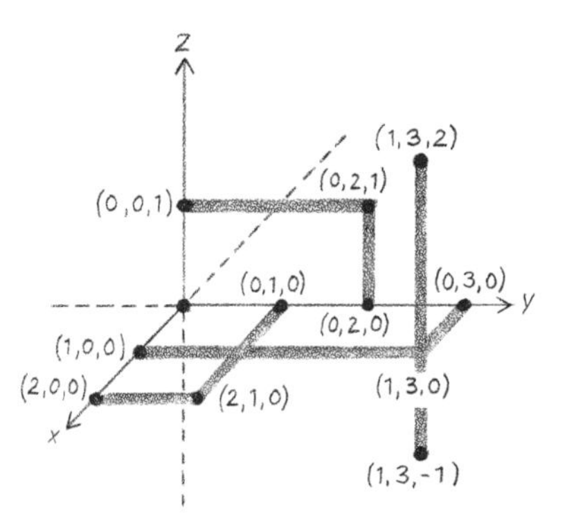

Figure 2.3

The distance from a point (x,y,z) to the origin is

$$\sqrt{x^2 + y^2 + z^2}\,. \tag{1}$$

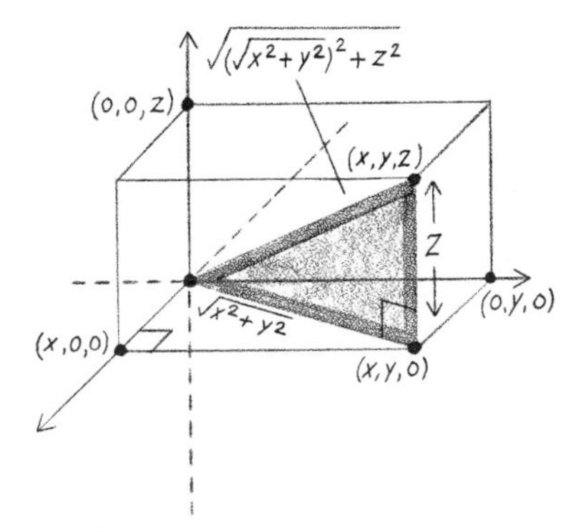

Figure 2.4

As in the case of the plane, this is based on the Pythagorean Theorem (Figure 2.4). More generally, the distance between two points (x_1,y_1,z_1) and (x_2,y_2,z_2) is

$$\sqrt{(x_2-x_1)^2+(y_2-y_1)^2+(z_2-z_1)^2}. \tag{2}$$

For example, the distance between (1,−2,0) and (2,4,3) is

$$\sqrt{(2-1)^2+(4+2)^2+(3-0)^2} = \sqrt{1+36+9} = \sqrt{46}\,.$$

In the plane the slope measures the inclination of a line. In three dimensions a single number like the slope is not sufficient to describe the direction of a line. The **direction numbers** of a line through the origin are the triple consisting of the coordinates of any point on the line: the direction numbers of the line through the origin and (x,y,z) are a triple of numbers kx, ky, kz where k is any constant not equal to 0. Thus, the direction numbers of a line are not a specific set of three numbers; it is any triple differing from (x,y,z) by a constant multiple. The **direction cosines** are the direction numbers divided by the distance of the point from the origin:

$$\frac{x}{\sqrt{x^2+y^2+z^2}},\ \frac{y}{\sqrt{x^2+y^2+z^2}},\ \frac{z}{\sqrt{x^2+y^2+z^2}}. \tag{3}$$

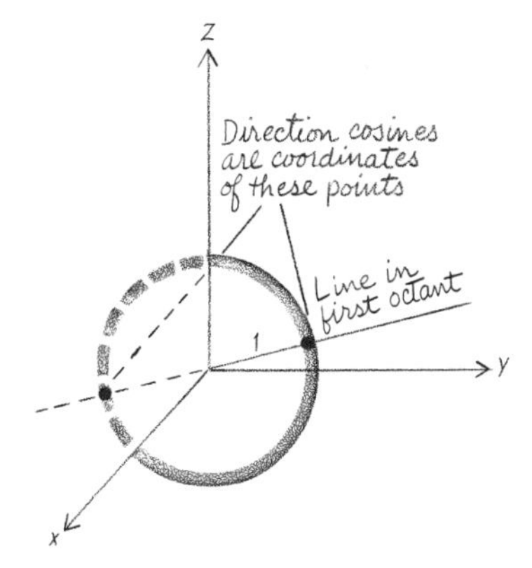

Figure 2.5

(These are called cosines because they are the cosines of the angles between the line and the coordinate axes. See Section 4.6.) These are the coordinates of the points on the line which are at a distance of 1 unit from the origin. There are two sets of direction cosines for each line; they differ in sign (Figure 2.5). The reader can easily verify that the sum of the squares of the coordinates in (3) is 1.

The **direction numbers** of a line through two points (x_1,y_1,z_1) and (x_2,y_2,z_2) are the differences of the coordinates:

$$(x_2-x_1, y_2-y_1, z_2-z_1). \tag{4}$$

The direction cosines are obtained by dividing each member of the triple (4) by the distance between the points.

EXAMPLE 1

The direction cosines of the line through the origin and (1,−2,0) are $(1/\sqrt{5},-2/\sqrt{5},0)$; indeed, the distance of the point from the origin is $\sqrt{5}$. The line through (1,−2,0) and (2,4,3) has the direction numbers (1,6,3). The direction cosines are obtained by division by the distance $\sqrt{1^2+6^2+3^2}=\sqrt{46}$. The other set of direction cosines for the first line is $(-1/\sqrt{5},2/\sqrt{5},0)$; for the second line it is (−1,−6, 3), divided by $\sqrt{46}$. ▲

The relations between lines in the plane are relatively simple: Two lines are either parallel or have a single point of intersection. Perpendicularity of lines is also very easily established. The relations between lines in space are more complex because nonparallel lines do not necessarily intersect. In the plane two lines are parallel if they have the same slope. In space two lines are called **parallel** if they have the same direction numbers.

Now we define perpendicularity for lines in space. The lines through the origin and (x_1,y_1,z_1) and (x_2,y_2,z_2), respectively, are called **perpendicular** if

$$x_1x_2+y_1y_2+z_1z_2=0. \tag{5}$$

More generally, the line through (x_1,y_1,z_1) and (x_2,y_2,z_2) is perpendicular to the line through (x_3,y_3,z_3) and (x_4,y_4,z_4) if

$$(x_2-x_1)(x_4-x_3)+(y_2-y_1)(y_4-y_3)+(z_2-z_1)(z_4-z_3)=0. \tag{6}$$

(Perpendicular lines do not necessarily intersect.) Observe that the criterion (5) is a generalization of the condition for perpendicularity in the plane: Two lines in the plane passing through the origin and the points (x_1,y_1) and (x_2,y_2) have slopes that are negative reciprocals of each other if and only if $x_1x_2+y_1y_2=0$.

The algebraic definition of perpendicularity, determined by Equation (5), has a geometric basis. If the lines from the origin through the two points (x_1,y_1,z_1) and (x_2,y_2,z_2) form a right angle at the origin, then by the Pythagorean Theorem, the sums of the squares of the distances of the two points from the origin is equal to the square of the distance between the two points:

$$(x_1^2+y_1^2+z_1^2)+(x_2^2+y_2^2+z_2^2)=(x_2-x_1)^2+(y_2-y_1)^2+(z_2-z_1)^2. \tag{7}$$

These two sums are equal if and only if the sum of the cross-product terms (5) is equal to 0.

EXAMPLE 2

Show that the line through $A(4,8,-11)$ and $B(-1,-7,-1)$ is perpendicular to the line through B and $C(4,-2,9)$:

$$(-1-4)(4+1)+(-7-8)(-2+7)+(-1+11)(9+1)=-25-75+100=0. \blacktriangle$$

2.1 EXERCISES

In Exercises 1 and 2 plot the positions of the points in space.

1 (a) $(2,0,0)$ (b) $(1,2,0)$ (c) $(-2,2,3)$ (d) $(0,-2,0)$

2 (a) $(0,1,0)$ (b) $(2,3,2)$ (c) $(3,2,-3)$ (d) $(0,0,3)$

In Exercises 3 and 4 find the direction cosines of the lines through the origin and the given points.

3 (a) $(1,0,2)$ (b) $(1,2,3)$ (c) $(-3,-2,2)$ (d) $(-2,3,-4)$

4 (a) $(-1,1,1)$ (b) $(-1,0,5)$ (c) $(2,-1,-2)$ (d) $(1,-1,2)$

In Exercises 5 and 6 compute the distances between the given points.

5 (a) $(0,0,0)$, $(2,4,3)$ (b) $(0,0,1)$, $(3,2,0)$ (c) $(-1,2,3)$, $(-3,4,4)$ (d) $(3,-2,-1)$, $(1,2,-3)$ (e) $(1,-2,-1)$, $(3,1,-2)$

6 (a) $(4,0,1)$, $(2,1,0)$ (b) $(-1,2,0)$, $(2,4,2)$ (c) $(4,0,-1)$, $(3,-1,0)$ (d) $(-1,2,1)$, $(-1,-2,3)$ (e) $(-3,-4,5)$, $(5,-4,-1)$

7 Show that the triangle with vertices $(4,3,1)$, $(7,1,2)$, and $(5,2,3)$ is isosceles.

8 Find the perimeter of the triangle with vertices $(1,-2,3)$, $(-4,2,1)$, and $(3,3,-1)$.

9 Show that the three points $(2,-3,4)$, $(3,4,-1)$, and $(4,11,-6)$ are on the same line. [*Hint:* Show that the lines through any pair have the same directions.]

10 Show that the line through $(3,5,-1)$ and $(4,3,-1)$ is perpendicular to the line through 0 and $(2,1,1)$.

11 By means of distance calculations, show that the three points $(5,3,-2)$, $(4,1,-1)$, and $(2,-3,1)$ are on the same line.

12 Show that $(0,1,3)$, $(-1,0,2)$, and $(-1,1,4)$ are the vertices of a right triangle.

13 The line from $(-3,2,5)$ to (x,y,z) has length 6 and the direction cosines $(-\frac{2}{3}, \frac{1}{3}, \frac{2}{3})$. What are the possible values of x, y, z?

14 Show that the line through $(-3,-2,-1)$ and $(-1,-4,13)$ is parallel to the line through $(6,1,4)$ and $(5,2,-3)$.

15 Show that $(5,1,5)$, $(4,3,2)$, and $(-3,-2,1)$ are the vertices of a right triangle.

16 Prove that the four points $(6,2,5)$, $(4,1,1)$, $(1,2,-1)$, and $(3,3,3)$ are the vertices of a parallelogram.

17 Verify that the lines from the origin to each of the two points are perpendicular:
(a) $(7,4,-5)$ and $(1,2,3)$ (b) $(2,9,-12)$ and $(-6,-16,-13)$

18 Show that there is one perpendicular bisector of the segment joining $(-5,2,0)$ and $(9,-4,6)$ which passes through $(5,1,-2)$ and $(-4,-5,13)$. [*Hint:* Find the coordinates of the midpoint.]

2.2 The Plane and Line in Space

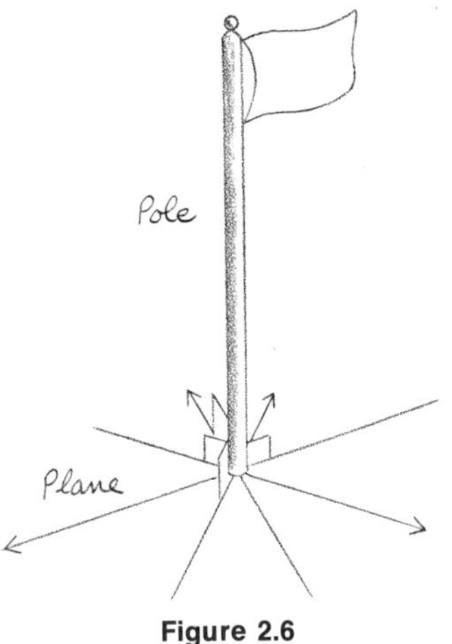

Figure 2.6

Given a line L and a plane, we say that L is perpendicular to the plane if it is perpendicular to every line in the plane. For example, an erect flagpole is perpendicular to the plane surface of the ground around it (Figure 2.6). It is evident that for any line L and point P on it, there is exactly one plane through P perpendicular to L. In other words, you cannot put two straight flagpoles at the same point on the ground. A line perpendicular to a plane is said to be **normal** to the plane.

Let L be a line through the origin with direction numbers (A,B,C); and let (x_0,y_0,z_0) be a point on this line. If (x,y,z) is a point in the plane passing through (x_0,y_0,z_0) perpendicular to L, then the line through these points is perpendicular to L. It follows from the definition of perpendicularity [Section 2.1, Equation (6)] that

$$A(x-x_0) + B(y-y_0) + C(z-z_0) = 0. \tag{1}$$

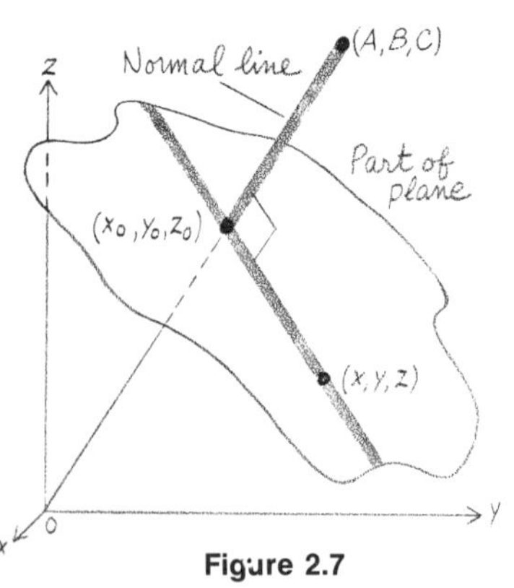

Figure 2.7

This is the **equation of the plane** because every point in the plane satisfies it, and, conversely, every point which satisfies it is in the plane (Figure 2.7). The numbers A, B, and C are called the **normal direction numbers** for the plane because the normal line has these direction numbers.

Another form of the equation of the plane is

$$Ax + By + Cz = D; \tag{2}$$

indeed, put $D = Ax_0 + By_0 + Cz_0$ in Equation (1). Thus, any linear equation in x, y, and z represents a plane in space.

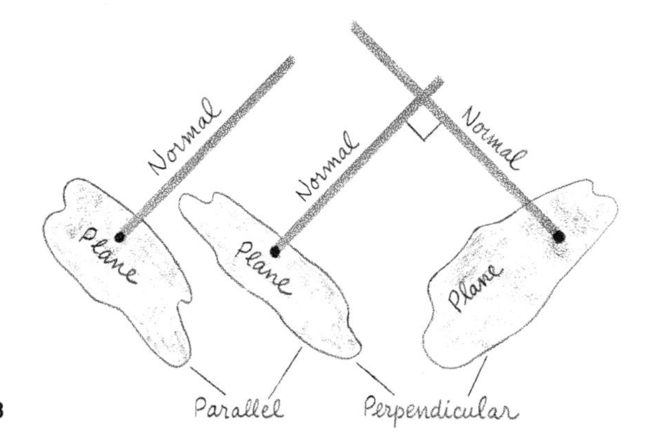

Figure 2.8

Any line parallel to the normal line is also perpendicular to the plane; in fact, the two lines have the same direction numbers. It follows that a line is perpendicular to a plane if the direction numbers of the line are equal to the normal direction numbers of the plane. Two **planes** are called **parallel** if their normal lines are. Two **planes** are called **perpendicular** if their normal lines are (Figure 2.8). There is a simple algebraic test to determine if two planes are parallel: If A_1, B_1, and C_1 are the normal direction numbers of one plane, and A_2, B_2, and C_2 those of the other plane, then they are parallel if and only if

$$\frac{A_1}{A_2} = \frac{B_1}{B_2} = \frac{C_1}{C_2}. \tag{3}$$

The planes are perpendicular if and only if

$$A_1A_2 + B_1B_2 + C_1C_2 = 0, \tag{4}$$

because this is just the condition for the perpendicularity of the normal lines.

EXAMPLE 1

Find the equation of the plane through $(1,-2,3)$ that is perpendicular to the line through the origin and $(4,-2,5)$. Here (A,B,C) is $(4,-2,5)$ and (x_0,y_0,z_0) is $(1,-2,3)$. From Equation (1) we obtain

$$4(x-1) - 2(y+2) + 5(z-3) = 0.$$

The equivalent form (2) is

$$4x - 2y + 5z = 23. \blacktriangle$$

EXAMPLE 2

Find the equation of the plane passing through $(1,-3,-5)$ and parallel to the plane $6x - 2y + 3z = 10$. The planes have the same normal direction numbers. It follows that the equation of the first plane must be

$$6(x-1)-2(y+3)+3(z+5)=0,$$

or

$$6x-2y+3z=-3. \blacktriangle$$

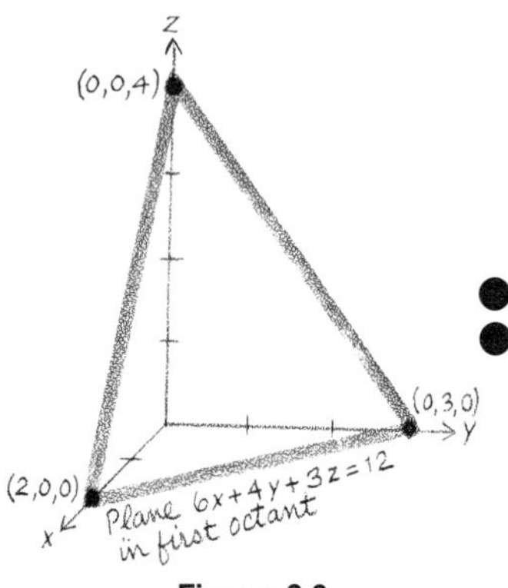

Figure 2.9

In order to sketch a plane in space, we find the points where it crosses each of the axes. Consider the plane of Equation (2). It cuts the x-axis at the point where $y=z=0$. Solve for x to obtain $x=D/A$. So the plane cuts the x-axis at the point $(D/A,0,0)$. Similarly, we find that the h- and z-axes are crossed at $(0,D/B,0)$ and $(0,0,D/C)$, respectively. In the above discussion it has been presumed that the numbers A, B, and C are not equal to 0, so that the division is possible.

EXAMPLE 3

The plane $6x+4y+3z=12$ is sketched in Figure 2.9. The axes are cut at the points (2,0,0), (0,3,0), and (0,0,4), respectively. Only the intersection of the plane with the **first** octant is shown. ▲

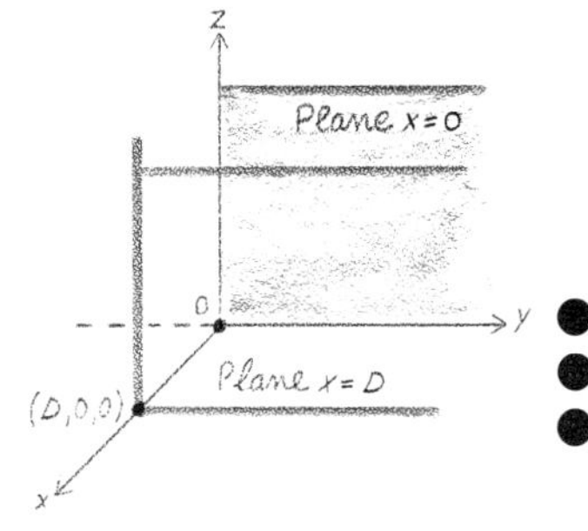

Figure 2.10

If one or more of the normal direction numbers of the plane is equal to 0, then the plane is perpendicular to one of the axes, or the normal is perpendicular to one of the axes. The simplest kind of such a plane has the equation $x=0$. This is special case of (2) with $A=1$, and $B=C=D=0$. The normal direction numbers are (1,0,0), which are the direction numbers of the x-axis. Hence the plane is perpendicular to this axis. It also contains (0,0,0). It follows that it is, in fact, the yz-plane. The plane with the equation $x=D$ passes through the point $(D,0,0)$ parallel to the yz-plane (Figure 2.10).

A plane with an equation $By+Cz=D$ has the normal direction numbers $(0,B,C)$. A line with these direction numbers is perpendicular to the x-axis because $1\cdot 0+B\cdot 0+C\cdot 0=0$. The plane is sketched in Figure 2.11

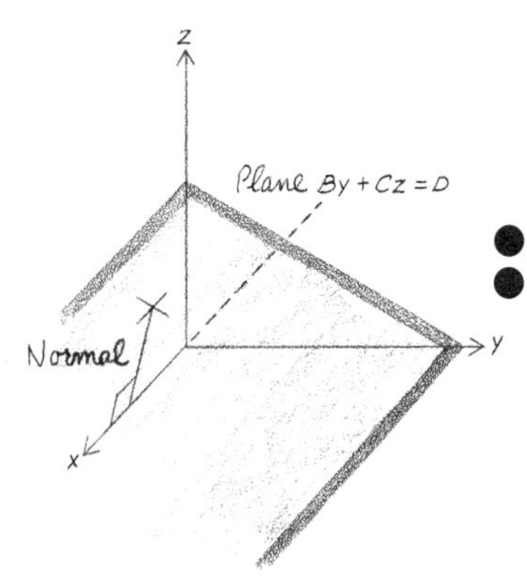

Figure 2.11

A straight line in space is the intersection of two nonparallel planes. Since the plane is represented by a single linear equation, the intersection is represented by a pair of linear equations,

$$Ax+By+Cz=D, \qquad A'x+B'y+C'z=D',$$

where each equation above is that of the corresponding plane. The system of two equations can be transformed to a more useful form. There are two equations in three unknowns. If we combine the two equations and eliminate one of the unknowns, we get a single equation in two unknowns. If we successively eliminate x, y, and z, we get **three linear equations each in two unknowns.** The most commonly used form of such a system is the **symmetric** form. If a line passes through the point (x_0,y_0,z_0) and has direction numbers (a,b,c), then every point on the line satisfies the relations

$$\frac{x-x_0}{a}=\frac{y-y_0}{b}=\frac{z-z_0}{c}. \tag{5}$$

(This represents a set of **two** equations.) The reason for this is that by definition the direction numbers of the line are $(x-x_0,y-y_0,z-z_0)$ and also (a,b,c); therefore, one set must be a multiple of the other, and so (5) must hold. Equations (5) represent the symmetric form. Since (x_0,y_0,z_0) is an arbitrary point on the line, the form (5) is not unique, and any other point on the line may be used in the equations. Here we have assumed that none of the direction numbers is 0. Equation (5) must be modified when this is not the case. If $a=0$, (5) is replaced by

$$x=x_0 \qquad \frac{y-y_0}{b}=\frac{z-z_0}{c}. \tag{6}$$

Similar modifications are made when b or c is equal to 0.

EXAMPLE 4

Find the symmetric form of the equations of the line formed by the intersection of the planes

$$2x-3y+2z=5, \qquad 3x-y-z=-6.$$

To eliminate the terms containing z, multiply the second equation by 2, and add:

$$\begin{array}{l} 2x-3y+2z=5 \\ 6x-2y-2z=-12 \\ \hline 8x-5y \qquad =-7, \end{array} \qquad \text{or} \qquad x=\frac{5y-7}{8}.$$

Next eliminate the terms containing y, and solve for x in terms of z. Multiply the second equation by -3 and add to obtain $-7x+5z=23$, or $x=(5z-23)/7$. The symmetric form of the equations of the line is now written as $x=(5y-7)/8=(5z-23)/7$, or

$$\frac{x-0}{1}=\frac{y-7/5}{8/5}=\frac{z-23/5}{7/5}.$$

It follows that the line passes through $(0,7/5,23/5)$ and has the direction numbers $(5,8,7)$. ▲

EXAMPLE 5

Find the equations of the line passing through (3,−1,2) and (4,−6,−5). The direction numbers are the differences of the respective coordinates, (1,−5,−7). It follows that the equations of the line are of the forms

$$\frac{x-3}{1}=\frac{y+1}{-5}=\frac{z-2}{-7}$$

(corresponding to the first point) or

$$\frac{x-4}{1}=\frac{y+6}{-5}=\frac{z+5}{-7}$$

(corresponding to the second point). ▲

EXAMPLE 6

Find the equations of the line passing through (2,−1,3) perpendicular to the plane $3x+2y+5z=3$. Then the line has direction numbers (3,2,5). It follows that the equations of the line are

$$\frac{x-2}{3}=\frac{y+1}{2}=\frac{z-3}{5}.\ ▲$$

2.2 EXERCISES

In Exercises 1 and 2 find the equation of the plane, in either form (1) or (2), satisfying the given conditions.

1 (a) containing point (4,2,−9), normal direction (1,1,1)
(b) containing (4,−3,8), normal direction (0,2,4)
(c) containing (−7,2,9), normal direction (−4,2,4)
(d) containing (2,5,5), parallel to $8x+3y+2z=1$
(e) containing (−2,8,1), parallel to $2x-8z=-2$

2 (a) containing (−1,4,−1), normal direction (5,−5,−1)
(b) containing (−3,−2,8), normal direction (−1,1,8)
(c) containing (2,−2,0), parallel to $7y-z=3$
(d) containing (6,0,−3), parallel to $3x+4y-6z=7$
(e) containing (4,−3,−3), parallel to $-7x+6y+7z=4$

3 Find the equation of the plane through P which is perpendicular to the line through A and B.
(a) $P(8,1,5)$, $A(8,2,-5)$, $B(-1,5,8)$
(b) $P(-1,4,7)$, $A(0,2,5)$, $B(8,-1,-3)$
(c) $P(-1,-6,-3)$, $A(-6,-8,0)$, $B(2,4,7)$
(d) $P(9,-1,3)$, $A(4,6,-8)$, $B(0,-1,-3)$
(e) $P(4,6,7)$, $A(-9,0,1)$, $B(-2,4,5)$

4 Sketch the portion of the plane in the first octant.

(a) $3x + 6y + 9z = 9$
(b) $x + 4y + 6z = 12$
(c) $3y + 6z = 10$
(d) $6x + 9y = 14$
(e) $4x + 3y + 6z = 24$

5 Find the symmetric form of the equations of the line formed by the intersection of the pair of planes.

(a) $3x - y - z = 4$; $x - y + z = 0$
(b) $x + y + z = 0$; $3x - y + z = -4$
(c) $2x - y - 2z = 5$; $3x - 2y - 2z = 5$
(d) $x - y - z = -1$; $x - 2y + 2z = 2$
(e) $2x + 3y - 3z = -8$; $x + y - 2z = -3$

6 Find the equations of the line passing through the given pair of points.

(a) $(0,1,2)$, $(-4,5,6)$
(b) $(6,-7,7)$, $(8,9,0)$
(c) $(0,-1,0)$, $(3,6,9)$
(d) $(9,-1,-4)$, $(-6,9,1)$
(e) $(4,-6,8)$, $(0,1,2)$

7 Find the equations of the line passing through the given point, perpendicular to the given plane.

(a) point $(1,5,8)$, plane $6x + 7y - 9z = 2$
(b) point $(4,-8,2)$, plane $6x - y + 6z = 1$
(c) point $(-7,4,1)$, plane $8x - 5y - 3z = -2$
(d) point $(0,9,9)$, planc $9x + 2z = 3$
(e) point $(-5,-7,-9)$, plane $2x - 5y - 8z = 1$

8 For each pair of planes determine whether or not they are parallel or perpendicular or neither.

(a) $6x + 4y - 2z = 1$; $3x + 2y - z = 5$
(b) $2x + y - 2z = 1$; $-7x + 4y - 5z = 1$
(c) $2x + 12y + 9z = 30$; $6x - 13y + 16z = -5$
(d) $4x - 3y - 6z = 1$; $2x - 3y - 3z = 4$

9 Find the equations of the line through $(1,-2,3)$ which is parallel to the intersections of the planes

$$2x - 4y + z = 3 \quad \text{and} \quad x + 2y - 6z = -4.$$

10 A line passes through $(0,1,-1)$ parallel to the line with the equations $y = 2x - 7$ and $z = 3x + 4$. Find the equations of the first line.

11 A plane contains the points $(1,0,0)$ and $(0,0,1)$, and is perpendicular to the plane $2x + y - z = -6$. Let A, B, and C be the normal direction numbers.

(a) Show that these satisfy the equations

$$2A + B - C = 0, \qquad A - C = 0.$$

(b) Find the equation of the plane.

12 Show that the line formed by the intersection of the planes $x - 2y = -2$ and $2y + z = -4$ is perpendicular to the line formed by the intersection of the planes $7x + 4y = 15$ and $y + 14z = -40$.

13 Find the equation of the line which is normal to the plane $x + 2y - z = 20$ and which passes through the point $(-1,2,1)$. At what point does this line meet the plane? [*Hint:* Solve the equations of the line and plane simultaneously.]

14 What is the length of the perpendicular segment from $(1,0,1)$ to the plane $x + y - z = 1$? [*Hint:* By the method of Exercise 13, find the point at which the perpendicular meets the plane.]

2.3 Surfaces in Space

The simplest surface in space is the plane. Another simple surface is the circular cylinder. If the axis of the cylinder coincides with the z-axis, then the cross section of the cylinder is a circle about the z-axis. The coordinates of the points on the cross section satisfy $x^2 + y^2 = r^2$, where r is the radius of the cylinder; thus, we take this to be the equation of the cylinder (Figure 2.12). In other words, the set of points in the **plane** which satisfy this equation is a circle, but the set of points in **space** satisfying the equation is a cylinder. If the axis of the cylinder is on the x-axis, then the equation is $y^2 + z^2 = r^2$. If it is on the y-axis, the equation is $x^2 + z^2 = r^2$ (Figure 2.13).

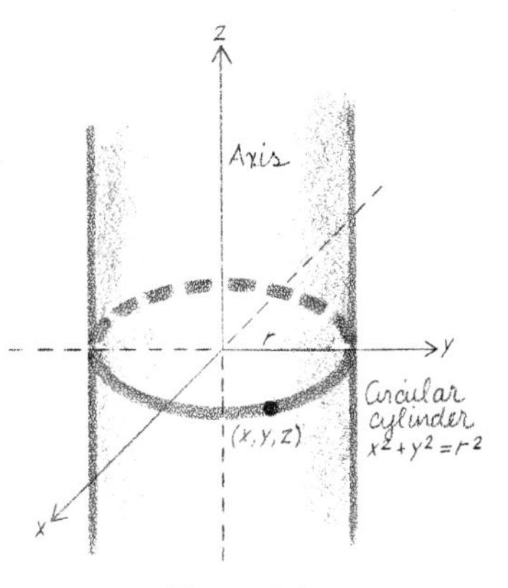

Figure 2.12

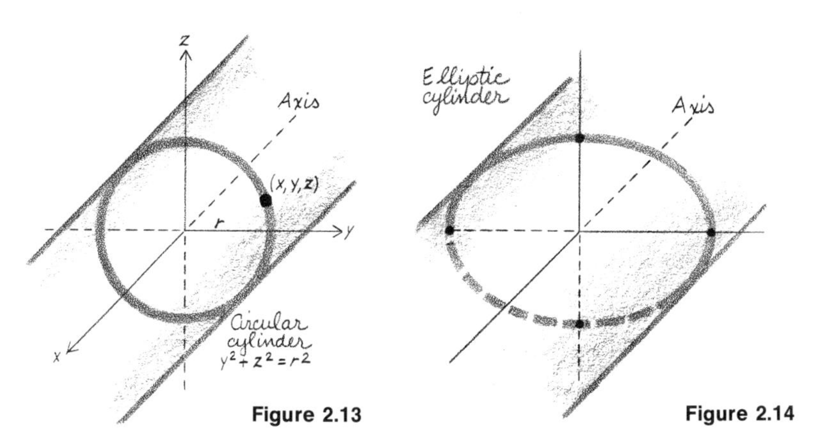

Figure 2.13

Figure 2.14

There are more general kinds of "cylinders" where the cross section is not necessarily a circle but may be any one of the second-degree curves we have studied. The elliptic cylinder has an ellipse as the cross section (Figure 2.14). For example, the elliptic cylinder with its axis along the z-axis has the equation of the plane ellipse in the variables x and y:

$$\frac{x^2}{a^2} + \frac{y^2}{b^2} = 1 \quad \text{or} \quad \frac{x^2}{b^2} + \frac{y^2}{a^2} = 1.$$

The line parallel to the z-axis and passing through the point $(x_0,y_0,0)$ has the set of equations $x = x_0$, $y = y_0$ (Figure 2.15). The equation of a cylinder with its axis along this line is similar to the equation for the axis on the z-axis, except that x and y are replaced by $x - x_0$ and $y - y_0$, respectively.

Other examples of cylinders are the parabolic cylinder with a parabolic cross section (Figure 2.16); and the hyperbolic cylinder with a hyperbolic cross section (Figure 2.17).

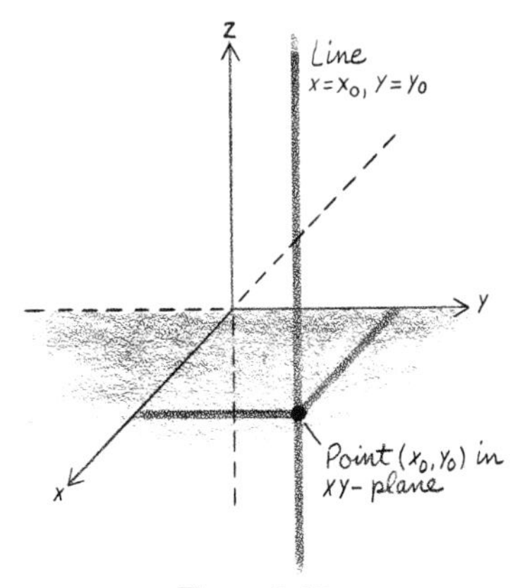

Figure 2.15

EXAMPLE 1

Describe and sketch the cylinder with the equation

$$\frac{(x-1)^2}{9} + \frac{(y-2)^2}{4} = 1.$$

This is an elliptic cylinder with the axis parallel to the z-axis, and cutting the xy-plane at (1,2,0). Its cross section in the plane is the ellipse with the equation above (Figure 2.18). ▲

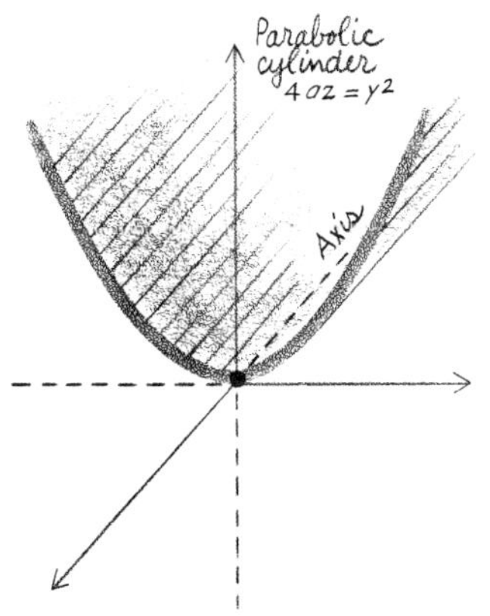

Figure 2.16

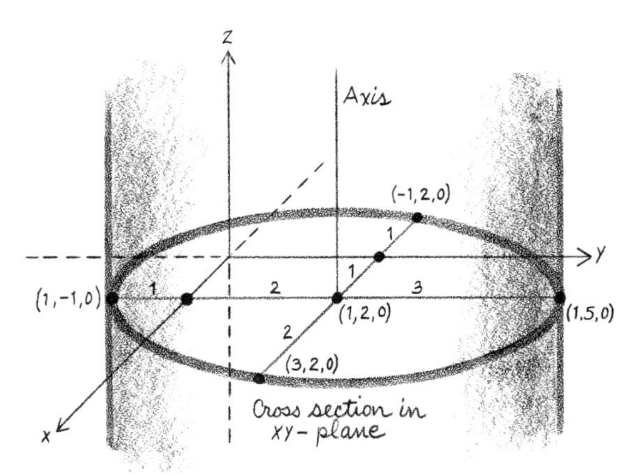

Figure 2.18

Now we shall study the **quadric surfaces** in space; these are further generalizations of second-degree curves. The standard forms of these surfaces are with the centers at the origin. They are the loci of points (x,y,z) in space satisfying equations of any of these three types:

$$\pm\frac{x^2}{a^2} \pm \frac{y^2}{b^2} \pm \frac{z^2}{c^2} = 1, \tag{1}$$

$$\pm\frac{x^2}{a^2} \pm \frac{y^2}{b^2} = 2cz, \tag{2}$$

$$\pm\frac{x^2}{a^2} \pm \frac{y^2}{b^2} \pm \frac{z^2}{c^2} = 0. \tag{3}$$

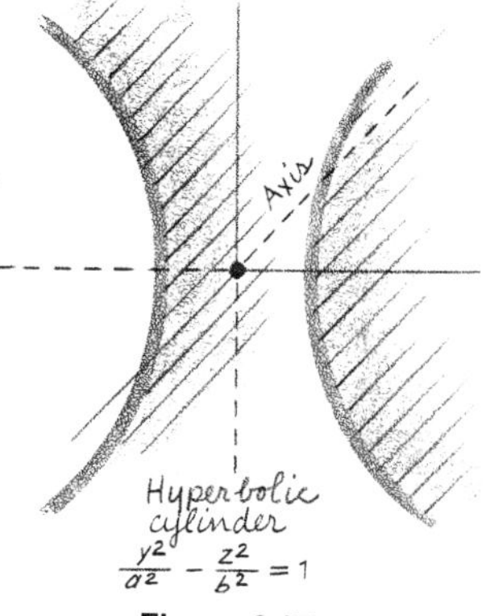

Figure 2.17

These surfaces have the property that their cross sections in any plane parallel to one of the coordinate planes is either a circle, ellipse, parabola, or hyperbola. This property is used in sketching the surface in three dimensions.

We examine the various classes of surfaces of this kind.

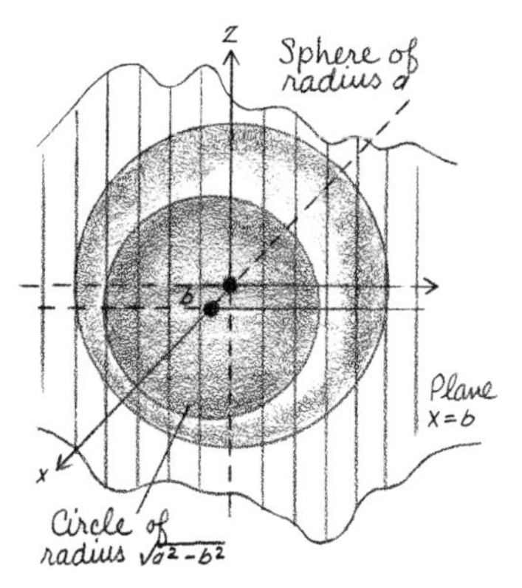

Figure 2.19

The sphere. If $a = b = c$ in (1), and all the signs are positive then the equation becomes

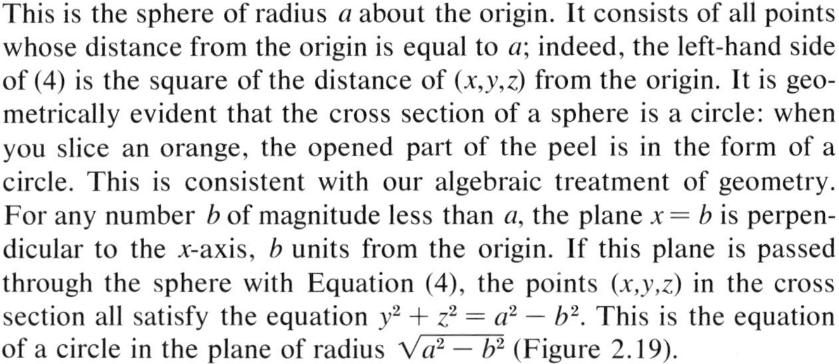

$$x^2 + y^2 + z^2 = a^2. \tag{4}$$

This is the sphere of radius a about the origin. It consists of all points whose distance from the origin is equal to a; indeed, the left-hand side of (4) is the square of the distance of (x,y,z) from the origin. It is geometrically evident that the cross section of a sphere is a circle: when you slice an orange, the opened part of the peel is in the form of a circle. This is consistent with our algebraic treatment of geometry. For any number b of magnitude less than a, the plane $x = b$ is perpendicular to the x-axis, b units from the origin. If this plane is passed through the sphere with Equation (4), the points (x,y,z) in the cross section all satisfy the equation $y^2 + z^2 = a^2 - b^2$. This is the equation of a circle in the plane of radius $\sqrt{a^2 - b^2}$ (Figure 2.19).

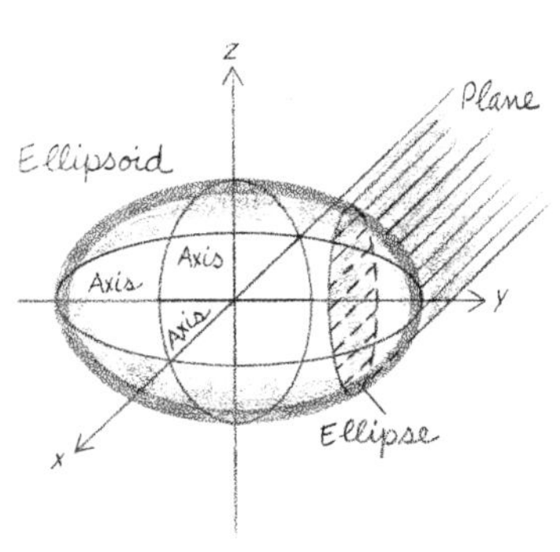

Figure 2.20

The ellipsoid. This is the surface represented by Equation (1) when the signs are all positive, but a, b, c not necessarily equal:

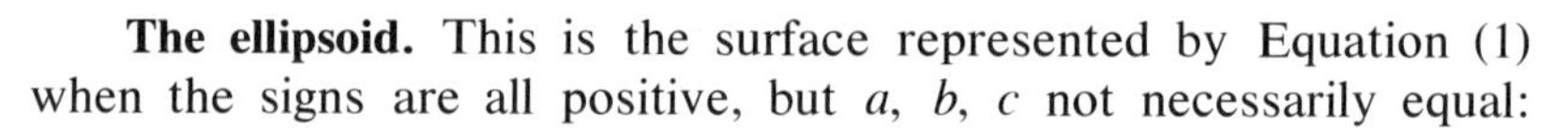

$$\frac{x^2}{a^2} + \frac{y^2}{b^2} + \frac{z^2}{c^2} = 1. \tag{5}$$

It has three axes of lengths $2a$, $2b$, and $2c$, respectively. The plane $x = d$, where d is any number of magnitude less than a, intersects the ellipsoid in the ellipse

$$\frac{y^2}{b^2} + \frac{z^2}{c^2} = 1 - \frac{d^2}{a^2} \quad \text{or} \quad \frac{y^2}{b^2(1 - d^2/a^2)} + \frac{z^2}{c^2(1 - d^2/a^2)} = 1.$$

The same is true of the intersection with any plane parallel to a coordinate plane (Figure 2.20).

Hyperboloid of one sheet. This is the surface represented by (1) when two signs are positive and one negative, as in

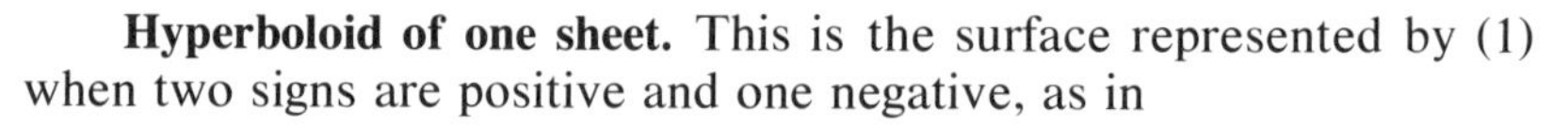

$$\frac{x^2}{a^2} + \frac{y^2}{b^2} - \frac{z^2}{c^2} = 1. \tag{6}$$

Intersections with planes $x = d$ or $y = d$ form hyperbolas; and those with $z = d$ are ellipses [Figures 2.21(a), 2.21(b)].

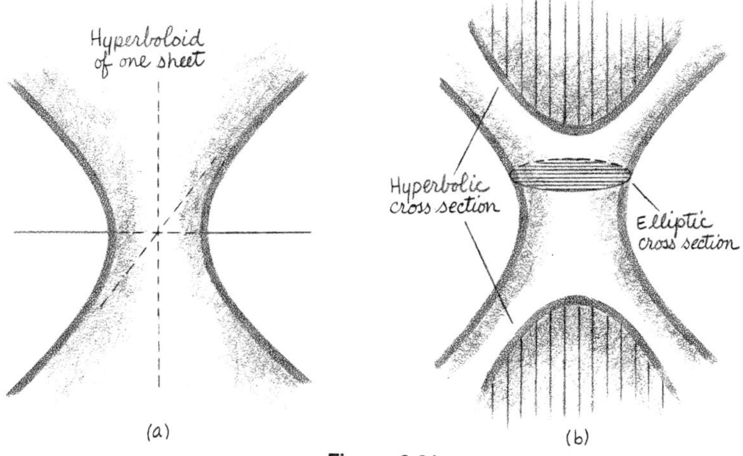

Figure 2.21

Hyperboloid of two sheets. When two signs in (1) are negative and one positive, as in

$$\frac{-x^2}{a^2} + \frac{y^2}{b^2} - \frac{z^2}{c^2} = 1, \tag{7}$$

the surface has two sheets [Figure 2.22(a)]. For d^2 larger than b^2, the intersection with the plane $y = d$ is an ellipse. The intersections with the planes $x = d$ and $y = d$ are hyperbolas [Figure 2.22(b)].

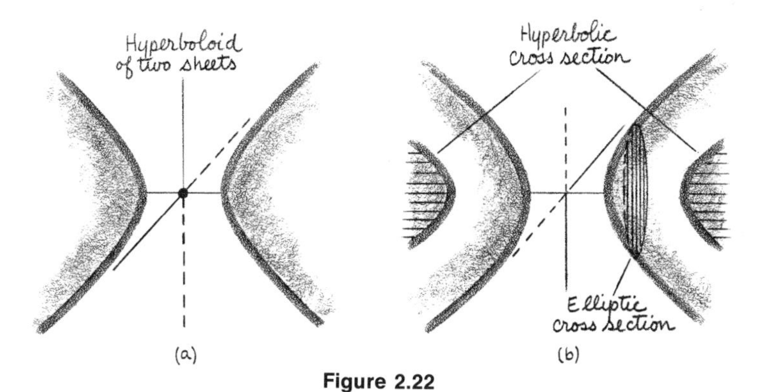

Figure 2.22

Elliptic paraboloid. This is the surface having Equation (2) with all positive signs:

$$\frac{x^2}{a^2} + \frac{y^2}{b^2} = 2cz. \tag{8}$$

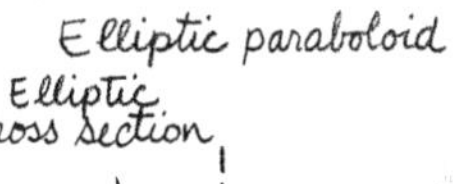

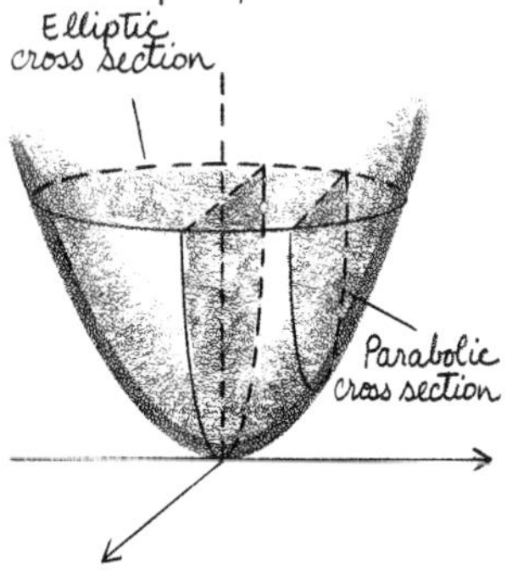

Figure 2.23

Intersections with planes perpendicular to the x- or y-axis are parabolas; and intersections with planes perpendicular to the z-axis are ellipses. If c is positive, then only positive values of z can satisfy Equation (7) (or $z = 0$). Therefore, the surface is contained in the half-space above the xy-plane. If c is negative, then the surface is reversed (Figure 2.23).

Hyperbolic paraboloid. This is the surface with the equation of the form

$$\frac{x^2}{a^2} - \frac{y^2}{b^2} = 2cz. \tag{9}$$

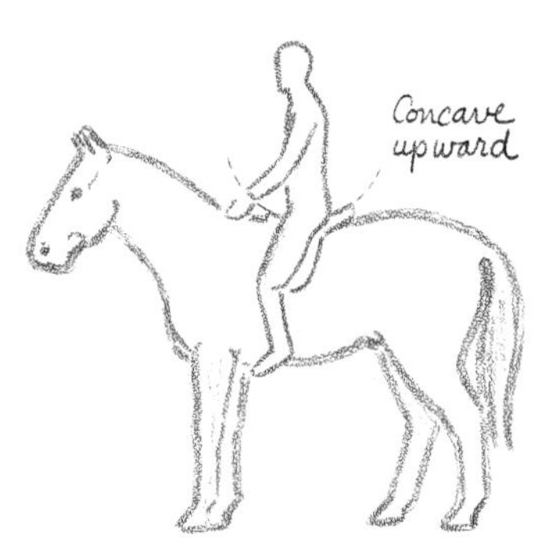

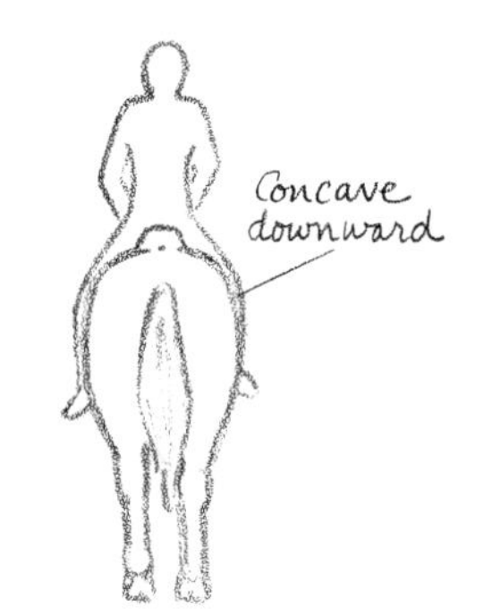

Figure 2.24

This is a saddle-shaped surface. The cross sections are easily described; however, the sketching is difficult because it is hard to put the pieces together. The surface of a saddle has the property that it is concave upward along the backbone of the horse, but concave downward across his back (Figure 2.24). The surface (9) is a saddle surface which extends indefinitely in all directions. Its cross sections are lines, hyperbolas, and parabolas. Let us consider the case where c is positive in (9). For simplicity consider the special case

$$x^2 - y^2 = z. \tag{10}$$

Here the horse is riding along the x-axis, and the rider's legs are placed on the y-axis. The view from the top of the rider appears in Figure 2.25: the cross-sectional curves cut horizontally are hyperbolas. The view from the front of the horse is in Figure 2.26: the vertical cross-sectional curves are parabolas which open downward. The view from the side of the horse is in Figure 2.27: The vertical cross-sectional curves are parabolas which open upward. The reader can verify all these statements in the following way: To get the view from a particular axis, substitute a number for the corresponding variable

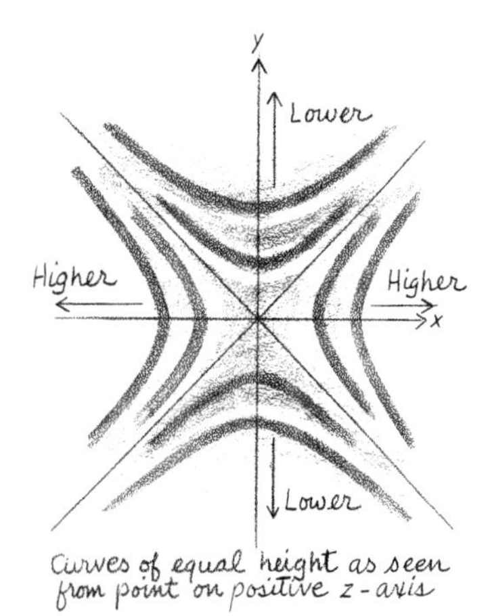

Figure 2.25

(x, y, or z) in Equation (10), and get the equation of the plane curve in the remaining two variables.

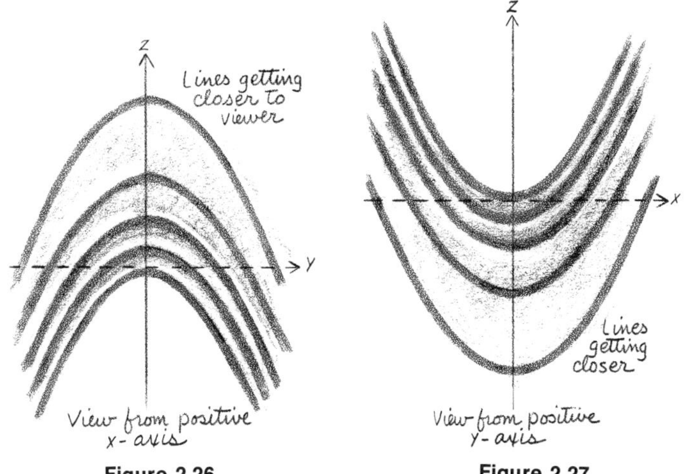

Figure 2.26

Figure 2.27

The following is our last surface.

Right circular cone. The equation is

$$x^2 + y^2 = c^2z^2. \tag{11}$$

It is of the form (3) with $a = b = 1$. The intersection with a plane perpendicular to the z-axis is a circle; and the intersection with a plane perpendicular to the x- or y-axis is a hyperbola (Figure 2.28).

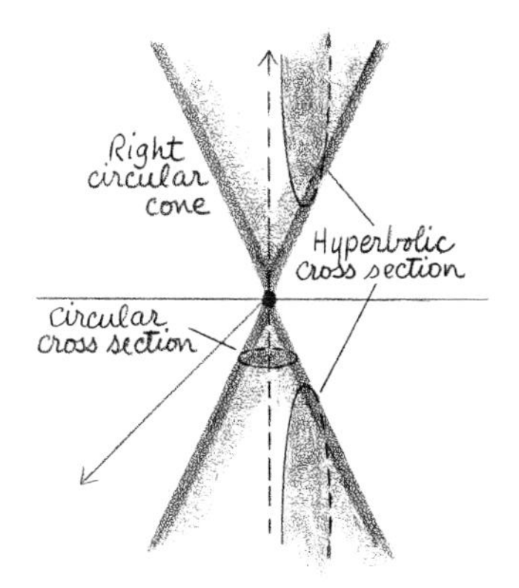

Figure 2.28

As we mentioned above these quadric surfaces have centers at the origin. If the center is moved to a point (h,k,l) then the equations are altered by replacing x, y, and z by $x - h$, $y - k$, and $z - l$, respectively.

EXAMPLE 2

Describe the surface with the equation

$$\frac{x^2}{9} + \frac{y^2}{4} - \frac{z^2}{16} = 1, \tag{12}$$

and its intersections with the planes $x = 2$, $y = \sqrt{3}$, and $z = 3$. This is the hyperboloid of one sheet with its axis along the z-axis. The intersection with the plane $x = 2$ is obtained by substituting $x = 2$ in Equation (12):

$$\frac{9y^2}{20} - \frac{9z^2}{80} = 1.$$

This is the equation of a hyperbola in the yz-plane with $a^2 = 20/9$ and $b^2 = 80/9$. The intersection with $y = \sqrt{3}$ is the hyperbola

$$\frac{4x^2}{9} - \frac{z^2}{4} = 1.$$

The intersection with $z = 3$ is the ellipse

$$\frac{16x^2}{225} + \frac{4y^2}{25} = 1. \blacktriangle$$

2.3 EXERCISES

In Exercises 1 and 2 describe the cross section of the cylinder and the direction of the axis, and, sketch it.

1 (a) $x^2 + y^2 = 16$
(b) $4x^2 + 9y^2 = 36$
(c) $-2x = z^2$
(d) $(y+1)^2/9 + (z-1)^2/4 = 1$
(e) $z^2 - 2x^2 = 4$

2 (a) $x^2 + z^2 = 25$
(b) $2y = z^2$
(c) $(x-2)^2 + (z+4)^2 = 16$
(d) $x^2 - 3y^2 = 1$
(e) $y^2/9 - z^2/25 = 1$

In Exercises 3 and 4 find the equation of the curve formed by the intersection of the surface and the plane. Identify the curve.

3 (a) $x^2 + y^2 + z^2 = 25, \quad x = 4$
(b) $x^2 + y^2 + z^2 = 169, \quad y = -12$
(c) $x^2 + 2y^2 + 3z^2 = 6, \quad y = \frac{1}{2}$
(d) $8x^2 + 2y^2 + 9z^2 = 144, \quad z = -3$
(e) $x^2/9 + y^2/4 - z^2/16 = 1, \quad z = 3$
(f) $x^2/16 + y^2/9 - z^2/36 = 1, \quad x = 2$
(g) $y^2 - x^2 - z^2 = 1, \quad z = 1$
(h) $y^2 - x^2/9 - z^2/4 = 1, \quad y = \sqrt{10}$
(i) $x^2 + y^2 = 4z, \quad x = -2$
(j) $x^2 - y^2/4 = -z, \quad z = 0$

4 (a) $x^2 + y^2 + z^2 = 52, \quad y = -6$
(b) $x^2 + y^2 + z^2 = 49, \quad x = 3$
(c) $x^2/9 + y^2/36 + z^2/4 = 1, \quad x = 2$
(d) $8x^2 + 2y^2 + z^2 = 144, \quad z = 10$
(e) $x^2 + y^2 - z^2 = 4, \quad x = -1$
(f) $x^2/36 + y^2/4 - z^2/16 = 1, \quad z = 0$
(g) $y^2 - x^2/4 - z^2/9 = 1, \quad y = \sqrt{5}$
(h) $y^2/16 - x^2/9 - z^2/36 = 1, \quad z = \sqrt{13}$
(i) $x^2/9 + y^2/4 = 16z, \quad x = -1$
(j) $x^2/36 - y^2/25 = -z, \quad z = 1$

5 Sketch the ellipsoid $x^2 + y^2 + z^2/4 = 1$. Describe the cross-sectional curves as seen from each of the three axes.

6 Sketch the hyperboloid of one sheet, $x^2 + y^2 - z^2 = 1$, and describe the cross sections.

7 Sketch the hyperboloid of two sheets, $y^2 - x^2 - z^2 = 4$, and describe the cross sections.

8 Sketch the elliptic paraboloid $x^2 + y^2 = -2z$, and describe the cross sections.

9 Sketch the hyperbolic paraboloid $x^2 - y^2 = 2z$ by drawing a rough outline of its intersections with each of the following planes: $x = 1$, $y = 1$, $z = 1$, $z = 0$, and $z = -1$.

10 A well-known classical result of geometry is that the four curves—circle, ellipse, parabola, and hyperbola—can be obtained as the intersection of planes with the right circular cone (Figure 2.28). Let the cone have Equation (11). Consider the intersection of the cone and the plane $Ax + Bz = D$, which is parallel to the y-axis.

(a) Derive the equation of the intersection by solving the equation of the plane for z, and then substituting this in the equation of the cone.

(b) Complete the square on the variable x in the equation in (a).

(c) Under what conditions on the constants A, B, c, and D does the equation in (b) represent a particular kind of curve? (These curves are called **conic sections.**)

2.4 Tangent Planes

In Section 1.7 we showed how to get the equation of the tangent line from that of the curve. In three dimensions, the curve is replaced by the surface, and the tangent line is replaced by the **tangent plane.** The simplest case is the plane tangent to the sphere. An example of this is the plane of the floor tangent to a spherical ball placed on it. Let $P_0(x_0,y_0,z_0)$ be an arbitrary point on the sphere. A plane through this point is tangent to the sphere if the plane is perpendicular to the line through the center (origin) and P_0. Since the direction numbers of this line are (x_0,y_0,z_0), these are also the normal direction numbers of the plane. It follows that the equation of the plane is

$$x_0(x - x_0) + y_0(y - y_0) + z_0(z - z_0) = 0. \tag{1}$$

This may be written as

$$x_0x + y_0y + z_0z = x_0^2 + y_0^2 + z_0^2.$$

Since P_0 is on the sphere, it follows that its coordinates must satisfy the equation of the sphere, namely, $x_0^2 + y_0^2 + z_0^2 = r^2$, where r is the radius. Therefore, Equation (1) takes the form

$$x_0x + y_0y + z_0z = r^2. \tag{2}$$

This is the equation of the plane tangent to the sphere of radius r at the point P_0.

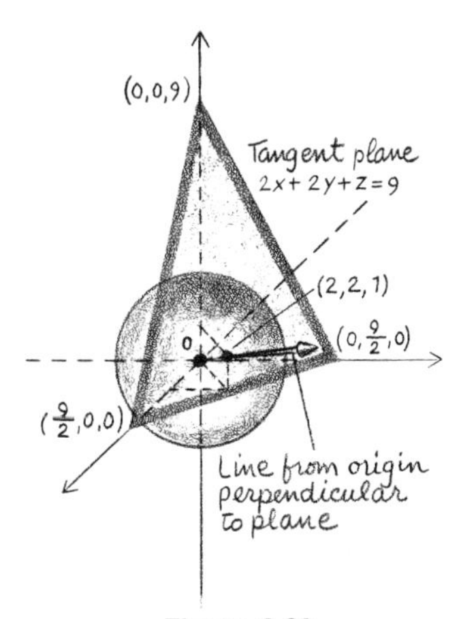

Figure 2.29

EXAMPLE 1

Find the equation of the plane tangent to the sphere of radius 3 at the point (2,2,1). Put the coordinates of this point equal to x_0, y_0, and z_0, respectively, in Equation (2):

$$2x + 2y + z = 9.$$

This plane cuts the axes at the points $(\frac{9}{2},0,0)$, $(0,\frac{9}{2},0)$ and $(0,0,9)$, respectively. The portion of the plane in the first octant appears in Figure 2.29. ▲

EXAMPLE 2

Given the plane $2x + y - z = 9$, find the sphere about the origin to which the plane is tangent, and the point of tangency. The numbers 2, 1, and -1 form a set of normal direction numbers (Section 2.2) for the plane; therefore, the normal line from the origin to the plane touches the plane at a point with coordinates of the form $(2c,c,-c)$, where c is some number. We have to find c. Since this point belongs to the plane, its coordinates must satisfy the equation of the plane:

$$2(2c) + c + c = 9, \qquad \text{or} \qquad c = \tfrac{3}{2}.$$

It follows that the point of tangency has the coordinates $(3,\frac{3}{2},-\frac{3}{2})$. Its distance from the origin is

$$\sqrt{3^2 + (\tfrac{3}{2})^2 + (-\tfrac{3}{2})^2} = 3\sqrt{1 + \tfrac{1}{4} + \tfrac{1}{4}} = \tfrac{3}{2}\sqrt{6}\,;$$

thus, the radius of the sphere is $\frac{3}{2}\sqrt{6}$. ▲

● Now we define the plane tangent to the ellipsoid

$$\frac{x^2}{a^2} + \frac{y^2}{b^2} + \frac{z^2}{c^2} = 1 \tag{3}$$

at the point $P_0(x_0,y_0,z_0)$. It is the plane

$$\frac{x_0x}{a^2} + \frac{y_0y}{b^2} + \frac{z_0z}{c^2} = 1 \tag{4}$$

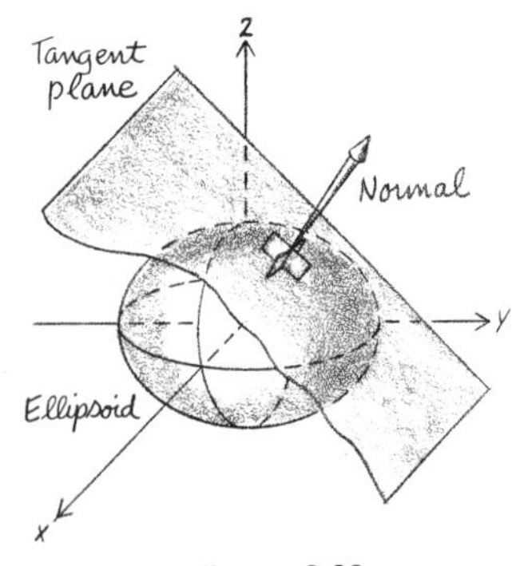

Figure 2.30

(Figure 2.30). The normal direction numbers are the coordinates of the points of tangency, divided by the corresponding factors a^2, b^2, and c^2.

We give a short explanation of Equation (4). Let $P(x,y,z)$ be another point on the ellipsoid; it satisfies Equation (3). Now write the latter equation in the equivalent form

$$\frac{(x - x_0 + x_0)^2}{a^2} + \frac{(y - y_0 + y_0)^2}{b^2} + \frac{(z - z_0 + z_0)^2}{c^2} = 1.$$

Put $A = x - x_0$ and $B = x_0$, and apply the formula for $(A + B)^2$; and do the same for the two other numerators in the equation above. After squaring and rearranging terms, we write the equation as

$$\left\{\frac{(x-x_0)^2}{a^2} + \frac{(y-y_0)^2}{b^2} + \frac{(z-z_0)^2}{c^2}\right\} + 2\left\{\frac{xx_0}{a^2} + \frac{yy_0}{b^2} + \frac{zz_0}{c^2}\right\} - \left\{\frac{x_0^2}{a^2} + \frac{y_0^2}{b^2} + \frac{z_0^2}{c^2}\right\} = 1.$$

The third term on the left-hand side is equal to 1 because P_0 is on the ellipse and so its coordinates satisfy Equation (3). The first term gets smaller and tends to 0 as P approaches P_0. Therefore, if P is very close to P_0, then the coordinates of P satisfy an equation which differs from (4) only in that there is a small extra term on the left-hand side. It follows from the reasoning above that **any** point P on the ellipse which is very close to P_0 satisfies an equation given approximately by (4). Since the latter is the equation of a plane we define it to be the equation of the **tangent** plane.

EXAMPLE 3

The equation of the plane tangent to the ellipsoid $x^2/9 + y^2/25 + z^2/16 = 1$ at the point $(-1, 3, \frac{4}{15}\sqrt{119})$ is

$$\frac{-x}{9} + \left(\frac{3}{25}\right) y + \left(\frac{\sqrt{119}}{60}\right) z = 1. \blacktriangle$$

t

Now we give the general rule for obtaining the tangent plane at a point $P_0(x_0, y_0, z_0)$ of a quadric surface of the form (1), (2), or (3) in Section 2.3. In each of the latter equations, replace the factors x^2, y^2, and z^2 by the corresponding factor $2x_0(x - x_0)$, $2y_0(y - y_0)$, and $2z_0(z - z_0)$, respectively; the factor z by $z - z_0$; and the terms 0 and 1 by 0. According to this rule the tangent planes are

$$\pm\frac{x_0(x-x_0)}{a^2} \pm \frac{y_0(y-y_0)}{b^2} \pm \frac{z_0(z-z_0)}{c^2} = 0; \tag{5}$$

$$\pm\frac{x_0(x-x_0)}{a^2} \pm \frac{y_0(y-y_0)}{b^2} = c(z-z_0); \tag{6}$$

$$\pm\frac{x_0(x-x_0)}{a^2} \pm \frac{y_0(y-y_0)}{b^2} \pm \frac{z_0(z-z_0)}{c^2} = 0, \tag{7}$$

corresponding to (1), (2), and (3), respectively. The reasoning is exactly the same as that outlined above for the special case of the ellipsoid. Write the equation of the surface in terms of x, y, and z; introduce the terms x_0 and $-x_0$, y_0, and $-y_0$, and z_0 and $-z_0$; expand the squares and rearrange terms; and apply the equation of the surface to the point P_0.

EXAMPLE 4

Find the plane tangent to the hyperboloid of one sheet

$$\frac{x^2}{4} + y^2 - \frac{z^2}{9} = 1$$

at the point $(2, \sqrt{7}/3, \sqrt{7})$. We apply Equation (5) with $a = 2$, $b = 1$, $c = 3$, and (x_0, y_0, z_0) the point of tangency:

$$\frac{x-2}{2} + \left(\frac{\sqrt{7}}{3}\right)\left(y - \frac{\sqrt{7}}{3}\right) - \left(\frac{\sqrt{7}}{9}\right)(z - \sqrt{7}) = 0,$$

or

$$\frac{x}{2} + (\sqrt{7})\frac{y}{3} + (\sqrt{7})\frac{z}{9} = 1.$$

[Compare Equation (4).] ▲

EXAMPLE 5

The plane tangent to the elliptic paraboloid

$$\frac{x^2}{9} + \frac{y^2}{4} = 4z$$

at the point (x_0, y_0, z_0) is, by Equation (6),

$$\frac{2x_0(x - x_0)}{9} + \frac{2y_0(y - y_0)}{4} = 4(z - z_0),$$

which, by the equation for the surface at (x_0, y_0, z_0) is

$$\frac{x_0 x}{9} + \frac{y_0 y}{4} = 2(z + z_0).$$

In the particular case in which the point of tangency is the origin, the equation reduces to $z = 0$, which is the equation of the xy-plane (Figure 2.31). ▲

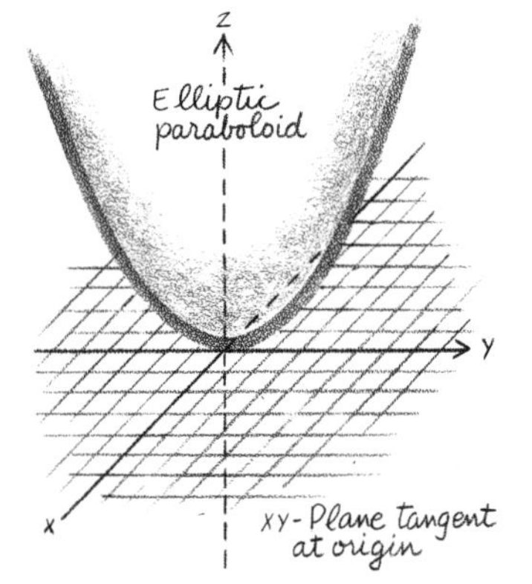

Figure 2.31

EXAMPLE 6

The plane tangent to a point of a cylinder is always parallel to the axis of the cylinder. Consider the circular cylinder $x^2 + y^2 = 16$. By the same method for deriving the tangent plane of the quadric surface, we find that the tangent **plane** of this cylinder has the same equation as the tangent line of the circular cross section of the cylinder. In this case the equation is $x_0 x + y_0 y = 16$ (Figure 2.32). This plane is parallel to the z-axis. For example, the plane tangent to the cylinder at the point $(3,\sqrt{7},-1)$ is $3x+\sqrt{7}y=16$. The z-coordinate has no role in the equation of this plane. ▲

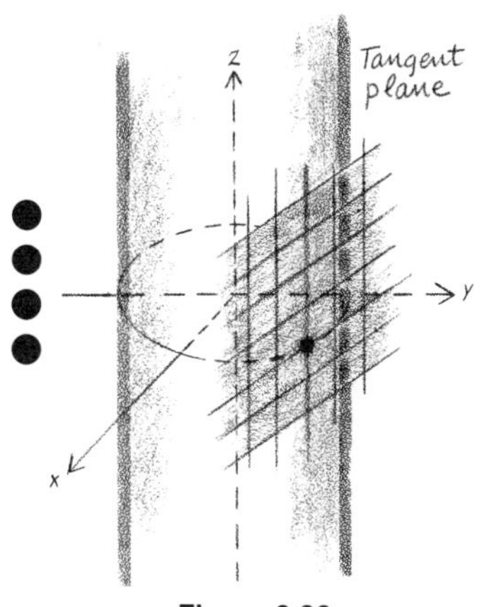

Figure 2.32

EXAMPLE 7

A surface has a **cap** at a point about which all neighboring points on the surface are **below** the former point. The surface has a **cup** at a point about which the neighboring points are all **above** the former point. For example, a sphere (in standard position) has a cap at the point where it cuts the positive z-axis, and a cup at the point where it cuts the negative z-axis. The elliptic paraboloid $x^2/a^2+y^2/b^2=2cz$ has a cup at the origin when c is positive, and a cap when c is negative. In general, the plane tangent to the surface at a cap or cup is perpendicular to the z-axis (Figure 2.33). The converse is not necessarily true; there does not have to be a cap or a cup whenever the tangent plane is perpendicular to the z-axis. An example of such a situation is that of the plane tangent to the hyperbolic paraboloid [Equation (9), Section 2.3] at the origin. According to formula (6) above for the tangent plane, the latter has the equation $z=0$, which is the xy-plane. However, the surface has a "saddlepoint" at the origin; it has a cap when viewed from the x-axis, and a cup when viewed from the y-axis. The relations between the cap and the cup and the tangent plane will be studied in more generality and detail in connection with functions of two variables (Section 7.3).

Other terms for the cap and the cup are "relative maximum and minimum," and "peak and pit." ▲

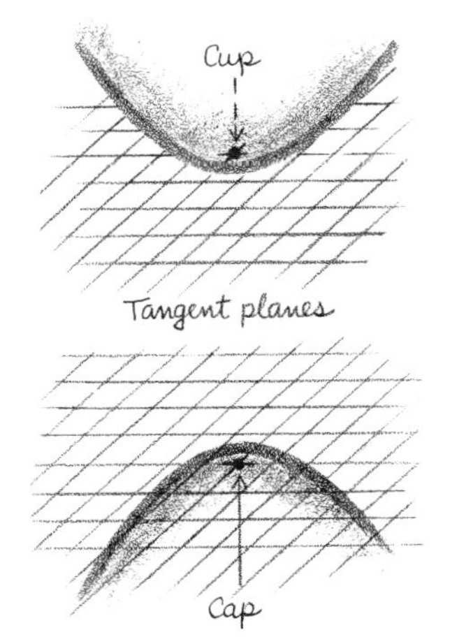

Figure 2.33

2.4 EXERCISES

In Exercises 1 and 2 find the equations of the tangent planes at the points of the given surfaces.

1 (a) $x^2+y^2+z^2=4$, at $(1,-1,\sqrt{2})$
(b) $x^2/4+y^2/9+z^2=1$, at $(-1,2,\sqrt{11}/6)$
(c) $2x^2+2y^2+z^2=9$, at $(1,\sqrt{3},-1)$
(d) $x^2/16+y^2/25=3z$, at $(4,-5,\frac{2}{3})$
(e) $x^2-y^2=6z$, at $(4,-3,\frac{7}{6})$.
(f) $\frac{1}{2}x^2-y^2-z^2=1$, at $(2\sqrt{3},1,2)$
(g) $x^2+y^2-4z^2=0$, at $(0,2,-1)$

2 (a) $x^2+y^2+z^2=16$, at $(2,3,-\sqrt{3})$
(b) $x^2+4y^2+z^2=9$, at $(-3,0,0)$
(c) $x^2/9+y^2/4=z$, at $(3,2,2)$
(d) $x^2/25-y^2/5=z$, at $(2\sqrt{5},2,0)$
(e) $x^2+2y^2-4z^2=1$, at $(1,-1,\sqrt{2}/2)$
(f) $x^2/4-y^2/2-z^2/3=1$, at $(2\sqrt{6},-2,3)$
(g) $x^2-4z^2=0$, at $(2,15,-1)$

3 Find the radius of the circle about the origin to which the given plane is tangent.
(a) $x + y + z = 4$
(b) $2x - y + 8z = 5$
(c) $x - 2y - 7z = 2$
(d) $3x - y = -7$
(e) $2x + 5z = 3$

4 The equation of the hyperboloid of one sheet rolled around the y-axis is $x^2/a^2 - y^2/b^2 + z^2/c^2 = 1$. Find the formula for the equation of the tangent plane at the point $(0,0,c)$. Is there a cap or cup? Consider the cases of both positive and negative c.

5 A hyperboloid of two sheets with its axis along the z-axis has the equation $z^2 - x^2 - y^2 = 1$. Find the formula for the equation of the tangent plane at the point (x_0,y_0,z_0). Find two points where the tangent plane is perpendicular to the z-axis. Is there a cap or cup at either of these points?

6 In Section 1.4 it was shown that the parabola has a reflection property which is used in mirrors. Explain why the circular paraboloid with an equation $x^2 + y^2 = 4cz$ is the **surface** that is appropriate for a headlight reflector. [*Hint:* Consider the cross section of the paraboloid on a plane through the z-axis. Any ray from the source at the focus travels along such a plane (Figure 2.34).]

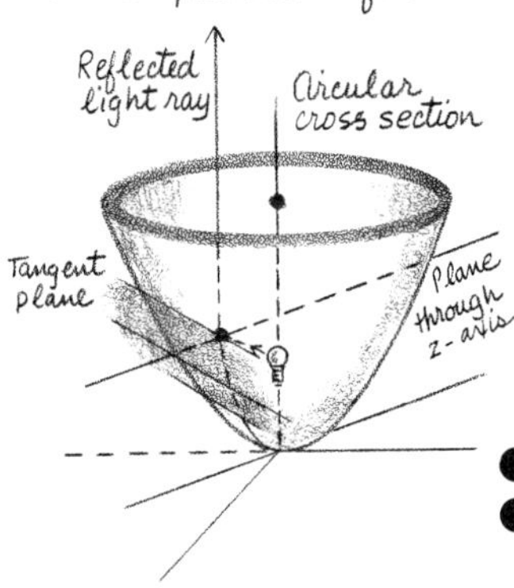

Figure 2.34

7 Consider a plane tangent to the sphere of radius 1 at a point P_0. Fill in the details of the proof of the following result: If P is a point on the plane distinct from P_0, then P is necessarily outside the sphere.
(a) The distance from $P(x,y,z)$ to P_0 is equal to $\sqrt{x^2 + y^2 + z^2 - 1}$.
(b) Use the result in (a) and the Pythagorean Theorem to deduce the asserted result.

8 The distance from the origin to a plane is defined as the radius of the sphere about the origin to which the plane is tangent. Let t represent this distance, and let A, B, and C be the normal direction numbers of the plane.
(a) Show that the coordinates of the point of tangency are represented by

$$\frac{At}{\sqrt{A^2 + B^2 + C^2}}, \qquad \frac{Bt}{\sqrt{A^2 + B^2 + C^2}}, \qquad \frac{Ct}{\sqrt{A^2 + B^2 + C^2}}.$$

(b) Let the equation of the plane be $Ax + By + Cz = D$. Show that the formula for the distance is

$$t = \frac{|D|}{\sqrt{A^2 + B^2 + C^2}},$$

[*Hint:* Substitute the coordinates in (a) in the plane equation, and solve for the unknown t.]

CHAPTER

3

Matrices and their Applications

3.1 Definitions of Matrix and Vector

A rectangular array of real numbers such as

$$\begin{bmatrix} 2 & -0.5 \\ 3.1 & 0 \end{bmatrix} \quad \text{or} \quad \begin{bmatrix} 2 & 1 & -7 \\ 3 & 8 & 15 \end{bmatrix}$$

is called a **matrix.** The **rows** of the matrix are the sets of numbers appearing in the same horizontal line; for example, the first row of the first matrix above consists of the numbers 2 and -0.5, in that order. The columns are those in the same vertical line; for example, the third column of the second matrix has the entries -7 and 15. A matrix with 2 rows and 3 columns, such as the second one above is called a "2-by-3" matrix. More generally, a matrix with m rows and n columns is called an "m-by-n" matrix. A matrix is **square** if it has the same number of rows and columns.

Capital letters represent matrices, and lower case letters represent the numbers appearing in them. These numbers are the **entries** or

elements of the matrix. They are indexed by a double subscript. For example, let the 2-by-3 matrix A be defined as

$$A = \begin{bmatrix} a_{11} & a_{12} & a_{13} \\ a_{21} & a_{22} & a_{23} \end{bmatrix}.$$

The first subscript on the entry is the number of the row to which it belongs, and the second subscript is the entry of the column; thus, a_{13} is the entry in the first row and the third column. In general, we denote an m-by-n matrix A by

$$A = \begin{bmatrix} a_{11} & a_{12} & \cdots & a_{1n} \\ a_{21} & a_{22} & \cdots & a_{2n} \\ & \vdots & & \\ a_{m1} & a_{m2} & \cdots & a_{mn} \end{bmatrix}.$$

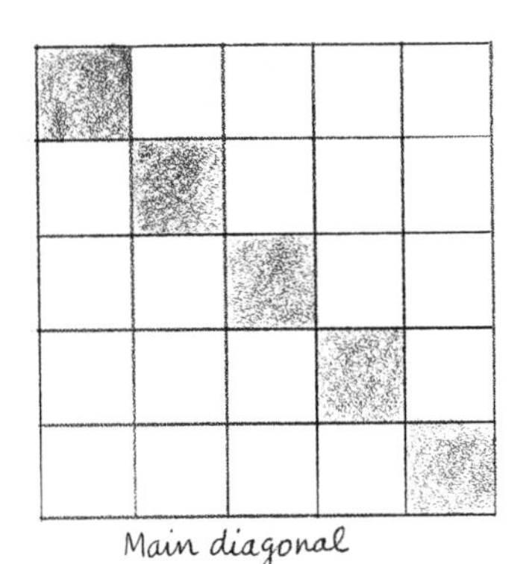

Figure 3.1

We refer to a_{ij} as the entry in the ith row and jth column. The indices i and j run from 1 to m and n, respectively. A is square if $m = n$. In this case the **main diagonal** is the set of entries consisting of all those with index i equal to index j (Figure 3.1).

Two matrices are defined as **equal** if they have the same numbers of rows and columns and if all the corresponding entries are equal. For example, although the two matrices

$$\begin{bmatrix} 2 & 1 & -7 \\ 3 & 8 & 15 \end{bmatrix} \quad \text{and} \quad \begin{bmatrix} 2 & 3 \\ 1 & 8 \\ -7 & 15 \end{bmatrix}$$

have the same populations of entries, they have different numbers of rows and columns, and therefore are not considered to be equal. Another example of a pair of unequal matrices is

$$\begin{bmatrix} 2 & 1 & -7 \\ 3 & 8 & 15 \end{bmatrix} \quad \text{and} \quad \begin{bmatrix} 2 & -7 & 1 \\ 3 & 8 & 15 \end{bmatrix};$$

here the last two numbers in the first rows do not appear in the same order.

Matrices with one row or one column are called **vectors.** A matrix with 1 row and n columns is called a row vector of order n. Similarly, a matrix with m rows and 1 column is called a column vector of order m. For example,

$$A = (1 \quad 3 \quad -1)$$

is a row vector of order 3, and

$$B = \begin{bmatrix} 7 \\ 2 \end{bmatrix}$$

is a column vector of order 2. The entries of a vector are often called **components,** and the order of the vector is the **dimension.** When we do not have to distinguish between a row and column vector with the same entries and of the same dimension, we shall refer to it simply as a vector.

Geometric interpretations may be omitted by those studying matrices before geometry.

EXAMPLE 1

There is a correspondence between points in the xy-plane and vectors of order 2. With each point we associate the vector whose first component is the abscissa and whose second is the ordinate. In three dimensions the points are associated with vectors of order 3 (Figure 3.2). ▲

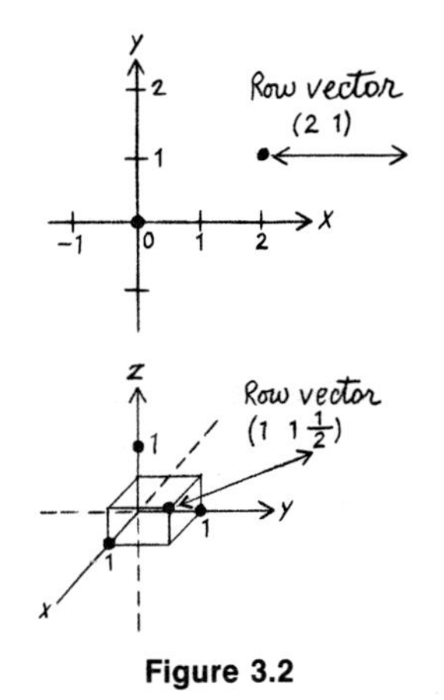

Figure 3.2

Two vectors can be added to form a third vector, as follows: Two vectors of the same dimension, which are both row vectors or both column vectors, have a **sum.** This is the vector formed by adding the corresponding components. For example, the sum of the two row vectors

$$X = (1 \quad 3 \quad -2) \qquad \text{and} \qquad Y = (0 \quad -4 \quad 1)$$

is the vector

$$(1+0 \quad 3-4 \quad -2+1) = (1 \quad -1 \quad -1);$$

it is denoted $X + Y$. Two vectors of different orders cannot be added.

EXAMPLE 2

There is a geometric interpretation for the addition of two-dimensional vectors in terms of their corresponding points in the plane. If X and Y are the vectors representing two points in the plane, then $X + Y$ is the vector which represents the corner of a parallelogram whose other three corners are X, Y, and the origin. For example, if $X = (2,1)$ and $Y = (1,3)$, then $X + Y$ is equal to (3,4), and is on the opposite corner of the parallelogram in Figure 3.3. ▲

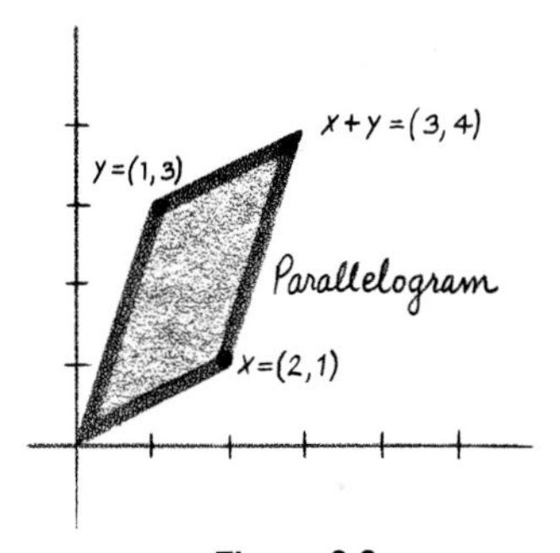

Figure 3.3

A vector can be multiplied by a number to form another vector. This multiplication is called **multiplication by a scalar;** a number is contrasted to a vector by referring to the former as a scalar. The product is obtained by multiplying each component of the vector by the given number. For example, the product of the number 3 and the vector $(1 \quad 0 \quad -4)$ is

$$3(1 \quad 0 \quad -4) = (3 \quad 0 \quad -12).$$

Next we define a product of two vectors. It differs from the usual concept of the product of two elements: The product of two numbers is a number; the sum of two vectors is a vector. However, the product of the two vectors, now to be defined, is a number, not a vector; that is,

vector **times** vector **equals** number.

This kind of product is called either **dot** product, **inner** product, or **scalar** product. For the sake of uniformity, we shall use the term inner product:

Let $A = (a_1, \ldots, a_n)$ and $B = (b_1, \ldots, b_n)$ be two vectors of the same dimension n. The inner product, written $A \cdot B$, is defined as

$$A \cdot B = a_1b_1 + a_2b_2 + \cdots + a_nb_n. \tag{1}$$

For example, if $A = (3 \quad 0 \quad 7)$ and $B = (-1 \quad 2 \quad 4)$, then $A \cdot B = 3(-1) + 0(2) + 7(4) = 25$.

A special kind of inner product is the product of A and itself:

$$A \cdot A = a_1^2 + \cdots + a_n^2.$$

The symbol $|A|$ stands for its square root:

$$|A| = \sqrt{a_1^2 + \cdots + a_n^2}.$$

It is called the **length** of the vector A. The reason for this is that in two and three dimensions the lengths of the vectors representing points are, by our definition, actually equal to the distances of the points from the origin. We think of the vector geometrically as a line pointing from the origin to the point in the plane or space, and the length of the vector as the length of the line connecting them (Figure 3.4).

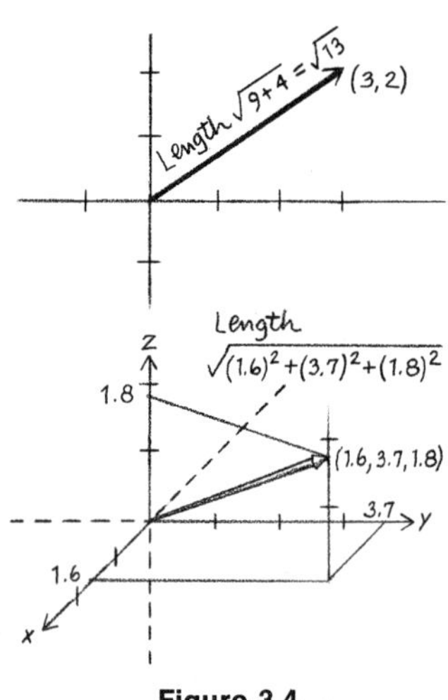

Figure 3.4

EXAMPLE 3

Formula (1) for the inner product of two vectors arises naturally in calculations relating unit prices to total cost in economics. Consider the list of prices for fruits in a supermarket on August 2, 1971, given in Figure 3.5. A housewife purchases 8, 5, 3, and 15 pounds of four kinds of fruit. The list of unit prices is taken as the vector A, the list of quantities as the vector B, and the total cost of the order as the inner product of the vectors, $A \cdot B$. ▲

The notions of addition of vectors and multiplication of vectors by scalars are now extended to matrices of any size, not just matrices that are vectors. Two matrices having the same numbers of rows and columns, respectively, have the **sum** which is the matrix obtained by adding the corresponding entries:

$$\begin{bmatrix} a_{11} & a_{12} & \cdots & a_{1n} \\ a_{21} & a_{22} & \cdots & a_{2n} \\ \cdot & & & \\ \cdot & & & \\ \cdot & & & \\ a_{m1} & a_{m2} & \cdots & a_{mn} \end{bmatrix} + \begin{bmatrix} b_{11} & b_{12} & \cdots & b_{1n} \\ b_{21} & b_{22} & \cdots & b_{2n} \\ \cdot & & & \\ \cdot & & & \\ \cdot & & & \\ b_{m1} & b_{m2} & \cdots & b_{mn} \end{bmatrix}$$

$$= \begin{bmatrix} a_{11} + b_{11} & a_{12} + b_{12} & \cdots & a_{1n} + b_{1n} \\ a_{21} + b_{21} & a_{22} + b_{22} & \cdots & a_{2n} + b_{2n} \\ \vdots & & & \\ a_{m1} + b_{m1} & a_{m2} + b_{m2} & \cdots & a_{mn} + b_{mn} \end{bmatrix}.$$

For example, if

$$A = \begin{bmatrix} 0 & -3 \\ 1 & 2 \end{bmatrix}, \qquad B = \begin{bmatrix} 1 & 2 \\ -2 & -3 \end{bmatrix},$$

then,

$$A + B = \begin{bmatrix} 0+1 & -3+2 \\ 1-2 & 2-3 \end{bmatrix} = \begin{bmatrix} 1 & -1 \\ -1 & -1 \end{bmatrix}.$$

Figure 3.5

	Vector A Price/lb	Vector B No. of lbs	Total price
Peaches 	$.29	8	$2.32
Bananas 	$.15	5	$.75
Grapes 	$.49	3	$1.47
Watermelon 	$.07	15	$1.05

Scalar product $A \cdot B =$ $5.59

If c is a number and A a matrix, then cA (c times A) is the matrix obtained by multiplying each entry of A by the same number c. For example, if $c = 3$ and $A = \begin{bmatrix} 1 & 3 \\ 5 & -2 \end{bmatrix}$, then

$$cA = \begin{bmatrix} 3 & 9 \\ 15 & -6 \end{bmatrix}.$$

Subtraction of vectors and matrices is defined by multiplying the subtrahend by -1, and proceeding as in addition; the difference $A - B$ is defined as A plus the matrix $(-1)B$. In other words, the difference is obtained by taking the difference of the corresponding entries of the matrices.

It is important to remember that the **sum is defined only for two matrices having exactly the same numbers of rows and columns.** The inner product has been defined only for two vectors, not matrices in general. This kind of multiplication will be extended to matrices of a general kind in Section 3.2.

3.1 EXERCISES

In Exercises 1 to 4 compute the sums and products whenever they are properly defined.

1 Put $U = (2 \quad -1)$, $V = (4 \quad 5)$, $W = (5 \quad -3 \quad 0)$, $R = (2 \quad -4 \quad 5)$, $S = (3 \quad 1 \quad 1)$, $Z = (0 \quad 0 \quad 0)$.
Compute:

(a) $U + V$
(b) $V + U$
(c) $W + Z$
(d) $Z + W$
(e) $R - S$
(f) $2U$
(g) $-5R$
(h) $V + W$
(i) $3(W + R) - S$

2 Put $A = (-3 \quad 5)$, $B = (7 \quad 10)$, $C = (1 \quad -2 \quad 3)$ $D = (4 \quad 6 \quad -1)$, $E = (0 \quad 0 \quad 0)$.
Compute:

(a) $B + A$
(b) $A + B$
(c) $C + D$
(d) $C + E$
(e) $B + C$
(f) $B + E$
(g) $2A + B$
(h) $C - \frac{1}{2}D$
(i) $4(A - B)$
(j) $2(E - C) + D$

3 Put $A = \begin{bmatrix} 1 & 2 \\ 3 & 4 \end{bmatrix}$, $B = \begin{bmatrix} 3 & 2 \\ 0 & 1 \end{bmatrix}$, $C = \begin{bmatrix} 1 & 0 \\ 0 & 1 \end{bmatrix}$,

$$D = \begin{bmatrix} 3 & 0 \\ 2 & 7 \\ 4 & 1 \end{bmatrix}, \quad E = \begin{bmatrix} -2 & 5 \\ -3 & 2 \\ -1 & 6 \end{bmatrix}.$$

Compute whenever defined:

(a) $A + B$
(b) $(A + B) + C$
(c) $(C + B) + A$
(d) $(3A - B) + 2C$
(e) $(2C - B) - 3A$
(f) $D + E$
(g) $D - 2E$
(h) $-2D + E$
(i) $3D - 2E$
(j) $(2A + B) + 7E$

4 Put $A = \begin{bmatrix} 0 & -1 \\ 1 & 0 \end{bmatrix}$, $B = \begin{bmatrix} 2 & 3 \\ 5 & 4 \end{bmatrix}$, $C = \begin{bmatrix} 2 & -1 \\ -3 & 1 \end{bmatrix}$,

$D = \begin{bmatrix} 2 & 4 & 2 \\ 6 & 2 & 4 \end{bmatrix}$, $E = \begin{bmatrix} 2 & -3 & 4 \\ 6 & 2 & 5 \end{bmatrix}$.

Compute whenever defined:

(a) $A + B$
(b) $A + C$
(c) $B + C$
(d) $(A + B) - C$
(e) $(B + C) - 2A$
(f) $-B + D$
(g) $3(B - C) + 5A$
(h) $2D - E$
(i) $2E - D$
(j) $-3D + 4E$

In Exercises 5 and 6 draw the pair of vectors and their sum as lines in the plane.

5 (a) $(3 \quad -1)$, $(4 \quad 2)$
(b) $(1 \quad 0)$, $(-4 \quad 3)$
(c) $(-2 \quad 2)$, $(1 \quad -7)$

6 (a) $(8 \quad 4)$, $(1 \quad -1)$
(b) $(2 \quad 2)$, $(-1 \quad 3)$
(c) $(-5 \quad 1)$, $(1 \quad -7)$

7 Calculate these inner products for $X = (1 \quad 2 \quad 3)$, $Y = (-2 \quad 0 \quad 2)$, $Z = (8 \quad 7 \quad 0)$:

(a) $X \cdot Y$
(b) $Z \cdot X$
(c) $Z \cdot Y$
(d) $X \cdot (Y - Z)$
(e) $X \cdot Y - X \cdot Z$
(f) $X \cdot (3Y - Z)$

8 Calculate these inner products for $X = (6 \quad 3 \quad 0)$, $Y = (2 \quad 5 \quad 4)$, $Z = (0 \quad 8 \quad 9)$:

(a) $Y \cdot X$
(b) $Z \cdot X$
(c) $Z \cdot Y$
(d) $X \cdot (Y - Z)$
(e) $X \cdot Y - X \cdot Z$
(f) $(X + Y) \cdot (Y - Z)$

9 If the vector $(x \quad y)$ is associated with the point (x,y) in the plane, how are the positions of the points associated with $3(x \quad y)$ and $(-1)(x \quad y)$ related to the original point?

10 If $X = (0 \quad 4 \quad -1)$ and $Y = (1 \quad 0 \quad 2)$, find the lengths of:

(a) X
(b) Y
(c) $X + Y$
(d) $X - Y$

11 Show that vector addition is associative and commutative; that is, if X, Y, and Z are vectors, then $(X + Y) + Z = X + (Y + Z)$ and $X + Y = Y + X$. Give the proof in the particular case of two-dimensional row vectors.

12 Show that the multiplication of a vector by a scalar is distributive with respect to vector addition: If c is a number and X and Y are vectors, then

$$c(X + Y) = cX + cY.$$

Give the proof for two-dimensional row vectors.

13 Extend the results of Exercises 11 and 12 from vectors to matrices. Give the proofs for the case of 2-by-2 matrices.

14 A mutual fund purchases 200 shares of telephone stock at \$45 per share, 50 shares of a computer stock at \$300 per share, 80 shares of automotive stock at \$125 per share, and 25 shares of telegraph stock at \$200 per share. Write the price and quantity vectors and express the total cost of the investment as an inner product.

15 Show directly from the definition of the length of a vector that for any number c and vector X, $|cX| = |c||X|$; that is, the length of cX is equal to the absolute value of c times the length of X.

16 Show that for any two vectors X and Y,

$$|X + Y|^2 = |X|^2 + 2X \cdot Y + |Y|^2.$$

Give the proof in the particular case of two-dimensional vectors.

Exercises 17 to 19 require elementary trigonometry.

17 Let $P(x,y)$ be a point in the plane, and θ the angle between the x-axis and the line from the origin to P. Express $\sin \theta$ and $\cos \theta$ in terms of the components and the length of the vector $(x \quad y)$.

18 Using the result of Exercise 17, prove that if $U = (x_1 \quad y_1)$ and $V = (x_2 \quad y_2)$ are vectors, then

$$U \cdot V = |U||V| \cos \phi,$$

where ϕ is the angle between the lines from the origin to $P(x_1,y_1)$ and $P(x_2,y_2)$, respectively (Figure 3.6). [*Hint:* Use the trigonometric identity $\cos(A - B) = \cos A \cos B + \sin A \sin B$.]

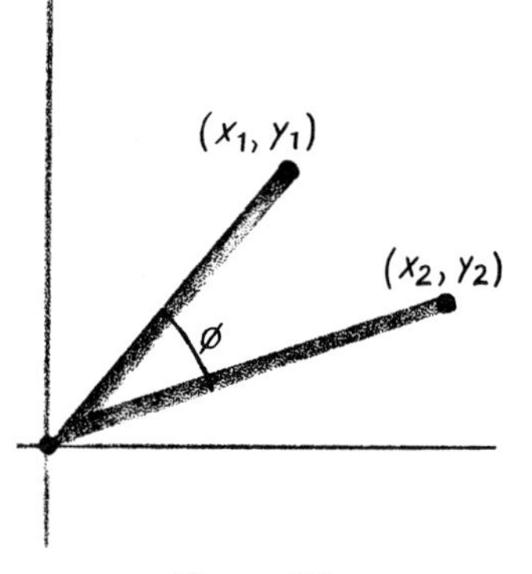

Figure 3.6

19 Why is $U \cdot V$ no greater in absolute value than the product $|U||V|$? Why is it **equal** in absolute value only when the origin and the points associated with the vectors are on one line? [*Hint:* Use the result of Exercise 18.]

3.2 Matrix Multiplication

Let A and B be matrices. If the number of columns of A is equal to the number of rows of B, then we define the product AB as follows:

The product $C = AB$ of the two matrices, multiplied in the given order, is the matrix whose (i,j)th entry is the inner product of the ith row of A and the jth column of B. The number of rows in the product matrix C is equal to the number of rows in A, and the number of columns in the product is equal to the number of columns in B.

EXAMPLE 1

Let A and B be the matrices

$$A = \begin{bmatrix} 1 & -2 \\ 3 & 1 \end{bmatrix} \qquad B = \begin{bmatrix} 3 & 0 & -2 \\ 5 & 1 & -1 \end{bmatrix}.$$

A has two columns and B two rows, so that the product C is well defined. Let A_1 and A_2 be the vectors representing the rows of A, and B_1, B_2, and B_3 the vectors representing the columns of B. The inner products are computed:

$$\begin{aligned}
A_1 \cdot B_1 &= 1(3) + (-2)5 = -7, \\
A_1 \cdot B_2 &= 1(0) + (-2)1 = -2, \\
A_1 \cdot B_3 &= 1(-2) + (-2)(-1) = 0, \\
A_2 \cdot B_1 &= 3(3) + 1(5) = 14, \\
A_2 \cdot B_2 &= 3(0) + 1(1) = 1, \\
A_2 \cdot B_3 &= 3(-2) + 1(-1) = -7.
\end{aligned}$$

Therefore, the product C of the two matrices is

$$C = \begin{bmatrix} -7 & -2 & 0 \\ 14 & 1 & -7 \end{bmatrix}.$$

The general characteristics of matrix multiplication are sketched in Figure 3.7.

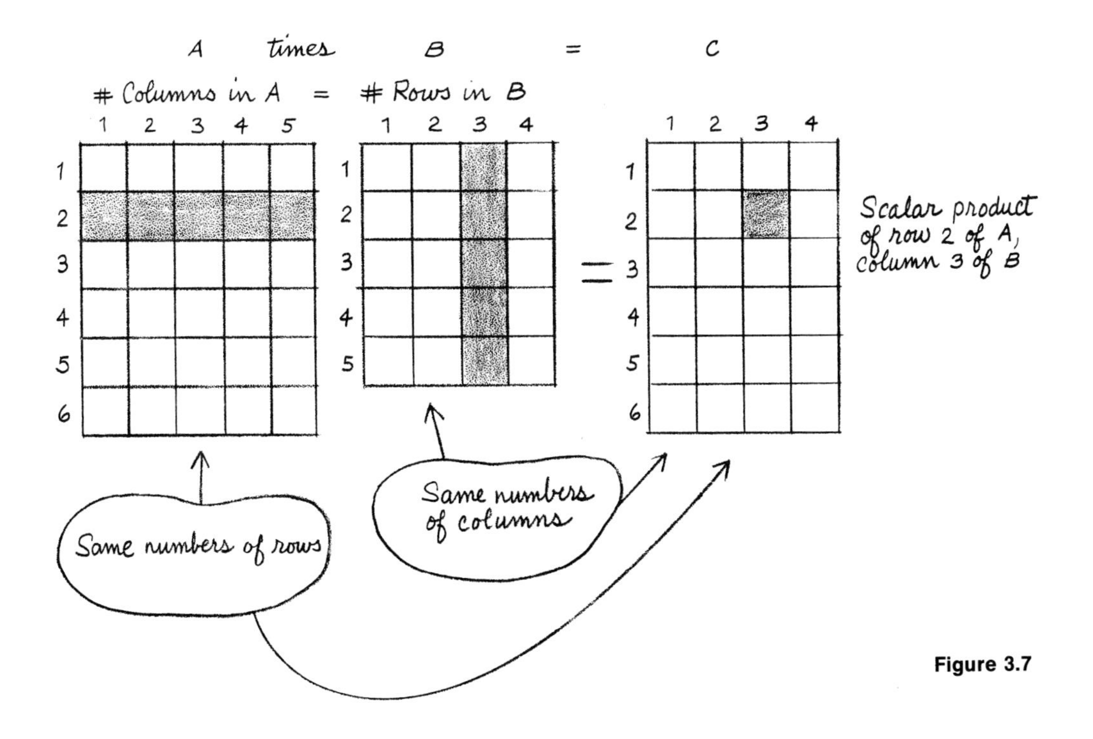

Figure 3.7

In forming the product AB, A is said to **premultiply** B, or B to **postmultiply** A. While in the ordinary multiplication of numbers a and b the product is **commutative,** that is, $ab = ba$, this is not generally true for matrices. First of all, while AB may be defined, the product in the reverse order may not necessarily be defined because the number of columns of B might not be equal to the number of rows of A. This is the case in Example 1, where B has 3 columns and A 2 rows. Even when the product **is** defined in both orders, the two product matrices are not always equal.

EXAMPLE 2

Put

$$A = \begin{bmatrix} 1 & 0 & 2 & 1 \\ 3 & -1 & 4 & -2 \end{bmatrix}, \qquad B = \begin{bmatrix} 1 & 0 \\ 0 & 2 \\ -3 & 1 \\ 8 & 1 \end{bmatrix};$$

then AB is the 2-by-2 matrix

$$\begin{bmatrix} 3 & 3 \\ -25 & 0 \end{bmatrix},$$

and BA is the 4-by-4 matrix

$$\begin{bmatrix} 1 & 0 & 2 & 1 \\ 6 & -2 & 8 & -4 \\ 0 & -1 & -2 & -5 \\ 11 & -1 & 20 & 6 \end{bmatrix}. \quad \blacktriangle$$

We mention that matrix multiplication is **associative;** that is, if A, B, and C are matrices for which the products AB and BC are defined, then $(AB)C = A(BC)$. In other words, the same product is obtained under the two orders of operation. The proof is left as an exercise (Exercise 5).

If A and B are square matrices of the same order, that is, the same numbers of rows and columns, then both AB and BA are defined but are not necessarily equal.

EXAMPLE 3

If

$$A = \begin{bmatrix} 1 & 2 \\ 0 & 1 \end{bmatrix} \quad \text{and} \quad B = \begin{bmatrix} -1 & 3 \\ 1 & 2 \end{bmatrix}$$

then

$$AB = \begin{bmatrix} 1 & 7 \\ 1 & 2 \end{bmatrix} \quad \text{and} \quad BA = \begin{bmatrix} -1 & 1 \\ 1 & 4 \end{bmatrix}. \quad \blacktriangle$$

A square matrix with 1's along the main diagonal and 0's in all places off the main diagonal is called an **identity matrix.** It serves for the multiplication of matrices the same role as the number 1 in the multiplication of numbers. If A is a matrix with n columns and if I_n is the n-by-n identity matrix, then AI is defined and is equal to A. For example, if

$$A = \begin{bmatrix} 1 & 2 & -1 \\ 1 & 0 & 3 \end{bmatrix} \quad \text{and} \quad I = \begin{bmatrix} 1 & 0 & 0 \\ 0 & 1 & 0 \\ 0 & 0 & 1 \end{bmatrix},$$

then

$$AI = \begin{bmatrix} 1 & 2 & -1 \\ 1 & 0 & 3 \end{bmatrix} = A.$$

Similarly, if B is a matrix with m rows, and I_m the identity of size m-by-m, then $IB = B$.

If A is a square matrix, then the product AA is defined; it is called A^2 ("A-square"). It is also a square matrix and has the same number of rows (and columns) as A. Then the product A^2A is also well defined; and, by the associative property of matrix multiplication, is equal to AA^2. This product is denoted A^3. The n-fold product $A \cdots A$ (n factors) is denoted A^n.

There are many interesting applications of matrix multiplication.

EXAMPLE 4

Matrix multiplication is used in calculating proportions of populations which change in accordance with a type of system known as a **Markov chain.** (It is named for A. A. Markov, 1856–1922.) Fish of a particular species live in a large lake. They move from one portion of the lake into another in accordance with the current, temperature, food, and chance factors. Suppose the lake is divided into 3 zones: A, B, and C. The fish are distributed throughout these zones. During a 24-hour period they migrate from zone to zone in the following manner:

(a) Of those in A, 25 percent remain, 40 percent move to B, and 35 percent to C.

(b) Of those in B, 50 percent remain, 25 percent move to A, and 25 percent to C.

(c) Of those in C, 40 percent remain, 20 percent move to A, and 40 percent to B. (See Figure 3.8.)

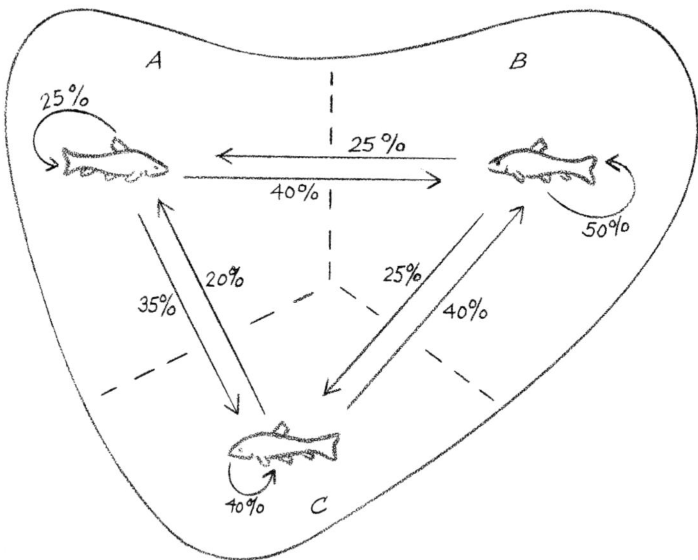

Figure 3.8

We arrange these proportions in a matrix P:

$$\begin{array}{l} \\ \text{from } A \\ \text{from } B \\ \text{from } C \end{array} \begin{array}{c} \begin{array}{ccc} \text{to } A & \text{to } B & \text{to } C \end{array} \\ \begin{bmatrix} 0.25 & 0.40 & 0.35 \\ 0.25 & 0.50 & 0.25 \\ 0.20 & 0.40 & 0.40 \end{bmatrix} \end{array} = P.$$

In each row, the numbers represent the proportions moving to the regions under the column heads. The sum of the entries in each row is 1. Such a matrix is called a **stochastic** matrix. We assume that the population remains fairly constant during the period under consideration.

Suppose on a given day that the fish are distributed in the zones A, B, and C in the proportions 0.30, 0.20, and 0.50, respectively. We express this distribution as a row vector $X = (0.30\ 0.20\ 0.50)$. We claim: The distribution of the fish after a 24-hour period is given by the components of the row vector XP, the matrix product of X and the "transition" matrix P above. To demonstrate this, let us calculate the proportion in zone A after the period. It consists of 25 percent of those who were in A at the beginning of the period, 25 percent of those who were in B, and 20 percent of those who were in C. Since the proportions in A, B, and C at the beginning of the period were $(0.30\ 0.20\ 0.50)$ it follows that the proportion in A at the end of the period is the sum of the products

$$(0.25)(0.30) + (0.25)(0.20) + (0.20)(0.50) = 0.225.$$

This is just the inner product of the vector X and the first column of P. Similarly, it is found that the proportion in B after the period is the inner product of X and the second column of P:

$$(0.40)(0.30) + (0.50)(0.20) + (0.40)(0.50) = 0.420.$$

The proportion in C is

$$(0.35)(0.30) + (0.25)(0.20) + (0.40)(0.50) = 0.355;$$

this is the inner product of X and the third column of P. It follows that the distribution after 1 period is

$$XP = (0.225 \quad 0.420 \quad 0.355).$$

To find the distribution after a 2-day period, we again apply the multiplication operation; however, this time the vector XP is the "present" distribution, so that it replaces the vector X, and the distribution after the next period is the product $(XP)\,P = XP^2$. Similarly, it can be shown that the distribution after 3 periods is XP^3; and, in general, the distribution after n periods is XP^n. Thus, the power P^n gives the transition proportions after n time periods. Although it is relatively easy to calculate small powers of the matrix, it may be rather difficult to directly calculate the higher powers. More powerful methods of calculation will be introduced later in connection with **characteristic vectors.** These will enable us to calculate the "long-term" distribution of the population in the various zones of the lake. ▲

EXAMPLE 5

In comparing the costs of living in various places, economists account not only for the prices but also for the quantities that are normally purchased. For example, a small rise in the cost of a universally used commodity (bread, rent) may have a more important effect on the population than a large rise in a rarely purchased item (shoe polish). Suppose a family of 4 consumes 3 dozen eggs per week, 10 quarts of milk, and 8 pounds of meat. At a particular time the prices in 4 American cities are

	Eggs	Milk	Meat
Chicago	60¢/doz	26¢/qt	85¢/lb
Los Angeles	80¢/doz	34¢/qt	94¢/lb
New York	65¢/doz	31¢/qt	93¢/lb
Philadelphia	70¢/doz	30¢/qt	91¢/lb

To compute the total cost of the 3 items in the various cities, we multiply the unit prices by the quantities purchased. This is done by writing the quantities as a column vector Q and the price schedule as a matrix P:

$$Q = \begin{bmatrix} 4 \\ 10 \\ 8 \end{bmatrix}, \qquad P = \begin{bmatrix} 0.60 & 0.26 & 0.85 \\ 0.80 & 0.34 & 0.94 \\ 0.65 & 0.31 & 0.93 \\ 0.70 & 0.30 & 0.91 \end{bmatrix};$$

then the total costs of the 3 items are given by the entries of the matrix product

$$PQ = \begin{bmatrix} 0.60 & 0.26 & 0.85 \\ 0.80 & 0.34 & 0.94 \\ 0.65 & 0.31 & 0.93 \\ 0.70 & 0.30 & 0.91 \end{bmatrix} \begin{bmatrix} 4 \\ 10 \\ 8 \end{bmatrix} = \begin{bmatrix} 11.80 \\ 14.12 \\ 13.14 \\ 13.08 \end{bmatrix}.$$

The entries give the costs for Chicago, Los Angeles, New York, and Philadelphia in that order. ▲

EXAMPLE 6

Sociologists and political scientists use matrices to analyze relations of **dominance** among individuals and groups. One is considered dominant over another if he is more powerful. In the theory of dominance relations, which we put into a mathematical form, we make the following assumptions:

(a) An individual does not dominate himself.

(b) For every pair of individuals under consideration, exactly one dominates the other, but not both.

Suppose that we are considering 3 politicians from a certain state, the governor, the senior senator, and the junior senator. The following power relations are assumed to exist among them: The senior senator dominates the junior senator, the latter dominates the governor, and the last one dominates the senior senator (Figure 3.9). These relations are summarized in the following matrix:

	Governor	Sr. Senator	Jr. Senator
Governor	0	1	0
Sr. Senator	0	0	1
Jr. Senator	1	0	0

The entry 1 in the ith row and the jth column signifies that the individual in row i dominates the one in column j. The entry 0 signifies that the individual does not dominate the other. We call this 3-by-3 matrix a **dominance matrix** $\boldsymbol{D}$**:**

$$D = \begin{bmatrix} 0 & 1 & 0 \\ 0 & 0 & 1 \\ 1 & 0 & 0 \end{bmatrix}.$$

Figure 3.9

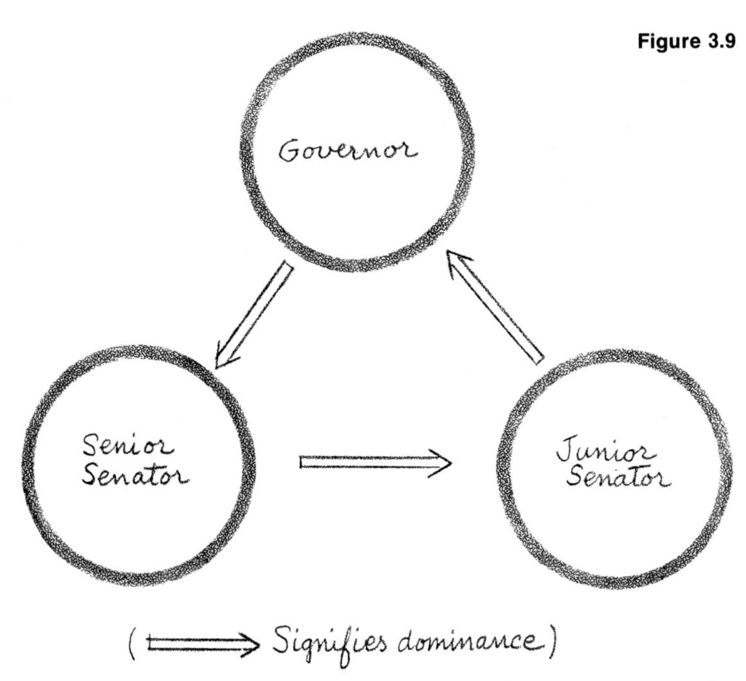

(⟹ Signifies dominance)

We say that the governor has *2-step dominance* over the junior senator because he dominates the senior senator, who, in turn, dominates the junior senator. Similarly, the junior senator has 2-step dominance over the senior senator; and the latter has 2-step dominance over the governor. The matrix for the 2-step dominance relations is

	Governor	Sr. Senator	Jr. Senator
Governor	0	0	1
Sr. Senator	1	0	0
Jr. Senator	0	1	0

This is the matrix obtained by multiplying the original matrix D itself:

$$DD = D^2 = \begin{bmatrix} 0 & 1 & 0 \\ 0 & 0 & 1 \\ 1 & 0 & 0 \end{bmatrix}\begin{bmatrix} 0 & 1 & 0 \\ 0 & 0 & 1 \\ 1 & 0 & 0 \end{bmatrix} = \begin{bmatrix} 0 & 0 & 1 \\ 1 & 0 & 0 \\ 0 & 1 & 0 \end{bmatrix}.$$

The power of an individual is often measured by the number of individuals whom he can dominate in either 1 or 2 steps. In terms of our matrix D, this is equal to the sum of the entries in the corresponding row of $D + D^2$. In this example, we have

$$D + D^2 = \begin{bmatrix} 0 & 1 & 1 \\ 1 & 0 & 1 \\ 1 & 1 & 0 \end{bmatrix}.$$

Each individual dominates two others in one or two stages so that they are considered equally powerful. In general D^2 may have positive entries larger than 1; these also signify dominance.

We mention an interesting mathematical result about dominance relations: Given a dominance relation on a set of n individuals, we can find at least one individual who dominates all others in 1 or 2 steps, and at least one individual who is dominated by all others in 1 or 2 steps. We omit the proof. ▲

EXAMPLE 7

As we shall see later, matrices are useful in solving simultaneous linear equations in several unknowns. Consider the pair of equations

$$\begin{aligned} x + 3y &= 6, \\ x + 2y &= 10. \end{aligned}$$

Let C be the matrix of coefficients:

$$C = \begin{bmatrix} 1 & 3 \\ 1 & 2 \end{bmatrix};$$

Then the system above may be expressed as a matrix equation,

$$\begin{bmatrix} 1 & 3 \\ 1 & 2 \end{bmatrix} \begin{bmatrix} x \\ y \end{bmatrix} = \begin{bmatrix} 6 \\ 10 \end{bmatrix} \quad \text{or} \quad C \begin{bmatrix} x \\ y \end{bmatrix} = \begin{bmatrix} 6 \\ 10 \end{bmatrix}.$$

The theory of matrices is used to solve for the unknown vector $\begin{bmatrix} x \\ y \end{bmatrix}$ in the most efficient way. ▲

3.2 EXERCISES

In Exercises 1 and 2 compute the products of the given matrices whenever possible. If it is not possible, state why.

1 Put $A = \begin{bmatrix} 3 \\ -1 \\ 2 \end{bmatrix}$, $B = (-5 \quad 3)$, $C = \begin{bmatrix} 2 & 3 \\ 1 & -1 \end{bmatrix}$, $D = \begin{bmatrix} 5 & 0 & -1 \\ 4 & 1 & 2 \\ 2 & -3 & -1 \end{bmatrix}$,

$$E = \begin{bmatrix} 6 & 0 \\ 3 & -2 \\ -2 & 1 \end{bmatrix}, \quad F = \begin{bmatrix} -3 & 1 & -2 \\ 5 & -4 & 0 \end{bmatrix}, \quad G = \begin{bmatrix} 2 & 3 & -2 \\ 1 & 0 & 0 \\ 1 & 2 & 1 \end{bmatrix}.$$

Compute:

(a) AD	(g) EG
(b) BC	(h) DG
(c) BF	(i) GD
(d) EF	(j) FA
(e) FE	(k) GDA
(f) GE	(l) BCF

2 Put $A = \begin{bmatrix} 5 \\ 4 \\ 2 \end{bmatrix}$, $B = (6 \quad 0)$, $C = \begin{bmatrix} 3 & -2 \\ -2 & 1 \end{bmatrix}$, $D = \begin{bmatrix} -3 & 1 & -2 \\ 0 & -4 & 5 \\ 0 & -2 & 1 \end{bmatrix}$,

$$E = \begin{bmatrix} 3 & 1 \\ 1 & 2 \\ -2 & -4 \end{bmatrix},\ F = \begin{bmatrix} 2 & -3 & 1 \\ 0 & 3 & -2 \end{bmatrix},\ G = \begin{bmatrix} 3 & 2 & -3 \\ 4 & 5 & 5 \\ 4 & 3 & 0 \end{bmatrix}.$$

Compute:

(a) DA	(g) CE
(b) AB	(h) FE
(c) FA	(i) FG
(d) BF	(j) GF
(e) BFE	(k) GD
(f) EC	(l) DG

3 Put

$$A = \begin{bmatrix} 1 & 4 & -1 \\ 2 & -2 & 0 \\ 3 & 0 & 1 \end{bmatrix} \quad \text{and} \quad I = \begin{bmatrix} 1 & 0 & 0 \\ 0 & 1 & 0 \\ 0 & 0 & 1 \end{bmatrix};$$

show that $AI = IA = A$.

4 Put

$$B = \begin{bmatrix} 1 & 0 & 1 \\ 3 & 1 & 0 \\ 4 & 3 & -1 \end{bmatrix};$$

compute B^2 and B^3.

5 Prove the associative property of matrix multiplication in the particular case of 2-by-2 matrices. Put

$$A = \begin{bmatrix} a_{11} & a_{12} \\ a_{21} & a_{22} \end{bmatrix}, \qquad B = \begin{bmatrix} b_{11} & b_{12} \\ b_{21} & b_{22} \end{bmatrix}, \qquad C = \begin{bmatrix} c_{11} & c_{12} \\ c_{21} & c_{22} \end{bmatrix},$$

and show that AB postmultiplied by C is equal to A postmultiplied by BC. Compute the products and compare.

6 Prove that matrix addition is distributive with respect to matrix multiplication in the 2-by-2 case: Using the matrices in Exercise 5, show that $A(B + C) = AB + AC$.

7 Voters select one candidate from the Democratic, Republican, and Third Parties. The voters shift from party to party in certain proportions: of those voting for the candidate of a given party in one election a fixed proportion will continue to vote for the same party in the next election, another proportion will switch to a second party, and another to a third party. Let the transition matrix P of proportions be

	to Democrat	to Republican	to Third
from Democrat	0.60	0.20	0.20
from Republican	0.30	0.60	0.10
from Third	0.30	0.20	0.50

Figure 3.10

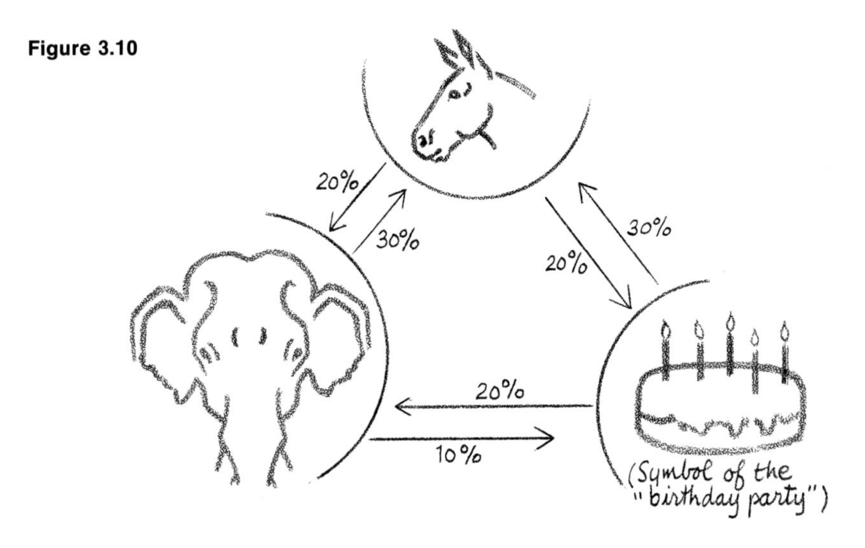

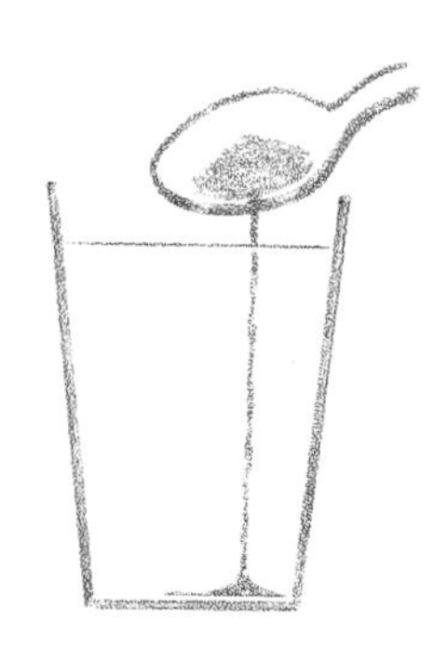

For example, of those who voted Republican in the election, 60 percent will continue to do so the next time, 30 percent will switch to the Democrats, and 10 percent to the Third Party (Figure 3.10). Let the distribution of voters in a particular year be

Democrats	Republicans	Third Party
50%	40%	10%

(a) Compute P^2, and P^3.

(b) Find the distributions of the voters in each of the next three elections. (See Example 4.)

8 A boy pours chocolate syrup into a glass of milk and stirs it with a spoon. Every time he completes a turn of the spoon, $\frac{1}{5}$ of the syrup in the bottom half of the glass moves to the top, and the other $\frac{4}{5}$ stays on the bottom. Similarly, $\frac{1}{5}$ of the syrup on the top moves to the bottom, and $\frac{4}{5}$ stays on the top (Figure 3.11).

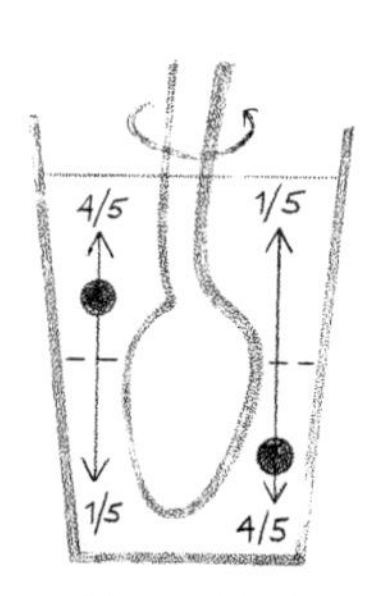

Figure 3.11

(a) Write the transition matrix P, and then compute P^2 and P^3.

(b) If all the syrup is at the bottom of the glass when the stirring begins, what proportion will have moved to the top after 3 turns?

(c) How does the stirring affect the distribution of syrup if it is already well mixed: that is, half the syrup is on the bottom and half on top? (See Example 4.)

9 Three countries trade gold among themselves in the international market. One of the 3 has a deliberate policy of hoarding the gold. Let the 3 countries be designated A, B, and C, respectively. The first 2 retain some of their gold every month, and freely sell the rest to the other 2. Country C sells none. Let the transition matrix for the movement of the gold be as follows:

	To		
From	A	B	C
A	0.80	0.10	0.10
B	0.20	0.60	0.20
C	0.00	0.00	1.00,

for example, A sells B and C 10 percent and 10 percent, respectively, of its gold supply, and keeps 80 percent. Suppose that A and B initially have all the gold, equally divided between them. What proportion will country C have after 2 months of trading? 3 months? (C is called an **absorbing** state.) (See Example 4.)

10 A medical experimenter observes the course of fever in mice infected with certain disease agents. The temperatures of the mice vary with time in the following way. Of those mice whose temperature is x degrees above normal on a certain day ($0 < x < 3$), 10 percent will have a temperature of $x + 1°$ on the following day, 40 percent a temperature of $x - 1°$, and 50 percent temperature unchanged. In other words, 10 percent gain 1 degree, 40 percent lose 1 degree and 50 percent remain unchanged. A mouse recovers when his temperature reaches the normal level, and dies when it goes 3 degrees above normal (Figure 3.12). A recovered mouse stays in that state. **Recovery** and **death** are states from which the mouse does not return to the "system"; they are **absorbing** states.

(a) Write the transition matrix for the proportions passing among the 4 states 0°, 1°, 2°, and 3° above normal.

(b) Calculate the distribution of temperatures on the third day for a group of mice with temperatures initially 1° above normal. (See Example 4.)

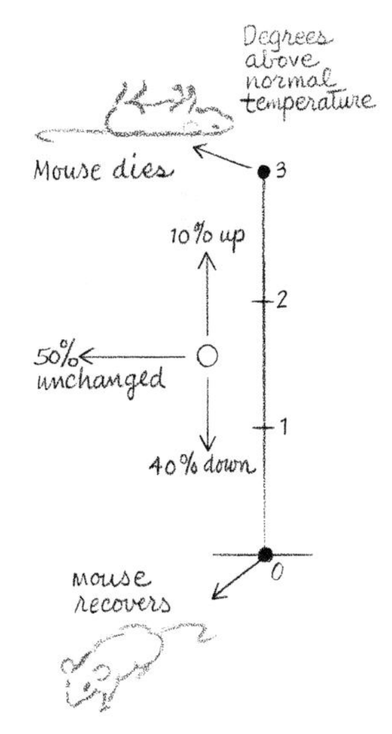

Figure 3.12

11 At the beginning of the year, an investor buys 200 shares of an industrial stock, 100 shares of a utility, and 50 shares of a railroad. He holds them for 4 quarters of the year. The prices at the beginning of each quarter are:

	Jan. 1	April 1	July 1	Oct. 1
Industrial	80	83	86	90
Utility	150	151	155	155
Railroad	60	64	61	63

Show, by means of matrix multiplication, how to calculate the total values of the investment at the beginning of each quarter. (See Example 5.)

12 As the first election results are counted, it appears on the basis of certain projections that 3 candidates A, B, and C will receive the following proportions of the total vote in each of the following 3 precincts:

	Candidate		
Precinct	A	B	C
1	0.55	0.40	0.05
2	0.45	0.45	0.10
3	0.30	0.65	0.05

Suppose that precincts 1, 2, and 3 have the estimated numbers of voters 2000, 1000, and 500, respectively. By means of matrix multiplication, estimate the total numbers of votes for each of the 3 candidates. (See Example 5.)

13 Consider 4 fund raisers for 4 different charitable organizations. One is considered to dominate the other if he can convince the latter to give \$100 to his organization. Two-step dominance is convincing by means of the intercession of a third person. Let the 4 persons be labeled A, B, C, and D, respectively. Suppose that A dominates B and D, B dominates C, C dominates A and D, and D dominates B.

(a) Write the 1-step dominance matrix d and calculate its square.

(b) How much money will each of the 4 charities receive through one- and two-step dominance? Assume each person gives at most once to each charity. (See Example 6.)

14 In ranking 4 football teams on the basis of their records, a sportswriter wishes to account not only for the raw won-and-lost figure, but also for the teams involved in each decision; for example, a victory over a good team should be assigned more credit than victory over a weak team. He assigns 1 point to team A for each victory over another team B, and also $\frac{1}{2}$ point to A for each victory of B; in other words, A is given credit for the victories of B. The 4 teams, A, B, C, and E, have the following records against each other: A defeated B and C. B defeated C. C defeated E. E defeated A and B.

(a) Write the dominance matrix D.

(b) Compute the number of points scored by each team in this method of ranking. [*Hint:* Find $D + \frac{1}{2}D^2$.] (See Example 6.)

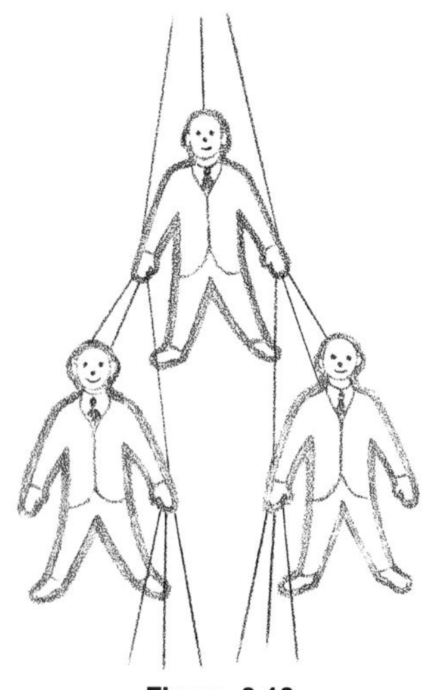

Figure 3.13

15 Hubert, John, George, and Ed are candidates for office at a nominating convention. Through certain political pressures, some can force others to make nominating speeches for them. This is a dominance relation. Suppose that Ed dominates Hubert and John, George dominates Ed, Hubert dominates George and John, and John dominates George.

(a) Under 1-step dominance how many speeches will be made for each of the 4 candidates?

(b) If the dominance extends to 2 steps, how many speeches will be made for each? Assume that one man can speak at most once for another. (Figure 3.13. See Example 6.)

16 Express each of the systems of linear equations as a matrix equation.

(a)
$$\begin{aligned} 4x + 2y + 3z &= 900 \\ 5x + y + 2z &= 600 \\ 3x + 2y + 5z &= 400 \end{aligned}$$

(b)
$$\begin{aligned} x + y + z &= 3 \\ 2y + z &= 2 \\ y + 2z &= 2 \end{aligned}$$

(c)
$$\begin{aligned} x - y &= 0 \\ y + z &= 2 \end{aligned}$$

(d)
$$\begin{aligned} x + 2y - 3z + w &= 0 \\ x + z - 7w &= 10 \\ y + 8w &= 1 \end{aligned}$$

3.3 Determinants

For every square matrix, there is a unique number called its **determinant.** First we consider the 2-by-2 matrix

$$A = \begin{bmatrix} a_{11} & a_{12} \\ a_{21} & a_{22} \end{bmatrix}.$$

Its determinant is denoted **det** A. When the entries are given as above, then **det** A is written as

$$\mathbf{det}\, A = \begin{vmatrix} a_{11} & a_{12} \\ a_{21} & a_{22} \end{vmatrix},$$

where the brackets of the matrix border are replaced by the vertical lines. The determinant is defined as the product of the main diagonal entries minus the product of the off-diagonal entries:

$$\begin{vmatrix} a_{11} & a_{12} \\ a_{21} & a_{22} \end{vmatrix} = a_{11}a_{22} - a_{12}a_{21}.$$

For example,

$$\begin{vmatrix} 1 & 2 \\ 4 & 3 \end{vmatrix} = 1(3) - 2(4) = -5$$

$$\begin{vmatrix} 1 & 2 \\ -4 & 3 \end{vmatrix} = 1(3) - 2(-4) = 11,$$

$$\begin{vmatrix} 3 & 2 \\ -6 & -4 \end{vmatrix} = 3(-4) - 2(-6) = 0.$$

Note that the determinant may be positive, negative, or zero.

When we turn to determinants of square matrices of order larger than 2, the definition of the determinant becomes more complicated. First we define the **minor** of each element of a square matrix. The minor of an entry of a 3-by-3 matrix is the determinant obtained in the following way:

1. Delete the row in which the entry appears.
2. Then delete the column in which the entry appears.

3. Form a 2-by-2 matrix from the remaining entries in the order in which they appear.
4. Calculate the determinant of the resulting 2-by-2 matrix, which is done according to the definition given above for such a determinant.

Using this definition of minor, let us find the forms for the minors in the 3-by-3 matrix

$$A = \begin{bmatrix} a_{11} & a_{12} & a_{13} \\ a_{21} & a_{22} & a_{23} \\ a_{31} & a_{32} & a_{33} \end{bmatrix}.$$

The minor of the entry a_{11} is the determinant of the 2-by-2 submatrix

$$\begin{bmatrix} a_{22} & a_{23} \\ a_{32} & a_{33} \end{bmatrix},$$

which is equal to $a_{22}a_{33} - a_{23}a_{32}$. Similarly, we find the minor of the entry a_{23}:

$$\begin{vmatrix} a_{11} & a_{12} & a_{13} \\ a_{21} & a_{22} & a_{23} \\ a_{31} & a_{32} & a_{33} \end{vmatrix} = \begin{vmatrix} a_{11} & a_{12} \\ a_{31} & a_{32} \end{vmatrix} = a_{11}a_{32} - a_{12}a_{31}.$$

EXAMPLE 1

Compute the minors of entries 1 and 7 of the matrix

$$\begin{bmatrix} 2 & 1 & 3 \\ 4 & 8 & 6 \\ 0 & 7 & 5 \end{bmatrix}.$$

The minor of entry 1 is

$$\begin{vmatrix} 4 & 6 \\ 0 & 5 \end{vmatrix} = 20.$$

The minor of entry 7 is

$$\begin{vmatrix} 2 & 3 \\ 4 & 6 \end{vmatrix} = 0. \blacktriangle$$

In order to complete our definition of the determinant of a matrix of order larger than 2, we define one more quantity associated with the entry of a matrix: the **cofactor.** The cofactor of the entry of a matrix is obtained from the minor of that entry by adjusting the algebraic sign (+) or (−) of the latter. The adjustment is as follows:

If the sum of the row number and the column number is **even,** the cofactor is equal to the minor. If the sum of the row and column numbers is **odd,** the cofactor is obtained from the minor by changing its sign. This pattern of sign adjustment is represented in the following diagram:

$$\begin{matrix} + & - & + \\ - & + & - \\ + & - & + \end{matrix}$$

This signifies that to get the cofactors from the minors the sign of the minors in the positions with + are unchanged, and the signs in positions with − are changed (Figure 3.14).

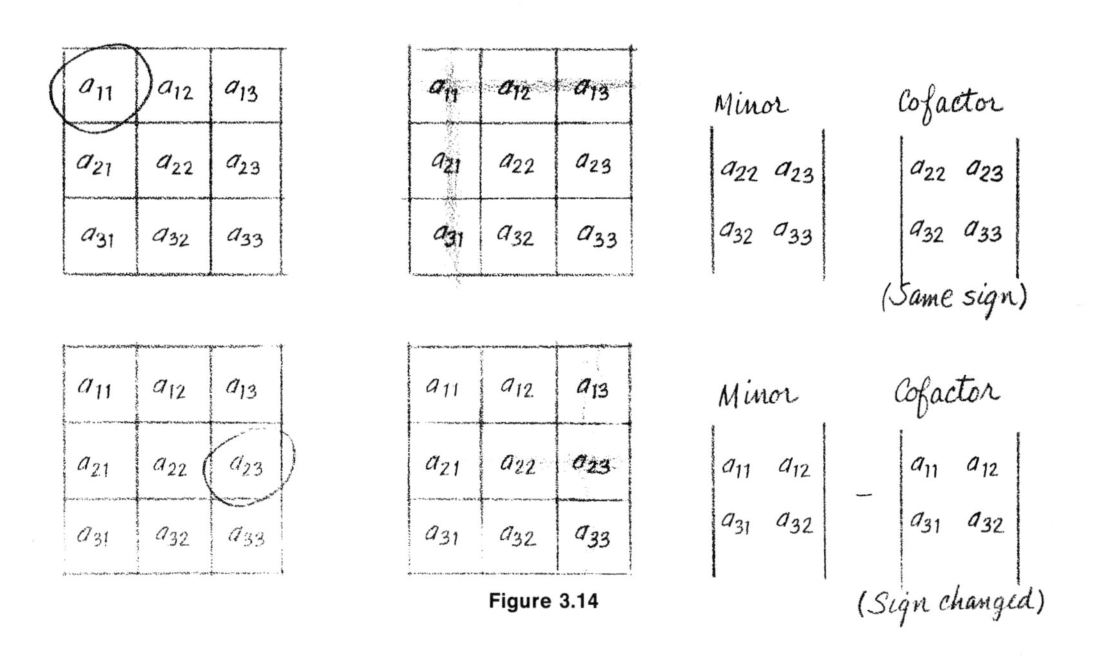

Figure 3.14

The determinant of a 3-by-3 matrix is the number obtained in these 2 steps:

1. Select any row or column of the matrix. Multiply each of its entries (3 of them) by the corresponding cofactor.
2. Add the 3 products.

It can be shown by actual computation that the number obtained in this way is the same for any initial choice of a row or column. This procedure for obtaining the determinant is called **expansion by cofactors.**

EXAMPLE 2

Consider the matrix in the previous example,

$$\begin{bmatrix} 2 & 1 & 3 \\ 4 & 8 & 6 \\ 0 & 7 & 5 \end{bmatrix}.$$

Let us expand by cofactors of entries of the first row.

Entry	Minor	Cofactor	Entry × cofactor
2	−2	−2	−4
1	20	−20	−20
3	28	28	84

The determinant is equal to the sum $(-4) + (-20) + 84 = 60$. The same number would have been obtained by choosing another row or column. Suppose, for example, that we choose the third column:

Entry	Minor	Cofactor	Entry × cofactor
3	28	28	84
6	14	−14	−84
5	12	12	60.

The determinant is again found to be 60. ▲

Finally we show how to calculate the determinant of a 4-by-4 matrix; this is the largest that we shall need. The minors and cofactors have definitions similar to the 3-by-3 case. The minor of an entry is the determinant of the 3-by-3 submatrix obtained by deleting from the original matrix the row and column to which the entry belongs. The cofactors are obtained from the minors by the same rule as above according to the sum of the row and column numbers. We get the following checkerboard pattern for the sign adjustment:

$$\begin{matrix} + & - & + & - \\ - & + & - & + \\ + & - & + & - \\ - & + & - & + \end{matrix}$$

The determinant is obtained as the sum of the products of the entries of a particular row or column and their cofactors.

EXAMPLE 3

Evaluate the determinant

$$\begin{vmatrix} 1 & 1 & 1 & 1 \\ 1 & 0 & -1 & 0 \\ 0 & 1 & 1 & -1 \\ 2 & 0 & -3 & -3 \end{vmatrix}.$$

To simplify the computations, we expand in the cofactors of the row or column having many 0 entries; for example, let us choose the second row. The entries, their minors, and their cofactors are:

Entry	Minor	Cofactor	Entry × cofactor
1	$\begin{vmatrix} 1 & 1 & 1 \\ 1 & 1 & -1 \\ 0 & -3 & -3 \end{vmatrix} = -6$	6	6
0	(No need to calculate.)		
-1	$\begin{vmatrix} 1 & 1 & 1 \\ 0 & 1 & -1 \\ 2 & 0 & -3 \end{vmatrix} = -7$	7	-7
0	(No need to calculate.)		

The determinant is equal to -1. ▲

3.3 EXERCISES

In Exercises 1 and 2 evaluate each determinant.

1 (a) $\begin{vmatrix} 5 & 3 \\ 0 & 2 \end{vmatrix}$ (c) $\begin{vmatrix} 2 & 5 \\ -1 & 3 \end{vmatrix}$

(b) $\begin{vmatrix} 2 & 2 \\ 7 & 6 \end{vmatrix}$ (d) $\begin{vmatrix} 2 & -2 \\ -1 & 1 \end{vmatrix}$

2 (a) $\begin{vmatrix} 1 & 3 \\ 9 & 6 \end{vmatrix}$ (c) $\begin{vmatrix} 9 & 2 \\ -6 & 5 \end{vmatrix}$

(b) $\begin{vmatrix} 2 & 7 \\ 0 & 9 \end{vmatrix}$ (d) $\begin{vmatrix} 1 & 7 \\ -2 & -2 \end{vmatrix}$

3 Evaluate by expansion of cofactors of (i) the first column; (ii) the second row:

(a) $\begin{vmatrix} 8 & 0 & 5 \\ 0 & -2 & 1 \\ 9 & -8 & 3 \end{vmatrix}$ (c) $\begin{vmatrix} 0 & 3 & 0 \\ -5 & 6 & 6 \\ 7 & 6 & -1 \end{vmatrix}$

(b) $\begin{vmatrix} -6 & 3 & 4 \\ 6 & 6 & 0 \\ -1 & 2 & 7 \end{vmatrix}$ (d) $\begin{vmatrix} 4 & -4 & 2 \\ 9 & 0 & -5 \\ 4 & 6 & 9 \end{vmatrix}$

4 Evaluate by expansion of cofactors of (i) the second column; (ii) the third row:

(a) $\begin{vmatrix} -1 & -7 & 2 \\ 3 & 0 & 1 \\ 1 & 6 & 4 \end{vmatrix}$ (c) $\begin{vmatrix} 5 & 5 & 0 \\ 5 & -6 & 8 \\ -3 & 1 & 8 \end{vmatrix}$

(b) $\begin{vmatrix} 1 & 4 & 3 \\ -5 & 4 & 8 \\ 3 & 0 & -4 \end{vmatrix}$ (d) $\begin{vmatrix} 4 & 0 & -3 \\ 2 & 6 & 1 \\ -8 & 9 & -1 \end{vmatrix}$

In Exercises 5 and 6, evaluate by expansion by cofactors in a row or column having the largest number of entries 0.

5 (a) $\begin{vmatrix} 0 & 0 & 1 & 2 \\ 1 & 1 & 2 & -1 \\ 2 & 0 & 0 & 2 \\ 0 & -1 & 1 & 2 \end{vmatrix}$ (c) $\begin{vmatrix} 0 & 0 & 1 & -1 \\ 2 & 0 & 1 & 0 \\ 1 & 1 & 0 & 1 \\ 0 & 1 & 2 & 0 \end{vmatrix}$

(b) $\begin{vmatrix} 0 & 1 & 0 & 2 \\ 0 & 0 & 0 & 0 \\ 2 & 0 & 0 & 0 \\ 2 & 0 & 1 & -1 \end{vmatrix}$ (d) $\begin{vmatrix} -2 & 0 & 0 & -2 \\ 1 & -1 & 2 & 0 \\ 1 & 0 & 0 & 0 \\ 0 & 0 & 2 & 2 \end{vmatrix}$

6 (a) $\begin{vmatrix} 2 & 0 & -2 & 0 \\ 0 & -2 & 0 & 1 \\ 0 & 0 & 2 & 2 \\ 1 & 0 & 0 & -1 \end{vmatrix}$ (c) $\begin{vmatrix} -1 & 0 & 1 & 0 \\ -2 & 2 & 2 & 2 \\ 0 & -2 & -1 & 2 \\ 0 & 0 & 0 & 2 \end{vmatrix}$

(b) $\begin{vmatrix} 1 & -1 & 2 & 1 \\ 1 & 0 & -1 & 0 \\ 2 & 0 & 1 & -1 \\ 2 & 1 & 2 & 1 \end{vmatrix}$ (d) $\begin{vmatrix} -1 & 1 & 1 & 1 \\ 2 & 0 & 0 & 1 \\ 2 & -2 & -1 & 1 \\ 1 & 0 & 0 & -2 \end{vmatrix}$

●●● **7** It was claimed just before Example 2 above that the value of the determinant is independent of the choice of the row or column chosen for the cofactor expansion. Give the proof in the case of a general 3-by-3 matrix A with entries a_{ij}, where i and j run from 1 through 3. Write the formulas for the cofactors of the first row, and the sum of their products with the entries; then repeat the same for any other row or column; for example, the second column. Show that the determinant is the same in both cases.

●●●● **8** A useful property of determinants is that if all the entries in a particular row or column have a common factor, then it can be factored out of the determinant before the evaluation. Give the proof in the case of a 3-by-3 matrix, showing that

$$\begin{vmatrix} ca_{11} & ca_{12} & ca_{13} \\ a_{21} & a_{22} & a_{23} \\ a_{31} & a_{32} & a_{33} \end{vmatrix} = c \begin{vmatrix} a_{11} & a_{12} & a_{13} \\ a_{21} & a_{22} & a_{23} \\ a_{31} & a_{32} & a_{33} \end{vmatrix}.$$

[*Hint:* Expand the determinant on the left by cofactors of the first row.] Use this result to evaluate:

(a) $\begin{vmatrix} 4 & 2 & 2 \\ 2 & -2 & 10 \\ 0 & 3 & -3 \end{vmatrix}$ (b) $\begin{vmatrix} 12 & 6 & 0 \\ 9 & -9 & 0 \\ 0 & 2 & 8 \end{vmatrix}$.

9 A determinant having two identical rows or two identical columns is always equal to 0. Give the proof for a 3-by-3 matrix; show that

$$\begin{vmatrix} a_{11} & a_{12} & a_{13} \\ a_{11} & a_{12} & a_{13} \\ a_{31} & a_{32} & a_{33} \end{vmatrix} = 0.$$

10 If 2 rows of a matrix, or 2 columns, are interchanged, then the sign of the corresponding determinant is changed. Give the proof for a 3-by-3 matrix; show that

$$\begin{vmatrix} a_{11} & a_{12} & a_{13} \\ a_{21} & a_{22} & a_{23} \\ a_{31} & a_{32} & a_{33} \end{vmatrix} = - \begin{vmatrix} a_{21} & a_{22} & a_{23} \\ a_{11} & a_{12} & a_{13} \\ a_{31} & a_{32} & a_{33} \end{vmatrix}.$$

[*Hint:* Expand by cofactors of the last row.]

11 If the entries above the main diagonal are all equal to 0, or those below are all equal to 0, then the determinant of the matrix is equal to the product of the entries on the main diagonal. Give the proof in the case of a 3-by-3 matrix; show that

$$\begin{vmatrix} a_{11} & 0 & 0 \\ a_{21} & a_{22} & 0 \\ a_{31} & a_{32} & a_{33} \end{vmatrix} = a_{11}a_{22}a_{33}.$$

3.4 Cramer's Rule

Determinants of 2-by-2 matrices naturally arise in the solution of a system of 2 linear equations in 2 unknowns. Let the linear equations be written as

$$\begin{aligned} a_{11}x_1 + a_{12}x_2 &= b_1, \\ a_{21}x_1 + a_{22}x_2 &= b_2. \end{aligned} \tag{1}$$

The a's and b's are given and we wish to solve for the unknowns x_1 and x_2. If we multiply the first equation by a_{22}, the second by $-a_{12}$, and add the resulting two equations, we eliminate x_2 and obtain a single equation for the unknown x_1:

$$(a_{11}a_{22} - a_{21}a_{12})x_1 = b_1a_{22} - b_2a_{12}.$$

This equation may be written in terms of two determinants as

$$\begin{vmatrix} a_{11} & a_{12} \\ a_{21} & a_{22} \end{vmatrix} x_1 = \begin{vmatrix} b_1 & a_{12} \\ b_2 & a_{22} \end{vmatrix}. \tag{2}$$

If we eliminate the variable x_1 from (1) by multiplying the first equation by $-a_{21}$ and the second by a_{11}, we get an equation analogous to (2) for x_2:

$$\begin{vmatrix} a_{11} & a_{12} \\ a_{21} & a_{22} \end{vmatrix} x_2 = \begin{vmatrix} a_{11} & b_1 \\ a_{21} & b_2 \end{vmatrix}. \tag{3}$$

If the determinant

$$D = \begin{vmatrix} a_{11} & a_{12} \\ a_{21} & a_{22} \end{vmatrix}$$

appearing in (2) and (3) is not equal to 0, then these equations can be solved for the unknowns x_1 and x_2 by division by D:

$$x_1 = \frac{\begin{vmatrix} b_1 & a_{12} \\ b_2 & a_{22} \end{vmatrix}}{\begin{vmatrix} a_{11} & a_{12} \\ a_{21} & a_{22} \end{vmatrix}}, \qquad x_2 = \frac{\begin{vmatrix} a_{11} & b_1 \\ a_{12} & b_2 \end{vmatrix}}{\begin{vmatrix} a_{11} & a_{12} \\ a_{21} & a_{22} \end{vmatrix}}. \tag{4}$$

This formula for the unknowns, given as ratios of determinants, is called **Cramer's Rule.** (G. Cramer was a Swiss mathematician of the eighteenth century.) Note that the denominator in both parts of (4) is equal to the determinant of the **coefficient matrix** of the system (1). The numerator in the expression for the solution x_1 is the determinant of the matrix obtained from the coefficient matrix by replacing the coefficients of x_1 by b_1 and b_2, respectively. Similarly, the numerator for x_2 is obtained by replacing the coefficients of x_2 by the same numbers b_1 and b_2.

EXAMPLE 1

Solve the system by Cramer's Rule:

$$\begin{aligned} 2x_1 - 3x_2 &= 4, \\ x_1 + 2x_2 &= 3. \end{aligned}$$

The coefficient matrix is $\begin{bmatrix} 2 & -3 \\ 1 & 2 \end{bmatrix}$, and its determinant is 7. x_1 is obtained by replacing the first column of the coefficient matrix by the column vector with the components 4 and 3, evaluating the determinant, and then dividing by the determinant of the coefficient matrix:

$$x_1 = \frac{\begin{vmatrix} 4 & -3 \\ 3 & 2 \end{vmatrix}}{7} = \frac{17}{7}.$$

Similarly, we obtain x_2 by replacing the second column in the coefficient matrix:

$$x_2 = \frac{\begin{vmatrix} 2 & 4 \\ 1 & 3 \end{vmatrix}}{7} = \frac{2}{7}. \blacktriangle$$

Cramer's Rule can be extended to 3 equations in 3 unknowns, and even to more than 3. Suppose we are given the system

$$\begin{aligned} a_{11}x_1 + a_{12}x_2 + a_{13}x_3 &= b_1, \\ a_{21}x_1 + a_{22}x_2 + a_{23}x_3 &= b_2, \\ a_{31}x_1 + a_{32}x_2 + a_{33}x_3 &= b_3. \end{aligned} \tag{5}$$

It may be written in matrix form as

$$\begin{bmatrix} a_{11} & a_{12} & a_{13} \\ a_{21} & a_{22} & a_{23} \\ a_{31} & a_{32} & a_{33} \end{bmatrix} \begin{bmatrix} x_1 \\ x_2 \\ x_3 \end{bmatrix} = \begin{bmatrix} b_1 \\ b_2 \\ b_3 \end{bmatrix}$$

or

$$AX = B,$$

where A is the coefficient matrix, X the vector of the unknowns, and B the vector of b's. Cramer's Rule states that if **det** $A \neq 0$, then the system has a unique solution for the unknowns in terms of quotients of determinants:

$$x_1 = \frac{1}{\det A} \begin{vmatrix} b_1 & a_{12} & a_{13} \\ b_2 & a_{22} & a_{23} \\ b_3 & a_{32} & a_{33} \end{vmatrix}, \tag{6}$$

$$x_2 = \frac{1}{\det A} \begin{vmatrix} a_{11} & b_1 & a_{13} \\ a_{21} & b_2 & a_{23} \\ a_{31} & b_3 & a_{33} \end{vmatrix},$$

$$x_3 = \frac{1}{\det A} \begin{vmatrix} a_{11} & a_{12} & b_1 \\ a_{21} & a_{22} & b_2 \\ a_{31} & a_{32} & b_3 \end{vmatrix}.$$

It is conventional to write $|A|$ for det A. The general procedure underlying formula (6) is outlined in Figure 3.15.

In numerical examples we shall ordinarily use the letters x and y for the unknowns in the place of the x's with subscripts; when there are 3 unknowns, we use z as the third.

EXAMPLE 2

Solve this system by Cramer's Rule:

$$\begin{aligned} x + y + z &= 2, \\ 2x - y + 2z &= 12, \\ 3x + y - z &= -2. \end{aligned}$$

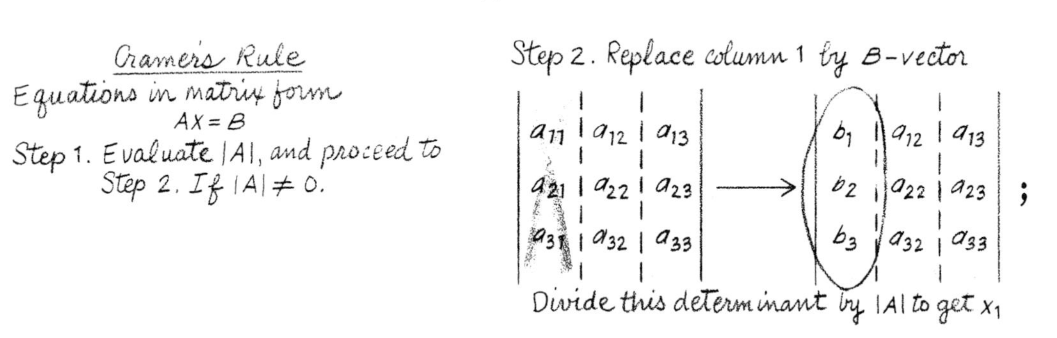

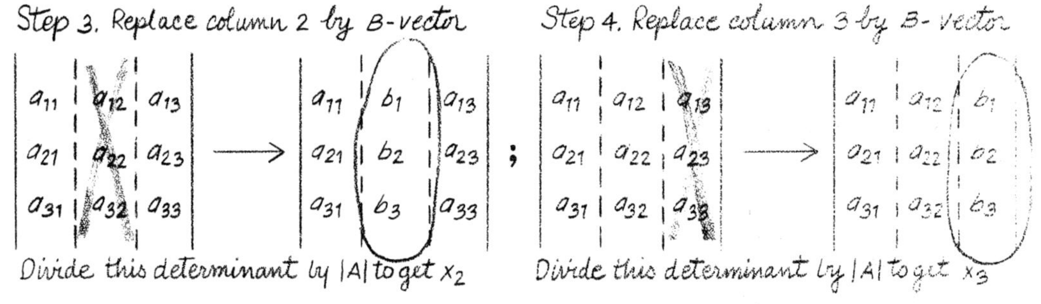

Figure 3.15

The determinant of the coefficient matrix is obtained by expansion in cofactors of the first row:

$$\begin{vmatrix} 1 & 1 & 1 \\ 2 & -1 & 2 \\ 3 & 1 & -1 \end{vmatrix} = (1-2) - (-2 \quad -6) + (2-(-3)) = 12.$$

The unknown x is obtained by substituting the b-vector for the first column of the coefficient matrix:

$$x = \frac{1}{12}\begin{vmatrix} 2 & 1 & 1 \\ 12 & -1 & 2 \\ -2 & 1 & -1 \end{vmatrix} = \frac{16}{12} = \frac{4}{3}.$$

Similarly, y and z are obtained by substituting the b-vector for the appropriate columns of the coefficient matrix:

$$y = \frac{1}{12}\begin{vmatrix} 1 & 2 & 1 \\ 2 & 12 & 2 \\ 3 & -2 & -1 \end{vmatrix} = -\frac{32}{12} = -\frac{8}{3};$$

$$z = \frac{1}{12}\begin{vmatrix} 1 & 1 & 2 \\ 2 & -1 & 12 \\ 3 & 1 & -2 \end{vmatrix} = \frac{40}{12} = \frac{10}{3}.$$

The solution of the system is

$$x = \frac{4}{3}, \qquad y = \frac{-8}{3}, \qquad z = \frac{10}{3}.$$

This can be checked by substituting these values in each of the 3 equations. ▲

Using Cramer's Rule is the most efficient way of solving a 2-by-2 system, but not a system with more than 2 unknowns. In the following sections, we shall develop more suitable ways of solving such systems; these methods depend more heavily on the theory of matrices.

The proof of Cramer's Rule in the 3-by-3 case is outlined in Exercise 5 below.

3.4 EXERCISES

Solve each of the systems by Cramer's Rule.

1 (a) $\begin{aligned} x - 8y &= 0 \\ 5x + 3y &= 5 \end{aligned}$ (c) $\begin{aligned} x - 2y &= 5 \\ 8x + 3y &= 5 \end{aligned}$

(b) $\begin{aligned} 5x + 3y &= 4 \\ 6x - 3y &= 2 \end{aligned}$ (d) $\begin{aligned} 7x + 2y &= -2 \\ 6x + 9y &= 3 \end{aligned}$

2 (a) $\begin{aligned} 6x + 8y &= 5 \\ 8x - 3y &= 1 \end{aligned}$ (c) $\begin{aligned} 9x + 7y &= -3 \\ 3x + y &= 1 \end{aligned}$

(b) $\begin{aligned} 3x - 2y &= -6 \\ 7x + 3y &= 2 \end{aligned}$ (d) $\begin{aligned} 3x + 5y &= 6 \\ 9x + y &= 5 \end{aligned}$

3 (a) $\begin{aligned} x + y + z &= 0 \\ -y + 2z &= 4 \\ 2x + 2y &= 1 \end{aligned}$ (c) $\begin{aligned} 2x + y + 3z &= 7 \\ x + 2y + 3z &= 5 \\ 7y + 5z &= 3 \end{aligned}$

(b) $\begin{aligned} x + y + z &= 2 \\ 2x + 3y + 4z &= 3 \\ x - 2y - z &= 1 \end{aligned}$ (d) $\begin{aligned} 2x - 7y + 4z &= 10 \\ 7x + 5y + 3z &= 13 \\ 5x + y + 3z &= 11 \end{aligned}$

4 (a) $\begin{aligned} x + y + z &= 9 \\ x + 2y + 3z &= 9 \\ x + 3y + 6z &= 3 \end{aligned}$ (c) $\begin{aligned} 3x - 7y + 7z &= -6 \\ x + 2y + 6z &= 6 \\ 5x + 6y - z &= -5 \end{aligned}$

(b) $\begin{aligned} x + 4y + 5z &= 9 \\ 2x + 3z &= 13 \\ 3x + 9z &= 33 \end{aligned}$ (d) $\begin{aligned} 2x + 5y + z &= 10 \\ x - 2y + 4z &= -11 \\ 2x + 3y &= 8 \end{aligned}$

5 Fill in the details of this outline of the proof of Cramer's Rule for a 3-by-3 system.

(a) Expand the determinants in (6) in cofactors of the columns with the entries b_1, b_2, and b_3.

(b) Substitute the expressions (6) for the unknowns in the left-hand side of the first equation in (5).

(c) In (b) collect the terms with factors b_1, b_2, and b_3.

(d) Show that the coefficient of b_1 is equal to 1, and those of b_2 and b_3 are equal to 0. This proves that the first equation is satisfied.

(e) Repeat the procedure above for the other two equations in (5). Then conclude that these are also satisfied by (6).

6 In 1967 the Premier of a major power happened to be at a meeting of the United Nations in New York. The President of another power was in Washington. They wanted to meet, but, for political reasons, neither would travel all the way to the other. They met at Glassboro, New Jersey, a point equidistant from the two big cities.

The recent admission of a third major power to the United Nations offers statesmen the more challenging mathematical problem of finding the point equidistant from 3 locations. Imagine a large diplomatic reception whose guests include a President, Premier, and Chairman; respective heads of 3 major powers. The President is on one side of the hall, examining a painting, the Premier is at the bar, and the Chairman is at the fruit bowl (Figure 3.16). Diplomatic officials are arranging a meeting of the 3 at a point equidistant from each. A map of the reception hall is drawn on rectangular coordinates. The painting is at the point (0,4), the bar at (2,−2), and the fruit at (−1,1). Where should they meet? [*Hint:* See the distance formula in Section 1.1. Let (x,y) be the meeting point. Then the squares of the distances from this point to the others must be equal.]

Figure 3.16

3.5 The Inverse Matrix

Let A be a square matrix. If there exists a square matrix A^{-1} of the same size such that

$$AA^{-1} = A^{-1}A = I,$$

where I is the identity matrix (see Section 3.2), then A^{-1} is called the **inverse of A** or **A inverse.** The equation above shows that the relation between a matrix and its inverse is similar to that between a number and its reciprocal.

Not every square matrix has an inverse. In arithmetic the only number without a reciprocal is 0. In matrix theory, a "zero matrix" is one with all 0 entries. It is similar to the number 0 in arithmetic: the product of any matrix with a zero matrix is also a zero matrix. From this it follows that the zero matrix does not have an inverse; indeed, the reason is the same as for the absence of a reciprocal of 0. **But there are matrices that are not zero matrices and do not have inverses.**

This fact is based on an important difference between matrix and numerical multiplication. Two nonzero matrices may have a product that is a zero matrix; for example,

$$\begin{bmatrix} 1 & -1 \\ -1 & 1 \end{bmatrix}\begin{bmatrix} 1 & 1 \\ 1 & 1 \end{bmatrix} = \begin{bmatrix} 0 & 0 \\ 0 & 0 \end{bmatrix}.$$

If the matrices on the left are denoted A and B, and the matrix on the right denoted 0, then the equation is $AB = 0$. We claim that neither A nor B has an inverse. Suppose, to the contrary, that A has an inverse, denoted A^{-1}; then we shall deduce a contradiction. (The proof that B^{-1} does not exist is similar.) By multiplying both sides of the equation $AB = 0$ by A^{-1}, and applying the associative law for matrix multiplication, we obtain

$$A^{-1}(AB) = 0, \qquad (A^{-1}A)B = 0, \qquad \text{or} \qquad IB = B = 0,$$

which is a contradiction.

We say that a square matrix A is **singular** if **det** $A = 0$; **nonsingular** if **det** $A \neq 0$. There is a fundamental relation between singularity and the existence of the inverse:

For a square matrix, the inverse exists if and only if the matrix is nonsingular.

The proof for 2-by-2 matrices will be given below; and the proof for the 3-by-3 will be outlined in the exercises (Exercises 14 and 15). First we show how to calculate the inverse of a 2-by-2 matrix,

$$A = \begin{bmatrix} a & b \\ c & d \end{bmatrix}.$$

Suppose that the determinant $ad - bc$ is not equal to 0. Perform the following three operations on the matrix (Figure 3.17):

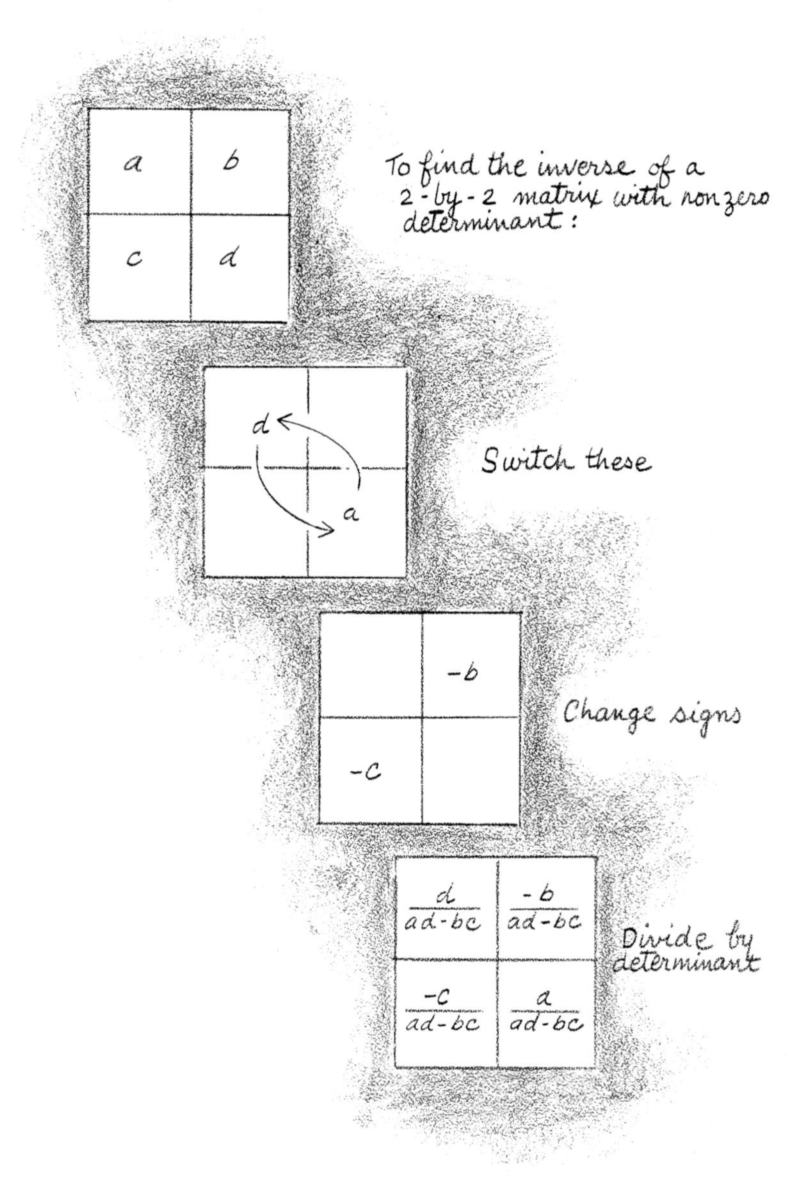

Figure 3.17

Interchange the positions of a and d, change the signs of b and c, and then divide each element of the resulting matrix by the determinant. We obtain

$$A^{-1} = \begin{bmatrix} \dfrac{d}{ad - bc} & \dfrac{-b}{ad - bc} \\ \dfrac{-c}{ad - bc} & \dfrac{a}{ad - bc} \end{bmatrix}.$$

This is really the inverse of A: to check it, we calculate the products AA^{-1} and $A^{-1}A$, and find them to be equal to the identity I of order 2.

EXAMPLE 1

Show that this matrix has an inverse, and then find it:

$$A = \begin{bmatrix} 1 & 3 \\ -2 & 4 \end{bmatrix}.$$

The inverse exists because the determinant is equal to 10, which is not 0. To calculate the inverse, switch the positions of the entries 1 and 4, change the signs of the other two entries, and then divide each entry of the resulting matrix by the determinant:

$$\begin{bmatrix} 1 & 3 \\ -2 & 4 \end{bmatrix} \longrightarrow \begin{bmatrix} 4 & 3 \\ -2 & 1 \end{bmatrix} \longrightarrow \begin{bmatrix} 4 & -3 \\ 2 & 1 \end{bmatrix} \longrightarrow \begin{bmatrix} 4/10 & -3/10 \\ 2/10 & 1/10 \end{bmatrix} = A^{-1}.$$

By matrix multiplication it can be checked that the products of A and A^{-1} are equal to the identity. ▲

Now we shall give the proof that a 2-by-2 matrix has an inverse if and only if it is nonsingular. It has already been shown that the inverse can be found **if** the determinant is not equal to zero; that is, when the matrix is nonsingular. It remains to be shown that if the determinant is equal to zero, then there does not exist an inverse. According to the argument at the beginning of this section, if A and B are nonzero matrices such that $AB = 0$, then A^{-1} does not exist. Let $A = \begin{bmatrix} a & b \\ c & d \end{bmatrix}$ be a singular matrix, and define the associated matrix

$$B = \begin{bmatrix} d & -b \\ -c & a \end{bmatrix};$$

then,

$$AB = \begin{bmatrix} a & b \\ c & d \end{bmatrix} \begin{bmatrix} d & -b \\ -c & a \end{bmatrix} = \begin{bmatrix} ad - bc & -ab + ab \\ dc - dc & ad - bc \end{bmatrix} = \begin{bmatrix} 0 & 0 \\ 0 & 0 \end{bmatrix}.$$

If A is a nonzero matrix, then so is B; thus, by the relation above, A^{-1} does not exist. On the other hand, if A **is** a zero matrix, then it certainly does not have an inverse.

The calculation of the inverse of the 3-by-3 matrix is more complicated. The main steps are (Figure 3.18):

1. Replace every entry in the matrix by its cofactor. This yields the **cofactor matrix.**
2. Interchange rows and corresponding columns in the cofactor matrix: replace the first, second, and third columns by the corresponding rows. The resulting matrix is called the **adjoint** matrix.
3. Multiply the adjoint matrix by the reciprocal of the determinant of A; in other words, divide each entry of the adjoint by **det** A.

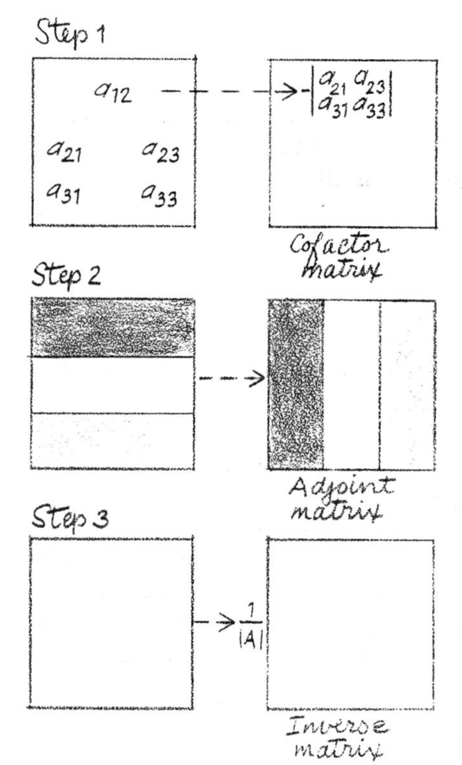

Figure 3.18

EXAMPLE 2

Calculate the inverse of

$$A = \begin{bmatrix} 1 & 6 & 0 \\ 2 & 4 & 1 \\ -3 & 0 & 2 \end{bmatrix}.$$

By expanding in cofactors of the first row, we find that the determinant is not 0, so that the inverse exists:

$$\begin{vmatrix} 4 & 1 \\ 0 & 2 \end{vmatrix} - 6 \begin{vmatrix} 2 & 1 \\ -3 & 2 \end{vmatrix} + 0 \begin{vmatrix} 2 & 4 \\ -3 & 0 \end{vmatrix} = 8 - 42 = -34.$$

The cofactor matrix is

$$\begin{bmatrix} \begin{vmatrix} 4 & 1 \\ 0 & 2 \end{vmatrix} & -\begin{vmatrix} 2 & 1 \\ -3 & 2 \end{vmatrix} & \begin{vmatrix} 2 & 4 \\ -3 & 0 \end{vmatrix} \\ -\begin{vmatrix} 6 & 0 \\ 0 & 2 \end{vmatrix} & \begin{vmatrix} 1 & 0 \\ -3 & 2 \end{vmatrix} & -\begin{vmatrix} 1 & 6 \\ -3 & 0 \end{vmatrix} \\ \begin{vmatrix} 6 & 0 \\ 4 & 1 \end{vmatrix} & -\begin{vmatrix} 1 & 0 \\ 2 & 1 \end{vmatrix} & \begin{vmatrix} 1 & 6 \\ 2 & 4 \end{vmatrix} \end{bmatrix} = \begin{bmatrix} 8 & -7 & 12 \\ -12 & 2 & -18 \\ 6 & -1 & -8 \end{bmatrix}.$$

[*Note:* The value and sign of the entry itself has no effect on the value of the corresponding cofactor.] Interchange rows and columns in the last matrix to get the adjoint:

$$\begin{bmatrix} 8 & -12 & 6 \\ -7 & 2 & -1 \\ 12 & -18 & -8 \end{bmatrix}.$$

Divide the matrix by the determinant -34 to get the inverse:

$$A^{-1} = -\frac{1}{34} \begin{bmatrix} 8 & -12 & 6 \\ -7 & 2 & -1 \\ 12 & -18 & -8 \end{bmatrix}.$$

By direct matrix multiplication we find that $AA^{-1} = A^{-1}A = I$, where I is the 3-by-3 identity. ▲

The inverse of a square matrix of order larger than 3 is calculated according to the same rules as for the 3-by-3: cofactor matrix, adjoint, and then division by the determinant.

The most important practical application of the inverse is in the solution of systems of linear equations. In Example 7 of Section 3.1 we showed that a system of equation in several unknowns can be represented as a matrix equation $AX = B$, where A is the matrix of coefficients, X the vector of unknowns, and B a given vector of constants. If the number of equations is the same as the number of unknowns, then the coefficient matrix is square. If A^{-1} exists, then we may multiply both sides of the matrix equation by A^{-1} to obtain the solution:

$$A^{-1}AX = A^{-1}B, \qquad IX = A^{-1}B, \qquad \text{or} \qquad X = A^{-1}B.$$

Thus the vector of unknowns is equal to the product of the inverse of the coefficient matrix and the vector of constants.

EXAMPLE 3

Solve the system

$$\begin{aligned} 2x + y &= 4, \\ x + 3y &= 7, \end{aligned}$$

by the method of the inverse. The inverse is found by the method above; here,

$$A = \begin{bmatrix} 2 & 1 \\ 1 & 3 \end{bmatrix} \quad \text{and} \quad A^{-1} = \frac{1}{5}\begin{bmatrix} 3 & -1 \\ -1 & 2 \end{bmatrix}.$$

It follows that the solution vector is

$$X = \begin{bmatrix} x \\ y \end{bmatrix} = \frac{1}{5}\begin{bmatrix} 3 & -1 \\ -1 & 2 \end{bmatrix}\begin{bmatrix} 4 \\ 7 \end{bmatrix} = \begin{bmatrix} 1 \\ 2 \end{bmatrix}.$$

This means that the solution of the system is $x = 1$ and $y = 2$. ▲

The solution of a 3-by-3 system is similar.

EXAMPLE 4

Solve the system

$$\begin{aligned} 5x + 7y + 6z &= -1, \\ 7x + 8y &= 7, \\ 2x + y - 3z &= 5. \end{aligned}$$

Here

$$A = \begin{bmatrix} 5 & 7 & 6 \\ 7 & 8 & 0 \\ 2 & 1 & -3 \end{bmatrix} \quad \text{and} \quad B = \begin{bmatrix} -1 \\ 7 \\ 5 \end{bmatrix}.$$

By expanding in cofactors of the third column, we find that

$$\textbf{det}\, A = 6\begin{bmatrix} 7 & 8 \\ 2 & 1 \end{bmatrix} - 0 - 3\begin{bmatrix} 5 & 7 \\ 7 & 8 \end{bmatrix} = 6(-9) - 3(-9) = -27.$$

It follows that A is nonsingular and so the inverse exists. The cofactor matrix is

$$\begin{bmatrix} -24 & 21 & -9 \\ 27 & -27 & 9 \\ -48 & 42 & -9 \end{bmatrix};$$

and the adjoint is

$$\begin{bmatrix} -24 & 27 & -48 \\ 21 & -27 & 42 \\ -9 & 9 & -9 \end{bmatrix};$$

thus, the inverse is

$$A^{-1} = -\frac{1}{27}\begin{bmatrix} -24 & 27 & -48 \\ 21 & -27 & 42 \\ -9 & 9 & -9 \end{bmatrix}.$$

By matrix multiplication we find the solution vector:

$$X = A^{-1}B = \begin{bmatrix} 1 \\ 0 \\ -1 \end{bmatrix}.$$

The solution of the system is $x = 1$, $y = 0$, $z = -1$. ▲

The reader may ask: Why study the inverse method for solving linear equations if we already have Cramer's Rule, which appears to be simpler? The answer is that the inverse method is very efficient when **several** sets of equations with the same coefficient matrices have to be solved. Instead of using Cramer's Rule for each set separately, we find A^{-1} and then premultiply each of the B-vectors to find the solutions, each in a single operation. This is particularly useful when the inverse is stored in a computer.

EXAMPLE 5

Sol and Ruth are interviewers at the college job placement office. Every applicant must have an interview with each of them. Sol's interview with each male applicant takes 10 minutes, and for obvious reasons, his interview with each female takes a bit more – 20 minutes. On the other hand, Ruth's interview with each female takes 10 minutes, and with each male, 15 minutes. If Sol works for 4 hours, and Ruth for 3, what is the maximum number of applicants, male and female, that can complete both interviews? Also, what happens if Sol and Ruth change their numbers of working hours?

We set up the problem as a system of two linear equations in two unknowns. Let x be the number of males interviewed, and y the number of females. Sol spends $x/6$ hours interviewing males, and $y/3$ for females. Since he works a total of 4 hours, x and y satisfy

$$\frac{x}{6} + \frac{y}{3} = 4.$$

Ruth spends $x/4$ hours on males, and $y/6$ on females, for a total of 3 hours; hence,

$$\frac{x}{4} + \frac{y}{6} = 3.$$

We solve the system

$$\begin{bmatrix} \frac{1}{6} & \frac{1}{3} \\ \frac{1}{4} & \frac{1}{6} \end{bmatrix} \begin{bmatrix} x \\ y \end{bmatrix} = \begin{bmatrix} 4 \\ 3 \end{bmatrix}.$$

The determinant of the coefficient matrix is $-1/18$; hence, the inverse exists. It is found to be

$$A^{-1} = \begin{bmatrix} -3 & 6 \\ \frac{9}{2} & -3 \end{bmatrix}.$$

The solution of the system is

$$\begin{bmatrix} -3 & 6 \\ \frac{9}{2} & -3 \end{bmatrix} \begin{bmatrix} 4 \\ 3 \end{bmatrix} = \begin{bmatrix} 6 \\ 9 \end{bmatrix};$$

thus, $x = 6$ and $y = 9$, so that together they can interview a maximum of 6 males and 9 females.

Suppose that the workload is increased so that Sol and Ruth work for 6 and 4 hours, respectively. The only change in the system of two equations is in the vector of constants on the right-hand side, as the coefficient matrix is unchanged:

$$\begin{bmatrix} \frac{1}{6} & \frac{1}{3} \\ \frac{1}{4} & \frac{1}{6} \end{bmatrix} \begin{bmatrix} x \\ y \end{bmatrix} = \begin{bmatrix} 6 \\ 4 \end{bmatrix}.$$

We use the same matrix A^{-1} to premultiply the vector on the right-hand side:

$$\begin{bmatrix} -3 & 6 \\ \frac{9}{2} & -3 \end{bmatrix} \begin{bmatrix} 6 \\ 4 \end{bmatrix} = \begin{bmatrix} 6 \\ 15 \end{bmatrix}.$$

Thus $x = 6$ (as before) and $y = 15$, and so together they can interview 6 males and 15 females.

The maximum number of applicants can be easily computed for any schedule of working hours by simply premultiplying the "working hours vector" by A^{-1}. The solution might not always be in terms of integers. In this case the solution has to be reduced to the nearest integer values. For example, if the two work 5 hours each, then the solution vector is

$$A^{-1} \begin{bmatrix} 5 \\ 5 \end{bmatrix} = \begin{bmatrix} 15 \\ 7\frac{1}{2} \end{bmatrix}.$$

The number of males is 15, but the number of females is 7 instead of $7\frac{1}{2}$. ▲

3.5 EXERCISES

In Exercises 1 to 4, find the inverses of the matrices.

1 (a) $\begin{bmatrix} 1 & -5 \\ 9 & 9 \end{bmatrix}$ (c) $\begin{bmatrix} -7 & 6 \\ 8 & -1 \end{bmatrix}$

(b) $\begin{bmatrix} 3 & 5 \\ -8 & 0 \end{bmatrix}$ (d) $\begin{bmatrix} 5 & 3 \\ -6 & -1 \end{bmatrix}$

2 (a) $\begin{bmatrix} 4 & 5 \\ 3 & 4 \end{bmatrix}$ (c) $\begin{bmatrix} 9 & 2 \\ 0 & -5 \end{bmatrix}$

(b) $\begin{bmatrix} 9 & 5 \\ -1 & 2 \end{bmatrix}$ (d) $\begin{bmatrix} 6 & -2 \\ 7 & -3 \end{bmatrix}$

3 (a) $\begin{bmatrix} 3 & 2 & -1 \\ 0 & 1 & 1 \\ 8 & 3 & 9 \end{bmatrix}$ (c) $\begin{bmatrix} 3 & 0 & 6 \\ 7 & 1 & -4 \\ -9 & 7 & 0 \end{bmatrix}$

(b) $\begin{bmatrix} 4 & 2 & 1 \\ 3 & -5 & -4 \\ 1 & 2 & 2 \end{bmatrix}$ (d) $\begin{bmatrix} 5 & 1 & 2 \\ -8 & 5 & 6 \\ 7 & 9 & 0 \end{bmatrix}$

4 (a) $\begin{bmatrix} 7 & 3 & 0 \\ -1 & 0 & 1 \\ 8 & 5 & 1 \end{bmatrix}$ (c) $\begin{bmatrix} 4 & -8 & 7 \\ 2 & 1 & 4 \\ 1 & 1 & 0 \end{bmatrix}$

(b) $\begin{bmatrix} 3 & 1 & 1 \\ 0 & 1 & -6 \\ 2 & 1 & 7 \end{bmatrix}$ (d) $\begin{bmatrix} 0 & 7 & -7 \\ -2 & 1 & 1 \\ 4 & 6 & 5 \end{bmatrix}$

In Exercises 5 to 8, solve the systems by the method of the inverse.

5 (a) $\begin{aligned} x + y &= 5 \\ 2x - y &= 7 \end{aligned}$ (c) $\begin{aligned} 7x + 2y &= 8 \\ 8x - 5y &= -20 \end{aligned}$

(b) $\begin{aligned} 3x + 4y &= 2 \\ 5x - 2y &= 12 \end{aligned}$ (d) $\begin{aligned} 5x - 6y &= -21 \\ 2x + 7y &= 1 \end{aligned}$

6 (a) $\begin{aligned} x + 7y &= 11 \\ 8x + y &= -22 \end{aligned}$ (c) $\begin{aligned} 3x - y &= 17 \\ 9x + 4y &= 37 \end{aligned}$

(b) $\begin{aligned} 8x + 5y &= 1 \\ x - 8y &= 26 \end{aligned}$ (d) $\begin{aligned} x - 2y &= 1 \\ 5x + 8y &= -31 \end{aligned}$

7 (a) $\begin{aligned} 7x + 4y - 5z &= 5 \\ x + 2y + z &= 7 \\ 5x - z &= 3 \end{aligned}$ (c) $\begin{aligned} 2x - 9y + 6z &= -11 \\ 5x + 6y + 4z &= 25 \\ 4y - 4z &= 0 \end{aligned}$

(b) $\begin{aligned} 3x - 4y + 3z &= 0 \\ 6x - 8z &= 14 \\ 2x - y &= 2 \end{aligned}$ (d) $\begin{aligned} 6x + y - z &= 2 \\ 4x + y - 4z &= 2 \\ 5y + 4z &= 10 \end{aligned}$

8 (a)
$$\begin{aligned} x + y + z &= 2 \\ 2x + 3y + 4z &= 3 \\ x - 2y - z &= 1 \end{aligned}$$

(b)
$$\begin{aligned} 2x - y + 5z &= -4 \\ -4x + 2y \quad &= -2 \\ x - y - z &= 1 \end{aligned}$$

(c)
$$\begin{aligned} x + 4y - 2z &= 5 \\ 2x - 9y + 3z &= -7 \\ 3x + y + z &= 4 \end{aligned}$$

(d)
$$\begin{aligned} 4x - y + 4z &= -17 \\ 2x + 2y - z &= 0 \\ -3x \quad + 7z &= -8 \end{aligned}$$

9 One kind of hen will yield a farmer 10 dozen eggs and 4 pounds of chicken meat. Another will yield 12 dozen eggs and 3 pounds of meat. The farmer has a market for 2700 dozen eggs and 900 pounds of meat. (a) How many hens of each type should he have to meet the demand of the market? [*Hint:* Let x and y be the numbers of the 2 kinds of hen. Solve a system of 2 equations in 2 unknowns.] (b) By the method of the inverse, find a general formula to use to satisfy a demand for b_1 dozen eggs and b_2 pounds of meat; in other words, express the numbers x and y in terms of the quantities demanded (Figure 3.19). (See Example 5.)

Figure 3.19

10 A small airline leases 2 kinds of passenger planes. One carries 30 first-class passengers and 80 coach passengers, and the other 40 first-class and 70 coach. It expects to carry at most 500 first-class and 1150 coach passengers. (a) How many planes of each kind should it lease for maximum efficiency? (b) What if it expects 700 first-class and 1500 coach passengers? (See Example 5.)

11 At a certain college the students are classified as either liberal arts, science, or engineering majors. In the freshman year, every liberal arts student takes 3 liberal arts courses and 1 science course; every science student takes 1 liberal arts course, 2 science courses, and 1 engineering course; and every engineer takes 1 liberal arts course, 1 science course, and 2 engineering courses. The course plan is summarized in the matrix:

	Students		
Courses	Liberal arts	Science	Engineering
Liberal arts	3	1	1
Science	1	2	1
Engineering	0	1	2

According to the composition of the faculty there are 1500 student places in liberal arts courses, 900 in science courses, and 400 in engineering courses. (Each student uses 4 places.) How many students should the college accept in each category in order to use the faculty resources as efficiently as possible? (Compare Example 5)

12 Cases of soda of assorted flavors are packed in groups of 20 bottles. There are 3 kinds of cases, with these flavor compositions:

	Flavors		
Cases	Cream soda	Root beer	Orange
A	10	5	5
B	5	10	5
C	0	10	10

The hostess at a large party wants to order 200 bottles of cream soda, 350 bottles of root beer, and 250 bottles of orange. (a) How many of each of the 3 cases should she order? (b) What if she wants 250 cream soda, 300 root beer, and 250 orange? (See Example 5.)

13 Men, women, and children are examined at a clinic. There is a standard 3-part medical examination: physical and interview, x-ray, and tests. The times consumed for each of the 3 categories of patients and for each part of the examination are as follows:

	Physical & interview	X-ray	Tests
Men	20 min	5	10
Women	15	5	15
Children	10	5	10

(Figure 3.20). Suppose that the physical examination section is in operation 15 hours per day, the x-ray section 6 hours, and the testing section 13 hours. (The patients have to wait for each of the parts of the examination.) How many men, women, and children, respectively, should be scheduled for appointments so that the 3 sections will be working at capacity? (See Example 5.)

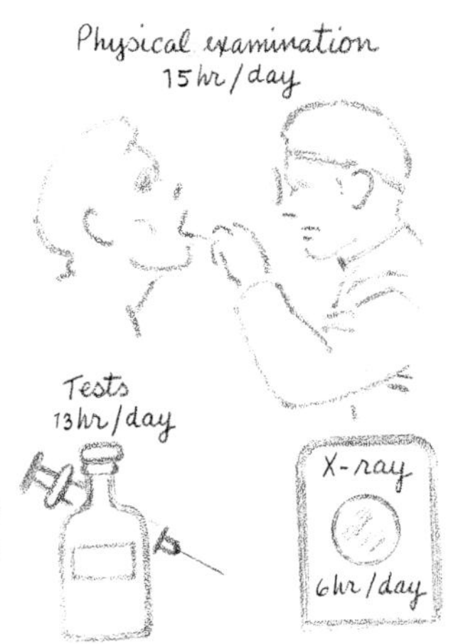

Figure 3.20

14 Let A be a square matrix and **cof** A be the corresponding cofactor matrix. A theorem in matrix theory states that the inner product of a row of A and a different row of **cof** A is equal to 0; for example, the inner product of the first row of A and the second row of **cof** A is 0. Give the proof in the 3-by-3 case. Use the fact that the determinant of a matrix with two identical rows is equal to 0. (See Exercise 9, Section 3.3.) ●●

15 Use the result of Exercise 14 to show that the inverse of A is equal to the adjoint multiplied by the reciprocal of the determinant. Give the proof in the 3-by-3 case. ●●

3.6 Elementary Operations on Matrices

We have already shown the reader 2 methods of solving a system of linear equations: Cramer's Rule and the method of the inverse. We shall now introduce a third method, that of **elementary operations** on the **augmented matrix** of the system. The advantages of this method are: it is more efficient than the others when the number of equations is large; it can be used even when the number of unknowns is not equal to the number of equations; it can be used to determine whether the system even has a solution; and finally whether the solution, if it exists, is unique. It is a generalization of the method of elimination used in high school algebra to solve simultaneous linear equations: the system is replaced by an equivalent but simpler system. The simplification is done by performing one or both of these two operations on the equations:

1. Both sides of the equation are multiplied by a number not equal to 0.
2. A nonzero multiple of one equation is added to another equation.

These operations correspond to certain "elementary operations" on the matrices involved in the matrix form of the system, $AX = B$. The augmented matrix of such a system is the coefficient matrix A with an added column consisting of the vector B.

EXAMPLE 1

Consider the system of 2 equations in 2 unknowns:

$$\begin{aligned} 3x + 2y &= 1, \\ x - y &= 2. \end{aligned} \tag{1}$$

Multiply both sides of the second equation by the constant -3, and add it to the first equation; we get the equivalent system

$$\begin{aligned} 5y &= -5, \\ x - y &= 2. \end{aligned}$$

Multiply the first equation on both sides by $\frac{1}{5}$:

$$\begin{aligned} y &= -1, \\ x - y &= 2. \end{aligned} \tag{2}$$

This is easily solved by substituting the value $y = -1$ in the second equation to get $x = 1$. Another way of writing system (2) is

$$\begin{aligned} x - y &= 2, \\ y &= -1. \end{aligned} \tag{3}$$

Observe that the operations leading from form (1) to (3) did not explicitly involve the unknowns x and y but only the numerical coeffients and the numbers on the right-hand sides of the equations. In other words, the operations involved only the matrix A and the vector B, but not the vector X. The **augmented** matrix of system (1) is the matrix A with the last column as B:

$$\begin{bmatrix} 3 & 2 & 1 \\ 1 & -1 & 2 \end{bmatrix}.$$

It is conventional to isolate the last column by placing a vertical broken line between it and the other columns:

$$\left[\begin{array}{cc|c} 3 & 2 & 1 \\ 1 & -1 & 2 \end{array}\right]. \tag{4}$$

Similarly, the augmented matrix of system (3) is

$$\left[\begin{array}{cc|c} 1 & -1 & 2 \\ 0 & 1 & -1 \end{array}\right]. \tag{5}$$

Matrix (5) is obtained from matrix (4) by means of the following operations: -3 times the second row is added to the first row, then the first row is multiplied by $\frac{1}{5}$, and finally, the rows are interchanged. The steps are

$$\left[\begin{array}{cc|c} 3 & 2 & 1 \\ 1 & -1 & 2 \end{array}\right] \longrightarrow \left[\begin{array}{cc|c} 0 & 5 & -5 \\ 1 & -1 & 2 \end{array}\right] \longrightarrow$$

$$\left[\begin{array}{cc|c} 0 & 1 & -1 \\ 1 & -1 & 2 \end{array}\right] \longrightarrow \left[\begin{array}{cc|c} 1 & -1 & 2 \\ 0 & 1 & -1 \end{array}\right]. \blacktriangle$$

The three elementary row operations on a matrix are:

1. interchange of rows
2. multiplication of the entries in a row by a common nonzero constant
3. addition of a multiple of one row to another row.

The example above illustrates the following important principle: Elementary row operations on the augmented matrix leave the system of equations essentially unchanged: the systems corresponding to the original augmented matrix and the transformed augmented matrix are equivalent, that is, have the same solution. As we indicated above, this method is justified because the elementary row operations are equivalent to operations on the system of equations which preserve the solution.

The message is that to solve a system of equations, simplify the augmented matrix by means of row operations, and then try to solve the corresponding system. In terms of Example 1 this means that the augmented matrix (4) should first be changed to (5), and then the corresponding system (3) is solved; the latter is very easy to do because the system is so simple.

What form should the augmented matrix assume to simplify the solution of the set of equations? The form of an **echelon** matrix. The French word **échelle** means **ladder.** An echelon matrix is one with the following properties (Figure 3.21):

1. The first several rows have some nonzero entries. Subsequent rows might have only zero entries. In any case the rows with nonzero entries appear higher.
2. In any row with a nonzero entry, the first such entry is always 1.
3. In any of the "nonzero" rows the number of zeros preceding the first nonzero entry is smaller than in the following row.

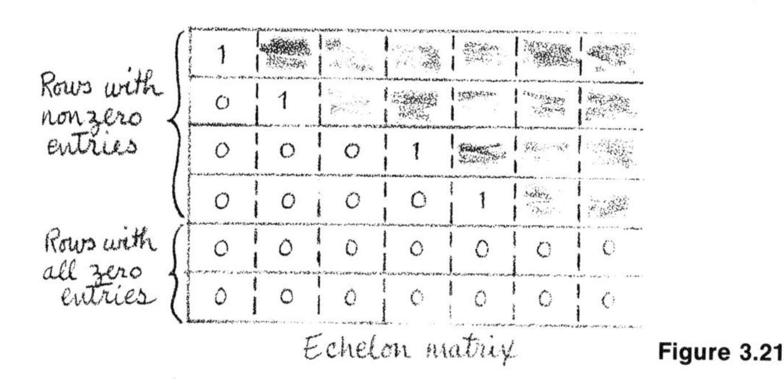

Figure 3.21

By means of elementary row operations any matrix with some nonzero entries can be transformed into an echelon matrix; or, as we shall say, into **echelon form.** The point is this: You can see that it is very easy to solve a system of equations whose augmented matrix is an echelon matrix; take, for example, system (3) with the echelon augmented matrix (5). The augmented matrix of **any** system can be put in echelon form; the corresponding equivalent system of equations is then very easily solved.

EXAMPLE 2

Consider the system of equations

$$
\begin{aligned}
x - 2y + 3z &= 4,\\
-x + y + z &= 0,\\
2x + 2y - z &= 1.
\end{aligned}
$$

The augmented matrix is

$$\left[\begin{array}{rrr|r} 1 & -2 & 3 & 4 \\ -1 & 1 & 1 & 0 \\ 2 & 2 & -1 & 1 \end{array}\right].$$

The first row is already in proper echelon form because it begins with 1. We shall eliminate the other nonzero entries in the first column: Add the entries of the first row to the second row, and -2 times the first row to the third. These operations are denoted $R_1 + R_2$ and $-2R_1 + R_3$, respectively; thus,

$$\left[\begin{array}{rrr|r} 1 & -2 & 3 & 4 \\ -1 & 1 & 1 & 0 \\ 2 & 2 & -1 & 1 \end{array}\right] \begin{array}{c} R_1 + R_2 \\ -2R_1 + R_3 \end{array} = \left[\begin{array}{rrr|r} 1 & -2 & 3 & 4 \\ 0 & -1 & 4 & 4 \\ 0 & 6 & -7 & -7 \end{array}\right].$$

The first nonzero entry in the second row is -1. It is changed to 1 by the row operation of multiplication by -1. This operation is denoted $(-1)R_2$:

$$\left[\begin{array}{rrr|r} 1 & -2 & 3 & 4 \\ 0 & -1 & 4 & 4 \\ 0 & 6 & -7 & -7 \end{array}\right] (-1)R_2 = \left[\begin{array}{rrr|r} 1 & -2 & 3 & 4 \\ 0 & 1 & -4 & -4 \\ 0 & 6 & -7 & -7 \end{array}\right].$$

Next we eliminate entry 6 under entry 1 in the second column; this is done by adding -6 times the second row to third:

$$\left[\begin{array}{rrr|r} 1 & -2 & 3 & 4 \\ 0 & 1 & -4 & -4 \\ 0 & 6 & -7 & -7 \end{array}\right] -6R_2 + R_3 = \left[\begin{array}{rrr|r} 1 & -2 & 3 & 4 \\ 0 & 1 & -4 & -4 \\ 0 & 0 & 17 & 17 \end{array}\right].$$

Finally, we multiply the third row by 1/17:

$$\left[\begin{array}{rrr|r} 1 & -2 & 3 & 4 \\ 0 & 1 & -4 & -4 \\ 0 & 0 & 1 & 1 \end{array}\right]. \tag{6}$$

This is in echelon form. The corresponding system of equations is

$$\begin{aligned} x - 2y + 3z &= 4, \\ y - 4z &= -4, \\ z &= 1. \end{aligned}$$

To solve the system, substitute $z = 1$ from the third equation into the second, and solve for y: $y = 0$. Then put $z = 1$ and $y = 0$ in the first equation and solve for x: $x = 1$. Therefore, the solution is $x = 1$, $y = 0$, and $z = 1$. The reader can check that this is also the solution of the original system. ▲

EXAMPLE 3

The particular echelon form of a given matrix is not unique; two different sequences of row operations may change a matrix into two different echelon forms. Take the augmented matrix appearing in Example 2:

$$\left[\begin{array}{rrr|r} 1 & -2 & 3 & 4 \\ -1 & 1 & 1 & 0 \\ 2 & 2 & -1 & 1 \end{array}\right].$$

Suppose we first interchange the first and second rows. This operation is denoted $R_1 \leftrightarrow R_2$:

$$\left[\begin{array}{rrr|r} 1 & -2 & 3 & 4 \\ -1 & 1 & 1 & 0 \\ 2 & 2 & -1 & 1 \end{array}\right] R_1 \longleftrightarrow R_2 = \left[\begin{array}{rrr|r} -1 & 1 & 1 & 0 \\ 1 & -2 & 3 & 4 \\ 2 & 2 & -1 & 1 \end{array}\right].$$

If we now perform the following operations in the given order,

$$R_1 + R_2 \quad 2R_1 + R_3 \quad (-1)R_1 \quad 4R_2 + R_3 \quad (-1)R_2 \quad (1/17)R_3,$$

we obtain the echelon matrix

$$\left[\begin{array}{rrr|r} 1 & -1 & -1 & 0 \\ 0 & 1 & -4 & -4 \\ 0 & 0 & 1 & 1 \end{array}\right].$$

The corresponding system of equations is equivalent to that in Example 2. ▲

If a matrix A is transformed into an echelon matrix A' by a series of row operations, and into another echelon matrix A'' by another series of row operations, then the echelon forms A' and A'' have the **same number of rows with nonzero entries.** This number is called the **rank** of the original matrix A. The reason for this is left to Exercise 13 below.

We remark that elementary **column** operations on matrices are defined analogously to row operations: interchange of columns, multiplication of the entries of a column by a common nonzero constant, and the addition of a constant multiple of one column to another. There is also a **column-echelon** form of a matrix. It can be shown that the "row rank" of a matrix (that defined by means of the row-echelon form) is always equal to the "column rank," defined in terms of the column-echelon form. It is only the set of row operations, not the column operations, that may be used on the augmented matrix in the solution of a system of linear equations.

3.6 EXERCISES

In Exercises 1 and 2 reduce the matrices to echelon form by means of row operations. (Recall that the form is not unique.)

1 (a) $\begin{bmatrix} 2 & 1 \\ 4 & 6 \end{bmatrix}$

(b) $\begin{bmatrix} 1 & 1 \\ 2 & 2 \end{bmatrix}$

(c) $\begin{bmatrix} 1 & -2 & 3 & 4 \\ 0 & 1 & 4 & 6 \\ 1 & 0 & 11 & 16 \end{bmatrix}$

(d) $\begin{bmatrix} 2 & 1 & -2 \\ 1 & 0 & 0 \\ 4 & 4 & 2 \end{bmatrix}$

(e) $\begin{bmatrix} 2 & 1 & 1 \\ 0 & 8 & 4 \\ 4 & 0 & 3 \\ 2 & 2 & 2 \end{bmatrix}$

2 (a) $\begin{bmatrix} 4 & -6 \\ 2 & 4 \end{bmatrix}$

(b) $\begin{bmatrix} 3 & 2 \\ -6 & -4 \end{bmatrix}$

(c) $\begin{bmatrix} 1 & 2 & 3 & 1 \\ 0 & -1 & -1 & -1 \\ 1 & -3 & -2 & -4 \end{bmatrix}$

(d) $\begin{bmatrix} 1 & -2 & 3 \\ -1 & 1 & 1 \\ 2 & 2 & -1 \end{bmatrix}$

(e) $\begin{bmatrix} 3 & -3 & 0 \\ 1 & 0 & 1 \\ 3 & 1 & 2 \\ 6 & 5 & -1 \end{bmatrix}$

3 Write equivalent forms of the following systems after changing the augmented matrices to echelon form. Do not solve.

(a) $\begin{aligned} x + 2y &= 10 \\ -3x + y &= 4 \end{aligned}$

(b) $\begin{aligned} x - 3y &= 4 \\ 8x - 5y &= 10 \end{aligned}$

(c) $\begin{aligned} 2x + z &= 2 \\ 2x + 2y &= 1 \\ x + y + 5z &= -2 \end{aligned}$

(d) $\begin{aligned} 3x + 6y + z &= 1 \\ x - 2y + 4z &= 0 \\ -x + 8y + 5z &= 2 \end{aligned}$

(e) $\begin{aligned} 5x + y - 5z &= 2 \\ 4y + 7z - 9w &= -1 \\ x - 8y + 2z - w &= 2 \end{aligned}$
(3 equations, 4 unknowns)

(f) $\begin{aligned} 2x + 6y &= -5 \\ 5x - y &= 5 \\ 7x + 5y &= 0 \end{aligned}$
(3 equations, 2 unknowns)

4 Follow the same instructions as for Exercise 3.

(a) $\begin{aligned} x - 3y &= 7 \\ 2x + y &= 5 \end{aligned}$

(b) $\begin{aligned} 2x - 9y &= 4 \\ -8x + y &= 9 \end{aligned}$

(c) $\begin{aligned} 2x - z &= 1 \\ 2x + 4y - z &= 1 \\ x - 8y - 3z &= -2 \end{aligned}$

(d) $\begin{aligned} x - y + 2z &= 2 \\ 2y - 3z &= -2 \\ 3x - 2y + 4z &= 5 \end{aligned}$

(e) $\begin{aligned} 2x + y + z + w &= 1 \\ x + 2y + z + w &= 0 \\ x + y + 2z + w &= 1 \end{aligned}$

(f) $\begin{aligned} x + 2y &= 4 \\ -x + 3y &= 1 \\ x + 12y &= 14 \end{aligned}$

5 A theorem of matrix theory states that an n-by-n matrix is of rank n if and only if it is nonsingular. Give the proof in the particular case $n = 2$.

6 Elementary row and column operations are powerful tools for calculating determinants. The determinant of a square matrix is unchanged if a multiple of one row is added to another row. Give the proof in the case of a 3-by-3 matrix.

7 Show that the multiplication of the entries of a row of a matrix by a nonzero constant does not affect the singularity or nonsingularity of the matrix.

8 Show that the interchange of two rows of a matrix does not affect the singularity or nonsingularity of the matrix. (See Exercise 10, Section 3.3.)

9 The rank of a matrix, defined above, is also equal to the number of rows (and columns) in the largest nonsingular square submatrix of the matrix. How does this follow from Exercises 6 to 8? (Note that the results about row operations are also valid for column operations.)

10 Elementary row operations can be carried out by premultiplying the matrix by an **elementary** matrix. This is a matrix obtained from the identity matrix by an elementary row (or column) operation.

(a) Let E be the matrix obtained from the identity by interchanging two rows such as the first and second rows:

$$E = \begin{bmatrix} 0 & 1 & 0 & 0 & 0 \\ 1 & 0 & 0 & 0 & 0 \\ 0 & 0 & 1 & 0 & 0 \\ 0 & 0 & 0 & 1 & 0 \\ & & & & \\ 0 & 0 & 0 & 0 & 1 \end{bmatrix}$$

If the number of rows of a matrix A is equal to the number of E, then the product EA is the matrix obtained from A by interchanging the first two rows. Give the proof for an arbitrary 3-by-3 matrix A: Show that

$$\begin{bmatrix} 0 & 1 & 0 \\ 1 & 0 & 0 \\ 0 & 0 & 1 \end{bmatrix} \begin{bmatrix} a_{11} & a_{12} & a_{13} \\ a_{21} & a_{22} & a_{23} \\ a_{31} & a_{32} & a_{33} \end{bmatrix} = \begin{bmatrix} a_{21} & a_{22} & a_{23} \\ a_{11} & a_{12} & a_{13} \\ a_{31} & a_{32} & a_{33} \end{bmatrix}.$$

(b) Let E be the elementary matrix obtained from the identity by multiplying a given row by a constant c. Show, in the 3-by-3 case, that EA is the matrix obtained from A by the same row operation.

(c) Let E be the elementary matrix obtained from the identity by adding a multiple of one row to another. Show, in the 3-by-3 case, that EA is the matrix obtained from A by the same row operation.

11 All of the results of Exercise 10 remain valid when column operations are used instead of row operations: postmultiplication by E takes the place of premultiplication. Give the details of the proof as in Exercise 10.

12 (Alternate method of computing the inverse.) Let A be a square matrix. Suppose that by a succession of row operations it can be transformed into the identity matrix. Let E_1, E_2, . . . be the corresponding elementary matrices, defined in Exercise 10. Show that the product of these matrices in the reverse order,

$$\ldots E_2 E_1,$$

is equal to A^{-1}. [*Hint:* Show that its product with A is equal to the identity.]

13 *Prove:* If A' and A'' are echelon matrices obtained from A by different sequences of row operations, then they have the same numbers of rows with nonzero entries. [*Hint:* First show that it is true if A itself is an echelon matrix.]

3.7 Existence and Uniqueness of Solutions of Linear Systems; Homogeneous Systems

A system of equations is like a **question** waiting for an **answer:** a solution. There are 3 kinds of systems corresponding to questions of 3 kinds: those with no answer, those with one answer, and those with more than one answer. Consider the question, "Zachary, what grade did you receive in mathematics as a high school senior?" If Zachary did not study mathematics in his senior year, then the question does not have an answer. If he took exactly one course in mathematics, then the question has one answer. Finally, if he took several courses, then the question has several answers.

Systems of equations are similar. A system may have no solution. It is then called **inconsistent.** Otherwise, it is called **consistent** (that is, when it has one or more solutions).

EXAMPLE 1

The system

$$\begin{aligned} x + y &= 1, \\ x + y &= 2, \end{aligned}$$

is inconsistent because there are no numbers x and y whose sum is equal to 1 and also equal to 2. ▲

One might wonder: What is our interest in such a system if it has no solution? The answer is that an inconsistent system may arise in a practical problem as an unknowing attempt to do that which is impossible. If the mathematics of the problem show that the system is inconsistent, then we infer that our effort is going to be unrewarded (see Exercise 21).

There is a simple criterion for determining whether or not a system is consistent:

A system of linear equations is consistent if and only if the rank of the coefficient matrix is equal to the rank of the augmented matrix.

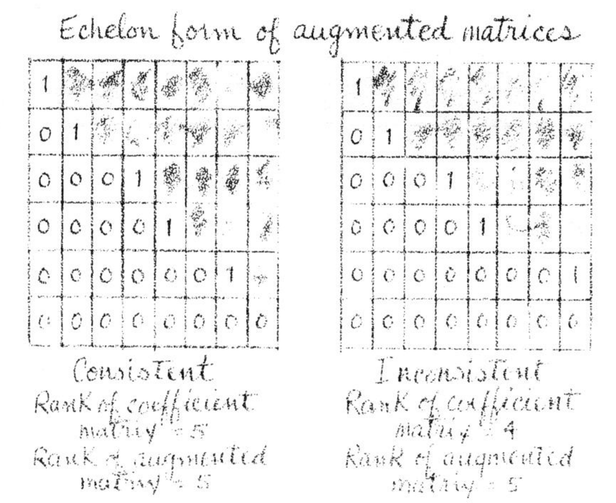

Figure 3.22

For the explanation for this criterion, we refer to Figure 3.22. The echelon forms of a consistent system and an inconsistent system are given. The system of equations corresponding to the first matrix is of the form

$$\begin{aligned} x + \cdots \qquad\qquad &= \cdots \\ y \quad + \cdots &= \cdots \\ z \quad &= \cdots \end{aligned}$$

The (unique) solution is obtained by substituting the value of z in the preceding equation, and then successively solving for each of the unknowns in each of the preceding equations. The last equation of the system corresponding to the second echelon matrix is of the form

$$(0)x + (0)y + \cdots + (0)z = 1.$$

There are no real numbers $x, y, \ldots, z$ satisfying such an equation; therefore, the system has no solution.

If a system is inconsistent, there is nothing more that we can do with it. On the other hand, if it is consistent, then we have to determine whether or not it has a unique solution. This can be determined by the following criterion:

If the common rank of the augmented and coefficient matrices is also equal to the number of unknowns, then the system has a unique

solution. If the rank is smaller than the number of unknowns, there are infinitely many solutions. [*Note:* The rank can never exceed the number of unknowns. Why?]

The proof is as follows. The number of unknowns is the number of columns of the coefficient matrix. If this is also equal to the rank of the coefficient matrix, then the latter matrix must be square, and its echelon form has 1's along the main diagonal and 0's below it. By further row operations, the coefficient matrix can be transformed to the identity. The corresponding equivalent system of equations is then of the form

$$\begin{aligned} x_1 \qquad\qquad &= b_1, \\ x_2 \qquad &= b_2, \\ \cdots \quad & \\ x_m &= b_m; \end{aligned}$$

therefore, the solution is obviously uniquely determined (Figure 3.23).

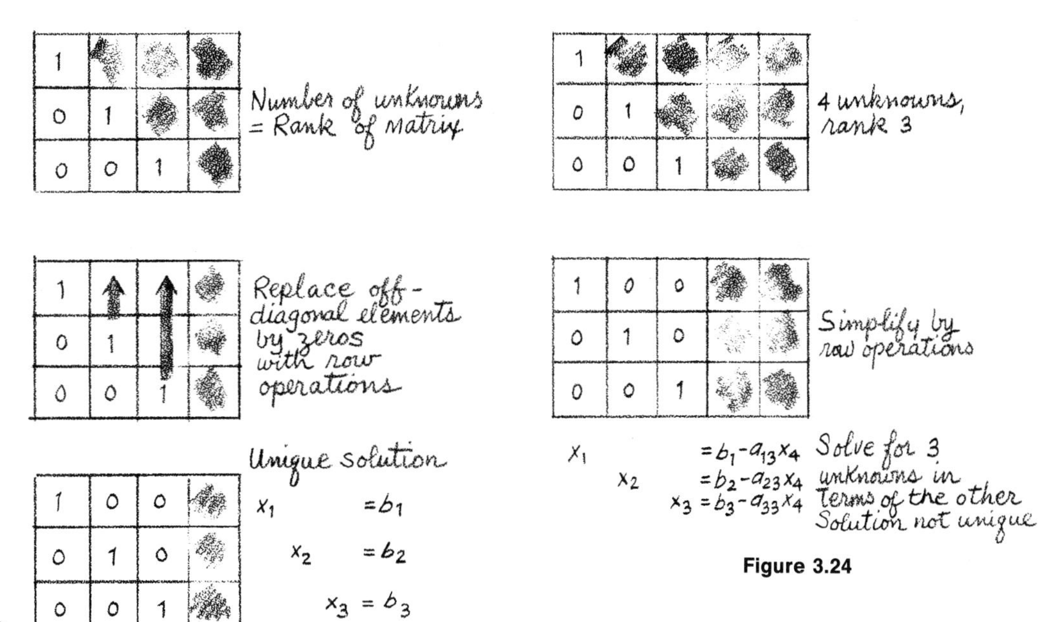

Figure 3.24

Figure 3.23

If the number of unknowns is larger than the rank, then we transpose the terms containing the "extra" unknowns and treat them as given constants. We solve for the other unknowns in terms of these (Figure 3.24). This method is known as the Gauss-Jordan elimination method. (Gauss was a German mathematician of the nineteenth century, and Jordan a Frenchman of the same century.) ●

EXAMPLE 2

Here is an example of a system of equations which we shall show to be inconsistent:

$$\begin{aligned} x - 2y + 3z &= 4, \\ -x + y + z &= 0, \\ 2x + 2y - 18z &= 0 \end{aligned}$$

The augmented matrix is

$$\left[\begin{array}{rrr|r} 1 & -2 & 3 & 4 \\ -1 & 1 & 1 & 0 \\ 2 & 2 & -18 & 0 \end{array}\right].$$

Reduce this to echelon form by row operations:

$$\left[\begin{array}{rrr|r} 1 & -2 & 3 & 4 \\ -1 & 1 & 1 & 0 \\ 2 & 2 & -18 & 0 \end{array}\right] \begin{array}{r} R_1 + R_2 \\ -2R_1 + R_3 \\ \end{array} = \left[\begin{array}{rrr|r} 1 & -2 & 3 & 4 \\ 0 & -1 & 4 & 4 \\ 0 & 6 & -24 & -8 \end{array}\right]$$

$$\left[\begin{array}{rrr|r} 1 & -2 & 3 & 4 \\ 0 & -1 & 4 & 4 \\ 0 & 6 & -24 & -8 \end{array}\right] \begin{array}{r} 6R_2 + R_3 \\ (-1)R_2 \\ (1/16)R_3 \end{array} = \left[\begin{array}{rrr|r} 1 & -2 & 3 & 4 \\ 0 & 1 & -4 & -4 \\ 0 & 0 & 0 & 1 \end{array}\right].$$

The rank of the coefficient matrix is 2. The rank of the augmented matrix is 3. Therefore, according to the criterion above, the system is inconsistent, and there is no solution. ▲

EXAMPLE 3

In this example the number of equations exceeds the number of unknowns, but the system will be shown to be consistent and have a unique solution:

$$\begin{aligned} x + y - 4z &= 2, \\ x - 2y + z &= 1, \\ x + y + z &= 0, \\ 2x - y + 2z &= 1. \end{aligned}$$

The augmented matrix is

$$\left[\begin{array}{rrr|r} 1 & 1 & -4 & 2 \\ 1 & -2 & 1 & 1 \\ 1 & 1 & 1 & 0 \\ 2 & -1 & 2 & 1 \end{array}\right].$$

The steps to echelon form are

Operations	Resulting matrix
$-R_1 + R_2$ $-R_1 + R_3$ $-2R_1 + R_4$	$\left[\begin{array}{rrr\|r} 1 & 1 & -4 & 2 \\ 0 & -3 & 5 & -1 \\ 0 & 0 & 5 & -2 \\ 0 & -3 & 10 & -3 \end{array}\right]$
$-R_2 + R_4$ $-R_3 + R_4$	$\left[\begin{array}{rrr\|r} 1 & 1 & -4 & 2 \\ 0 & -3 & 5 & -1 \\ 0 & 0 & 5 & -2 \\ 0 & 0 & 0 & 0 \end{array}\right]$
$(-1/3)R_2$ $(1/5)R_3$	$\left[\begin{array}{ccc\|c} 1 & 1 & -4 & 2 \\ 0 & 1 & -5/3 & 1/3 \\ 0 & 0 & 1 & -2/5 \\ 0 & 0 & 0 & 0 \end{array}\right]$

The number of unknowns and the rank are both equal to 3, and so the system is consistent. The coefficient matrix can be reduced to the identity by performing the row operations

$$(5/3)R_3 + R_2, \qquad 4R_3 + R_1, \qquad -R_2 + R_1.$$

As a result we obtain the augmented matrix

$$\left[\begin{array}{ccc|c} 1 & 0 & 0 & 11/15 \\ 0 & 1 & 0 & -1/3 \\ 0 & 0 & 1 & -2/5 \\ 0 & 0 & 0 & 0 \end{array}\right].$$

The corresponding system of equations yields the unique solution $x = 11/15$, $y = -1/3$, $z = -2/5$. This can be checked by substitution in each of the four original equations. ▲

EXAMPLE 4

In this example we have a consistent system with infinitely many solutions:

$$\begin{aligned} x + y + z + w &= 2, \\ x + 2z &= 4, \\ x - y + z - w &= 1, \\ x + 2y + 2z + 2w &= 5. \end{aligned}$$

The augmented matrix is

$$\left[\begin{array}{rrrr|r} 1 & 1 & 1 & 1 & 2 \\ 1 & 0 & 2 & 0 & 4 \\ 1 & -1 & 1 & -1 & 1 \\ 1 & 2 & 2 & 2 & 5 \end{array}\right].$$

The steps in the reduction to echelon form are

Operations	Resulting matrix
$-R_1 + R_2$ $-R_1 + R_3$ $-R_1 + R_4$	$\left[\begin{array}{cccc|c} 1 & 1 & 1 & 1 & 2 \\ 0 & -1 & 1 & -1 & 2 \\ 0 & -2 & 0 & -2 & -1 \\ 0 & 1 & 1 & 1 & 3 \end{array}\right]$
$-2R_2 + R_3$ $R_2 + R_4$ $(-1)R_2$	$\left[\begin{array}{cccc|c} 1 & 1 & 1 & 1 & 2 \\ 0 & 1 & -1 & 1 & -2 \\ 0 & 0 & -2 & 0 & -5 \\ 0 & 0 & 2 & 0 & 5 \end{array}\right]$
$R_3 + R_4$ $(-\frac{1}{2})R_3$	$\left[\begin{array}{cccc|c} 1 & 1 & 1 & 1 & 2 \\ 0 & 1 & -1 & 1 & -2 \\ 0 & 0 & 1 & 0 & \frac{5}{2} \\ 0 & 0 & 0 & 0 & 0 \end{array}\right]$

The ranks of the coefficient and augmented matrices are equal to 3; hence, the system is consistent. The number of unknowns is 4, larger than the rank; therefore, the solution is not unique. We solve for the unknowns x, y, and z in terms of the remaining unknown w by means of the Gauss-Jordan procedure. Transform the first part of the augmented matrix to the 3-by-3 identity by these operations:

$$R_3 + R_2, \qquad -R_3 + R_1, \qquad -R_2 + R_1.$$

The augmented matrix becomes

$$\left[\begin{array}{cccc|c} 1 & 0 & 0 & 0 & -1 \\ 0 & 1 & 0 & 1 & \frac{1}{2} \\ 0 & 0 & 1 & 0 & \frac{5}{2} \\ 0 & 0 & 0 & 0 & 0 \end{array}\right]$$

This is the augmented matrix of the system

$$\begin{aligned} x \qquad\qquad\qquad &= -1, \\ y \qquad + w &= \tfrac{1}{2}, \\ z \qquad\quad &= \tfrac{5}{2}; \end{aligned}$$

hence, the solution is $x = -1$, $y = \frac{1}{2} - w$, $z = \frac{5}{2}$. This is called the **complete** solution of the system. There are infinitely many solutions because for each value of the variable w there is a solution. **Particular** solutions are obtained by assigning numerical values to w; for example, if $w = 0$, then the particular solution is $x = -1$, $y = \frac{1}{2}$, and $z = \frac{5}{2}$. ▲

A system of equations is called **homogeneous** if the constants on the right-hand sides of the equations are all equal to 0. Our examples will usually involve systems of 2 homogeneous equations in 2 unknowns, and also 3 equations in 3 unknowns:

$$\begin{aligned} a_{11}x_1 + a_{12}x_2 &= 0, \\ a_{21}x_1 + a_{22}x_2 &= 0, \end{aligned} \qquad \begin{aligned} a_{11}x_1 + a_{12}x_2 + a_{13}x_3 &= 0, \\ a_{21}x_1 + a_{22}x_2 + a_{23}x_3 &= 0, \\ a_{31}x_1 + a_{32}x_2 + a_{33}x_3 &= 0. \end{aligned}$$

Such a system is always consistent; indeed, the rank of the augmented matrix is always the same as the coefficient matrix since the last column of the former matrix contains only zeros. In addition, the reader can see that the system always has the trivial solution $x_1 = x_2 = x_3 = 0$. The primary question about a homogeneous system is the uniqueness of the solution. The only solution is the trivial "zero solution" if the rank of the coefficient matrix is equal to the number of unknowns; and there is a solution other than the zero solution if the rank is less than the number of unknowns.

EXAMPLE 5

Consider the homogeneous system

$$\begin{aligned} x + 2y - z &= 0, \\ 3x - 4y + 2z &= 0, \\ x + 12y - 6z &= 0, \end{aligned}$$

with the augmented matrix

$$\left[\begin{array}{rrr|r} 1 & 2 & -1 & 0 \\ 3 & -4 & 2 & 0 \\ 1 & 12 & -6 & 0 \end{array}\right].$$

By means of the operations $(-3)R_1 + R_2$, $-R_1 + R_3$, $R_2 + R_3$, $(-1/10)R_2$, the matrix is transformed into

$$\left[\begin{array}{rrr|r} 1 & 2 & -1 & 0 \\ 0 & 1 & -\frac{1}{2} & 0 \\ 0 & 0 & 0 & 0 \end{array}\right].$$

The rank is 2 and there are 3 unknowns; hence, there is a solution other than the zero solution. It is obtained by the Gauss-Jordan method, by performing the row operation $(-2)R_2 + R_1$; the augmented matrix becomes

$$\left[\begin{array}{rrr|r} 1 & 0 & 0 & 0 \\ 0 & 1 & -\frac{1}{2} & 0 \\ 0 & 0 & 0 & 0 \end{array}\right].$$

We solve for x and y in terms of z: $x = 0$ and $y = \frac{1}{2}z$. This is the complete solution. ▲

The theory of homogeneous systems is used for Markov chains (compare Section 3.2). Let X be a row vector representing the distribution of the proportions of a population over a set of states. Let P be the transition matrix for the proportions; then the distribution of the proportions after one transition was shown in Section 3.2 to be the row vector XP. The distribution X is called **stationary** if it unchanged by the transition operation, that is, if $XP = X$. Given a transition matrix P, we can find a stationary distribution by solving the system of equations represented by the equation $XP = X$. Stationary distributions are of great importance in problems involving the long-term properties of Markov chains; such problems will be studied in Section 3.11.

EXAMPLE 6

We consider a simpler version of the fish-distribution problem discussed in Example 4 of Section 3.2. A lake contains a population of fish. We imagine that the lake is divided into 2 parts: that part within 6 feet of the surface, and that below the 6-foot level. The fish move above and below the level according to the following rule: During a 5-minute time period, $\frac{1}{3}$ of the fish that were above the 6-foot level move below the level, and the remaining $\frac{2}{3}$ stay above it; and, of those who were below the level, $\frac{1}{4}$ move above it and $\frac{3}{4}$ stay below. The transition matrix P is

	To	
From	Above level	Below level
Above level	$\frac{2}{3}$	$\frac{1}{3}$
Below level	$\frac{1}{4}$	$\frac{3}{4}$

A stationary distribution for this system is a distribution of fish above and below the level such that the **proportions** remain unchanged after a transition; however, particular fish may change their positions. A stationary distribution is found by solving the equation $XP = X$, which in this case is the system

$$(x \quad y)\begin{bmatrix} \frac{2}{3} & \frac{1}{3} \\ \frac{1}{4} & \frac{3}{4} \end{bmatrix} = (x \quad y) \qquad \text{or} \qquad \begin{aligned} \tfrac{2}{3}x + \tfrac{1}{4}y &= x, \\ \tfrac{1}{3}x + \tfrac{3}{4}y &= y. \end{aligned}$$

Transpose the x and y terms to the left-hand side of the equations, and obtain the homogeneous system

$$\begin{aligned} (-\tfrac{1}{3})x + (\tfrac{1}{4})y &= 0, \\ (\tfrac{1}{3})x + (-\tfrac{1}{4})y &= 0. \end{aligned}$$

Solve for x in terms of y: $x = \frac{3}{4}y$. There are infinitely many solutions corresponding to values of y; however, there is only one solution that is satisfactory for our problem, namely, the one with nonnegative x and y whose sum is 1. The reason for this is that **proportions** are nonnegative and add to 1. So we put $y = 1 - x$, and substitute for y in the equation $x = \frac{3}{4}y$; we obtain

$$x = \tfrac{3}{4}(1 - x) \qquad \text{or} \qquad x = \tfrac{3}{7}.$$

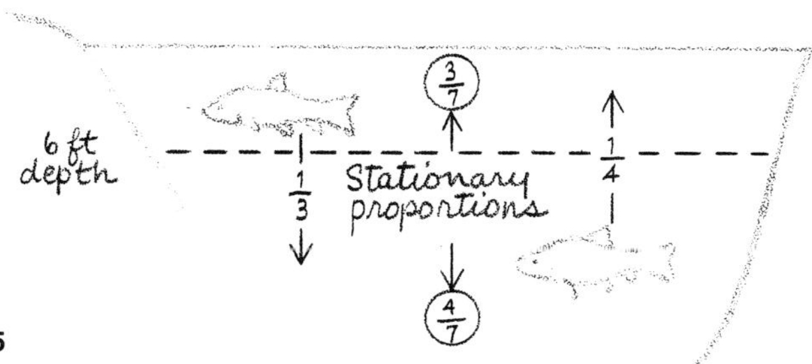

Figure 3.25

It follows that $y = \frac{4}{7}$. Therefore, the stationary distribution is found to be the pair of proportions $(\frac{3}{7},\frac{4}{7})$ (Figure 3.25). If the distribution of fish is this stationary one, then the distribution will remain the same after the transition period; similarly, it will remain the same after a second transition period, and so on forever. Later, in Section 3.11, we shall see that Markov chains tend to drift **into** a stationary distribution from whatever distribution they had at the beginning. It follows that the chain "forgets" its original distribution, whatever it was, and moves into a stationary one. ▲

We close this section by summarizing the procedure for analyzing a system of linear equations (Figure 3.26):

1. Transform the augmented matrix to echelon form and determine whether or not the system is consistent according to the ranks of the coefficient and augmented matrices.
2. If the system is consistent, determine whether or not the solution is unique according to the rank and the number of unknowns.
3. Solve the system by the Gauss-Jordan method.

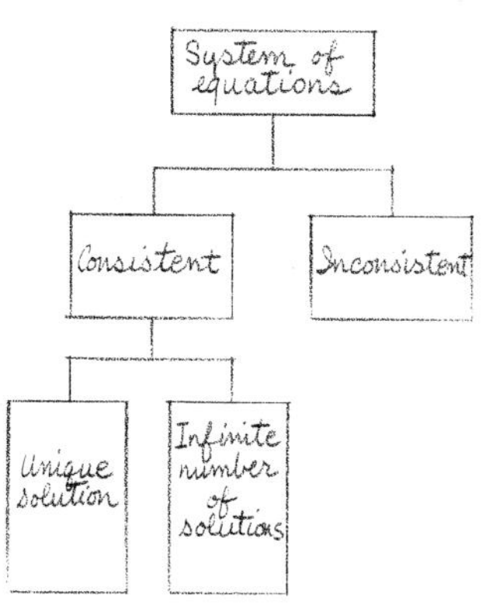

Figure 3.26

3.7 EXERCISES

In Exercises 1 to 20 determine whether the systems are consistent. If they are, find the unique solutions or the complete solutions accordingly.

1 $\begin{aligned} x + 3y + z &= 4 \\ x + y - z &= 1 \\ x + 2y \phantom{{}-z} &= 0 \end{aligned}$

2 $\begin{aligned} x - y &= -6 \\ x + 2y &= 5 \\ 3x + 3y &= 4 \end{aligned}$

3
$$\begin{aligned} x + 2y + z &= 6 \\ x - y &= 2 \\ x - y + z &= 0 \end{aligned}$$

4
$$\begin{aligned} x + y + z &= 5 \\ x - y - z &= 4 \end{aligned}$$

5
$$\begin{aligned} x + 3y - z &= 4 \\ x + 2y + z &= 2 \\ 3x + 7y + z &= 9 \end{aligned}$$

6
$$\begin{aligned} x - y + z + w &= 1 \\ 2x + y + w &= 2 \\ -x + z + w &= 3 \\ 2x + 2z + 3w &= 6 \end{aligned}$$

7
$$\begin{aligned} 2x + y - z &= -4 \\ x - y + 3z &= 3 \\ x + 2y - 4z &= 1 \end{aligned}$$

8
$$\begin{aligned} x + 2y - z &= 6 \\ 2x - y + 3z &= -13 \\ 3x - 2y + 3z &= -16 \end{aligned}$$

9
$$\begin{aligned} 6x + 3y + 2z &= 1 \\ 3x - 4z &= 4 \\ 5x - y &= 14 \end{aligned}$$

10
$$\begin{aligned} x - y - z - w &= 5 \\ x + 2y + 3z + w &= -2 \\ 2x + 2z + 3w &= 3 \\ 3x + y + 2w &= 1 \end{aligned}$$

11
$$\begin{aligned} x + 2y + z &= 2 \\ 2x - 2z + w &= 6 \\ 4y + 3z + 2w &= -1 \\ -x + 6y - z - w &= 2 \end{aligned}$$

12
$$\begin{aligned} x + 2y + 3z &= 0 \\ 2x - y + z &= 0 \\ 2x + 3y + 4z &= 0 \end{aligned}$$

13
$$\begin{aligned} x + y + z &= 0 \\ x - y - 2z &= 0 \\ 2x - z &= 0 \end{aligned}$$

14
$$\begin{aligned} x + y - z &= 4 \\ x - y + z &= 2 \end{aligned}$$

15
$$\begin{aligned} 3x - y + z &= 4 \\ x - y &= 0 \\ 2x + z &= 4 \\ 4x - 2y + z &= 4 \end{aligned}$$

16
$$\begin{aligned} x - 2y + z - w &= 3 \\ 2x - 3y &= 3 \\ x - y - z + w &= 0 \end{aligned}$$

17
$$\begin{aligned} x + z &= 2 \\ y + z &= 6 \\ y + w &= 0 \\ x + y + z + w &= 2 \end{aligned}$$

18
$$\begin{aligned} x + y + z &= 4 \\ -x - y &= 2 \\ 2x + 2y + 2z &= 8 \end{aligned}$$

19
$$\begin{aligned} x + 2y + z - 4w + v &= 0 \\ x + 2y - z + 2w + v &= 0 \\ 2x + 4y + z - 5w &= 0 \\ x + 2y + 3z - 10w + v &= 0 \end{aligned}$$

20
$$\begin{aligned} x + 2y + w &= 1 \\ 2x + 4y + z + 4w + 3v &= 6 \\ x + 2y + 2z + 4w - 2v &= 1 \\ -x - 2y + 3z + 5w + 4v &= 6 \end{aligned}$$

21 A small airline leases 2 kinds of passenger planes. The first kind carries 30 first-class and 90 coach passengers, and the second 40 first-class and 120 coach passengers. The airline usually carries about 2000 first-class and 4800 coach passengers at any given time. How many planes of each kind should the airline lease in order to fill all seats? Does this problem have a solution, and why? [*Hint:* Let x and y be the numbers of the 2 kinds of planes.] (See Section 3.5, Example 5.)

22 Give a geometric interpretation of the inconsistency of a system of 3 equations in 3 unknowns. [*Hint:* What is the set of points (x,y,z) in space satisfying 1, 2, and 3 linear equations in the variables x, y, and z?]

23 Ice cream, soy flour, and carrots have the following quantities (in grams) of calcium and phosphorus for each 100 grams of edible portions of food:

	Calcium	Phosphorus
Ice cream	0.1	0.1
Soy flour	0.2	0.6
Carrots	0.2	0.1

What combination or combinations of the 3 foods will yield 10 grams of calcium and 5 grams of phosphorus? Is the solution unique? (See Section 3.5, Example 5.)

24 Three brands of a can of vegetables contain the following numbers of units of protein, carbohydrates, and minerals:

	Protein	Carbohydrates	Minerals
Brand *A*	4	3	1
Brand *B*	2	4	3
Brand *C*	4	4	2

How many cans of each brand should be bought to get 100 units of protein, 200 units of carbohydrates, and 100 units of minerals? (See Section 3.5, Example 5.)

25 Recall the milk-and-chocolate problem in Exercise 8, Section 3.2: Each stirring of the spoon sends $\frac{1}{5}$ of the syrup on the bottom to the top, and $\frac{1}{5}$ of the syrup on the top to the bottom. Find the stationary distribution for the transition matrix. (See Example 6.)

26 Suppose that instead of mixing syrup with milk, as in Exercise 25, the boy opens a raw egg and mixes it with the milk. Each turn of the spoon sends only $\frac{1}{8}$ of the egg from the bottom half to the top half (and leaves the other $\frac{7}{8}$ on the bottom) and sends $\frac{5}{8}$ of the egg on the top to the bottom. Write the transition matrix and find the stationary distribution. (See Example 6.)

27 A tree is considered to be in exactly 1 of 3 states: healthy, diseased, or dead. Consider a forest with a fixed number of trees. Suppose in 1 year that 8 percent of all healthy trees become diseased, 2 percent actually die, and the remaining 90 percent remain healthy. Eighty percent of diseased trees remain diseased, 15 percent die, and 5 percent recover and become healthy. (Naturally, all dead trees remain dead.) Write the transition matrix, and then find the stationary distribution. What is the nature of the stationary distribution? (Figure 3.27) (See Example 6.)

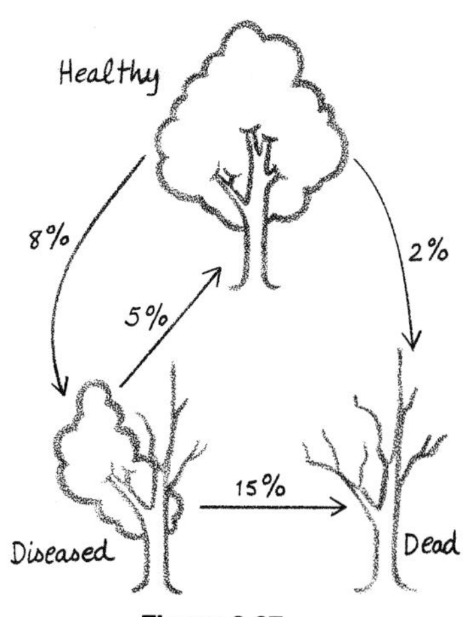

Figure 3.27

28 Consider the voting problem in Exercise 7, Section 3.2, where the voters shift among the 3 parties according to the transition matrix P

	To Dems	To Reps	To Third
From Dems	0.60	0.20	0.20
From Reps	0.30	0.60	0.10
From Third	0.30	0.20	0.50

Find the stationary distribution. To simplify the computations, multiply the homogeneous equations by the factor 10 to remove the decimals. (See Example 6.)

29 The following mathematical model is called a **random walk with reflecting barriers.** Consider a system of particles (electrons, molecules, viruses, microbes, blood corpuscles, and so on) which move up and down a numbered line segment in the following manner. During a given interval of time, and at any point on the segment except the endpoints, half the particles originally at the point at the beginning of the time interval move up a short distance, and half move down. All particles that reach the endpoints are then reflected back into the segment. Consider the simple case of particles located at the states 0, 1, 2, 3, and 4. Half the particles at the interior states 1, 2, and 3 move up and half move down 1 step, but those at 0 move up to 1, and those at 4 move back to 3 (Figure 3.28). The states 0 and 4 are the **reflecting barriers.** The 5-state system has the transition matrix:

From state	To state 0	1	2	3	4
0	0	1	0	0	0
1	$\frac{1}{2}$	0	$\frac{1}{2}$	0	0
2	0	$\frac{1}{2}$	0	$\frac{1}{2}$	0
3	0	0	$\frac{1}{2}$	0	$\frac{1}{2}$
4	0	0	0	1	0

Find the stationary distribution. (See Example 6.)

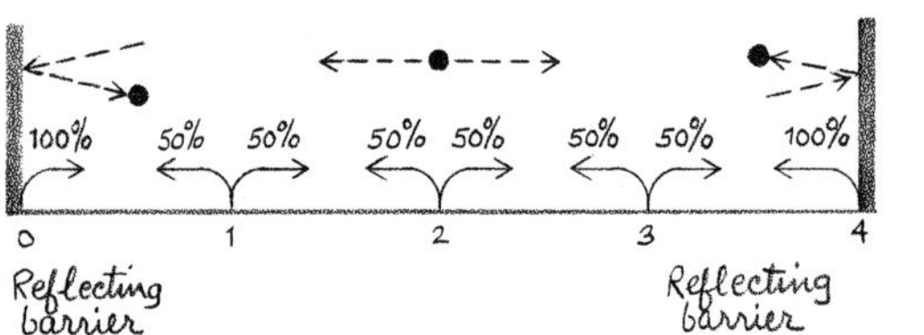

Figure 3.28

30 Here is a variant of the random walk with reflecting barriers applied to a problem in economic currency control. A government wants the price of its monetary unit to cost no more than \$10 and no less than \$9. Assume for simplicity that the exchange rate varies in units of one-quarter of a dollar. Whenever the price reaches \$10, the government itself buys dollars with its own currency to drive down the price; and whenever it falls to \$9, the government buys its currency with dollars to push up the price. Suppose that the price moves according to the following general pattern. If, on one day, the price is either \$9.25, \$9.50, or \$9.75, then on 25 percent of the following days it will drop by \$0.25, on 25 percent of the days it will rise by \$0.25, and on 50 percent of the days it will remain unchanged. If it is at \$9 on one day, then on half of the following days it will remain unchanged, and on half will rise by \$0.25. Similarly, if it is at \$10, then half the time it remains unchanged, and half the time it falls by

\$0.25 (Figure 3.29). Write the matrix of 1-day transition proportions, and find the stationary distribution. (See Example 6.)

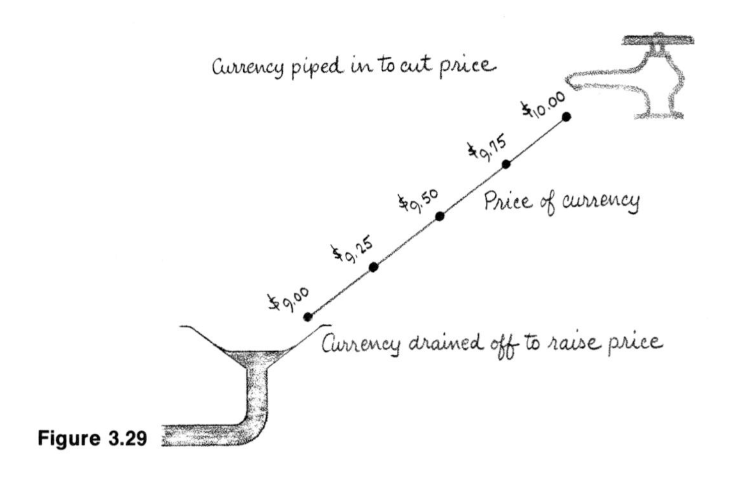

Figure 3.29

31 Let

$$P = \begin{bmatrix} p_{11} & p_{12} \\ p_{21} & p_{22} \end{bmatrix}$$

be an arbitrary stochastic matrix, that is, with nonnegative entries with row-sums equal to 1. Show that there always exists at least one stationary distribution. Then find the conditions on the matrix under which the stationary distribution is unique. [Find the formula for the solution of the homogeneous system

$$(x \quad y)P = (x \quad y).]$$

3.8 Linear Inequalities and Convex Sets

In this section and the next, we introduce **linear programming.** This branch of mathematics was developed during the 1930s and 1940s by mathematical economists in various countries of Europe and in the United States. It is a blend of analytic geometry and systems of linear equations. The practical problems solved by linear programming are concerned with allocation and optimization of economic resources and outputs, respectively. The theory and its applications are introduced in the next section. Here we begin with the mathematical basis of the subject.

According to our earlier work in analytic geometry, every line in the plane is characterized by a linear equation, which, in general form, is

$$Ax + By + C = 0,$$

or, equivalently,

$$Ax + By = -C. \tag{1}$$

A line cuts the plane into two half-planes and serves as the boundary between them. If a point does not lie on the line (1), then it does not satisfy that equation, and so must satisfy either

$$Ax + By < -C, \qquad \text{or} \qquad Ax + By > -C.$$

Each of these inequalities characterizes a half-plane. When the boundary is added to the half-plane, it becomes a **closed** half-plane. When the boundary is added to the first half-plane above, it becomes the closed half-plane with the inequality

$$Ax + By \leq -C. \tag{2}$$

Similarly, the second half-plane becomes

$$Ax + By \geq -C. \tag{3}$$

In problems of linear programming our half-planes usually include the boundary; hence, we shall usually refer to Equations (2) and (3) as those of half-planes, without having to specify that they are closed.

EXAMPLE 1

The line $x + 2y = 2$ cuts the plane into two half-planes, characterized by the inequalities

$$x + 2y \leq 2, \tag{4}$$

and

$$x + 2y \geq 2. \tag{5}$$

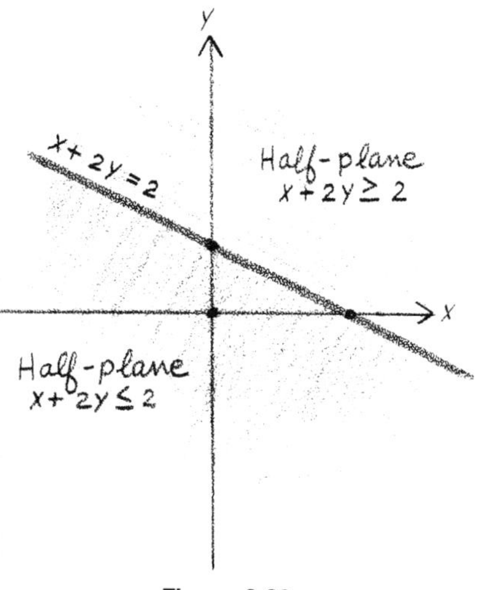

Figure 3.30

The first half-plane lies below and to the left of the bounding line; and the second one is above and to the right (Figure 3.30). A simple way to see which half-plane is on a given side of the line is to determine the half-plane containing the origin. Put $x = 0$ and $y = 0$ in the left-hand side of (4); then the origin (0,0) satisfies inequality (4). Since the origin is below and to the left of the line $x + 2y = 2$, so is the entire half-plane (4). In a case where the origin is itself on the boundary line, we can use another simple point such as (1,0) or (0,1) for the test. ▲

A set formed by the intersection of two or more closed half-planes is called a **polygonally convex set.** For simplicity, we shall refer to it as just a **convex** set.

EXAMPLE 2

The intersection of the half-planes $x + 3y \leq 6$ and $x - y \geq 7$ appears in Figure 3.31. The first half-plane is below and to the left of the line $x + 3y = 6$, and the second below and to the right of $x - y = 7$. Note that the origin belongs to the first half-plane but not to the second. This convex set is **unbounded** because it extends indefinitely to the right. ▲

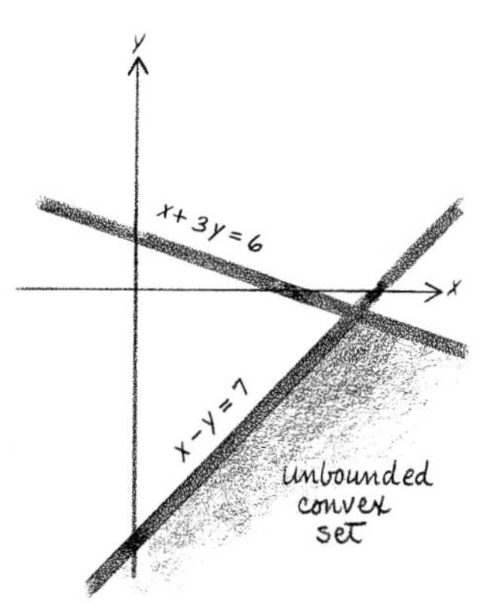

Figure 3.31

EXAMPLE 3

Consider the intersection of the 3 half-planes, $x \geq -1, x + y \leq 2$, and $x - y \leq 4$. The origin is contained in each of these. The convex set appears in Figure 3.32. This convex set is **bounded** because it is confined to a finite portion of the plane. ▲

The **extreme points** of a convex set are those points of the set formed by the intersections of pairs of the bounding lines. Note that there may be points in the intersection of the bounding lines which are outside the convex set; these are not extreme points.

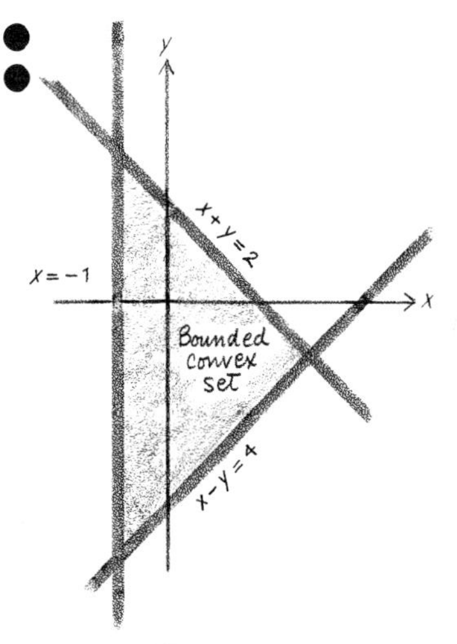

Figure 3.32

EXAMPLE 4

Consider the bounded convex set consisting of the intersections of the half-planes $2x - y \leq 2$, $x + y \geq -4$, $x + 4y \leq -2$. The extreme points are found by solving pairs of the 3 equations

$$\text{(a) } 2x - y = 2, \qquad \text{(b) } x + y = -4, \qquad \text{(c) } x + 4y = -2.$$

Solving (a) and (b) simultaneously (for, example, by Cramer's Rule) we find $x = -2/3$, $y = -10/3$. Similarly, solving (a) and (c) we get $x = 2/3$, $y = -2/3$. Solving (b) and (c) we get $x = -14/3$, $y = 2/3$. The convex set and the extreme points appear in Figure 3.33. ▲

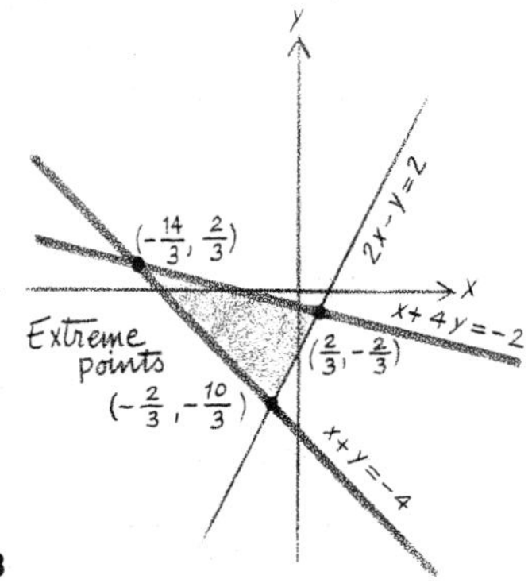

Figure 3.33

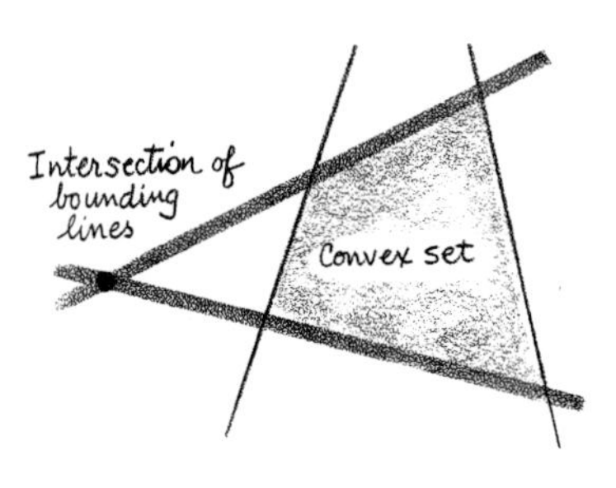

Figure 3.34

Figure 3.34 contains an example of a convex set whose bounding lines intersect at points outside the set, so that they are not extreme points.

The theory of convex sets in space—three dimensions—is algebraically similar to that in the plane, but is a bit more difficult to visualize. According to our work in solid analytic geometry, an equation

$$Ax + By + Cz = D \tag{6}$$

represents the set of points (x,y,z) forming a plane in space. This divides the space into two closed half-spaces: the set of points satisfying

$$Ax + By + Cz \leq D, \tag{7}$$

and the set satisfying

$$Ax + By + Cz \geq D. \tag{8}$$

The intersection of several closed half-spaces is, as in the case of the plane, a polygonally convex set. The extreme points are the points of the set formed by the intersections of at least three bounding planes of the form (6). These are determined by solving the corresponding systems of 3 equations.

EXAMPLE 5

We draw the picture of the intersection of the 5 half-spaces characterized by the inequalities

$$\begin{aligned} x + y + 2z \leq 2, \qquad x + 2y + z \leq 2, \\ x \geq 0, \qquad y \geq 0, \qquad z \geq 0. \end{aligned} \tag{9}$$

The intersection of the last 3 is the first octant; thus the intersection of all 5 is that part of the intersection of the first 2 half-spaces which lie in the first octant. These are bounded by the planes $x + y + 2z = 2$ and $x + 2y + z = 2$ [Figures 3.35(a) and 3.35(b)]. The convex set appears in Figure 3.35(c). The extreme points are found by solving sets of 3 equations out of the 5:

$$\begin{array}{ll} \text{(i)} & x + \;\; y + 2z = 2, \\ \text{(ii)} & x + 2y + \quad\;\; = 2, \\ \text{(iii)} & \qquad\quad x \quad\;\; = 0, \\ \text{(iv)} & \qquad y \qquad\;\; = 0, \\ \text{(v)} & \qquad\qquad\;\; z = 0. \end{array}$$

After solving each set of 3, we check that the solution actually belongs to the convex set, that is, it satisfies the original inequalities (9). The solution does not always belong to the set because the bounding planes might intersect outside the set.

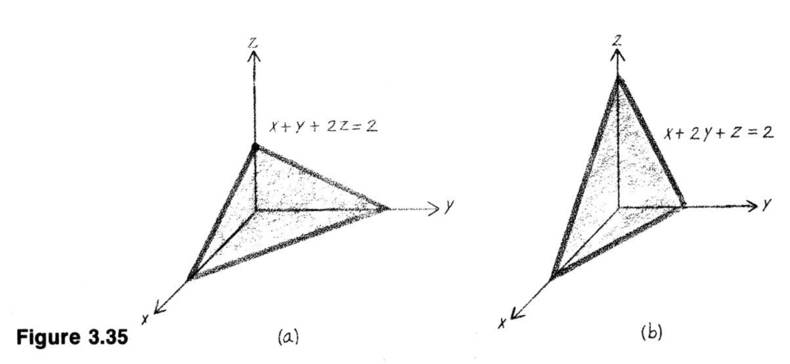

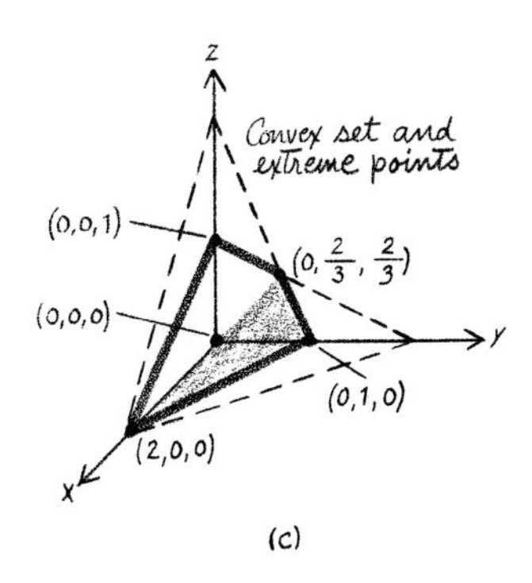

Figure 3.35

First let us solve the set consisting of (i), (ii), and (iii). Substituting from (iii) into the first 2 equations, we obtain the system of 2 equations

$$\begin{aligned} y + 2z &= 2, \\ 2y + \ \ z &= 2. \end{aligned}$$

By using any of the methods of solving this 2-by-2 system, we find that the solution is $y = \frac{2}{3}$ and $z = \frac{2}{3}$. This yields the point $(0,\frac{2}{3},\frac{2}{3})$ as the solution. This point satisfies each of the inequalities in (9):

$$0 + \tfrac{2}{3} + 2(\tfrac{2}{3}) \leq 2, \qquad 0 + 2(\tfrac{2}{3}) + \tfrac{2}{3} \leq 2,$$
$$0 \geq 0, \qquad \tfrac{2}{3} \geq 0, \qquad \tfrac{2}{3} \geq 0,$$

hence, it is an extreme point of the set.

Next we solve (i), (ii), and (iv). Substituting from the last equation into the first 2, we obtain the system

$$x + 2z = 2, \qquad x + z = 2,$$

which has the solution $x = 2$, $z = 0$. It follows that the solution of the set of 3 equations is the point (2,0,0). The latter also satisfies all the inequalities in (9), so that it is an extreme point.

We list the sets of the other equations and their solutions; not all of them yield extreme points.

Equations	Point	(9) satisfied?
(i), (ii), (v)	(2,0,0)	Yes
(i), (iii), (iv)	(0,0,1)	Yes
(i), (iii), (v)	(0,2,0)	No
(i), (iv), (v)	(2,0,0)	Yes
(ii), (iii), (iv)	(0,0,2)	No
(ii), (iii), (v)	(0,1,0)	Yes
(ii), (iv), (v)	(2,0,0)	Yes
(iii), (iv), (v)	(0,0,0)	Yes

These are all extreme points except (0,2,0) and (0,0,2). The extreme points appear in Figure 3.35(c). ▲

In dimensions greater than three, convex sets and extreme points are algebraically similar to those in the lower dimensions; however, there is no visual interpretation. In a space of n dimensions, the "points" correspond to n-tuples of real numbers $(x_1, \ldots, x_n)$. A hyperplane is a set of points whose coordinates satisfy a linear equation $A_1x_1 + \cdots + A_nx_n = d$. Closed half-spaces are defined by replacing the equations by inequalities. Intersections of closed half-spaces form convex sets. The extreme points are those points of the set which are points of intersection of n bounding hyperplanes.

EXAMPLE 6

In four-dimensional space, a convex set is formed by the intersection of the 6 half-spaces

$$x + y + 2z + w \leq 2, \qquad x + y + z + 2w \leq 2,$$
$$x \geq 0, \qquad y \geq 0, \qquad z \geq 0, \qquad w \geq 0.$$

(Here we have used the variables x, y, z, w in place of the x's with subscripts.) We find the extreme points by solving sets of 4 of the 6 equations:

$$\begin{array}{llll}
\text{(i)} & x + y + 2z + w = 2, & \text{(iv)} & y = 0, \\
\text{(ii)} & x + y + z + 2w = 2, & \text{(v)} & z = 0, \\
\text{(iii)} & x = 0, & \text{(vi)} & w = 0.
\end{array}$$

In the table below we list the solutions of the sets of 4 equations, and indicate which of the solutions are really extreme points, that is, which satisfy all the inequalities defining the convex set.

Equations	Point	Extreme Point?
(i), (ii), (iii), (iv)	$(0,0,\frac{2}{3},\frac{2}{3})$	Yes
(i), (ii), (iii), (v)	(0,2,0,0)	Yes
(i), (ii), (iv), (v)	(2,0,0,0)	Yes
(i), (ii), (iii), (vi)	(0,2,0,0)	Yes
(i), (ii), (iv), (vi)	(2,0,0,0)	Yes
(i), (ii), (v), (vi)	$(x, 2-x, 0, 0)$	Yes, for $0 \leq x \leq 2$
(i), (iii), (iv), (vi)	(0,0,0,2)	No
(i), (iii), (iv), (vi)	(0,0,1,0)	Yes
(i), (iii), (v), (vi)	(0,2,0,0)	Yes
(i), (iv), (v), (vi)	(2,0,0,0)	Yes
(ii), (iii), (iv), (v)	(0,0,0,1)	Yes
(ii), (iii), (iv), (vi)	(0,0,2,0)	No
(ii), (iii), (v), (vi)	(0,2,0,0)	Yes
(ii), (iv), (v), (vi)	(2,0,0,0)	Yes
(iii), (iv), (v), (vi)	(0,0,0,0)	Yes

The extreme points include $(0,0,,\frac{2}{3})$, $(0,2,0,0)$, $(2,0,0,0)$, $(0,0,0,1)$, and $(0,0,0,0)$, as well as all points with $x+y=2$, $0 \leq x \leq 2$, $z=w=0$. ▲

If 2 half-spaces are disjoint, that is, have no points of intersection, then the resulting convex set is called **empty.** Naturally there are no extreme points. An example of 2 disjoint half-spaces is the pair $x+y \leq 0$ and $x+y \geq 1$.

3.8 EXERCISES

In Exercises 1 to 4 draw a picture of the convex set and determine the extreme points.

1 (a) $x+3y \leq 9, \quad x \geq 0, \quad y \geq 0$
(b) $y \geq 0, \quad x \geq 0, \quad x \leq 4$
(c) $3x+2y \geq 0, \quad x \leq 1, \quad y \leq 2$
(d) $2x+3y \leq 12, \quad 2x+y \leq 8$
(e) $x+2y \geq 20, \quad 3x+y \geq 15, \quad x \geq 0, \quad y \geq 0$
(f) $x-y \leq 1, \quad x+y \geq 1$
(g) $-x+y \geq -3, \quad x+5y \geq 21, \quad x+y \geq 5$
(h) $x+2y \geq 4, \quad 2x-y \leq 4, \quad -3x+2y \leq 6$
(i) $2x-y \leq 10, \quad x+y \leq 8, \quad 2x+3y \leq 20, \quad x-y \geq -5$

2 (a) $x-y \geq 0, \quad y \geq 0, x \leq 2$
(b) $x \leq 0, \quad y \leq 0, \quad x \geq -1$
(c) $-5x+y \leq 0, \quad -2x+y \geq 0$
(d) $x-2y \geq -6, \quad x+y \geq 2$
(e) $x+y \leq 2, \quad x+y \geq -1, \quad x \geq 0, \quad y \geq 0$
(f) $x+2y \geq 4, \quad 2x+y \geq 4, \quad x \geq 0, \quad y \geq 0$
(g) $-4x+y \leq 1, \quad -x+y \geq 2, \quad x+y \leq 4$
(h) $2x+y \geq -1, \quad -x+y \leq 4, \quad x \leq 2$
(i) $x-y \leq 1, \quad x+2y \geq 4, \quad 2x+3y \leq 17, \quad -2x+3y \leq 13$

3 (a) $x+y+z \leq 4, \quad x \geq 0, \quad y \geq 0, \quad z \geq 0$
(b) $x+3y+z \leq 6, \quad x \geq 0, \quad y \geq 0, \quad z \geq 0$
(c) $x+y+z \leq 4, \quad x \leq 2, \quad x \geq 0, \quad y \geq 0, \quad z \geq 0$
(d) $x+y+3z \leq 6, \quad 3x+y+z \leq 6. \quad x \geq 0, \quad y \geq 0, \quad z \geq 0$
(e) $2x+3y+z \leq 6, \quad 2x+z \leq 2, \quad x \geq 0, \quad y \geq 0, \quad z \geq 0$

4 (a) $x+y+z \leq 2, \quad x \geq 0, \quad y \geq 0, \quad z \geq 0$
(b) $x+4y+2z \leq 4, \quad x \geq 0, \quad y \geq 0, \quad z \geq 0$
(c) $x+2y+z \leq 4, \quad x \leq 1, \quad x \geq 0, \quad y \geq 0, \quad z \geq 0$
(d) $x+2y+4z \leq 4, \quad 2x+y+z \leq 2, \quad x \geq 0, \quad y \geq 0, \quad z \geq 0$
(e) $x+2y+3z \leq 6, \quad 3x+2y+z \leq 6, \quad x \geq 0, \quad y \geq 0, \quad z \geq 0$

Find the extreme points of the convex sets in Exercises 5 and 6.

5 (a) $x+2y+2z+4w \leq 4, \quad x-y-2z+w \leq 2.$
$x \geq 0, \quad y \geq 0, \quad z \geq 0, \quad w \geq 0$
(b) $x-2y+2z \leq 2, \quad y+z+w \leq 1, \quad x \geq 0, \quad y \geq 0, \quad z \geq 0, \quad w \geq 0$

6 (a) $x + y - z \leq 2,\ \ y + 2z + w \leq 2,\ \ x \geq 0,\ \ y \geq 0,$
$z \geq 0,\ \ w \geq 0$

(b) $2x + y + 3z + 3w \leq 6,\ \ x - y - 3z + w \leq 2,$
$x \geq 0,\ \ y \geq 0,\ \ z \geq 0,\ \ w \geq 0$

3.9 Linear Programming

Linear programming is a branch of applied mathematics concerned with problems of allocation and optimization. The basis of this subject is an elementary problem involving **products** and **packages.** Each package contains several products of various kinds. The same kinds of product may appear in each package but in different amounts. In this way we form several kinds of packages, classified according to their contents. The packages vary in price. The basic problem of linear programming has two forms, one called the **consumer's problem** and the other the **producer's problem:**

Consumer's Problem: If you want specified minimum quantities of each product, how many packages of each kind should you buy to minimize the total cost? (Figure 3.36).

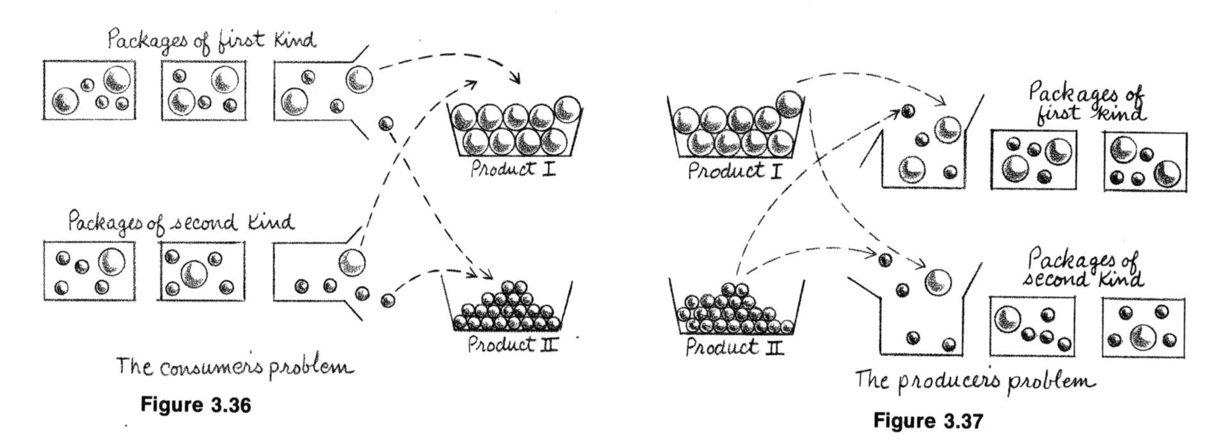

Figure 3.36

Figure 3.37

Producer's Problem: If you have given amounts of the various products, how many packages of each kind should you prepare for sale to maximize the total market value of the production? (Figure 3.37).

EXAMPLE 1

A package of 3 fine and 2 broad point pens costs 15¢. A second kind of package has 1 fine and 4 broad point pens; it costs 20¢. If you want at least 36 fine and 44 broad point pens, how many packages of each kind should you buy to minimize the cost? The data are summarized in the following table:

	Package I	Package II	Requirements
Fine Point	3	1	36
Broad point	2	4	44
Cost	15¢	20¢	

Let x and y be the numbers of packages I and II, respectively, that you purchase; then

$$\begin{aligned} \text{No. of fine point pens} &= 3x + y, \\ \text{No. of broad point pens} &= 2x + 4y, \\ \text{Total cost} &= \$0.15x + \$0.20y. \end{aligned}$$

The problem is now to determine the values of the variables x and y which minimize the total cost $15x + 20y$ (in cents) and which also satisfy the minimum quantity requirements of the 2 kinds of pens. The mathematical statement of the problem is: Find the values x and y which minimize the linear function $15x + 20y$ subject to the following inequalities:

$$3x + y \geq 36, \quad 2x + 4y \geq 44, \quad x \geq 0, \quad y \geq 0. \qquad (1)$$

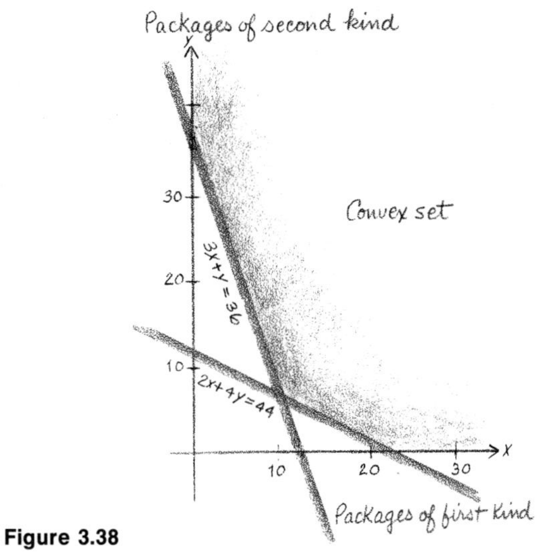

Figure 3.38

The first 2 inequalities reflect the quantity requirements, and the second 2 inequalities state that the quantities have to be nonnegative. Every pair (x,y) satisfying the inequalities in (1) represents a possible combination order for the packages of the 2 kinds. The corresponding set of points in the plane is an intersection of half-spaces, so that it is a convex set (Figure 3.38). At every point

(x,y) in this set the linear function $15x + 20y$ represents the cost of the order. The problem is to find that point in the convex set for which the expression $15x + 20y$ is a minimum. This example illustrates the consumer's problem. We shall solve it later. ▲

EXAMPLE 2

Let us look at a similar problem from the point of view of the distributor who packages the pens. Suppose he has 36 fine and 44 broad point pens. The market price of each package is the same as before (Example 1). How many packages of each kind should he assemble to maximize the market value of the production? Let x and y be the number of the 2 kinds of packages. Then, as in Example 1, the number of fine point pens required is $3x + y$, and the number of broad point pens is $2x + 4y$: and the market value is $15x + 20y$ (in cents). But this time the producer is trying to **maximize** the market value $15x + 20y$ under the restrictions of the numbers of pens that he has, namely,

$$3x + y \leq 36, \qquad 2x + 4y \leq 44, \qquad x \geq 0, \qquad y \geq 0. \qquad (2)$$

The inequality signs in the first 2 above have been reversed; and the second pair of inequalities is the same as in (1). Every pair (x,y) represents a production schedule for the packages of the 2 kinds. The corresponding set of points in the plane is a convex set (Figure 3.39). At every point in this set, the linear function $15x + 20y$ represents the market value of the production. The problem is to find that point (x,y) in the convex set for which the expression $15x + 20y$ is a maximum. ▲

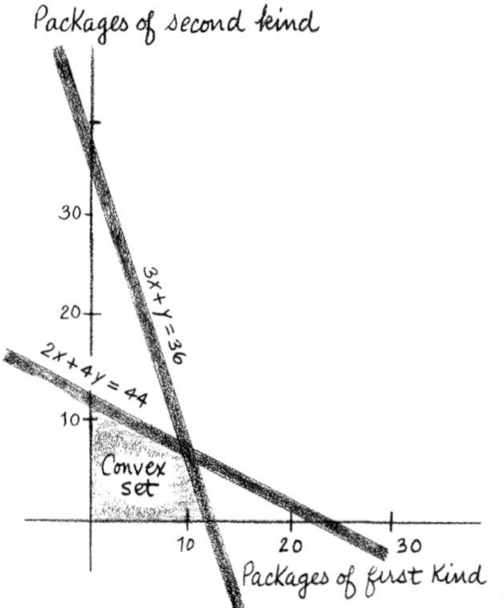

Figure 3.39

The fundamental theorem underlying the solution of the linear programming problem is this:

The maximum and/or the minimum of a linear function on a convex set is always assumed at an extreme point of the set.

The significance of this theorem is that in our search for the maximum or minimum of the linear function we have only to examine the values of the function at the extreme points of the set; indeed, the maximum or the minimum can occur only at such points and no others. Before showing how this powerful theorem is used, we give a graphic explanation of it. Let c and d be constants, and $cx + dy$ a linear function. For each x and y, let z be the value such that $z = cx + dy$. The set of points (x,y,z) in space which satisfy the equation above is a plane (with the equation $cx + dy - z = 0$. Scc Scction 2.2). The value of z is the height of the plane directly above the point (x,y) in the xy-plane below (Figure 3.40). The plane forms a flat roof over the convex set. It is evident from the picture in Figure 3.41 that the lowest and highest points of the roof, respectively, necessarily occur at directly above "corner points" of the wall, that is, at the extreme points. A second possibility is high (or low) points all along the seam between the roof and the wall. However, in either case there is a corner point no lower than the rest of the roof. This is the explanation for bounded convex sets in the plane.

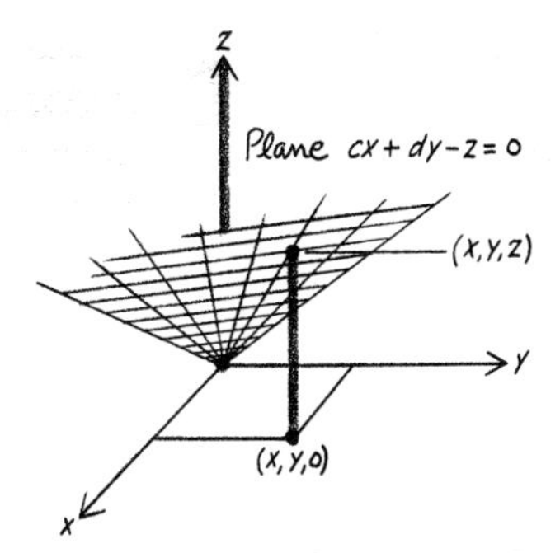

Figure 3.40

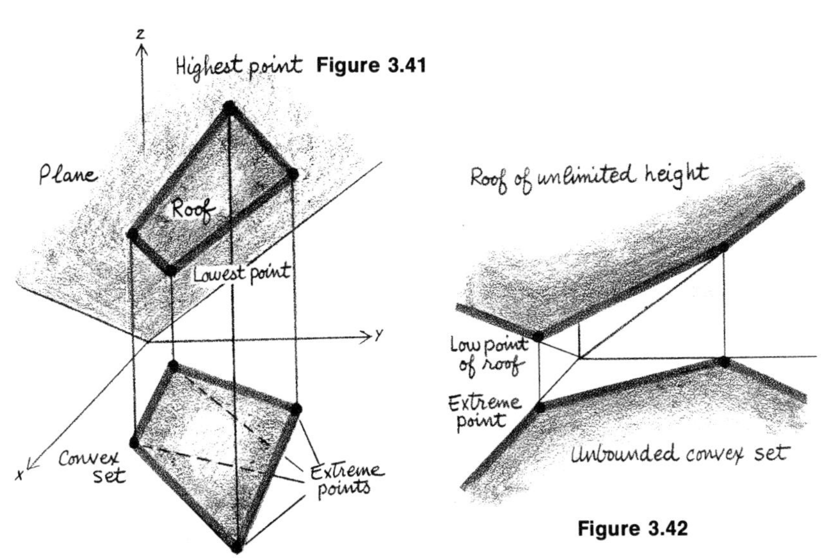

Figure 3.41

Figure 3.42

When the convex set is unbounded, then the linear function may fail to have a finite maximum because its values may "tend to infinity"; in other words, the roof may become of unlimited height (Figure 3.42). When the linear function assumes negative values, the plane is below the xy-plane and forms a slanted floor below the convex set. If the set is unbounded, the floor may go indefinitely down and down (Figure 3.43). If the linear function has no maximum but only a finite minimum on the convex set, then the minimum is necessarily assumed at one of the extreme points. Similarly, if only a maximum exists, then it is assumed at an extreme point.

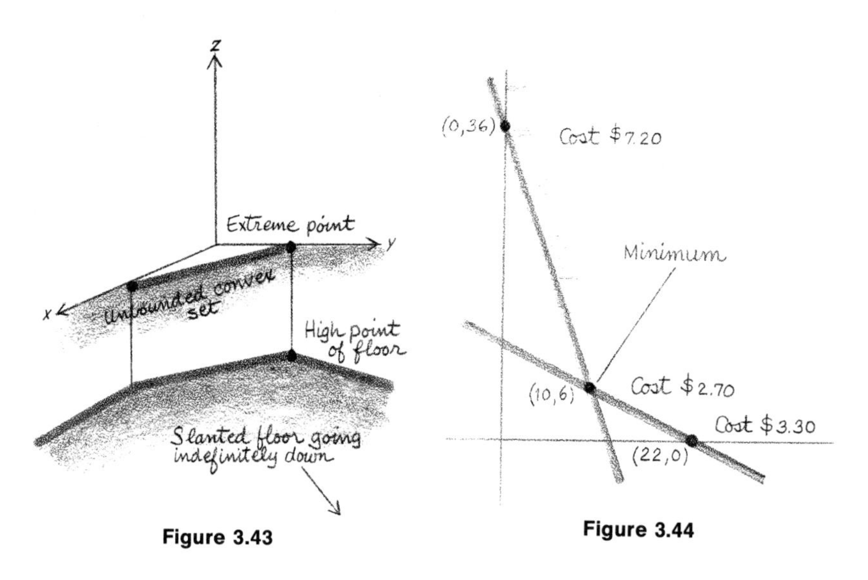

Figure 3.43

Figure 3.44

EXAMPLE 3

We use the fundamental theorem on the extreme points to solve the consumer problem in Example 1 above. Convert the inequalities in (1) into equations, and then find the extreme points of the convex set by solving pairs of the four equations,

$$3x + y = 36, \qquad 2x + 4y = 44, \qquad x = 0, \qquad y = 0.$$

Following the method of Section 3.8, we find the extreme points to be (0,36), (22,0), and (10,6). The values of the linear function at these 3 points are \$7.20, \$3.30, and \$2.70, respectively. The smallest of these is the third, which is assumed at the point with $x = 10$ and $y = 6$. It follows that the latter combination gives the minimum value of the linear function over the whole convex set. The most economical combination of packages is 10 of the first kind and 6 of the second (Figure 3.44). ▲

EXAMPLE 4

Next we solve the producer problem in Example 2 above. The extreme points of the convex set (2) are (0,0), (0,11), (10,6), and (12,0). The corresponding values of the linear function are

$$\$0.00, \qquad \$2.20, \qquad \$2.70, \qquad \text{and} \qquad \$1.80.$$

The maximum of these is the third, which occurs at the point with $x = 10$ and $y = 6$. This is the maximum for the whole convex set (Figure 3.45). This is the combination with the largest market value. ▲

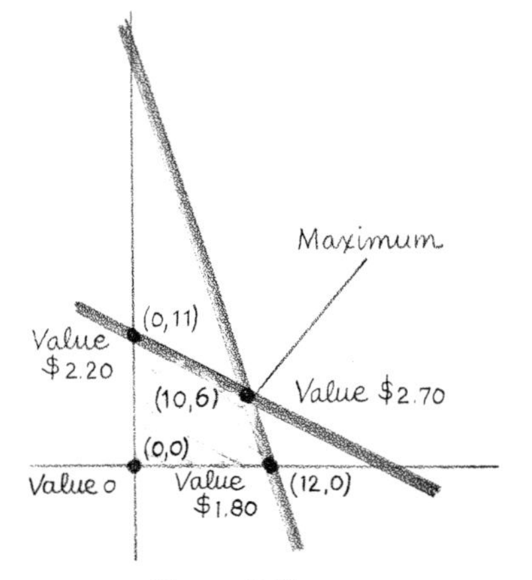

Figure 3.45

Now we consider an example involving three variables.

EXAMPLE 5

The governor of a state is building political support for his legislative program. It consists of 3 major bills, A, B, and C. The needed votes can be obtained from the legislators through the influence of selective appointments to positions in the state government. The state consists of urban, suburban, and rural areas with representation in the legislature (Figure 3.46). For each urban resident appointed to a position, the governor gets 2 urban votes for Bill A, and 2 for C. For each suburban appointment he gets 2 votes for A, 1 for B, and 1 for C. For each rural position, he gets 2 for B and 2 for C. Suppose that he needs at least 30 votes for A, 15 for B, and 35 for C. How should he allocate the appointments among the 3 areas to minimize the total number of such appointments and still obtain the votes he needs? In this problem the appointments are the packages and the votes are the products; each appointment carries several votes of various kinds. The data is summarized as follows:

	Urban	Suburban	Rural	Votes needed
Bill A	2	2	0	30
Bill B	0	1	2	15
Bill C	2	1	2	35
No. of positions	1	1	1	

Let x, y, and z be the numbers of appointments from the 3 areas, respectively; then

$$\begin{aligned} \text{No. of votes for } A &= 2x + 2y, \\ \text{No. of votes for } B &= \phantom{2x+{}} y + 2z, \\ \text{No. of votes for } C &= 2x + y + 2z, \\ \text{No. of appointments} &= x + y + z. \end{aligned}$$

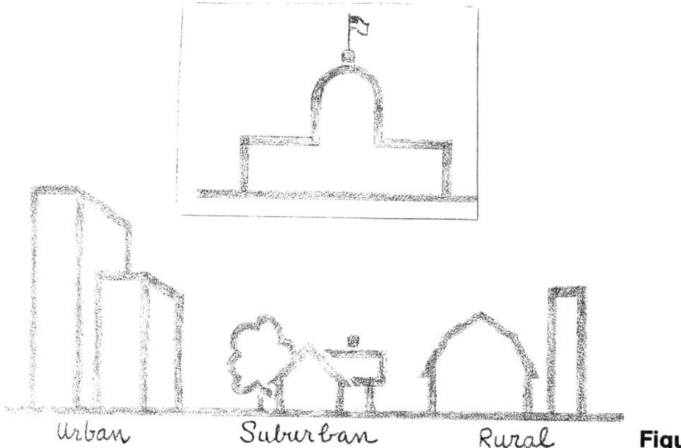

Figure 3.46

The governor's problem is to minimize the cost function $x+y+z$ subject to these conditions dictated by the vote requirements:

$$\begin{array}{ccc} 2x+2y\geq 30, & y+2z\geq 15, & 2x+y+2z\geq 35, \\ x\geq 0, & y\geq 0, & z\geq 0. \end{array} \tag{3}$$

It is not difficult to draw this convex set in a three-dimensional diagram, but neither is it necessary. The problem can be solved purely algebraically as before. We just have to find the extreme points of the convex set (3), and then evaluate the linear cost function at each of these. Recall that the extreme points of a set in three dimensions are the points of the set where 3 of the 6 bounding planes intersect. We solve these equations in groups of 3:

$$\begin{array}{lrl} \text{(i)} & x+y & =15, \\ \text{(ii)} & y+2z & =15, \\ \text{(iii)} & 2x+y+2z & =35, \\ \text{(iv)} & x & =0, \\ \text{(v)} & & =0, \\ \text{(vi)} & z & =0. \end{array} \tag{4}$$

The solutions are listed below, and the extreme points are identified:

Equations	Point	Extreme point?
(i), (ii), (iii)	(10,5,5)	Yes
(i), (ii), (iv) (i), (ii), (vi) (i), (iii), (iv) (i), (iv), (vi) (ii), (iv), (vi)	(0,15,0)	No

Equations	Point	Extreme point?
(i), (ii), (v)	(15,0,15/2)	Yes
(i), (iii), (v)	(15,0,5/2)	No
(i), (iii), (vi)	(20,−5,0)	No
(ii), (iii), (iv)	inconsistent	...
(ii), (iii), (v)	(10,0,15/2)	No
(ii), (iii), (vi)	(10,15,0)	Yes
(i), (v), (vi)	(15,0,0)	No
(ii), (iv), (v)	(0,0,15/2)	No
(i), (iv), (v) } (ii), (v), (vi) }	inconsistent	. . .
(iv), (v), (vi)	(0,0,0)	No
(iii), (iv), (v)	(0,0,35/2)	No
(iii), (iv), (vi)	(0,35,0)	Yes
(iii), (v), (vi)	(35/2,0,0)	No

The 4 extreme points and the corresponding costs are

Extreme point	(10,5,5)	(15,0,15/2)	(10,15,0)	(0,35,0)
Cost $x + y + z$	20	$22\frac{1}{2}$	25	35

The minimum number of appointments is 20, and the optimal allocation among the 3 areas is 10, 5, and 5. ▲

3.9 EXERCISES

In Exercises 1 and 2 find the maximum and the minimum of the given linear function over the given convex set. Sketch the set and indicate the extreme points.

	Convex set	Linear function
1 (a)	$3x + 2y \leqslant 8$, $2x + 4y \leqslant 8$, $x \geqslant 0,\ y \geqslant 0$	$7x + 6y$
(b)	$x + 2y \leqslant 2$, $x \geqslant 0,\ y \geqslant 0$	$x + y$
(c)	$3x + 4y \leqslant 10$, $x - y \leqslant 1$, $x \geq 0,\ y \geq 0$	$2x + y$
(d)	$x + y \geqslant 1$, $-x + 2y \leqslant 1$, $x \leqslant 1$	$x + 3y$

	Convex set	Linear function
2 (a)	$\left.\begin{aligned} 2x + 3y &\leqslant 15 \\ 3x + y &\leqslant 12 \\ x \geqslant 0,\quad y &\geqslant 0 \end{aligned}\right\}$	$5x - 7y$
(b)	$\left.\begin{aligned} 2x + y &\leqslant 7 \\ 3x + y &\leqslant 8 \\ x \geqslant 0,\quad y &\geqslant 0 \end{aligned}\right\}$	$x + y$
(c)	$\left.\begin{aligned} 4x + y &\leqslant 11 \\ x + 3y &\leqslant 11 \\ x \geqslant 0,\quad y &\geqslant 0 \end{aligned}\right\}$	$x - 5y$
(d)	$\left.\begin{aligned} x - y &\leqslant 4 \\ x + y &\leqslant 5 \\ x - y &\geqslant 2 \\ x \geqslant 0,\quad y &\geqslant 0 \end{aligned}\right\}$	$x + 2y$

3 Find the maximum of the linear function over the given convex set.

	Convex set	Linear Function
(a)	$\left.\begin{aligned} &x + 3y + z \leqslant 4 \\ &x + z \leqslant 5,\quad y + z \leqslant 2 \\ &x,\ y, \text{ and } z \text{ nonnegative} \end{aligned}\right\}$	$2x + y + 6z$
(b)	$\left.\begin{aligned} x + y + 2z &\leqslant 6 \\ x + y \qquad &\leqslant 4 \\ 2x + y + z &\leqslant 9 \\ x,\ y, \text{ and } z &\text{ nonnegative} \end{aligned}\right\}$	$4x + 2y + z$
(c)	$\left.\begin{aligned} &y + 2z \leqslant 4 \\ &2x + z \leqslant 6,\quad 2y + z \leqslant 1, \\ &x,\ y, \text{ and } z \text{ nonnegative} \end{aligned}\right\}$	$4x - 6y + 5z$

4 Find the minimum of the linear function over the given convex set.

	Convex set	Linear function
(a)	$\left.\begin{aligned} x + y - z &\leqslant -2 \\ x - y - 2z &\leqslant -1 \\ x,\ y, \text{ and } z &\text{ nonnegative} \end{aligned}\right\}$	$x + y + 4z$
(b)	$\left.\begin{aligned} x + y + 3z &\geqslant 4 \\ x + 2y \qquad &\geqslant 2 \\ x + y + z &\geqslant 6 \\ x,\ y, \text{ and } z &\text{ nonnegative} \end{aligned}\right\}$	$3x + 5y + 2z$
(c)	$\left.\begin{aligned} 4x + y &\geqslant 8 \\ x + y &\geqslant 5 \\ 2x + 7y &\geqslant 20 \\ x,\ y &\text{ nonnegative} \end{aligned}\right\}$	$x + 2y$

5 A fuel refinery has 2 processes for producing 3 grades of gasoline, A, B, and C. The first process produces 1 tank of A, 3 of B, and 3 of C in 10 hours. The second yields 3 of A, 4 of B, and 1 of C in 5 hours. How many times should each process be used to produce 8 tanks of A, 19 of B, and 7 of C in minimum time? [*Hint:* The grades of fuel are the products, and the processes are the packages. The cost is time of production.] (See Example 3.)

6 Two types of tables are made from a combination of oak and mahogany lumber. The first table requires 10 feet of oak and 5 of mahogany and yields a profit of $30 per table. The second requires 6 feet of oak and 4 of mahogany and brings a profit of $20. From a supply of 1000 feet of oak and 600 feet of mahogany, how many tables of each type should be made to maximize the total profit? [*Hint:* The lumber types are the products and the tables are the packages.] (See Example 4.)

7 Potatoes contain approximately 2 percent protein and 20 percent carbohydrates, but no fat. Lean meat contains about 20 percent protein, 8 percent fat, and virtually no carbohydrates. Suppose you want a combination of meat and potatoes to yield 1 pound of protein, 0.1 pound of fat, and 4 pounds of carbohydrates. Potatoes cost 10¢/lb and meat $2/lb What combination of meat and potatoes is the most economical? (The solution does not have to be in whole units of pounds.) (See Example 3.)

8 Jeremy is scheduled to have examinations in math, Latin, and history. There are many days before the exams will take place. He finds that when he spends 2 hours on math, then on the same day, he can study 1 hour of Latin and 1 hour of history. When he spends 1 hour on math, he can do 2 hours more of history and 2 hours more of Latin. When, on a particular day, he studies no math, he can spend 2 hours on Latin and 3 on history. He needs at least 20 hours of math study, 22 hours of Latin, and 20 hours of history. How many days should he spend on each type of study schedule to complete his preparation in a minimum number of days? (See Example 5.)

9 A small coal company owns 2 mines. In 1 hour the mines produce the following quantities of coal of various grades:

	Mine I	Mine II
High grade	2 tons	1 ton
Medium grade	3 tons	2 tons
Low grade	3 tons	6 tons

The hourly cost of running the mines is $250 for the first mine and $275 for the second. If the company has an order for at least 100, 180, and 240 tons of high, medium, and low grade coal, respectively, how many hours should each mine operate in order to minimize the cost? (See Example 3.)

10 A computer-dating agency buys advertising space in 3 college newspapers: a men's college, a women's college, and a coed college. A $25 ad in the men's college paper brings 8 new male customers; a $20 ad in the

women's college paper brings 6 females; and a \$20 ad in the paper of the coed college brings 4 male and 2 females. The agency wants to build a list of 360 males and 360 females. How many ads should it put in each of the papers to minimize the cost of advertising? (See Example 3.)

11 A group of 32 children and 24 adults is to have a picnic on an island in the middle of a lake. There are 2 motor boats (with one driver) available to transport them to the island. The smaller motorboat can carry 4 children and 2 adults, and can make the trip in 2 minutes. The larger can carry 6 children and 4 adults and can make the trip in 3 minutes. How many times should each boat be used to minimize the travel time for the whole group? (See Example 3.)

12 A utility company has found that operations at its 2 power plants emit the following amounts of air pollutants per hour:

	Plant I	Plant II
Sulphur dioxide	1 unit	1 unit
Smokeshade	1 unit	4 units
Carbon monoxide	0 units	2 units

The law specifies that the total amounts emitted by the 2 plants in 1 day should not exceed 20 units of sulphur dioxide, 40 units of smokeshade, and 30 units of carbon monoxide. If the utility follows the law, how many hours should each plant run to maximize the total number of hours of operation? (See Example 4.)

13 The Controllers of the Brave New World have two methods for the artificial production of 3 types of humans, the Alphas, Betas, and Gammas. Each method produces the following numbers in the given times:

	Method I	Method II
Alphas	1	3
Betas	1	0
Gammas	1	1
Production time	3 hr	4 hr

An order is placed for 6 Alphas, 2 Betas, and 4 Gammas. How many times should each method be used to minimize the production time? (See Example 3.)

3.10 Characteristic Values and Vectors

Let A be a square matrix. Consider the following problem: Does there exist a number λ, real or complex, and a column vector X such that

$$AX = \lambda X \tag{1}$$

holds? There is a trivial answer to this: Yes, just let X be the vector with all entries equal to 0. In this case the equation reduces to the zero vector on both sides. Our real interest is in the existence of a **nonzero** vector X such that (1) holds. So we restate our question as: Does there exist a nonzero X such that (1) holds for some λ?

We transform Equation (1). There is no loss in writing λX as λIX, where I is the identity. Then (1) is equivalent to

$$(A - \lambda I)X = 0, \tag{2}$$

where 0 represents the zero vector. This equation represents a system of homogeneous equations in which the number of unknowns is equal to the number of equations. According to the theory of linear equations (Section 3.7), there is a nontrivial solution of the system, other than the zero solution, if and only if the rank of the coefficient matrix, $A - \lambda I$, is smaller than the number of unknowns. This is true if and only if the matrix is **singular,** that is, its determinant is equal to 0 (Section 3.6, Exercise 5). Thus, Equation (2) has a nonzero solution if and only if

$$\det(A - \lambda I) = 0. \tag{3}$$

This is called the **characteristic equation** of the matrix A. The existence of a number λ satisfying (3) implies the existence of a nonzero vector X satisfying (1) and (2). When such a solution λ exists it is called a **characteristic value** of A: and the vector X is called a **characteristic vector.** (Alternate terminology in use is **eigenvalue** and **eigenvector, proper value** and **proper vector,** and **latent root** and **latent vector.**)

The left-hand side of (3) is a polynomial in the variable λ. Therefore, the equation may have several roots, not necessarily all distinct. Corresponding to every root λ, there is an X satisfying (1). It is obtained by solving the homogeneous system (2).

We devote this section to the calculation of characteristic values and vectors. Several important applications will be given in the next section.

EXAMPLE 1

Put

$$A = \begin{bmatrix} 1 & 2 \\ 2 & 4 \end{bmatrix}.$$

Equation (3) takes the form

$$\det \begin{bmatrix} 1-\lambda & 2 \\ 2 & 4-\lambda \end{bmatrix} = 0 \quad \text{or} \quad \lambda^2 - 5\lambda = 0.$$

The roots are $\lambda_1 = 0$ and $\lambda_2 = 5$. To find the corresponding characteristic vectors, we solve system (2), first with one value of λ, and then with the other. For $\lambda = 0$ the homogeneous system (2) becomes

$$\begin{bmatrix} 1 & 2 \\ 2 & 4 \end{bmatrix} \begin{bmatrix} x \\ y \end{bmatrix} = \begin{bmatrix} 0 \\ 0 \end{bmatrix} \qquad \text{or} \qquad \begin{aligned} x + 2y &= 0, \\ 2x + 4y &= 0; \end{aligned}$$

here X is the column vector with components x and y. The complete solution is $x = -2y$; thus, the characteristic vector corresponding to the characteristic value 0 is any vector with the first component equal to (-2) times the second,

$$X_1 = \begin{bmatrix} x \\ -\frac{1}{2}x \end{bmatrix}.$$

The characteristic vector is not unique. Particular ones are obtained by substituting specific values of x (or y); for example, if $x = -2$, then the particular characteristic vector is

$$\begin{bmatrix} -2 \\ 1 \end{bmatrix}.$$

The characteristic vector X_2 corresponding to the other characteristic root is obtained by solving the system (2) with $\lambda = 5$:

$$\begin{bmatrix} -4 & 2 \\ 2 & -1 \end{bmatrix} \begin{bmatrix} x \\ y \end{bmatrix} = \begin{bmatrix} 0 \\ 0 \end{bmatrix} \qquad \text{or} \qquad \begin{aligned} -4x + 2y &= 0, \\ 2x - y &= 0. \end{aligned}$$

The complete solution is

$$X_2 = \begin{bmatrix} x \\ 2x \end{bmatrix}. \quad \blacktriangle$$

Computation of the characteristic vectors becomes more complicated when the roots of the characteristic equation are irrational or complex.

EXAMPLE 2

Let A be the matrix

$$A = \begin{bmatrix} 2 & -1 \\ 3 & 2 \end{bmatrix}.$$

The characteristic equation is

$$\det \begin{bmatrix} 2 - \lambda & -1 \\ 3 & 2 - \lambda \end{bmatrix} = 0 \qquad \text{or} \qquad (2 - \lambda)^2 + 3 = 0,$$

which is equivalent to $\lambda^2 - 4\lambda + 7 = 0$. By the quadratic formula, the roots of this equation are

$$\lambda_1 = 2 - i\sqrt{3}, \qquad \lambda_2 = 2 + i\sqrt{3},$$

where $i = \sqrt{-1}$ is the imaginary unit. (The quadratic formula is given at the end of this section for reference.) The characteristic vector corresponding to the first of these numbers is the solution of the system

$$(i\sqrt{3})x - y = 0, \qquad 3x + (i\sqrt{3})y = 0.$$

The solution is $y = (i\sqrt{3})x$, so that

$$X_1 = \begin{bmatrix} x \\ (i\sqrt{3})x \end{bmatrix}.$$

The other characteristic vector is similarly found to be

$$X_2 = \begin{bmatrix} x \\ (\ i\sqrt{3})x \end{bmatrix}. \ \blacktriangle$$

Although it is relatively easy to get the characteristic values and vectors of a 2-by-2 matrix, it is generally more difficult to find the characteristic values of a larger matrix because the characteristic equation is then of degree higher than 2. When the equation is reducible by factoring to a quadratic, the roots can be found by the same method as for the 2-by-2 matrix.

EXAMPLE 3

Put

$$A = \begin{bmatrix} 2 & 1 & 1 \\ 1 & 1 & 0 \\ 1 & 0 & 1 \end{bmatrix}.$$

The characteristic equation is

$$\det \begin{bmatrix} 2-\lambda & 1 & 1 \\ 1 & 1-\lambda & 0 \\ 1 & 0 & 1-\lambda \end{bmatrix} = 0.$$

Expand by cofactors of the first row:

$$(2-\lambda)(1-\lambda)^2 - (1-\lambda) - (1-\lambda) = 0$$

or

$$\lambda(1-\lambda)(\lambda-3) = 0.$$

The roots are 1, 0, and 3. X_1 is found by solving the system

$$\begin{bmatrix} 1 & 1 & 1 \\ 1 & 0 & 0 \\ 1 & 0 & 0 \end{bmatrix} \begin{bmatrix} x \\ y \\ z \end{bmatrix} = \begin{bmatrix} 0 \\ 0 \\ 0 \end{bmatrix}.$$

By the method of Section 3.7, we solve for 2 of the unknowns in terms of the third; the solution is $x = 0$, $y = -z$. Thus,

$$X_1 = \begin{bmatrix} 0 \\ -z \\ z \end{bmatrix}.$$

X_2 is found by solving the system

$$\begin{bmatrix} 2 & 1 & 1 \\ 1 & 1 & 0 \\ 1 & 0 & 1 \end{bmatrix}\begin{bmatrix} x \\ y \\ z \end{bmatrix} = \begin{bmatrix} 0 \\ 0 \\ 0 \end{bmatrix}.$$

Here the complete solution is $x = -z$ and $y = z$; thus,

$$X_2 = \begin{bmatrix} -z \\ z \\ z \end{bmatrix}.$$

X_3 is found by solving

$$\begin{bmatrix} 1 & 1 & 1 \\ 1 & -2 & 0 \\ 1 & 0 & -2 \end{bmatrix}\begin{bmatrix} x \\ y \\ z \end{bmatrix} = \begin{bmatrix} 0 \\ 0 \\ 0 \end{bmatrix},$$

and so

$$X_3 = \begin{bmatrix} 2z \\ z \\ \end{bmatrix}. \blacktriangle$$

A square matrix is called **symmetric** if its entries are symmetric about the main diagonal: the matrix is unchanged if the columns are interchanged with the corresponding rows. For example, the matrices

$$\begin{bmatrix} 2 & 1 \\ 1 & 0 \end{bmatrix}, \begin{bmatrix} 2 & 2 \\ 2 & 4 \end{bmatrix}, \begin{bmatrix} 1 & \frac{1}{2} & \frac{1}{3} \\ \frac{1}{2} & 1 & -\frac{1}{2} \\ \frac{1}{3} & -\frac{1}{2} & 1 \end{bmatrix}$$

are symmetric (Figure 3.47). A classical result of the theory of characteristic values is that

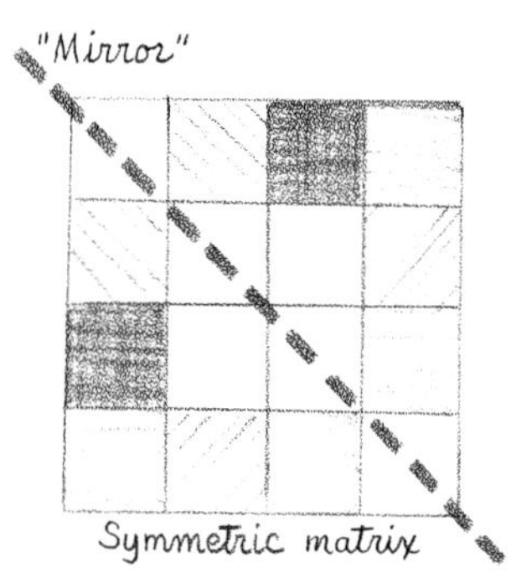

Figure 3.47

1. The roots of the characteristic equation of a symmetric matrix are always real numbers, not complex. (The proof for the 2-by-2 matrix is left as Exercise 5.)

2. The characteristic vectors corresponding to distinct characteristic roots are "perpendicular." In the cases of two and three dimensions this means that the lines through the origin and the points representing two characteristic vectors are perpendicular (see Section 2.1). The proof is outlined in Exercise 8.

EXAMPLE 4

Consider the symmetric matrix A in Example 1 whose characteristic vectors are

$$X_1 = \begin{bmatrix} x \\ -\frac{1}{2}x \end{bmatrix} \quad \text{and} \quad X_2 = \begin{bmatrix} x \\ 2x \end{bmatrix}.$$

These represent the lines with the equations

$$y = -\tfrac{1}{2}x \quad \text{and} \quad y = 2x,$$

respectively. These lines are perpendicular because the slope of one is the negative reciprocal of that of the other (Figure 3.48). ▲

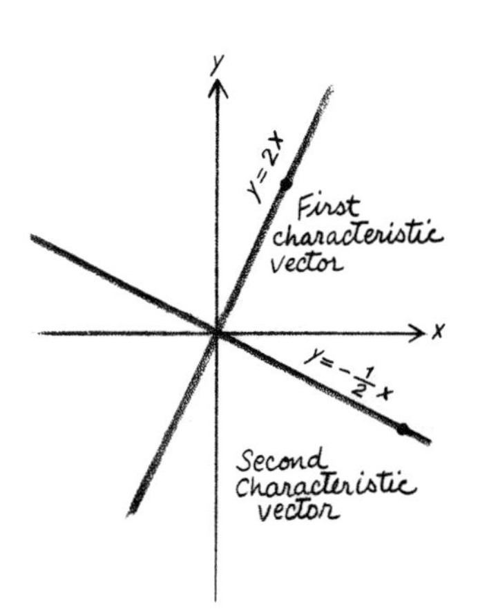

Figure 3.48

EXAMPLE 5

The 3-by-3 matrix A in Example 3 above is symmetric, and the characteristic vectors are

$$X_1 = \begin{bmatrix} 0 \\ -z \\ z \end{bmatrix}, \quad X_2 = \begin{bmatrix} -z \\ z \\ z \end{bmatrix}, \quad X_3 = \begin{bmatrix} 2z \\ z \\ z \end{bmatrix}.$$

The lines through the origin and X_1 and X_2 are perpendicular because the inner product of the two vectors is equal to 0:

$$0(-z) + (-z)z + z(z) = 0.$$

Similarly, the inner products of X_1 and X_3, and of X_2 and X_3 are equal to 0 (compare Section 2.1). ▲

3.10 EXERCISES

Find the characteristic values and vectors of the matrices in Exercises 1 to 4. Show that the characteristic vectors are perpendicular when the matrix is symmetric.

1 (a) $\begin{bmatrix} 5 & 3 \\ 1 & 3 \end{bmatrix}$ (b) $\begin{bmatrix} 5 & 0 \\ 0 & 1 \end{bmatrix}$ (c) $\begin{bmatrix} 1 & 2 \\ -1 & 4 \end{bmatrix}$ (d) $\begin{bmatrix} 5 & 4 \\ 1 & 2 \end{bmatrix}$ (e) $\begin{bmatrix} 4 & -3 \\ -3 & 5 \end{bmatrix}$ (f) $\begin{bmatrix} 1 & -5 \\ 2 & -2 \end{bmatrix}$

2 (a) $\begin{bmatrix} 1 & -1 \\ 3 & 5 \end{bmatrix}$ (b) $\begin{bmatrix} 5 & -1 \\ 3 & 1 \end{bmatrix}$ (c) $\begin{bmatrix} -1 & 0 \\ 10 & 2 \end{bmatrix}$ (d) $\begin{bmatrix} 1 & 2 \\ 5 & 4 \end{bmatrix}$ (e) $\begin{bmatrix} 1 & 2 \\ 2 & 5 \end{bmatrix}$ (f) $\begin{bmatrix} 1 & -2 \\ 3 & -1 \end{bmatrix}$

3 (a) $\begin{bmatrix} 1 & 2 & 0 \\ 2 & -2 & 0 \\ 0 & 0 & 3 \end{bmatrix}$

(b) $\begin{bmatrix} 1 & 0 & 0 \\ 3 & -2 & 0 \\ -2 & 3 & 1 \end{bmatrix}$

(c) $\begin{bmatrix} 1 & 0 & 0 & 0 \\ 3 & 4 & 0 & 0 \\ 0 & 0 & -1 & 2 \\ 0 & 0 & -2 & 3 \end{bmatrix}$

4 (a) $\begin{bmatrix} 4 & -5 & 0 \\ -4 & 3 & 0 \\ 0 & 0 & 7 \end{bmatrix}$

(b) $\begin{bmatrix} 2 & 0 & 0 \\ -1 & -3 & 0 \\ 4 & -2 & 1 \end{bmatrix}$

(c) $\begin{bmatrix} 3 & 0 & 0 & 0 \\ 0 & 1 & 0 & 0 \\ 2 & 0 & 4 & 0 \\ 0 & 0 & 0 & -1 \end{bmatrix}$

5 Let A be an arbitrary symmetric 2-by-2 matrix,

$$A = \begin{bmatrix} a & b \\ b & c \end{bmatrix}.$$

Verify that the characteristic equation has only real solutions. [*Hint:* Refer to the quadratic formula in the paragraph following this section.]

● **6** *Prove:* If λ is a characteristic value of A with the corresponding characteristic vector X, then λ^2 is a characteristic value of the matrix A^2 with the same characteristic vector X, that is, $A^2X = \lambda^2 X$.

● **7** Let A be a square symmetric matrix, and X and Y vectors of the dimension of the order of A. Then the inner product of the vector AX and the vector Y is equal to the inner product of X and AY. [In the notation of Section 3.1: $(AX) \cdot Y = X \cdot (AY)$.] Give the proof for the 2-by-2 matrix A and the vectors X and Y defined as

$$A = \begin{bmatrix} a & b \\ b & c \end{bmatrix}, \qquad X = \begin{bmatrix} x_1 \\ x_2 \end{bmatrix}, \qquad Y = \begin{bmatrix} y_1 \\ y_2 \end{bmatrix}.$$

● **8** If X and Y are characteristic vectors corresponding to distinct characteristic values of a symmetric matrix, then their inner product is equal to 0. Fill in the proofs of these steps of the proof of this theorem. Here λ_1 and λ_2 are characteristic values of A, and X and Y their corresponding characteristic vectors.

(a) For any number λ: $(\lambda X) \cdot Y = \lambda(X \cdot Y)$. Here the dot stands for inner product.

(b) $(AX) \cdot (AY) = \lambda_1\lambda_2(X \cdot Y)$.

(c) $(AX) \cdot (AY) = (A^2X) \cdot Y = X \cdot (A^2Y)$ (compare Exercise 7)

(d) $\lambda_1^2(X \cdot Y) = \lambda_2^2(X \cdot Y) = \lambda_1\lambda_2(X \cdot Y)$
(e) Why does it follow from the last set of equations that $X \cdot Y = 0$ if $\lambda_1 \neq \lambda_2$?

9 Let A be the 2-by-2 matrix, $\begin{bmatrix} a & b \\ c & d \end{bmatrix}$. The characteristic equation assumes the form

$$\lambda^2 - \lambda(a + d) + ad - bc = 0.$$

Substitute the matrix A for λ, and the matrix $(ad - bc)I$ for the number $ad - bc$; then the equation becomes the matrix equation

$$A^2 - (a + d)A + (ad - bc)I = 0,$$

where 0 is the 2-by-2 matrix with all entries 0. Verify that this equation actually holds by comparing the entries of the matrices on both sides. This is a special case of the classical Cayley-Hamilton Theorem which states that every square matrix satisfies its own characteristic equation.

Note on the Quadratic Formula. The equation

$$Ax^2 + Bx + C = 0$$

has the solutions

$$x = \frac{-B \pm \sqrt{B^2 - 4AC}}{2A}.$$

These are real if $B^2 - 4AC$ is not negative, complex if it is negative.

3.11 Powers of Square Matrices and Expansion in Characteristic Vectors

In this section we use the characteristic values and vectors of a square matrix A to get explicit expressions for the successive powers $A^2, A^3, \ldots$. Although it is possible to write the expressions for these powers based directly on the definition of matrix multiplication, these are usually too complicated to be of practical use. The method given in this section furnishes convenient algebraic formulas for such powers. It is based on two facts:

1. If λ is the characteristic value of A with the corresponding characteristic vector X, then, for every positive integer n, λ^n is the characteristic value of A^n with the same characteristic vector X; that is,

$$A^nX = \lambda^nX. \tag{1}$$

[The proof for $n = 2$ is given as Exercise 6 of Section 3.10; and the proof for the general case is similar.]

2. The characteristic vectors of a matrix often **span** their space. This means: If $X_1, \ldots, X_n$ are n-dimensional vectors which are the characteristic vectors of a matrix of size n-by-n, then any n-dimensional vector Y may be expressed as a **linear combination** of the characteristic vectors, that is, it can be expressed as

$$Y = c_1X_1 + \cdots + c_nX_n, \tag{2}$$

where the c's are suitable numbers. Although not true of all square matrices, this property is valid for all symmetric matrices. The numbers $c_1, \ldots, c_n$ are determined as the solution of a system of n equations in n unknowns, which in matrix form is equivalent to (2) above, where the X's and Y are given and the c's are unknown. The geometric interpretation is that an arbitrary vector can be expressed as the vector sum of characteristic vectors after the latter have been suitably squeezed or stretched (Figure 3.49).

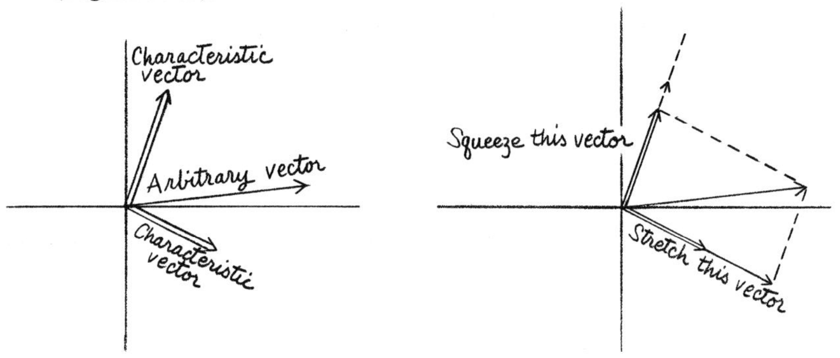

Figure 3.49

EXAMPLE 1

Let A be the matrix $\begin{bmatrix} 5 & -2 \\ -2 & 2 \end{bmatrix}$. The characteristic values and vectors are

$$\lambda_1 = 1, \quad X_1 = \begin{bmatrix} x \\ 2x \end{bmatrix}, \quad \text{and} \quad \lambda_2 = 6, \quad X_2 = \begin{bmatrix} x \\ -\frac{1}{2}x \end{bmatrix}.$$

Several powers of A are

$$A^2 = \begin{bmatrix} 29 & -14 \\ -14 & 8 \end{bmatrix}, \quad A^3 = \begin{bmatrix} 173 & -86 \\ -86 & 44 \end{bmatrix}.$$

By direct matrix multiplication, we find

$$A^2X_1 = \begin{bmatrix} 29 & -14 \\ -14 & 8 \end{bmatrix} \begin{bmatrix} x \\ 2x \end{bmatrix} = 1^2 \begin{bmatrix} x \\ 2x \end{bmatrix} = \lambda_1^2 X_1.$$

Similarly, we find

$$A^3X_1 = X_1, \qquad A^2X_2 = 6^2X_2, \qquad A^3X_2 = 6^3X_2.$$

This illustrates formula (1).

Next we show how to express a vector Y as a linear combination of the X's. Choose particular versions of the characteristic vectors; for example, put $x = 1$ in X_1 and $x = 2$ in X_2:

$$X_1 = \begin{bmatrix} 1 \\ 2 \end{bmatrix}, \qquad X_2 = \begin{bmatrix} 2 \\ -1 \end{bmatrix}.$$

Let the vector Y be $\begin{bmatrix} 5 \\ -2 \end{bmatrix}$; then system (2) takes the form

$$\begin{aligned} 5 &= c_1 + 2c_2, \\ -2 &= 2c_1 - c_2. \end{aligned}$$

Solving by Cramer's Rule (Section 3.4), we find

$$c_1 = \frac{1}{5}, \qquad c_2 = \frac{12}{5};$$

hence,

$$Y = \left(\frac{1}{5}\right) X_1 + \left(\frac{12}{5}\right) X_2$$

(Figure 3.50). ▲

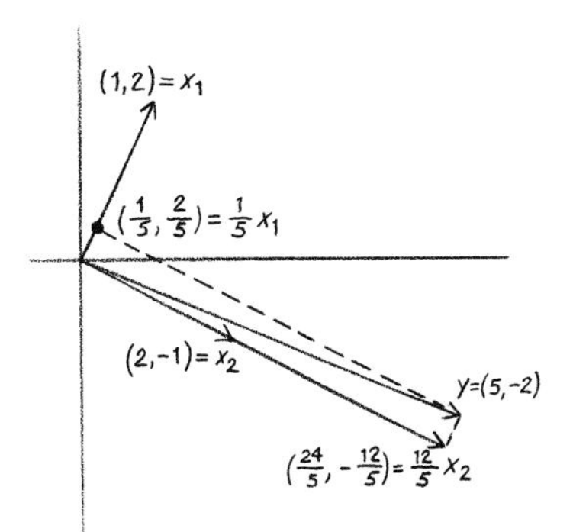

Figure 3.50

Now let X_1 and X_2 be characteristic vectors of A corresponding to characteristic values λ_1 and λ_2. Let $c_1X_1 + c_2X_2$ be an arbitrary linear combination of the X's. Let us premultiply this combination by A:

$$\begin{aligned} A(c_1X_1 + c_2X_2) &= A(c_1X_1) + A(c_2X_2) \\ &\quad \text{(by the distributive property of multiplication).} \\ &= c_1AX_1 + c_2AX_2 \\ &\quad \text{(by the definition of } \textbf{number} \times \textbf{matrix}) \\ &= \lambda_1c_1X_1 + \lambda_2c_2X_2 \\ &\quad \text{(since the } X\text{'s are characteristic vectors).} \end{aligned}$$

If we premultiply again by A and apply the same argument, we obtain

$$A^2(c_1X_1 + c_2X_2) = \lambda_1^2c_1X_1 + \lambda_2^2c_2X_2.$$

This may be extended to A^3 and then to any power of A:

$$A^k(c_1X_1 + c_2X_2) = \lambda_1^kc_1X_1 + \lambda_2^kc_2X_2.$$

Furthermore, this is valid also for a combination of more than two X's:

$$A^k(c_1X_1 + c_2X_2 + c_3X_3 + \cdot\cdot\cdot) = \lambda_1^k c_1X_1 + \lambda_2^k c_2X_2 + \lambda_3^k c_3X_3 + \cdot\cdot\cdot . \quad (3)$$

In other words, the effect of the premultiplication by A^k is to multiply the c's by the kth powers of the λ's.

EXAMPLE 2

Let A be the matrix in Example 1 and Y the same vector, so that

$$Y = \left(\frac{1}{5}\right) X_1 + \left(\frac{12}{5}\right) X_2.$$

Recall that the characteristic values are 1 and 6; thus, by formula (3),

$$A^kY = 1^k \left(\frac{1}{5}\right) X_1 + 6^k \left(\frac{12}{5}\right) X_2$$

$$= \begin{bmatrix} \frac{1}{5} + 6^k \left(\frac{24}{5}\right) \\ \frac{2}{5} - 6^k \left(\frac{12}{5}\right) \end{bmatrix}. \quad \blacktriangle$$

By means of formula (3) we can find the entries of A^k itself, not only those of A^kY: We express the columns of A as linear combinations of the characteristic vectors. Let $Y_1, Y_2, \ldots$ be the vectors representing the columns of A. By the definition of matrix multiplication, the columns of A^2 are the vectors $AY_1, AY_2, \ldots$. More generally, the columns of A^{k+1} are $A^kY_1, A^kY_2, \ldots$. Each of the Y's is first expressed as a linear combination of the X's; then the columns of A^{k+1} are found by formula (3).

EXAMPLE 3

The columns of matrix A in Examples 1 and 2 are

$$Y_1 = \begin{bmatrix} 5 \\ -2 \end{bmatrix} \quad \text{and} \quad Y_2 = \begin{bmatrix} -2 \\ 2 \end{bmatrix}.$$

We have already shown that $Y_1 = (1/5)X_1 + (12/5)X_2$. By the same method, we find

$$Y_2 = \left(\frac{2}{5}\right) X_1 - \left(\frac{6}{5}\right) X_2.$$

The columns of A^{k+1} are, according to formula (3),

$$\left(\frac{1}{5}\right) X_1 + 6^k \left(\frac{12}{5}\right) X_2 \qquad \text{and} \qquad \left(\frac{2}{5}\right) X_1 - 6^k \left(\frac{6}{5}\right) X_2.$$

By substituting the explicit forms of the X's we find the formula for the matrix A^{k+1}:

$$\left(\frac{1}{5}\right) \begin{bmatrix} 1 + 24(6)^k & 2 - 12(6)^k \\ 2 - 12(6)^k & 4 + 6^{k+1} \end{bmatrix}.$$

The advantage of using this formula over the iterative method of matrix multiplication is that the power of any magnitude can be found directly without first computing the lower powers. ▲

Before turning to a variety of applications, we review the steps in the method of raising a matrix to an arbitrary power:

1. Find the characteristic values of the matrix by solving the characteristic equation. Then find the characteristic vectors.
2. Express the columns of the matrix as linear combinations of the characteristic vectors by solving the system of equations (2).
3. Apply formula (3) to the linear combinations in step (2).

In the following examples applications of this method to Markov chains and to **difference equations** are given.

EXAMPLE 4

Consider the 2-by-2 stochastic matrix

$$P = \begin{bmatrix} 0.6 & 0.4 \\ 0.2 & 0.8 \end{bmatrix}.$$

Let us find the expression for P^n where n is an arbitrary positive integer. The characteristic equation is

$$\begin{vmatrix} 0.6 - \lambda & 0.4 \\ 0.2 & 0.8 - \lambda \end{vmatrix} = 0 \qquad \text{or} \qquad \lambda^2 - (1.4)\lambda + 0.4 = 0.$$

The solutions are $\lambda_1 = 1$ and $\lambda_2 = 0.4$. The characteristic vectors are

$$X_1 = \begin{bmatrix} x \\ x \end{bmatrix}, \qquad X_2 = \begin{bmatrix} x \\ -\frac{1}{2}x \end{bmatrix}.$$

Particular characteristic vectors are obtained by putting $x = 1$ in the first vector and $x = 2$ in the second:

$$X_1 = \begin{bmatrix} 1 \\ 1 \end{bmatrix}, \qquad X_2 = \begin{bmatrix} 2 \\ -1 \end{bmatrix}.$$

Next we express the columns of P as linear combinations of the characteristic vectors. The coefficients for the first column are obtained by solving the system

$$\begin{aligned} 0.6 &= c_1 + 2c_2, \\ 0.2 &= c_1 - 2c_2, \end{aligned}$$

so that $c_1 = 1/3$ and $c_2 = 0.4/3$. Therefore, the first column Y_1 is the linear combination

$$Y_1 = \left(\frac{1}{3}\right) X_1 + \left(\frac{0.4}{3}\right) X_2.$$

Similarly, by solving the system

$$\begin{aligned} 0.4 &= c_1 + 2c_2, \\ 0.8 &= c_1 - c_2, \end{aligned}$$

we find that the second column Y_2 is the combination

$$Y_2 = \left(\frac{2}{3}\right) X_1 - \left(\frac{0.4}{3}\right) X_2.$$

Now we apply formula (3) with n in place of $k + 1$, or $n - 1$ in place of k:

$$A^{n-1}Y_1 = \left(\frac{1}{3}\right) X_1 + \left(\frac{0.4}{3}\right) (0.4)^{n-1}X_2$$

$$A^{n-1}Y_2 = \left(\frac{2}{3}\right) X_1 - \left(\frac{0.4}{3}\right) (0.4)^{n-1}X_2.$$

By substituting the explicit forms of the X's, we obtain

$$P^n = \begin{bmatrix} \frac{1}{3} + \frac{2(0.4)^n}{3} & \frac{2}{3} - \frac{2(0.4)^n}{3} \\ \frac{1}{3} - \frac{(0.4)^n}{3} & \frac{2}{3} + \frac{(0.4)^n}{3} \end{bmatrix}.$$

As n increases, the factor $(0.4)^n$ becomes smaller and smaller; or, as we shall say in the terminology of calculus in a later chapter, **it approaches 0 as a limit.** It follows that the matrix P^n approaches the matrix

$$Q = \begin{bmatrix} \frac{1}{3} & \frac{2}{3} \\ \frac{1}{3} & \frac{2}{3} \end{bmatrix}.$$

For example, the first several powers of P are

$$P^2 = \begin{bmatrix} 0.440 & 0.560 \\ 0.280 & 0.720 \end{bmatrix}, \qquad P^3 = \begin{bmatrix} 0.376 & 0.624 \\ 0.312 & 0.688 \end{bmatrix},$$

$$P^4 = \begin{bmatrix} 0.3504 & 0.6496 \\ 0.3248 & 0.6752 \end{bmatrix}, \qquad P^5 = \begin{bmatrix} 0.34016 & 0.65984 \\ 0.32992 & 0.67008 \end{bmatrix}.$$

The two rows of the matrix Q are identical. Each represents the stationary distribution:

$$\left(\tfrac{1}{3} \ \ \tfrac{2}{3}\right)\begin{bmatrix} 0.6 & 0.4 \\ 0.2 & 0.8 \end{bmatrix} = \left(\tfrac{1}{3} \ \ \tfrac{2}{3}\right).$$

It follows from this that for **any** initial distribution of proportions over the states, the long-term proportions ultimately approach the stationary distribution; indeed, if X is any initial distribution over the 2 states, then

$$XP^n \quad \text{approaches} \quad XQ = (x \ \ y)\begin{bmatrix} \tfrac{1}{3} & \tfrac{2}{3} \\ \tfrac{1}{3} & \tfrac{2}{3} \end{bmatrix} = (x \ \ y).$$

(Recall that $x + y = 1$.) This is a special case of the **ergodic** theorem for Markov chains. ▲

EXAMPLE 5

Now we consider an application involving a 3-by-3 stochastic matrix. A large engineering firm has 2 categories of employees: administrative and technical. The company observes that of all the administrative employees present during a given year, $\frac{9}{10}$ remain the following year and $\frac{1}{10}$ leave. Among technical employees $\frac{6}{10}$ remain for the following year as technical employees, $\frac{1}{10}$ are promoted to administrative positions, and $\frac{3}{10}$ leave the firm (Figure 3.51). This is a Markov chain with 3 states: administrative, technical, and "out." The transition matrix P for the 3 states is

	Administrative	Technical	Out
Administrative	$\frac{9}{10}$	0	$\frac{1}{10}$
Technical	$\frac{1}{10}$	$\frac{6}{10}$	$\frac{3}{10}$
Out	0	0	1

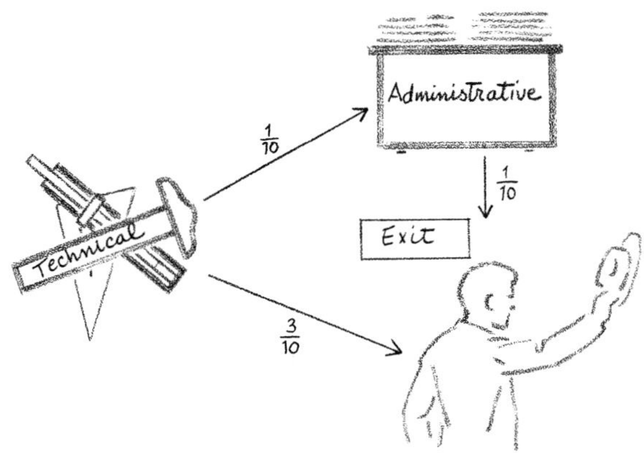

Figure 3.51

Entry 1 in the third row and column signifies that an employee who goes out never returns; this is an **absorbing** state of the chain (compare Section 3.2, Exercise 9). Of all the employees with the company on January 1 of a given year, what proportions will be in each of the 3 categories after n years? We have to find the formula for P^n, where

$$P = \begin{bmatrix} 0.9 & 0 & 0.1 \\ 0.1 & 0.6 & 0.3 \\ 0 & 0 & 1 \end{bmatrix}.$$

The characteristic equation is

$$\begin{vmatrix} 0.9-\lambda & 0 & 0.1 \\ 0.1 & 0.6-\lambda & 0.3 \\ 0 & 0 & 1-\lambda \end{vmatrix} = 0,$$

or

$$(0.9-\lambda)(0.6-\lambda)(1-\lambda) = 0.$$

Thus, the characteristic values are $\lambda_1 = 1$, $\lambda_2 = 0.9$, and $\lambda_3 = 0.6$. The characteristic vectors are found to be

$$X_1 = \begin{bmatrix} x \\ x \\ x \end{bmatrix}, \qquad X_2 = \begin{bmatrix} 3y \\ y \\ 0 \end{bmatrix}, \qquad X_3 = \begin{bmatrix} 0 \\ y \\ 0 \end{bmatrix}.$$

As special cases, we take

$$X_1 = \begin{bmatrix} 1 \\ 1 \\ 1 \end{bmatrix}, \qquad X_2 = \begin{bmatrix} 3 \\ 1 \\ 0 \end{bmatrix}, \qquad X_3 = \begin{bmatrix} 0 \\ 1 \\ 0 \end{bmatrix}.$$

Next we express the columns of the matrix P as linear combinations of the characteristic vectors:

$$Y_1 = \begin{bmatrix} 0.9 \\ 0.1 \\ 0 \end{bmatrix} = 0.3 \begin{bmatrix} 3 \\ 1 \\ 0 \end{bmatrix} - 0.2 \begin{bmatrix} 0 \\ 1 \\ 0 \end{bmatrix} = 0.3X_2 - 0.2X_3.$$

$$Y_2 = \begin{bmatrix} 0 \\ 0.6 \\ 0 \end{bmatrix} = 0.6 \begin{bmatrix} 0 \\ 1 \\ 0 \end{bmatrix} = 0.6X_3.$$

$$Y_3 = \begin{bmatrix} 0.1 \\ 0.3 \\ 1 \end{bmatrix} = \begin{bmatrix} 1 \\ 1 \\ 1 \end{bmatrix} - 0.3 \begin{bmatrix} 3 \\ 1 \\ 0 \end{bmatrix} - 0.4 \begin{bmatrix} 0 \\ 1 \\ 0 \end{bmatrix}.$$

Our final step is the application of formula (3), with $k = n - 1$. The first column of P^n is

$$0.3(0.9)^{n-1}X_2 - 0.2(0.6)^{n-1}X_3 = \begin{bmatrix} (0.9)^n \\ 0.3(0.9)^{n-1} - 0.2(0.6)^{n-1} \\ 0 \end{bmatrix}.$$

The second column is

$$0.6(0.6)^{n-1}X_3 = \begin{bmatrix} 0 \\ (0.6)^n \\ 0 \end{bmatrix}.$$

The third column is

$$X_1 - 0.3(0.9)^{n-1}X_2 - 0.4(0.6)^{n-1}X_3 = \begin{bmatrix} 1 - (0.9)^n \\ 1 - 0.3(0.9)^{n-1} - 0.4(0.6)^{n-1} \\ 1 \end{bmatrix}.$$

In summary the matrix P^n is

$$\begin{bmatrix} (0.9)^n & 0 & 1 - (0.9)^n \\ 0.3(0.9)^{n-1} - 0.2(0.6)^{n-1} & (0.6)^n & 1 - 0.3(0.9)^{n-1} - 0.4(0.6)^{n-1} \\ 0 & 0 & 1 \end{bmatrix}.$$

For example

$$P^2 = \begin{bmatrix} 0.81 & 0 & 0.19 \\ 0.15 & 0.36 & 0.49 \\ 0 & 0 & 1 \end{bmatrix}, \qquad P^3 = \begin{bmatrix} 0.729 & 0 & 0.271 \\ 0.171 & 0.216 & 0.613 \\ 0 & 0 & 1 \end{bmatrix},$$

$$P^5 = \begin{bmatrix} 0.59049 & 0 & 0.40951 \\ 0.17091 & 0.07776 & 0.75133 \\ 0 & 0 & 1 \end{bmatrix}.$$

From these matrices we can see what happens to the various employees over the years; for example, about 41 percent of the original administrative employees will have left after 5 years, and, of the original technical employees, about 8 percent will have moved to administrative positions, and 75 percent will have left. ▲

EXAMPLE 6

Here is an application of the theory to a biological problem involving a **difference equation.** A one-celled organism reproduces periodically in accordance with the following law: A new organism (just born) reproduces after 2 time units (seconds, minutes, or hours), and thereafter at every time unit; in other words, it reproduces at times 2, 3, 4, ... Each of its offspring follows the same pattern of reproduction. None of the organisms dies. If a single organism is born at time 0, what is the size of the population he generates after n time units? The growth of the population is illustrated in Figure 3.52; there the organisms are numbered according to their order of birth. By direct counting we find the sizes of the population over the first 10 time units:

Time	0	1	2	3	4	5	6	7	8	9	10
Number	1	1	2	3	5	8	13	21	34	55	89.

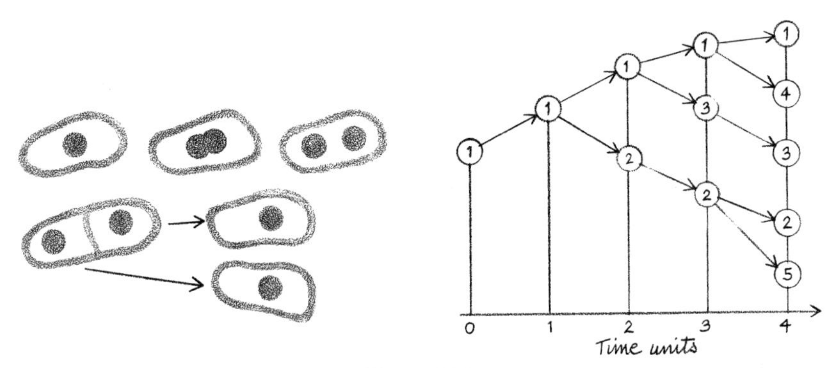

Figure 3.52

(These are the famous Fibonacci numbers first discussed in the thirteenth century in connection with rabbits. These numbers have many interesting properties.) In order to count the population by this method at a later time, we would have to calculate the sizes for all the preceding times. The method of **difference equations** furnishes a formula for the exact size of the population at any time n. The data given above for the first 10 time units follow this pattern: The size of the population at any given time is equal to the sum of the sizes at the previous 2 times; for example, the size at time 8 is the sum of the sizes at times 6 and 7: $34 = 13 + 21$. We express this relation in this way: Let x_n be the population size at time n; then

$$x_0 = x_1 = 1 \quad \text{and} \quad x_n = x_{n-1} + x_{n-2}, \qquad \text{for } n \geq 2. \quad (4)$$

This formula is explained in the diagram in Figure 3.53; the newborn are called **infants** and those at least 1 unit old are **adults.**

Equation (4) implies that the information on the size at a given time can be deduced from the sizes at the previous 2 times. We summarize the information at 2 successive times $n-1$ and n by means of a vector with components x_n and x_{n-1}; then the successive vectors

$$\begin{bmatrix} x_1 \\ x_0 \end{bmatrix}\begin{bmatrix} x_2 \\ x_1 \end{bmatrix}, \ldots, \begin{bmatrix} x_n \\ x_{n-1} \end{bmatrix}$$

are related by [compare (4) above]

$$\begin{aligned} x_{n+1} &= x_n + x_{n-1}, \\ x_n &= x_n. \end{aligned}$$

In matrix form this is

$$\begin{bmatrix} 1 & 1 \\ 1 & 0 \end{bmatrix}\begin{bmatrix} x_n \\ x_{n-1} \end{bmatrix} = \begin{bmatrix} x_{n+1} \\ x_n \end{bmatrix}.$$

Let A be the matrix $\begin{bmatrix} 1 & 1 \\ 1 & 0 \end{bmatrix}$; it follows that

$$\begin{bmatrix} x_2 \\ x_1 \end{bmatrix} = A\begin{bmatrix} x_1 \\ x_0 \end{bmatrix} = A\begin{bmatrix} 1 \\ 1 \end{bmatrix}; \qquad \begin{bmatrix} x_3 \\ x_2 \end{bmatrix} = A\begin{bmatrix} x_2 \\ x_1 \end{bmatrix} = AA\begin{bmatrix} x_1 \\ x_0 \end{bmatrix} = A^2\begin{bmatrix} 1 \\ 1 \end{bmatrix}.$$

In general, we have

$$\begin{bmatrix} x_{n+1} \\ x_n \end{bmatrix} = A^n\begin{bmatrix} 1 \\ 1 \end{bmatrix}. \tag{5}$$

An explicit expression for the x-vector can be obtained from this and the formula for A^n.

We follow the 3-step procedure for computing the power of the matrix. The characteristic equation is

$$\begin{vmatrix} 1-\lambda & 1 \\ 1 & -\lambda \end{vmatrix} = 0 \qquad \text{or} \qquad \lambda^2 - \lambda - 1 = 0.$$

By the quadratic formula the solutions are

$$\lambda_1 = \frac{1+\sqrt{5}}{2}, \qquad \lambda_2 = \frac{1-\sqrt{5}}{2}.$$

The characteristic vectors are

$$X_1 = \begin{bmatrix} x \\ \frac{1}{2}x(\sqrt{5}-1) \end{bmatrix}, \qquad X_2 = \begin{bmatrix} x \\ -\frac{1}{2}x(\sqrt{5}+1) \end{bmatrix}.$$

For simplicity we pick the special cases with $x = 2$:

$$X_1 = \begin{bmatrix} 2 \\ \sqrt{5}-1 \end{bmatrix}, \qquad X_2 = \begin{bmatrix} 2 \\ -\sqrt{5}-1 \end{bmatrix}.$$

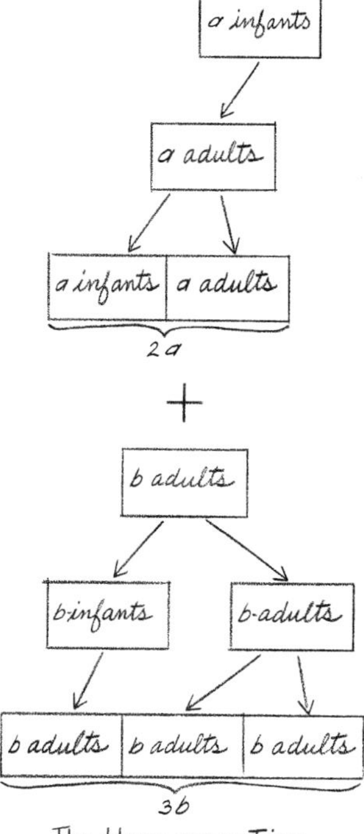

Figure 3.53

Next we express the vector $\begin{bmatrix}1\\1\end{bmatrix}$ as a linear combination of the X's:

$$\begin{bmatrix}1\\1\end{bmatrix} = \frac{\sqrt{5}+3}{4\sqrt{5}}\begin{bmatrix}2\\\sqrt{5}-1\end{bmatrix} + \frac{\sqrt{5}-3}{4\sqrt{5}}\begin{bmatrix}2\\-\sqrt{5}-1\end{bmatrix}$$
$$= \frac{\sqrt{5}+3}{4\sqrt{5}}X_1 + \frac{\sqrt{5}-3}{4\sqrt{5}}X_2.$$

Apply formula (3) with $n = k$:

$$A^n\begin{bmatrix}1\\1\end{bmatrix} = \frac{\sqrt{5}+3}{4\sqrt{5}}\lambda_1{}^n X_1 + \frac{\sqrt{5}-3}{4\sqrt{5}}\lambda_2{}^n X_2$$
$$= \frac{\sqrt{5}+3}{4\sqrt{5}}\left(\frac{1+\sqrt{5}}{2}\right)^n\begin{bmatrix}2\\\sqrt{5}-1\end{bmatrix}$$
$$+ \frac{\sqrt{5}-3}{4\sqrt{5}}\left(\frac{1-\sqrt{5}}{2}\right)^n\begin{bmatrix}2\\-1-\sqrt{5}\end{bmatrix}.$$

This can be simplified to

$$\begin{bmatrix}\dfrac{\sqrt{5}+3}{2\sqrt{5}}\left(\dfrac{1+\sqrt{5}}{2}\right)^n + \dfrac{\sqrt{5}-3}{2\sqrt{5}}\left(\dfrac{1-\sqrt{5}}{2}\right)^n\\[2ex] \dfrac{1+\sqrt{5}}{2\sqrt{5}}\left(\dfrac{1+\sqrt{5}}{2}\right)^n - \dfrac{1-\sqrt{5}}{2\sqrt{5}}\left(\dfrac{1-\sqrt{5}}{2}\right)^n\end{bmatrix}.$$

The second component of this vector, which is x_n, may be simplified and written as

$$x_n = \frac{1}{\sqrt{5}}\left[\left(\frac{1+\sqrt{5}}{2}\right)^{n+1} - \left(\frac{1-\sqrt{5}}{2}\right)^{n+1}\right] \tag{6}$$

(See Exercise 16). This furnishes the explicit formula for the size of the population at time n. We do not have to express the second column of A as a linear combination of the X's because we already have our answer. ▲

3.11. EXERCISES

In Exercises 1 and 2 find the characteristic values λ and verify by actual computation that λ^2 is the characteristic value of the square of the matrix.

1 (a) $\begin{bmatrix}1 & 2\\5 & -2\end{bmatrix}$

(b) $\begin{bmatrix}2 & \sqrt{3}\\\sqrt{3} & 4\end{bmatrix}$

(c) $\begin{bmatrix}3 & 1\\11/4 & -2\end{bmatrix}$

(d) $\begin{bmatrix}3 & 0 & 0\\0 & -2 & 2\\0 & 2 & 1\end{bmatrix}$

2 (a) $\begin{bmatrix} 2 & -1 \\ 0 & 1 \end{bmatrix}$

(b) $\begin{bmatrix} 5 & \frac{3}{2} \\ \frac{3}{2} & 1 \end{bmatrix}$

(c) $\begin{bmatrix} -3 & 11 \\ 1 & 7 \end{bmatrix}$

(d) $\begin{bmatrix} 1 & 0 & 0 \\ 0 & 3 & 2 \\ 0 & -1 & 0 \end{bmatrix}$

In Exercises 3 and 4 express the vector Y *as a linear combination of the* X*'s.*

3 (a) $Y = \begin{bmatrix} 3 \\ 1 \end{bmatrix}, \quad X_1 = \begin{bmatrix} -6 \\ 5 \end{bmatrix}, \quad X_2 = \begin{bmatrix} 3 \\ -3 \end{bmatrix}$

(b) $Y = \begin{bmatrix} 2 \\ -1 \end{bmatrix}, \quad X_1 = \begin{bmatrix} 2 \\ 0 \end{bmatrix}, \quad X_2 = \begin{bmatrix} -2 \\ 5 \end{bmatrix}$

(c) $Y = \begin{bmatrix} 6 \\ 3 \end{bmatrix}, \quad X_1 = \begin{bmatrix} -1 \\ 5 \end{bmatrix}, \quad X_2 = \begin{bmatrix} 2 \\ 1 \end{bmatrix}$

(d) $Y = \begin{bmatrix} 1 \\ 5 \\ 6 \end{bmatrix}, \quad X_1 = \begin{bmatrix} 1 \\ 3 \\ 1 \end{bmatrix}, \quad X_2 = \begin{bmatrix} 2 \\ 0 \\ -1 \end{bmatrix}, \quad X_3 = \begin{bmatrix} 1 \\ 1 \\ 2 \end{bmatrix}$

(e) $Y = \begin{bmatrix} 0 \\ 1 \\ 3 \end{bmatrix}, \quad X_1 = \begin{bmatrix} 1 \\ 0 \\ 5 \end{bmatrix}, \quad X_2 = \begin{bmatrix} 0 \\ 2 \\ -1 \end{bmatrix}, \quad X_3 = \begin{bmatrix} 1 \\ 1 \\ 1 \end{bmatrix}$

(f) $Y = \begin{bmatrix} 9 \\ -7 \\ 5 \end{bmatrix}, \quad X_1 = \begin{bmatrix} 2 \\ -3 \\ 0 \end{bmatrix}, \quad X_2 = \begin{bmatrix} 3 \\ 0 \\ -2 \end{bmatrix}, \quad X_3 = \begin{bmatrix} 2 \\ -1 \\ -1 \end{bmatrix}$

4 (a) $Y = \begin{bmatrix} 1 \\ 1 \end{bmatrix}, \quad X_1 = \begin{bmatrix} 1 \\ 0 \end{bmatrix}, \quad X_2 = \begin{bmatrix} 1 \\ -1 \end{bmatrix}$

(b) $Y = \begin{bmatrix} -2 \\ -23 \end{bmatrix}, \quad X_1 = \begin{bmatrix} 2 \\ -3 \end{bmatrix}, \quad X_2 = \begin{bmatrix} 1 \\ 5 \end{bmatrix}$

(c) $Y = \begin{bmatrix} 1 \\ 0 \end{bmatrix}, \quad X_1 = \begin{bmatrix} 4 \\ -6 \end{bmatrix}, \quad X_2 = \begin{bmatrix} 3 \\ -1 \end{bmatrix}$

(d) $Y = \begin{bmatrix} -1 \\ -9 \\ -3 \end{bmatrix}, \quad X_1 = \begin{bmatrix} 1 \\ 0 \\ 0 \end{bmatrix}, \quad X_2 = \begin{bmatrix} 1 \\ 2 \\ 1 \end{bmatrix}, \quad X_3 = \begin{bmatrix} 1 \\ 1 \\ 2 \end{bmatrix}$

(e) $Y = \begin{bmatrix} 4 \\ 10 \\ 2 \end{bmatrix}, \quad X_1 = \begin{bmatrix} 1 \\ -5 \\ 0 \end{bmatrix}, \quad X_2 = \begin{bmatrix} 3 \\ 0 \\ 1 \end{bmatrix}, \quad X_3 = \begin{bmatrix} -1 \\ 2 \\ 1 \end{bmatrix}$

(f) $Y = \begin{bmatrix} 2 \\ 15 \\ 10 \end{bmatrix}, \quad X_1 = \begin{bmatrix} 1 \\ 1 \\ 1 \end{bmatrix}, \quad X_2 = \begin{bmatrix} 1 \\ -1 \\ 0 \end{bmatrix}, \quad X_3 = \begin{bmatrix} 1 \\ -2 \\ -1 \end{bmatrix}$

In Exercises 5 and 6 find the expressions for the kth powers of the following matrices.

5 (a) $\begin{bmatrix} 4 & 1 \\ 0 & 2 \end{bmatrix}$

(b) $\begin{bmatrix} 5 & 1 \\ 4 & 8 \end{bmatrix}$

(c) $\begin{bmatrix} 2 & 1 \\ 1 & 2 \end{bmatrix}$

(d) $\begin{bmatrix} 1 & 4 & -2 \\ 0 & 2 & 5 \\ 0 & 0 & -1 \end{bmatrix}$

(e) $\begin{bmatrix} 2 & 0 & 0 \\ -1 & 1 & 4 \\ 0 & 0 & -1 \end{bmatrix}$

6 (a) $\begin{bmatrix} -5 & 4 \\ 1 & -2 \end{bmatrix}$

(b) $\begin{bmatrix} 3 & 1 \\ 0 & -2 \end{bmatrix}$

(c) $\begin{bmatrix} -2 & 2 \\ 2 & 1 \end{bmatrix}$

(d) $\begin{bmatrix} 3 & -2 & 1 \\ 0 & 1 & 1 \\ 0 & 0 & -2 \end{bmatrix}$

(e) $\begin{bmatrix} 4 & 0 & 0 \\ 1 & 2 & 1 \\ 0 & 0 & 3 \end{bmatrix}$

7 In a certain town there is a public issue of water fluoridation. Every voter has an opinion, either **for** or **against** the proposal. Suppose that opinion in the town shifts from year to year according to the following rule: Of those who oppose fluoridation one year, 80 percent will continue to oppose it the following year, but 20 percent will shift to support it. Of those who now support fluoridation, 90 percent will continue to support it the following year, but 10 percent will shift to oppose it. Suppose that the supporters of the proposal initially have 25 percent of the town in their favor; how much will they have in 5 years? (See Example 4).

8 On an amendment proposed for the State Constitution the voters are either **for, against,** or **undecided.** The first two categories of voters maintain the same opinion from week to week before the ballot. The undecided voters drift to one side or the other in the following manner: $\frac{1}{3}$ of the undecided voters go **for,** $\frac{1}{6}$ go **against,** and $\frac{1}{2}$ remain undecided. Suppose that 30 percent of the voters are initially **for,** 35 percent **against,** and the remaining 35 percent undecided. What is the distribution of opinion after 4 weeks? After 5 weeks? (See Example 5.) [*Hint:* There are 2 characteristic vectors

$$\begin{pmatrix} 1 \\ 1 \\ 1 \end{pmatrix} \quad \begin{pmatrix} 1 \\ 0 \\ -2 \end{pmatrix}$$

for $\lambda = 1$.]

9 In a school system, the employees are classified as teachers, supervisors, or "out of the system." In a 1-year period the employees shift among the various states in the proportions given in the matrix P below:

	Teacher	Supervisor	Out
Teacher	0.7	0.1	0.2
Supervisor	0	0.9	0.1
Out	0	0.0	1.0

Find the transition proportions for a 4-year period, that is, the matrix P^4 (See Example 5).

10 A university has the following experience in the hiring and promotion of professors. At each 3-year interval 40 percent of the assistant professors have been promoted to associate professor, and 60 percent have left (because of termination of contract, resignation, death, and so on). Among associate professors, 20 percent have been promoted to professor, 10 percent have left, and 70 percent remain at their rank. Finally, 90 percent of professors remain, and 10 percent leave. What proportion of assistant professors will be (a) associate professors; (b) professors at the end of 12 years of service? (See Example 5.)

11 Consider the 4 states of a cigarette smoker: "quit smoking," "mild smoker," "strong smoker," and "incurable smoker." Suppose that the transition matrix P for passage among these states in a 1-year period is

	Quit	Mild	Strong	Incurable
Quit	1	0	0	0
Mild	0.1	0.7	0.1	0.1
Strong	0.1	0.1	0.7	0.1
Incurable	0	0	0	1

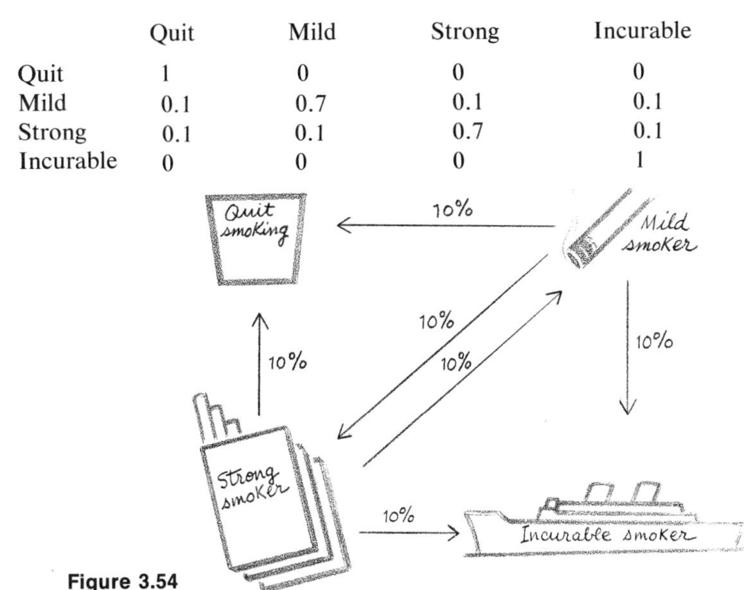

Figure 3.54

(Figure 3.54). Find the general expression P^k. What proportion of (a) mild smokers, (b) strong smokers, eventually quit? (Remember that there are 4 characteristic vectors.) See Example 5 and Exercise 8; the characteristic vectors of $\lambda = 1$ are

$$\begin{bmatrix} 1 \\ 1 \\ 1 \\ 1 \end{bmatrix} \text{ and } \begin{bmatrix} 1 \\ 0 \\ 0 \\ -1 \end{bmatrix}$$

12 The size of a population at time n is equal to the size at time $n-1$ **plus** $\frac{1}{4}$ of the size at time $n-2$. If x_n stands for the size at time n, then it satisfies the difference equation

$$x_n = x_{n-1} + \tfrac{1}{4}x_{n-2}.$$

Suppose also that $x_0 = x_1 = 4$.

(a) What matrix carries the vector

$$\begin{bmatrix} x_n \\ x_{n-1} \end{bmatrix} \quad \text{into} \quad \begin{bmatrix} x_{n+1} \\ x_n \end{bmatrix}?$$

(b) Using the method of Example 6, find the general expression for x_n.

13 An economic theory for growth of national income states that the national income x_n at year n is equal to $(k+A)x_{n-1} - Ax_{n-2}$, where k is the **multiplier constant** and A the **acceleration constant.** Assume that

$$A = 2.5, \qquad k = 0.5, \qquad x_0 = 300, \qquad \text{and} \qquad x_1 = 450.$$

Find the general expression for x_n. Note that the characteristic values are complex.

14 The price of a commodity on January 1 of a given year is equal to a weighted average of its prices on the same dates of the two previous years: $\frac{8}{10}$ of the price of last year plus $\frac{2}{10}$ of the price of the preceding year. Letting x_n be the price at time n, we have the difference equation

$$x_n = 8x_{n-1} + 0.2x_{n-2}.$$

Assume that $x_0 = 5$ and $x_1 = 6$. Find the general expression for x_n. What is the long-term behavior of the price as n increases? (See Example 6.)

15 Using the argument in Exercise 6 of Section 3.10, show that $A^3X = \lambda^3X$, $A^4X = \lambda^4X$, and so on.

16 In this exercise we outline the proof of a special case of the **ergodic theorem** for Markov chains, which states that the successive powers of the transition matrix approach a matrix whose rows are identical with the stationary distribution. (This is not valid for all chains, but a more general theorem covers all.) Let P be an arbitrary 2-by-2 stochastic matrix

$$P = \begin{bmatrix} 1-p & p \\ q & 1-q \end{bmatrix},$$

where p and q are arbitrary numbers between 0 and 1, inclusive, but with $0 < p + q < 2$.

(a) Show that the characteristic values are 1 and $1 - p - q$.

(b) Show that the corresponding characteristic vectors are

$$X_1 = \begin{bmatrix} 1 \\ 1 \end{bmatrix}, \qquad X_2 = \begin{bmatrix} p \\ -q \end{bmatrix}.$$

(c) Express the columns of P as linear combinations of the X's by solving the systems

$$\begin{aligned} 1-p &= c_1 + pc_2, & p &= c_1 + pc_2, \\ q &= c_1 - qc_2, & 1-q &= c_1 - qc_2, \end{aligned}$$

for the c's in terms of p and q.

(d) Using formula (3) find the expression for P^n.

(e) Show that each row of P^n approaches the same vector

$$\left(\frac{q}{q+p} \quad \frac{p}{q+p}\right).$$

(f) Verify that the latter vector is the stationary distribution. Why is it the only stationary distribution?

(g) What if $p = q = 0$?

17 Verify by algebra that x_n is given by expression (6).

CHAPTER

4

The Functions of Calculus

4.1 Elementary Functions

Calculus is the analysis of certain classes of **functions** by means of the **limit** concept. The idea of **function** is present throughout modern mathematics. Since it is of particular importance in calculus, we devote this entire chapter to the description, construction, and applications of many functions arising in the calculus. Much of this material is ordinarily included in calculus texts after the introduction of **limits** and **derivatives.** Our approach is different: We acquaint the reader with functions and their applications before the other concepts of the calculus; and it is our hope that the applications will stir the reader's interest in the latter concepts. In the usual calculus texts complex examples of functions are saved until the student has to do a maximization-minimization problem, so that he has to learn two difficult things at the same time. Here we introduce the two separately. Our study of the calculus actually begins in this chapter.

We begin with the formal definition of a function of a single variable:

A real-valued function of a real variable consists of two elements:

1. **A set of real numbers, called the *domain*.**
2. **A rule for associating one and only one real number with each number of the domain: if x is a number in the domain, then, according to the rule, there is a unique number y associated with it.**

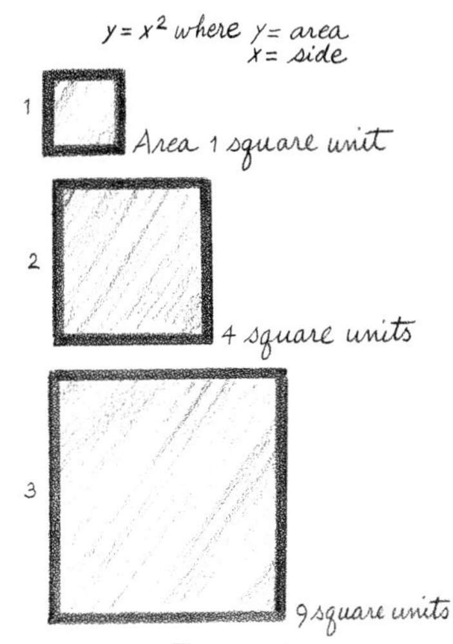

Figure 4.1

The set of **assigned** numbers is the **range** of the function. The domain is often called the set of values of the **independent variable,** and the range is the set of values of the **dependent variable.** A typical number in the domain is denoted x, and is called **the independent variable;** and the typical number in the range is denoted y, and is called **the dependent variable.** We say that the function associates with each x in the domain a unique (that is, uniquely determined by x) y in the range; y is said to **correspond** to x.

The **graph** of a function is the set of points (x,y) in the plane such that x is a number in the domain and y is the corresponding number in the range. In many modern texts the function itself is formally defined as the set of **ordered pairs** (x,y), where x belongs to the domain and y is the corresponding range number; however, the definition we have chosen is more suitable for the purposes of our applications.

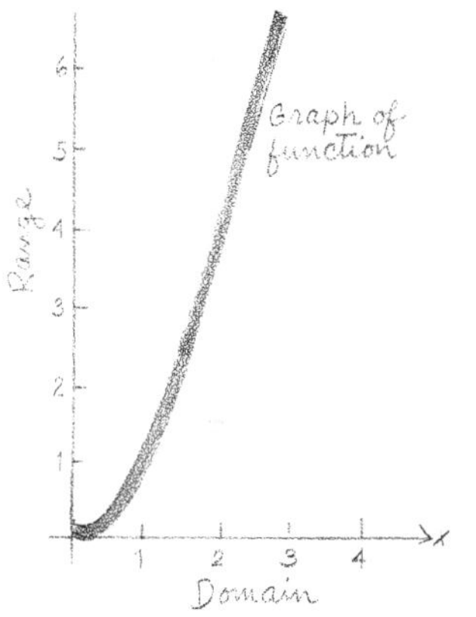

Figure 4.2

EXAMPLE 1

The area of a square is equal to the square of the length of the side. If y represents the area and x the side, then the mathematical relation between the variables is given by the equation $y = x^2$. It is natural to consider the side as the independent variable, and the area as the dependent variable (Figure 4.1); however, we can easily reverse their roles if we desire. Since length and area are nonnegative quantities, the domain and the range of the function both consist of the nonnegative real numbers.

The graph of this function is the set of points (x,y), where $y = x^2$, and $x \geq 0$: the set of points whose first coordinate is nonnegative, and whose second coordinate is the square of the first. This is simply the portion of the parabola $y = x^2$ over the nonnegative axis (Figure 4.2). ▲

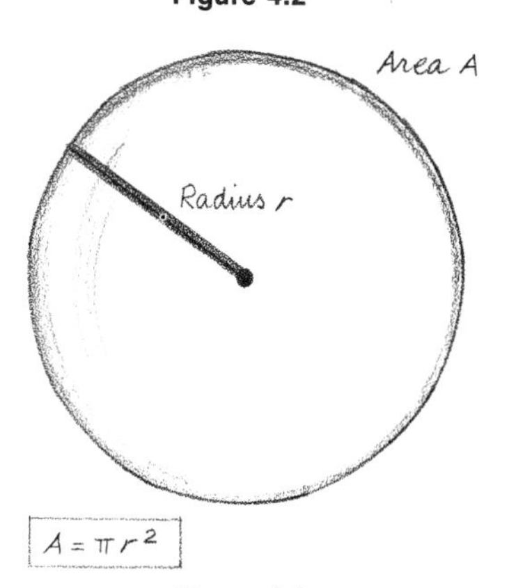

Figure 4.3

Sometimes it is convenient to use letters other than x and y for the variables. For example, more suggestive symbols for the variables **area** and **length-of-side** are A and s, respectively. Then the relation between them is denoted $A = s^2$. A similar function is the one describing the relation between the area A of a circle and its radius r; here $A = \pi r^2$ (Figure 4.3).

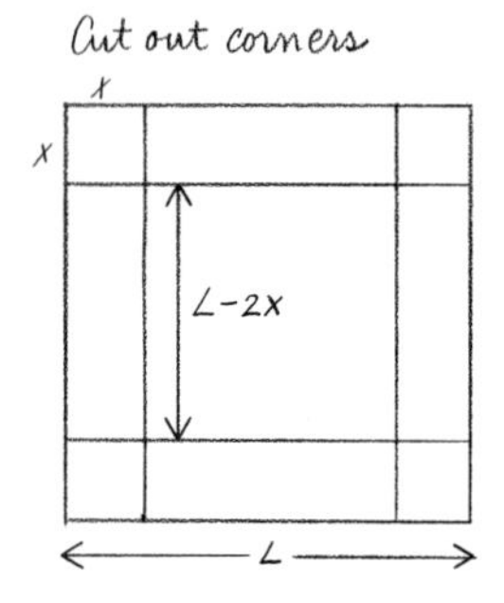

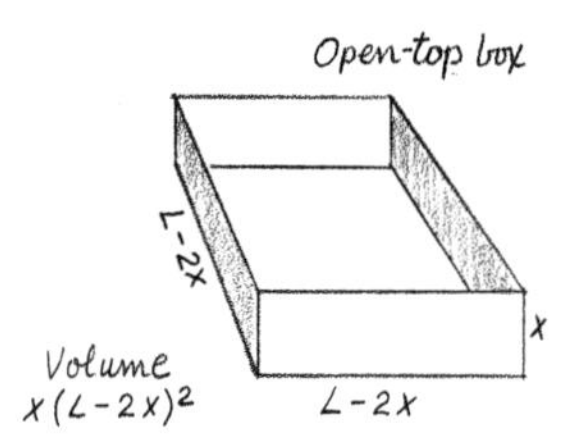

Figure 4.4

EXAMPLE 2

A box is to be made from a square piece of sheetmetal by cutting out squares from the corners and bending up the sides. Suppose that the quantity L is the given length of the edge of the original piece of metal; L is a fixed quantity. Let x be the length of the edge of the cut corner. When the corners are cut out, the box has a square base of edge $L-2x$, which is the original length L minus a corner from each end. The height of the box is equal to the length of the cut corner x. The volume of the box is the product of the length, height, and width:

$$\text{Volume } V = x(L-2x)(L-2x) = x(L-2x)^2$$

(Figure 4.4). The volume is a function of the edge of the cut corners. The dependence of V upon x is portrayed in Figure 4.5.

The domain of the function V consists of all real numbers between 0 and $\frac{1}{2}L$. The reason for this is that, on the one hand, the number x is positive; on the other hand, it must remain smaller than half the length of the side L. However, it is convenient to include the endpoints $x=0$ and $x=\frac{1}{2}L$ in the domain; these are the extreme cases in which the volume of the box is equal to 0.

The range of the function is not evident from the formula for V. We get some idea of the range in special cases of values L by substituting values of x. For example, when $L=10$ units, we have the following values of V for the given values of x:

x	0	0.5	1.0	1.5	2.0	2.5	3.0	3.5	4.0	4.5	5.0
V	0	40.5	64.0	73.5	72.0	62.5	48.0	31.5	16.0	4.5	0.0

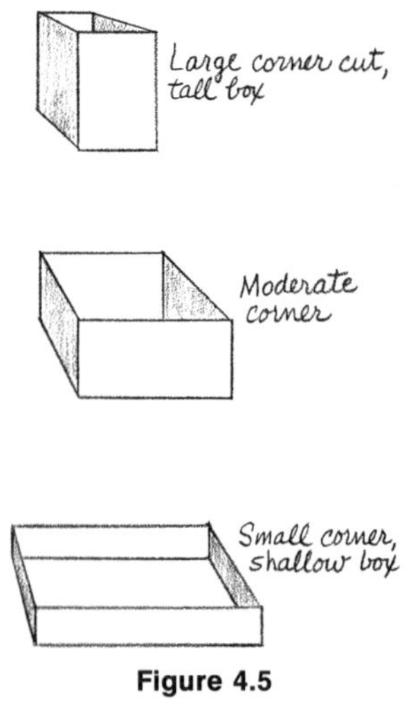

Figure 4.5

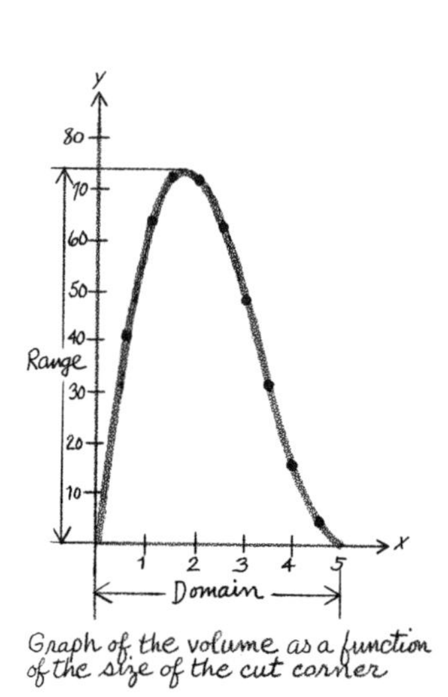

Figure 4.6

The largest of these values is 73.5. The graph of the function is sketched in Figure 4.6; it is obtained from the formula $y = x(10 - 2x)^2$. Since the complete graph appears to be smooth, and the tabled values of x are relatively close to each other, we expect the largest value of the function over the entire domain to be not much greater than the largest value in the table above. In Figure 4.6 the scale of the range is approximately $\frac{1}{10}$ that of the domain. Later we shall use the calculus to derive the exact maximum value. ▲

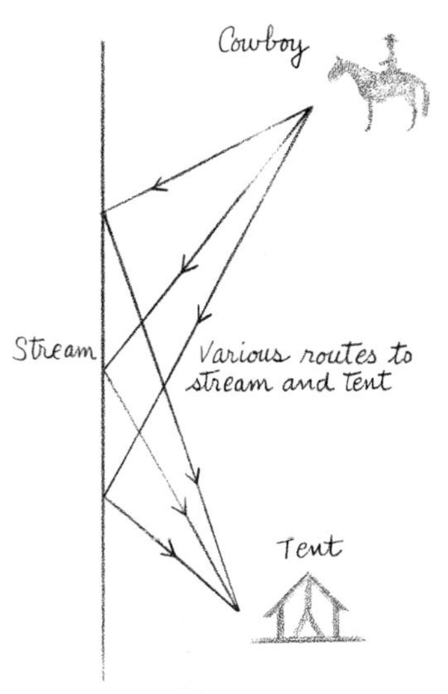

Figure 4.7

EXAMPLE 3

A cowboy has a tent not far from a stream, in the middle of a field. He approaches the stream and the tent from a distance. His plan is first to ride the horse to the stream for a drink, and then to return to the tent. We assume that he will travel on a straight line to a point along the stream, and from there on a straight line to the tent (Figure 4.7). The length of the entire trip, to the stream and then to the tent, will vary with the position which he chooses along the stream. We shall show how to describe the total length of the trip as a function of this position.

We refer to Figure 4.8. Let a be the perpendicular distance of the cowboy from the stream, b the perpendicular distance of the tent from the stream, and c the distance **along the stream** from the cowboy to the tent. Let x be the downstream distance from the cowboy to the point along the stream where the horse will drink; then $c - x$ is the remaining downstream distance to the tent. The variable x is evidently at least equal to 0 and at most equal to c.

According to the Pythagorean Theorem (see Figure 4.8), the distance from the cowboy to the point on the stream is the square root of the sum of the squares of a and x, $\sqrt{a^2 + x^2}$. Similarly, the remaining distance to the tent is $\sqrt{b^2 + (c - x)^2}$. If D is the total distance from the cowboy to the stream and to the tent, then

$$D = \sqrt{a^2 + x^2} + \sqrt{b^2 + (c - x)^2}.$$

D is a function of x; the quantities a, b, and c are assumed to be fixed. The independent variable x takes on values between 0 and c inclusive.

Suppose that $a = 100$ yards, $b = 50$, and $c = 300$; then

$$D = \sqrt{10{,}000 + x^2} + \sqrt{2500 + (300 - x)^2}.$$

The values of D for various values of x are

x	0	50	100	150	200	250	300
D	404	367	347	338	336	340	366

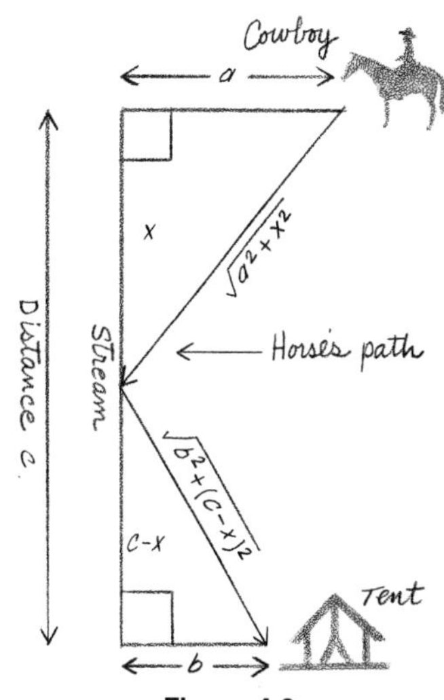

Figure 4.8

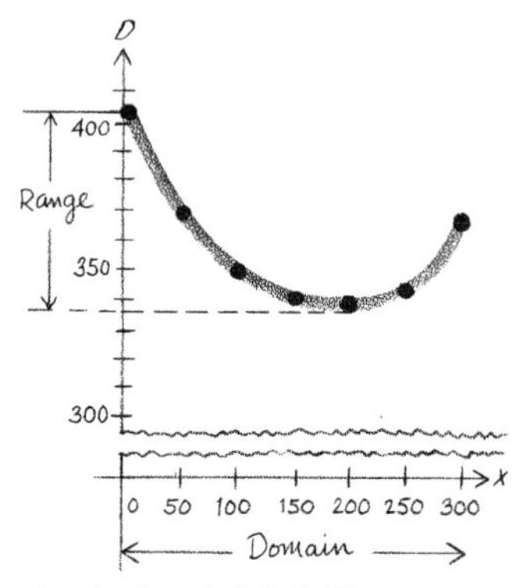

Figure 4.9

It appears from the table that the distance D varies with x in the following manner: It decreases as x goes from 0 to about 200, and then increases. The range of the function D is the set of numbers approximately between 335 and 404. The graph appears in Figure 4.9. Note that the vertical scale ranges from 300 to slightly more than 400, and the horizontal scale from 0 to 300. ▲

EXAMPLE 4

We now describe an **average cost function** (AC function) in economic theory. Our particular example is about the construction of an office building.

We begin with a figurative example. A child has a set of wooden blocks. He wants to pile the blocks in a large triangle so that each block in the triangle rests on half of each of the 2 blocks immediately below it (Figure 4.10). A pile 2 blocks high requires a total of 3 blocks; a pile 3 high requires $1 + 2 + 3 = 6$ blocks; a pile 4 high requires $1 + 2 + 3 + 4 = 10$. In general, a pile n blocks high requires $1 + 2 + \cdots + n$ blocks. The general formula for this sum is given by

$$1 + 2 + \cdots + n = \frac{n(n+1)}{2} \tag{1}$$

(Exercise 19).

This simple model is useful in analyzing the cost of constructing a building with a number of floors. There is a certain initial cost for a building of any size: site purchase, clearing, digging, and so on. After that the cost depends on the number of floors planned. The cost of a building with 2 floors is much more than double the cost of a building with just 1 floor: the foundation has to be stronger, and so does the first floor. The costs of each additional floor tend to pile up like the child's blocks. Suppose that the costs of building the successive floors are

Site purchase, clearing, etc.	\$800,000
First floor	\$50,000
Second floor	\$100,000
Third floor	\$150,000
. . .	
nth floor	\$50,000n

Figure 4.10

It follows that the cost of the building depends on the number of floors in this way:

Number of floors	Total cost
1	$\$800{,}000 + \$50{,}000$
2	$\$800{,}000 + \$50{,}000(1 + 2)$
3	$\$800{,}000 + \$50{,}000(1 + 2 + 3)$
.	.
.	.
.	.
n	$\$800{,}000 + \$50{,}000(1 + \cdots + n)$

By Equation (1) the cost of an n-floor building may be put in the form

$$\$800{,}000 + \frac{\$50{,}000n(n+1)}{2}.$$

The **average cost** is this quantity divided by the number n of floors:

$$\text{Average cost } (AC) = \frac{\$800{,}000}{n} + \$25{,}000(n+1).$$

The importance of the AC function is that if the floor space is rented at a uniform price, then the most profitable plan for the building is to minimize the average cost of each floor. The AC function (in thousands of dollars) is tabled for several values of n:

n	1	2	3	4	5	6	7	8	9	10
AC	850	475	367	325	311	308	314	325	339	355

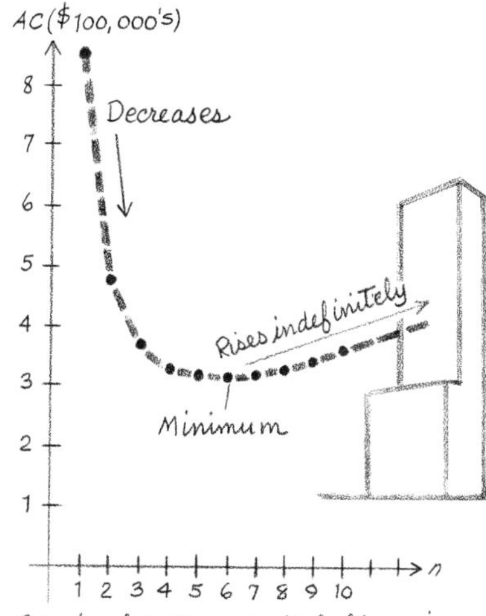

Figure 4.11 Graph of average cost of floor in building with n floors

Note that the average cost is very high at the beginning, falls to a minimum at 6, and then increases as it goes to 7 and up. After that it rises without limit (Figure 4.11). The domain of the AC function consists of the positive integers. The range consists of the set of numbers obtained from the AC formula by substituting integer values of n. ▲

EXAMPLE 5

This is a case of the **inventory problem,** which arises in the area known as **operations research.** The director of the laboratory of a medical school has found that approximately 500 white mice are consumed every year. He must decide how much "inventory" he should stock. If he keeps a small number in stock, then he has to reorder often, and pay the additional costs of ordering and shipping. If he keeps too many, then he has the additional costs of feeding and caring. The total **cost of inventory** is the sum of the cost of ordering and shipping **plus** the cost of maintenence (feeding and caring). We shall show how the cost of inventory varies with the size of the shipment ordered each time.

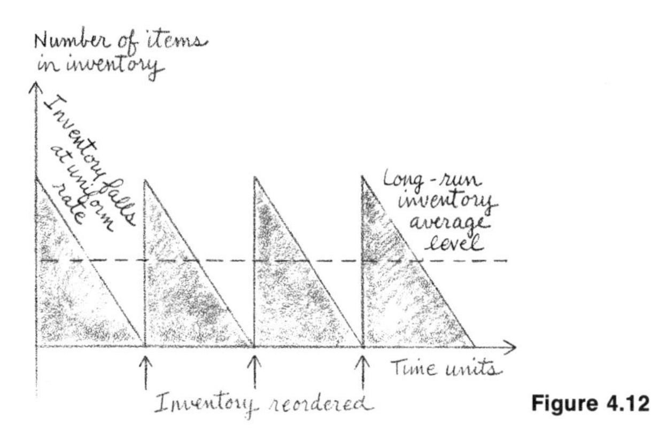

Figure 4.12

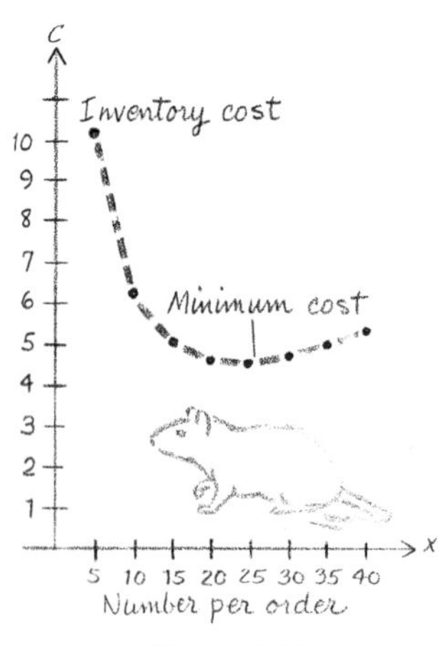

Figure 4.13

Let x stand for the number of mice in each shipment; then $500/x$ is the number of shipments per year. (We are ignoring the possibility that $500/x$ might not be an integer.) Suppose that the cost of maintaining 1 mouse is 5¢ per day. The cost of ordering is \$10 for each batch of mice (handling and shipping) **plus** 50¢ for each mouse. The stock is assumed to be consumed uniformly over time so that it decreases steadily from x to 0 over the period of the inventory (Figure 4.12). (This is a simplified version of what is done in practice because it is undesirable to have an empty stock several times per year.) Then the average number of mice in stock over the whole year is $x/2$; thus, the cost of maintenence is

Daily cost × no. of days per year × average stock size

$$= 0.05 \times 365 \times \left(\frac{x}{2}\right) = \tfrac{1}{2}(18.25)x.$$

The annual cost of ordering is

Cost of ordering x mice **times** no. of orders

$$= (10 + 0.05x)\left(\frac{500}{x}\right) = \frac{5000}{x} + 25.$$

The total cost of inventory is the sum of these:

$$C = \frac{5000}{x} + 25 + \tfrac{1}{2}(18.25)x.$$

The values of the cost function C for corresponding values of x (in multiples of 5) are

x	5	10	15	20	25	30	35	40
C	1071	616	495	457	453	466	487	515

If shipments are made only in batches of 5, then the shipment size with minimum cost of inventory is 25. The graph of the cost function appears in Figure 4.13. ▲

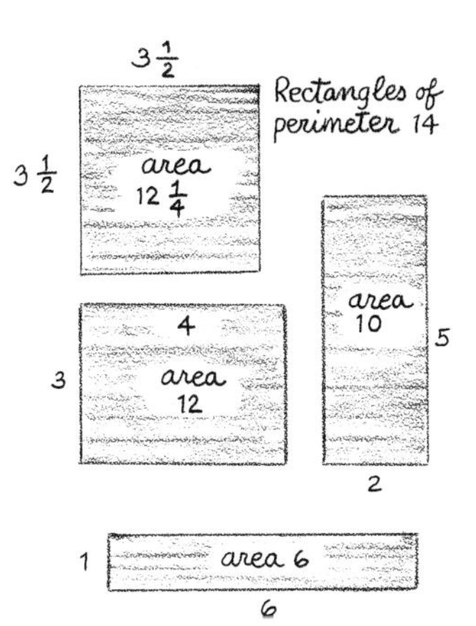

Figure 4.14

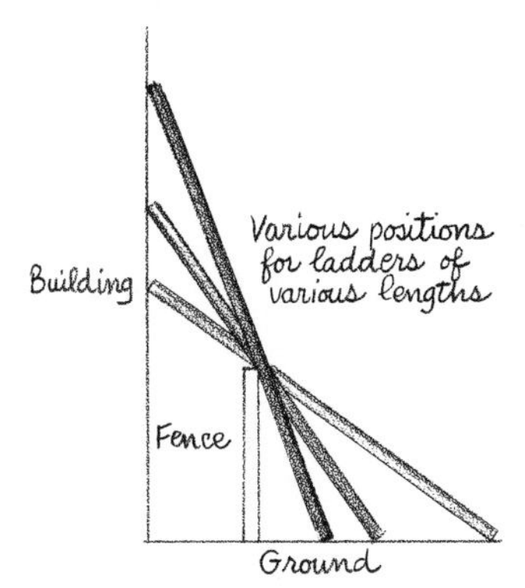

Figure 4.15

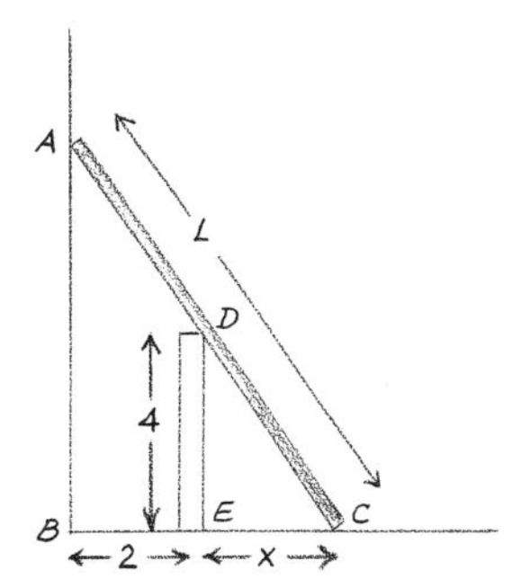

Figure 4.16

4.1 EXERCISES

1 Rectangles of various areas can be enclosed by a 14-foot rope (Figure 4.14). Let x be the bottom horizontal portion of the boundary of such a rectangle and A the area of the rectangle.

(a) Express each side as a function of x.

(b) Using the formula for the area of a rectangle (length times width) express A as a function of x.

(c) What is the domain of the function?

(d) Draw a graph for values $x = 0, 1, 2, \ldots$ in the domain. When does the area appear to be largest?

2 A 4-foot fence is located 2 feet from the base of a building. An extension ladder is to be placed to lean against the house in such a way that the ladder rests on the fence (Figure 4.15). How does the length of the ladder vary with the distance from the foot of the ladder to the foot of the fence? In other words, express the length L as a function of this distance x. [*Hint:* Let x be the distance from the fence to the ladder (Figure 4.16). Note that triangles ABC and DEC are similar, and then use the Pythagorean Theorem.] Find the values of L for $x = 1, 2, 5$.

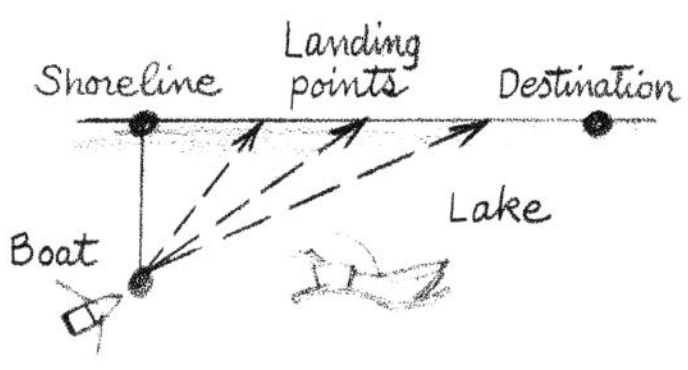

Figure 4.17

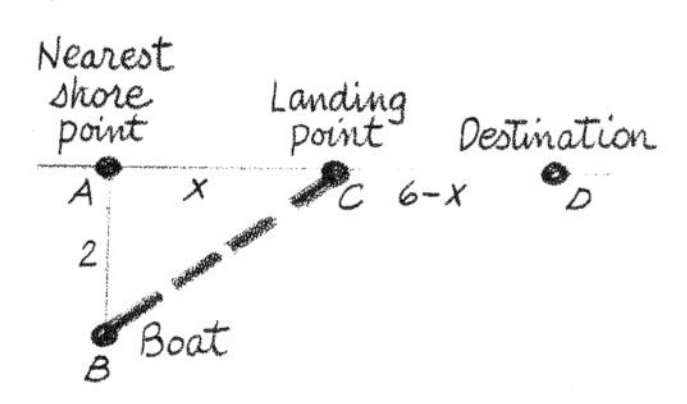

Figure 4.18

3 A man is out in a rowboat on a large lake, 2 miles from the nearest point on a straight shore. He is fishing and listening to his transistor radio. Suddenly he hears that his name is announced as a winner in the state lottery. He wants to get back to the dock as soon as he can. The dock is 6 miles east of the nearest point on the shore. He can row 3 miles per hour and walk 4 miles per hour. He plans to row the boat on a straight line to a point somewhere between the nearest shore point and his destination, the dock, and walk the rest of the way (Figure 4.17). Let x be the number of miles from the nearest shore point and the landing point (Figure 4.18).

(a) Express the distance from the initial position of the boat to the landing point in terms of x.

(b) Express the remaining distance in terms of x.

(c) Using the formula "rate times time equals distance," express the time spent rowing and walking in terms of x.

(d) What is the domain of the function $T =$ time spent rowing and walking?

(e) Compute the values of T for $x = 0, 1, \ldots, 6$.

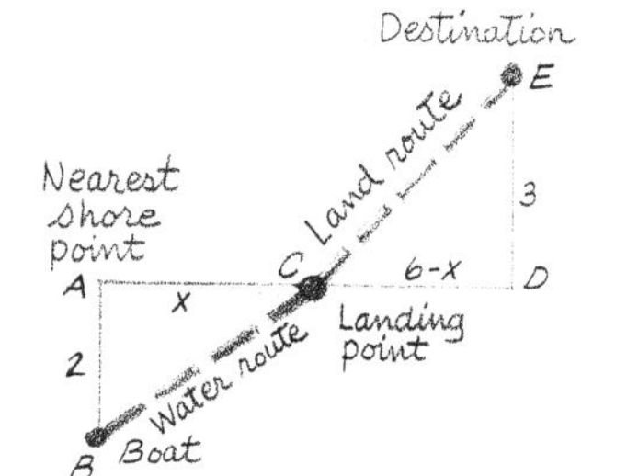

Figure 4.19

4 Here is a more general version of the rowboat problem considered in Exercise 3. We suppose this time that the man's destination is 3 miles directly inland from the dock. He is to walk there on a straight line from the landing point on the shore (Figure 4.19). Assuming the same speeds as before, express the total time for the trip as a function of x.

5 A batch of particles is released from a source and they move out uniformly in space along straight lines (sound waves from a noise source, photons from a light source, seeds from a flower, mucous droplets from a sneezing nose, and so on). A circular disk of unit radius is placed 1 unit away from the source; it catches the particles coming in its direction. The **density** of particles on the disk is the ratio

$$\frac{\text{Number of particles hitting the disk}}{\text{Area of disk}}.$$

Since the area of the disk is $\pi 1^2 = \pi$, the density is (no. of particles)/π. Suppose the disk is removed and replaced by another disk at a distance x from the source (Figure 4.20).

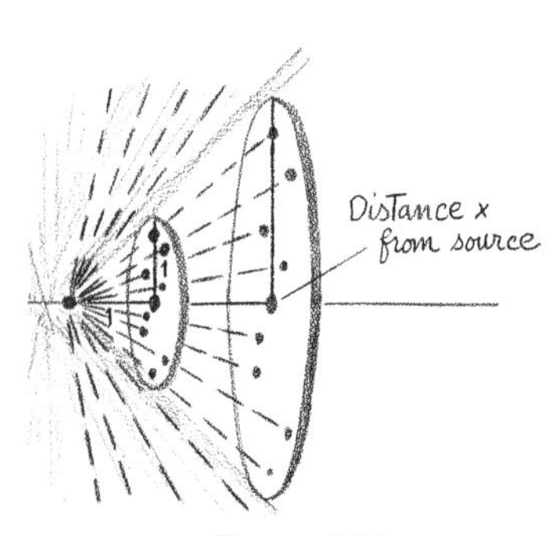

Figure 4.20

(a) How large should the disk be in order for it to catch exactly the same number of particles as the first disk? (Find the radius and the area.)

(b) Express the density of particles on the second disk in terms of (i) the density on the first disk, and (ii) the distance x. If 25 particles hit the first disk, what is the particle density on the second disk as a function of x?

(c) Draw the graph of this function for $x = 1, 2, \ldots, 10$. (This exercise is a derivation of the well-known **inverse square law** of physics.)

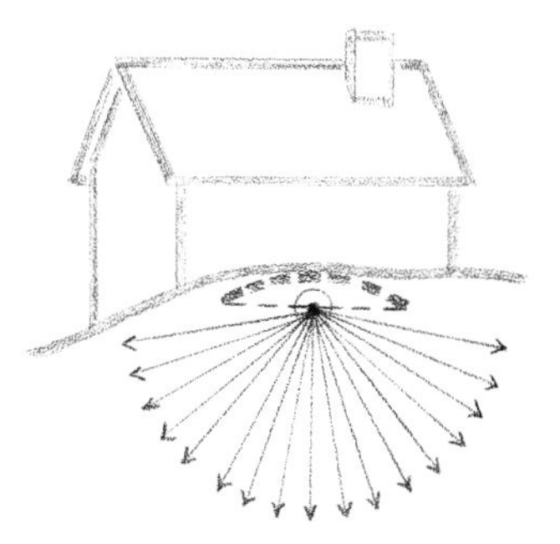
Figure 4.21

6 A cesspool is under a suburban home. Phosphates from washing detergents are emptied into the cesspool. From there they diffuse uniformly outward through the soil (Figure 4.21). Assume that the density of phosphate particles varies with the inverse square of the distance from the cesspool (compare Exercise 5). If the density is 0.001 mg per cubic foot of soil at a distance of 25 feet, what is the density at (a) 50 ft, (b) 100 ft, (c) 200 ft, and (d) x ft?

7 The density of carbon monoxide in the air in the vicinity of a large bus terminal varies inversely as the square of the distance from the terminal (compare Exercise 5). The air contains 4 parts of carbon monoxide per million at a distance of 100 feet. What is the general expression for the number of parts per million at x feet?

8 Two coal-burning generating plants are located in an open countryside 10 miles apart. The concentration of sulphur dioxide in the surrounding air due to a particular plant varies according to the inverse square of the distance from the plant. At a distance of 1 mile from plant A, the concentration of the chemical in the air coming from A is 0.1 parts per million. At a distance of 1 mile from B, the chemical concentration due to B is 0.3 parts per million. Let x be the number of miles from A on a straight line from A to B (Figure 4.22). Express the total concentration from the two sources at x as a function of x. Draw a graph. What is the domain of the function?

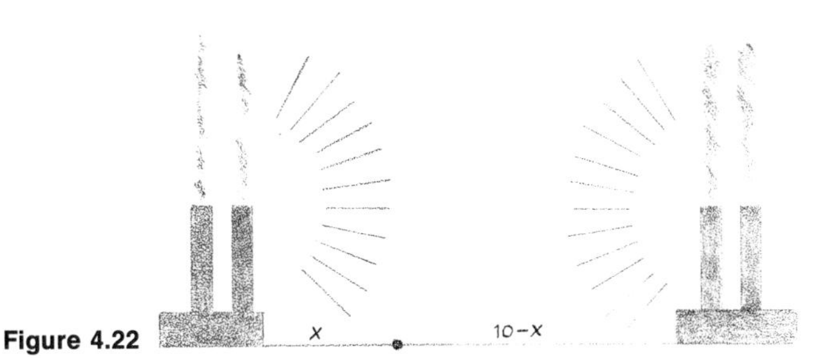

Figure 4.22

9 The owner of a dress factory finds that a lone worker can produce 100 dresses per week. If a second worker is hired, then the production of each is 98 dresses because the two workers tend to spend time talking to each other. In general, for each additional worker hired, the average number of dresses produced falls by 2. The factory capacity is 50 workers. Let x be the number of workers hired, for x between 1 and 50, inclusive. Express the number of dresses produced as a function of x. Draw a graph of the function.

10 Every day Daniel jogs 5 laps around the track. If he runs more than that, then the number of calories consumed on an individual lap increases with the number of laps after the first 5. (A tired body burns the fuel less efficiently.) Suppose that the first 5 laps consume 5 calories each; and the number increases by 2 per lap after that: 7 calories for the sixth lap, 9 calories for the seventh lap, and so on. Put $x =$ no. of laps. Express the average number of calories consumed per lap (no. of calories/no. of laps) as a function of x, and draw the graph for $x \geq 5$, (see Example 4).

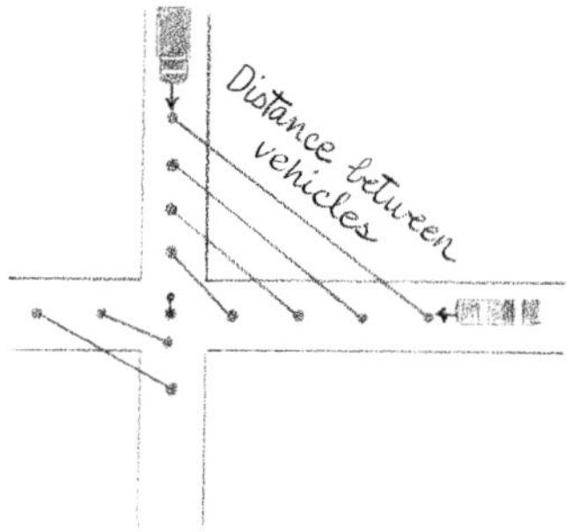

11 A truck is moving south on a straight road at 40 miles per hour (mph) and is approaching an intersection. A bus is moving west on the intersecting road at 60 mph, approaching the same intersection. At a particular moment, the truck is 3 miles north of the intersection, and the bus is 4 miles east of it (Figure 4.23). Let D stand for the distance between the two vehicles t minutes later.

(a) Express D as a function of t. [*Hint:* Let the bus run along the x-axis and the truck along the y-axis. After t minutes how far are the vehicles from the origin?]

(b) Compute the values of D for $t = 0, 1, 2, \ldots$, up to 5.

(c) Which one gets to the intersection first?

(d) Draw a graph of the function for $t = 0, 1, \ldots, 5$.

Figure 4.23

12 Situated among several hills is a shallow valley in the shape of an inverted cone. The depth of the cone is $\frac{1}{8}$ of its radius (Figure 4.24). Let x be the volume of water (in thousands of gallons) falling into the cone.

(a) Express the radius R of the resulting lake as a function of x; also, the depth D of the water over the point of the inverted cone. [*Hint:* The formula for the volume of a cone is $V = (\frac{1}{3})\ \pi\ (\text{radius})^2 \cdot \text{height}$ (Figure 4.25).]

(b) Sketch the graph of R as a function of x by picking a few convenient points.

(c) Express the size of the flooded area (in square units) as a function of x.

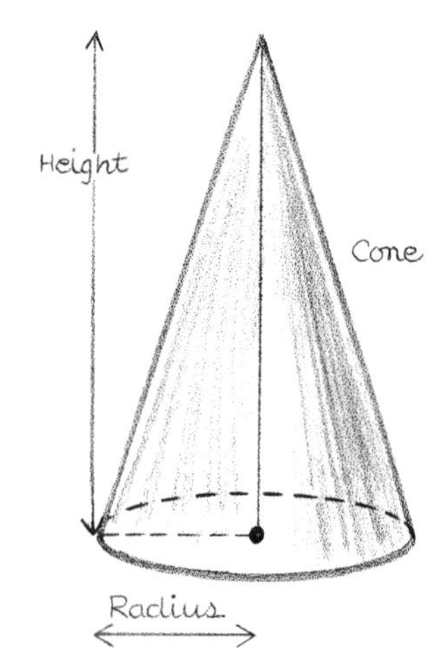

Figure 4.25

Figure 4.24

13 Cars travel along a road in a single lane at a uniform rate of 50 mph. Let d be the distance (in feet) between successive cars; and let N be the number of cars passing a fixed point on the road every minute. Express N as a function of d, and plot the graph.

14 The convention on the road is that there should be at least 1 car length separating moving cars for every 10 mph in speed:

$$\frac{\text{Minimum distance (car lengths)}}{\text{Car speed}} = 0.1.$$

On a crowded road the drivers become impatient and the ratio of the actual distance between cars to the speed decreases. Suppose that it decreases by 0.001 of the car speed:

$$\frac{\text{Distance between cars}}{\text{Car speed}} = 0.1 - 0.001 \text{ car speed}.$$

Put $x =$ car speed and $D =$ distance between cars, so that, from the previous equation,

$$D = 0.1x - 0.001x^2$$

where D is measured in car lengths. Suppose that each car is 15 feet long.

(a) What is the formula for the distance between cars in feet at speed x?

(b) Let N be the number of cars that pass a fixed point on the road every minute (see Exercise 13). Express N as a function of x, and draw its graph. Ignore the fact that N assumes only positive integer values.

15 A department store sells portable television sets at a rate of approximately 100 sets per year. The manager must decide how much inventory to stock. The average cost of carrying 1 set in the store for 1 year (cost of space, insurance, and so on) is \$5. The cost of ordering a shipment of sets is \$10 plus \$1 for each set. Let x be the number of sets in each order shipped to the store during the year. Let C be the sum of the inventory costs (see Example 5).

(a) Express C as a function of x.

(b) Compute the values of C for $x = 5, 10, 15, 20, 25, 30$.

(c) Plot the graph of C.

16 A pumping station is to be built between two neighboring towns A and B, on the bank of a river. A and B are 4 and 5 miles, respectively, from the nearest points C and D on the edge of the river. C and D are 12 miles apart. Let x be the distance from C to the point P where the station is to be built. Express the total length of the pipelines from P to the two towns as a function of x. [*Hint:* This is like the cowboy problem of Example 3; the pumping station is analogous to the point on the stream where the horse is to drink.]

17 How does the surface area of a homogeneous body vary with the volume of the body? For example, consider a cube of edge x.

(a) Let S be the surface area; express it as a function of x.

(b) Express the volume V as a function of x.

(c) Using the results of (a) and (b), find V as a function of S, and then S as a function of V.

(d) Find S for $V = 1, 8, 27$, and 64; then draw a rough sketch of the graph.

18 Instead of the cube in Exercise 17, consider a sphere of radius r. The formula for the volume is $(\frac{4}{3})\pi r^3$, and the surface area is $4\pi r^2$.

(a) Express r in terms of the volume V, and then the surface area S in terms of V.

(b) In what way is the result for the sphere similar to that for the cube in Exercise 17?

(c) Why do these results for the cube and sphere suggest a "two-thirds power law" for the relation between the volume of air inhaled and the surface area of the lungs?

19 Here is an outline of the proof justifying formula (1) which states that the sum of the first n integers is $n(n+1)/2$.

(a) Write the sum forward and then backward:

$$\begin{array}{c} 1 + 2 + \cdots + n \\ n + (n-1) + \cdots + 1. \end{array}$$

(b) What is the sum of each column; for example,

$$\frac{1}{n}, \quad \frac{2}{n-1}, \quad \frac{3}{n-2}, \ldots ?$$

(c) Sum by columns and then add all the summands.

(d) The right-hand side of (1) is the sum in (c) divided by 2.

4.2 Empirical Functions and Least-Squares Approximation

The functions considered in Section 4.1 were deduced from the known relations between the dependent and independent variables; then they were summarized in formulas. There are many functions that are specified not by formulas but by a complete listing of the points in the domain and the corresponding points in the range. Such a listing consists of ordered pairs (x,y) where x and y stand for the values of the independent and dependent variables, respectively. **Empirical** functions are those arising from recorded data.

EXAMPLE 1

Our first several examples are those of **time series.** Here the independent variable represents time units; in this case, years. The annual death rate per 100,000 population per year for a given mortal disease is defined as the ratio

$$\frac{\text{Annual number of deaths} \times 100{,}000}{\text{Number of persons in population}}.$$

Here are the death rates from tuberculosis in the United States for the years 1950–1964 (*Source:* U.S. Department of Health, Education and Welfare):

1950	20.6	1955	8.3	1960	5.6
1951	18.5	1956	7.8	1961	5.0
1952	14.4	1957	7.3	1962	4.7
1953	11.3	1958	6.6	1963	4.6
1954	9.3	1959	6.0	1964	4.0

The domain consists of the numbers of the years from 1950 through 1964. The function is plotted as a set of points (x,y) in the plane; the abscissa of each point is a year and the ordinate is the corresponding death rate (Figure 4.26). ▲

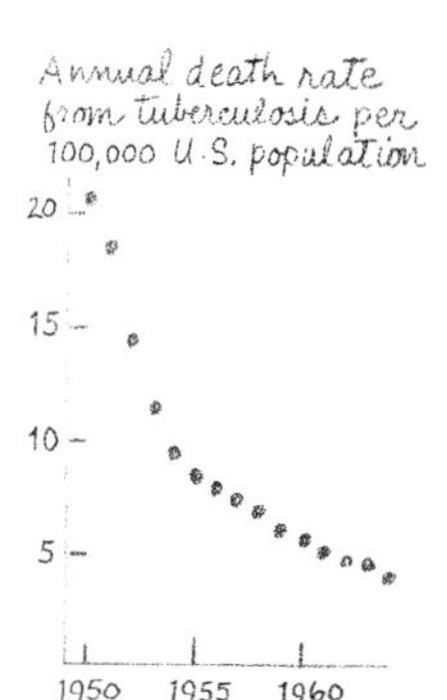

Figure 4.26

EXAMPLE 2

In the table below are given the annual death rates per 100,000 population in the United States for cancer of the respiratory system for the years 1950–1964 (*Source:* U.S. Department of Health, Education and Welfare).

1950	14.1	1955	18.2	1960	22.2
1951	14.7	1956	19.2	1961	23.1
1952	15.6	1957	19.9	1962	24.0
1953	16.7	1958	20.4	1963	24.9
1954	17.1	1959	21.2	1964	25.7

These points are plotted in Figure 4.27. ▲

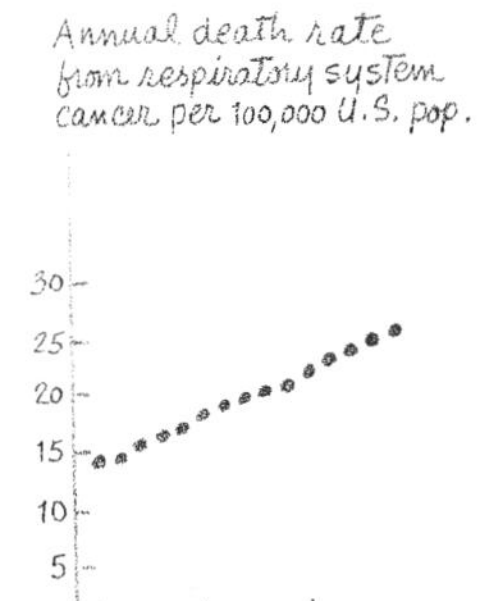

Figure 4.27

EXAMPLE 3

The marriage rates per 1000 U.S. population are given below for 5-year intervals for the period 1900–1965 (*Source:* U.S. Department of Health, Education and Welfare).

1900	9.3	1925	10.3	1950	11.1
1905	10.0	1930	9.2	1955	9.3
1910	10.3	1935	10.4	1960	8.5
1915	10.0	1940	12.1	1965	9.3
1920	12.0	1945	12.2		

The points are plotted in Figure 4.28. ▲

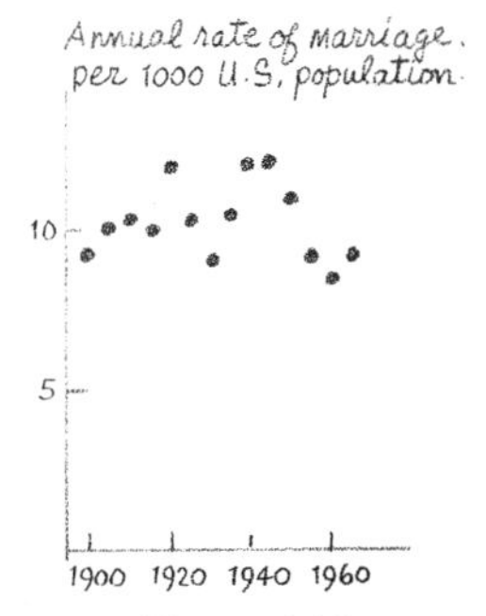

Figure 4.28

Let us compare the 3 functions observed in Examples 1, 2, and 3. In Figure 4.26 we see that the death rate from tuberculosis dropped sharply and steadily from 1950 (when the disease was first treated with modern drugs) to 1954, and then continued a slower but still steady decline to 1964. The points for 1950 to 1954 seem to lie close to a straight line; and those for the subsequent years lie close to another straight line. The data plotted in Figure 4.27 also appear to

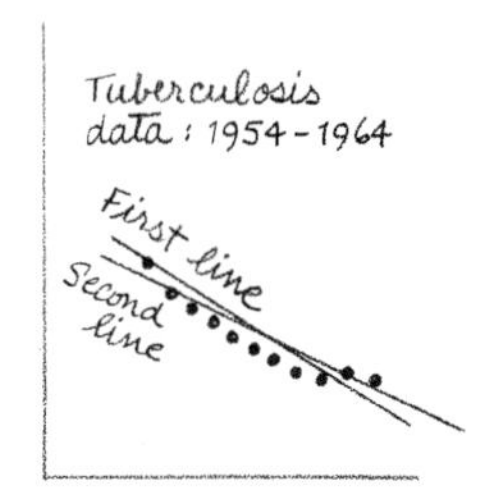

Figure 4.29

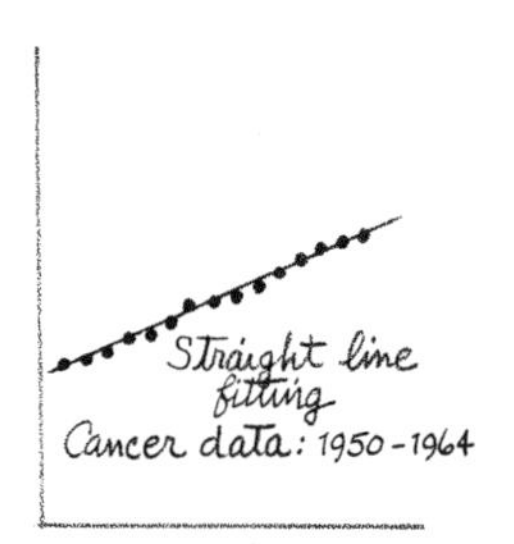

Figure 4.30

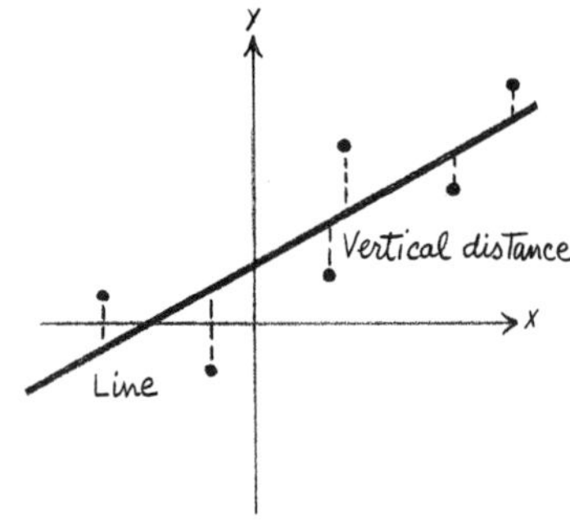

Figure 4.31

run along a straight line, at least very close to it. The data in Figure 4.28 show no movement along a line.

If data were exactly on a straight line, then the function would have an explicit formula, namely, the equation of the line. In the above examples, as with many others, the points often lie close to a straight line, but not exactly on it. The line is said to "fit" the data. More than one line can fit a given set of data; for example, two lines that fit the data of Figure 4.26 are drawn in Figure 4.29; however, it is not obvious which line fits "better." The line fitting the cancer data in Figure 4.27 is fairly well determined, and is drawn in Figure 4.30. In the case of the marriage data the points do not fit any line.

The "best fitting" line is obtained from the data by the method of **least squares.** We put the adjective in quotation marks because it is not best in a universal sense, but best according to a mathematical principle which we now describe. Consider a finite set of points in the plane and an arbitrary line. It is not necessary at the moment to assume that the points lie close to the line. Take the vertical distance between each point and the line, square it, and form the sum of squares (Figure 4.31). This is called the **sum of the squares of the vertical deviations of the points from the given line.** For a given set of points, there is a sum of squares associated with each line; it measures how well the line fits the points. If the sum of squares is "relatively" large, then the fit is poor; if relatively small, then the fit is good. A line is considered to be a **best fitting line** if the associated sum of squares is the smallest possible for the set of points. This line is called the **least squares line.** It might seem arbitrary to crown this line with the title "best." But this is justified by several facts: (i) The line is easily computable; (ii) a given set of data has one and only one least squares line (unless the x's are identical, which is a case of no practical interest; (iii) the line fits well when the data are almost linear.

Now we present the formulas for the slope and the y-intercept of the least squares line fitted to a set of points $(x_1,y_1), \ldots, (x_n,y_n)$. Define the 4 sums:

Σx = Sum of all abscissas (read **sigma x**),
Σy = Sum of all ordinates (read **sigma y**),
Σx^2 = Sum of squares of abscissas (read **sigma x²**),
Σxy = Sum of products of abscissas and ordinates (read **sigma xy**).

Then the slope m and the y-intercept b of the least squares line are

$$m = \frac{n\Sigma xy - (\Sigma x)(\Sigma y)}{n\Sigma x^2 - (\Sigma x)^2}, \qquad b = \frac{(\Sigma y)(\Sigma x^2) - (\Sigma x)(\Sigma xy)}{n\Sigma x^2 - (\Sigma x)^2}. \quad (1)$$

In the case when the sum of the abscissas is equal to 0, that is, $\Sigma x = 0$, the formulas are simpler:

$$m = \frac{\Sigma xy}{\Sigma x^2}, \qquad b = \frac{\Sigma y}{n}. \tag{2}$$

In practice, formulas (1) are rarely used: the abscissas are coded so that $\Sigma x = 0$, and then the simpler formulas (2) are used.

The proof of formulas (1) and (2)—that is, the proof that the line with the slope and intercept (1) minimizes the sum of squares of vertical deviations—is outlined in Exercise 13. The proof is elementary, requiring only elementary algebra; however, it involves notation and computations that are not directly relevant to applications, and so it is left as an exercise.

EXAMPLE 4

Calculate the least squares line to fit the data given by these points:

x	0	1	2	3	4
y	−3	−1	0	2	4

The points are plotted in Figure 4.32. We calculate the 4 sums appearing in formula (1), and then find m and b. Construct the table:

	x	y	x^2	xy	
	0	−3	0	0	
	1	−1	1	1	
	2	0	4	0	
	3	2	9	6	
	4	4	16	16	
Sums:	10	2	30	21	$n = 5$.

The slope and y-intercept are obtained by applying (1):

$$m = \frac{5(21) - 10(21)}{5(30) - (10)^2} = 1.7,$$

$$b = \frac{2(30) - 10(21)}{5(30) - (10)^2} = -3.$$

It follows that the least squares line has the formula

$$y = (1.7)x - 3.$$

It appears with the 5 points in Figure 4.32.

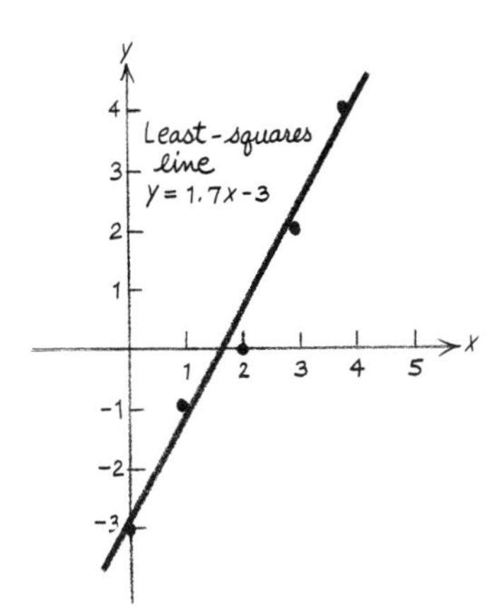

Figure 4.32

Now we show how to compute the least squares line by coding the abscissa and using formula (2). The average value of the x's is

$$\frac{0+1+2+3+4}{5}=2.$$

Subtract 2 from each of the x's so that the new table of values is

x	-2	-1	0	1	2
y	-3	-1	0	2	4

Now the sum of the x's is equal to 0. We compute the other 3 sums:

	x	y	x^2	xy	
	-2	-3	4	6	
	-1	-1	1	1	
	0	0	0	0	
	1	2	1	2	
	2	4	4	8	
Sums:	0	2	10	17	$n=5.$

By formula (2),

$$m=\frac{17}{10}=1.7; \qquad b=\frac{2}{5}=0.4.$$

In terms of the coded abscissa, the formula of the least squares line is $y=(1.7)x+0.4$. In order to get the formula for the original abscissa, we replace x by $x-2$:

$$y=1.7(x-2)+0.4=(1.7)x-3.$$

This is the same as the line obtained above by formula (1). ▲

EXAMPLE 5

Find the least squares line to fit the portion of the tuberculosis death rate data in Example 1 for the years from 1954 through 1964. First we code the abscissa so that the sum of the values of x is equal to 0. The average of 1954 through 1964 is 1959; hence, we code 1959 as 0, and 1954, 1955, . . . , 1964 as -5, -4, . . . , 5. The 4 sums are computed:

x	y	x^2	xy
-5	9.3	25	-46.5
-4	8.3	16	-33.2
-3	7.8	9	-23.4

	−2	7.3	4	−14.6	
	−1	6.6	1	−6.6	
	0	6.0	0	0.0	
	1	5.6	1	5.6	
	2	5.0	4	10.0	
	3	4.7	9	14.1	
	4	4.6	16	18.4	
	5	4.0	25	20.0	
Sums:	0	69.2	110	−56.2	$n = 11.$

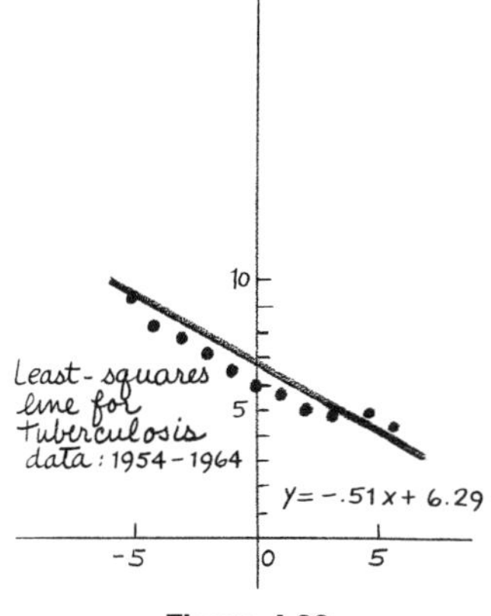

Figure 4.33

By formula (2) we have

$$m = \frac{-56.2}{110} = -0.51; \qquad b = \frac{69.2}{11} = 6.29.$$

The least squares line is drawn with the data in Figure 4.33. The observed death rates are very close to those values computed from the formula for the least squares line:

x	−5	−4	−3	−2	−1	0	1	2	3	4	5
Observed y	9.3	8.3	7.8	7.3	6.6	6.0	5.6	5.0	4.7	4.6	4 0
Computed y	8.8	8.3	7.8	7.3	6.8	6.3	5.8	5.3	4.8	4.2	3.7 ▲

EXAMPLE 6

The average weights of American men in the age group 20–24 vary with height as follows (*Source:* Society of Actuaries, 1959):

Height	*Weight*	*Height*	*Weight*
5′4″	141	5′11″	167
5′5″	144	6′	172
5′6″	148	6′1″	177
5′7″	151	6′2″	182
5′8″	155	6′3″	186
5′9″	159	6′4″	190
5′10″	163		

In order to compute the least squares line with the simpler formula (2), we code the data so that the middle abscissa value 5′10″ is taken as 0, and the inches are replaced by units. Then the independent variable runs from −6 to 6. Since the y-values are also very large, it is convenient to code these also. The range of y is from 141 to 190, and the middle value is approximately 165. So we subtract 165 from each of the y-values; the dependent variable then runs from −24 to 25. The sums are now computed:

	x	y	x^2	xy
	-6	-24	36	144
	-5	-21	25	105
	-4	-17	16	68
	-3	-14	9	42
	-2	-10	4	20
	-1	-6	1	7
	0	-2	0	0
	1	2	1	2
	2	7	4	14
	3	12	9	36
	4	17	16	68
	5	21	25	105
	6	25	36	150
Sums:	0	-10	182	761

$n = 13$.

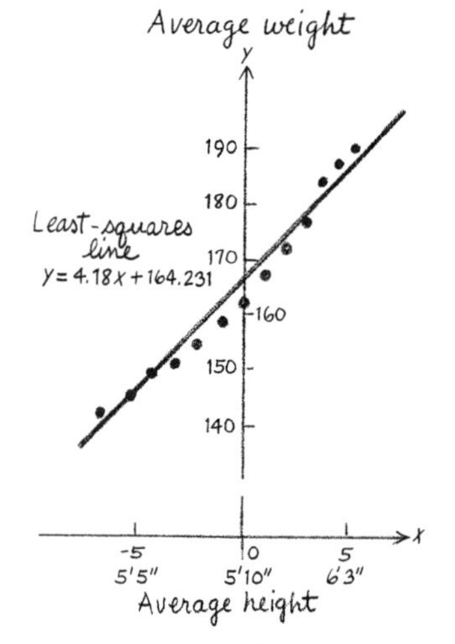

Figure 4.34

The slope and y-intercept are

$$m = \frac{761}{182} = 4.18, \qquad b = \frac{-10}{13} = -0.769.$$

The equation of the line in coded coordinates is

$$y = 4.18x - 0.769,$$

where x is measured in inches above 5′ 10″. The original y-values are obtained from the coded ones by replacing y in the equation above by $y - 165$; the equation becomes

$$y = 4.18x + 164.231.$$

It is plotted against the data in Figure 4.34. ▲

4.2 EXERCISES

In Exercises 1 and 2 find the equation of the least squares line for each of the given sets of data. Code whenever necessary. Plot the data and the line.

1 (a)

x	-2	0	2
y	4	3	1

(b)

x	10	11	12
y	98	99	103

(c)

x	21	22	23	24	25
y	-1	-2	0	2	1

(d)

x	16	17	18	19
y	3	3	4	5

[In (d) code the x's by subtracting their average.]

2 (a)

x	-1	0	1
y	-4	-2	1

(b)

x	51	53	55
y	34	37	41

(c)

x	17	19	21	23	25
y	2	3	1	0	-1

(d)

x	20	22	24	26
y	1	1	2	4

3 The **expectation of life** for a particular age group is the average number of years that a person of the given age lives after that age. For white American males of ages 16–20, the life expectancies are

Age	16	17	18	19	20
Years	53.9	53.0	52.1	51.1	50.2

(*Source:* U.S. Public Health Service, 1967). Plot the data and compute the least squares line.

4 Repeat Exercise 3 for the following data for nonwhite American males:

Age	16	17	18	19	20
Years	48.4	47.5	46.6	45.7	44.8

5 The percentage of the American labor force listed as unemployed for the years 1964–1969 was

1964	1965	1966	1967	1968	1969
5.2	4.5	3.8	3.6	3.6	3.5

(*Source:* U.S. Department of Labor). Plot the data and the least squares line. Code the years as $-5, -3, -1, 1, 3, 5$.

6 At a given temperature, exposed flesh feels colder in the presence of wind. The **wind chill index** is the temperature equivalent in cooling power in the absence of wind; for example, when the temperature is 35°F and the wind is blowing at 10 mph, the flesh loses as much heat as when the wind is calm and the temperature is 21°F. The index for 35°F is given below for various wind speeds in mph (*Source:* Environmental Science Services Administration):

Wind	5	10	15	20	25
Temperature equivalent	33	21	16	12	7

Plot the data and the least squares line.

7 The number of quarts of milk that could be purchased by the average American with the wages of 1 hour of his labor is given for various years (*Source:* U.S. Department of Agriculture):

1929	1939	1949	1959	1969
3.9	5.1	6.5	8.7	10.2

Plot the data and the least squares line.

8 The rate of twin births increases with the number of children previously born to a mother. The rates per 1000 deliveries in the United States in 1964 are given below (*Source:* U.S. Department of Health, Education and Welfare):

Previous live births	0	1	2	3	4	5	6
Rate of twins	6.8	8.9	10.5	11.8	13.6	14.4	16.1

Plot the data and the least squares line.

9 Expenditures for public school education in the United States (in billions of dollars) for various years are given below (*Source:* U.S. Office of Education):

School years	1959–1960	1961–1962	1963–1964	1965–1966	1967–1968
Expenditure	16	18	21	26	31

Plot the data and the least squares line.

10 The labor force of the United States increased over the 5-year period 1965–1969 as follows (*Source:* U.S. Department of Labor):

Year	1965	1966	1967	1968	1969
Millions of workers	77.1	78.9	80.8	82.3	84.2

Plot the data and the least squares line.

11 In the United States the numbers of minors (under 18) arrested by police for selected years are (in thousands) (*Source:* U.S. Department of Justice):

1953	1954	1955	1956	1957
149	164	195	235	254

Plot the data and the least squares line.

12 The amount of money (millions of dollars) paid for medical care and hospitalization in the United States under workmen's compensation for selected years is (*Source:* U.S. Department of Health, Education and Welfare):

1952	1953	1954	1955	1956
260	280	308	325	350

Plot the data and the least squares line.

13 In this exercise we outline the proof of the fact that formulas (1) and (2) actually are the lines whose sum of squares deviation is the smallest. The proof is given just for the case of formula (2), under the assumption that $\Sigma x = 0$. Let the coordinates of the n points be $(x_1,y_1), \ldots, (x_n,y_n)$. Let $y = mx + b$ be an arbitrary line.

(a) Show that the sum of the squares of the vertical deviations is $\Sigma(y - mx - b)^2$, which stands for

$$(y_1 - mx_1 - b)^2 + \cdots + (y_n - mx_n - b)^2.$$

(b) Verify that the sum of the squares above may be expressed in terms of the 4 sums appearing in (1):

$$\Sigma y^2 + m^2\Sigma x^2 + nb^2 - 2m\Sigma xy - 2b\Sigma y.$$

(c) Show that the sum above is equal to

$$(\Sigma x^2)\left(m - \frac{\Sigma xy}{\Sigma x^2}\right)^2 + n\left(b - \left(\frac{1}{n}\right)\Sigma y\right)^2 + \Sigma y^2 - \frac{(\Sigma xy)^2}{(\Sigma x^2)} - \left(\frac{1}{n}\right)(\Sigma y)^2.$$

[*Hint:* Complete the square on m and then on b.]

(d) Show that the expression in (c) is minimized when m and b are chosen according to formula (2).

4.3 The Exponential Function

The exponential function describes the change of quantities over time in applications involving growth and decay. It also arises in the theory of probability, as well as in many other areas of mathematics. The logarithm function, to be studied in Section 4.4, is the "inverse" of the exponential.

We review some elementary properties of exponents. Let b be an arbitrary nonzero real number; then, for any positive integer n, b^n stands for the n-fold product

$$b^n = b \cdot b \cdot \cdots \cdot b \qquad (n \text{ factors}).$$

The negative integer exponents are defined as reciprocals:

$$b^{-1} = \frac{1}{b}, \qquad b^{-2} = \frac{1}{b^2}, \qquad \ldots, \qquad b^{-n} = \frac{1}{b^n}.$$

The exponent 0 satisfies

$$b^0 = 1.$$

Exponents that are reciprocals of integers are defined as roots:

$$b^{1/2} = \sqrt{b}, \qquad b^{1/3} = \sqrt[3]{b}, \qquad b^{-1/4} = \frac{1}{\sqrt[4]{b}};$$

and, in general,

$$b^{1/n} = \sqrt[n]{b}, \qquad b^{-1/n} = \frac{1}{\sqrt[n]{b}}.$$

Fractional exponents are defined as multiples of reciprocal exponents:

$$b^{2/3} = (b^{1/3})^2 \qquad \text{or} \qquad (\sqrt[3]{b})^2;$$

or as reciprocals of multiples:

$$b^{2/3} = (b^2)^{1/3} \qquad \text{or} \qquad \sqrt[3]{b^2}.$$

In general, if m and n are integers, then

$$b^{m/n} = \sqrt[n]{b^m} \qquad \text{or} \qquad (\sqrt[n]{b})^m.$$

Note that the power $b^{m/n}$ above may be imaginary if b is negative and n even; for example, $(-2)^{1/2} = i(2)^{1/2}$, where i is the imaginary unit. However, in our applications, b will usually be positive, so that the powers of b are real numbers.

EXAMPLE 1

$$4^{1/2} = \sqrt{4} = 2; \qquad 4^{-1/2} = \frac{1}{\sqrt{4}} = \frac{1}{2};$$
$$4^{3/2} = (\sqrt{4})^3 = 2^3 = 8. \blacktriangle$$

Recall that a rational number is one that can be expressed as the ratio of two integers. If x is a rational number and b any positive real number, then b^x has a meaning defined as above in terms of a power of a root. Now we extend this definition so that it holds for all real numbers x including the irrational numbers; for example, $x = \sqrt{2}$.

Every irrational number has a nonrepeating and nonterminating decimal expansion. It can be computed to within an arbitrary degree of accuracy by taking a finite part of the decimal expansion. For example, the irrational number $\sqrt{2}$ has a decimal expansion 1.41421. . . . It can be calculated with as much accuracy as desired by taking sufficiently many decimal places. We say that the successive terms 1., 1.4, 1.41, . . . "converge to the limit $\sqrt{2}$" as the number of decimal places increases.

The number b^x, with x irrational, is defined in a similar way. Suppose again that $x = \sqrt{2}$. We take the successive decimal approximations to the power of b:

$$b^1, \quad b^{1.4}, \quad b^{1.41}, \quad b^{1.414}, \; . \; . \; . \; .$$

It can be shown that these numbers "converge to a limit" whose value is denoted $b^{\sqrt{2}}$. Observe that the successive powers do not substantially vary as the number of decimal places gets large; for example, the difference between the two numbers $b^{1.4142\ldots}$ and $b^{1.4142}$ (with a finite decimal of 4 places) is

$$b^{1.4142}(b^{0.0000\cdots} - 1),$$

which is certainly not more than $b^{1.4142}|b^{0.0001} - 1|$. Therefore, it seems that the successive terms tend to cluster about a value which we call $b^{\sqrt{2}}$.

In general, for any real, irrational number x, the expression b^x is defined as the "limit" of the corresponding sequence b^r of powers where the exponent r is a finite piece of the infinite decimal expansion. This defines b^x for every real number x. For the moment, we omit a discussion of **limit;** this is left to the section on calculus.

The expression b^x with a fixed positive b and variable x defines an **exponential function.** It inherits the properties of the original **laws of exponents:**

$$b^{x+y} = b^x b^y, \qquad b^{x-y} = b^x/b^y,$$
$$b^0 = 1, \qquad (b^x)^y = b^{xy}.$$

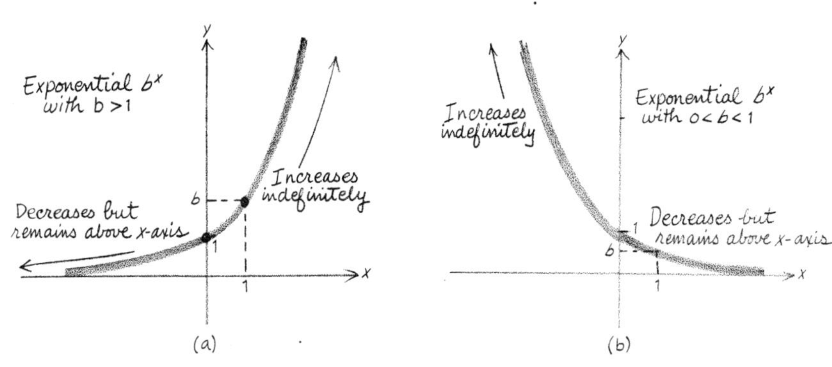

Figure 4.35

If $b > 1$, then b^x increases as x increases; the graph appears in Figure 4.35(a). If $b = 1$, then $b^x = 1^x = 1$ for all x, and the graph is the horizontal line $y = 1$. If $0 < b < 1$, then b^x decreases as x increases [Figure 4.35(b)]. Note that the graph passes through the y-axis 1 unit above the origin because $b^0 = 1$. When $x = 1$, the value of the function is equal to b itself because $b^1 = b$.

In applications we shall usually consider exponential functions with the domain restricted to the nonnegative real numbers or nonnegative integers. The independent variable is usually considered as a **time** variable. We extend the class of exponential functions to include those of the form cb^x, where c is a positive number; the latter number is called the initial value of the exponential function because it is the value of the function when $x = 0$: $cb^0 = c1 = c$. The number b is the **base** of the function.

The exponential function with $b > 1$ arises in problems of growth: that of populations, incomes, organisms, and so on. The variable x represents time, and the function cb^x is the size at time x. First we consider growth over integer time units: 1, 2, ... Let c be the size of the population at time $x = 0$. Suppose that it increases by $100p$ percent of its current size at every time unit. Then the size after x time units is given by the exponential function with $b = 1 + p$:

$$c(1+p)^x. \tag{1}$$

Let us check this in the simplest cases. After 1 unit of time the size is c plus $100p$ percent of c:

$$c + cp = c(1+p).$$

After the next time unit the size is $c(1 + p)$ plus $100p$ percent of it:

$$\begin{aligned} c(1+p) + c(1+p)p &= c(1+p)(1+p) \qquad \text{[by factoring } c(1+p)\text{]} \\ &= c(1+p)^2. \end{aligned}$$

Similarly, after the third time unit the size is

$$c(1+p)^3.$$

Thus, it seems plausible that the general formula for the size after x time units is given by (1).

EXAMPLE 2

Suppose that the national income of a country grows by 5 percent every year, and that the present national income (in billions of dollars) is 100. The size after x years is given by (1) with $c = 100$ and $p = 0.05$:

$$\text{National income} = 100(1.05)^x.$$

The values of the function for $x = 0, 1, \ldots, 10$ are given in the table below, and the graph appears in Figure 4.36:

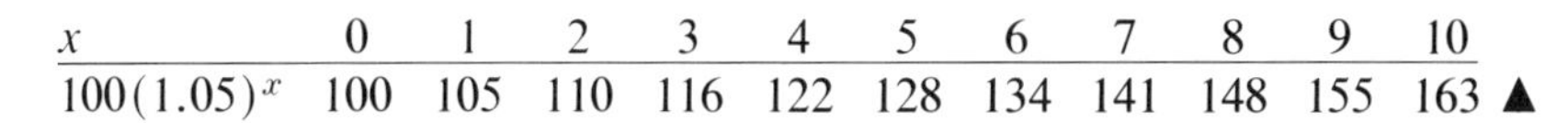

x	0	1	2	3	4	5	6	7	8	9	10
$100(1.05)^x$	100	105	110	116	122	128	134	141	148	155	163 ▲

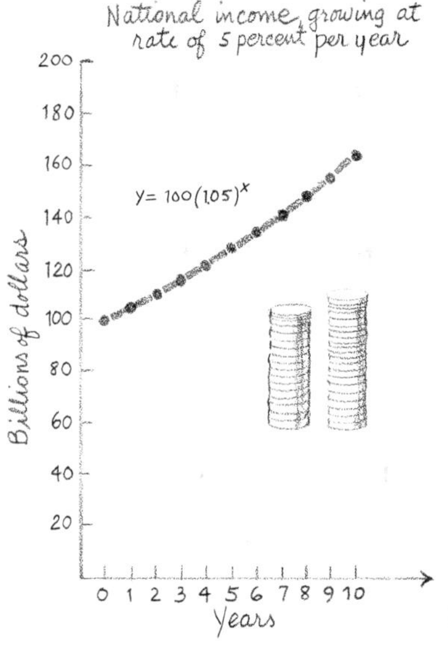

Figure 4.36

EXAMPLE 3

Suppose that the radius of the cross-section of a tree trunk increases by 20 percent every year. The present radius is 0.5 inches. The radius after x years is given by formula (1) with $c = 0.5$ and $p = 0.2$:

$$\text{Trunk radius} = 0.5(1.2)^x.$$

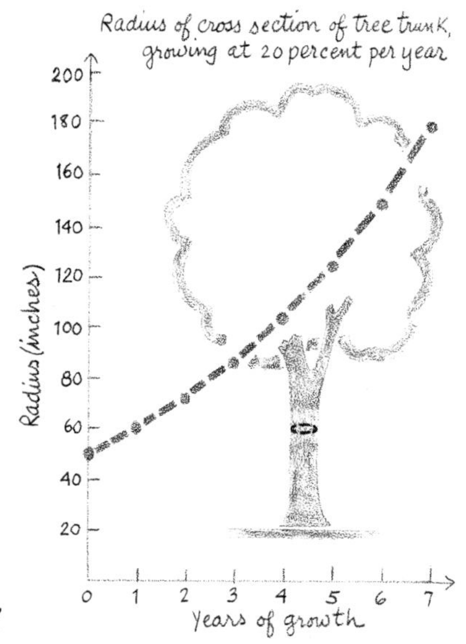

Figure 4.37

The values for the first 7 years are given in the table below, and the graph is in Figure 4.37:

x	0	1	2	3	4	5	6	7
$0.5(1.2)^x$	0.50	0.60	0.72	0.86	1.04	1.24	1.49	1.79 ▲

EXAMPLE 4

In 1790 the population of the United States was approximately 3.9 million. It increased on the average by 35 percent per decade for the first 7 decades. According to formula (1), the population after x decades can also be described by the formula

$$3.9(1.35)^x.$$

In the table below, we compare the actual census figures (*Source:* Bureau of the Census) for the population (in millions) with the values of the exponential function given above. Note that the two sets of figures are practically identical:

Year	x	Census data	$3.9(1.35)^x$
1790	0	3.9	3.9
1800	1	5.3	5.3
1810	2	7.2	7.1
1820	3	9.6	9.6
1830	4	12.9	13.0
1840	5	17.1	17.5
1850	6	23.2	23.6
1860	7	31.4	31.9

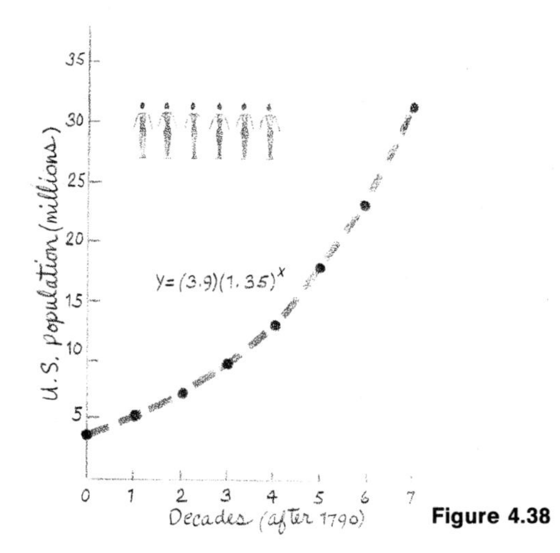

Figure 4.38

The function is plotted in Figure 4.38. ▲

The exponential function also occurs in applications involving decay; that is, negative growth. Suppose that a population which initially consists of c members decreases by $100p$ percent every time unit. After 1 time unit the population size is $c - cp = c(1 - p)$. After 2 time units it is

$$c(1-p) - c(1-p)p = c(1-p)^2.$$

Similarly, after x time units the size is

$$c(1-p)^x, \tag{2}$$

which is the same as (1) except that the sign preceding p has been changed to negative. This corresponds to the exponential function cb^x where $0 < b < 1$.

EXAMPLE 5

Suppose we take a glass of whiskey, and water it down in the following manner. Pour out $\frac{1}{5}$ of the pure whiskey, replace it with water, and stir it to a uniform mixture. Then pour out $\frac{1}{5}$ of the resulting mixture, replace it with water, and stir; and so on. The proportion of pure whiskey remaining in the mixture after x waterings is given by formula (2) with $c = 1$ and $p = 0.2$:

$$\text{Proportion whiskey} = (0.8)^x.$$

The proportions for values $x = 0, 1, \ldots, 8$ are given in the table below, and the function is plotted in Figure 4.39.

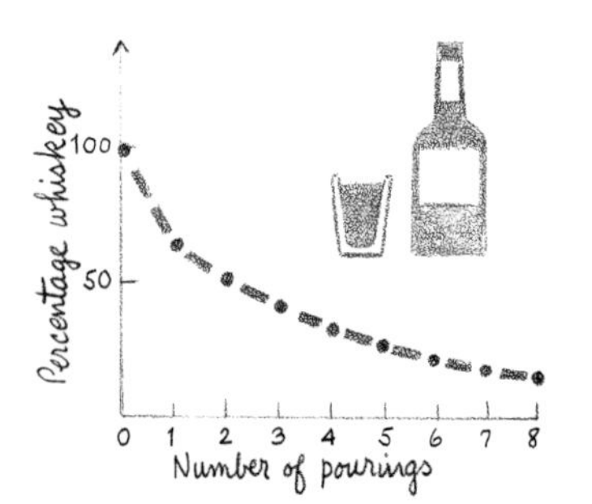

Figure 4.39

x	0	1	2	3	4	5	6	7	8
$(0.8)^x$	0.8	0.64	0.51	0.41	0.33	0.26	0.21	0.17	0.15 ▲

In the examples above the growth and decay were measured at equally spaced time points. In many examples of practical interest, these changes occur continuously over time. Here a natural standard base arises for the exponential function: It is a particular irrational number called e which lies between 2 and 3. Its value to 5 decimal places is

$$e = 2.71828. . . .$$

The growth of a quantity of money drawing compound interest at a fixed rate is typical of many growth processes: the growth is continuous over time and the rate of growth at any time is proportional to the attained size of the growing quantity. Let us explain the meaning of **continuously compounded interest.** If 100 percent annual interest is paid on \$1 and it is compounded twice per year, then the sum of the principle and interest at the end of 1 year is the same as that resulting from ordinary 50 percent interest for 2 years. By formula (1) this is

$$(1 + \tfrac{1}{2})^2.$$

If the compounding is done 4 times per year the interest paid at the end of the year is equal to that of ordinary 25 percent interest for 4 years:

$$(1 + \tfrac{1}{4})^4.$$

In general, if the compounding is done m times per year, then the sum of the principle and interest is

$$\left(1 + \frac{1}{m}\right)^m.$$

The values for various numbers m are

m	1	2	3	4	5	6	100
$(1 + 1/m)^m$	2	2.25	2.37	2.44	2.49	2.52	2.70

If we let m "become infinite," then the compounding becomes a continuous process; and the total value of the principle plus interest gets closer and closer to the number e mentioned above. Thus, e represents the value of 1 monetary unit compounded continuously for 1 year at an annual rate of 100 percent. In a rigorous study of calculus it is defined as the "limit" of $(1 + 1/m)^m$ as "m becomes infinite"; it is denoted

$$e = \lim_{m\to\infty} \left(1 + \frac{1}{m}\right)^m.$$

The sum of the principle and interest is modified when the period of continuous compounding is not exactly in whole years, and when the annual interest rate is not 100 percent. If the annual rate is $100k$ percent per year and the compounding period is t years, not necessarily an integer, then the sum of the principle and interest compounded m times per year is

$$\left(1+\frac{k}{m}\right)^{mt}.$$

(It is like ordinary $100k/m$ percent interest for a period of mt years.) As "m becomes infinite," this approaches the value e^{kt}. The total value for c dollars is

$$ce^{kt}. \qquad (3)$$

This is the formula for the size of a quantity growing for t years like continuously compounded interest paid at an annual rate of $100k$ percent per year. The number c is the initial size and the number k is called the **nominal** rate of growth.

The definition of continuous decay or dilution is similar. If c units decay continuously at an annual rate of $100k$ percent per year for t years, then the quantity left after t years is

$$ce^{-kt}. \qquad (4)$$

This differs from (3) by the sign in the exponent; note that (2) differed from (1) by the sign of p.

A table of values of the exponential functions e^x and e^{-x} are in Table II of the Appendix.

EXAMPLE 6

If \$1 is compounded continuously for t years at an annual rate of 6 percent per year, then its final value is $e^{0.06t}$. Values of this function for various t can be obtained from Table II (Appendix):

t	0.5	1.0	2.0	3.0	4.0
$x = 0.06t$	0.03	0.06	0.12	0.18	0.24
e^x	1.0305	1.0618	1.1275	1.1972	1.2712

In problems where money is replaced by a growing population or organism, the "annual interest rate" is called the **nominal rate of growth;** for example, the size of a population growing continuously at a nominal rate of 4 percent per year is $e^{0.04t}$ after t years. ▲

EXAMPLE 7

One way to start cleaning latex paint from a brush is to hold it under running warm water. Suppose that the amount of paint on the brush decreases continuously at a nominal rate of 25 percent per minute. If the brush originally has 3 ounces of paint on it, then, by formula (4), after t minutes in the water it has

$$3e^{-0.25t} \text{ ounces}$$

left on it. The graph of this function appears in Figure 4.40; in particular, for $t = 1$, 2, and 5 minutes, the amounts left are 2.33640, 1.81959, and 0.85950, respectively. ▲

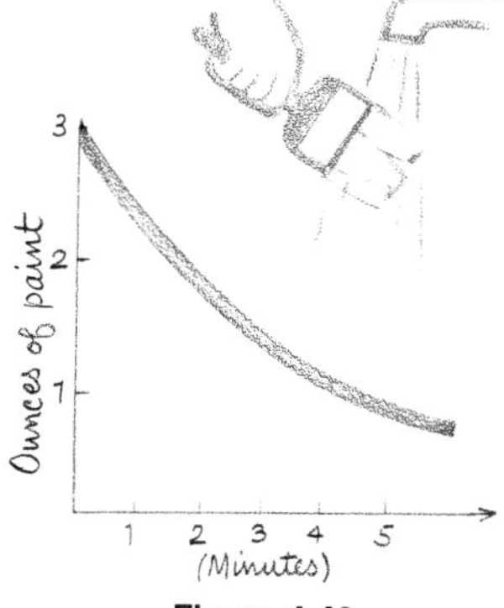

Figure 4.40

4.3 EXERCISES

1 Evaluate:
(a) $(1+\frac{1}{6})^3$ (b) $(1+\frac{3}{4})^4$ (c) $(1-\frac{1}{4})^4$
How do these compare with $e^{1/2}$, e^3, and e^{-1}, respectively?

2 Evaluate:
(a) $(1+\frac{1}{8})^4$ (b) $(1-\frac{1}{9})^3$ (c) $(1+\frac{1}{5})^5$
Compare these to the values of $e^{0.5}$, $e^{-1/3}$, and e, respectively.

3 The assets of a company grow at an annual rate of 10 percent. If it started with \$10 million, what is the value after $t = 1, 2, 3, 4$ years? (See Example 2.)

4 A roach-infested house is sprayed daily. Each spraying reduces the roach population by 15 percent. What percentages of the roaches remain after 3, 4, and 5 days? (See Example 5.)

5 The number of crimes in a city increase by 8 percent per year. If 50,000 crimes were reported for 1971, how many were there in 1974? (See Example 2.)

6 The population of the United States increased by approximately 26 percent per decade for the 3 decades beginning with 1860. The population that year was 31.4 million. Using formula (1), compute the population size for 1890. (The actual population then was 62.9 million.) (See Example 4.)

7 An unfortunate political candidate loses 20 percent of his supporters after each major speech. If he starts with 1 million enthusiasts, how many will be left after the fifth oration? (See Example 5.)

8 The population of a state is 2 million. If it increases by 5 percent annually, what will be the population after 5 years? What if the growth is considered to be continuous, at a nominal annual rate of 5 percent? (See Examples 2 and 5.)

9 The number of bacteria in a culture increases continuously at a nominal rate of 25 percent per day. The original culture had 6000 bacteria. What is the formula for the number present after t days? Evaluate for $t = 2$, 5, and 8. (See Example 6.)

10 Upon reaching his fiftieth birthday, a man is told that he will continuously lose his hair at a nominal rate of 5 percent per year. What percentage of his hair will remain at his 65th birthday? (See Example 7.)

11 A tank containing a solution of water and alcohol is simultaneously drained and refilled with pure water. This is done by a continuous process. The original solution contained 70 percent alcohol. It decreases at a nominal rate of 15 percent per minute. What is the remaining percentage of alcohol after $t = 1, 2, 3, 4$ minutes? (See Example 7.)

12 The population of a town is continuously growing at an annual nominal rate of 3 percent. If, on January 1, there were 40,325 residents, approximately how many were there on September 1 of the same year? (See Example 6.)

13 The oil necessary to heat a certain apartment house in the winter continuously increases at a nominal rate of 2 percent per degree for every degree of heat maintained in the house above 65°F. If 50 gallons per day are needed to maintain a temperature of 65°F, how many gallons are needed to keep a temperature of 70°F; 72°F? (See Example 6.)

14 The age-specific death rate in a population tends to increase exponentially with the age group. In the table below the rates per 1000 U.S. population, 1963, for the various age groups are given (*Source:* U.S. Department of Health, Education and Welfare):

Age group	Rate	Age group	Rate
5–14	0.4	45–54	7.5
15–24	1.1	55–64	17.3
25–34	1.5	65–74	38.8
35–44	3.0	75–84	85.2

(a) Plot these on a graph, using the midyear of the age group as the abscissa, and the rate as the ordinate; for example, for the group 35–44, we put $x = 40$ and $y = 3.0$.

(b) Let x represent the number of decades after the year 10, that is, $x = 0, 1, 2, \ldots$ correspond to ages 10, 20, 30, . . . , respectively. Compare the values of the function

$$y = \tfrac{1}{2}(2^x - 1)$$

with the corresponding values of the rates above. Draw the graph of this function on the same sheet as that in (a).

4.4 The Logarithm Function

The logarithm function is obtained from the exponential by interchanging the roles of the independent and dependent variables. In the exponential function b^x, x is the independent variable, and $y = b^x$ the corresponding value of the dependent variable. The logarithm function is defined, when $b > 0$ and $b \neq 1$, by the reversed relation between the variables: $x = b^y$. In more advanced mathematics it can be shown that for every positive number x there exists a real number y satisfying the latter equation. The exponent y is called the **logarithm of x to the base b,** and is denoted

$$y = \log_b x.$$

This equation expresses the same relation between x and y as the equation $x = b^y$.

The existence of a number y corresponding to the given x can also be inferred from the graph of the exponential function: For every positive ordinate value y on the graph of the exponential, there exists a unique abscissa value x such that $y = b^x$ (Figure 4.41). The logarithm function is then obtained by interchanging the labels x and y.

By definition, the graph of the logarithm function is the set of points (x,y) in the plane such that $y = \log_b x$, or, equivalently, $x = b^y$. Therefore, the set is obtained from that of the exponential function simply by interchanging the ordinate and the abscissa. This is equivalent to the graph obtained from the exponential by reflecting the latter through the diagonal line $y = x$: the graph of the logarithm is the mirror image of the graph of the exponential (Figure 4.42).

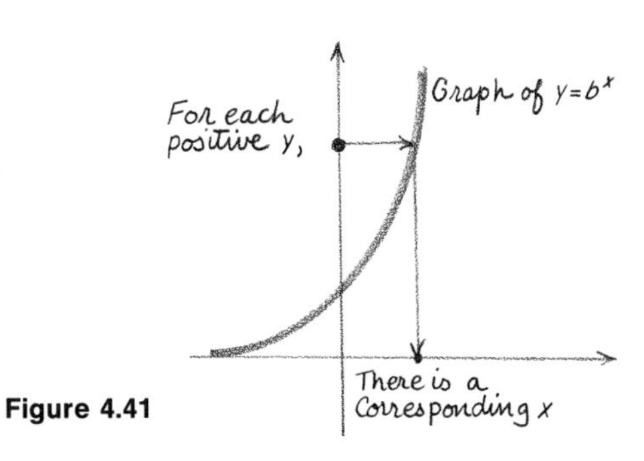

Figure 4.41

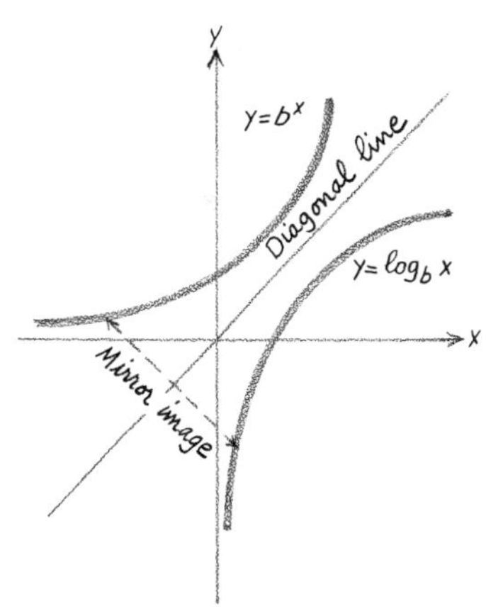

Figure 4.42

The fundamental properties of the logarithm are derived from the corresponding laws of exponents:

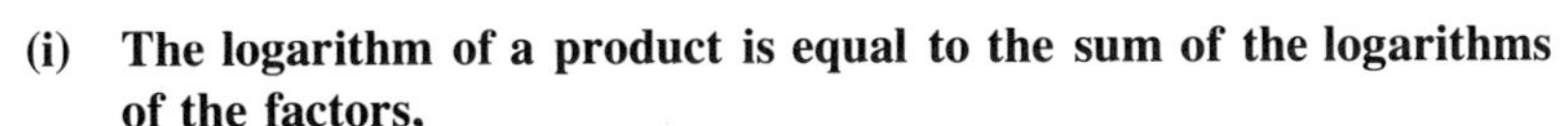

(i) The logarithm of a product is equal to the sum of the logarithms of the factors,

$$\log_b xy = \log_b x + \log_b y.$$

(ii) The logarithm of a quotient is equal to the logarithm of the numerator minus the logarithm of the denominator,

$$\log_b \left(\frac{x}{y}\right) = \log_b x - \log_b y.$$

(iii) The logarithm of the number 1 is 0,

$$\log_b 1 = 0.$$

(iv) The logarithm of the power of a number is equal to the power times the logarithm of the number,

$$\log_b x^p = p \log_b x.$$

The student is likely to be familiar with logarithms to the base $b = 10$; these are called **common** logarithms. The logarithms appearing throughout this book are those to the base e, where e is the number defined in Section 4.3. The reason for this is the importance of the number e in the calculus. Instead of writing $\log_e x$, we shall omit the subscript e with the understanding that e is the base. Many other texts use $\log x$ to stand for the common logarithm. The base e logarithm is called the **natural** logarithm; other texts use the notation **ln** x for this logarithm.

The natural logarithm function is useful in problems of growth and decay when the nominal rate is unknown, but has to be found from the given information. The values of the natural logarithm are given in Table III of the Appendix.

EXAMPLE 1

A culture of bacteria grows like continuously compounded interest. If the colony increases from 5000 to 10,000 cells in 3 days, what will be the size in 5 days? According to formula (3) of Section 4.3, the size of the culture after t days is

$$y = 5000\, e^{kt}, \tag{1}$$

where k is the unknown nominal rate of growth. We have to find k. Since the size is 10,000 after 3 days, we substitute $t = 3$ and $y = 10{,}000$ in Equation (1) and solve for k:

$$10{,}000 = 5000\, e^{3k} \qquad \text{or} \qquad 2 = e^{3k}.$$

This equation is the same as

$$3k = \log 2 \qquad \text{or} \qquad k = \tfrac{1}{3} \log 2.$$

From Table III we get $\log 2 = 0.6931$; hence,

$$k = (\tfrac{1}{3})(0.6931) = 0.2310.$$

It follows that the formula for the size of the culture after t days is

$$y = 5000\, e^{0.2310t}.$$

In particular, after 5 days, the size is

$$y = 5000\ e^{1.15} = 5000(3.16) = 15{,}800 \qquad \text{(approximately)}. \ \blacktriangle$$

EXAMPLE 2

If a quantity grows like compound interest at a nominal rate of k, then the time it takes for the population to double in size is given by the formula (Exercise 11)

$$t = \left(\frac{1}{k}\right) \log 2 = \left(\frac{1}{k}\right) 0.6931. \tag{2}$$

The doubling time is inversely proportional to the nominal rate of growth. Suppose the radius of a tree grows continuously, and increases from 1 to 3 inches in 10 years. How long will it take to grow to 6 inches? First we find the nominal rate of growth. Since the radius grows from 1 to 3 inches in 10 years, we have

$$3 = (1)e^{10k}, \qquad \text{or} \qquad \log 3 = 10k,$$

$$\text{or} \qquad k = 0.1 \log 3 = 0.10986.$$

It follows from (2) that the number of years after the tenth to double is

$$\frac{0.6931}{0.10986} = 6.309 \qquad \text{(approximately)}. \ \blacktriangle$$

EXAMPLE 3

The atmospheric pressure (in pounds per square inch) decreases continuously relative to the altitude above sea level: it falls at a constant nominal rate. The pressure at sea level is 14.7 pounds per square inch; at 3000 feet it is 13.0. What is the pressure at 1000 feet? We start with the decay equation

$$y = 14.7\ e^{-kt},$$

and substitute $t = 3000$ and $y = 13.0$; then, by property (ii) of the logarithm function,

$$\log 13.0 = \log 14.7 - 3000k.$$

In order to use the table for these logarithms, we have to convert them to logarithms of numbers between 1 and 10. Write

$$\log 13.0 = \log(1.3 \times 10) = \log 1.30 + \log 10,$$

$$\log 14.7 = \log(1.47 \times 10) = \log 1.47 + \log 10.$$

Then the equation for k becomes

$$\log 1.30 = \log 1.47 - 3000k,$$

which by Table III is equivalent to

$$0.2624 = 0.3853 - 3000k.$$

This linear equation is solved for k: $k = 0.0410 \times 10^{-3}$; thus, the decay equation takes the form

$$y = 14.7e^{-(0.0410\times 10^{-3})t}.$$

If $t = 1000$, then $y = 14.7e^{-0.0410} = 14.7(0.9608) = 14.1$ (approximately). We conclude that the pressure at 1000 feet is 14.1 pounds. ▲

4.4 EXERCISES

1 The population of a country increased from 10 to 10.4 million in 2 years. If the growth is like compound interest, how long will it take for the population to reach 18 million? (See Example 1.)

2 If the number of bacteria in a culture triples in 48 hours, how long does it take for the number to double? (See Example 2.)

3 After exposure to a toxic substance, all the members of a group of experimental mice eventually show harmful reactions. The proportion of unaffected mice declines at a constant rate. Suppose that half the group showed no reaction by the time 3 days had passed. How long will it take for 90 percent of the original group to react? (See Example 3.)

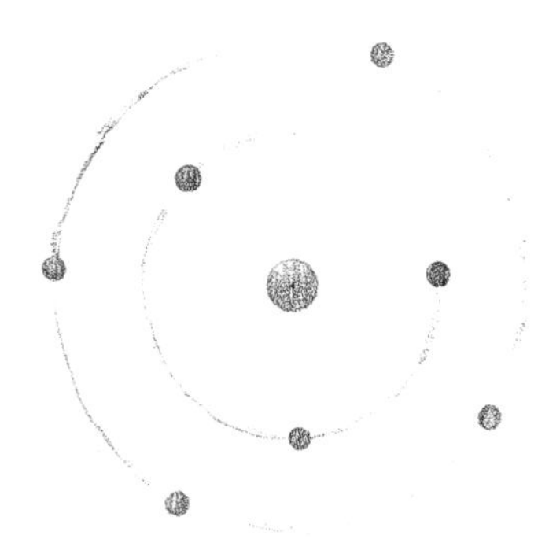

Figure 4.43

4 Between April 1, 1930 and April 1, 1940, the population of the United States grew from 123 million to 132 million. If the growth was like continuously compounded interest, when did the population reach 130 million? (See Example 1.)

5 A radioactive substance decays at a constant rate. The **half-life** is the number of years it takes for half the substance to decay. If 1 gram decays to 0.8 in 100 years, what is the half-life of the substance? (Figure 4.43.) (See Example 3.)

6 Dead organic objects are dated in the following way. A living organism contains radioactive and stable (nonradioactive) carbon in a fixed ratio. When the organism dies, the radioactive carbon decays at a constant rate. Radioactive carbon has a half-life of 550 years. The proportion of radioactive carbon in the organism is measured; from this the time since death is inferred. If a piece of charcoal has only 60 percent of its original radioactive carbon, how many years ago did the tree, from which it came, die? (See Example 3.)

7 The proportion of television sets of a given year and model in operation declines by a continuous process of decay at a constant rate. If 40 percent of the sets are still in operation after 6 years, how long will it take for only 10 percent to be in use? (See Example 3.)

8 A short time after a drug is introduced into the body, it is carried by the blood to various organs and reaches a certain level of concentration. After that the body dilutes the solution by a continuous process, and the level falls at a constant rate. If 50 percent of the drug content of a particular organ is gone after 10 hours, how much longer will it take for 90 percent to be eliminated? (Figure 4.44) (See Example 3).

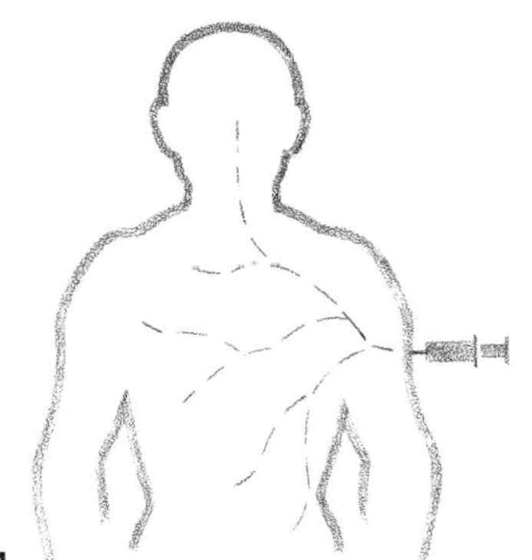

Figure 4.44

9 If the population of a fishtank doubles in 6 months, how long will it take to quadruple? (See Example 2.)

10 Shortly after the abolition of graduate school deferrments in 1969, the number of full-time mathematics graduate students declined by 8 percent per year for 2 successive years. If we assume a continuous process of decay at a nominal rate of 8 percent per year, when will the number drop to 30 percent of the original number? (See Example 3.)

11 Verify formula (2) for the doubling time in a continuous compound interest growth process. [*Hint:* Why must T satisfy $2 = e^{kT}$?]

12 Let T be the half-life of a substance decaying continuously at the nominal rate of $100k$ percent. Show that T is also equal to the expression in formula (2).

13 Prove the following generalization of formula (2): The time it takes for the population to multiply by a factor m is $T = (1/k) \log m$.

14 When \$1 is invested at $100k$ percent annual interest compounded continuously, it grows to e^{kt} dollars in t years. Suppose a man invests \$1 on January 1 of each year. Find the formula for the total value of the investment on December 31, m years later. [*Hint:* Use the well-known formula for the sum of a finite geometric series: $x + x^2 + \cdots + x^m = x(x^m - 1)/(x - 1)$.] (See Example 1.)

15 Using the formula obtained in Exercise 14, find the number of years it will take for the investment to grow to \$50 if the nominal interest rate is 6 percent.

4.5 Motion, Velocity, and Acceleration

Figure 4.45

Sir Isaac Newton was a famous English scientist and mathematician of the seventeenth century. The following is usually called Newton's First Law of Motion:

I. **If an object is at rest, then it will remain at rest unless a force acts on it.**

II. **If an object is in motion, and if there is no force acting on it, then it will continue in motion, moving in a straight line and at constant speed.**

The first part of the law is simple. However, the second part is unusual because on our earth there are no situations in which an object moves without a force acting on it. There are always the forces of ground friction and air resistance (Figure 4.45). But whenever these forces are relatively small, then the motion of the moving object closely follows what is described in the second part of the law; in other words, the latter part describes an ideal situation which is never quite attained. For example, the motion of a baseball after leaving the pitcher's hand is, for a few feet, nearly uniform in speed and straight in direction: air friction and gravity have not had sufficient time to bend its course and slow its speed (Figure 4.46). Another example is the motion of a space missile on its way to the moon, far from the effective range of the fields of gravitation of the earth and the moon (Figure 4.47).

Figure 4.46

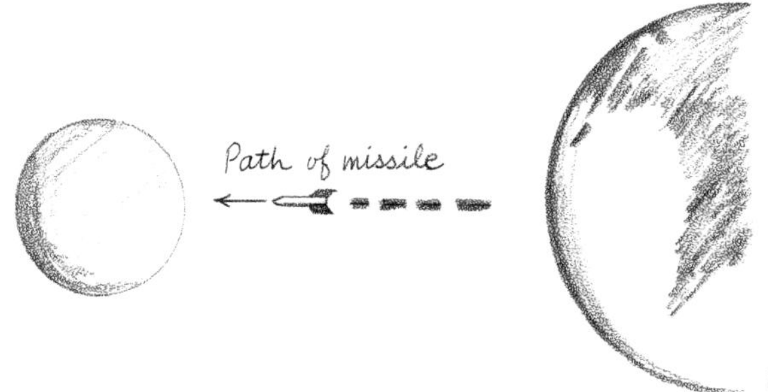

Figure 4.47

Problems of force, speed, and motion led Sir Isaac Newton and his German contemporary, Gottfried Leibniz, to invention of calculus. While such problems of physics are not of direct relevance to those specializing in the biological, behavioral, social, and business sciences, they are helpful in understanding the quantitative concepts of these sciences. Economists refer to the forces of **inflation** or the forces of **recession.** We may think of the level of an economic indicator as the path of an object in motion, and the economic acts of the government or individual as a force acting on the moving object (Figure 4.48). Sociologists deal with forces of family decay (Example 7 below). The medical treatment of an infection is a force acting on the moving level of fever in a patient.

Figure 4.48

The **speed** of an object moving uniformly and on a straight line is defined as the distance traveled divided by the time taken; it is measured in miles per hour (mph) or feet per second (ft/sec). A force acting on a moving object changes the speed or the direction, or both. The rate of change of the speed is called the **acceleration** of the force; it is measured in mph per hour (mph/hr) or ft/sec per second (ft/sec^2). When the force increases the speed, the acceleration is positive; when it decreases the speed, the acceleration is negative. (Negative acceleration is often called **deceleration.**) For example, when the speed of an object changes from 30 ft/sec to 50 ft/sec in 4 seconds, then the average acceleration over this time period is 5 ft/sec^2: the change in speed is divided by the change in the time. If the speed changes in the other direction, from 50 to 30 ft/sec, then the acceleration is -5 ft/sec^2.

A continuous force acting a moving object is conceived in a manner similar to that of continuous compound interest (Section 4.3). There the nominal interest rate was divided into many equal "pieces" and distributed over the year; then the number of pieces was increased indefinitely. We shall do the same for a force: we break up a quantity of acceleration and distribute it at equally spaced time points. Let the quantity of acceleration be proportional to the length of the interval considered: the acceleration for an interval of length t is denoted gt, where g is a fixed number called the **acceleration constant.** The interval from 0 to t is marked with the $m+1$ equally spaced points $0, t/m, 2t/m, \ldots, t$; here m is an arbitrary positive integer. The acceleration is divided into m equal parts gt/m and applied to the object at the first m time points. The speed increases by gt/m at each point (Figure 4.49). The distance covered in each time interval is the product of the speed in that interval and the length of the time interval: $(gt/m)(t/m)$ in the first interval, $(gt/m)(2t/m)$ in the second, . . . , $(gt/m)(t)$ in the mth interval. It follows that the total distance traveled is the sum (Figure 4.49)

$$\frac{gt}{m}\left(\frac{t}{m}+\frac{2t}{m}+\cdots+\frac{mt}{m}\right)=\frac{gt^2(1+2+\cdots+m)}{m^2}.$$

By the formula for the sum of the first m integers [formula (1), Section 4.1], the expression on the right-hand side above is

$$\frac{gt^2\, m(m+1)}{m^2}.$$

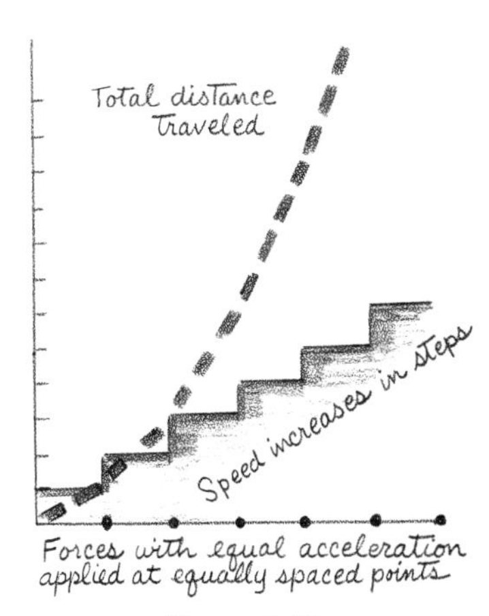

Figure 4.49

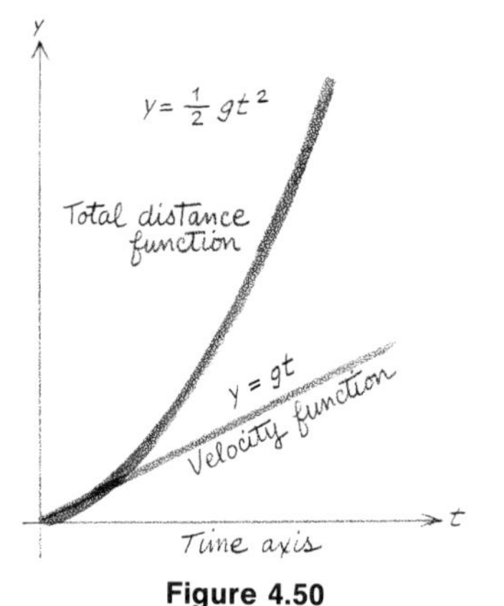

Figure 4.50

This is approximately equal to $\frac{1}{2}gt^2$ when m is very large because the factor $m(m+1)/m^2$ is approximately 1. It follows that as we let m increase indefinitely, that is, we distribute the acceleration more and more thinly over the time interval, the speed at time t is

$$gt \tag{1}$$

and the distance covered up to time t is

$$\tfrac{1}{2}gt^2. \tag{2}$$

The "staircase function" of the speed in Figure 4.49 is smoothed into the straight line $y = gt$ in Figure 4.50; and the "piecewise linear" distance function in Figure 4.49 becomes the smooth parabola $y = \frac{1}{2}gt^2$ in Figure 4.50. This limiting form of the speed function is conventionally called the **velocity function;** note that it changes with time at a constant rate. Observe also that the distance covered in t time units varies directly as the square of the time.

EXAMPLE 1

If an object is released to fall from a height to the ground, then the constant force of gravity acts on the object during its flight to the ground. The constant of acceleration due to the force of gravity is known to be

$$g = 32 \text{ ft/sec}^2. \tag{3}$$

If we neglect the effect of air resistance, then the velocity of the object after t seconds in flight is

$$32 \text{ ft/sec} \tag{4}$$

and the distance fallen is

$$16t^2 \text{ ft} \tag{5}$$

Thus, the velocity is 32 after 1 second, 64 after 2 seconds, and so on. The distance fallen increases rapidly: 16 feet after 1 second, 64 feet after 2 seconds, 144 feet after 3 seconds, and so on. ▲

EXAMPLE 2

An object is thrown directly upward from the ground with an initial speed of 80 ft/sec. According to the second part of the law of motion, if there were no gravity, the object would travel forever in a straight line and at the same speed of 80 ft/sec. In reality the force of gravity pulls down the object at the same time; the height lost because of the force of gravity is, by formula (5), $16t^2$ ft in t seconds. The actual height of the object t seconds after being thrown up is the difference between what the height would have been in the absence of gravity and the height lost because of the gravity:

$$80t - 16t^2.$$

In general, the formula for the height of the object thrown up with an initial speed of v ft/sec is

$$vt - 16t^2 \qquad (6)$$

after t seconds in flight (Figure 4.51). Similarly, the velocity of the object after t seconds is the difference between what the velocity would have been in the absence of gravity—80 ft/sec—and the velocity lost because of the force of gravity, $32t$ ft/sec; hence, the velocity is

$$80 - 32t.$$

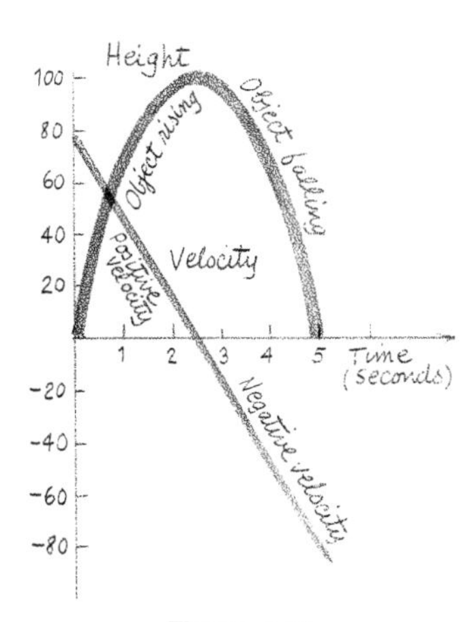

Figure 4.51

More generally, the formula for the velocity after t seconds in flight is

$$v - 32t. \qquad (7)$$

The graph of the function (6) is a parabola. The velocity is a positive quantity as the object rises, and negative as it falls (Figure 4.51). ▲

EXAMPLE 3

When a buoyant object is placed below the surface of the water, the force of buoyancy causes the object to accelerate upward. Suppose that after diving into a pool, a swimmer floats (without self-propulsion) to the surface from a depth of 8 feet in 3 seconds (Figure 4.52). What is the constant of acceleration? How long would it take the swimmer to rise from 15 feet?

Let g be the unknown constant of acceleration. Since the distance traveled is 8 feet and the time is 3 seconds, we substitute these values in formula (2) and solve for g:

$$8 = \tfrac{1}{2}g(3)^2 \qquad \text{or} \qquad g = \frac{16}{9} \text{ ft/sec}^2$$

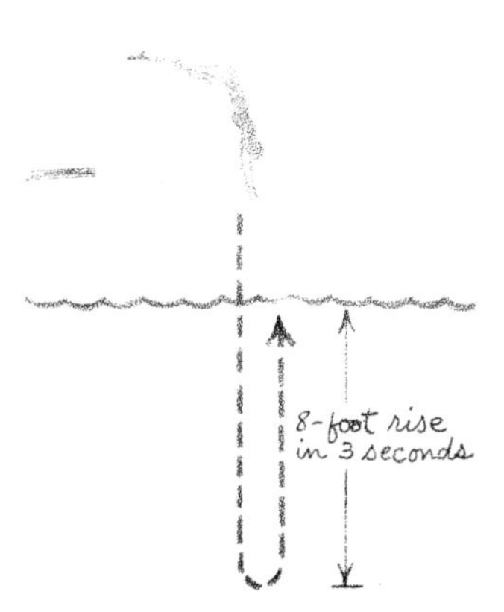

Figure 4.52

In order to find the time it takes to rise to the surface from 15 feet, we substitute this value of g in formula (2) and equate the latter to 15:

$$15 = \tfrac{1}{2}\left(\frac{16}{9}\right) t^2.$$

Solve this for t, using the table of square roots (Table I of the Appendix):

$$t = \sqrt{\frac{(15)(9)}{8}} = \left(\frac{3}{4}\right)\sqrt{30} = 4.1. \ \blacktriangle$$

Now we apply the notions of acceleration and velocity to examples in the nonphysical sciences. There are quantities in these areas which "snowball" with time, that is, grow with increasing velocity. This motion can often be explained in terms of continuous social or economic forces. Although these forces are usually not constant over long time periods, they often are fairly constant over short periods.

EXAMPLE 4

The rate of divorce in the United States increased during World War II. We may think of the disruptive force of war, with its sudden separations of husbands and wives, as accelerating the rate of family dissolution. In the table below the divorce rates for the United States for 1939–1946 are given; the figures are numbers of divorces per 1000 married women of age 15 and over (*Source:* U.S. Department of Health, Education and Welfare):

Year	1939 Rate	Increase over 1939
1939	8.5	0
1940	8.8	0.3
1941	9.4	0.9
1942	10.1	1.6
1943	11.0	3.5
1944	12.0	4.5
1945	14.4	5.9
1946	17.9	9.4

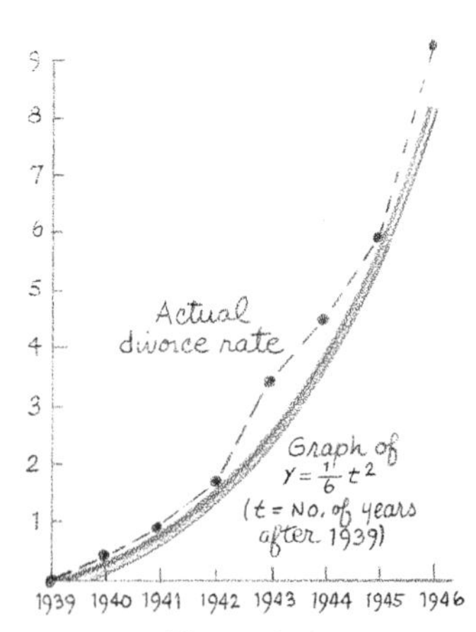

Figure 4.53

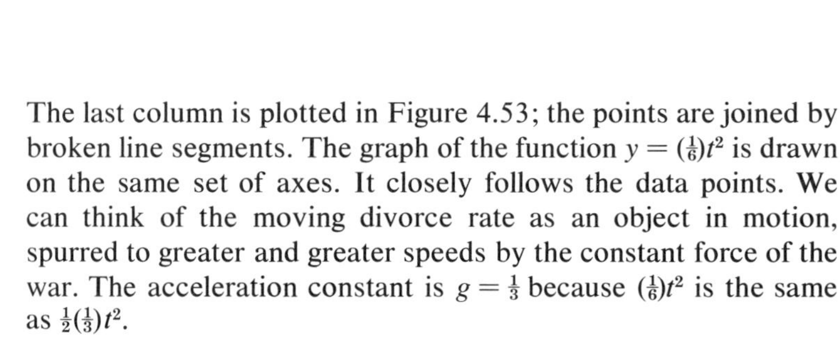

The last column is plotted in Figure 4.53; the points are joined by broken line segments. The graph of the function $y = (\frac{1}{6})t^2$ is drawn on the same set of axes. It closely follows the data points. We can think of the moving divorce rate as an object in motion, spurred to greater and greater speeds by the constant force of the war. The acceleration constant is $g = \frac{1}{3}$ because $(\frac{1}{6})t^2$ is the same as $\frac{1}{2}(\frac{1}{3})t^2$.

What would have been the rate of divorce in 1947 if the war had continued? According to our theoretical formula, it would have been

$$(\tfrac{1}{6})8^2 = 10.7$$

points higher than in 1939, namely, 19.2 per 1000 married women. ▲

4.5 EXERCISES

1 A stone is dropped from a skyscraper 400 feet above the ground. How many seconds will it take to reach the ground? At that time, what will be its velocity in ft/sec? (See Example 1.)

2 A bomb is dropped from a plane flying at an altitude of 2000 feet. How many seconds does it take for the bomb to hit the ground? (See Example 1.)

3 The velocity of an object thrown upward from the ground is equal to 0 at the moment when it attains the high point of its path (Figure 4.51). If it is thrown up with an initial velocity of 96 ft/sec, when (in seconds) will it reach the high point? What is the height at that time? (See Example 2.)

4 Using formulas (6) and (7), prove: If the object is sent up with an initial speed of v ft/sec, then the high point is attained after $v/32$ sec and the height at that time is $v^2/64$. (See Example 2.)

5 An object is thrown up with an initial speed of 100 ft/sec. After how many seconds will it return to the ground? [*Hint:* Put the expression for the height (6) equal to 0, and solve for t.] (See Example 2.)

6 Generalize the result of Exercise 5 to the case where the initial speed is an arbitrary number v; derive the formula for the time to return. (See Example 2.)

7 At the end of a 30-second jump, a parachutist lands with a speed of 25 mph. From what height did he jump? (Assume a constant force of downward acceleration, and determine it. Note 1 mile = 5280 feet.) (Figure 4.54.) (See Example 3.)

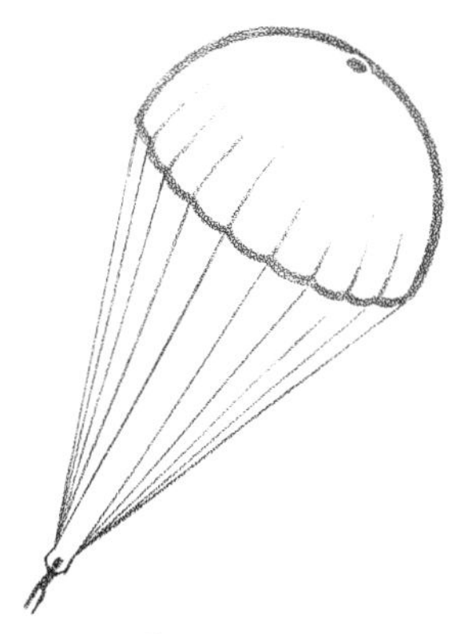

Figure 4.54

8 If the skydiver in Exercise 7 wanted to practice landing with a velocity of 25 mph **without** a parachute, from what height should he jump? (See Example 3.)

9 The wind acts like a continuous force on a relatively dense object, such as a baseball or a stone. As a ball moves in the wind the air molecules continuously pound it, causing a constant acceleration in the direction of the wind. If you have watched baseball fielders catching fly balls on a windy day, you must have noticed that the ball takes a sharp turn in the direction of the wind after spending several seconds in flight. Suppose a ball is in the air for 3 seconds and is displaced by the wind 18 feet from its original landing position. Find the constant of acceleration due to the wind. (See Example 3.)

Figure 4.55

10 A pitcher's curve ball follows a path with an increasing "bend" (Figure 4.55). This is caused by the spin that he puts on the ball: the constant force of the spin accelerates the ball in the direction of the spin. If the curve ball moves 2 feet off its straight line path in its 60-foot trip to the plate from the pitcher's mound, what is the constant of acceleration due to the spin? Let the independent and dependent variables be the distances traveled forward and laterally, respectively. (See Example 3.)

11 A motorist is traveling at a uniform rate of 40 mph. Suddenly he sees an obstruction on the road and applies the brakes. Suppose that this puts a constant negative acceleration on the moving car, equal to -15 mph/hr.

(a) What is the formula for the number of feet the car moved t seconds after the brakes were put on? [*Hint:* Convert mph to ft/sec by the relation 1 mile $=$ 5280 feet; then find the difference between the distance the car would have moved if the brakes had not been used, and the distance lost because the brakes were used.]

(b) What is the formula for the velocity of the car t seconds after the brakes were applied?

(c) Using the result in (b) find how long it takes for the car to come to a stop after the brakes are applied. [*Hint:* Find when the velocity is equal to 0.] (See Example 3.)

12 A car traveling on an expressway at 70 mph is brought to a stop in 400 feet by a constant negative acceleration. How many seconds does it take for the car to stop? (See Example 3.)

13 A series of good corporate earnings reports arrive at the floor of a stock exchange, acting like a constant continuous force on the stock price index. If the index gains 1.5 points in 1 hour of trading, how much should it gain, with the continued good news, after 2 hours? (See Example 3.)

14 One of the economic forces behind the recovery of the American economy from the depression of the 1930s was the demand for the construction of housing. Below are given the numbers of dwelling units (in thousands) constructed annually during the period 1936–1941 (*Source:* U.S. Department of Commerce):

1936	1937	1938	1939	1940	1941
319	336	406	515	603	715.

(World War II ended the rise in 1942.) Let x be the number of years after 1936, and y the number of units of housing in excess of 319.

(a) Plot the points (x,y).

(b) Suppose that the acceleration of the level of construction was a constant equal to 32,000 units per year, per year. What formula would describe the annual number of units above 319 for the given years? Compare the graph and values of this function with the given data. (See Example 4.)

15 An organism is attacked by a fever-inducing infection. If left unchecked, the fever will rise by 1°F every 3 hours until the organism dies. However, as soon as the infection begins, the antigens of the body produce a constant force which decelerates the level of fever. What rate of deceleration (negative acceleration) is required to halt the rise in fever in exactly 12 hours? (Deceleration is measured in degrees per hour per hour.) What is the maximum fever above normal? [*Hint:* Compare the infection to the object being thrown up, and the antigenic action to the force of gravity, as in Example 2, and Exercises 3 to 6.]

16 In Exercise 15, how long will the fever last?

4.6 Trigonometric Functions

The next three sections contain those elements of trigonometry needed for our study of calculus and its applications. These sections are not intended as an introduction to the main subject material of trigonometry, as many topics have been omitted. Our specific interest is in the trigonometric functions: their definitions, properties, and applications.

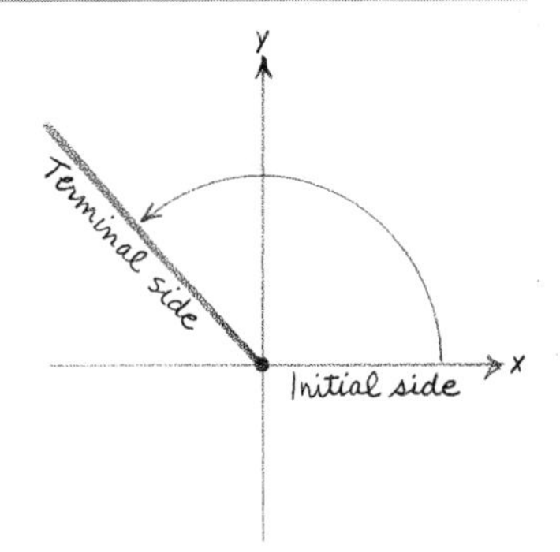

Figure 4.56

First we discuss the notion of an **angle,** which has been studied in high school plane geometry. Consider a half-line in a given position in the plane. Suppose that the "finite" endpoint is held fixed, and the half-line is rotated to a new position. This rotation generates an angle. The original position of the half-line is the **initial** side of the angle, and the new position is the **terminal** side. The angle is said to be in **standard position** if the vertex is at the origin and the initial side is along the positive x-axis (Figure 4.56). The angles to be considered will be in standard position.

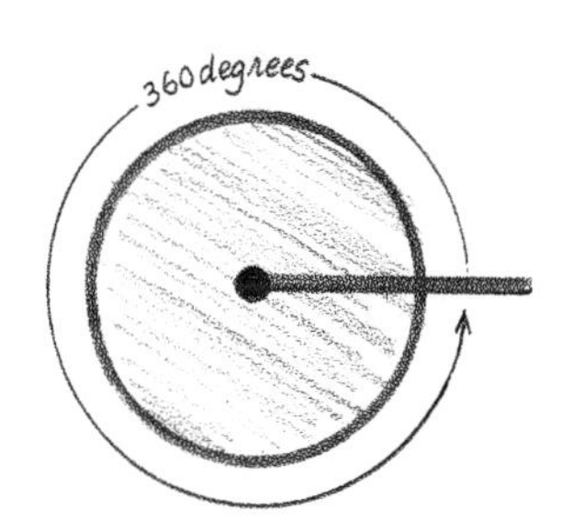

There are two systems of units for measuring angles. The first is in **degrees.** These are defined in terms of the circular arc generated by the angle. Let P be a point on the initial side. As the side is rotated into the terminal position, P sketches the arc of a circle. The angle is measured according to the length of this arc. The complete circle is divided into 360° (degrees). An angle formed by a complete counterclockwise rotation is defined to be an angle of 360°. In general, the number of degrees in an angle is defined to be the corresponding number in the circular arc generated by the counterclockwise rotation. For example, an angle formed by one-quarter of a complete rotation counterclockwise has 90° (Figure 4.57); an angle of one-half of a complete rotation has 180°; and an angle formed by two complete rotations has 720°.

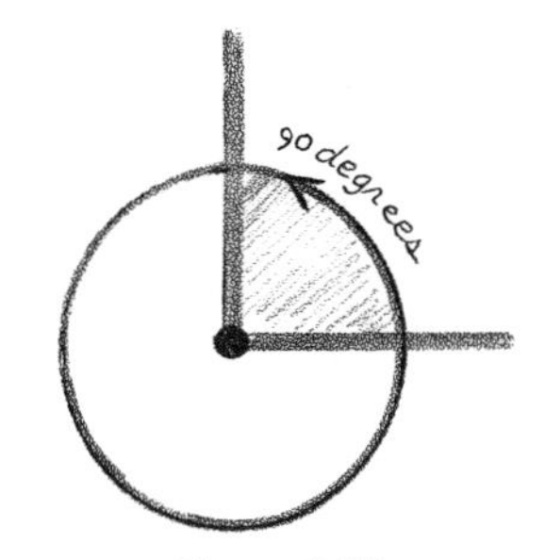

Figure 4.57

Clockwise rotations are measured in a similar way except that the angles are defined to be **negative.** For example, the angle gen-

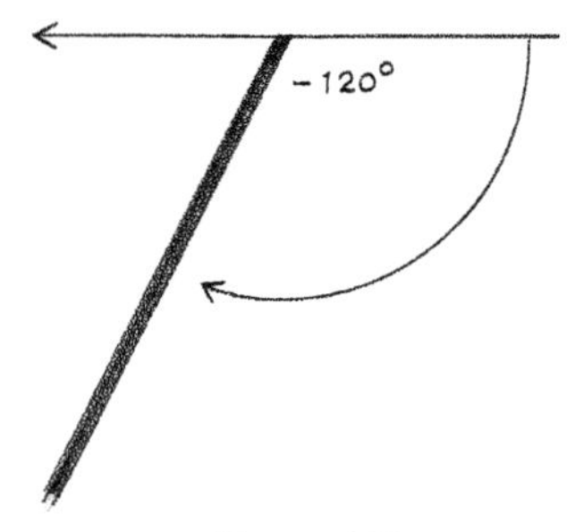

Figure 4.58

erated by one-third of a complete clockwise rotation has $-120°$ (Figure 4.58).

Another unit of angle measurement is the **radian.** In plane geometry we learned that the length of the circumference of a circle is π times the diameter, or, equivalently, 2π times the radius of the circle. Thus, we say that the circumference of the circle has the length "2π **radians**"; that is, the circumference is a 2π-multiple of the radius. This defines a unit for measuring the length of an arc of the circle: the latter is divided into 2π **radians. The radian measure of an angle sketched in the counterclockwise sense** is defined to be the length of the corresponding circular arc, measured in radians. For example, an angle of 360° corresponds to 2π radians, or approximately 6.2832 radians. The relation between degrees and radians is given by the equation

$$2\pi \text{ radians} = 360°,$$

or

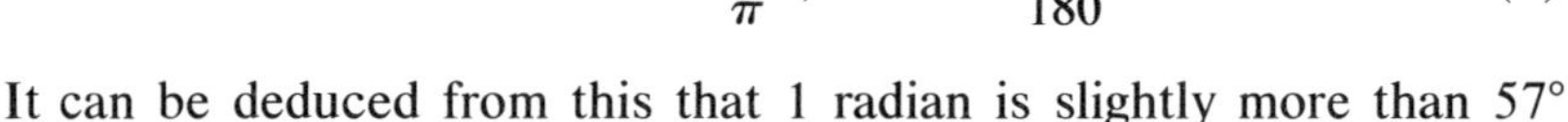

$$1 \text{ radian} = \frac{180°}{\pi}, \qquad 1° = \frac{\pi}{180}. \tag{1}$$

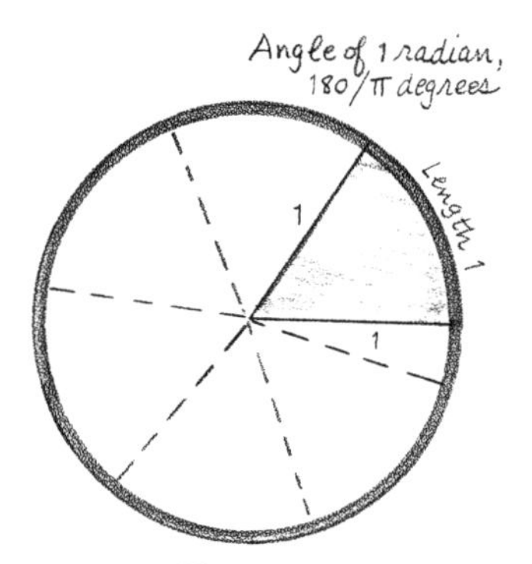

Figure 4.59

It can be deduced from this that 1 radian is slightly more than 57° (Figure 4.59).

It is customary to leave the factor π in angles measured in radians, and not to convert it to the approximate numerical value. Also, the angle is written simply as a multiple of π, and the word **radians** is left out; for example, an angle of 180° is written in terms of radians as π.

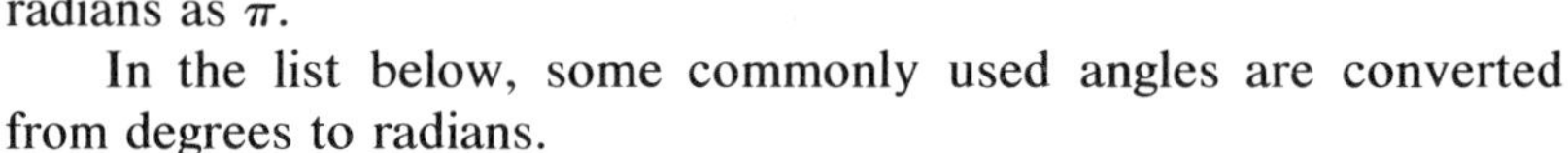

In the list below, some commonly used angles are converted from degrees to radians.

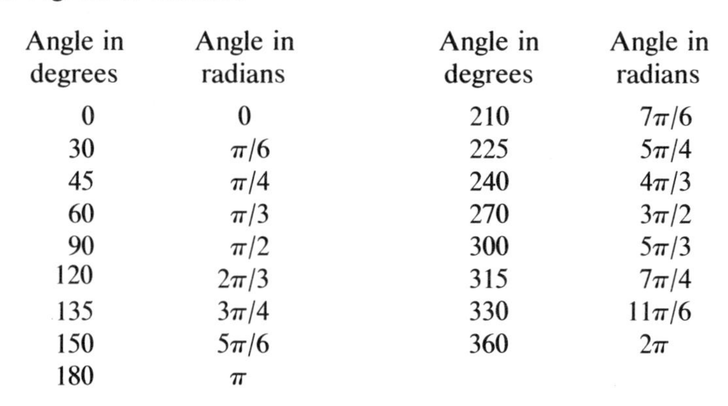

Angle in degrees	Angle in radians	Angle in degrees	Angle in radians
0	0	210	$7\pi/6$
30	$\pi/6$	225	$5\pi/4$
45	$\pi/4$	240	$4\pi/3$
60	$\pi/3$	270	$3\pi/2$
90	$\pi/2$	300	$5\pi/3$
120	$2\pi/3$	315	$7\pi/4$
135	$3\pi/4$	330	$11\pi/6$
150	$5\pi/6$	360	2π
180	π		

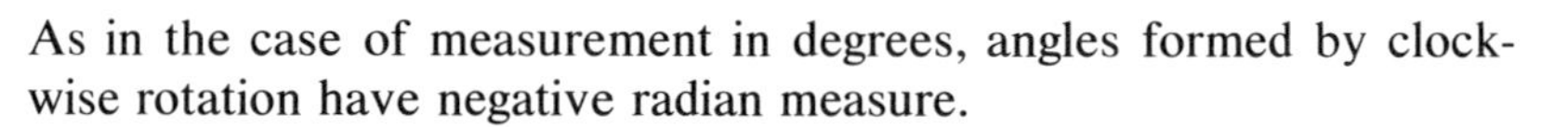

As in the case of measurement in degrees, angles formed by clockwise rotation have negative radian measure.

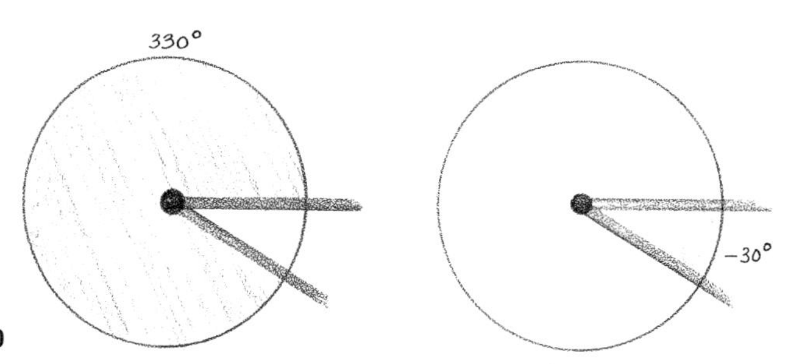

Figure 4.60

The position of the terminal side of an angle is called the **terminal position.** Every angle (in standard position) has exactly one terminal position; however, different angles may share the same terminal position. In particular, two angles that differ by 360° or an integer multiple of it (or by 2π or an integer multiple of it) have the same terminal position; such angles are **coterminal.** For example, an angle generated by $1\frac{1}{2}$ turns in the counterclockwise sense is coterminal with an angle of $\frac{1}{2}$ turn in either sense. Another example of a set of coterminal angles is: −30°, 330°, 690°, . . . (Figure 4.60).

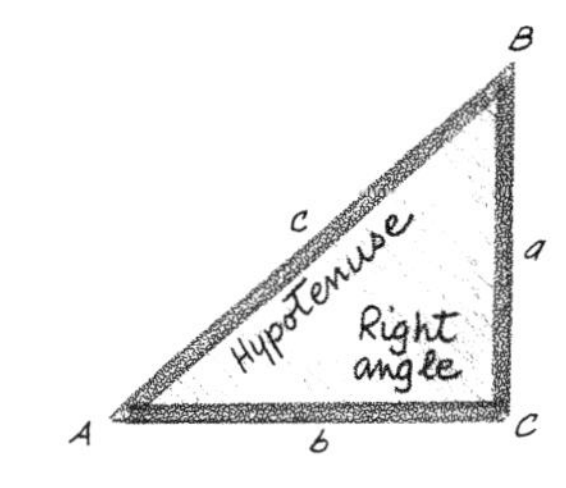

The **trigonometric** functions are certain functions whose domain is the set of values of angles, measured in degrees or radians. We begin with the classical definitions of these functions. Let ABC be a right triangle with right angle at C, and sides of lengths a, b, and c, respectively. The **sine** of angle A, denoted sin A, is the ratio of the opposite side a to the hypotenuse c. The **cosine** of angle A, denoted cos A, is the ratio of the adjacent side b to the hypotenuse c. The **tangent** of angle A, denoted tan A, is the ratio of the opposite side a to the adjacent side b (Figure 4.61). Thus,

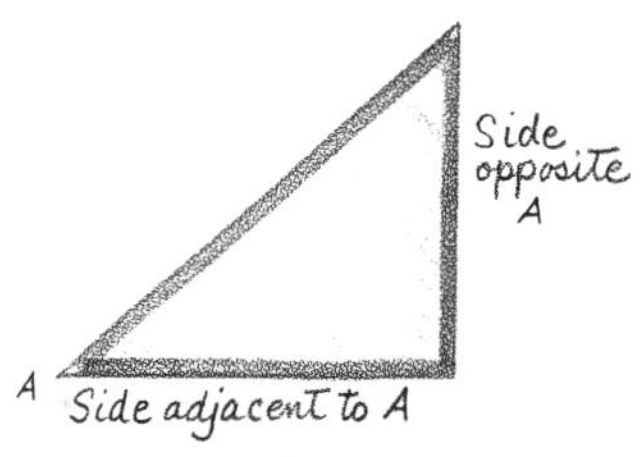

Figure 4.61

$$\sin A = \frac{a}{c}, \qquad \cos A = \frac{b}{c}, \qquad \tan A = \frac{a}{b}. \tag{2}$$

If another right triangle has sides proportional to the first triangle, then the two triangles are similar and the corresponding angles are equal. It follows that the sine, cosine, and tangent are the same for the corresponding angles of both triangles; therefore, these functions are independent of the particular triangles and depend only on the angles involved (Figure 4.62). For each angle A between 0° and 90° there is a well-defined number associated with it and is called sin A; thus, sin A represents a function with the independent variable A. Similarly, the cosine and tangent are functions of A. The three functions have the domain consisting of angles which, measured in degrees, are between 0° and 90°, or, measured in radians, are between 0 and $\pi/2$.

[*Note:* We omit the definitions of three other associated trigonometric functions: the secant, cosecant and cotangent; these will not be used here.]

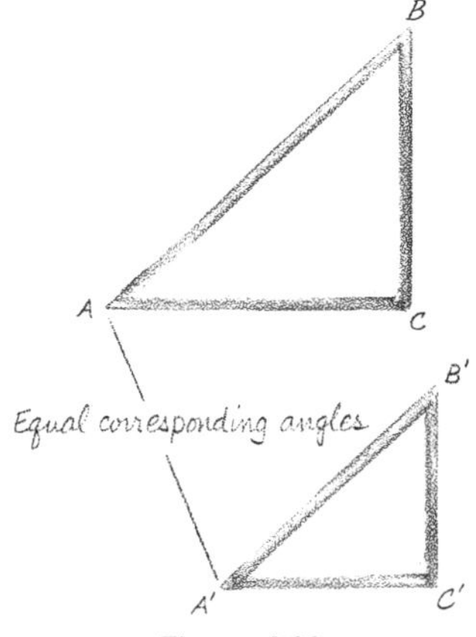

Figure 4.62

We now extend the definitions of these functions. The domain is extended to all angles (in degrees or radians), not only angles between 0° and 90°. Let θ be an angle in standard position, and (x,y) a point on the terminal side; then we define

$$\sin\theta = \frac{y}{\sqrt{x^2+y^2}}, \qquad \cos\theta = \frac{x}{\sqrt{x^2+y^2}}, \tag{3}$$

$$\text{and, for } x \neq 0,\ \tan\theta = y/x$$

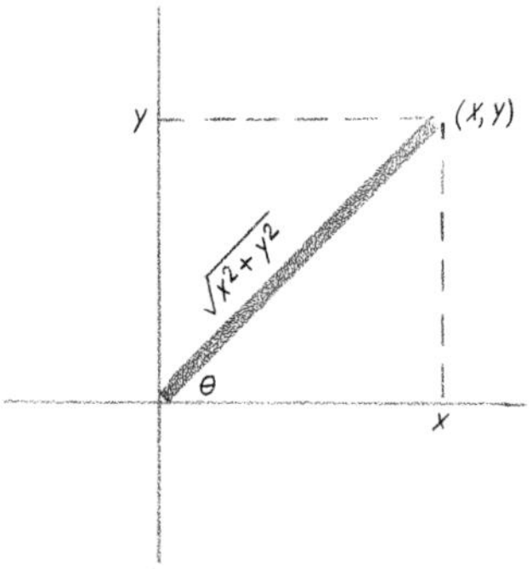

Figure 4.63

(Figure 4.63). Note that $\sqrt{(x^2+y^2)}$ is the distance from the point (x,y) to the origin. If θ terminates in the first quadrant, then the definitions in (3) coincide with those in (2). These functions depend only on the terminal position of the angle: coterminal angles are assigned the same values. The tangent function is undefined for angles of 90°, 270°, 450° and so on; or, equivalently, for angles of $\pi/2$, $3\pi/2$, $5\pi/2$, ...

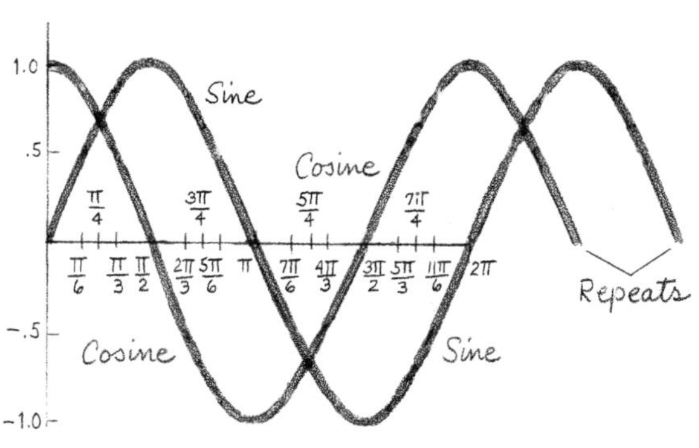

Figure 4.64

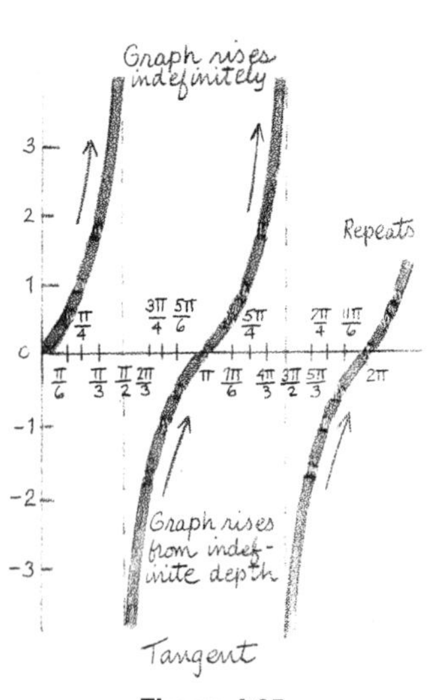

Figure 4.65

The graphs of the sine and cosine for angles between 0 and 2π, inclusive, appear in Figure 4.64; and that of the tangent in Figure 4.65. Tables IV and V of the Appendix contain the values of these functions for angles in terms of radians from 0 to $\pi/2$, and in terms of degrees from 0° to 90°, respectively. These angles are called the **reference angles;** the value of a trigonometric function of any angle can be obtained by appropriately referring to the value of the function at one of the reference angles. The latter angles terminate in the first quadrant. The use of the reference angles is based on the fact that for any point (x,y) in the plane there exists a point (x',y') in the first quadrant such that the ratios (3) are the same for both points or else several may differ in sign.

EXAMPLE 1

Let us find the trigonometric functions of an angle of 45°, or, equivalently $\pi/4$. This angle is generated by $\frac{1}{8}$ of a complete counterclockwise rotation, so that the terminal side is along the line $x = y$ (Figure 4.66). When we put $x = y$ in the formulas in (3), we obtain

$$\sin 45° = \frac{x}{\sqrt{x^2 + x^2}} = \frac{1}{\sqrt{2}} \quad \text{or} \quad \frac{\sqrt{2}}{2};$$

also, $\cos 45° = \sqrt{2}/2$; and $\tan 45° = 1$.

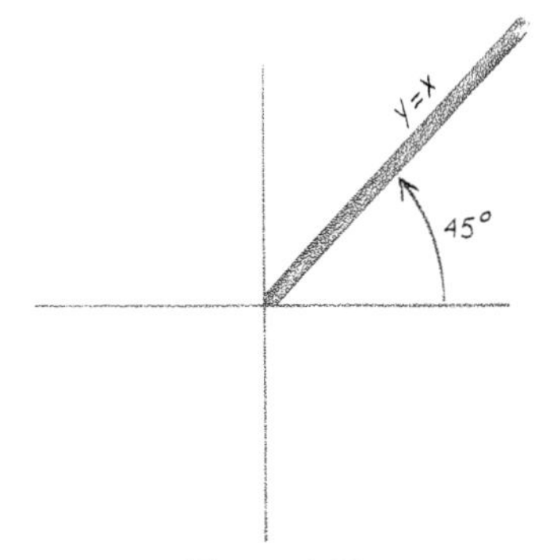

Figure 4.66

From these values we can deduce those for angles of 135°, 225°, and 315°. The angle of 135° is obtained by $\frac{3}{8}$ of a complete rotation since $135/360 = \frac{3}{8}$; therefore, the terminal side is along the line $y = -x$ (Figure 4.67). It follows that the cosine and the tangent of 135° are obtained from those of 45° by simply changing their signs, while the sine remains the same:

$$\sin 135° = \frac{\sqrt{2}}{2}, \qquad \cos 135° = \frac{-\sqrt{2}}{2}, \qquad \tan 135° = -1.$$

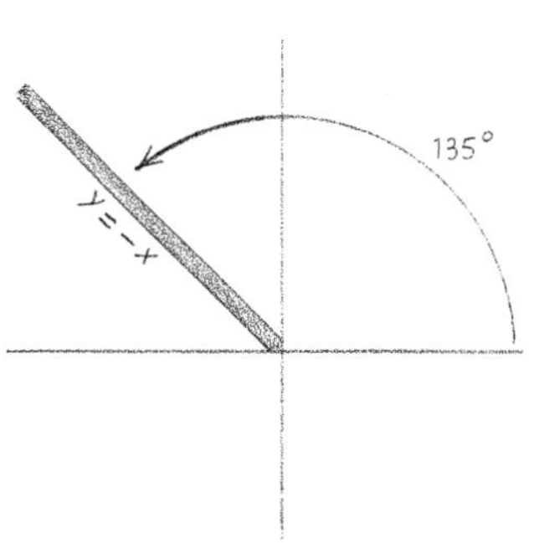

Figure 4.67

Similarly, we find that the terminal side of the angle of 225° is along the line $y = x$, in the third quadrant, and

$$\sin 225° = \frac{-\sqrt{2}}{2}, \qquad \cos 225° = \frac{-\sqrt{2}}{2}, \qquad \tan 225° = 1.$$

Finally, we find that the terminal side of an angle of 315° is in the fourth quadrant, along the line $y = -x$, and

$$\sin 315° = \frac{-\sqrt{2}}{2}, \qquad \cos 315° = \frac{\sqrt{2}}{2}, \qquad \tan 315° = -1. \blacktriangle$$

EXAMPLE 2

Let us find the functions of 30°, or equivalently, $\pi/6$ radians. Such an angle is generated by $\frac{1}{12}$ of a complete rotation. To evaluate the functions we use a result from plane geometry: In a 30°–60°–90° triangle, the smaller leg is necessarily one-half the hypotenuse. (The reason is outlined in Figure 4.68.) Such a right triangle is similar to the triangle ABC, with $a = 1$ and $c = 2$, and by the Pythagorean Theorem, $b = \sqrt{3}$. It follows from (2) that

$$\sin 30° = \frac{a}{c} = \frac{1}{2}, \qquad \cos 30° = \frac{b}{c} = \frac{\sqrt{3}}{2}, \qquad \tan 30° = \frac{\sqrt{3}}{3}.$$

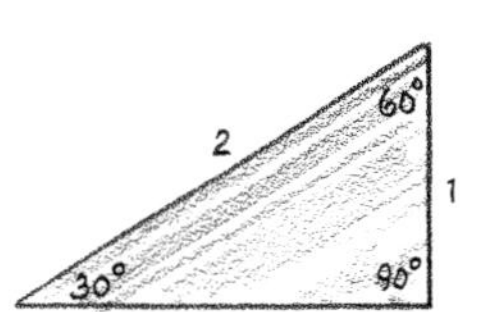

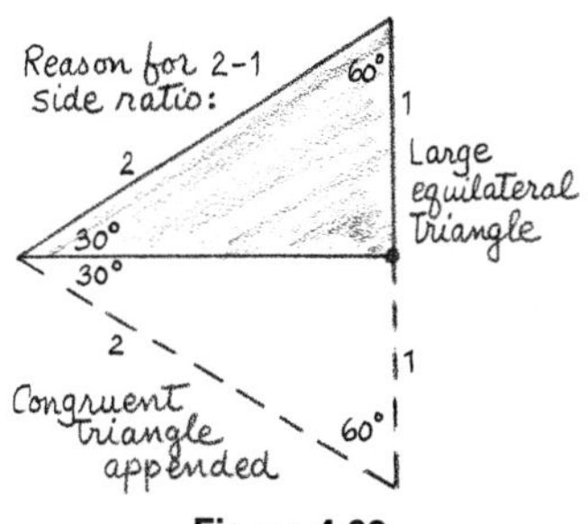

Figure 4.68

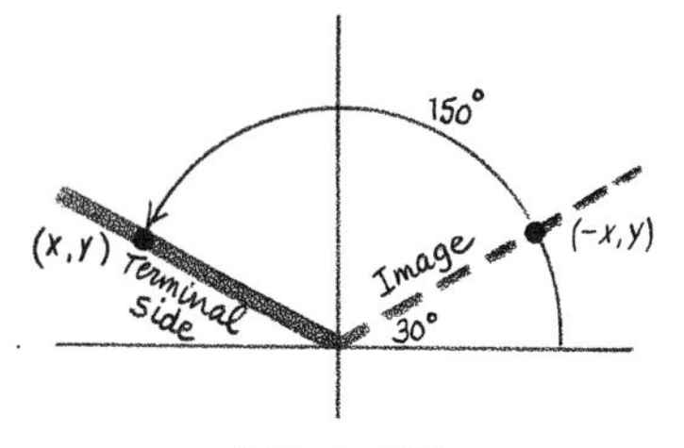

Figure 4.69

From these we can deduce the functions of 150°, 210°, and 330°. The terminal side of the angle of 150° is the "mirror image" through the y-axis of the terminal side of the angle of 30°: If (x,y) is a point on the terminal side, then its "image" is the point $(-x,y)$ (Figure 4.69). It now follows from the definitions (3) that the trigonometric functions of 30° and 150° are related as follows:

$$\sin 150° = \sin 30°, \qquad \cos 150° = -\cos 30°,$$
$$\tan 150° = -\tan 150°.$$

The terminal side of an angle of 210° is related to the terminal side of an angle of 30°: If (x,y) is a point on the former, then $(-x,-y)$ is on the latter. It follows from the definitions (3) that trigonometric functions are related as follows:

$$\sin 210° = -\sin 30°, \qquad \cos 210° = -\cos 30°,$$
$$\tan 210° = \tan 30°.$$

The terminal side of an angle of 330° is also related to that of 30°: If (x,y) is a point on the first, then $(x,-y)$ is on the second, and the trigonometric functions are related as follows:

$$\sin 330° = -\sin 30°, \qquad \cos 330° = \cos 30°,$$
$$\tan 330° = -\tan 30°$$

(Figure 4.70). ▲

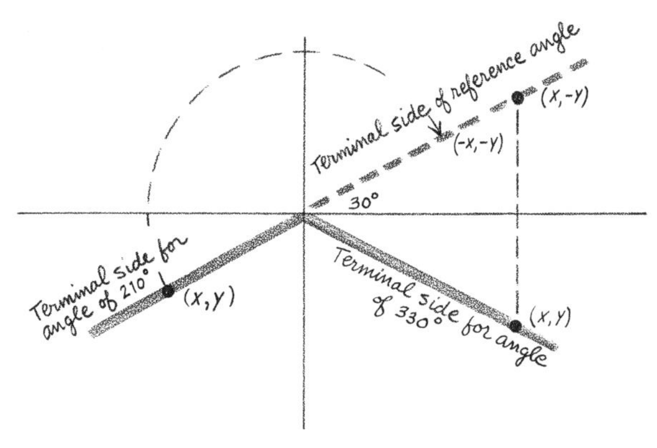

Figure 4.70

EXAMPLE 3

Find the sine, cosine, and tangent of an angle of 592°. The first step is to find the angle between 0° and 360° which is coterminal with the given angle. It is obtained by subtracting 360° from the given 592°: 592° − 360° = 232°. The latter terminates in the third quadrant. The reference angle is obtained by subtracting

180° to get 52°: if (x,y) is a point on the terminal side of the angle of 232°, then $(-x,-y)$ is the corresponding point on the terminal side of the angle of 52°. As for the functions of 210° in Example 2, the functions of 232° and 52° are related as follows:

$$\sin 232° = -\sin 52°, \qquad \cos 232° = -\cos 52°,$$
$$\tan 232° = \tan 52°.$$

From Table V we get the numerical values:

$$\sin 52° = 0.7880, \qquad \cos 52° = 0.6157, \qquad \tan 52° = 1.280.$$

From these we finally obtain

$$\sin 592° = -0.7880, \qquad \cos 592° = -0.6157,$$
$$\tan 592° = 1.280. \quad ▲$$

In the table below we list the values of the trigonometric functions for selected angles given in radians.

Angle	Sine	Cosine	Tangent
0	0	1	0
$\pi/6$	$1/2$	$\sqrt{3}/2$	$\sqrt{3}/3$
$\pi/4$	$\sqrt{2}/2$	$\sqrt{2}/2$	1
$\pi/3$	$\sqrt{3}/2$	$1/2$	$\sqrt{3}$
$\pi/2$	1	0	...
$2\pi/3$	$\sqrt{3}/2$	$-1/2$	$-\sqrt{3}$
$3\pi/4$	$\sqrt{2}/2$	$-\sqrt{2}/2$	-1
$5\pi/6$	$1/2$	$-\sqrt{3}/2$	$-\sqrt{3}/3$
π	0	-1	0
$7\pi/6$	$-1/2$	$-\sqrt{3}/2$	$\sqrt{3}/3$
$5\pi/4$	$-\sqrt{2}/2$	$-\sqrt{2}/2$	1
$4\pi/3$	$-\sqrt{3}/2$	$-1/2$	$\sqrt{3}$
$3\pi/2$	-1	0	...
$5\pi/3$	$-\sqrt{3}/2$	$1/2$	$-\sqrt{3}$
$7\pi/4$	$-\sqrt{2}/2$	$\sqrt{2}/2$	-1
$11\pi/6$	$-1/2$	$\sqrt{3}/2$	$-\sqrt{3}/3$
2π	0	1	0

The cosine is $\pi/2$ units "ahead" of the sine:

$$\cos \theta = \sin\left(\theta + \frac{\pi}{2}\right). \tag{4}$$

This is reflected in the relation between their graphs (Figure 4.64): The graph of the cosine "lags" $\pi/2$ units behind that of the sine.

4.6 EXERCISES

1 Convert to radians in multiples of π:
(a) $20°$ (b) $225°$ (c) $75°$ (d) $150°$ (e) $312°$

2 Convert to radians in multiples of π:
(a) $24°$ (b) $230°$ (c) $80°$ (d) $160°$ (e) $315°$

3 Convert to degrees:
(a) $7\pi/8$ (b) $9\pi/8$ (c) $5\pi/12$ (d) $12\pi/7$ (e) $\pi/5$

4 Convert to degrees:
(a) $3\pi/8$ (b) $5\pi/8$ (c) $11\pi/10$ (d) $7\pi/4$ (e) $\pi/8$

In Exercises 5 and 6 confirm the values of the trigonometric functions for each of the given angles; the values are given in the table above.

5 (a) $\pi/4$ (b) $\pi/2$ (c) $3\pi/4$ (d) π (e) $5\pi/4$ (f) $3\pi/2$ (g) $7\pi/4$ (h) 2π

6 (a) $\pi/6$ (b) $\pi/3$ (c) $2\pi/3$ (d) $5\pi/6$ (e) $7\pi/6$ (f) $4\pi/3$ (g) $5\pi/3$ (h) $11\pi/4$

In Exercises 7 and 8 find the trigonometric functions of the angles.

7 (a) $290°$ (b) $187°$ (c) $130°$ (d) $219°$ (e) $112°$ (f) $175°$

8 (a) $182°$ (b) $94°$ (c) $322°$ (d) $127°$ (e) $218°$ (f) $316°$

In Exercises 9 and 12 express each trigonometric function as a corresponding function of an angle in the first quadrant with + or − preceding it.

9 (a) $\sin 705°$ (b) $\sin 490°$ (c) $\sin(-400°)$ (d) $\sin(-510°)$ (e) $\cos 387°$ (f) $\cos 680°$ (g) $\cos 870°$ (h) $\tan 460°$ (i) $\tan(-390°)$ (j) $\tan 800°$

10 (a) $\sin 463°$ (b) $\sin 620°$ (c) $\sin(-819°)$ (d) $\sin(-546°)$ (e) $\cos 369°$ (f) $\cos 624°$ (g) $\cos(-572°)$ (h) $\tan 703°$ (i) $\tan 884°$ (j) $\tan(-403°)$

11 (a) $\sin 7\pi$ (b) $\sin 9\pi/2$ (c) $\sin(-8\pi/3)$ (d) $\sin(-7\pi/3)$ (e) $\cos 9\pi/4$ (f) $\cos 10\pi$ (g) $\cos(-6\pi/5)$ (h) $\tan 7\pi/2$ (i) $\tan 5\pi$ (j) $\tan(-5\pi/3)$

12 (a) $\sin 8\pi/3$ (b) $\sin 5\pi$ (c) $\sin(-4\pi/3)$ (d) $\sin(-11\pi/2)$ (e) $\cos 6\pi$ (f) $\cos(11\pi/3)$ (g) $\cos(5\pi/3)$ (h) $\tan 8\pi$ (i) $\tan(12\pi/5)$ (j) $\tan(-4\pi/3)$

4.7 The Inverse Trigonometric Functions

The **inverse** trigonometric functions are related to the corresponding trigonometric functions (sine, cosine, and tangent) as the logarithm function is related to the exponential (Section 4.4). The sine and the cosine assume values from -1 to 1, inclusive, and their (common) domain is the entire set of real numbers (in degrees or radians). The domain of the tangent function is the set of the real numbers except for integer multiples of 90° (or $\pi/2$ radians). The inverse functions are obtained by "interchanging" the roles of the independent and dependent variables, respectively. The process is not as direct as in the case of the logarithm because we havc to take a careful account of the domains and the ranges.

The main problem in the definition of the inverse trigonometric functions is that the trigonometric functions repeat their values infinitely often. For example, the sine assumes the value $\frac{1}{2}$ at 30°, 150°, 390°, 510°, and so on. In general, for every value in the domain there corresponds a unique value in the range, as for any function (Section 4.1). However, the converse is not true: A single value in the range corresponds to many in the domain.

The set of **principal values** of a trigonometric function is a portion of the domain which generates a complete "copy" of the range of the function, and such that every value in the range is assumed exactly **once** over this portion. A set of principal values for the sine function is the interval of real numbers from $-\pi/2$ through $\pi/2$, including the endpoints. (Here the independent variable of the trigonometric function is taken in radians.) This is verified by noting that the dependent variable of the sine function ranges from -1 through 1 as the independent variable goes from $-\pi/2$ through $\pi/2$. This is the standard choice for the set of principal values, but not the only possible one: any interval of length π, with endpoints which are integer multiples of $\pi/2$, may serve as a set of principal values for the sine. The standard set of principal values for the cosine function is the interval from 0 through π. The principal values of the tangent function are formed by the interval from $-\pi/2$ to $\pi/2$.

The inverse trigonometric functions are obtained by restricting the trigonometric functions to the set of principal values, and then interchanging the roles of the independent and dependent variables. The domain of the inverse function is the range of the original function; and the range of the inverse is the set of principal values of the original function.

Let us define the inverse sine function, $y = \sin x$. As x varies from $-\pi/2$ to $\pi/2$, y assumes each particular value y between -1 and

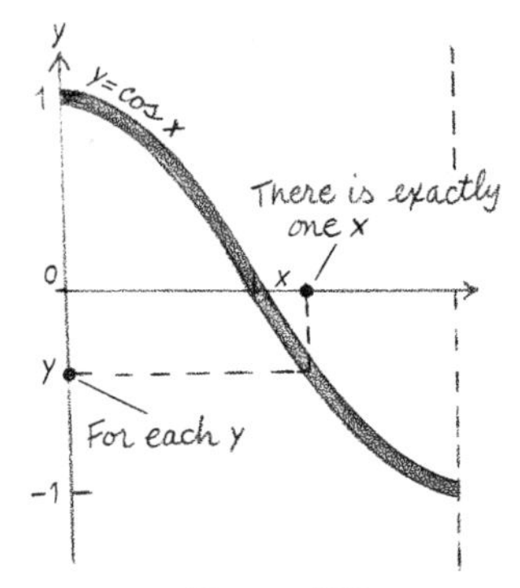

Figure 4.73

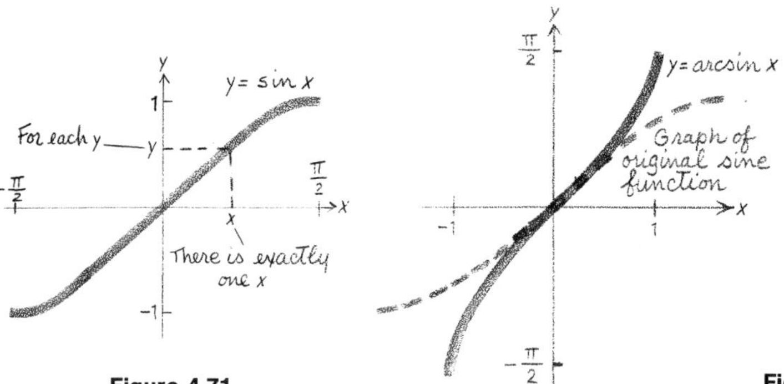

Figure 4.71

Figure 4.72

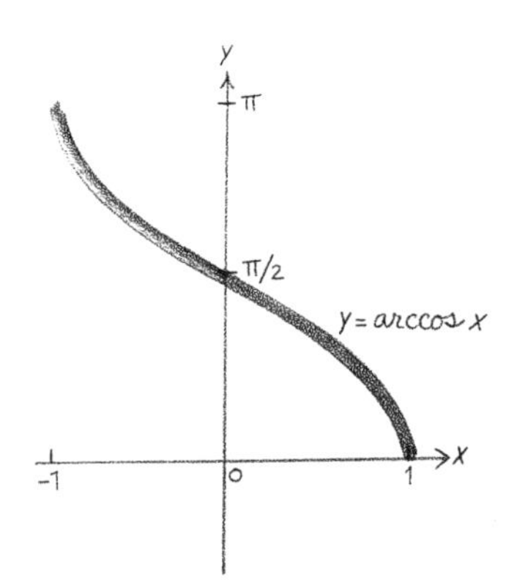

Figure 4.74

1 exactly once (Figure 4.71). When we reverse the roles of x and y, the resulting function is known as the **inverse sine function,** or **arcsine** function. The relation $y = \arcsin x$ means that y is the angle in the set of principal values whose sine is equal to x; for example, $\arcsin \frac{1}{2} = \pi/6$, since $\sin(\pi/6) = \frac{1}{2}$. The graph of the arcsine function is obtained from that of the sine by reflecting the latter through the diagonal line (Figure 4.72). This is how the graph of the logarithm function was obtained from that of the exponential (Section 4.4).

Next we define the **inverse cosine** function, or arccosine function. For every x in the set of principal values from 0 to π there corresponds exactly one y in the range from -1 to 1; and conversely, for every y in the range, there corresponds exactly one x in the set of principal values such that $y = \cos x$ (Figure 4.73). Now we interchange the roles of x and y. The relation $y = \arccos x$ means that y is the angle whose cosine is x; for example, $\arccos 1 = 0$. The graph of this function appears in Figure 4.74; it is obtained by reflecting the graph of the cosine function.

We have a similar "one-to-one" correspondence between the principal values of the tangent in the interval from $-\pi/2$ to $\pi/2$ (excluding the endpoints) (Figure 4.75). The **inverse tangent** function, or **arctangent** function, is obtained by inverting the relation between the angle and its tangent: $y = \arctan x$ means that y is the angle whose tangent is x. The graph appears in Figure 4.76.

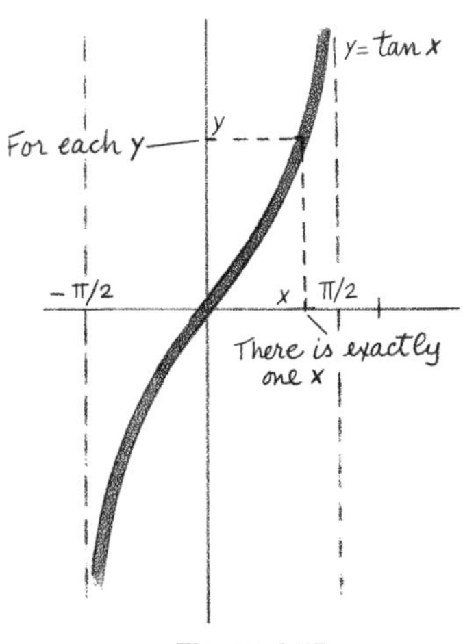

Figure 4.75

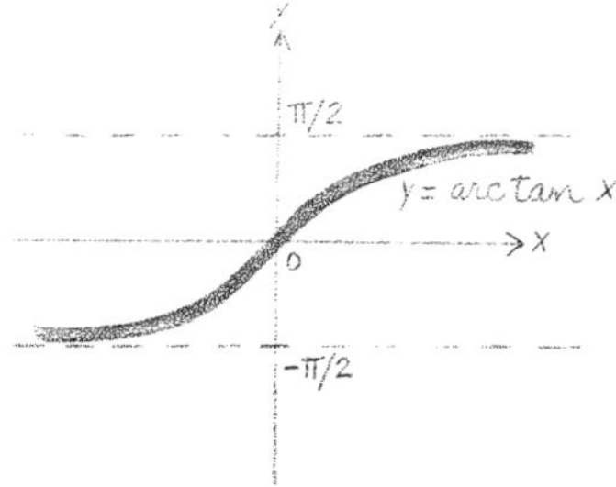

Figure 4.76

EXAMPLE 1

Find the arcsine and arccosine of $-\frac{1}{2}$. We have to find the principal value "angle" whose sine is equal to $-\frac{1}{2}$. We claim that this is equal to $-\pi/6$. In order to confirm this we verify that

$$\sin\left(-\frac{\pi}{6}\right) = -\frac{1}{2}.$$

The latter follows from the fact that $\pi/6$ is the "reference angle" for $-\pi/6$, and that the sine function is negative for angles which terminate in the fourth quadrant (Section 4.6).

Similarly, it can be shown that

$$\cos\left(\frac{2\pi}{3}\right) = -\frac{1}{2},$$

so that $\arccos(-\frac{1}{2}) = 2\pi/3$. ▲

We think of arcsin x as the radian measure of an angle: If $y = \arcsin x$, it is the measure of the angle such that $\sin y = x$. It follows that the sine of the angle "arcsin x" is x (for $-1 \leq x \leq 1$). Similarly, we have $\cos(\arccos x) = x$, and $\tan(\arctan x) = x$ for all x in the respective domains of the inverse functions. We can also find the formula for the "composition" of a trigonometric function with the inverse of another trigonometric function. This is done by a simple geometric analysis of a right triangle with a leg of length x, and a hypotenuse of length 1.

EXAMPLE 2

Find sin(arccos x), that is, the sine of the angle whose cosine is x. Draw a right triangle ABC, with right angle at C, and let the angle arccos x be at A (Figure 4.77). For simplicity take the hypotenuse as 1; then the leg AC must have length x (by the definition of the cosine function for the right triangle). By the Pythagorean Theorem the other leg must have length $\sqrt{1 - x^2}$. It now follows from the definition of the sine function for the right triangle that the sine of the angle at A is $\sqrt{1 - x^2}$, the ratio of the opposite leg to the hypotenuse. Therefore, $\sin(\arccos x) = \sqrt{1 - x^2}$. ▲

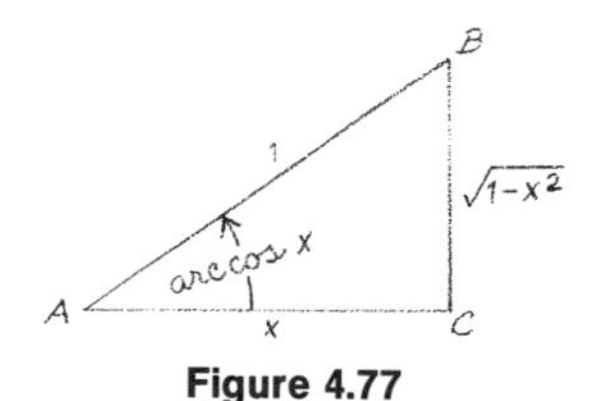

Figure 4.77

Next we consider the "composition" of the inverse function with a trigonometric function, done in the opposite order, for example, arcsin(cos x). This is done by the analysis of a right triangle with an **angle** of x radians, and hypotenuse 1.

EXAMPLE 3

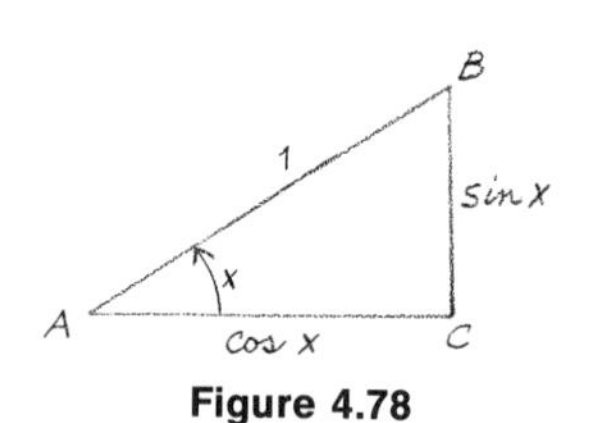

Figure 4.78

Find arcsin(cos x) for $0 \leq x \leq \pi/2$. In the right triangle ABC, the angle at A is taken to have x radians, and the hypotenuse to be 1. It follows from the definitions of the sine and the cosine that the length of the leg AC is cos x and the length of the leg BC is sin x (Figure 4.78). It also follows that the **sine** of the angle at B is equal to cos x. It follows from the elementary property of the triangle that the sum of the angles is π, and that the radian measure of the angle at B is $\pi/2 - x$; therefore,

$$\arcsin(\cos x) = \frac{\pi}{2} - x.$$

This formula can be extended to all values of x. ▲

Let us now record a simple but extremely useful trigonometric identity:

$$\sin^2 x + \cos^2 x = 1, \tag{1}$$

where $\sin^2 x$ is the symbol for the square of the sine of x,$(\sin x)^2$, and $\cos^2 x$ is the symbol for $(\cos x)^2$. This formula is a direct consequence of the definitions of the sine and cosine functions [formula (3), Section 4.6].

4.7 EXERCISES

In Exercises 1 and 2 find the values of the inverse functions in radians.

1 (a) $\arcsin \frac{1}{2}$ (b) $\arcsin(-\sqrt{2}/2)$ (c) $\arccos(-\sqrt{3}/2)$ (d) $\arccos \frac{1}{2}$ (e) $\arctan(-1)$ (f) $\arctan \sqrt{3}$

2 (a) $\arcsin(\sqrt{2}/2)$ (b) $\arcsin(-\sqrt{3}/2)$ (c) $\arccos(-\sqrt{2}/2)$ (d) $\arccos 1$ (e) $\arctan 0$ (f) $\arctan(-\sqrt{3}/3)$

In Exercises 3 and 4 find the given functions in terms of x.

3 (a) $\arccos(\sin x)$ (b) $\cos(\arctan x)$ (c) $\cos(\arcsin x)$,

4 (a) $\tan(\arccos x)$ (b) $\tan(\arcsin x)$, (c) $\sin(\arctan x)$

5 Verify the result of Example 2 by using the identity $\cos(\arccos x) = x$ together with Equation (1).

4.8 Frequency, Amplitude, Phase, and Mean of Trigonometric Functions

Throughout this section the independent variable of the trigonometric functions is measured in radians; it is denoted by x. The sine and cosine functions repeat their values at intervals of 2π:

$$\sin x = \sin(x \pm 2\pi) = \sin(x \pm 4\pi) = \ldots,$$
$$\cos x = \cos(x \pm 2\pi) = \cos(x \pm 4\pi) = \ldots.$$

These functions are described on a domain of length 2π and then copied and "pasted together" on the other intervals. The interval from 0 to 2π, inclusive, is the **fundamental** domain. Its length, 2π, is the **period** of the function. The reciprocal, $1/2\pi$, is the **frequency.** The fundamental domain for the tangent function is the interval from $-\pi/2$ to $\pi/2$; its period is π, and frequency $1/\pi$.

Instead of the function $\sin x$, consider the function $\sin 2x$; here the independent variable x is replaced by its multiple $2x$. As x goes from 0 to π, $2x$ goes over the whole fundamental interval from 0 to 2π; thus, the interval from 0 to π produces a complete copy of the function $\sin 2x$. The values are repeated every π units. The graphs of $\sin x$ and $\sin 2x$ are compared in Figure 4.79. Note that the graph of the latter oscillates twice as rapidly as the former. The function $\sin 2x$ is said to have the period π, half that of the function $\sin x$, and the frequency $1/\pi$.

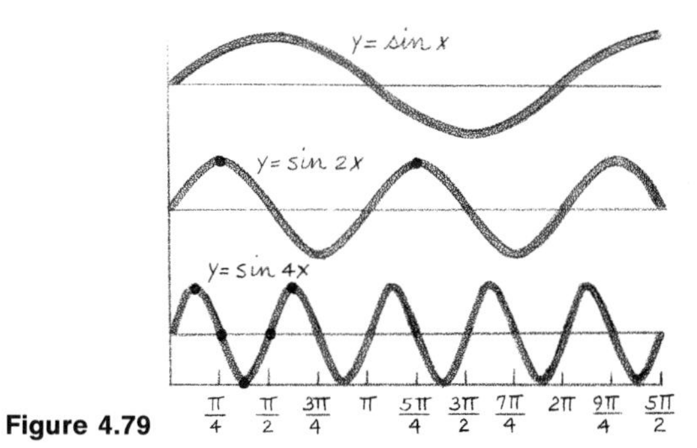

Figure 4.79

The function $\sin 4x$ has period $\pi/2$, one-quarter of $\sin x$, and the frequency $2/\pi$. It makes four complete cycles for every cycle of $\sin x$ (Figure 4.79). In general, the function $\sin mx$, where m is any real number, is said to have period $2\pi/m$ and frequency $m/2\pi$. The discussion for the cosine is similar: $\cos mx$ has the same period and frequency. The period of the corresponding tangent function is always half that of the sine function, and the frequency is twice.

Let A be an arbitrary positive real number. While the ordinary sine function assumes values from -1 to 1, the multiple of the sine function, $A \sin x$, ranges from $-A$ to A. The number A is the **amplitude:** it is half the difference between the largest and smallest value in the range. The graphs of the sine function for $A = 3$ and $A = \frac{1}{2}$ are compared to that of the standard sine function ($A = 1$) in Figure 4.80. When A is larger, the function has a larger oscillation. The amplitude for the cosine is similar: the cosine is just a shifted sine. The tangent does not have an amplitude because its values range over the whole line.

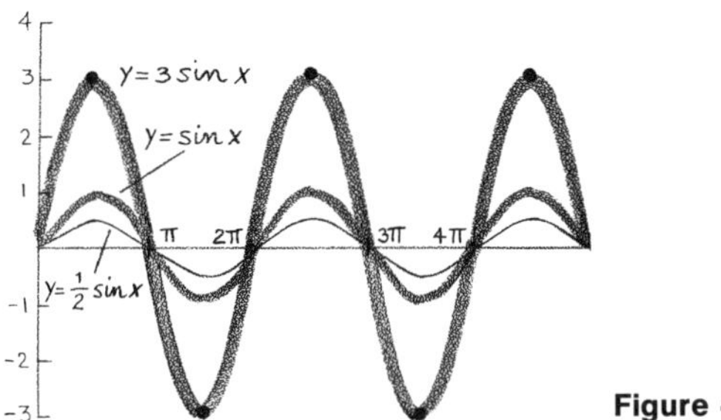

Figure 4.80

Next we consider a transformation of the sine function called a **phase shift.** Instead of $\sin x$, consider the function $\sin(x + c)$, where c is a fixed real number. This function is c units "ahead" of the sine: the graph has the same shape, but is displaced c units to the left (Figure 4.81). In particular, the sine function displaced by $\pi/2$ units, $\sin(x + \pi/2)$, coincides with the cosine. (See Figure 4.64.) The shift of phase is the same for the cosine and tangent.

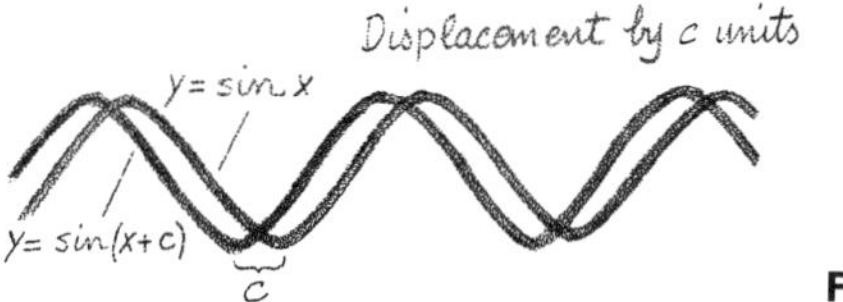

Figure 4.81

Finally we consider the **mean shift.** The sine oscillates evenly about the value 0, going from -1 to 1, and back to -1. If instead of $\sin x$, we consider the function $b + \sin x$, where b is some real number, then the latter varies from $b - 1$ to $b + 1$, and the **mean** is b. The mean of the standard sine function is 0. The graphs of the standard sine and the sine shifted by 3 units are compared in Figure 4.82. The mean shift for the cosine is similarly defined as $b + \cos x$; the same for the tangent.

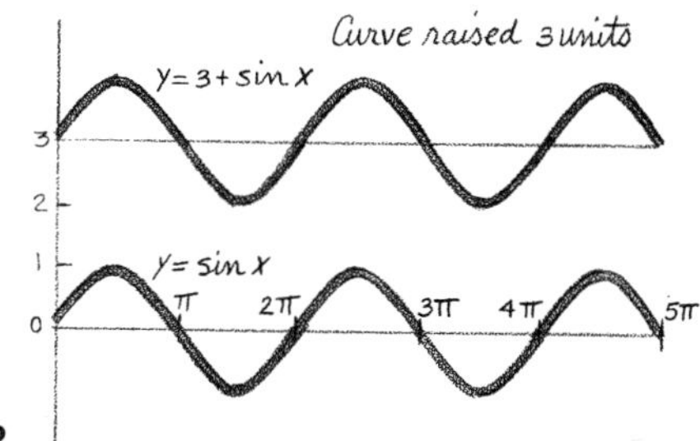

Figure 4.82

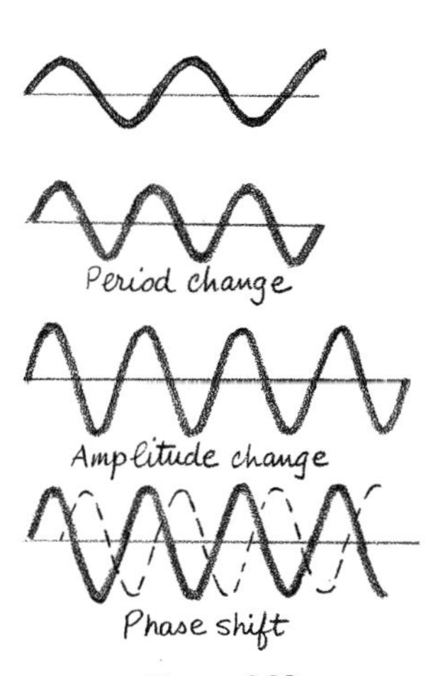

Figure 4.83

In the applications to follow we shall use the sine and cosine but not the tangent; four kinds of transformations have been defined (Figure 4.83):

1. change of period
2. change of amplitude
3. phase shift
4. mean shift

These can be applied to the sine and cosine individually or in succession; for example, we may first change the period from 2π to $2\pi/m$,

$$\sin mx;$$

then the amplitude to A,

$$A \sin mx;$$

then the phase to c,

$$A \sin[m(x+c)];$$

and, finally, the mean to b:

$$b + A \sin[m(x+c)]. \qquad (1)$$

This is the general form of the transformed sine; it is similar for the cosine.

EXAMPLE 1

Find the equation of the transformed sine function which has a high value of 10, a low of 2, a period of 3, and which attains its high value 10 at $x = 0$.

The mean is the average of the high and low values; thus, $b = \frac{1}{2}(10 + 2) = 6$.

The amplitude is half the difference between the largest and smallest values; thus, $A = \frac{1}{2}(10 - 2) = 4$.

The period is given as 3. Since the period is defined above as

$$\text{Period} = \frac{2\pi}{\text{Coefficient of } x}, \tag{2}$$

it follows that the coefficient of x is $2\pi/3$; hence, this is the value of m in (1). It follows that the formula for the function we seek is

$$y = 6 + 4 \sin\left[\left(\frac{2\pi}{3}\right)(x + c)\right],$$

where c has to be determined.

Now we determine the phase c. We have been given the fact that $y = 10$ when $x = 0$; therefore, the sine function above, the coefficient of 4, must have the value 1 at $x = 0$. Then the entire variable of the sine must be equal to $\pi/2$:

$$\left(\frac{2\pi}{3}\right)(0 + c) = \frac{\pi}{2};$$

therefore, c must be equal to $\frac{3}{4}$, and so the phase is $\frac{3}{4}$. The function has the formula

$$y = 6 + 4 \sin\left[\left(\frac{2\pi}{3}\right)\left(x + \frac{3}{4}\right)\right].$$

This function may also be written as a cosine:

$$y = 6 + 4 \cos\left(\frac{2\pi x}{3}\right),$$

because the cosine is obtained from the sine by a phase shift of $\pi/2$. ▲

Now we discuss two simple physical models which involve the sine function; these have applications to weather and air pollution study, and to business cycles.

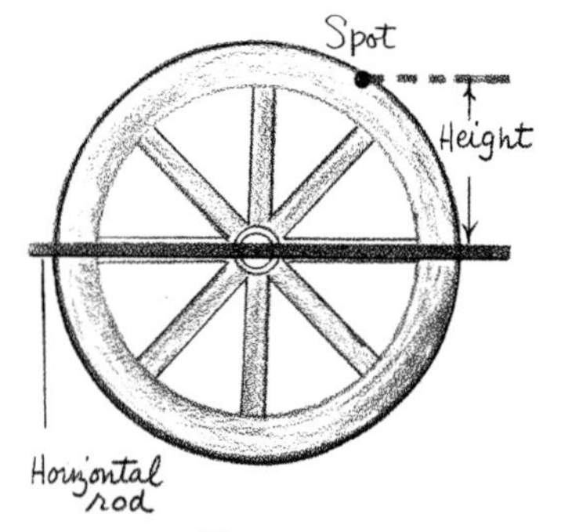

Figure 4.84

EXAMPLE 2

A wheel of radius r rotates about an axle at constant speed. A point on the rim of the wheel is painted with a spot of red. A rod is placed horizontally through the axle. How does the height of the red spot above the rod vary as a function of time (Figure 4.84)? The height is considered positive or negative accordingly as the spot is above or below the rod.

Let θ be the angle formed by the line from the center of the wheel to the spot and the rod. The wheel starts when $\theta = 0$. Let

m be the **angular speed** of the wheel in radians per second: it turns m radians in 1 second. After t seconds the wheel has turned mt radians, so that $\theta = mt$. It follows that the height of the spot above the rod after t seconds is given by the formula $r \sin mt$ (Figure 4.85). The period of this function is, by formula (2), $2\pi/m$, and the amplitude is r. If the wheel had started at an angle of c radians, then the function would have been $r \sin(mt + c)$, which is equivalent to

$$r \sin\left[m\left(t + \frac{c}{m}\right)\right];$$

hence, the phase shift is c/m radians. ▲

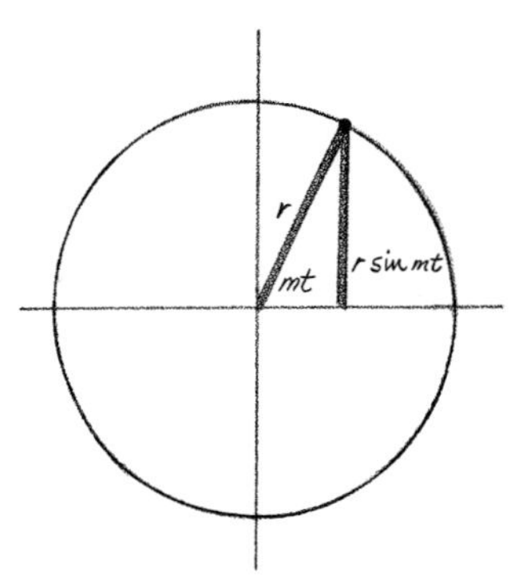

Figure 4.85

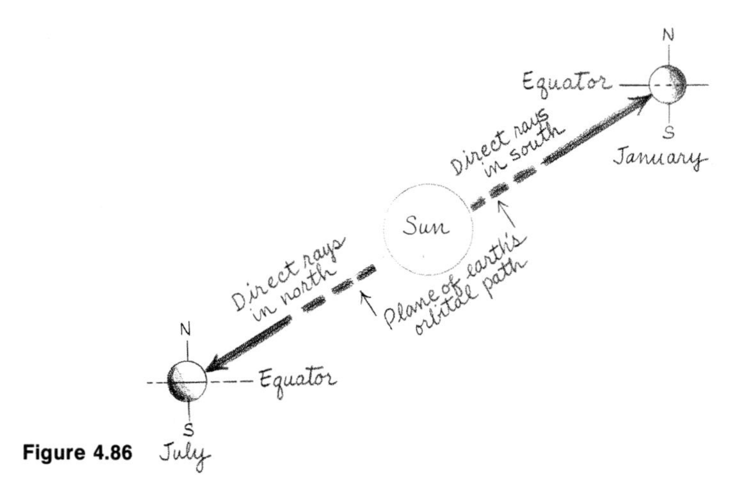

Figure 4.86

This simple example of the height of a painted spot on the rim of a rotating wheel explains the pattern of our seasons on earth, with all its biological, economic, and social implications. The earth revolves about the sun in a nearly circular path and at constant speed. Its axis is tilted at a slight angle. As it goes around the sun, the points of equal latitude that are struck most directly by the rays of the sun are the warmest. This level of latitude varies with the position of the earth in its orbit (Figure 4.86): as the earth moves around the sun, this latitude varies like the height of the spot on the rim of the wheel. This accounts for the "sine wave" characteristic of seasonal records: monthly mean temperatures, sunset times, sunrise times, and air pollution data.

EXAMPLE 3

The monthly average temperatures for New York City (in degrees Fahrenheit) are (*Source:* U.S. National Weather Records Center):

January	33.2	July	76.8
February	33.4	August	75.1
March	40.5	September	68.5
April	51.4	October	58.3
May	62.4	November	47.0
June	71.4	December	35.9

These are plotted in Figure 4.87. Its shape is similar to that of the sine curve over the portion of the domain from $-\pi/2$ to $3\pi/2$. We shall transform the sine function to fit the temperature record, and refer to the general formula (1).

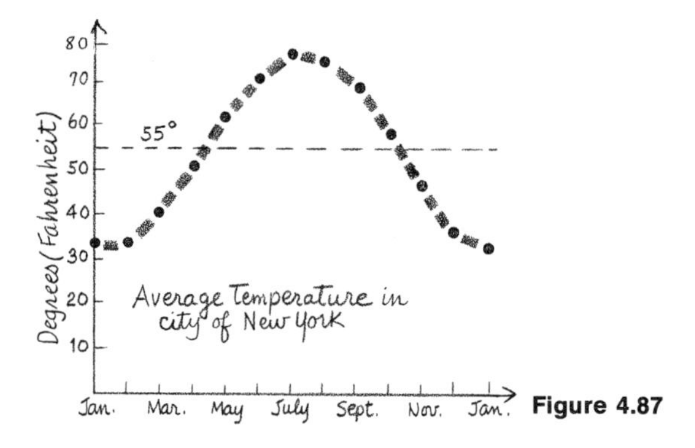

Figure 4.87

The temperature range is from a low of 33.2 to a high of 76.8; therefore, the mean is the average,

$$b = 55.0.$$

and the amplitude is half the difference,

$$A = 21.8.$$

If we take the month as the unit of time, then 12 months form a complete cycle, and so the period of our sine function is 12; therefore, by formula (2), the coefficient of x in the sine function is $\pi/6$. The transformed sine function now has the form

$$55.0 + 21.8 \sin\left[\left(\frac{\pi}{6}\right)(x + c)\right], \tag{3}$$

where the phase shift c has to be determined.

We label January as 0, February as 1, and so on. The low point of the function is at $x = 0$; here the sine in (3) is equal to -1:

$$\text{sine}\left(\frac{\pi}{6}\right)(0 + c) = -1,$$

or

$$\frac{\pi c}{6} = \frac{-\pi}{2},$$

so that $c = -3$. The function (3) takes the specific form

$$55.0 + 21.8 \sin\left[\left(\frac{\pi}{6}\right)(x - 3),\right], \qquad \text{for } 0 \leq x \leq 12. \tag{4}$$

Using the relation

$$\left(\frac{\pi}{6}\right)(x - 3) \text{ radians} = 30(x - 3) \text{ degrees},$$

we find the values of the function (4):

x	0	1	2	3	4	5	
Function (4)	33.2	36.1	44.1	55.0	65.9	73.9	
	6	7	8	9	10	11	12
	76.8	73.9	65.9	55.0	44.1	36.1	33.2

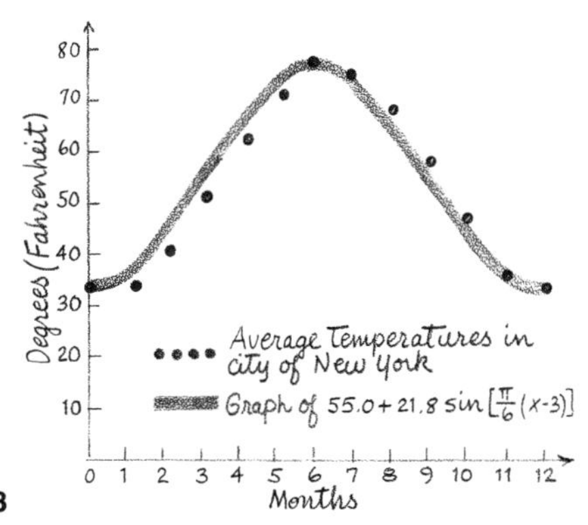

Figure 4.88

This adjusted sine function is plotted with the temperature data in Figure 4.88. Note that the temperature lags slightly behind the sine curve in the spring, and a bit above it in the autumn. This is due to the warm-up time in the spring, and the cooling-off period in the autumn. ▲

Our second physical model is of the science-fiction type. (Or should we say math-fiction.)

EXAMPLE 4

Imagine that the earth is flattened by a gigantic rolling pin, like a ball of dough, and then has a hole cut in it. An object is placed directly over this hole, and then propelled directly upward. The object will rise to a certain height, and then fall back toward the surface of the earth. But it will fall **into** the hole, and not be stopped by the ground. It will pass through the hole and to the other side of the flat surface of this earth. But then the force of gravity will slow it down and pull it back toward the "ground level" from the other side. It will then pass through the hole, and back up again (Figure 4.89).

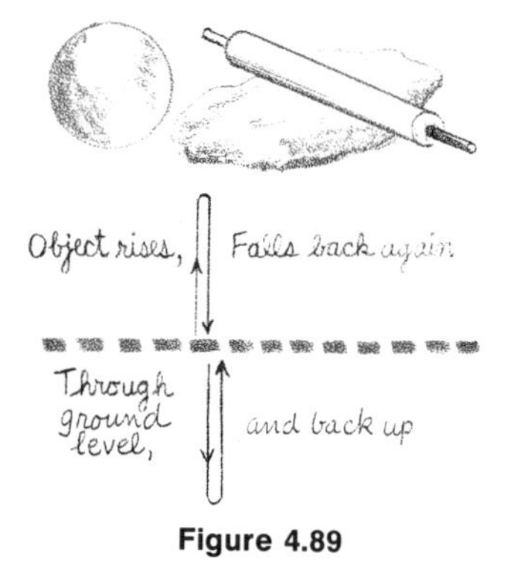

Figure 4.89

Let us deduce the formula for the height of the object t seconds after being shot up. If v is the initial speed of the rising object, and g the acceleration constant for this flat earth (see Section 4.5), then the height of the object after t seconds of its flight above the surface of the earth is given by the formula $vt - \frac{1}{2}gt^2$. The flight curve, as a function of time, is a parabola. The object reaches the ground with the same velocity v with which it started: this is the initial speed for the flight in the direction below the surface. Therefore the flight below the surface is the mirror-image of the flight above it: its function is the inverted parabola. The third part of the journey is described by a parabola above the surface. The flight curve for the entire series of excursions above and below the surface is drawn in Figure 4.90. It is a sequence of parts of parabolas, one opening up and the next down. **It looks very much like the sine function**! ▲

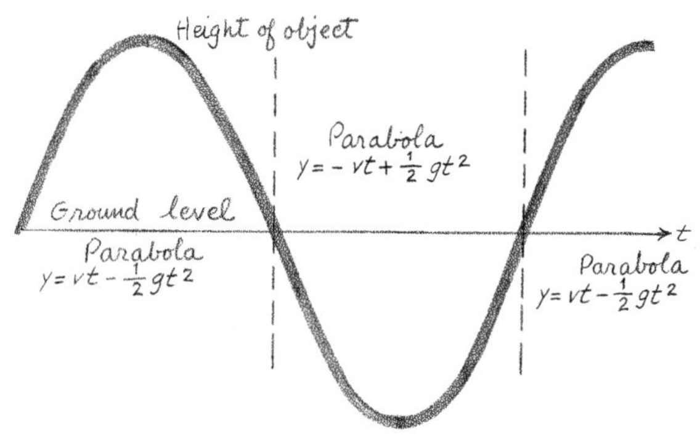

Figure 4.90

Is there any relevance of this fictitious flat-earth model to the "real" world? Yes, we can use it to explain business cycles.

EXAMPLE 5

Economic activity seems to run in cycles: expansion, retrenchment, recession, and recovery. A typical economic indicator following such a pattern often has a graph which looks like a sine curve (Figure 4.91). The early part of the expansion is rapid, but then it slows down as the peak is reached. As the peak passes, the retrenchment is gradual at first, and then picks up speed as recession approaches. Recovery from the low point is slow at first, but then moves more quickly.

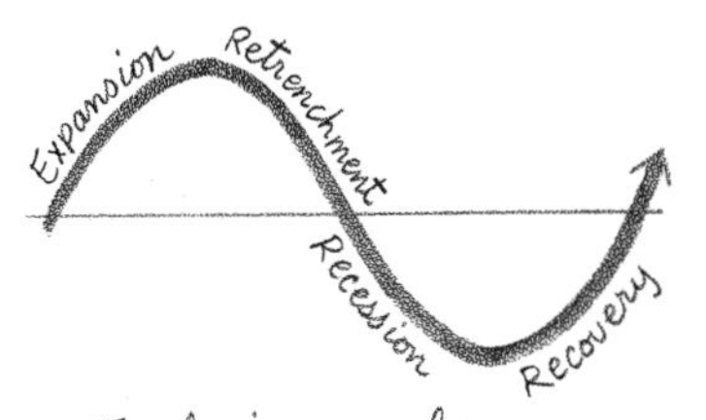

Figure 4.91

Let us use the model in Example 4 to explain this sine variation of the business cycle. Consider the price of a commodity as a function of time. Let the "equilibrium" price of the commodity correspond to the surface of the flat earth. If the price of the commodity is suddenly driven up, like an object thrown into the air, then, under free market conditions, more of the commodity will be put on the market as the sellers try to gain on the abnormally high price. This acts like gravity: it forces down the price. But as the price falls quickly to the equilibrium level, the market has been "flooded" and so the price continues to fall past the level until the supply is used up. As the sellers keep the product from the market, the price moves back toward the equilibrium level. And on it goes.

The world of business is really more complicated than this example presumes; however, we have introduced it to show the reader why trigonometric functions have been so valuable in the theory of econometrics—which is the application of mathematical methods to economic problems. ▲

4.8 EXERCISES

In Exercises 1 and 2 find the mean, amplitude, and period of each function.

1 (a) $\sin x$
(b) $-4 \cos x$
(c) $\sin 3x$
(d) $\sin 4\pi x$
(e) $1 - \sin 2x$
(f) $\sin 5x - 2$

2 (a) $4 \sin x$
(b) $3 \cos(x - \pi)$
(c) $\cos 7x$
(d) $3 + \sin \pi x$
(e) $8 - 2 \sin 3x$
(f) $-1 - 7 \sin 2x$

In Exercises 3 and 4 find the formulas for the sine curves with the stated properties.

3 (a) maximum value 2, minimum -1, period 1, maximum value assumed at $x = 0$.
(b) maximum 14, minimum 8, period 9, mean value assumed at $x = 0$.

(c) maximum 5, minimum 0, period 4, minimum value assumed at $x = 0$.

(d) maximum -1, minimum -5, period 6π, maximum value assumed at $x = 0$.

4 (a) maximum 3, minimum -2, period 1, minimum value assumed at $x = 0$.

(b) maximum 7, minimum 1, period 8, mean value assumed at $x = 0$.

(c) maximum 6, minimum 0, period 3, maximum value assumed at $x = 0$.

(d) maximum -2, minimum -6, period 3π, maximum value assumed at $x = 0$.

5 As the earth revolves around the sun, the moon revolves about the earth approximately 12 or 13 times per year. The moon moves in the plane perpendicular to the path of the earth (Figure 4.92).

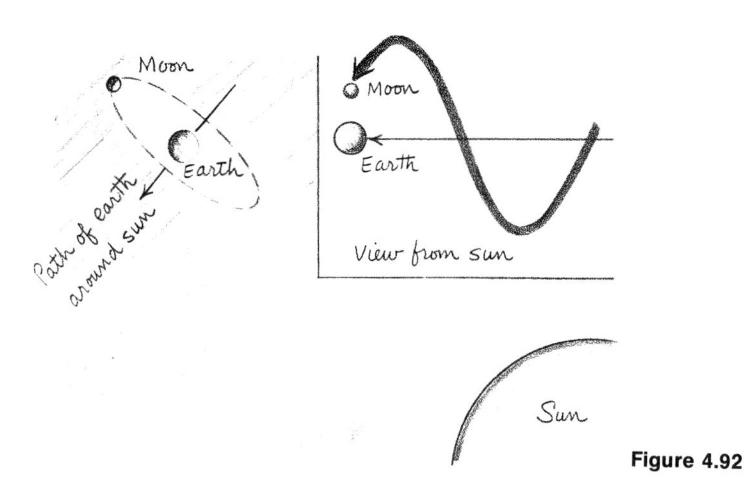

Figure 4.92

(a) If an observer on the sun watched the height of the moon above the path of the earth, why would the height trace a sine curve?

(b) Assume, for simplicity, that there are 360 days per year, and that the moon revolves about the earth 12 times per year. What is the period of the sine function in (a)?

(c) What is the formula for the height of the moon above the orbital path of the earth as a function of the time in units of months? (The distance between the two bodies is 240,000 miles.) (See Example 2.)

Figure 4.93

6 Why does the height of a horse on a "merry-go-round" vary with time like a sine function (Figure 4.93)? (See Example 2.)

7 Here are the average monthly temperatures for the city of Atlanta, Georgia (*Source:* U.S. National Weather Records Center):

January	44.7
February	46.1
March	51.4
April	60.2
May	69.1
June	76.6
July	78.9
August	78.2
September	73.1
October	62.4
November	51.2
December	44.8

(a) Number the months 0, 1, . . . , 11, and plot the data.
(b) Find a sine function which well fits the data. Compare its values for $x = 0, 1, \ldots, 11$ to the given data. (See Example 3.)

8 Repeat the procedure of Exercise 7 for the following data for the average monthly temperatures of Sioux Falls, South Dakota:

January	15.2
February	19.1
March	30.1
April	45.9
May	58.3
June	68.1
July	74.3
August	71.8
September	61.8
October	50.3
November	32.6
December	21.1

(See Example 3.)

9 These are the times of sunset (in local standard time) at 40° North Latitude in the year 1969:

Jan. 1	4:45	May 1	6:54	Sept. 1	6:32
Feb. 1	5:19	June 1	7:22	Oct. 1	5:43
Mar. 1	5:52	July 1	7:33	Nov. 1	4:58
April 1	6:24	Aug. 1	7:14	Dec. 1	4:35

(a) Label the months January, February, . . . as 0, 1, ... Let $y =$ number of minutes after 6:00. Plot the graph of y as a function of the month x.
(b) Find a sine function which well fits the data, and compute its values for the given x. (See Example 3.)

10 Repeat the procedure of Exercise 9 for the sunrise times at 30° North Latitude:

Jan. 1	6:56	May 1	5:17	Sept. 1	5:37
Feb. 1	6:51	June 1	5:00	Oct. 1	5:53
Mar. 1	6:26	July 1	5:02	Nov. 1	6:14
April 1	5:50	Aug. 1	5:19	Dec. 1	6:38

(See Example 3.)

11 The "normal" price of a utility stock is $50 per share. If the price rises over that level, it is subject to a downward force having an acceleration of $−2 per day, per day. If it falls below the normal level, it is subject to an upward force with an acceleration of $2 per day, per day. At time $t = 0$, the price rises from the normal level with an initial speed of $4 per day.

(a) Find the formula for the price of the stock at time t, before it falls back to the normal level of $50.

(b) What is the maximum value that the stock will assume during this surge?

(c) What is the minimum value assumed by the stock after returning to the normal level and falling below it?

(d) What is the period of the stock as a function of the number of days t?

(e) Find the formula of the sine function which best fits this price function. (See Example 4.)

12 Work out the more general version of Exercise 11, with a normal price level p, an acceleration constant g, and an initial velocity v. (See Example 4.)

13 The major source of sulphur dioxide pollution of the air is the burning of coal and oil. Since there is a direct relation between the temperature and the amount of heating fuel burned, there is an inverse relation between the level of pollution and the temperature. Since the average monthly temperature follows a sine curve, so does the pollution level; indeed, when the sine curve is flipped over it retains its characteristic shape. The average monthly concentrations of sulphur dioxide in the air over New York (in parts per million) are given below (*Source:* Annual Data Report, Department of Air Resources, City of New York):

January	0.19	July	0.06
February	0.16	August	0.07
March	0.15	September	0.07
April	0.14	October	0.11
May	0.09	November	0.12
June	0.08	December	0.12

Let x be the number of months after January. Fit a sine curve to the data as a function of x. (See Example 3.)

CHAPTER

5

Differential Calculus

5.1 Functions and Their Derivatives

In the previous chapter we presented many examples of functions. They are typical of the kinds of functions studied in calculus, and are of two major types:

(i) **Algebraic** functions. These include the functions studied in Sections 4.1, 4.2, and 4.5: polynomials and ratios of polynomials.

(ii) **Transcendental** functions. These include the exponential, logarithm, and trigonometric functions.

These functions are usually defined by a formula and an associated domain. We introduce the general notation for the formula of a function. If x is a value in the domain, and y the corresponding value in the range, then the relation between them is expressed by means of the equation

$$y = f(x). \tag{1}$$

This is read "y is equal to f of x." It signifies that the value y is obtained by putting x in the expression on the right-hand side of Equation (1). For example, if the function is the rule for associating the area of a circle (dependent variable) with the radius (independent variable), then the particular form of f is $f(x) = \pi x^2$. The values of the function at particular points in the domain are obtained by substituting the values in the domain for x; for example, $f(2) = \pi(2)^2 = 4\pi$ is the area of a circle of radius 2. The **graph** of the function is the set of all points (x,y) in the plane satisfying Equation (1).

In Section 1.7 we defined the notion of the tangent line for the curves studied in analytic geometry. Now we extend the definition to the graph of a general function f. The procedure is similar to that for the previous curves. Let (x,y) be a fixed point on the graph and (x^*,y^*) another point on the graph, near the first. Draw a secant line through the two points, and then let the second point approach the first. If the secant line approaches a **final nonvertical position,** and the **same** one for all possible ways in which (x^*,y^*) can approach (x,y), whether from the right or the left or a combination of both, then the line in this final position is called the **tangent** to the curve at (x,y) (Figure 5.1). In this case the function is said to be **differentiable** at the (abscissa) point x.

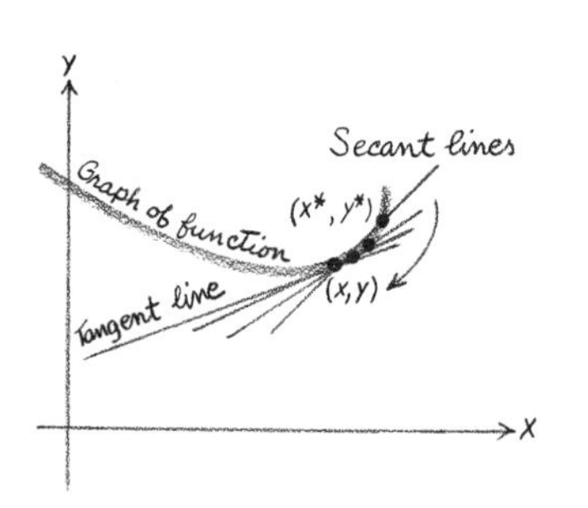

Figure 5.1

Our main interest is in the slope of the tangent line. It is calculated in the following way. The slope of the secant line is the difference quotient

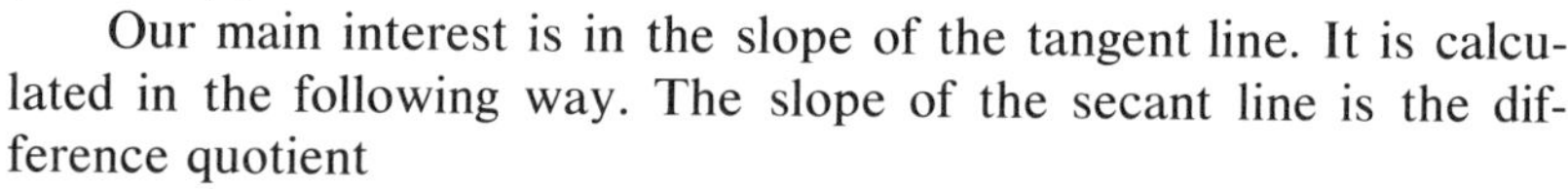

$$\frac{y - y^*}{x - x^*}, \qquad \text{or equivalently,} \qquad \frac{y^* - y}{x^* - x}. \tag{2}$$

As the second point approaches the first, the difference quotient approaches the slope of the tangent line. We say that

$$\lim_{x^* \to x} \frac{y - y^*}{x - x^*} = \text{Slope of the tangent line.} \tag{3}$$

This is read, "The limit of $(y - y^*)/(x - x^*)$ as x^* approaches x is equal to the slope of the tangent line." We shall say more about **limits** in Section 5.3.

In this section we study the slopes of the tangent lines of algebraic functions. Transcendental functions are studied in the next section.

EXAMPLE 1

If f is a linear function, $f(x) = mx + b$, then the graph is the straight line with the equation $y = mx + b$. The tangent line at any point is the line itself. Its slope is m. This agrees with formula (3) for the limit of the difference quotient; indeed, for the linear function, the slope is, by definition, equal to the difference quotient (2). ▲

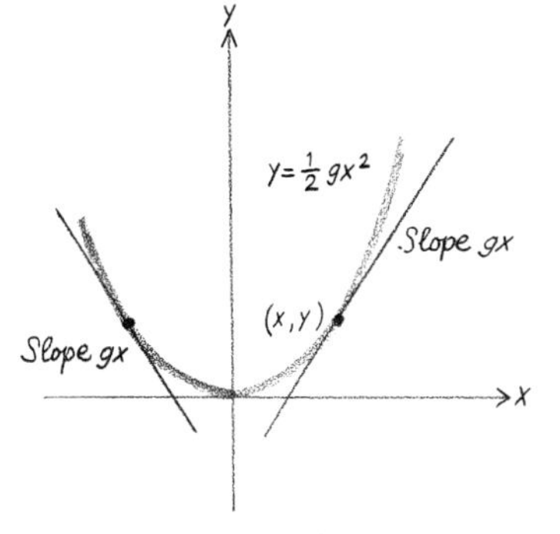

Figure 5.2

EXAMPLE 2

In Section 1.7, we found that the slope of the line tangent to the parabola $4ay = x^2$ at the point with the abscissa x is given by $x/2a$. It follows that the slope of the line tangent to the graph of the function $f(x) = \frac{1}{2}gx^2$ is gx; in fact, we put $g = 1/2a$. The slope is positive when x is positive, negative when x is negative (Figure 5.2). (According to our previous notation, we are assuming that the constant g is positive.) ▲

EXAMPLE 3

Find the formula for the slope of the line tangent to the graph of the function $f(x) = x^3$. The graph is the locus of points (x,y) with $y = x^3$. The difference quotient (2) takes the specific form

$$\frac{x^3 - x^{*3}}{x - x^*}.$$

Now let x^* approach x. We appear to have some trouble: Both the numerator and the denominator get smaller and smaller, so that the ratio becomes 0/0, which is undefined. We cannot estimate the limit of the quotient without first simplifying it. So first we shall divide the numerator by the denominator:

$$(x^3 - x^{*3})/(x - x^*) = x^2 + xx^* + x^{*2}.$$

If now x^* gets closer and closer to x, then the right-hand side approaches $x^2 + x^2 + x^2 = 3x^2$, so that

$$\lim_{x^* \to x} \frac{x^3 - x^{*3}}{x - x^*} = 3x^2.$$

This is the formula for the slope of the tangent line (Figure 5.3). For example, when $x = 2$, the slope of the tangent line is $3(4) = 12$. ▲

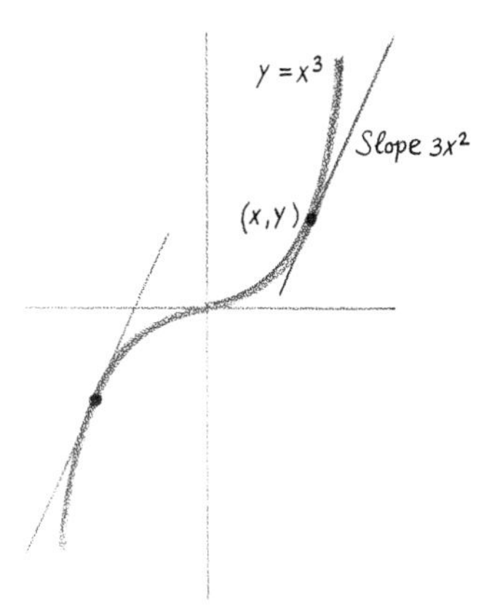

Figure 5.3

The examples above suggest a general result (which is, in fact, valid): For any positive integer n, the slope of the line tangent to the curve $y = x^n$ is given by the formula

$$nx^{n-1} \tag{4}$$

(Exercise 9). For example, the tangent to $y = x^5$ at $x = \frac{4}{3}$ is equal to $5(\frac{4}{3})^4 = 1280/81$.

Formula (4) for the slope can be extended from positive integer exponents n to any real exponent p: If p is any real number not equal to 0, and if $y = x^p$ for $x > 0$, then the slope of the tangent line at a point with the abscissa x is

$$px^{p-1}. \tag{5}$$

We shall not give a complete proof of the general case but only in typical particular cases in the following examples.

EXAMPLE 4

Let us find the formula for the slope of the tangent for the function $f(x) = x^{-2}$. The difference quotient is

$$\frac{x^{-2} - x^{*-2}}{x - x^*}.$$

Factor the numerator and rewrite the ratio:

$$\begin{aligned}\frac{(x^{-1} + x^{*-1})(x^{-1} - x^{*-1})}{x - x^*} &= \frac{x^{-1} + x^{*-1}}{x - x^*}\left(\frac{1}{x} - \frac{1}{x^*}\right)\\ &= \frac{x^{-1} + x^{*-1}}{x - x^*} \cdot \frac{x^* - x}{xx^*}\\ &= -\frac{x^{-1} + x^{*-1}}{xx^*}.\end{aligned}$$

As x^* tends to x, the last ratio above becomes

$$\frac{-2x^{-1}}{x^2} \qquad \text{or} \qquad -2x^{-3}.$$

This agrees with formula (5) for $p = -2$. The method used here is typical of that used for negative integer powers of x. ▲

EXAMPLE 5

Let us find the formula for the slope of the line tangent to the graph of $f(x) = x^{2/3}$. The difference quotient (2) takes the form

$$\frac{x^{2/3} - x^{*2/3}}{x - x^*}.$$

Our object is to divide the numerator by the denominator and then let x^* approach x. This would be relatively easy if the numerator contained only positive integer powers of x and x^*. We change the quotient to this more desirable form by changing the independent variable from x to $u = x^{1/3}$; the 1/3 power "cancels" the root because $x = u^3$. The difference quotient assumes the form

$$\frac{u^2 - u^{*2}}{u^3 - u^{*3}};$$

here we have replaced x and x^* by u^3 and u^{*3}, respectively. We divide the numerator and denominator by $u - u^*$ (using long division); the ratio becomes

$$\frac{u + u^*}{u^2 + uu^* + u^{*2}}.$$

As x^* tends to x, u^* also tends to u, and the ratio tends to

$$\frac{2u}{u^2 + u^2 + u^2} = \frac{2u}{3u^2} \qquad \text{or} \qquad \frac{2}{3u}.$$

Now replace u by $x^{1/3}$; the slope is

$$\frac{2}{3x^{1/3}} \qquad \text{or} \qquad \tfrac{2}{3}x^{-1/3}.$$

This is consistent with formula (5) for $p = \frac{2}{3}$. ▲

The **derivative** of a function is a new function whose domain consists of those points where the former function is differentiable, and whose values are the slopes of tangent lines at the corresponding points. The derivative of a function $f(x)$ is denoted $f'(x)$; its value at the point x is the slope of the line tangent to f at the point x. The formula for the derivative is exactly the same as the formula for the slope of the tangent line. The term **derivative** is based on the fact that the new function is derived from the original one.

A function is not necessarily differentiable at every point of its domain, and so the derivative is not necessarily defined at every point of the domain of the former function.

EXAMPLE 6

Consider the function

$$f(x) = \sqrt{x^2} \qquad \text{(positive square root)}$$

defined for all x. It is equal to x for positive x, $-x$ for negative x, and 0 at the origin (Figure 5.4). We shall show that it is not differentiable at the origin. Let x^* be an arbitrary positive number, and let (x^*, y^*) be the corresponding point on the graph of the function. Put $x = 0$, $y = 0$, and form the difference quotient: Since $y^* = x^*$, the difference quotient is

$$\frac{y - y^*}{x - x^*} \qquad \text{or} \qquad \frac{y^*}{x^*} = 1.$$

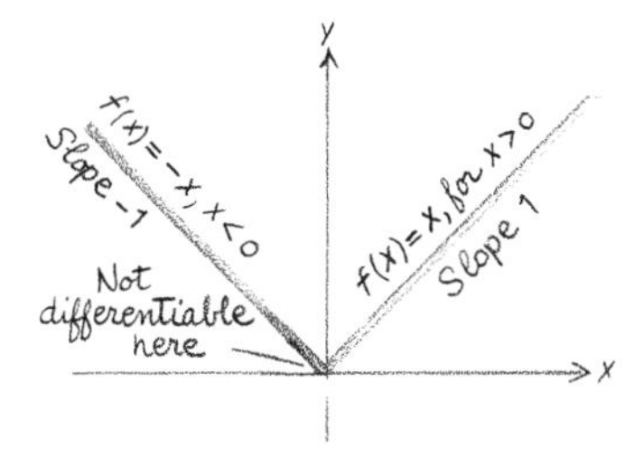

Figure 5.4

As (x^*,y^*) moves toward the origin, the difference quotient remains equal to 1, so that the "tangent line from the right" has the slope 1. Similarly, we find that the "tangent line from the left" has the slope -1. Therefore, the function is not differentiable at the origin; indeed, the difference quotients approach different limits from the right and from the left, respectively. ▲

By formula (5), the derivative of the function $f(x) = x^p$ is $f'(x) = px^{p-1}$. This requires the qualification that the expressions x^p and x^{p-1} are real numbers (see Section 4.3); thus $x \geq 0$ in our applications.

EXAMPLE 7

Let $f(x)$ be x^4; find $f(0)$, $f(1)$, $f(2)$, and $f(-2)$; and $f'(0)$, $f'(1)$, $f'(2)$, and $f'(-2)$. According to formulas (4) or (5), we have $f'(x) = 4x^3$ and so

$$f(0) = 0^4 = 0, \qquad f'(0) = 4(0)^3 = 0,$$
$$f(1) = 1^4 = 1, \qquad f'(1) = 4(1)^3 = 4,$$
$$f(2) = 2^4 = 16, \qquad f'(2) = 4(2)^3 = 32,$$
$$f(-2) = (-2)^4 = 16, \qquad f'(-2) = 4(-2)^3 = -32. \ ▲$$

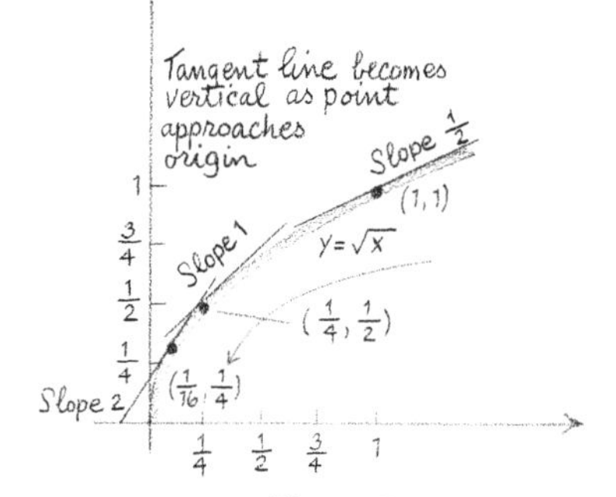

Figure 5.5

EXAMPLE 8

Consider $f(x) = x^{1/2}$, defined for $x \geq 0$. The derivative is $f'(x) = \frac{1}{2}x^{-1/2}$. The derivative is not defined for $x = 0$ because the tangent lines near the origin approach a vertical position (Figure 5.5). Note that the value of the derivative increases as the point on the graph moves down to the origin:

$$f(1) = 1, \quad f'(1) = \tfrac{1}{2}; \qquad f(\tfrac{1}{4}) = \tfrac{1}{2}, \quad f'(\tfrac{1}{4}) = 1;$$
$$f(\tfrac{1}{16}) = \tfrac{1}{4}, \quad f'(\tfrac{1}{16}) = 2; \qquad f(\tfrac{1}{64}) = \tfrac{1}{8}, \quad f'(\tfrac{1}{64}) = 4. \ ▲$$

Derivatives have many applications; these are given throughout this chapter. At this point we mention and illustrate just one application: curve sketching.

EXAMPLE 9

Let us use the derivative to sketch the graph of the function $f(x) = x^{3/2}$ over the domain $0 \leq x \leq 4$. By formula (5), the derivative is $f'(x) = \frac{3}{2}x^{1/2}$. Here are the values of the function and the derivative for $x = 0$, 1, and 4:

x	0	1	4
$f(x)$	0	1	8
$f'(x)$	0	$\frac{3}{2}$	3

Plot the points (0,0), (1,1), and (4,8), and then pass through them lines of slopes 0, $\frac{3}{2}$, and 3, respectively (Figure 5.6). The graph of the function is then filled in, guided by the points and the tangent lines (Figure 5.7). ▲

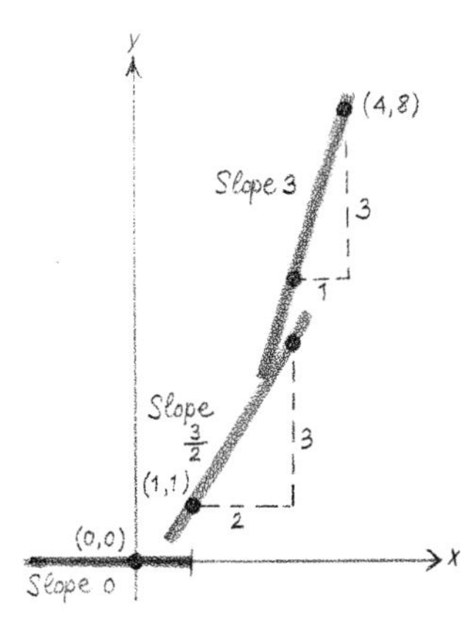

Figure 5.6

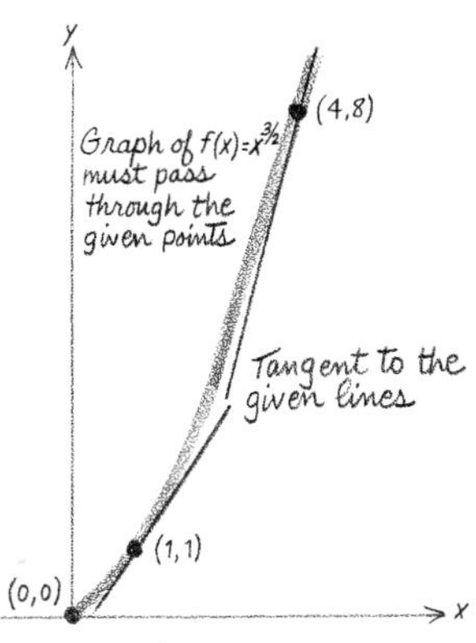

Figure 5.7

Using the derivative, we can find not only the slope of the tangent line, but also the **equation** of the line.

EXAMPLE 10

Find the equation of the line tangent to the curve $y = x^3$ at the point with the abscissa $x = \frac{1}{2}$. By formula (4), the derivative of $f(x) = x^3$ is $f'(x) = 3x^2$. The ordinate on the curve corresponding to the abscissa $\frac{1}{2}$ is $y = (\frac{1}{2})^3 = \frac{1}{8}$. The slope of the tangent line is $3(\frac{1}{2})^2 = \frac{3}{4}$. Therefore, the tangent line passes through the point $(\frac{1}{2}, \frac{1}{8})$ with slope $\frac{3}{4}$. By the point-slope formula (Section 1.2), the equation of the line is

$$y - \tfrac{1}{8} = (\tfrac{3}{4})(x - \tfrac{1}{2}). \ ▲$$

5.1 EXERCISES

In Exercises 1 and 2 find the formula for the derivative of the function using the **methods** *of Examples 1 to 3. Do not use the given formulas.*

1 (a) $f(x) = 4$
(b) $f(x) = 2x + 3$
(c) $f(x) = 10x^2$
(d) $f(x) = x^2 + x$
(e) $f(x) = 7x - x^2$
(f) $f(x) = x^3 - 2x^2$
(g) $f(x) = x^4$
(h) $f(x) = -4x^7$

2 (a) $f(x) = -2$
(b) $f(x) = 3x + 4$
(c) $f(x) = 5x^2$
(d) $f(x) = 2x^2 + x$
(e) $f(x) = 5x - 2x^2$
(f) $f(x) = 2x^3 - x^2$
(g) $f(x) = x^5$
(h) $f(x) = x^6$

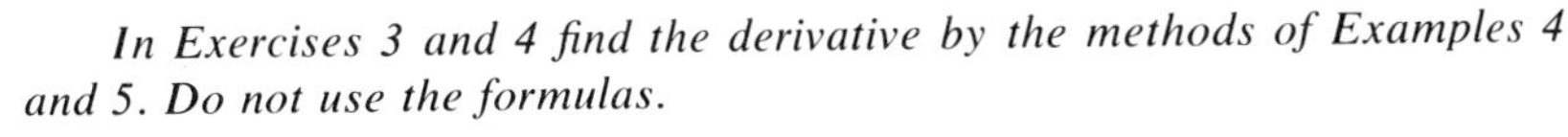

In Exercises 3 and 4 find the derivative by the methods of Examples 4 and 5. Do not use the formulas.

3 (a) $f(x) = 1/x$
(b) $f(x) = -2/x$
(c) $f(x) = x^{1/4}$
(d) $f(x) = x^{1/2} + x^{1/3}$
(e) $f(x) = x + x^{-1}$
(f) $f(x) = x^{-3}$
(g) $f(x) = x^{-1/2}$
(h) $f(x) = x^2 - 2/x^2$

4 (a) $f(x) = 3x^{1/2}$
(b) $f(x) = -1/x^3$
(c) $f(x) = 2x^{4/3}$
(d) $f(x) = x + x^{1/4}$
(e) $f(x) = 2x - x^{-2}$
(f) $f(x) = x^{-4}$
(g) $f(x) = x^{-1/3}$
(h) $f(x) = 8x + x^2 - 2/x$

In Exercises 5 and 6 find the derivatives by means of the general formula (5). Then sketch the graph of the function over the indicated domain by the method of Example 9. Use the values of the derivative at the given abscissas x.

5 (a) $f(x) = x^2$ on $-1 \leq x \leq 2$, $x = -1, 0, 1$
(b) $f(x) = -x^3$ on $-1 \leq x \leq 1$, $x = -1, 0, 1$
(c) $f(x) = x^{-1}$ on $0 < x \leq 1$, $x = \frac{1}{2}, \frac{3}{4}, 1$
(d) $f(x) = \frac{1}{2}x^{-2}$ on $1 \leq x \leq 2$, $x = 1, 2$
(e) $f(x) = x^{2/3}$ on $0 \leq x \leq 27$, $x = 1, 9, 27$
(f) $f(x) = x^{-2/3}$ on $0 < x \leq 27$, $x = 1, 9, 27$

6 (a) $f(x) = \frac{1}{2}x^2$ on $-2 \leq x \leq 1$, $x = -2, -1, 0, 1$
(b) $f(x) = x^{-3}$ on $0 < x \leq 1$, $x = \frac{1}{4}, \frac{1}{2}, \frac{3}{4}, 1$
(c) $f(x) = x^{-2}$ on $\frac{1}{2} \leq x \leq 2$, $x = \frac{1}{2}, 1, 2$
(d) $f(x) = \frac{1}{4}x^4$ on $0 \leq x \leq 2$, $x = 0, 1, 2$
(e) $f(x) = x^{1/3}$ on $0 \leq x \leq 27$, $x = 1, 9, 27$
(f) $f(x) = x^{-1/2}$ on $0 < x \leq 4$, $x = 1, 4$

In Exercises 7 and 8 find the equation of the line tangent to the curve at the indicated abscissa value.

7 (a) $y = 1 - 2x^2$ at $x = 2$
(b) $y = 4x^2 - 10x + 1$ at $x = 3$
(c) $y = x^3$ at $x = 1$
(d) $y = x + x^{-1}$ at $x = -2$
(e) $y = x^{2/3}$ at $x = 8$
(f) $y = x^{-4/5}$ at $x = 1$

8 (a) $y = 2x^2$ at $x = 2$
(b) $y = 4x - x^2$ at $x = 2$
(c) $y = 4 - x^2$ at $x = 1$
(d) $y = 2x - 3/x$ at $x = 2$
(e) $y = x^{3/4}$ at $x = 16$
(f) $y = x^{-3/5}$ at $x = -32$

9 Let A and B be two numbers that are not equal. Show by long division that

$$\frac{A^n - B^n}{A - B} = A^{n-1} + A^{n-2}B + A^{n-3}B^2 + \cdots + B^{n-1} \qquad (n \text{ terms}).$$

Use this formula to derive the assertion that the slope of the tangent line for $y = x^n$ is nx^{n-1}. [*Hint:* What becomes of the right side of the equation above when $A = B$?]

5.2 Derivatives of the Exponential, Logarithm, and Trigonometric Functions

In this section, we derive the formulas for the derivatives of the transcendental functions:

Function	Derivative	
$\log x$ (base e)	$1/x$	(1)
e^x	e^x	(2)
$\sin x$	$\cos x$	(3)
$\cos x$	$-\sin x$	(4)

arcsin x	$(1-x^2)^{-1/2}$	(5)
arccos x	$-(1-x^2)^{-1/2}$	(6)

By numerical calculation let us illustrate the fact that the derivative of log x is $1/x$. This is meant not as a proof but as an indication of the consistency of the formula with our geometric concept of the derivative.

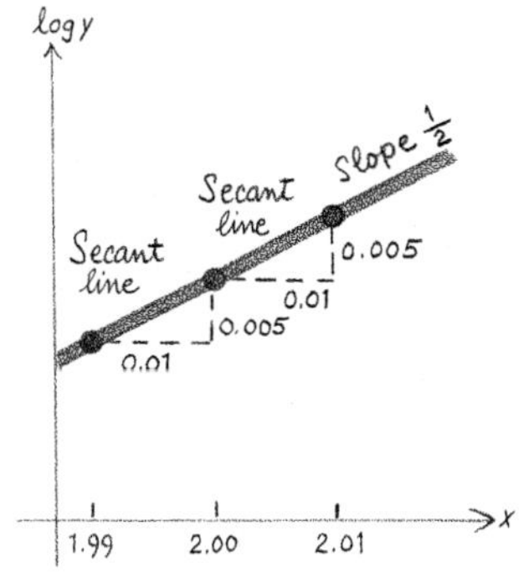

Figure 5.8

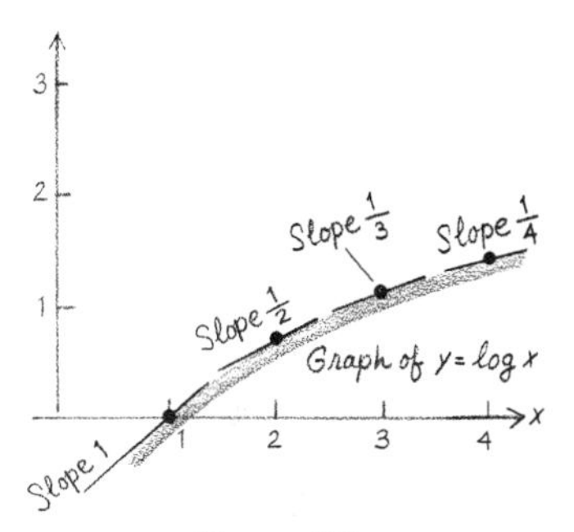

Figure 5.9

EXAMPLE 1

Using the values in Table III of the Appendix, we shall find the slope of the secant line drawn to the curve of the log function. Let $x = 2$ and $x^* = 2.01$; then the slope of the secant line through the corresponding points on the curve is the difference quotient

$$\frac{y^* - y}{x^* - x} = \frac{\log 2.01 - \log 2}{2.01 - 2}$$

$$= \frac{0.6981 - 0.6931}{0.01} = 0.5000 \text{ (approximately)}.$$

We get the same slope for the secant drawn from $x^* = 1.99$:

$$\frac{y - y^*}{x - x^*} = \frac{\log 2 - \log 1.99}{2 - 1.99}$$

$$= \frac{0.6931 - 0.6881}{0.01} = 0.5000 \qquad \text{(approximately)}$$

(Figure 5.8). Since the points (x^*,y^*) and (x,y) are so close to each other, the slope of the secant line is practically the same as that of the tangent line; thus, the derivative of the logarithm appears to be equal to $\frac{1}{2}$ (approximately). The general formula (1) for the derivative of log x states that the derivative is the reciprocal of the abscissa (Figure 5.9). The reader can verify by means of the same calculation that (1) is consistent with other values of the logarithm (Exercise 1). ▲

While the numerical calculation indicated in Example 1 might be sufficient to convince the reader of the validity of (1), we will also present the direct mathematical argument for (1). We indicate the proof in the particular case $x = 1$, and leave the general case (for all $x > 0$) as an exercise (Exercise 7). Recall (Section 4.3) that $(1 + 1/n)^n$ is approximately equal to the number e if n is large; this is equivalent to the fact that $(1 + h)^{1/h}$ is approximately equal to e if h is small. Converting this relation to that between the respective logarithms, we find

$$\log(1 + h)^{1/h} \text{ approximately equal to } \log e$$

for small h. By the "power property" of logarithms,

$$\log(1 + h)^{1/h} = (1/h)\log(1 + h),$$

and the relation $\log e = 1$, we deduce that

$$\frac{\log(1+h)}{h} \text{ is approximately equal to 1}$$

for small h. But the ratio above is the difference quotient of the log function at $x = 1$:

$$\frac{\log(1+h) - \log 1}{h}$$

(because $\log 1 = 0$). We conclude that the difference quotient of the logarithm function tends to 1 at $x = 1$.

The formula for the derivative of e^x is obtained from the formula for the derivative of the logarithm, and the relation between e^x and the logarithm. Recall (Section 4.4) that one function is obtained from the other by interchanging x and y, or, equivalently, reflecting the graph through the diagonal line. If (x,y) and (x^*,y^*) are two points on the graph of e^x, then (y,x) and (y^*,x^*) are the corresponding points on the graph of the log function. It follows that the difference quotient of the first function is the reciprocal of the difference quotient for the second function; hence, by the "limiting process" for the tangent line, the slope of the line tangent to the graph of the first function at (x,y) is the reciprocal of the slope of the line tangent to the graph of the second function at the corresponding point (y, x) (Figure 5.10). Now a point on the graph of the exponential function $y = e^x$ is of the form (x,e^x), and so the corresponding point on the graph of the log function is of the form (e^x, x). Formula (1) states that the slope of the line tangent to the graph of the log function at a given point is the reciprocal of the abscissa of that point; thus, the slope of the line through (e^x, x) is $1/e^x$. By the previous remarks about the relation between the slopes of the exponential and logarithm functions, it follows that the slope of the line tangent to the graph of $y = e^x$ at the point (x,e^x) is the reciprocal of $1/e^x$; namely, it is equal to e^x.

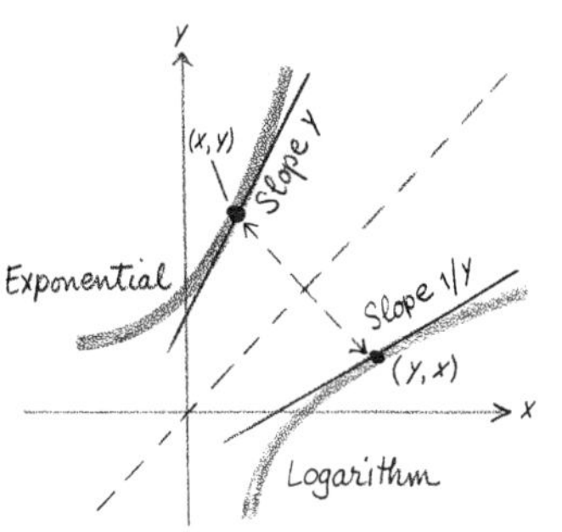

Figure 5.10

Next we illustrate the relation (3) for a particular value of x.

EXAMPLE 2

We will calculate the slope of the line tangent to the sine function by means of the numerical values of the function given in Table IV; here the independent variable is in radians. Let us numerically estimate the derivative of $\sin x$ at $x = 0.30$. We have $\sin 0.31 = 0.3051$ and $\sin 0.30 = 0.2955$; hence, the difference quotient for $x = 0.30$ and $x^* = 0.31$ is

$$\frac{\sin 0.31 - \sin 0.30}{0.31 - 0.30} = \frac{0.3051 - 0.2955}{0.01} = 0.96.$$

This agrees, to 2 decimal places, with the value of cos 0.30, which is 0.9553 (Table IV). So the line tangent to the graph of the sine function has a slope approximately equal to the value of the cosine function at that point. The reader can verify that this relation holds at other values of x. This illustrates formula (3). ▲

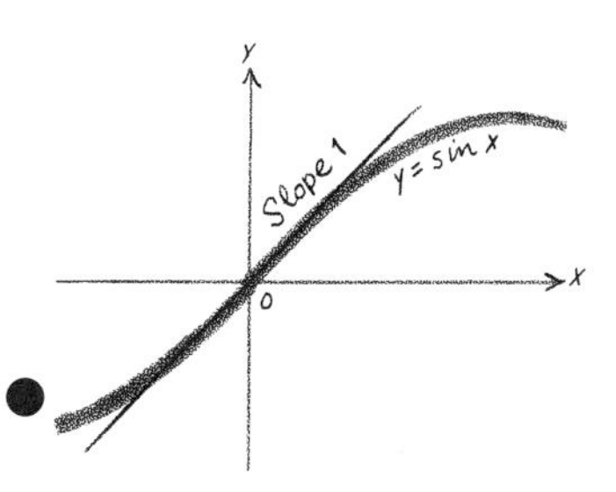

Figure 5.11

We also give the direct mathematical argument for (3). First we give the geometric basis of the proof for the particular case $x=0$: we will show that the derivative of the sine is equal to cos 0 (which is 1) at $x=0$ (Figure 5.11). The slope of the secant line through the origin and a point (x^*,y^*) on the graph of $y=\sin x$ is the difference quotient

$$\frac{y^*-0}{x^*-0}=\frac{y^*}{x^*}=\frac{\sin x^*}{x^*}.$$

The numerator $\sin x^*$ represents the ordinate on an arc cut out on a circle of radius 1 by an angle of x^* radians. The denominator x^* is, by definition of radian measure, the length of the arc. As x^* becomes smaller and smaller, the ratio of the ordinate to the length of the arc tends to 1; indeed, the vertical line of length $\sin x^*$ becomes "indistinguishable" from the arc (Figure 5.12). A complete algebraic proof of this is outlined in Exercise 6. This finishes the case $x=0$.

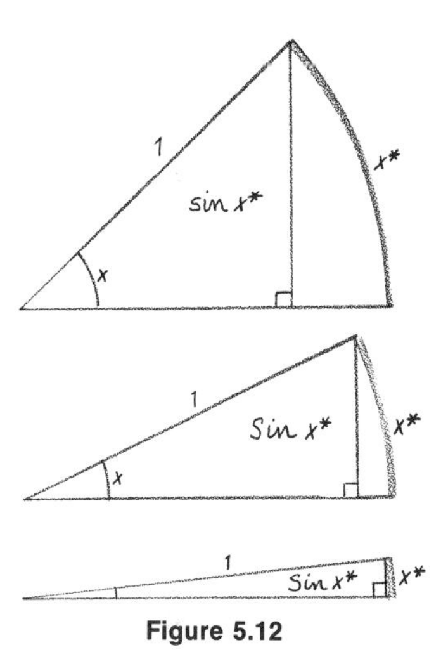

Figure 5.12

The proof of the general case is similar. Let (x,y) and (x^*,y^*) be two points on the curve $y=\sin x$. The absolute difference of $x-x^*$ is the length of a small arc, and the absolute difference of $\sin x-\sin x^*$ is the side of a right triangle, adjacent to an angle of $x-x^*$ radians. The arc is very nearly the same as the hypotenuse of the right triangle when the angle is small. It follows from the formula defining the cosine in terms of ratios of sides of a right triangle (Section 4.6) that

$$\textbf{cos } x \textbf{ is approximately } \frac{\sin x^*-\sin x}{x^*-x}$$

(Figure 5.13).

The relation (4) can be deduced from (3). Recall that the cosine is the same as the sine except for a phase shift of $\pi/2$ radians: $\cos x=\sin(x+\pi/2)$ (Section 4.8). The relation (3) states that the differentiation operation shifts the sine $\pi/2$ units ahead:

$$\textbf{derivative of } \sin x=\cos x=\sin\left(x+\frac{\pi}{2}\right);$$

therefore,

$$\textbf{derivative of } \cos x=\textbf{derivative of } \sin\left(x+\frac{\pi}{2}\right)$$

$$=\sin\left(x+\frac{\pi}{2}+\frac{\pi}{2}\right)=\sin(x+\pi).$$

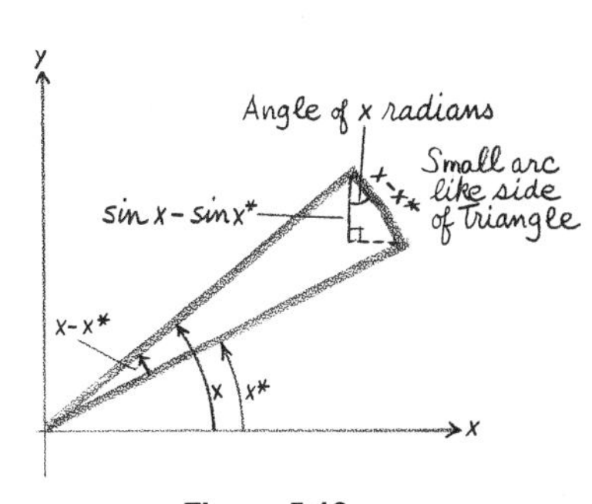

Figure 5.13

The last expression, $\sin(x + \pi)$, is equal to $-\sin x$ because the phase shift by π radians "flips over" the graph of the sine function:

$$\sin(x + \pi) = -\sin x. \tag{7}$$

It follows that the derivative of $\cos x$ is $-\sin x$ (Exercise 8).

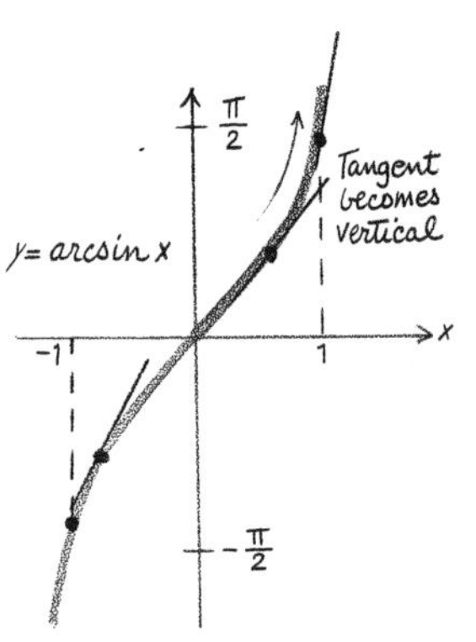

Figure 5.14

Formula (5) for the derivative of the arcsine function is obtained from the sine just as the derivative of e^x was obtained from that of the log. Let $(x, \arcsin x)$ be a point on the graph of the arcsine function; then $(\arcsin x, x)$ is the mirror-image point on the graph of the sine function. The derivative of the latter function at the point $(\arcsin x, x)$ on the graph is equal to the cosine of the abscissa, namely, $\cos(\arcsin x)$. The latter is equal to $\sqrt{1 - x^2}$ (Section 4.7). The derivative of the arcsine function at the point $(x, \arcsin x)$ is therefore equal to the reciprocal, $1/\sqrt{1 - x^2}$. This function is not differentiable at $x = \pm 1$ (Figure 5.14).

The derivation of formula (6) for the arccosine is similar. The derivatives of the tangent and arctangent will be obtained in Section 5.4.

5.2 EXERCISES

1 Compare the derivative of $\log x$ with the approximation obtained by the method of Example 1 at the following values of x:
(a) $x = 3$ (b) $x = 4.5$ (c) $x = 20$ (d) $x = 36$

2 Compare the derivative of e^x with the approximation obtained by computing the difference quotient at these values of x:
(a) $x = 0.15$ (b) $x = 0.30$ (c) $x = 0.60$ (d) $x = 2.05$ (e) $x = 4.10$

In Exercises 3 and 4 find the derivative at the given point by applying formulas (3) or (4) for the derivative and also by computing the difference quotients using Table IV.

3 (a) $\sin x$ at $x = 0.1$ (b) $\cos x$ at $x = 0.1$ (c) $\sin x$ at $x = 0.5$ (d) $\cos x$ at $x = 0.5$ (e) $\sin x$ at $x = 0.9$ (f) $\cos x$ at $x = 0.8$

4 (a) $\cos x$ at $x = 0.25$ (b) $\sin x$ at $x = 0.14$ (c) $\cos x$ at $x = 0.43$ (d) $\sin x$ at $x = 0.38$ (e) $\cos x$ at $x = 1.00$ (f) $\sin x$ at $x = 1.25$

5 Calculate the values of the ratio of $(\sin x)/x$ for $x = 0.10, 0.09, 0.06, 0.03, 0.01$. What does this indicate about the "limiting value" of the ratio as x tends to 0?

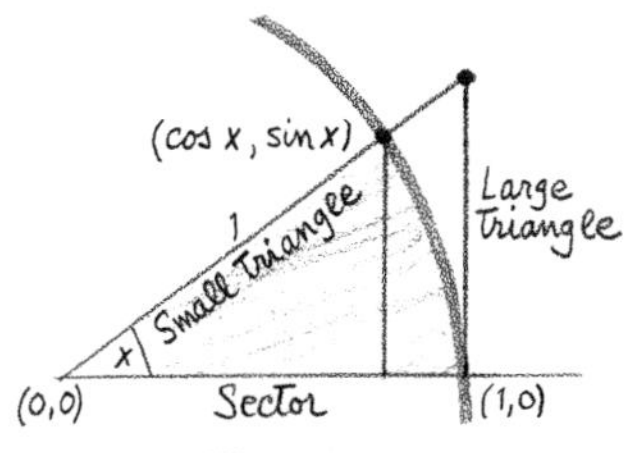

Figure 5.15

6 Here is the outline of the proof that $(\sin x)/x$ tends to 1 as x tends to 0; refer to Figure 5.15.

(a) Show that the area of a sector of a circle of radius r and central angle of x radians is equal to $\frac{1}{2}xr^2$. [*Hint:* What is the area for $x = 2\pi$?]

(b) Show that the area of the shaded sector is $x/2$.

(c) Show that the area of the larger triangle is

$$\frac{\sin x}{2\cos x}.$$

[*Hint:* The side opposite x is of length $\tan x$.]

(d) Show that the area of the smaller triangle is

$$\tfrac{1}{2}\cos x \sin x.$$

(e) Show how the double inequality

$$\cos x \sin x < x < \frac{\sin x}{\cos x}$$

follows from (b), (c), and (d).

(f) Deduce that $\cos x < (\sin x)/x < 1/\cos x$. What happens to $(\sin x)/x$ when x approaches 0?

7 By using the addition property of the logarithm (Section 4.4) show that the difference quotient

$$\frac{\log(x+h) - \log x}{h}$$

may be written in the form

$$\frac{1}{x}\,\frac{\log(1 + h/x)}{h/x}.$$

Show that this tends to $1/x$ as h tends to 0.

8 Verify identity (7) by comparing the graphs of the sine and its shift by π units.

5.3 The Derivative as an Instantaneous Rate of Change; Limits

There are several places where we have used, either implicitly or explicitly, the concept of **limit:**

1. The slope of the tangent line is the limit of the slope of the secant line as the point (x^*,y^*) approaches (x,y) (Sections 1.7, 5.1, and 5.2).
2. The powers P^n of certain stochastic matrices have limiting forms (Section 3.11).

3. The exponential b^x for real x is the limiting form of the same expression for rational x. The base e of the natural logarithm is the limit of $(1 + 1/n)^n$ for n increasing to infinity (Section 4.3).
4. The distance traveled by an object under a continuous force is the limit of the distance it travels when the force is broken up into pieces and distributed over the time interval (Section 4.5).

The limit concept is the basis of calculus. We shall formally define it. The relation between independent and dependent variables can be viewed as **static:** For each x in the domain, there is a unique y in the range such that $y = f(x)$. But this relation also has a **dynamic** aspect: **Changes** in the independent variable determine **corresponding changes** in the dependent variable. The dependent variable is the puppet, and the independent variable is the hand of the puppeteer. The static aspect of the relation between them is the correspondence between the limbs of the puppet and the fingers of the puppeteer. The dynamic aspect is the correspondence of the **movements** of the limbs and the fingers (Figure 5.16). For example, a movement of the puppet's head might correspond to that of the forefinger of the puppeteer.

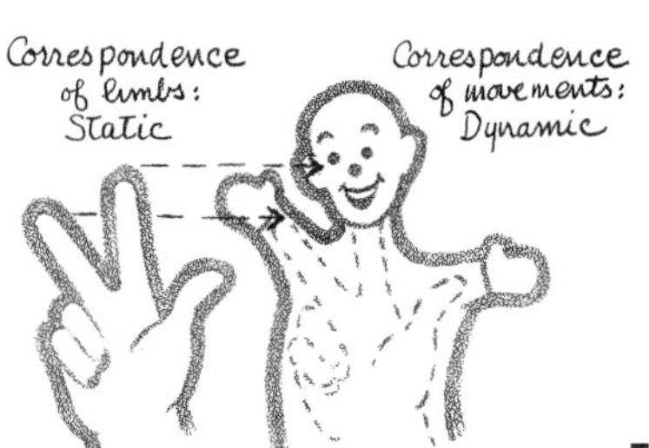

Figure 5.16

The limit concept involves a "local" dynamic relation: The "closeness" of the independent variable to a certain point implies the "closeness" of the dependent variable to a corresponding point. Let x be a fixed abscissa point and y an ordinate, and let the domain of the function f include values arbitrarily close to but not necessarily including x; and, finally, let x^* be an arbitrary point in the domain and $y^* = f(x^*)$ the corresponding point in the range. We say that y^* approaches y as x^* approaches x if y^* can be made arbitrarily close to y by choosing x^* sufficiently close to, but not equal to, x. In this case y is called the **limit of y^* as x^* tends to x,** and is denoted

$$y = \lim_{x^* \to x} y^*.$$

This signifies that the movement of the independent variable toward x is reflected in the corresponding movement of the dependent variable toward y. In the case of the puppet, small movements of the fingers cause small movements of the puppet's limbs; therefore, if the finger is moved a little bit from a given position, the puppet's limb will move by a correspondingly small amount.

The limit concept has a more precise mathematical definition. We say that y^* approaches y as x^* approaches x if:

For every positive real number ϵ, no matter how small, there exists a corresponding positive number δ, such that

y^* is within ϵ units of y

whenever

x^* is within δ units of x

(Figure 5.17). In other words, if we are given an **arbitrary** "tolerance" for differences from y, then we can find a **sufficiently** small "tolerance" for x such that all x^* within the given tolerance of x correspond to y^* which are within the given tolerance of y.

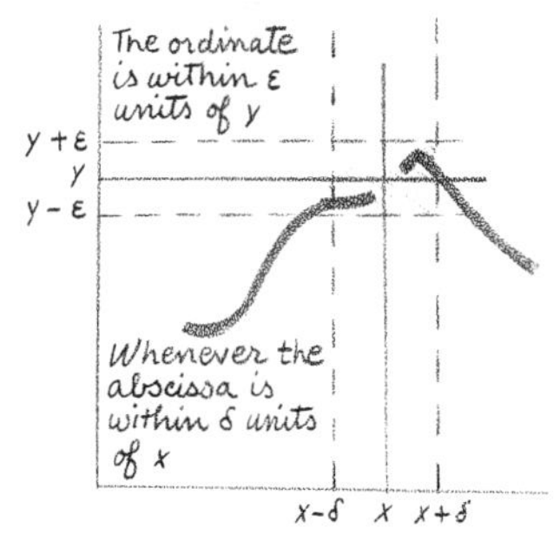

Figure 5.17

The formal mathematical definition of limit, just given, is known as the "epsilon-delta" definition of limit. It is used extensively in higher mathematics. We will not refer to it in this book because our main interest is in the intuitive concepts of the calculus and their applications. We have mentioned it only because it is the central concept in the theoretical development of calculus. But limits in various forms will arise throughout the book and the reader should have at least some precise information about them.

In the previous definition of **limit** the abscissa x and ordinate y were finite. The definition can be extended to the case where either or both are infinite; that is, $x = +\infty$ or $-\infty$, or $y = +\infty$ or $-\infty$. Suppose first that $x = \infty$ and y is finite, and that the domain of f includes arbitrarily large positive values x^*. We say that y^* approaches y as x^* tends to infinity if y^* can be brought arbitrarily close to y by choosing x^* sufficiently large; and we write

$$\lim_{x^* \to \infty} y^* = y$$

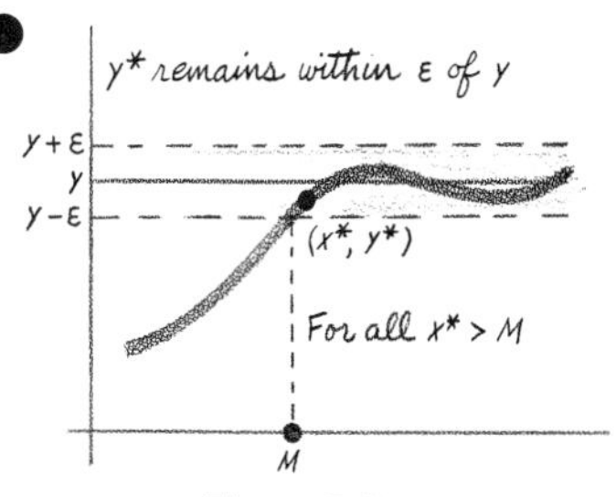

Figure 5.18

(Figure 5.18). (The definition of the limit for $x^* \to -\infty$ is similar.) More precisely, we say that **y^* approaches y as x^* approaches ∞ if for every $\epsilon > 0$ there exists a positive number M, sufficiently large, such that y^* is within ϵ units of y whenever x^* exceeds M (Figure 5.18).**

An example of such a limit is the **limit of a sequence** of real numbers. A **sequence** is an infinite set of real numbers labeled by the integers; for example, the set

$$(1 + 1), \qquad (1 + 1/2)^2, \qquad (1 + 1/3)^3, \ldots, \qquad (1 + 1/n)^n, \ldots$$

is a sequence. A sequence can be viewed as a function whose domain consists of an infinite set of integers, usually the positive integers or the nonnegative integers. The independent variable is denoted by n in place of x, and the corresponding values of the dependent variable are denoted $\{a_n\}$ (with a subscript) in place of $f(n)$. The sequence is said to approach the limit L if the dependent variable a_n tends to the limit L as $n \to \infty$ in the sense of the definition of limit for $x = \infty$. In other words, **the sequence approaches the limit L if the terms of the sequence are arbitrarily close to L whenever their indices are sufficiently large.** The limit relation is denoted

$$\lim_{n \to \infty} a_n = L.$$

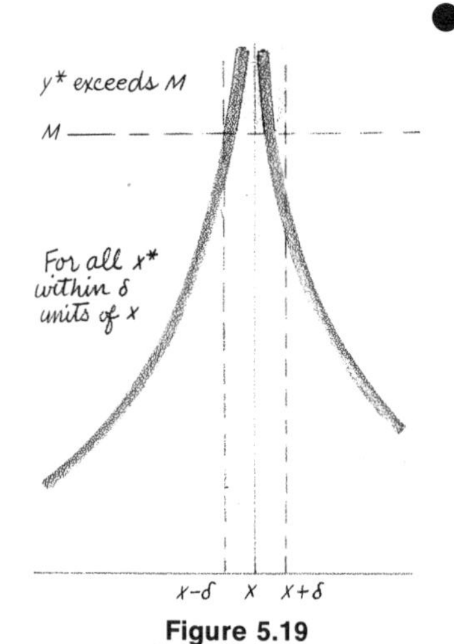

Figure 5.19

Finally we define **limit** for $y = \infty$. We say that y^* tends to ∞ as x^* tends to x if y^* can be made arbitrarily large (in the positive sense) by choosing x^* sufficiently close to, but not equal to, x. In terms of the precise definition, this means that **for every positive number M, no matter how large, there exists a positive δ such that $y^* > M$ whenever x^* is within δ units of x (Figure 5.19).** (The definition for $y = -\infty$ is similar.)

Another kind of limit is for x and y both infinite; we omit the details.

Let us point out the significance of the definition of limit in the examples mentioned above:

1. The slope of the secant line can be brought arbitrarily close to that of the tangent by choosing x^* sufficiently close to x.
2. The power P^n of a stochastic matrix can be brought arbitrarily close to the limiting matrix by choosing the power n sufficiently large.
3. The quantity $(1 + 1/n)^n$ can be made arbitrarily close to e by choosing n sufficiently large.

A question often asked in practice is: If we are given the "tolerance" for y, the number ϵ, how do we find the tolerance for x, the number δ? This is, in general, a challenging and difficult question. At our level we will be concerned only with the **existence** of limits but not the calculation of the tolerances.

In Sections 5.1 and 5.2 our definition of the derivative was based on geometry: it is the slope of the tangent line. Now we show the meaning of the derivative in terms of the dynamic relation between the independent and dependent variables.

The derivative measures the rate of change of the dependent variable with respect to the independent variable. Let $y = f(x)$ be a given function. The difference quotient $(y - y^*)/(x - x^*)$ is equal to

$$\frac{f(x) - f(x^*)}{x - x^*}. \tag{1}$$

By formula (3) of Section 5.1, the derivative is the limit

$$f'(x) = \lim_{x^* \to x} \frac{f(x) - f(x^*)}{x - x^*}. \tag{2}$$

The numerator in (1) is the change in the dependent variable as the independent variable changes from x^* to x; therefore, the ratio (1) is the **rate** of change of the latter with respect to the former. The derivative is the limit of the rate as the interval considered is taken smaller and smaller, and finally crystallizes at the point x. The derivative is called the **instantaneous** rate of change of the function. It measures the "local" change.

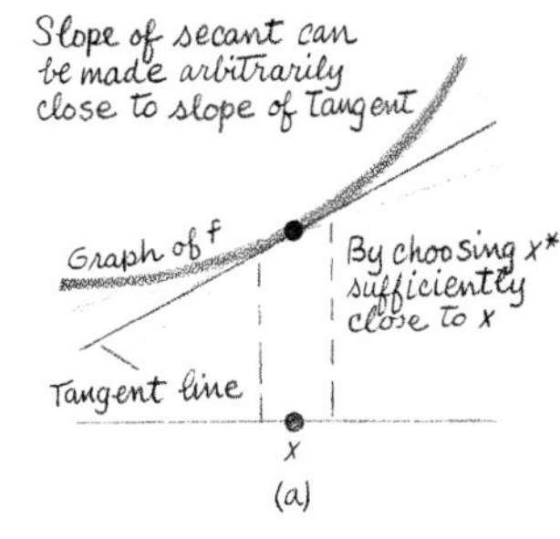

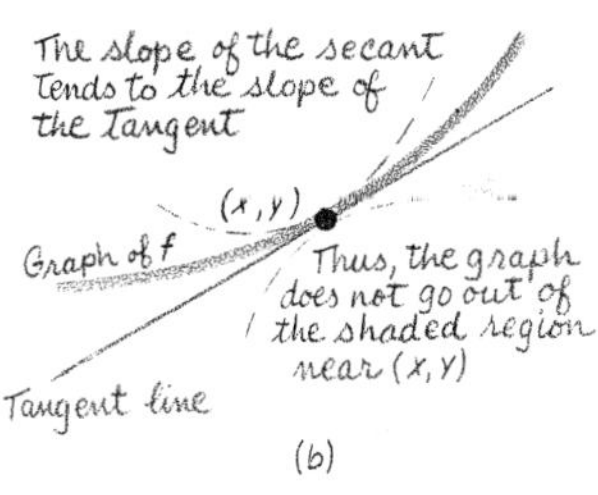

Figure 5.20

Let us consider the geometric interpretation of the relation (2) in terms of the graph of the function near the point x. The difference quotient (1) is the slope of the secant line through (x,y) and (x^*,y^*). Equation (2) states that the slope of this line can be made arbitrarily close to that of the tangent line by choosing the second point sufficiently close to the first [Figure 5.20(a)]. This implies that the graph of the function cannot "wiggle" too much near the point x [Figure 5.20(b)]. It appears in the latter figure that the function is "approximated" near x by the linear function whose graph is the line tangent to the graph at x. This is the **local linear approximation** of a differentiable function.

A number x is said to be a **point of increase** for a function f if, in a small interval to the left of x, the values of the function are less than $f(x)$, and, in a small interval to the right of x, the values of f are greater than $f(x)$ (Figure 5.21). When f is differentiable at x there is a simple criterion to determine whether x is such a point:

If $f'(x) > 0$, then x is a point of increase.

The geometric basis of the proof is found in the local linear approximation in Figure 5.20(b) above: the graph of the function is bound to stay close to the tangent line near the point x. The direct proof is: Since the difference quotient $(f(x) - f(x^*))/(x - x^*)$ tends to a positive limit, $f'(x)$, as x^* tends to x, it is **positive** for x^* near x; therefore, $f(x) - f(x^*)$ has the same sign as $x - x^*$.

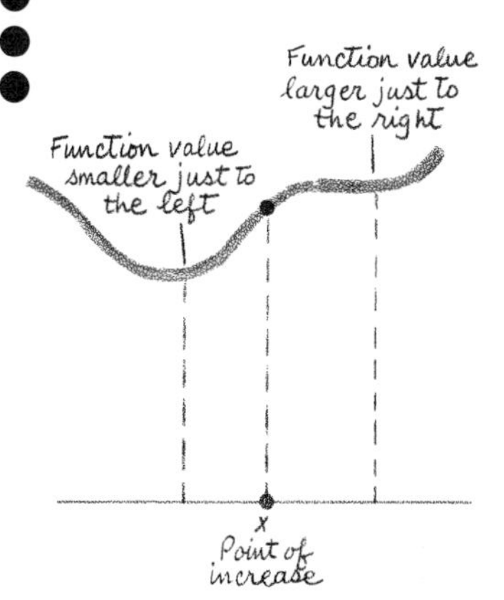

Figure 5.21

The graph is "rising" at a point of increase. For example, the derivative of the function $f(x) = x^2$ is $f'(x) = 2x$. The latter is positive for every $x > 0$; therefore, every such x is a point of increase for the function, and the graph "rises" at every $x > 0$.

A **point of decrease** is similar to a point of increase except that the inequalities are reversed: If $x^* < x$, then $f(x^*) > f(x)$, and if $x^* < x$, then $f(x^*) > f(x)$, for all x^* close to x. When f is differentiable at x, then $\boldsymbol{x}$ **is a point of decrease if** $\boldsymbol{f'(x) < 0}$ (Figure 5.22).

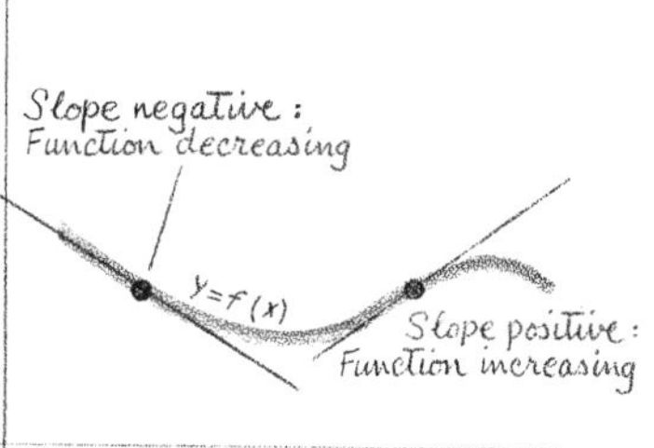

Figure 5.22

There is another commonly used symbol for the derivative:

$$\frac{dy}{dx}. \tag{3}$$

This is read "dy over dx," or, more simply, "dy dx." The letter d stands for **difference** or **differential.** The ratio (3) represents the limit of the difference quotient of y with respect to x. It is not to be interpreted as an algebraic fraction involving factors d, x, and y; it is an indecomposable symbol for the derivative, interchangeable with $f'(x)$.

We mention two applications of the **concept** of the derivative: to motion in physics, and marginals in economics.

EXAMPLE 1

An object is placed in a field of continuous constant force; for example, a ball is released from the top of a tower and placed in the earth's field of gravitation. Let g be the constant of acceleration. Then the formula for the distance covered in x seconds is $f(x) = \frac{1}{2}gx^2$, and the velocity after x seconds is gx (Section 4.5). Note that the velocity function is the derivative of the distance function:

$$f'(x) = gx.$$

The velocity is the instantaneous rate of change of the distance with respect to time, or the **instantaneous speed.** It tells the speed of the object at any particular point in time. Similarly, the acceleration function, which is the constant g, is the derivative of the velocity function; it is the instantaneous rate of change of the velocity with respect to time. ▲

EXAMPLE 2

Let the independent variable x represent the output of a firm measured in items produced; and let $f(x)$ be the cost of producing x items. f is the **cost function.** In some industries, the variable x assumes only integer values; for example, in the aircraft industry, the output of a company is measured in whole planes produced, and there is a significant weight attached to each plane produced. However, in most industries, the number of items produced is hundreds or thousands; for example, a steel company produces thousands of tons of steel, and a mine produces hundreds of tons of coal. In the latter case, the output may be considered to be a variable which is measured continuously, not in whole units; for example, we may report a production of 3.2 hundreds of tons of coal. Here the cost function has a domain given by the set of positive real numbers, not just integers. In mathematical economics, the **marginal cost** of producing x units is defined as the derivative $f'(x)$ of the cost function. It is the instantaneous rate of change of the cost with respect to the output (Figure 5.23). It is interpreted as the cost of producing an additional unit of output when the output has already attained the level x. In this case $f'(x)$ is called the **marginal cost function.**

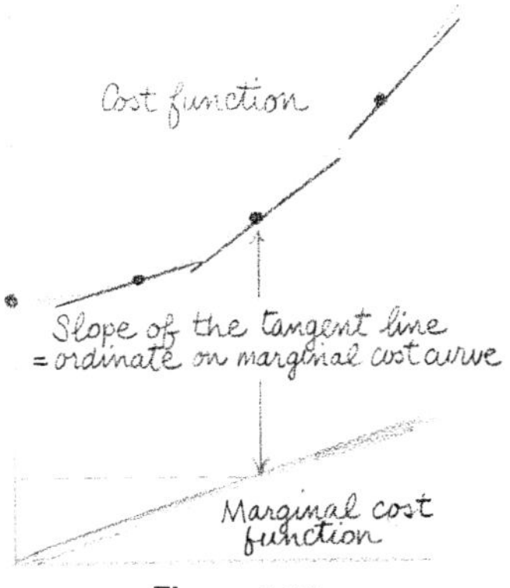

Figure 5.23

The **revenue function** $f(x)$ is the amount earned (in monetary units) from an output of x units. The **marginal revenue function** is the derivative $f'(x)$: it represents the revenue derived from an additional unit of output at the level x. The revenue of a firm typically increases with the output, but at an increasingly slower rate; thus, the marginal revenue tends to decrease with the level of output (Figure 5.24). For example, the automobile industry of the United States might find a relatively large difference in the revenue derived from annual productions of 10 million cars and 11 million; however, it would find a much smaller difference between the revenues from the production of 210 million and 211 million cars.

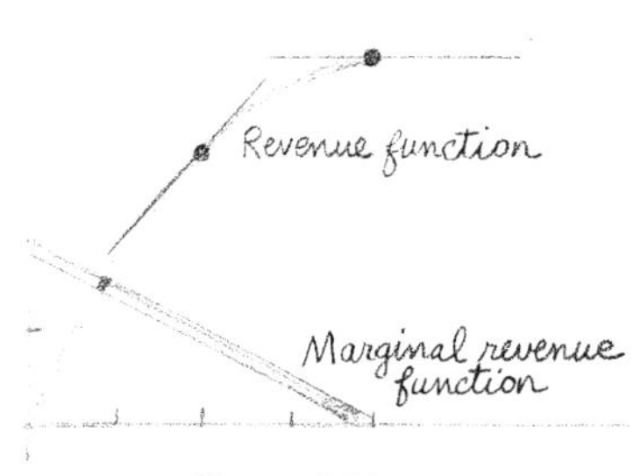

Figure 5.24

If we think of the cost and revenue functions as "distance" and the level of output as "time," then the marginal functions are like the "velocity." ▲

The purpose of this section is to describe the fundamental idea of the calculus and motivate the reader for its study. No exercises are assigned.

5.4 Techniques of Differentiation

The process of computing the derivative of a function is called differentiation. Recall that a function is differentiable at x if the derivative exists there (Section 5.1). In the following we will suppose that all functions under consideration are differentiable at the points involved. The formulas for the derivatives of certain basic functions are stated in formula (5) of Section 5.1, and formulas (1) to (6) of Section 5.2. In this section we show how to differentiate more complicated functions built up from the seven basic ones.

The symbol f has been used to represent a general function. Now there will often be two or more functions involved in the same formula. In order to distinguish one from another, we shall use letters other than f to represent functions: $g(x)$, $h(x)$, $u(x)$, $v(x)$, and $w(x)$. The corresponding derivatives are denoted with primes: $g'(x)$, $h'(x)$, and so on.

In addition to the two symbols we have already used for the derivative, $f'(x)$ and dy/dx, there is another:

$$\frac{d}{dx} f(x), \quad \text{or} \quad \left(\frac{d}{dx}\right) f(x), \quad \text{or} \quad \frac{df(x)}{dx}. \tag{1}$$

The symbol d/dx, placed before f, is an instruction to differentiate the function.

We state 4 general rules for differentiating certain combinations of functions:

Rule 1. If C is a constant and $f(x)$ differentiable, then the derivative of $Cf(x)$ exists and is equal to $Cf'(x)$:

$$\left(\frac{d}{dx}\right) Cf(x) = Cf'(x).$$

In other words, the derivative of a constant times a function is the constant times the derivative of the function.

The reason is this: Since the difference quotient of $Cf(x)$ is equal to C times the difference quotient of $f(x)$,

$$\frac{Cf(x) - Cf(x^*)}{x - x^*} = C\,\frac{f(x) - f(x^*)}{x - x^*},$$

the same relation holds for the derivative.

EXAMPLE 1

Consider the function $2x^3$. This is of the form $Cf(x)$, where $C = 2$ and $f(x) = x^3$. By formula (5) of Section 5.1, $f'(x)$ is equal to $3x^2$. By Rule 1, the derivative of the product $2x^3$ is $(2)3x^2$, or $6x^2$. ▲

Rule 2. Let $f(x)$ and $g(x)$ be two differentiable functions, and $f(x) + g(x)$ their sum; then the latter is also differentiable, and its derivative is the sum of the derivatives of the summands:

$$\left(\frac{d}{dx}\right)(f(x) + g(x)) = \left(\frac{d}{dx}\right)f(x) + \left(\frac{d}{dx}\right)g(x).$$

We shall give an intuitive, geometric explanation for this rule, leaving the direct proof based on the difference quotient as an exercise for the reader (Exercise 16). The intuitive idea is that a differentiable function is "locally linearly approximable" (Section 5.3). The derivative of a function at a given point depends on the values of the function only in a very small interval around the point; therefore, since the function can be approximated by a linear function near the point, we are going to suppose that the function is itself actually linear. We can easily see that Rule 2 is valid for linear functions: If $f(x) = ax + b$ and $g(x) = cx + d$, where a, b, c, and d are constants, then $f + g$ is also linear:

$$f(x) + g(x) = (a + c)x + (b + d);$$

and the derivative of the sum, $a + c$, is equal to the sum of the derivatives a and c. It is now a reasonable conclusion (although we have not given a conclusive proof) that Rule 2 holds for all f and g differentiable at x. This kind of geometric explanation for the properties of differentiation will be used several times in this chapter.

EXAMPLE 2

Find the derivative of $x^3 + 4x$. This function is the sum of $f(x) = x^3$ and $g(x) = 4x$. Its derivative is the sum of $f'(x) = 3x^2$ and $g'(x) = 4$; that is, $(d/dx)(x^3 + 4x) = 3x^2 + 4$. ▲

Rule 2 can be extended to differences of functions, and to sums of 3 functions or more:

$$\left(\frac{d}{dx}\right)(f(x) - g(x)) = \left(\frac{d}{dx}\right)f(x) - \left(\frac{d}{dx}\right)g(x),$$

$$\left(\frac{d}{dx}\right)(f(x) + g(x)) + h(x) + \cdots)$$
$$= \left(\frac{d}{dx}\right)f(x) + \left(\frac{d}{dx}\right)g(x) + \left(\frac{d}{dx}\right)h(x) + \cdots$$

For example, the derivative of $x^3 + 4x - 2/x$ is $3x^2 + 4 + 2/x^2$.

Rule 3. The product of two differentiable functions $f(x)$ and $g(x)$, $f(x)g(x)$, is differentiable, and the derivative of the product is equal to the derivative of f times g, plus f times the derivative of g:

$$\left(\frac{d}{dx}\right) f(x)g(x) = \left[\left(\frac{d}{dx}\right) f(x)\right] g(x) + f(x) \left[\left(\frac{d}{dx}\right) g(x)\right].$$

As in the explanation of the previous rule, we consider only the special case of linear functions: $f(x) = ax + b$ and $g(x) = cx + d$. On the one hand, by Rules 1 and 2, we find

$$\left(\frac{d}{dx}\right) f(x)g(x) = \left(\frac{d}{dx}\right) [acx^2 + (bc + ad)x + bd]$$
$$= 2acx + bc + ad;$$

and, on the other hand,

$$\left[\left(\frac{d}{dx}\right) f(x)\right] g(x) + f(x) \left[\left(\frac{d}{dx}\right) g(x)\right] = a(cx + d) + c(ax + b)$$
$$= 2acx + bc + ad.$$

EXAMPLE 3

The function $x^2 \sin x$ is the product of $f(x) = x^2$ and $g(x) = \sin x$. It follows that the derivative of the product is

$$\left[\left(\frac{d}{dx}\right) x^2\right] \sin x + x^2 \cdot \left(\frac{d}{dx}\right) \sin x = (2x) \sin x + x^2 \cos x.$$

Similarly, the function $4e^x x^6$ is the product of $f(x) = 4e^x$ and $g(x) = x^6$; therefore, the derivative is

$$\left[\left(\frac{d}{dx}\right) (4e^x)\right] x^6 + 4e^x \cdot \left(\frac{d}{dx}\right) x^6 = 4e^x x^6 + 24e^x x^5. \ \blacktriangle$$

Rule 4. Let $f(x)$ and $g(x)$ be differentiable, and $g(x)$ not equal to 0. Then the quotient $f(x)/g(x)$ is also differentiable, and its derivative is given by the formula

$$\frac{g(x) \cdot (d/dx)f(x) - f(x) \cdot (d/dx)g(x)}{(g(x))^2}$$

In words: The derivative of the quotient is equal to:

the denominator times the derivative of the numerator

minus

the numerator times the derivative of the denominator

all divided by

the square of the denominator.

Note that the condition $g(x)$ not equal to 0 is necessary for the definition of the ratio representing the derivative. The proof of this rule for linear functions is outlined in Exercise 17.

EXAMPLE 4

The function

$$\frac{x^3 + x}{x^2 + 1}$$

is the quotient of $f(x) = x^3 + x$ and $g(x) = x^2 + 1$. Their derivatives are $3x^2 + 1$ and $2x$, respectively. By Rule 4, the derivative of the quotient is

$$\frac{(x^2+1)(d/dx)(x^3+x) - (x^3+x)(d/dx)(x^2+1)}{(x^2+1)^2}$$
$$= \frac{(x^2+1)(3x^2+1) - (x^3+x)(2x)}{(x^2+1)^2} = \frac{x^4+2x^2}{(x^2+1)^2}. \quad ▲$$

EXAMPLE 5

We use Rule 4 to derive the expression for the derivative of tan x and arctan x. The function tan x is the quotient of the sine and cosine; hence, by Rule 4 and the formulas for the derivatives of the sine and cosine, $(d/dx)\sin x = \cos x$, $(d/dx)\cos x = -\sin x$, we obtain

$$\left(\frac{d}{dx}\right)\frac{\sin x}{\cos x} = \frac{\cos x\left(\dfrac{d}{dx}\sin x\right) - \sin x\left(\dfrac{d}{dx}\cos x\right)}{\cos^2 x}$$
$$= \frac{\cos^2 x + \sin^2 x}{\cos^2 x},$$

which, by the formula $\cos^2 x + \sin^2 x = 1$, is $1/\cos^2 x$. From this we get the formula

$$\left(\frac{d}{dx}\right)\tan x = \frac{1}{\cos^2 x}. \tag{2}$$

Next we find the derivative of the function arctan x. We use the same procedure as for the exponential function in Section 5.2. Let (x,y) be a point on the graph of $y = \arctan x$ and let (y,x) be the corresponding "image point" on the graph of the tan function. By the previous result, the derivative of the tan function at (y,x) is $1/\cos^2 y$ (because the coordinates have been reversed). It follows that the derivative of the arctan function at the original point (x,y) is the reciprocal $\cos^2 y$. Since the latter is equal to $\cos^2$

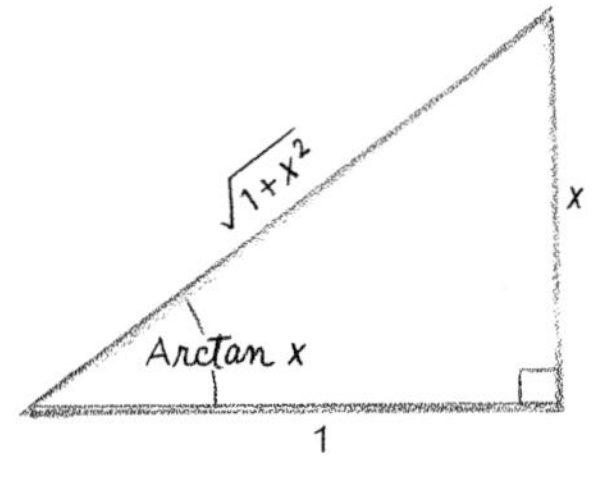

Figure 5.25

(arctan x), which is equal to $1/(1+x^2)$ (Figure 5.25), it follows that we get

$$\left(\frac{d}{dx}\right)\arctan x = \frac{1}{1+x^2}. \qquad (3) \ ▲$$

5.4 EXERCISES

Find the derivatives of the functions in Exercises 1 to 6.

1 (a) $12x^3 - 5x + 3$
(b) $5x^4 + 12x^2$
(c) $x^3 - 5x^2 + 6x^{-10}$
(d) 5 (constant)
(e) $\sqrt{x} - 1/\sqrt{x}$
(f) $x^{1/3} - x^{-2}$
(g) $\sin x - 2\cos x$
(h) $x - \log x$
(i) $e^x - 1 - x$
(j) $x^{-3/2} + 2\log x$

2 (a) $x^5 + 5x^4 + 20x^3$
(b) $3x^{1/2} - x^{3/2} + 2x^{-1/2}$
(c) 1 (constant)
(d) $3x^3 - 7x^2 + 2x - 1$
(e) $x^4 + \log x - \sin x$
(f) $e^x - 30x^2$
(g) $\sin x - x - \frac{1}{3}x^3$
(h) $\cos x - 1 - \frac{1}{2}x^2$
(i) $\log x + \cos x$
(j) $-e^x + x^{-1} + \log x$

3 (a) $x\cos x$
(b) $x\log x$
(c) xe^x
(d) $\sin x\cos x$
(e) $x^3\sin x$
(f) $(1+x^2)e^x$
(g) $e^x\log x$
(h) $(\log x)^2$
(i) $(e^x + x)(1+\log x)$
(j) $(x+1)(x^2+2)$

4 (a) $(x^2+3x+5)(x^2-7x+3)$
(b) $(x^2+3)^2$
(c) $x^{-2}\sin x$
(d) $\sin^2 x$
(e) $x^4\arctan x$
(f) $\cos^2 x$
(g) $x^3 e^x$
(h) $x^3\cos x$
(i) $x\arccos x$
(j) $\log x \cdot \tan x$

5 (a) $\dfrac{x^2+7x+4}{x+2}$
(b) $\dfrac{2x}{x^2+3x+1}$
(c) $\dfrac{e^x}{x^3}$
(d) $\dfrac{\sin x}{1+\log x}$
(e) $\dfrac{\log x}{x}$
(f) $\dfrac{1}{\tan x}$
(g) $\dfrac{\arcsin x}{x}$
(h) $\dfrac{\arcsin x}{\arccos x}$

6 (a) $\dfrac{2x}{x^2+1}$
(b) $\dfrac{1-2x}{4x^2+x-2}$
(c) $\dfrac{1}{\arctan x}$
(d) $\dfrac{\sin x}{e^x-1}$
(e) $\dfrac{\cos x - 1}{\frac{1}{2}x^2}$
(f) $\dfrac{\log x}{\cos x}$
(g) $\dfrac{e^x}{1+e^x}$
(h) $\dfrac{\log x}{1-e^x}$

7 Write x^5 as $x^2 \cdot x^3$. Find the derivative by applying the product rule (Rule 3) to the functions $f(x) = x^2$ and $g(x) = x^3$.

8 Write x^3 as the quotient of $f(x) = x^{10}$ and $g(x) = x^7$. Find the derivative by applying Rule 4.

9 Find the derivative of $(x^2 + 2x + 1)^2$ by two methods:

(a) Expand the square and apply Rules 1 and 2 to the sum.

(b) Consider the square as the product of identical functions $x^2 + 2x + 1$, and apply Rule 3.

10 Repeat Exercise 9 with the function $(x + 1/x)^2$.

11 Using Method (b) of Exercise 9, derive the formula for the derivative of the square of a function:

$$\frac{d}{dx}(f(x))^2 = 2f(x)\,f'(x).$$

12 Find the derivative:

(a) $(x^3 + 2x + 3)\ \sin x\ e^x$

(b) $(x^3 + 2x - 1)^2(x^2 - 3)$
[*Hint:* Let f be the first factor and g the product of the last two factors in (a); then apply Rule 3.]

13 Using the definition of the derivative as the limit of a difference quotient, indicate the steps in the proof of this formula for the derivative of the reciprocal of a function:

$$\frac{d}{dx}\left(\frac{1}{g(x)}\right) = -\frac{(d/dx)g(x)}{(g(x))^2}.$$

[*Hint:* Simplify the difference quotient $\dfrac{1/g(x) - 1/g(x^*)}{x - x^*}$ and then let x^* tend to x.] This formula is a special case of Rule 4 with a numerator of 1.

14 Find the derivative of $(1 - 2x)/(1 + x^2)$ by two methods:

(a) Apply Rule 3 to the product $(1 - 2x) \cdot \dfrac{1}{1 + x^2}$.

(b) Use Rule 4.

15 Generalize the result of Exercise 14: show that the quotient rule can be derived from the product rule by writing the quotient as a product

$$f(x)\left(\frac{1}{g(x)}\right),$$

and then applying the formula of Exercise 13 to the product rule.

16 Show that the difference quotient of the sum $f + g$ is equal to the sum of the difference quotients of f and g, respectively. How does this prove the validity of Rule 2?

17 We outline the proof of Rule 4 for linear functions $f(x) = ax + b$ and $g(x) = cx + d$.

(a) Show that the difference quotient of $f(x)/g(x)$ is

$$\frac{\dfrac{ax+b}{cx+d} - \dfrac{ax^*+b}{cx^*+d}}{x - x^*}.$$

(b) Reduce the complex fraction to a simple one in lowest terms. Show that it tends to $(ad - bc)/(cx + d)^2$ as x^* tends to x. Is this consistent with the statement of Rule 4?

5.5 The Chain Rule for Composite Functions

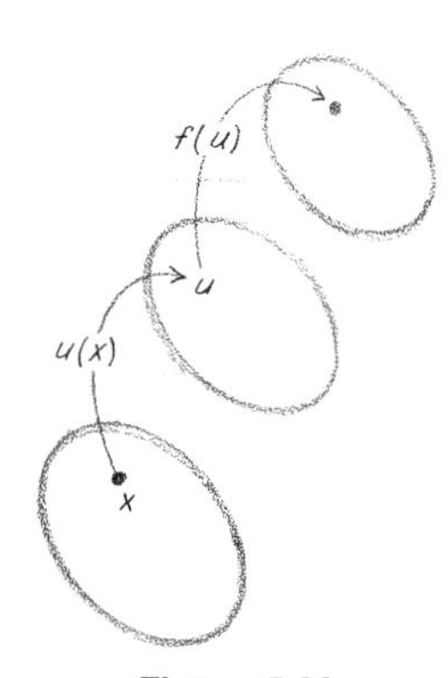

Figure 5.26

Many functions arising in calculus are **composite.** For example, take the function $g(x) = (x^2 + x - 1)^4$. It can be considered as having been composed of two functions,

$$u(x) = x^2 + x - 1 \qquad \text{and} \qquad f(u) = u^4,$$

in the following way. The value $g(x)$ is obtained by first taking the value $u(x)$, and then substituting this as the independent variable of the function $f(u) = u^4$:

$$g(x) = f(u(x)) = (u(x))^4 = (x^2 + x - 1)^4.$$

The range of the function $u(x)$ is contained in the domain of the function $f(u)$ (Figure 5.26). Thus, the u-values serve as the dependent variable for the first function, but as the independent variable for the second.

Another example is $g(x) = e^{-(1/2)x^2}$; this is the composition of $u(x) = -\frac{1}{2}x^2$ and $f(u) = e^u$. This representation as a composition is not unique; indeed, the function is also the composition of the function $u(x) = \frac{1}{2}x^2$ and $f(u) = e^{-u}$.

In either case the composite function is expressed in terms of the first independent variable x. The second independent variable has only an auxiliary role. The **chain rule** is a formula for differentiating the composite function with respect to x, where the derivative is expressed in terms of the derivatives of the component functions f and u.

To differentiate $(x^2 + x - 1)^4$ by the techniques of Section 5.4, we would first expand $x^2 + x - 1$ to the fourth power, obtain a polynomial containing 9 terms, and then find the derivative term-by-term (Rule 2). That is a long calculation. By contrast, the derivative can be simply found by the chain rule. The latter is useful not only in simplifying long calculations with composite functions, but also in getting the derivatives of many functions that cannot be obtained by the methods of Section 5.4.

The chain rule states:

If $u(x)$ and $f(u)$ are differentiable functions, then the composite function $f(u(x))$ is differentiable, and its derivative is the derivative of f at $u = u(x)$ times the derivative of the function $u(x)$:

$$\frac{d}{dx} f(u(x)) = \left\{\frac{d}{du} f(u) \,(\text{evaluated at } u = u(x))\right\} \frac{du}{dx}.$$

This is expressed more compactly as

$$g'(x) = f'(u(x))u'(x), \tag{1}$$

Here the prime over f stands for the derivative with respect to u. It is easy to check the validity of the rule when f and u are linear functions (Exercise 11). As in Section 5.4, it is then plausible that the rule holds for all **differentiable** f and u.

EXAMPLE 1

Let us obtain the derivative of the first function considered above, $g(x) = (x^2 + x - 1)^4$. We have $f'(u) = 4u^3$; this is the same as the formula for the derivative of x^4 except that the independent variable is denoted u in place of x. The derivative of $u(x)$ is $u'(x) = 2x + 1$. By the chain rule, formula (1), we have

$$g'(x) = 4(u(x))^3 u'(x) = 4(x^2 + x - 1)^3(2x + 1). \ \blacktriangle$$

EXAMPLE 2

If $g(x) = e^{-(1/2)x^2}$, then $u(x) = -\frac{1}{2}x^2$ and $f(u) = e^u$; then

$$u'(x) = -x, \qquad f'(u) = e^u;$$

therefore,

$$g'(x) = e^{u(x)}u'(x) = e^{-(1/2)x^2}(-x) = -xe^{-(1/2)x^2}. \ \blacktriangle$$

EXAMPLE 3

Find the derivative of $(x^2 + 4)^2(2x^3 - 1)^3$. First apply the product rule to the product of the functions $(x^2 + 4)^2$ and $(2x^3 - 1)^3$:

$$\left(\frac{d}{dx}\right)(x^2+4)^2(2x^3-1)^3 = \left[\left(\frac{d}{dx}\right)(x^2+4)^2\right](2x^3-1)^3 + (x^2+4)^2\left[\left(\frac{d}{dx}\right)(2x^3-1)^3\right].$$

Apply the chain rule to the derivatives of the powers appearing in each of the terms above:

$$\left(\frac{d}{dx}\right)(x^2+4)^2 = 2(x^2+4)(2x) = 4x(x^2+4),$$

$$\left(\frac{d}{dx}\right)(2x^3-1)^3 = 3(2x^3-1)^2(6x) = 18x(2x^3-1)^2.$$

Substitute these in the previously displayed expression:

$$\left(\frac{d}{dx}\right)[(x^2+4)^2(2x^3-1)^3]$$
$$= 4x(x^2+4)(2x^3-1)^3 + 18x(x^2+4)^2(2x^3-1)^2.$$

It may be left in this form. ▲

EXAMPLE 4

Find the derivative of

$$\frac{x^2}{\sqrt{4-x^2}}, \qquad \text{or equivalently,} \qquad x^2(4-x^2)^{-1/2}.$$

This can be found in either of two ways.

1. Apply the quotient rule to the first form with $f(x) = x^2$ and $g(x) = (4-x^2)^{1/2}$; the derivative is

$$\frac{(4-x^2)^{1/2}(d/dx)x^2 - x^2(d/dx)(4-x^2)^{1/2}}{4-x^2}.$$

   We have $(d/dx)x^2 = 2x$; and by the chain rule,

$$\left(\frac{d}{dx}\right)(4-x^2)^{1/2} = \tfrac{1}{2}(4-x^2)^{-1/2}(-2x) = -x(4-x^2)^{-1/2}.$$

   It follows that the derivative of the quotient is

$$\frac{2x(4-x^2)^{1/2} + x^3(4-x^2)^{-1/2}}{4-x^2} \qquad \text{or}$$
$$2x(4-x^2)^{-1/2} + x^3(4-x^2)^{-3/2}.$$

2. The derivative can also be obtained by applying the product rule to the second form of the function with $f(x) = x^2$ and $g(x) = (4-x^2)^{-1/2}$:

$$\frac{d}{dx}x^2(4-x^2)^{-1/2} = \frac{d}{dx}x^2 \cdot (4-x^2)^{-1/2} + x^2\frac{d}{dx}(4-x^2)^{-1/2}$$

$$= 2x(4-x^2)^{-1/2} + x^2[-\tfrac{1}{2}(4-x^2)^{-3/2}(-2x)]$$

$$= 2x(4-x^2)^{-1/2} + x^3(4-x^2)^{-3/2}. \ ▲$$

EXAMPLE 5

Find the derivative of $\log(x^2-1)$. This is a composite function with $f(u) = \log u$ and $u(x) = x^2 - 1$; hence, $f'(u) = 1/u$ and $u'(x) = 2x$, and by the chain rule,

$$\left(\frac{d}{dx}\right)\log(x^2-1) = \frac{u'(x)}{u(x)} = \frac{2x}{x^2-1}. \quad \blacktriangle$$

EXAMPLE 6

Find the derivative of $\sin(1/x)$. Here $f(u) = \sin u$ and $u(x) = x^{-1}$; hence, $f'(u) = \cos u$ and $u'(x) = -x^{-2}$, and, by the chain rule,

$$\left(\frac{d}{dx}\right)\sin\left(\frac{1}{x}\right) = [\cos u(x)]\, u'(x) = \left[\cos\left(\frac{1}{x}\right)\right]\left(\frac{-1}{x^2}\right). \quad \blacktriangle$$

EXAMPLE 7

Find the derivative of $\arccos(e^x)$. Here $f(u) = \arccos u$ and $u(x) = e^x$; hence,

$$f'(u) = \frac{-1}{\sqrt{1-u^2}} \qquad \text{and} \qquad u'(x) = e^x,$$

and, by the chain rule,

$$\left(\frac{d}{dx}\right)\arccos e^x = \frac{-u'(x)}{\sqrt{1-u^2}} = \frac{-e^x}{\sqrt{1-e^{2x}}}. \quad \blacktriangle$$

We summarize the formulas for the compositions of the basic functions f with a function $u(x)$:

Function	Derivative
$(u(x))^p, \quad p \neq 0$	$p(u(x))^{p-1}u'(x)$
$\log(u(x))$	$u'(x)/u(x)$
$e^{u(x)}$	$e^{u(x)}u'(x)$
$\sin(u(x))$	$u'(x)\cos(u(x))$
$\cos(u(x))$	$-u'(x)\sin(u(x))$
$\tan(u(x))$	$\dfrac{u'(x)}{\cos^2 u(x)}$
$\arcsin(u(x))$	$\dfrac{u'(x)}{\sqrt{1-(u(x))^2}}$
$\arccos(u(x))$	$\dfrac{-u'(x)}{\sqrt{1-(u(x))^2}}$
$\arctan(u(x))$	$\dfrac{u'(x)}{1+(u(x))^2}$

(2)

We summarize the results of Sections 5.4 and 5.5. In Sections 5.1 and 5.2 we are given the formulas for the derivatives of certain basic functions, x^p, $\log x$, and so on. Most of the formulas arising in calculus can be formed from these functions by

1. rational binary operations on these functions, that is, forming the sum, difference, quotient, or product of two or more functions; and by
2. forming the composition of two or more functions. We have shown how the derivative of the "built up" function is obtained in terms of the derivatives of the component functions (Figure 5.27). By means of the rules of Sections 5.4 and 5.5 we can find the derivative of almost any function arising in a practical problem.

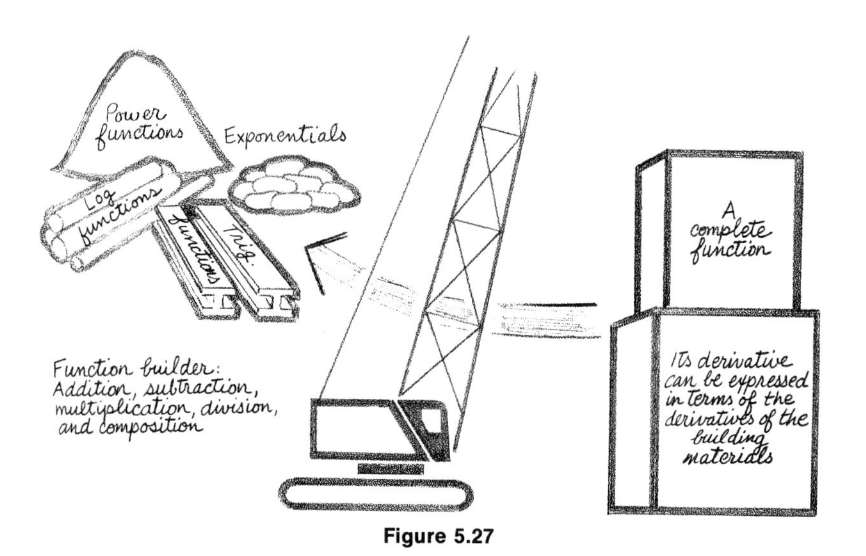

Figure 5.27

5.5 EXERCISES

In Exercises 1 to 10 find the derivative by means of the chain rule, and wherever appropriate, one of the rules of Sections 5.1 and 5.2.

1 (a) $(2 + 3x^2)^4$
(b) $2(4x^2 - 1)^5$
(c) $\sqrt{4 - 3x^2}$
(d) $\sqrt{x^3 + 8}$
(e) $(3x - x^3 + 1)^4$
(f) $(3 + 4x - x^2)^{3/2}$
(g) $(x/(1 + x))^5$
(h) $(x - 1)\sqrt{x^2 - 2x + 2}$

2 (a) $(2x^3 - 5)^3$
(b) $(1 - x^3)^4$
(c) $(2x^2 - 5)^{2/3}$
(d) $3\sqrt{2x + 3x^2}$
(e) $x(9 - x^2)^{2/3}$
(f) $\frac{1}{2}(x^2 - 3x + 5)^{-2}$
(g) $(x^3 - 1)^{-5/2}$
(h) $(x + 1)\sqrt{x^4 + 2}$

3 (a) $(x^2+1)^3(9-x^2)^{1/2}$ (e) $\dfrac{x}{(x+1)(2x-3)^2}$

(b) $(x^2-1)^2\sqrt{x-1}$ (f) $\sqrt{\dfrac{x^2-1}{x^2+1}}$

(c) $\dfrac{1-2x}{4+x^2}$ (g) $\dfrac{x-1}{\sqrt{x^2-2x+2}}$

(d) $\dfrac{x+2}{\sqrt{1-x^3}}$ (h) $\left(\dfrac{x^3-1}{2x^3+1}\right)^4$

4 (a) $(2x+1)^4\sqrt{3x-2}$ (e) $\dfrac{3x}{(x^2+1)^5(4x+5)}$

(b) $\sqrt{x+2}\,(x^3+3)^3$ (f) $\dfrac{x}{\sqrt{x^2-1}}$

(c) $\dfrac{x^2+1}{x^2-1}$ (g) $\dfrac{(x^2+3)^4}{(2x^3+5)^3}$

(d) $\dfrac{x-1}{\sqrt{x^2-1}}$ (h) $\left(\dfrac{x^2+2}{3-x^2}\right)^3$

5 (a) $5\log(x^2+2)$ (d) $[\log(x+1)]^2$
(b) $\log(2x-x^2)$ (e) $\log(\tan x)$
(c) $\log(\sin x)$ (f) $\log(\log x)$

6 (a) $7\log(1+2x)$ (d) $[\log(x^2+2)]^3$
(b) $-\log(3x^2+x-1)$ (e) $\log(1+e^x)$
(c) $\log(\cos x)$ (f) $\log(1+\log x)$

7 (a) e^{x^2-3x} (d) $e^{\sin x}$
(b) $e^{1/x}$ (e) $e^{\sqrt{1+x^2}}$
(c) $e^{\sqrt{x}}$ (f) $e^{\sin x+\cos x}$

8 (a) e^{2x-x^2} (d) $e^{\cos x}$
(b) e^{-1/x^2} (e) $e^{x\log x}$
(c) $e^{(x/1+x)}$ (f) $e^{x^2/\sqrt{1-x^2}}$

9 (a) $5\cos(3x+2)$ (d) $\sin\sqrt{1-x^2}$
(b) $\cos(2x+1)^2$ (e) $\arcsin\sqrt{x}$
(c) $\sin^5 x$ (f) $\arctan\left(\dfrac{x-1}{x+1}\right)$

10 (a) $2\sin(7x-5)$
(b) $\sin\sqrt{x}$
(c) $\cos^4 x$
(d) $\sin^3 x\cos x$
(e) $\arccos\sqrt{1-x}$
(f) $\tan\dfrac{1-x^2}{1+x^2}$

11 Put $u(x)=ax+b$ and $f(u)=cu+d$. Express $f(u(x))$ as a function of x, and verify the chain rule formula (1).

5.6 Approximation by Differentials

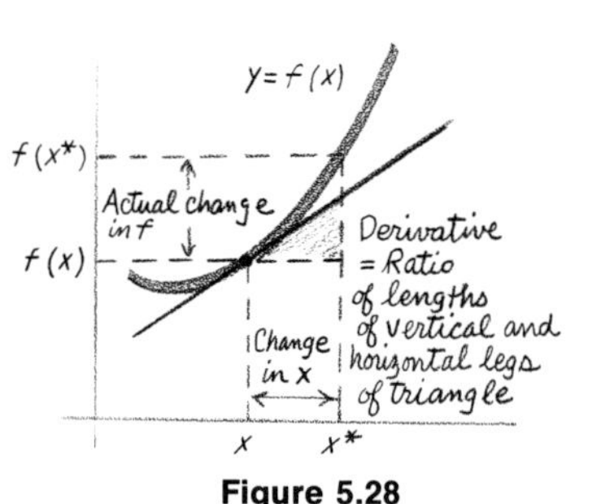

Figure 5.28

Now we turn to the many applications of the derivative.

In Section 5.3 we defined the derivative as the limit of the change of the dependent variable divided by the change of the independent variable: the limit of the ratio of the differences [see formulas (1) and (2) of Section 5.3]. It follows that if the differences are small, then their ratio is a good approximation to the derivative (Figure 5.28):

$$\frac{\text{Change in dependent variable}}{\text{Change in independent variable}} = \text{Derivative (approximately)}, \quad (1)$$

or

$$\frac{f(x^*) - f(x)}{x^* - x} \quad \text{approximately equal to } f'(x).$$

We will use the above relation in the form of an "equation"

$$f(x^*) - f(x) \approx f'(x)(x^* - x). \quad (2)$$

This is not meant to be an equation in the sense that the two sides are equal; they are **approximately** equal.

Formulas (1) and (2) are the basis of the method of **differentials,** which is used to relate small changes in one variable to corresponding changes in the other. In a typical problem we are given one of the changes and have to find the other. Our first application is to the calculation of roots of real numbers.

EXAMPLE 1

We approximate $\sqrt{101}$ by means of formula (2). Let $f(x)$ be the function $\sqrt{x}$. We know its value at $x = 100$: $f(100) = \sqrt{100} = 10$. We seek the value of the function at the point $x^* = 101$, which is relatively close to 100.

The derivative f' is

$$f'(x) = \frac{1}{2\sqrt{x}},$$

and so $f'(100) = 1/20$. Substitute $x = 100$ and $x^* = 101$ in Equation (2) and solve for $f(101)$:

$$f(101) = 10 + (1)f'(100) = 10 + \frac{1}{20} = 10.05.$$

This approximation is accurate to the given number of decimal places (see Table I). ▲

The method of differentials can also be used for interpolation in tables.

EXAMPLE 2

In Table IV the sine of 0.60 radians is given as 0.56464. Suppose we want the sine of 0.604. We put $f(x) = \sin x$, $x = 0.60$, and $x^* = 0.604$; then

$$f'(x) = \cos x = \cos 0.60 = 0.82534.$$

Apply formula (2) and solve for $f(x^*)$:

$$\begin{aligned} f(x^*) &= f(x) + (x^* - x)f^*(x) \\ &= 0.56464 + 0.004(0.82534) = 0.56794. \ \blacktriangle \end{aligned}$$

Here are further applications.

EXAMPLE 3

A small leak in a spherical balloon reduces the radius from 4 to 3.80 feet. Find the change in the surface area of the balloon (Figure 5.29). The formula for the area of the surface of a sphere of radius r is $f(r) = 4\pi r^2$. The derivative is $f'(r) = 8\pi r$. By formula (1), the change in the surface area is approximately equal to the change in the radius times the derivative:

$$f(4) - f(3.80) = 0.20\, f'(4) = 6.4\pi = 20.1. \ \blacktriangle$$

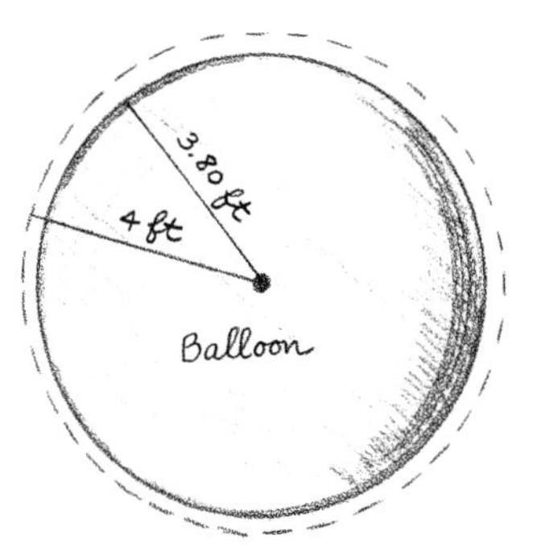

Figure 5.29

EXAMPLE 4

The acceleration due to gravity was stated in Section 4.5 to be 32 ft/sec^2. This is really valid only for objects relatively close to sea level. Newton's Law of Gravitation asserts that the acceleration varies inversely as the square of the distance of the object from the center of the earth. Furthermore, Newton's Second Law of Motion states that the weight of an object is equal to the product of the acceleration of gravity and the mass of the object. As the astronauts have found, an object loses weight as it rises from the surface of the earth (Figure 5.30); indeed the weight varies as

$$\text{Weight} = \frac{\text{Constant}}{(\text{Distance})^2}.$$

If $f(x)$ is the weight (pounds) at distance x (miles), then

$$f(x) = \frac{k}{x^2},$$

where k is a certain constant.

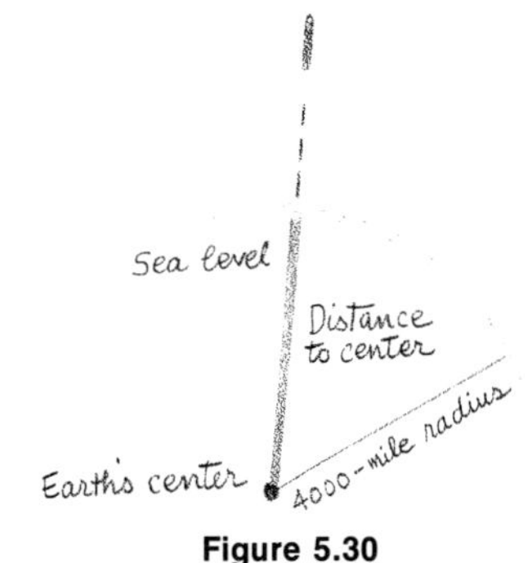

Figure 5.30

If an astronaut weighs 180 pounds at sea level, how much weight does he lose as his rocket climbs to an altitude of 100 miles? The radius of the earth is about 4000 miles; thus, $f(4000) = k/(4000)^2 = 180$, and so

$$k = 180(4000)^2.$$

The derivative of f is $f'(x) = -2k/x^3$, and so

$$f'(4000) = -\frac{2k}{(4000)^3} = -\frac{2(180)}{4000}.$$

By Equation (1) the change in weight is equal to the change in altitude times $f'(4000)$:

$$100\,\frac{-360}{4000} = -9;$$

thus the loss is 9 pounds. ▲

EXAMPLE 5

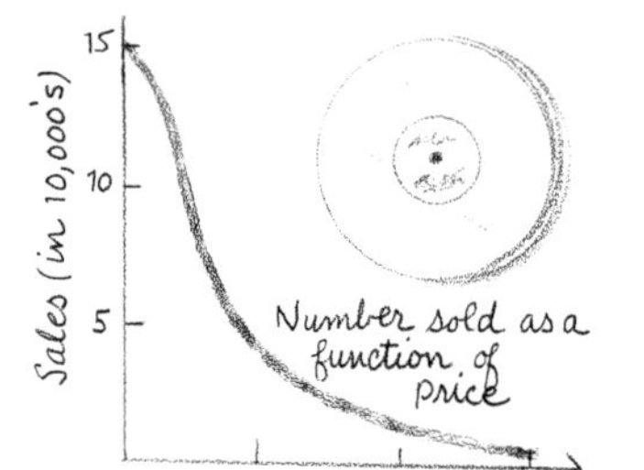

Figure 5.31

According to classical economic theory, the demand for a commodity in a free market decreases as the price increases. Suppose that the sales of a popular record album (in numbers sold per month) is given by the function

$$f(x) = \frac{150{,}000}{10 + x^2},$$

where x is the price of the album in dollars (Figure 5.31). In reality, price is a small factor in determining the sales of records. In our formula, the other factors—advertising, popularity, and so on—are considered to be fixed; the sales are given as a function of price alone. The graph of the function appears in Figure 5.31.

If the current price is \$4, what decrease in the price will increase the sales by 3 percent? According to formula (1), the change in sales is equal to the change in price times the derivative of f. The latter is found by the chain rule:

$$f'(x) = 150{,}000(-2x)(10 + x^2)^{-2}.$$

It follows that the change in price corresponding to a change of $0.03f(4)$ in sales is

$$\frac{0.03f(4)}{f'(4)} = \frac{(0.03)\,150{,}000/(26)}{-(8)\,150{,}000/(26)^2} = \frac{-(0.03)(26)}{8} = -0.095.$$

This is \$0.095, or $9\frac{1}{2}$¢; such a decrease in price will raise sales by 3 percent. Such a decrease is approximately 2.5 percent of the price. In economic theory, the ratio of the percentage increase in sales to the percentage decrease in price is called the **elasticity** of the demand with respect to price. It is a measure of the responsiveness of the market to price changes. In this example the elasticity at $x = 4$ is the ratio $3/2.5 = 1.2$. ▲

EXAMPLE 6

An engineer has to measure the volume of a large steel ball. His measurement of the radius is 2.3 feet; however, there is a possible error of 0.05 feet in this measurement, and the actual radius is somewhere between 2.25 and 2.35. The **relative error** is the ratio of the error to the measurement. Find the relative error in the measured volume of the ball based on the error in measuring the radius.

The volume of a ball of radius r is

$$f(r) = (\tfrac{4}{3})\pi r^3;$$

the derivative is $f'(r) = 4\pi r^2$. By Equation (1),

$$\frac{\text{Error in measurement of volume}}{\text{Error in measurement of radius}} = f'(r);$$

thus, the error in volume is $0.05\,f'(2.3) = 4\pi(2.3)^2(0.05)$. We divide this by the measured volume, $f(2.3)$, to get the relative error:

$$\text{Relative error} = \frac{\text{Error in volume}}{\text{Measured volume}} = \frac{0.05f'(2.3)}{f(2.3)}$$

$$= \frac{4\pi(2.3)^2(0.05)}{(\frac{4}{3})\pi(2.3)^3} = \frac{3(0.05)}{2.3} = 0.065.$$

The relative error is approximately $6\frac{1}{2}$ percent. ▲

EXAMPLE 7

Sewage is dumped at sea several miles from a beach (Figure 5.32). The level of sewage pollution in the bathing water varies inversely as the square of the distance between the beach and the dumping site. If the latter is 20 miles from the beach, how much further should it be moved to reduce the beach pollution by 5 percent?

Let x be the distance to the dumping site and $f(x)$ the level of pollution; then

$$f(x) = \frac{k}{x^2},$$

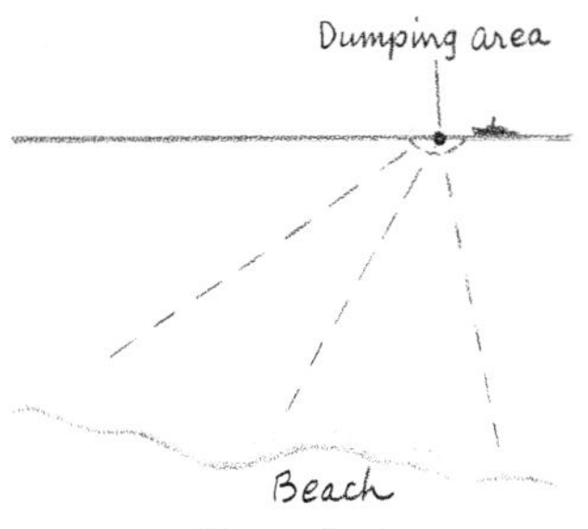

Figure 5.32

where k is an appropriate constant. The change in the level of pollution is 5 percent of $-f(20)$; and the change in the distance of the dumping site is unknown, and denoted d. The derivative of f is

$$f'(x) = \frac{-2k}{x^3}.$$

By Equation (1),

$$\frac{\text{Change in pollution}}{\text{Change in distance}} = \frac{-0.05f(20)}{d} = f'(20),$$

or

$$d = \frac{-0.05f(20)}{f'(20)} = \frac{-0.05k/(20)^2}{-2k/(20)^3} = (\tfrac{1}{2})20(0.05) = \tfrac{1}{2}.$$

We conclude that if the sewage is carried out $\frac{1}{2}$ mile more, then the pollution at the beach will be reduced by 5 percent. ▲

5.6 EXERCISES

In Exercises 1 to 4 find the approximate values using differentials.

1 (a) $\sqrt{37}$ (b) $\sqrt{98}$ (c) $\sqrt[3]{29}$ (d) $(80)^{-1/4}$ (e) $\sqrt{35.23}$ (f) $(220)^{4/3}$

2 (a) $\sqrt{50}$ (b) $\sqrt{63}$ (c) $\sqrt[3]{997}$ (d) $(623)^{-1/4}$ (e) $\sqrt[3]{27.50}$ (f) $(260)^{-3/4}$

3 (Use the tables where necessary.)
(a) $\sin 0.095$ (b) $\cos 0.785$ (c) $\tan 0.333$ (d) $\arccos 0.01$ (e) $\log 7.501$ (f) $1/(1+e^{0.01})$ (g) $e^{1.554}$ (h) $\arcsin 0.49$

4 Compute without the use of tables. Convert degrees to radians.
(a) $\cos 61°$ (b) $\sin^4 59°$ (c) $1/(1+\sin 44°)$ (d) $\arctan(\sqrt{3}-0.01)$ (e) $\log 1.03$ (f) $e^{0.03}-1$ (g) $e^{\sqrt{1.01}}$ (h) $(\arccos 0.51)^2$

5 A square metal sheet expands so that its edge increases from 20 to 20.002 inches. What is the change in the area? Approximately what change in the edge would increase the area by 0.6? (See Example 3.)

6 A metal cube contracts so that its edge decreases from 10 to 9.98 inches. What is the approximate change in (a) the surface area; (b) the volume? What change in the edge will decrease the volume by 0.5 cubic inches? (See Example 3.)

7 Find the approximate changes in the volume and surface area, respectively, of a ball bearing if its radius wears down from 6 to 5.96 millimeters. (See Example 3.)

8 In Example 3 of Section 4.8 the average monthly temperature for New York was given by $55.0 + 21.8 \sin(\pi/6(x - 3))$, where x is the number of months after January. Assume that the year is divided into 12 months, each of 30 days; and that x is a continuous variable, measured in months after January 15. Find the approximate change in the average temperature from October 15 to October 17; and from April 13 to April 15. (See Example 2.)

9 The side of a square is measured as 15 feet with a possible error of 0.03 foot. Find the relative error in the computed area of the square. (See Example 6.)

10 If the population of the earth grows like continuously compounded interest, and will double in the next 25 years, what will be the approximate percentage increase over (a) the first year; (b) the last of the 25 years? (See Example 6.)

11 In Exercise 18 of Section 4.1 we showed that the surface area of simple solids like the cube and the sphere vary proportionally to the $\frac{2}{3}$ power of the volume. What relative increase in the volume of air inhaled will increase the surface area of the lungs by 10 percent? (See Example 6.)

12 If the relative error in the calculated volume of a ball is to be no more than 0.004, what is the maximum allowable relative error in the measurement of the radius? (See Example 6.)

13 The sale of peaches in August is found to be a constant divided by $(1 + x)^{4/3}$, where x is the price per pound. If the actual price is 26¢/lb, how much should it be lowered to increase the sales by 8 percent? Find the elasticity. (See Example 5.)

14 A high voltage electrical machine gives off a small amount of radiation as it operates. The amount of radiation absorbed by a body in the vicinity of the machine is inversely proportional to the square of the distance between them. If a person at a distance of 10 feet absorbs 3 units of radiation per hour, what is the approximate rate of absorption at 11 feet? (See Examples 4 and 7.)

15 An observer is on the ground, 10 miles from the launching pad of a rocket. As the rocket rises, on a vertical line, he measures the angle he sees between the ground and a straight line to the rocket. If the angle increases from $\pi/6$ to $\pi/6 + 0.3$ in $\frac{1}{2}$ second, how quickly is the rocket rising at that time? The angle is measured in **radians.** [*Hint:* How far did the rocket move?] (See Example 2.)

16 In a baseball game a long drive is hit over the outfield fence. How much further did it travel beyond the fence before striking the ground? Or, if it hit the fence, how much further would it have gone if it did not hit the fence? The author has proposed a simple formula for estimating the additional distance (*Sports Illustrated,* August 16, 1971, p. 56):

$$\text{Additional distance} = \frac{dh}{16s^2},$$

where d = distance from the plate (where the ball was hit) to the fence, h = height of the ball at the fence, and s = no. of seconds for the flight of the ball to the fence (Figure 5.33). The effects of air resistance and wind have been ignored in the derivation of this formula. Suppose that a ball was hit over the fence at a distance of 400 feet from the plate, the height of the ball passing over the fence was 50 feet, and s was measured (by stopwatch) to be 4 seconds.

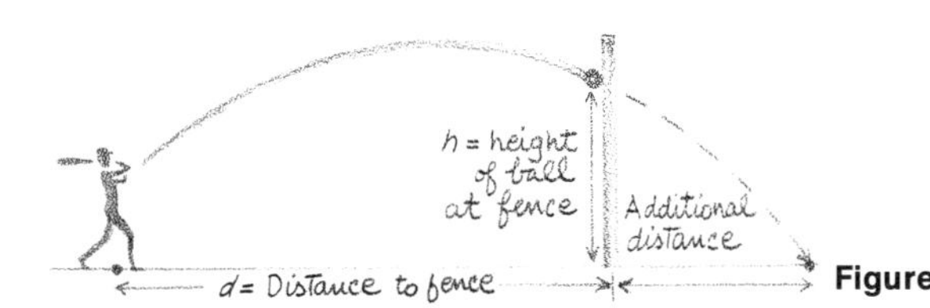

Figure 5.33

(a) How much further did the ball travel?

(b) If the stopwatch measures in tenths of a second, there is a maximum error of 0.05 in s. What is the maximum error in the estimated distance in (a)? (See Examples 4 and 6.)

17 Let $f(x)$ be the sales of a commodity at price x. The price increases slightly to x^*.

(a) What is the formula for the **relative** increase in price?

(b) What is the corresponding expression for the relative decrease in sales?

(c) Show that the elasticity is given approximately by the formula

$$\frac{-xf'(x)}{f(x)}.$$

(See Example 5.)

5.7 Related Rates

In the previous section, we studied the change in one variable relative to that in another variable; for example, the change in the area of a circle or volume of a ball caused by a change in the radius. In this section we show how the **rates** of change are related; for example, how the rate of change in the area of a circle is related to the rate of change of the radius. The method is to:

1. write the equation relating the original variables;
2. consider the two variables to be functions of a time variable t; and differentiate both members of the equation with respect to t;

3. finally solve for the unknown rate on the basis of the information given.
The chain rule is frequently used in the step of differentiation.

EXAMPLE 1

The edge of a square is increasing at a rate of 2 inches per hour at the moment when the edge is 10 inches. Find the rate of change of the area at that moment.

Let y be the area of the square and x the length of the edge; then $y = x^2$. Now we consider x and y to be functions of the time variable t and express them as $x(t)$ and $y(t)$, respectively. They are related by

$$y(t) = (x(t))^2.$$

Differentiate both sides with respect to t. The derivative on the left-hand side is dy/dt, the rate of change of the area. Apply the chain rule to the right-hand side, taking x as a function of t (instead of u as a function of x): according to the first formula on the list in the display (2) of Section 5.5, the derivative of $(x(t))^2$ with respect to t is $2x(t)\ dx/dt$; thus,

$$\frac{dy}{dt} = 2x(t)\ \frac{dx}{dt}.$$

By definition, dx/dt is the rate of change of the edge, which is given as 2 at the moment that $x(t) = 10$. We substitute this pair of values on the right-hand side of the equation above, and obtain

$$\frac{dy}{dt} = 2(10)(2) = 40.$$

Thus, the rate of change in the area is 40 square inches per hour. Note that the numerical values are substituted in the equation involving the **derivatives.** ▲

EXAMPLE 2

A passenger liner is sailing south at 16 mph. A freighter, 32 miles directly south in the path of the liner, is sailing east at 12 mph. What is the rate of change of the straight line distance between them at the end of 1 hour? 2 hours? Are they approaching or separating at these times?

We imagine the liner as moving down the y-axis, and the freighter moving up the x-axis. When the liner is 32 miles up the y-axis, the freighter is at the origin. Let x be the number of miles traveled by the freighter after t hours, y the number of miles

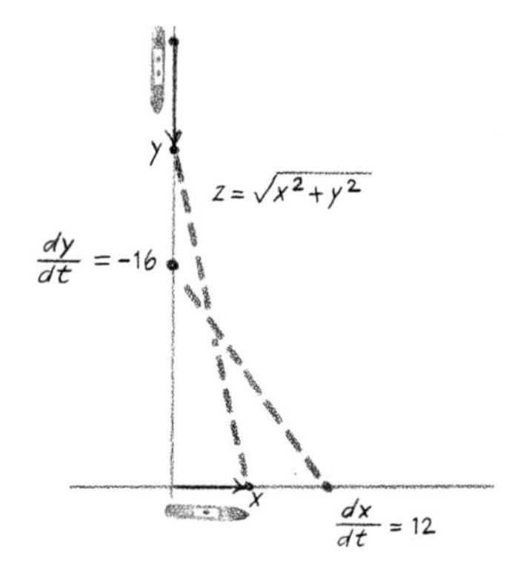

Figure 5.34

between the liner and the origin after t hours, and z the distance between the 2 crafts. These variables are related by the equation (Figure 5.34)

$$x^2 + y^2 = z^2. \tag{1}$$

We consider these to be functions of t, and differentiate according to the chain rule. As in Example 1, we have

$$\left(\frac{d}{dt}\right)(x(t))^2 = 2x(t)\frac{dx}{dt}$$
$$\left(\frac{d}{dt}\right)(y(t))^2 = 2y(t)\frac{dy}{dt},$$
$$\left(\frac{d}{dt}\right)(z(t))^2 = 2z(t)\frac{dz}{dt}.$$

We may omit the variable t in writing the functions x, y, and z; thus the relations above may be written simply as

$$\left(\frac{d}{dt}\right)x^2 = 2x\left(\frac{dx}{dt}\right),$$
$$\left(\frac{d}{dt}\right)y^2 = 2y\left(\frac{dy}{dt}\right),$$
$$\left(\frac{d}{dt}\right)z^2 = 2z\left(\frac{dz}{dt}\right).$$

When Equation (1) is differentiated on both sides, it becomes $2x(dx/dt) + 2y(dy/dt) = 2z(dz/dt)$, or

$$x\left(\frac{dx}{dt}\right) + y\left(\frac{dy}{dt}\right) = z\left(\frac{dz}{dt}\right). \tag{2}$$

At the end of 1 hour, the liner has moved 16 miles south to $y = 16$, and the freighter 12 miles east to $x = 12$. By Equation (1), the distance between them is $z = \sqrt{x^2 + y^2} = 20$. The rates of change of the boats are

$$\frac{dy}{dt} = -16 \quad \text{and} \quad \frac{dx}{dt} = 12,$$

for the liner and the freighter, respectively. The first rate is negative because y is decreasing with time. We substitute $x = 12$, $dx/dt = 12$, $y = 16$, $dy/dt = -16$, and $z = 20$ in Equation (2), and solve for the unknown dz/dt:

$$(12)(12) + (16)(-16) = 20\frac{dz}{dt} \quad \text{or} \quad \frac{dz}{dt} = \frac{-28}{5}.$$

Thus, the rate of change of the straight line distance between them is -5.6 mph. The negative sign means that the distance is decreasing; therefore, the ships are approaching each other.

After 2 hours, when $t = 2$, the liner has moved to the origin, and the freighter to $x = 24$. The distance between them is 24. We put $x = 24$, $dx/dt = 12$, $y = 0$, $dy/dt = -16$, and $z = 24$ in Equation (2), and solve for dz/dt:

$$(24)(12) + (0)(-16) = 24\frac{dz}{dt} \qquad \text{or} \qquad \frac{dz}{dt} = 12.$$

This means that the distance is growing at a rate of 12 mph, so that the ships are separating.

The process of finding dz/dt by direct differentiation of both sides of Equation (1) is called **implicit** differentiation. We were able to find dz/dt without first finding the explicit formula for z itself as a function of t. The latter is relatively complicated (see Section 4.1, Exercise 11). In Exercise 17 below, the reader is asked to solve this problem by direct differentiation of z as a function of t. ▲

EXAMPLE 3

The weekly revenue y (in thousands of dollars) of a manufacturer is given by the formula

$$y = 200x - x^2, \tag{3}$$

where x is the number of units produced per week (see Section 5.3, Example 2). If 75 units are now being produced each week, and if production is now increasing at the rate of 2 units per week, at what rate is the revenue increasing?

We differentiate both sides of (3) with respect to t:

$$\frac{dy}{dt} = 200\frac{dx}{dt} - 2x\frac{dx}{dt},$$

and substitute $dx/dt = 2$, $x = 75$, to obtain $dy/dt = 100$. The revenue is increasing at a rate of \$100,000. ▲

5.7 EXERCISES

In Exercises 1 and 2, the variables x, y, *and* z *are all functions of* t. *From the following equations, deduce the equations for the derivatives.* [Hint: *Use the form of the chain rule: If* $y = f(x)$, *then* $dy/dt = (d/dx)f(x)\,dx/dt$.]

1 (a) $y = x^{3/2}$
(b) $z^2 = 2x^2 + 3y^2$
(c) $z = xy$
(d) $y = \sin x$
(e) $y = x/(x^2 + 1)$
(f) $y = \log x$
(g) $z = \tan x$
(h) $y = 1/(1 + e^{-x})$

2 (a) $y = 1/x$
(b) $z^2 = x^3 - 2y^4$
(c) $z = x/y$
(d) $y = \cos x$
(e) $y = \sin^2 x$
(f) $y = e^x$
(g) $z = \cos x \sin x$
(h) $y = \sin x/(1 + \cos x)$

3 The radius of a spherical tumor on the lung of a laboratory rat is 5 millimeters and growing at a rate of 0.2 millimeters per week. Find the rate of growth of the volume. [Volume of a ball: $(\frac{4}{3})\pi(\text{radius})^3$.] (See Example 1.)

4 In order to escape the gravitational field of the earth, a rocket must take off from the surface of the earth at the speed of 7 miles per second; this is called the **escape velocity.** The weight of an object in the earth's gravitational field is inversely proportional to the square of the distance from the center of the earth (see Section 5.6, Example 4). An astronaut weighs 180 pounds at sea level, 4000 miles from the center of the earth. At what rate is he losing weight as his rocket lifts off the ground at the escape velocity? (See Example 1.)

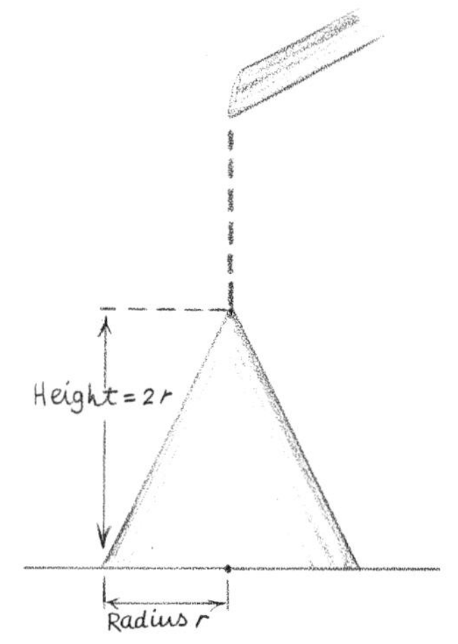

Figure 5.35

5 Sand falls from a chute and forms a pile in the shape of a cone. The altitude of the cone is always twice the radius of the base circle (Figure 5.35). If the pile is 6 feet high and the sand is falling at the rate of 5 cu ft/sec, how fast is the height of the pile increasing? [Volume of cone = $(\frac{1}{3})\pi$ height $(\text{radius})^2$.] (See Example 1.)

6 A baseball player can run 30 ft/sec. The catcher on the opposing team can throw a ball at a speed of 125 ft/sec. If the runner is attempting to steal third base and is 20 feet from it, and the catcher throws the ball to the third baseman, what is the rate of change of the distance between the ball and the runner at the instant the ball is thrown (Figure 5.36)? Note that the distance from the catcher to the third baseman is 90 feet. (See Example 2.)

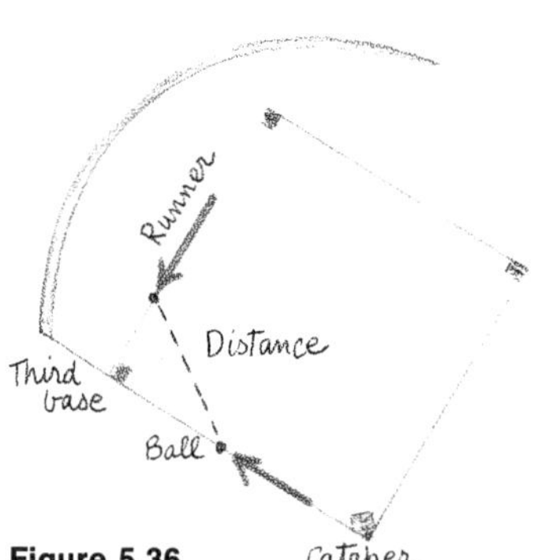

Figure 5.36

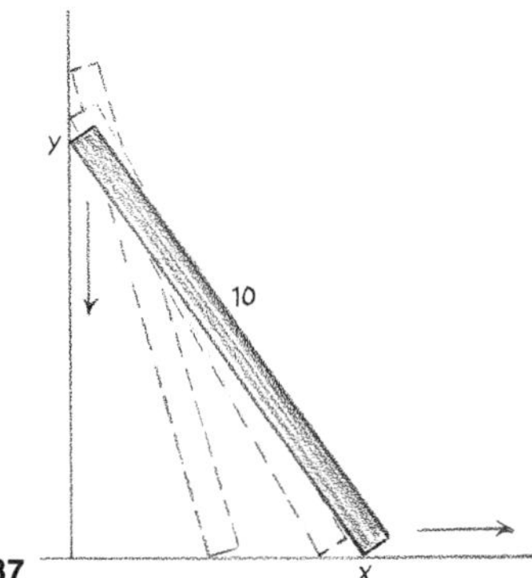

Figure 5.37

7 A 10-foot ladder leans against the side of a house. The top of the ladder slides down the side as the bottom of the ladder moves away from the house (Figure 5.37). Find the rate at which the top of the ladder is moving downward if the bottom is 6 feet from the house and is moving at 2 ft/sec. (See Example 2.)

8 A fighter-bomber dives toward its target along a parabolic path. The parabola is $y = 1 + \frac{1}{2}x^2$, where

y = height of the plane above the ground (100's of feet)
x = horizontal distance from target (100's of feet)

(Figure 5.38). How quickly is the plane losing altitude at the moment its horizontal distance to the target is 150 feet and its horizontal speed is 300 ft/sec? (See Example 1.)

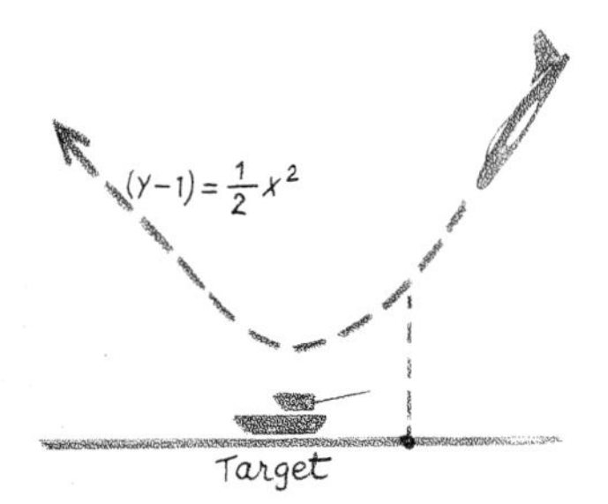

Figure 5.38

9 A hydrant on the curb of a city street is opened during the hot weather and a spray cap is placed on the mouth. The water is sent in a semicircular spray onto the roadway. The density of water hitting an object in the road is inversely proportional to the square of the distance from the hydrant and is given (in cubic feet of water hitting 1 square foot of area in 1 second) by $1/(\text{distance})^2$. A car travels on a straight line along a path parallel to the curb. The distance from the hydrant to the nearest point P on the path of the car is 10 feet (Figure 5.39).

(a) Find the formula for the density of water hitting the car when it is x feet from the point P.
(b) The car is moving at 25 mph. How quickly is the density rising when the car is 8 feet before the point P? (See Example 1.)

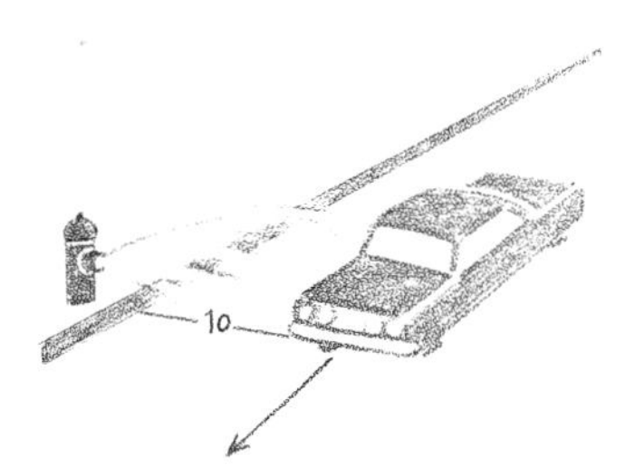

Figure 5.39

10 The cost (in \$100's) of producing x items is given by the formula $y = \frac{1}{2}x + (1/50)x^2$. If the production is increasing at a rate of 3 units per unit time, how quickly is the cost rising when the level of production is 25 items? (See Examples 1 and 3.)

11 A point on the rim of a bicycle tire of radius R is painted white. As the tire rolls along the ground, the height of the white point varies with the angle of rotation (in radians), measured from the horizontal through the center of the wheel and the line to the white point (Figure 5.40). Let x be the angle of rotation.

(a) Find the formula relating x to the height of the point above the ground.
(b) If x is $\pi/3$ radians and is increasing at a rate of 2 radians per second, how quickly is the height of the point changing? What if $x = 7\pi/6$? (See Example 1.)

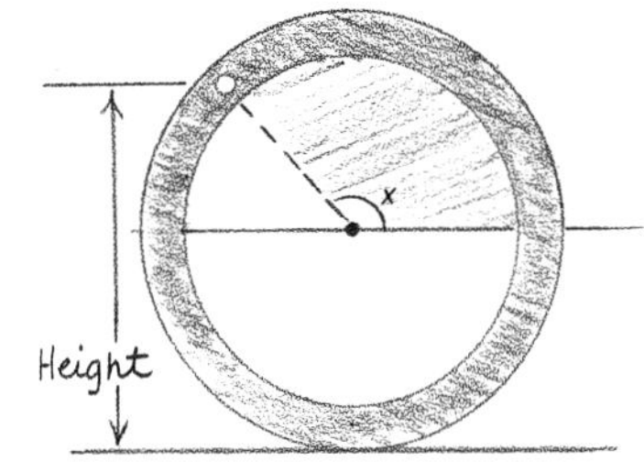

Figure 5.40

12 The top of a fishing pole is 8 feet above the surface of a lake. The line is pulled in at a rate of $\frac{3}{4}$ foot line per second. How quickly is the end of the line moving toward the shore when 24 feet of the line is still out (Figure 5.41)? (See Example 2.)

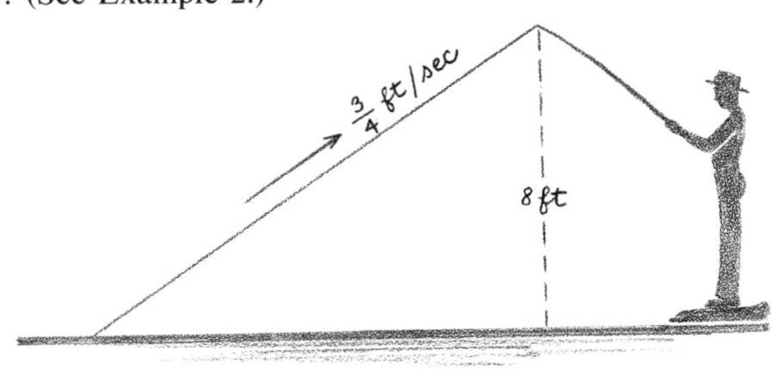

Figure 5.41

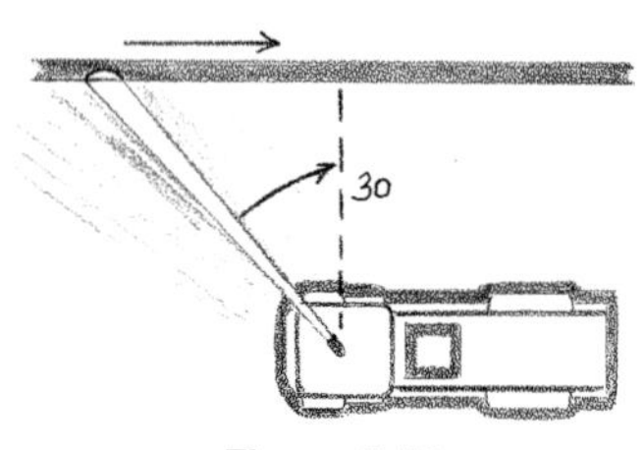

Figure 5.42

13 A beacon on the top of a parked fire engine rotates, casting a moving spot of light on the faces of the houses along the street. The beacon is 30 feet from the nearest point P on the line of the facades of the houses. The beacon makes 1 complete turn every second. Find the rate at which the spot of light moves along the part of a building 100 feet from P (Figure 5.42). (See Example 1.)

14 An icicle forms under the edge of the roof of a house when the snow melts on the heated roof, but refreezes upon dropping off the edge of the roof. The ice forms in the shape of an inverted cone, whose length is 12 times the radius of the base circle. If the volume of the icicle grows 2 cu cm/hr, how quickly is the length growing when the icicle is 10 cm long? (See Exercise 5.)

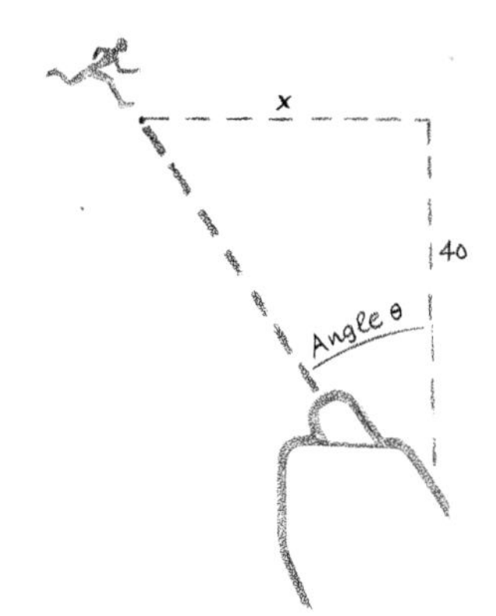

Figure 5.43

15 A television cameraman at a track meet is following a runner with his camera, which is located at the side of the track, 40 feet from the finish line. The distance of the runner from the line is x, and the angle between the runner and the line is θ (Figure 5.43).

(a) What equation relates x and θ?

(b) The camera is turning at a rate of $\pi/12$ radians per second at the moment the runner is 30 feet from the line. What is the speed of the runner? (See Example 1.)

16 A piston engine is moved by a shaft connected to a flywheel (Figure 5.44). The wheel turns clockwise. The radius of the wheel is denoted R, the shaft is of length l, and θ is the angle of rotation.

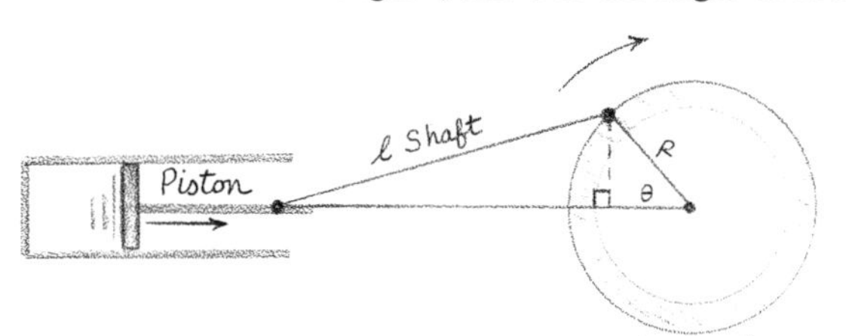

Figure 5.44

(a) Drop a perpendicular from the point on the wheel connected to the shaft, down to the horizontal line through the center of the wheel. Express the length of this line in terms of R and θ; consider the two cases $-\pi/2 \leq \theta \leq \pi/2$ and $\pi/2 \leq \theta \leq 3\pi/2$.

(b) Using the result in (a) express the distance D from the end of the shaft to the center of the wheel in terms of l, R, and θ.

(c) Find the formula for the relation between the rates of change of D and θ with respect to time.

(d) If $R = 1$, $l = 2$, $\theta = \pi/2$, and the wheel makes 2 complete revolutions every second, how quickly is the end of the shaft moving? (See Example 1.)

17 We refer to Example 2. Find the formula for the distance between the boats after t hours (see Section 4.1, Exercise 11). By direct differentiation find the rate of change of the distance at $t = 1$ and $t = 2$. (The distance is a composite function of t.)

5.8 Higher Derivatives

From a given function $f(x)$ we derive another function $f'(x)$ by differentiation. If we perform the same operation on $f'(x)$, we get the derivative of the derivative. It is called the **second derivative** of f, and is denoted $f''(x)$. The derivative of the latter is called the **third derivative** of f, and denoted $f'''(x)$. The fourth, . . . , nth derivatives are similarly defined, and are denoted $f^{(4)}(x)$, . . . , $f^{(n)}(x)$, respectively.

The higher derivatives also have the "dy over dx" notation. Since the second derivative is obtained by differentiation of the derivative, it is denoted

$$\frac{d}{dx}\left(\frac{dy}{dx}\right) \quad \text{or simply} \quad \frac{d^2y}{dx^2}.$$

As in the case of the symbol dy/dx, the letters are not joined by algebraic operations; the symbol for the second derivative is not "reducible" by cancellation. The third derivative is the derivative of the second:

$$\frac{d}{dx}\left(\frac{d^2y}{dx^2}\right) \quad \text{or simply} \quad \frac{d^3y}{dx^3}.$$

The nth derivative is denoted

$$\frac{d^ny}{dx^n}.$$

The higher derivatives are obtained by step-by-step differentiation.

EXAMPLE 1

Find the higher derivatives of

$$y = x^4 - 3x^2 - 6x + 4.$$

We have

$$\frac{dy}{dx} = 4x^3 - 6x - 6, \qquad \frac{d^3y}{dx^3} = 24x,$$

$$\frac{d^2y}{dx^2} = 12x^2 - 6, \qquad \frac{d^4y}{dx^4} = 24,$$

and d^5y/dx^5 and all subsequent derivatives are equal to 0. ▲

EXAMPLE 2

Find the higher derivatives of $\sin x$.

$$\frac{dy}{dx} = \cos x, \qquad \frac{d^3y}{dx^3} = -\cos x,$$

$$\frac{d^2y}{dx^2} = -\sin x, \qquad \frac{d^4y}{dx^4} = \sin x.$$

The subsequent derivatives follow the same pattern:

$$\cos x, \quad -\sin x, \quad -\cos x, \quad \sin x, \quad \text{and so on. } \blacktriangle$$

EXAMPLE 3

Find the first two derivatives of

$$f(x) = \frac{x}{\sqrt{1-x^2}} \quad \text{or} \quad \frac{x}{(1-x^2)^{1/2}}.$$

We find $f'(x)$ by the quotient and chain rules:

$$f'(x) = \frac{(1-x^2)^{1/2}(d/dx)(x) - x(d/dx)(1-x^2)^{1/2}}{((1-x^2)^{1/2})^2}$$

$$= \frac{(1-x^2)^{1/2} + x^2(1-x^2)^{-1/2}}{1-x^2}.$$

To simplify this, multiply the numerator and denominator by $(1-x^2)^{1/2}$:

$$\frac{1-x^2+x^2}{(1-x^2)^{3/2}} = (1-x^2)^{-3/2}.$$

The second derivative is obtained by differentiating the latter:

$$f''(x) = (-\tfrac{3}{2})(1-x^2)^{-5/2}(-2x) = 3x(1-x^2)^{-5/2}. \ \blacktriangle$$

EXAMPLE 4

Find the first three derivatives of $y = e^{-x^2}$. By formula (2) of Section 5.5 (the chain rule),

$$\frac{dy}{dx} = (-2x)e^{-x^2}.$$

The second derivative is obtained by the product and chain rules:

$$\frac{d^2y}{dx^2} = \left(\frac{d}{dx}\right)(-2x) \cdot e^{-x^2} + (-2x)\left(\frac{d}{dx}\right)e^{-x^2}$$

$$= -2e^{-x^2} + (-2x)(-2x)e^{-x^2} = -2e^{-x^2}(1-2x^2).$$

The third derivative is similarly obtained:

$$\frac{d^3y}{dx^3} = \left(\frac{d}{dx}\right)(-2e^{-x^2}) \cdot (1 - 2x^2) + (-2e^{-x^2})\left(\frac{d}{dx}\right)(1 - 2x^2)$$

$$= (4xe^{-x^2})(1 - 2x^2) + (-2e^{-x^2})(-4x)$$

$$= 4e^{-x^2}[x(1 - 2x^2) + 2x] = 4e^{-x^2}(3x - 2x^3). \ \blacktriangle$$

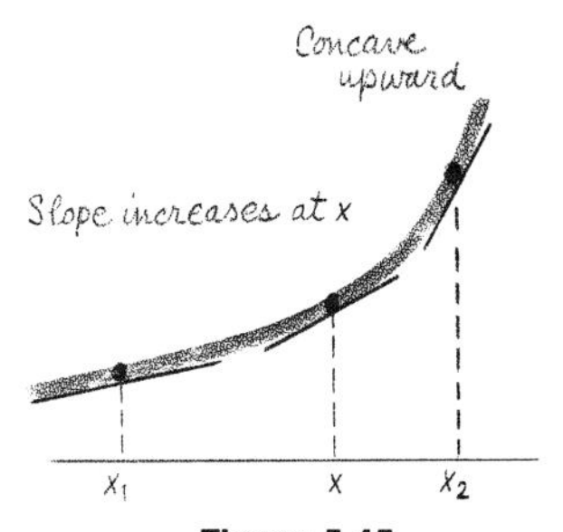

Figure 5.45

In most applications we need at most the first and second derivatives, not the higher ones. Just as f' measures the rate of change of f, so f'' measures that of f'. We will show how f'' influences the shape of thc graph of thc original function f.

Suppose, at a given point x, that $f''(x)$ is positive. Then f' has a point of increase at x (Section 5.3): If x_1 is a point just to the left of x and x_2 a point just to the right, then

$$f'(x_1) < f'(x) < f'(x_2).$$

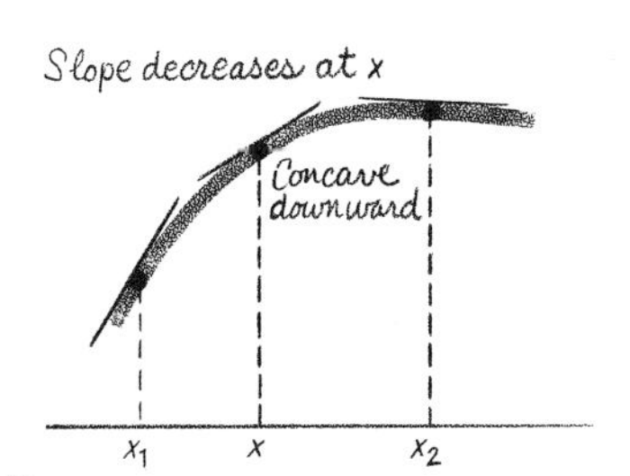

Figure 5.46

Thus, the slope of the tangent line increases at x, and the graph of the function is **concave upward** at x (Figure 5.45).

If $f''(x)$ is negative, then the slope of the tangent line decreases at the point x, and the graph is **concave downward** (Figure 5.46).

If $f''(x) = 0$, then the curve may have one of various shapes near x.

1. If f'' changes sign at x, from positive to negative or negative to positive, then x is called a point of **inflection.** The direction of concavity changes at this point (Figure 5.47)
2. If f'' is positive just to the right and to the left of x, then the curve is concave upward both to the right and to the left of x [Figure 5.48(a)].
3. If f'' is negative just to the right and left of x, then the curve is concave downward to the right and left [Figure 5.48(b)].
4. If $f'' = 0$ throughout an interval containing x, then it is a linear function in this interval because the slope (f') is constant.
5. There are several other more complicated possibilities for the sign of f'' around x, but these do not occur in our applications so that we omit them.

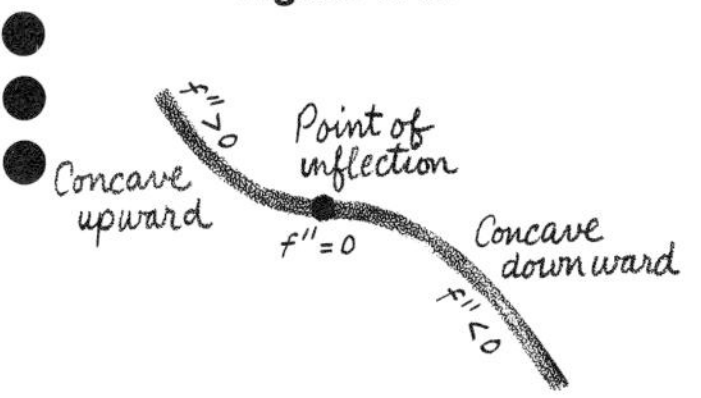

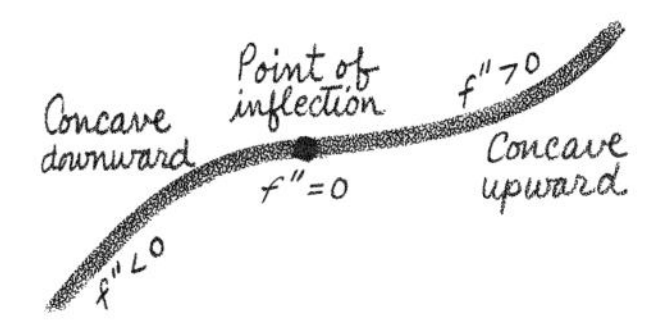

Figure 5.47

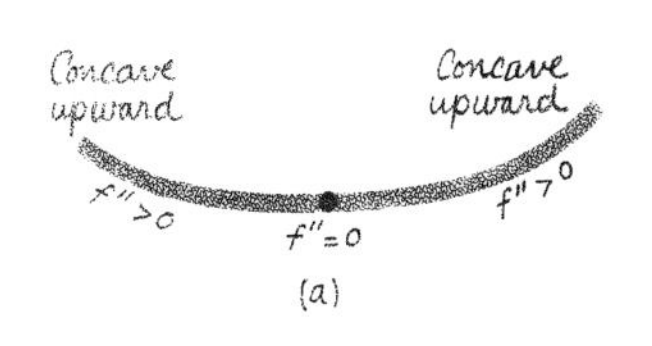

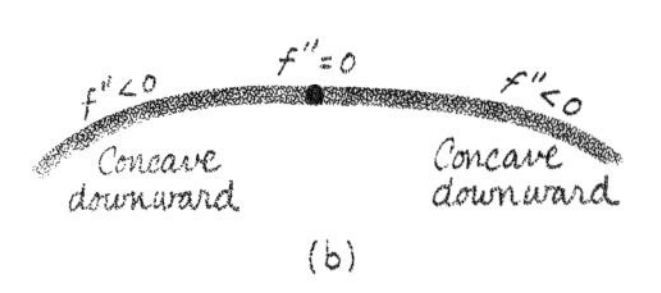

Figure 5.48

EXAMPLE 5

Consider the function $f(x) = (\frac{1}{3})x^3$. The first two derivatives are

$$f'(x) = x^2 \qquad \text{and} \qquad f''(x) = 2x.$$

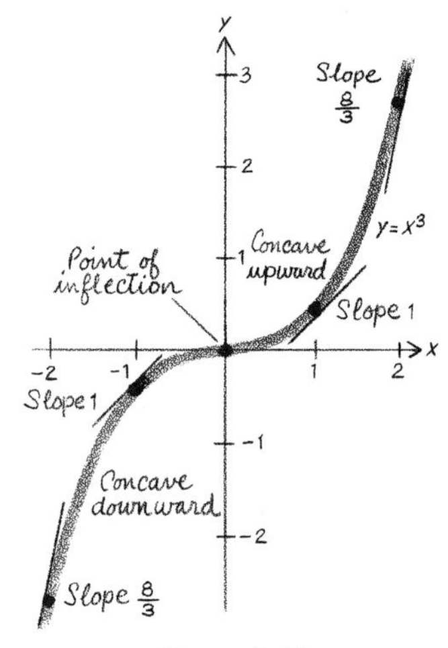

Figure 5.49

The graph is shown in Figure 5.49. The first derivative is always positive except at the origin, where it is equal to 0; therefore, the slope of the tangent line is positive except at the origin. The second derivative is negative to the left of the origin, where the curve is concave downward; and it is positive to the right of the origin, where it is concave upward. There is a point of inflection at the origin, where the second derivative is equal to 0. ▲

In the context of the motion problem, the second derivative represents acceleration. Let $f(t)$ be the distance an object travels up to time t. The derivative $f'(t)$ is the velocity. Its derivative $f''(t)$ is the acceleration, the rate of change of the velocity. In the case of a field of constant force, such as the earth's gravitational field near sea level, the acceleration is a constant; however, in general, the acceleration does not have to be a constant but may vary as a function of time.

EXAMPLE 6

Let us first consider, as a special case, motion in a field of constant force (Section 4.5); for example, the distance (in feet) an object falls in t seconds. The distance function is $f(t) = 16t^2$. The velocity is the first derivative, $f'(t) = 32t$. The acceleration is $f''(t) = 32$. ▲

EXAMPLE 7

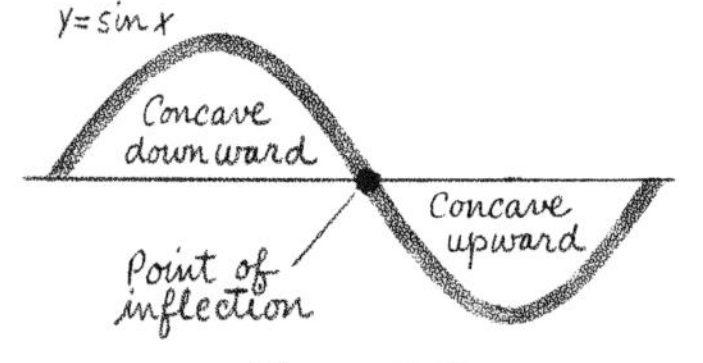

Figure 5.50

In Example 2 above, the second derivative of the sine function was shown to be the negative of the function itself. The graph is concave downward whenever the graph is above the x-axis, and concave upward whenever the graph is below the x-axis (Figure 5.50). This is consistent with the representation of the business cycle as a sine curve (Example 5, Section 4.8). When the economic indicator falls below its normal level, forces with positive acceleration are generated to drive the indicator up to the normal level. When the indicator rises above the normal level, forces of negative acceleration drive it back to normal. ▲

EXAMPLE 8

A 24 foot ladder leans against the wall of a building. The top of the ladder is lowered by a rope down the side of the wall, and the foot of the ladder slides horizontally, away from the wall. The top of the ladder moves down at a constant rate of 2 ft/sec. Find the acceleration of the foot of the ladder when (a) the ladder is just starting to move down; (b) the foot of the ladder is halfway out; (c) the top of the ladder just reaches the ground.

Let y be the height of the top of the ladder above the ground, and x the distance of the foot of the ladder from the base of the wall (Figure 5.51); then, by the Pythagorean Theorem,

$$x^2 + y^2 = (24)^2. \tag{1}$$

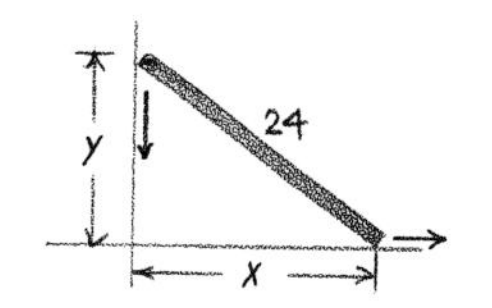

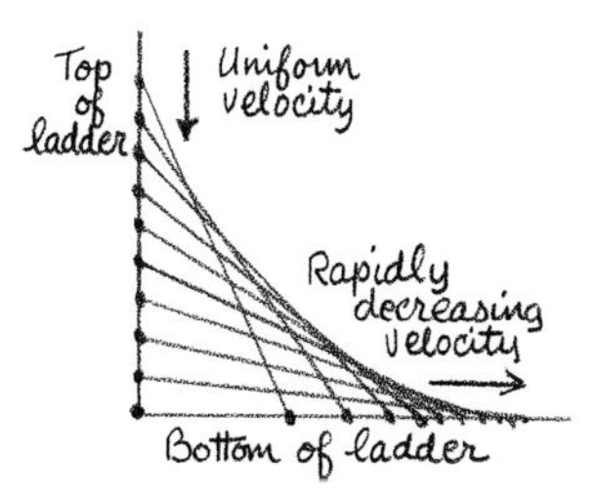

Figure 5.51

Take the derivative of each side with respect to t:

$$2x\left(\frac{dx}{dt}\right) + 2y\left(\frac{dy}{dt}\right) = \left(\frac{d}{dx}\right)(24)^2 = 0,$$

or

$$x\left(\frac{dx}{dt}\right) + y\left(\frac{dy}{dt}\right) = 0. \tag{2}$$

We seek the second derivative because it is the acceleration; hence, we differentiate each side of (2):

$$\frac{d}{dt}\left(x\,\frac{dx}{dt}\right) + \frac{d}{dt}\left(y\,\frac{dy}{dt}\right) = 0,$$

or

$$\left(\frac{dx}{dt}\right)^2 + x\left(\frac{d^2x}{dt^2}\right) + \left(\frac{dy}{dt}\right)^2 + y\left(\frac{d^2y}{dt^2}\right) = 0. \tag{3}$$

Since the top of the ladder is moving down at a constant rate, dy/dt is constant and equal to -2; therefore, the acceleration down is 0, that is, $d^2y/dt^2 = 0$. Equations (1), (2), and (3) are reduced to

$$\begin{aligned} x^2 + y^2 &= (24)^2, \\ x\left(\frac{dx}{dt}\right) &= 2y, \\ \left(\frac{dx}{dt}\right)^2 + x\left(\frac{d^2x}{dt^2}\right) + 4 &= 0. \end{aligned} \tag{4}$$

Solve the second equation in (4) for dx/dt:

$$\frac{dx}{dt} = \frac{2y}{x}.$$

Substitute this in the third equation in (4) and solve for the second derivative:

$$\frac{4y^2}{x^2} + x\left(\frac{d^2x}{dt^2}\right) + 4 = 0,$$

or

$$\frac{d^2x}{dt^2} = -4\,\frac{(x^2 + y^2)}{x^3} = \frac{-4(24)^2}{x^3}.$$

Thus, the acceleration varies inversely as the negative cube of the distance of the foot of the ladder from the wall.

The acceleration is undefined at $x = 0$ because the denominator of the acceleration is 0. When the foot of the ladder moves out just a bit, the acceleration is large and negative; for example, when $x = 1$, it is $-4(24)^2 = -2304$ ft/sec². This means that the bottom of the ladder starts out with a very large negative acceleration: it loses speed very rapidly at the beginning. When the foot of the ladder is halfway out, then $x = 12$ and the acceleration is

$$\frac{d^2x}{dt^2} = \frac{-4(24)^2}{12^3} = \frac{-4}{3} \text{ ft/sec}^2.$$

The speed is still decreasing but at a slower rate. When the foot of the ladder is almost all the way out, that is, the top just reaches the ground, then $x = 24$ and the acceleration is

$$\frac{d^2x}{dt^2} = \frac{-4(24)^2}{(24)^3} = \frac{-1}{6} \text{ ft/sec}^2. \blacktriangle$$

EXAMPLE 9

The **logistic function** describes certain growth processes which are **bounded.** The compound interest growth process described in Section 4.3 is an unbounded process because it will grow without limit if unchecked. A bounded growth process has a built-in limit. The number of cases in a flu epidemic in a fixed population is bounded by the size of the population. The growth of the population of a developing country is limited, in the long run, by space and economic resources. Such a growth process has the following character: The growth is very rapid at the beginning, and the rate of growth increases; this means that the growth has positive acceleration. As the population approaches its natural limit, the rate of growth decreases, so that the acceleration is negative.

The size of a population growing in this manner is often well described by the general logistic function

$$f(x) = \frac{A}{1 + e^{-k(x-c)}},$$

where A and k are arbitrary positive constants and c is an arbitrary constant. The constant A stands for the limit of the population size and c is the time at which the growth begins its slowdown. The constant k is a scale factor for the units of time x. The graph of the **standard** logistic, with $A = k = 1$ and $c = 0$,

$$y = \frac{1}{1 + e^{-x}},$$

appears in Figure 5.52. The second derivative is found to be

$$\frac{d^2y}{dx^2} = \frac{e^{-2x}(1 - e^x)}{(1 + e^{-x})^3}.$$

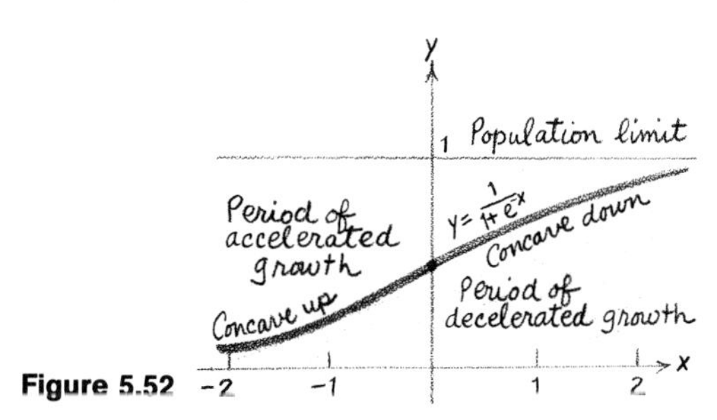

Figure 5.52

It is positive when x is negative, and negative when x is positive. This signifies that the acceleration of the population size is positive in the early stages, and negative in the later stages. The formula for the logistic function can be derived from certain reasonable assumptions about the growth process. This will be done in Section 6.10.

The population of the United States for the decades from 1790 to 1910 was discovered to be remarkably well described by the logistic function with

$$A = 197{,}273{,}000, \qquad k = 0.3134,$$
$$c = 12.4, \qquad \text{and} \qquad x = \text{No. of decades after 1790.}$$

The actual census figures (Bureau of the Census) are compared, in the table below, to the corresponding values of the logistic function:

Year	U.S. population (thousands)	x	$f(x)$
1790	3929	0	4031
1800	5308	1	5402
1810	7240	2	7293
1820	9638	3	9180
1830	12,866	4	13,258
1840	17,069	5	17,645
1850	23,192	6	23,318
1860	31,443	7	30,728
1870	39,818	8	39,693
1880	50,156	9	50,325
1890	62,948	10	63,228
1900	75,995	11	77,362
1910	91,972	12	92,184

▲

5.8 EXERCISES

In Exercises 1 to 6 compute the first two derivatives of each function.

1 (a) $f(x) = x^3 - 6x^2 + 9x + 1$
(b) $f(x) = 2x^6 + 3x^5 + 10x$
(c) $f(x) = 3x^4 - 12x^3 + 12x^2 - 4$
(d) $f(x) = 1/(x+1)$
(e) $f(x) = x\sqrt{x^2+1}$
(f) $f(x) = x/(1+x^2)$
(g) $f(x) = (x+1)^2(x+2)$
(h) $f(x) = (x^2+2x+1)^3$

2 (a) $f(x) = x^4 - 8x^3 + 64x + 8$
(b) $f(x) = 3x^4 - 2x^3 - 12x^2$
(c) $f(x) = x^6 - 16x^5 - x^4$
(d) $f(x) = (2+x)^{-1/2}$
(e) $f(x) = (x^2+1)\sqrt{1-x^2}$
(f) $f(x) = x/\sqrt{1+x^2}$
(g) $f(x) = (x+3)^4x^5$
(h) $f(x) = (x^2-x+9)^4$

3 (a) $f(x) = \sin^4 x$
(b) $f(x) = \dfrac{\sin x}{1-\sin x}$
(c) $f(x) = \dfrac{1}{\sin x} + \dfrac{1}{\cos x}$
(d) $f(x) = \cos^2 x \sin x$
(e) $f(x) = (\sin x)/x$
(f) $f(x) = \dfrac{1-\cos x}{x^2}$
(g) $f(x) = \arcsin x$
(h) $f(x) = x^4 \sin^2 x$

4 (a) $f(x) = \cos^3(2x)$
(b) $f(x) = \dfrac{\cos x + 1}{\cos x - 1}$
(c) $f(x) = \dfrac{1}{\cos x} - \dfrac{1}{\sin x}$
(d) $f(x) = \sin^2 x \cos x$
(e) $f(x) = \tan x$
(f) $f(x) = x^2 \cos x$
(g) $f(x) = \arccos x$
(h) $f(x) = \arctan x$

5 (a) $f(x) = \log x$
(b) $f(x) = x^2 \log x$
(c) $f(x) = \log(1-x^2)$
(d) $f(x) = e^{1/x}$
(e) $f(x) = xe^{-x}$
(f) $f(x) = \log(\log x)$
(g) $f(x) = e^{e^x}$
(h) $f(x) = e^{x^2} \log \sqrt{x}$

6 (a) $f(x) = e^x$
(b) $f(x) = (\log x)/x$
(c) $f(x) = \log(\sin^2 x)$
(d) $f(x) = e^{\sqrt{x}}$
(e) $f(x) = x^2e^{-(1/2)x^2}$
(f) $f(x) = e^x + e^{-x}$
(g) $f(x) = \log(\tan x)$
(h) $f(x) = e^{\sin x}$

7 The area of the surface of a spherical balloon is increasing at a constant rate of 2 square ft/min; thus, a point on the surface of the balloon is moving outward from the center on a straight line. What is the acceleration of a point at the moment that the radius of the balloon is 8 inches ($\frac{2}{3}$ foot)? [*Hint:* Surface area $= 4\pi(\text{radius})^2$; differentiate each member of the equation twice with respect to t.] (See Example 6.)

8 Generalize the ladder problem of Example 8. Let L be the length of the ladder and $-v$ the constant velocity of the top of the ladder.

(a) Find the formula (in terms of L, x, and v) for the acceleration of the foot of the ladder when it is x feet from the wall.

(b) Using the formula obtained in (a), find the acceleration when the bottom of the ladder is halfway out.
(c) What is the acceleration of the bottom as the top just reaches the ground?

9 A 15-foot ladder rests vertically against the side of a house. The ladder suddenly slips because of an outside force, and the top of the ladder falls directly downward with a constant acceleration of 20 ft/sec^2. Find the acceleration of the foot of the ladder at the moment the top of the ladder has fallen 10 feet. [*Hint:* First derive the formulas for the velocity and distance fallen for the top of the ladder.] (See Example 8.)

10 A white spot is painted on the rim of a bicycle wheel of radius 2 feet. The wheel rolls over the ground making 3 revolutions per second. Find the general formula for the acceleration of the height of the spot above the ground in terms of the actual height at the particular moment. (See Example 7.)

11 A colony of bacteria grows like compound interest, doubling in size every two hours. If the colony started with 10 bacteria, what is the acceleration of the colony size after 3 hours? (See Section 4.3)

12 A radioactive substance decays continuously at a constant rate. One gram decays to 0.8 gram in 100 years. What is the acceleration of the amount of material remaining after 50 years? x years? (See Section 4.4.)

5.9 Curve Sketching with the Aid of Derivatives

Suppose you want to draw a rough sketch of the graph of a function $y=f(x)$. The most direct way is to substitute successive values for x, compute the corresponding values of y, and then plot the points (x,y). This method uses only the **static** properties of the relation between x and y (Section 5.3). The curve can be sketched more accurately and efficiently by using the derivatives of the function; these give much information about the **dynamic** properties of the graph: its bends, rises, and falls.

Consider a function whose domain is the entire x-axis, and which has derivatives everywhere, at least up to the third order. Then the first and second derivatives vary "smoothly" with x: the slope of the tangent line and the concavity of the curve change gradually, not suddenly, with x. The first step in sketching the graph of such a function is to locate the points where the derivative is equal to 0; these are the **critical** points. The slope of the tangent is equal to 0 at these points. The second step is to determine where the second derivative is equal to 0: the direction of concavity may change at such points.

EXAMPLE 1

Sketch the graph of $y = x^3 - 3x + 4$. The first and second derivatives are

$$\frac{dy}{dx} = 3x^2 - 3 \qquad \frac{d^2y}{dx^2} = 6x.$$

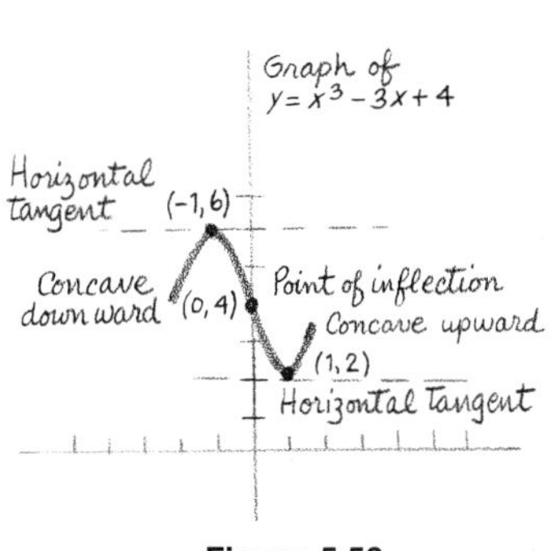

Figure 5.53

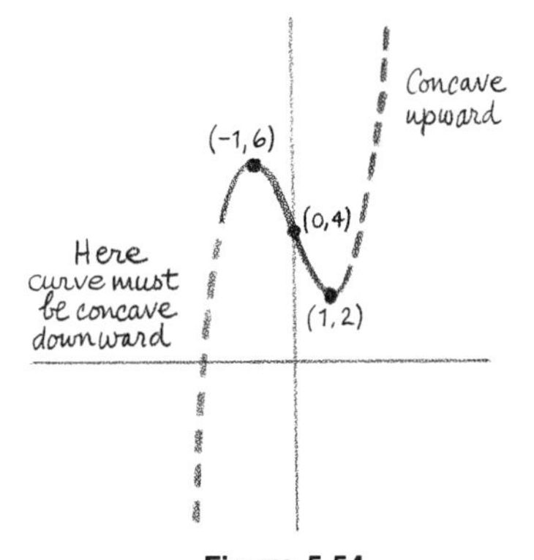

Figure 5.54

The critical points are obtained by setting the first derivative equal to 0 and solving for x:

$$3x^2 - 3 = 0 \qquad \text{so that} \qquad x = \pm 1.$$

The corresponding values of y are obtained by substituting these values of x in the formula for y:

$$y = 1^3 - 3(1) + 4 = 2, \qquad y = (-1)^3 - (-3) + 4 = 6.$$

So the horizontal tangent lines on the graph pass through the points $(-1,6)$ and $(1,2)$.

The only possible point of inflection is where the second derivative, $6x$, is equal to 0; namely, at $x = 0$. The second derivative is positive to the right of the origin, and so the curve is concave upward; and it is negative to the left of the origin, and the curve is concave downward. The ordinate of the point of inflection is obtained by putting $x = 0$ in the expression for $y: y = 0^3 - 3(0) + 4 = 4$ (Figure 5.53).

This information is sufficient for the sketch of the portion of the curve in Figure 5.53. The slope of the tangent line does not change sign at any point with abscissa larger than 1, or smaller than -1. The concavity is always upward to the right of the origin, and downward to the left. It follows that the rest of the graph must roughly follow the dotted lines attached to the curve in Figure 5.54. ▲

In the next example we consider a function whose behavior for large positive and large negative x has to be determined without reference to the concavity of the graph.

EXAMPLE 2

Sketch the graph of $y = x/(x^2 + 1)$. We compute the first and second derivatives by the quotient and chain rules:

$$\begin{aligned}\frac{dy}{dx} &= \frac{(x^2+1)(d/dx)(x) - x(d/dx)(x^2+1)}{(x^2+1)^2} \\ &= \frac{x^2+1-2x^2}{(x^2+1)^2} = \frac{1-x^2}{(x^2+1)^2}.\end{aligned}$$

$$\frac{d^2y}{dx^2} = \frac{(x+1)^2(d/dx)(1-x^2) - (1-x^2)(d/dx)(x^2+1)^2}{(x^2+1)^4}$$

$$= \frac{(x^2+1)^2(-2x) - (1-x^2)4x(x^2+1)}{(x^2+1)^4}$$

$$= \frac{-2x(3-x^2)}{(x^2+1)^3}.$$

Set the first derivative equal to 0 and solve for x:

$$\frac{1-x^2}{(x^2+1)^2} = 0, \qquad 1-x^2 = 0, \qquad \text{or} \qquad x = \pm 1.$$

The corresponding y-values are

$$y = \frac{1}{1^2+1} = \tfrac{1}{2}, \qquad y = \frac{-1}{(-1)^2+1} = -\tfrac{1}{2};$$

thus, the critical points are at $(1,\frac{1}{2})$ and $(-1,-\frac{1}{2})$.

Set the second derivative equal to 0, and solve for x:

$$\frac{-2x(3-x^2)}{(x^2+1)^3} = 0 \qquad \text{or} \qquad -2x(3-x^2) = 0.$$

The solutions are found by setting each of the two factors equal to 0: $x=0$, $x=\sqrt{3}$, and $x=-\sqrt{3}$. It follows that the second derivative is equal to 0 at the points $(0,0)$, $(\sqrt{3},\sqrt{3}/4)$, and $(-\sqrt{3},\sqrt{3}/4)$. The value of $\sqrt{3}$ is approximately 1.7 and that of $\sqrt{3}/4$ approximately 0.4.

Next we determine the sign of the second derivative between the points where it is equal to 0; we shall evaluate it at $x=-2, -1, 1,$ and 2:

at $x=-2$, $d^2y/dx^2 = -4/5^3$ (negative),

at $x=-1$, $d^2y/dx^2 = 4/2^3$ (positive),

at $x=1$, $d^2y/dx^2 = -4/2^3$ (negative),

at $x=2$, $d^2y/dx^2 = 4/2^3$ (positive).

Finally, we determine the nature of the graph as x gets larger in either the negative or positive senses. Write the function in the equivalent form

$$y = \frac{1}{x+x^{-2}}.$$

As x gets larger and larger, the term x^{-2} in the denominator tends to 0, and so y is approximately $1/x$. This approaches 0 as x increases.

The information given above is sufficient for sketching the curve (Figure 5.55). ▲

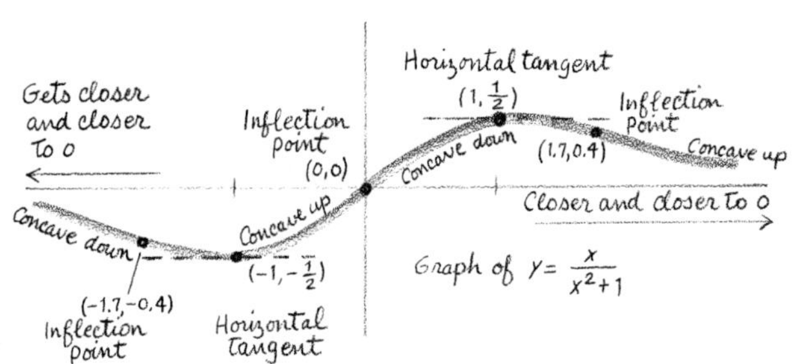

Figure 5.55

Next we consider a function whose domain is the entire axis except for a single point where the function is not defined. The graph has to be carefully analyzed around that point.

EXAMPLE 3

Sketch the graph of $y = x + 1/x$, which is defined for all $x \neq 0$. The derivatives are

$$\frac{dy}{dx} = 1 - \frac{1}{x^2}, \qquad \frac{d^2y}{dx^2} = \frac{2}{x^3}.$$

The first derivative is equal to 0 when $x = \pm 1$; the corresponding y-values are ± 2; hence, the critical points are (1,2) and $(-1,-2)$. The second derivative is positive to the right of the origin and negative to the left; it is never equal to 0. At $x = 0$ the function is not defined. When x is close to 0, the first derivative is large in the negative sense; for example, when $x = 0.1$, $dy/dx = 1 - 100 = -99$. This signifies that the curve dives down the negative y-axis as it approaches the origin from the left, and climbs sharply up the positive y-axis as it approaches the origin from the right.

For large values of x, in either the positive or the negative sense, the term $1/x$ is very small; therefore, the function $y = x + 1/x$ is like the straight line $y = x$.

Using the information above, we can get a rough sketch of the curve (Figure 5.56). It has two branches: one in the first quadrant, and one in the third. ▲

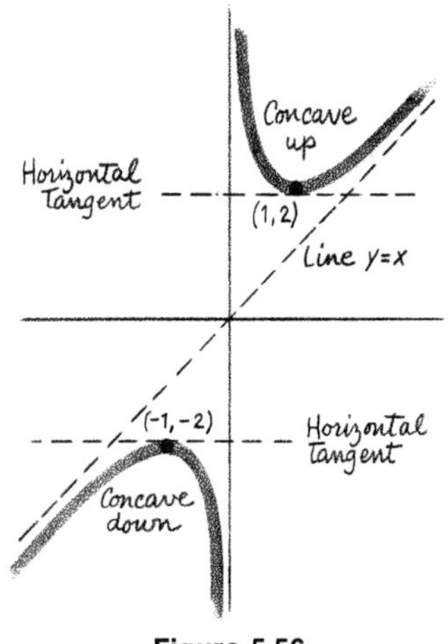

Figure 5.56

Finally we consider a function whose domain is a finite interval.

EXAMPLE 4

Sketch $y = \sin x + \cos x$, for $-\pi \le x \le \pi$. The derivatives are

$$\frac{dy}{dx} = \cos x - \sin x, \qquad \frac{d^2y}{dx^2} = -\sin x - \cos x.$$

The critical points are the solutions of

$$\cos x - \sin x = 0,$$

or $x = \pi/4$ and $x = -3\pi/4$. (Recall that we are now considering only solutions in the interval from $-\pi$ to π.) The corresponding y-values are

$$y = \sin\left(\frac{\pi}{4}\right) + \cos\left(\frac{\pi}{4}\right) = \tfrac{1}{2}(\sqrt{2} + \sqrt{2}) = \sqrt{2},$$

$$y = \sin\left(\frac{-3\pi}{4}\right) + \cos\left(\frac{-3\pi}{4}\right) = -\tfrac{1}{2}(\sqrt{2} + \sqrt{2}) = -\sqrt{2}.$$

The critical points are found to be $(\pi/4, \sqrt{2})$ and $(-3\pi/4, -\sqrt{2})$.

The second derivative is equal to 0 when $\sin x = -\cos x$, or $x = 3\pi/4$ and $x = -\pi/4$. The corresponding y-values are

$$y = \sin\left(\frac{3\pi}{4}\right) + \cos\left(\frac{3\pi}{4}\right) = \frac{\sqrt{2}}{2} - \frac{\sqrt{2}}{2} = 0,$$

$$y = \sin\left(\frac{-\pi}{4}\right) + \cos\left(\frac{-\pi}{4}\right) = \frac{-\sqrt{2}}{2} + \frac{\sqrt{2}}{2} = 0.$$

Thus, the second derivative is equal to 0 at the abscissa values $3\pi/4$ and $-\pi/4$ along the x-axis.

In order to determine the direction of the concavity for the various portions of the curve, we evaluate the second derivative at three conveniently chosen points whose abscissas are between the successive pairs $-\pi$ and $-\pi/4$, $-\pi/4$ and $3\pi/4$, and $3\pi/4$ and π. Take $-\pi/2$, 0, and $5\pi/6$. At $x = -\pi/2$:

$$\frac{d^2y}{dx^2} = -\sin\left(\frac{-\pi}{2}\right) - \cos\left(\frac{-\pi}{2}\right) = 1,$$

so that the curve is concave upward between $-\pi$ and $-\pi/4$. At $x = 0$,

$$\frac{d^2y}{dx^2} = -\sin 0 - \cos 0 = -1,$$

so that the curve is concave downward from $-\pi/4$ to $3\pi/4$. At $x = 5\pi/6$,

$$\frac{d^2y}{dx^2} = -\sin\left(\frac{5\pi}{6}\right) - \cos\left(\frac{5\pi}{6}\right) = \frac{-1}{2} + \frac{\sqrt{3}}{2}.$$

This is positive; hence, the curve is concave upward from $3\pi/4$ to π.

The value of y at the endpoints $-\pi$ and π is -1.

With this information, we sketch the curve (Figure 5.57). ▲

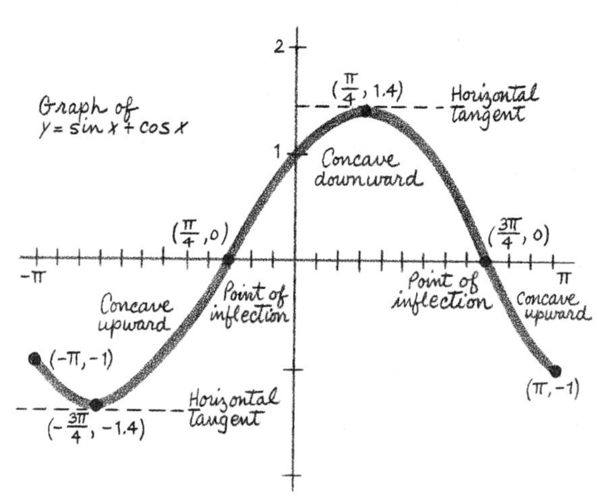

Figure 5.57

5.9 EXERCISES

Sketch the graphs of the functions, indicating critical points, inflection points, and the direction of concavity (wherever possible). When the domain is unspecified, it is understood to be the entire x*-axis; in this case, determine the behavior of the graph for large values of* x *in the positive and negative senses.*

1 (a) $y = x^2 - x - 6$
(b) $y = x^2 - x$
(c) $y = x^3 - 3x^2$
(d) $y = x^3 + x - 5$
(e) $y = x^4 + 4x^3 - 1$
(f) $y = (x+1)^3(x-1)$
(g) $y = 1/(x^2+1)$
(h) $y = x^2 + 1/x$, for $x \neq 0$

2 (a) $y = x^2 - 8x + 7$
(b) $y = x^2 - 10x + 11$
(c) $y = 4x^3 - 6x^2$
(d) $y = x^3 - 6x^2 + 9x - 8$
(e) $y = x^4 + 2x^2 + 1$
(f) $y = (x-2)^2(x+3)$
(g) $y = x^2/(x+1)^2$
(h) $y = x/(x+2)$, for $x \neq -2$

3 (a) $y = e^{-(1/2)x^2}$
(b) $y = xe^{-(1/2)x^2}$
(c) $y = x \log x$, for $x > 0$
(d) $y = (\log x)^2$, for $x > 0$

4 (a) $y = xe^{-x}$, for $x \geq 0$
(b) $y = \dfrac{1}{1+e^{-x}}$
(c) $y = \log \sqrt{1+x^2}$
(d) $y = \log(1 + \sin x)$, for $0 \leq x \leq \pi$

5 (a) $y = \sin^2 x$, for $0 \leq x \leq 2\pi$
(b) $y = e^{-x} \sin x$, for $x \geq 0$
(c) $y = 2 \cos x + \cos^2 x$, for $0 \leq x \leq 2\pi$
(d) $y = \sin x (\cos x)$, for $-\pi \leq x \leq \pi$

6 (a) $y = \cos^2 x$, for $0 \leq x \leq 2\pi$
(b) $y = x - \sin x$, for $-\pi/4 \leq x \leq 3\pi/4$
(c) $y = 2 \sin x + \sin^2 x$, for $0 \leq x \leq 2\pi$
(d) $y = \cos^3 x + 3 \cos x$, for $-\pi \leq x \leq \pi$

5.10 Maxima and Minima of Functions; Continuity

One of the major applications of differential calculus is a method of solving maximization and minimization problems. In a typical problem of this kind, the independent variable x is under the control of an operator. A function $y = f(x)$ determines a dependent variable of interest to him. In the maximization problem he seeks the value of x which makes y as large as possible; and in the minimization problem, the value which makes it as small as possible.

EXAMPLE 1

The operator is a botanist who wants to measure the effect of frequency of watering on the growth of a young plant. Let x be the number of times that the plant is watered each month; and let $y = f(x)$ be the height of the plant after 1 year of growth. The botanist seeks that value of x which will make y as large as possible. If x is too small, then the plant is not expected to grow well; hence y will be small. If x is too large, then the plant may suffer from too much water, and y will also be small. We deduce that there is 1 or possibly several intermediate values of x for which $f(x)$ takes on its largest value; this is sought by the botanist (Figure 5.58). ▲

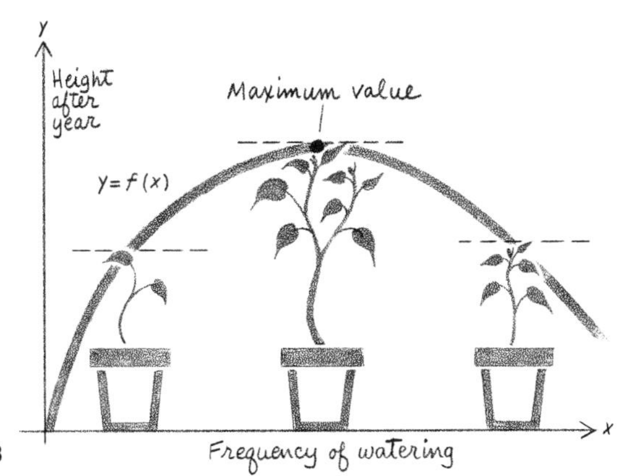

Figure 5.58

EXAMPLE 2

Let $y = f(x)$ be the average cost per floor in the construction of an office building of x floors. If x is too small, then the overhead costs dominate, and the average cost will be large; on the other hand, the average cost will also be too large if the number of floors is too large. There is some middle value x such that $f(x)$ is as small as possible. This is sought by the operator: the builder of the building (see Section 4.1, Example 4). ▲

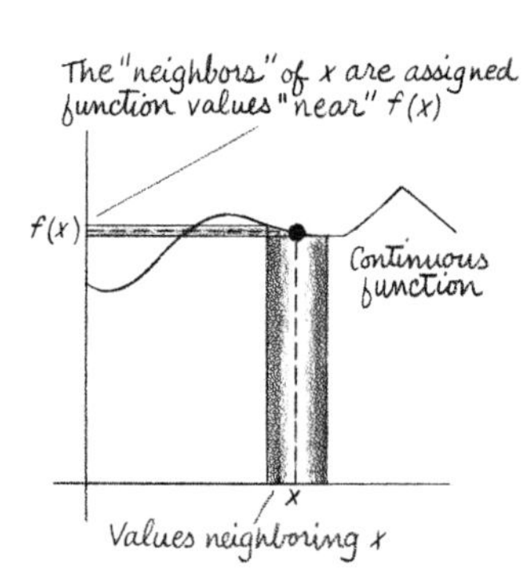

Figure 5.59

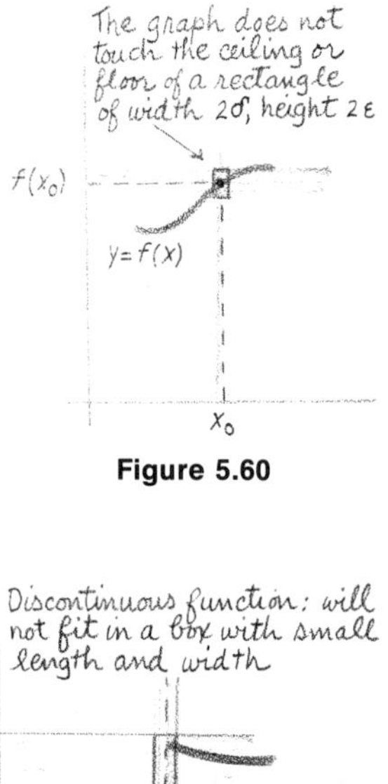

Figure 5.60

One of the major ideas associated with the concept of the function is **continuity.** Roughly speaking, we call a function **continuous** if it assigns values in such a way as to preserve "nearness": points that are "sufficiently close" in the domain have correspondingly "close" values in the range. It is convenient to think of a continuous function as one whose graph can be drawn over the domain "without lifting the pencil from the paper (Figure 5.59)." This is not the strict mathematical definition of continuity, but is useful in intuitive explanations of properties of continuous functions.

Let us also give the strict mathematical definition. The function f is **continuous at the point x_0** of its domain if

$$\lim_{x \to x_0} f(x) = f(x_0),$$

where x tends to x_0 through the domain of f (Section 5.3). According to the "epsilon-delta" definition of limit, this means that for every $\epsilon > 0$ there exists $\delta > 0$ such that the portion of the graph corresponding to abscissa values within δ units of x_0 is within ϵ units of $f(x_0)$ (Figure 5.60). An example of a function which is not continuous—**discontinuous**—at a point x_0 appears in Figure 5.61: There is a break in the graph at that point.

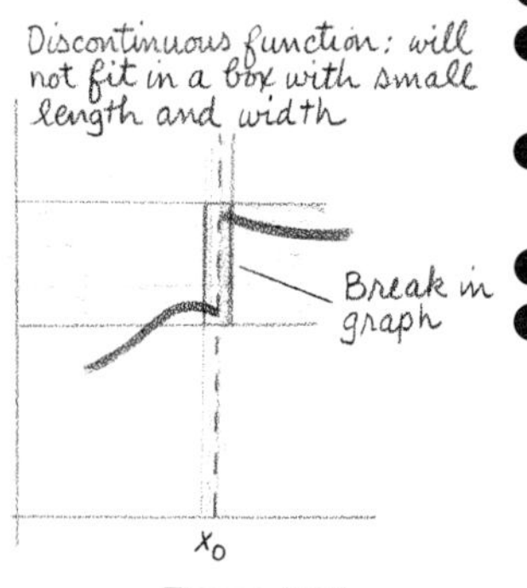

Figure 5.61

In most problems of practical interest the domain is formed by sets of the following kinds:

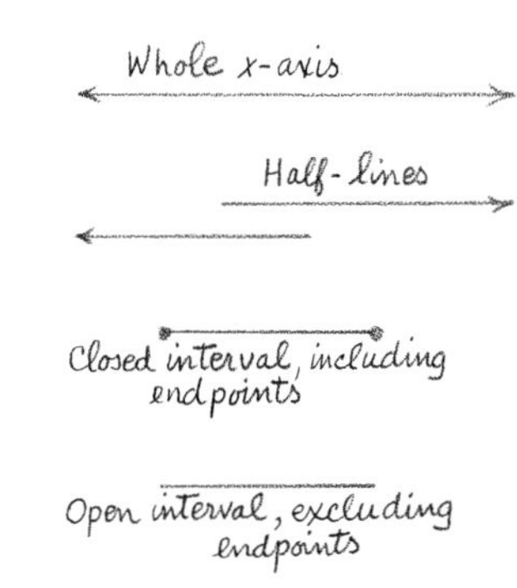

Figure 5.62

1. **The entire set of real numbers.** That is, the whole x-axis.
2. **A half-line.** Examples are: the set of positive real numbers, the set of real numbers at least equal to 2, or at least equal to any specified number c. The half-line may or may not include the finite endpoint.
3. **A finite interval.** This may or may not include one or both endpoints. A closed interval includes both endpoints. An open interval contains neither endpoint (Figure 5.62).

A function is called **continuous on an interval** if it is continuous at every point of the interval, in the sense of the definition given above.

In many problems the domain of a function is a closed interval, and the function is continuous on the interval. Here is an important theorem of the calculus concerning a continuous function on such a domain:

EXTREME VALUE THEOREM. Let f be a continuous function whose domain is a closed finite interval. Then there is at least one value x_0 in the domain such that the corresponding function value $f(x_0)$ is at least as large as $f(x)$ for any other x in the domain; that is,

$$f(x_0) \geq f(x) \quad \text{for all } x \text{ in the domain} \quad \text{[Figure 5.63(a)].}$$

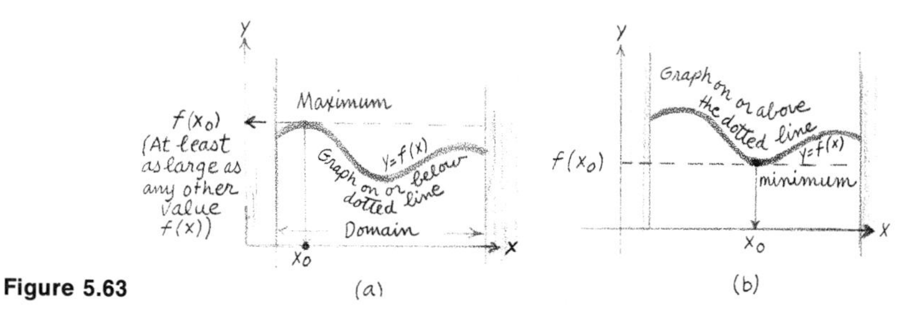

Figure 5.63

In this case we say that f has its maximum at x_0. Similarly, there exists a value x_0 such that $f(x_0)$ is at most as large as $f(x)$ for every other x in the domain; that is,

$$f(x_0) \leq f(x) \quad \text{for all } x \text{ in the domain} \quad \text{[Figure 5.63(b)].}$$

We say that f has its minimum at x_0.

The maximum and the minimum values of the function are called the **extreme values** or **extrema.**

A rigorous mathematical proof is contained in more advanced books on calculus. However, let us give an intuitive explanation based on the "pencil-and-paper" description of continuity. If you draw the graph of a function without lifting the pencil from the paper and start at one endpoint and end at the other, then it seems plausible that at **some** point (or points) the curve will be at its highest and at some point (or points) it will be at its lowest.

The hypothesis of the Extreme Value Theorem, that the interval be closed (and finite), is essential for its conclusion. If the function is not defined at an endpoint, then the conclusion might fail.

EXAMPLE 3

The function $f(x) = 1/x$ has an unbroken graph over the open interval from 0 to 1 (Figure 5.64); however, it is not defined at $x = 0$ and has no maximum on the interval. Similarly, the function has no minimum over the interval from -1 to 0. There is no maximum on the positive side of the origin because there is no x_0 such that $f(x_0)$ dominates all other function values $f(x)$ for all $x > 0$. Similarly, there is no minimum on the negative side of the origin because there is no x_0 such that $f(x_0)$ is dominated by the other function values $f(x)$ for $x < 0$. ▲

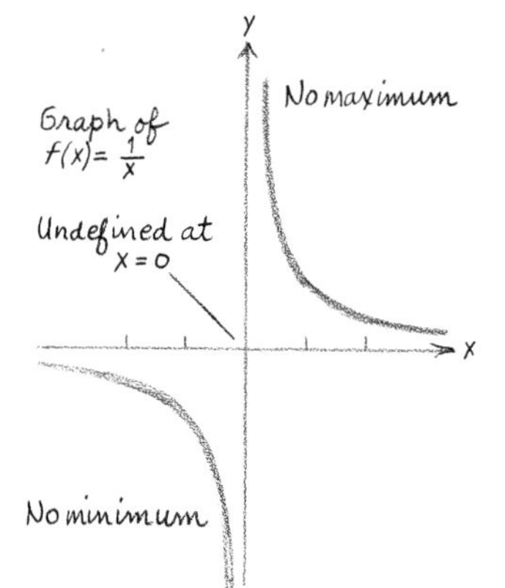

Figure 5.64

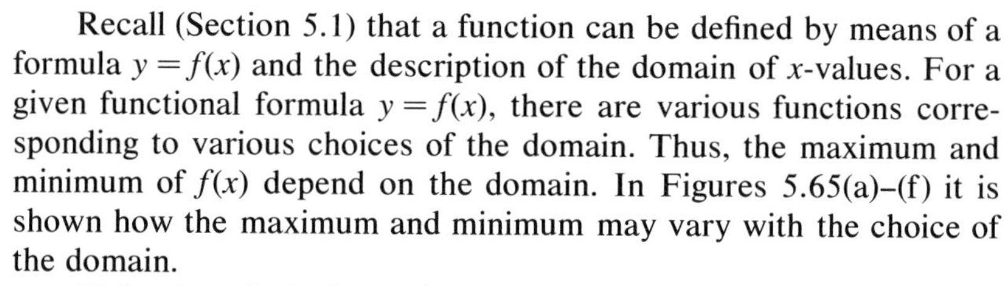

Recall (Section 5.1) that a function can be defined by means of a formula $y = f(x)$ and the description of the domain of x-values. For a given functional formula $y = f(x)$, there are various functions corresponding to various choices of the domain. Thus, the maximum and minimum of $f(x)$ depend on the domain. In Figures 5.65(a)–(f) it is shown how the maximum and minimum may vary with the choice of the domain.

If the domain is the entire real line, then the function in Figure 5.65(a) has no maximum or minimum.

When the domain is restricted to the positive half-line, then there is a minimum but no maximum [Figure 5.65(b)].

If the domain is restricted to the negative half-line, there is a maximum but no minimum [Figure 5.65(c)].

If the domain is taken as in Figure 5.65(d), the minimum is at the left endpoint and the maximum is at the right endpoint.

In the domain in Figure 5.65(e), the maximum and the minimum are located in the interior of the interval.

In Figure 5.65(f), the minimum is inside the interval, and the maximum is at the right endpoint.

The function in Figures 5.65(a)–(f) is typical of all functions with smooth graphs: the maximum and the minimum necessarily occur at

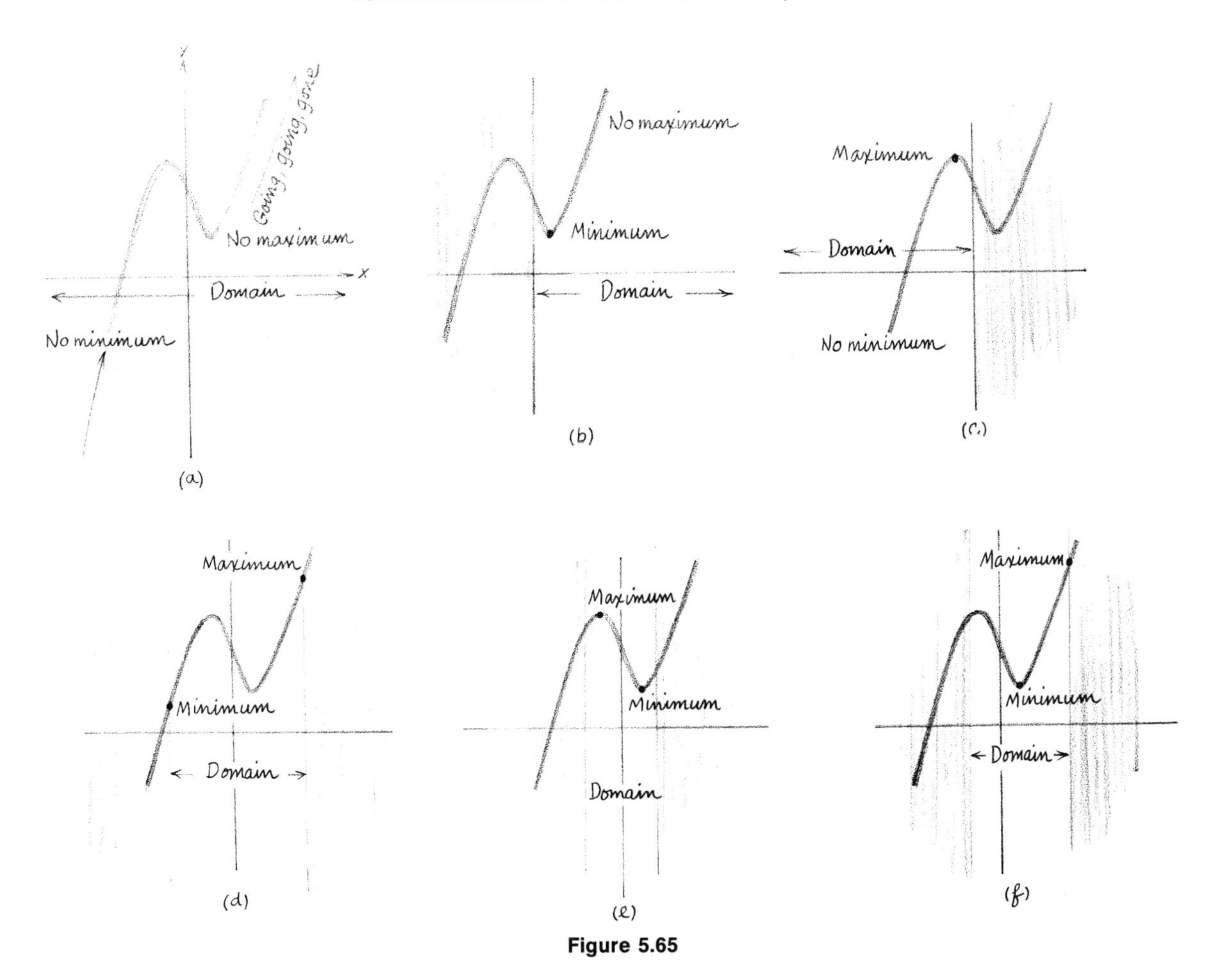

Figure 5.65

(a) at critical points, that is, where the tangent is horizontal, or

(b) at endpoints of the interval, or

(c) at a point where the derivative is undefined. For example, the function $f(x)$ defined as

$$\begin{aligned} f(x) &= \sqrt{x}, && \text{for positive } x, \\ &= 0, && \text{for } x = 0, \\ &= \sqrt{-x}, && \text{for negative } x, \end{aligned}$$

is a continuous function; however, its derivative is not defined at $x = 0$, where the minimum value $f(0) = 0$ occurs (Figure 5.66).

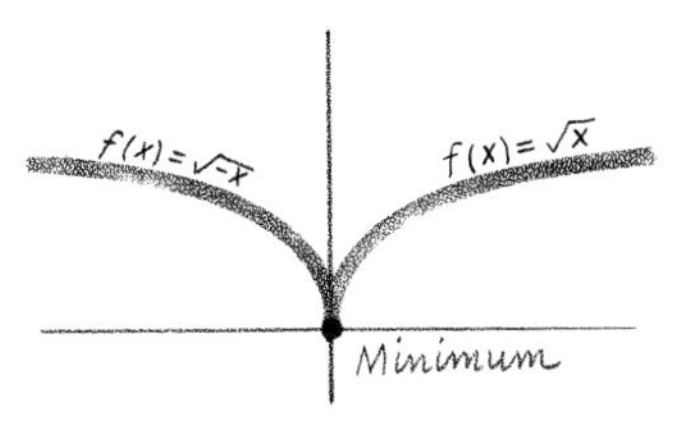

Figure 5.66

EXAMPLE 4

Find the extrema of the function $f(x) = x^4 - 2x^2 + 5$ on the closed interval from $x = -1.2$ to $x = 1$. The critical points are found by solving

$$f'(x) = 4x^3 - 4x = 0 \qquad \text{or} \qquad 4x(x-1)(x+1) = 0.$$

The solutions are $x = 0$, 1, and -1. The endpoints of the domain are -1.2 and 1. The maximum and the minimum necessarily occur at 1 of the 4 points where $x = -1.2, -1, 0$, and 1 because the derivative is defined everywhere on the domain. We have

$$f(-1.2) = (-1.2)^2((1.2)^2 - 2) + 5 = 4.2$$

$$f(-1) = 1 - 2 + 5 = 4,$$

$$f(0) = 5,$$

$$f(1) = 4.$$

It follows that the maximum value, 5, is assumed at $x = 0$; the minimum, 4, is assumed at both $x = 1$ and $x = -1$. ▲

The method of Example 4 suffices when the domain is a closed finite interval: the values at the endpoints are compared to the values at the critical points. When the domain is a half-line or the entire x-axis, the maximum or the minimum may not exist. In this case we can often determine which extreme value exists by analyzing the shape of the graph near the critical points. A function f is said to have a **relative maximum** (or **local maximum**) at a point x_0 if $f(x)$ is at most equal to $f(x_0)$ for all x near x_0. If the graph is concave downward at a critical point, then the function necessarily has a relative maximum at that point. This is the basis of a criterion for a relative maximum:

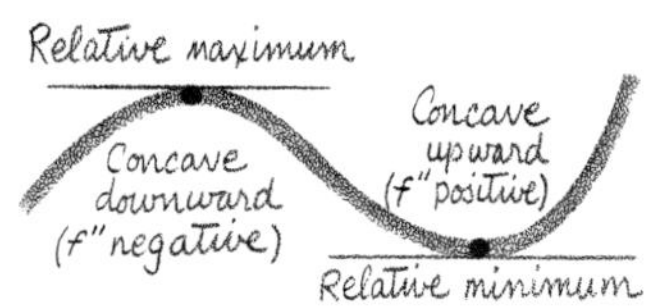

Figure 5.67

If the second derivative is negative at a critical point, that is, if

$$f'(x_0) = 0 \qquad \textbf{and} \qquad f''(x_0) \textbf{ is negative,}$$

then there is a relative maximum at x_0 (Figure 5.67).

In many cases we can then establish that the **relative** maximum is actually the maximum of the function over the entire domain. (The latter maximum is called the **absolute** or **global** maximum of the function.)

Similarly, a function f is said to have a **relative minimum** (or **local minimum**) at a point x_0 if $f(x)$ is at least equal to $f(x_0)$ for all x near x_0. If the graph is concave upward at a critical point, then the function has a relative minimum at the point:

If the second derivative is positive at a critical point, that is, if

$$f'(x_0) = 0 \quad \textbf{and} \quad f''(x_0) \textbf{ is positive,}$$

then there is a relative minimum at x_0 (Figure 5.67).

EXAMPLE 5

Find the extrema of the curve

$$y = \frac{1}{1 + x^2}$$

on the domain consisting of the entire x-axis. Since there are no endpoints, the extrema can occur only at the critical points. Solve the equation

$$\frac{dy}{dx} = \frac{-2x}{(1 + x^2)^2} = 0.$$

The only solution is $x = 0$. The second derivative is

$$\frac{d^2y}{dx^2} = \frac{(1 + x^2)^2(-2) + 2x(4x)(1 + x^2)}{(1 + x^2)^4} = \frac{2(3x^2 - 1)}{(1 + x^2)^3}$$

The second derivative has the value -2 at the critical point $x = 0$. This implies that the curve is concave downward. There are no other points where the curve can turn back up because $x = 0$ is the only critical point (Figure 5.68). It follows that a maximum occurs at $x = 0$ and nowhere else; the maximum value is

$$\frac{1}{1 + 0^2} = 1. \ \blacktriangle$$

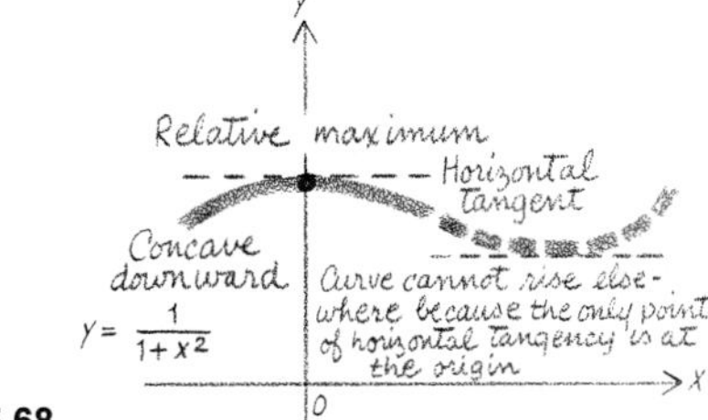

Figure 5.68

The second derivative criterion for a maximum or a minimum fails when the second derivative is equal to 0 at a critical point. In this case we examine the values of the second derivative **near** the critical point.

EXAMPLE 6

Consider the function $y = (x-2)^4$, with the domain consisting of the entire real axis. The first derivative is $dy/dx = 4(x-2)^3$; therefore, the only critical point is at $x = 2$. The second derivative is $12(x-2)^2$; it is equal to 0 at the critical point $x = 2$. We cannot tell from the value of the second derivative whether the graph is concave upward or downward because it is equal to 0; however, it is positive at all other points, and so must be concave upward around $x = 2$. It follows that there is a **minimum** at $x = 2$; the minimum value is 0. ▲

5.10 EXERCISES

Find the values of x *in each of the given domains for which the function is a maximum or a minimum. The domain consisting of the closed finite interval from* a *to* b *is denoted* $a \le x \le b$*; the half-line of numbers at least equal to* a *is denoted* $x \ge a$*; the half-line of numbers greater than* a *is denoted* $x > a$*, and so on. Indicate whenever the maximum or minimum does not exist.*

Function	Domains
1 $f(x) = x^2$	(a) $-1 \le x \le 2$ (b) $x \le -1$ (c) $x \ge 2$
2 $f(x) = x^3 - 3x^2$	(a) $-1 \le x \le 1$ (b) $x \ge 1$ (c) $x \le -1$
3 $f(x) = \frac{1}{3}x^3 + \frac{1}{2}x^2 - 6x + 8$	(a) $x \le -2$ (b) $-2 \le x \le 0$ (c) $x \ge 0$
4 $f(x) = 2x^3 - 8x^2 - 6x + 2$	(a) $x \le 0$ (b) $0 \le x \le 1$ (c) $x \ge 0$
5 $f(x) = x^4 - 8x^2$	(a) $x \le 2$ (b) $-1 \le x \le 5$ (c) $x \ge 1$
6 $f(x) = x^3 - 6x^2 + 9x + 1$	(a) $x \le 2$ (b) $x \ge 2$ (c) $0 \le x \le 4$
7 $f(x) = \dfrac{6x}{x^2+1}$	(a) $x \le 0$ (b) $x \ge 0$ (c) $-2 \le x \le 2$
8 $f(x) = x^{1/3}$	(a) $x \le -8$ (b) $-1 \le x \le 8$ (c) $x \ge -1$

9 $f(x) = \dfrac{x}{\sqrt{x^2+1}}$ (a) $x \le 0$ (b) $-\sqrt{8} \le x \le \sqrt{8}$ (c) $x \ge 1$

10 $f(x) = \dfrac{\log x}{x}$ (a) $0 < x \le 1$ (b) $x \ge 1$ (c) $x \ge e^2$

11 $f(x) = \sin x \cos x$ (a) $0 \le x \le \pi/2$ (b) $0 \le x \le \pi/6$ (c) $\pi/2 \le x \le \pi$

12 $f(x) = \sin^2 x$ (a) $-\pi/4 \le x \le \pi/4$ (b) $0 \le x \le \pi/2$ (c) $\pi/2 \le x \le 5\pi/6$

13 $f(x) = e^{-(1/2)x^2}$ (a) $x \ge -1$ (b) $x \le -1$ (c) $2 \le x \le 5$

14 $f(x) = xe^{-(1/2)x^2}$ (a) $x \ge 0$ (b) $x \le 0$ (c) $\frac{1}{2} \le x \le \frac{3}{4}$

15 $f(x) = (1 + \cos^2 x)^{-1/2}$ (a) $-\pi \le x \le \pi$ (b) $-\pi \le x \le 0$ (c) $\pi/4 \le x \le 5\pi/6$

16 $f(x) = \dfrac{1}{1 + \sin^2 x}$ (a) $-\pi \le x \le \pi$ (b) $-\pi/3 \le x \le \pi/4$ (c) $\pi/6 \le x \le 3\pi/4$

17 If a function is **differentiable** at a point x_0, why is it also **continuous** at x_0? (Indicate why the range values are close to each other if the corresponding domain values are.)

18 The Intermediate Value Theorem of calculus states that if f is continuous on an interval and assumes two values a and b, with $a \neq b$, then it necessarily assumes every value between a and b. Give a geometric proof based on the pencil-and-paper definition of continuity.

19 Let $f(x)$ have a "smooth" graph over the interval from a to b. By an intuitive geometric argument, show that there is at least one point c between a and b such that the tangent to the curve at c is parallel to the line connecting the endpoints of the graph (Figure 5.69). (The point c might coincide with a or b.) State this result in terms of f and its derivative f'. This result is known as the Law of the Mean.

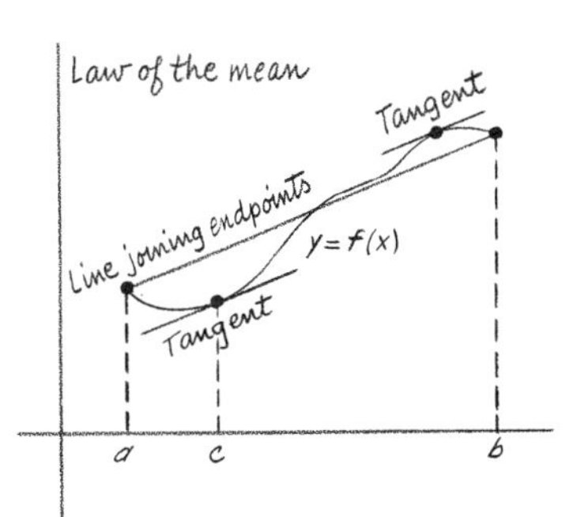

Figure 5.69

5.11 Applications with Maxima and Minima

In this section, we apply the method of determining the extreme values to various situations.

EXAMPLE 1

A box with an open top is to be made from a square piece of sheetmetal 10 inches long by cutting out square corners and bending up the sides. Let x be the length of the edges of the cut corners; then x varies from 0 to 5. The volume of such a box is given by the function

$$f(x) = x(10 - 2x)^2, \qquad \text{for } 0 \leq x \leq 5.$$

This was derived in Example 2, Section 4.1. There it was shown, by direct evaluation of the function at various points, that a corner of length $x = 1.5$ produced a box of volume 73.5 sq in., and that this is the largest volume over all x-values which are multiples of 0.5. Now we are going to show that this is close to the maximum over all values of x, the volume of the largest box that can be so constructed.

Our task is to find the maximum of the function f over the given domain. The critical points are found as the solutions of the equation $f'(x) = 0$:

$$f'(x) = -4x(10 - 2x) + (10 - 2x)^2 = 0,$$

or

$$(10 - 2x)(10 - 6x) = 0.$$

The solutions are $x = 5/3$ and $x = 5$. The endpoints of the domain are $x = 0$ and $x = 5$. It follows that the maximum of the function is the largest of the values $f(0)$, $f(5/3)$, and $f(5)$. These are 0, 74.1, and 0, respectively. It follows that $x = 5/3$ yields the largest volume: 74.1. ▲

EXAMPLE 2

We refer to the cowboy-and-his-horse problem in Example 3, Section 4.1. The total distance from the cowboy to the stream and then to his tent is given by the function

$$f(x) = (10{,}000 + x^2)^{1/2} + (2500 + (300 - x)^2)^{1/2}, \qquad \text{for } 0 \leq x \leq 300.$$

The minimum distance was shown to be approximately 336 feet. Now we shall determine this exactly.

Our problem is to minimize f over the given interval. To find the critical points, we first compute f' by means of the chain rule:

$$f'(x) = \frac{x}{(10{,}000 + x^2)^{1/2}} - \frac{300 - x}{(2500 - (300 - x)^2)^{1/2}}.$$

Set this equal to 0 and solve for x:

$$\frac{x}{(10{,}000 + x^2)^{1/2}} - \frac{300 - x}{(2500 - (300 - x)^2)^{1/2}} = 0.$$

Transpose the second term and square both sides of the resulting equation:

$$\frac{x^2}{10{,}000 + x^2} = \frac{(300 - x)^2}{2500 + (300 - x)^2}.$$

Divide the numerator and denominator of each fraction by the corresponding numerator:

$$\frac{1}{\dfrac{10{,}000}{x^2} + 1} = \frac{1}{\dfrac{2500}{(300 - x)^2} + 1}.$$

It follows that

$$\frac{x^2}{10{,}000} = \frac{(300 - x)^2}{2500},$$

and so, by taking the square root, we find

$$\frac{x}{100} = \frac{300 - x}{50} \qquad \text{or} \qquad x = 200.$$

The only critical point is at $x = 200$. Since the endpoints are 0 and 300, the minimum of the function over the entire domain is the minimum of $f(0)$, $f(200)$, and $f(300)$. By the table in the example cited above, the minimum is at $x = 200$, where $f(200) = 336$. ▲

EXAMPLE 3

An engineer is designing a large cylinder to hold 1 cu ft of commercial cake mix. The side of the can is constructed from cardboard, and the top and bottom from metal. The metal costs twice as much as the cardboard. What dimensions should be used to minimize the cost of material of an individual can? If too little metal is used in the top and the bottom, then the can will have to be very long to hold the required volume; this will increase the amount of cardboard that has to be used. If too little cardboard is used for the sides, then the can will be wide and shallow and will require much metal for the top and the bottom (Figure 5.70). The engineer seeks the right combination to minimize the cost.

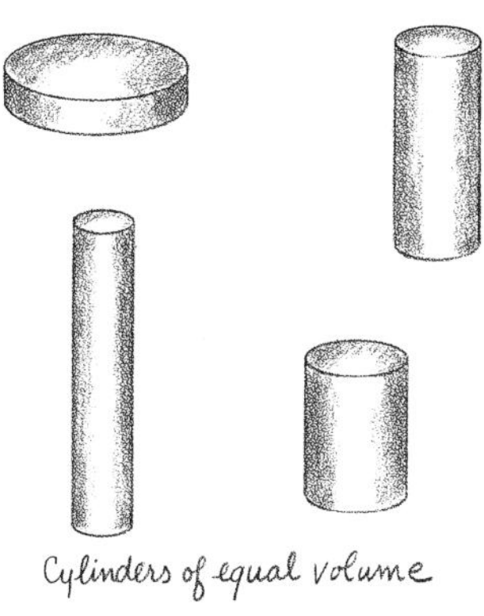

Figure 5.70

Let R be the radius of the base circle of the cylinder; then the areas of the top and the bottom are equal to πR^2. Let H be the height of the cylinder; then the area of the side of the cylinder is $2\pi RH$, which is the length of the circumference of the base circle multiplied by the height.

The variables R and H are related through the condition that the volume is 1. The formula for the volume is $\pi R^2 H$; hence, since the volume is 1, $\pi R^2 H = 1$, or

$$H = \frac{1}{\pi R^2}. \tag{1}$$

Next we choose an independent variable x, and express the cost of the cylinder as a function of x. Let the cost of the cardboard be c cents per square foot. The actual amount is unimportant so that we take $c = 1$. Since the metal costs twice as much as the cardboard, it costs 2 cents per square foot. Let x be the variable representing the radius of the base circle: $x = R$. Then the amount of cardboard in a single can is the lateral area of the cylinder, $2\pi RH$, which, by (1), is $2/R$, or, in terms of x, $2/x$. This is also the **cost** of the cardboard in a can. The area of the top is πR^2, or πx^2; therefore, the sum of the areas of the top and the bottom is twice that amount, $2\pi x^2$; hence, the **cost** of the metal in the can is $2(2\pi x^2)$, or $4\pi x^2$. It follows that the total cost of a can is

$$f(x) = 4\pi x^2 + \frac{2}{x}, \qquad \text{for } x > 0.$$

We seek the minimum of this function. The domain is the half-line consisting of the positive real numbers; the endpoint 0 is not included. The minimum, if it exists, is at a critical point. Set the first derivative equal to 0:

$$f'(x) = 8\pi x - \frac{2}{x^2} = 0 \qquad \text{or} \qquad \frac{4x^3 - 1}{x^2} = 0.$$

The solution is that of the equation $4\pi x^3 = 1$, namely, $x = (4\pi)^{-1/3}$, which is approximately 0.43. In order to show that the function is really minimized at this point, we demonstrate that the second derivative is positive, so that the graph is concave upward:

$$f''(x) = 8 + \frac{4}{x^3};$$

and this is positive for all $x > 0$; in particular, at the critical point. It follows that the minimum cost is achieved with a radius $(4\pi)^{-1/3}$. By Equation (1) the height is then $1/\pi R^2$, or $(4\pi)^{2/3}/\pi$. The cost is then

$$f((4\pi)^{-1/3}) = 3\sqrt[3]{4\pi} = 6.9 \qquad \text{(approximately)}.$$

Note that the ratio of the height to the radius is

$$\frac{H}{R}=\frac{(4\pi)^{2/3}/\pi}{1/(4\pi)^{1/3}}$$

thus, the height should be taken as 4 times the radius, or twice the diameter. ▲

EXAMPLE 4

A retail chain has a branch store in each of 3 cities, A, B, and C. A and B are 320 miles apart, and C is 200 miles from each of them. The warehouse is to be built equidistant from A and B. In addition, in order to minimize the time and costs of transportation, it should be located so that the sum of the 3 distances from the warehouse to each of the cities is as small as possible. Where should it be built? (Figure 5.71.)

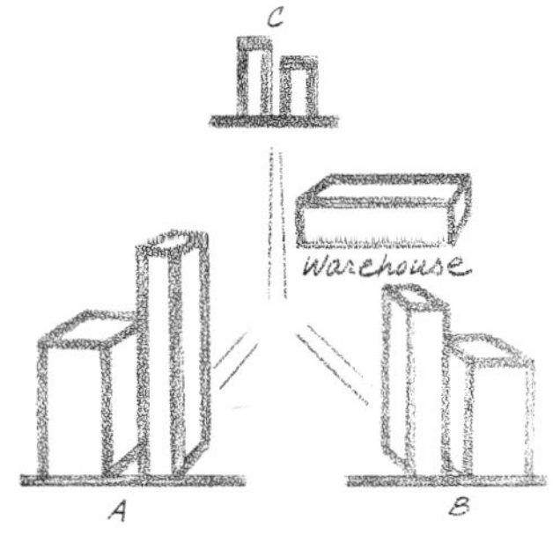

Figure 5.71

A, B, and C are the corners of an isosceles triangle. By a simple application of the Pythagorean Theorem, the distance from C to the midpoint of the line joining A and B is 120 miles [Figure 5.72(a)]. The warehouse is to be built at a point P along the line from C to the midpoint. The sum of the 3 distances to the respective cities is the sum of the lengths of the lines from P to A, B, and C, respectively. We shall express this sum as a function of an independent variable x, and then find the minimum of the function. Let x be the radian measure of the angle formed by the lines from A to P and A to B; then the 3 distances are [Figure 5.72(b)]

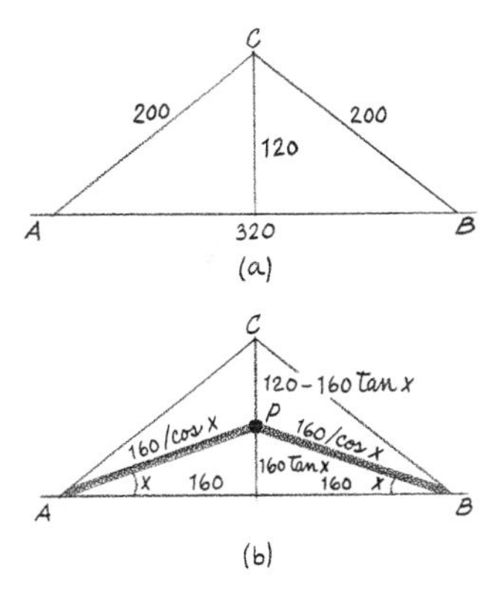

Figure 5.72

from P to A	$160/\cos x$,
from P to B	$160/\cos x$,
from P to C	$120-160\tan x$.

The sum of the 3 distances is

$$f(x)=\frac{320}{\cos x}+120-160\tan x.$$

The domain of the function is $0\leq x\leq\arctan(\frac{3}{4})$; indeed, the tangent of the angle cannot exceed that of the angle at A, which is $120/160=\frac{3}{4}$. Now we find the critical points:

$$f'(x)=\frac{320\sin x}{\cos^2 x}-\frac{160}{\cos^2 x}=0,$$

or

$$2\sin x=1,\qquad\text{so that}\qquad x=\frac{\pi}{6}.$$

(This is the only solution for the given domain of x.) The value of the function at this point is

$$\begin{aligned} f\left(\frac{\pi}{6}\right) &= \frac{320}{\cos(\pi/6)} + 120 - 160\tan\left(\frac{\pi}{6}\right) \\ &= \frac{320(2)}{\sqrt{3}} + 120 - \frac{160}{\sqrt{3}} \\ &= 120 + 160\sqrt{3} = 397 \qquad \text{(approximately)}. \end{aligned}$$

The minimum of the function is the minimum of

$$f(0), \qquad f\left(\frac{\pi}{6}\right), \qquad \text{and} \qquad f(\arctan(\tfrac{3}{4})).$$

When $x = 0$, the warehouse is situated right at the bisector of the line from A to B, and the sum of the 3 distances is

$$f(0) = 160 + 160 + 120 = 440.$$

When $x = \arctan(\frac{3}{4})$, then the warehouse is at C, and the sum of the distances is $200 + 200 + 0 = 400$; thus,

$$f(\arctan(\tfrac{3}{4})) = 400.$$

Since $f(\pi/6) = 397$ is smaller than either of these, it is the minimum value of the function on the given domain. The warehouse should be located where $x = \pi/6$. ▲

EXAMPLE 5

Let us reconsider the "rowboat problem" of Exercises 3 and 4 of Section 4.1. A man is in a rowboat at a point A on the lake, and wants to get as quickly as possible to a point B on the land. He can row at v_1 mph and walk v_2 mph. If he rows on a straight line to a point on the shore, and then walks on a straight line over land to his destination, where should he land the boat to minimize the total time traveling?

This is really a masked version of a celebrated scientific problem first considered by the Alexandrian Greeks 2000 years ago, and finally solved by Snell and Descartes in the seventeenth century: the problem of the refraction of light. The reader is surely familiar with the bending of light rays as they pass from water to air or through a lens (Figure 5.73). The principle of light refraction is the basis of man's accomplishments in optics. It rests on **Fermat's Principle** (he was a mathematician of the seventeenth century): When light travels at various speeds through various media, it takes the path requiring the least time. For ex-

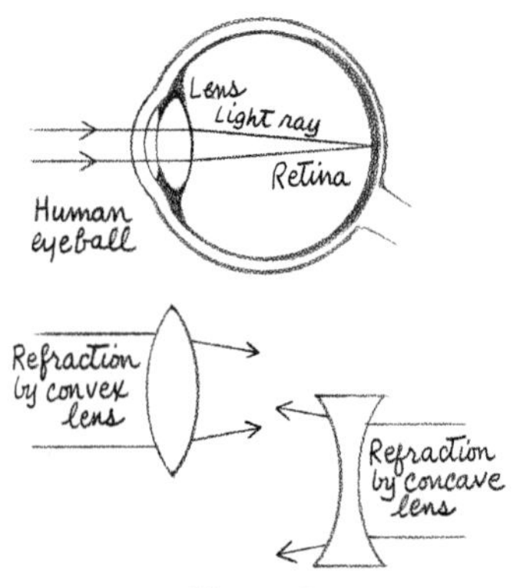

Figure 5.73

ample, when light travels from a point in the water to a point above the surface of the water, it goes on a straight line to the surface, and then on another straight line to the point in the air (Figure 5.74). The point where it hits the surface is chosen to make the trip as brief as possible, in units of time. The path of light is like the path of the man in the rowboat who takes the quickest route to his destination.

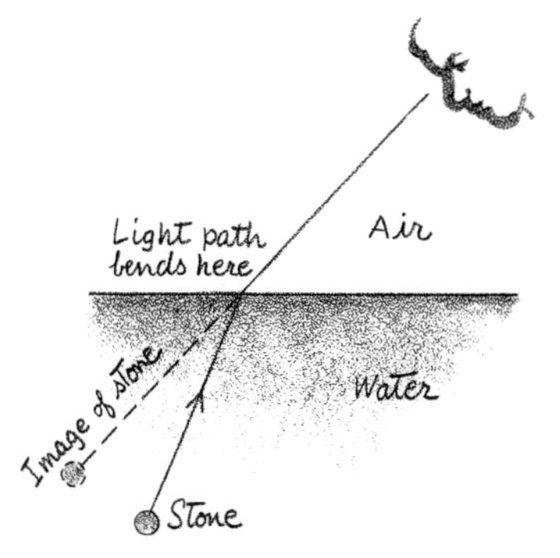

Figure 5.74

To make the solution of the problem as simple as possible, we suppose that the light source is at a point A 1 unit below the surface of the water, and the receiver is at a point B 1 unit above the surface; finally, the horizontal distance from A to B is taken as 2 units. Let v_1 and v_2 be the velocities of light in water and air, respectively. At what point should the ray pass through the surface of the water to minimize the travel time? We shall express the travel time as a function of an independent variable x: Let x be the **horizontal** distance from the source at A to the point on the surface struck by the ray. The distance from A to the surface, along a straight line, is $\sqrt{1+x^2}$; the velocity of the light is v_1; therefore, the time spent traveling through the water is $\sqrt{1+x^2}/v_1$. The remaining distance to B is $\sqrt{1+(2-x)^2}$, and the velocity is v_2; therefore, the time spent is $\sqrt{1+(2-x)^2}/v_2$ (Figure 5.75). The total time for the trip is

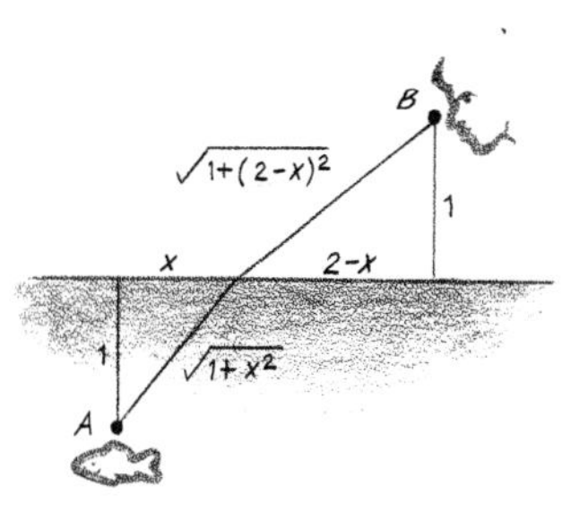

Figure 5.75

$$f(x) = \frac{\sqrt{1+x^2}}{v_1} + \frac{\sqrt{1+(2-x)^2}}{v_2}, \qquad \text{for } 0 \le x \le 2.$$

We wish to minimize this function on the interval.

The derivatives are

$$f'(x) = \frac{x}{v_1\sqrt{1+x^2}} - \frac{2-x}{v_2\sqrt{1+(2-x)^2}},$$

$$f''(x) = \frac{1}{v_1(1+x^2)^{3/2}} + \frac{1}{v_2(1+(2-x)^2)^{3/2}}.$$

Since v_1 and v_2 are positive quantities, $f''(x)$ is positive for all x, and so the curve is concave upward for all x; therefore, it will have a **minimum** at any critical point. When we write the equation $f'(x) = 0$, we get

$$\frac{x}{v_1\sqrt{1+x^2}} = \frac{2-x}{v_2\sqrt{1+(2-x)^2}},$$

or

$$\frac{v_1}{v_2} = \frac{x/\sqrt{1+x^2}}{(2-x)/\sqrt{1+(2-x)^2}}.$$

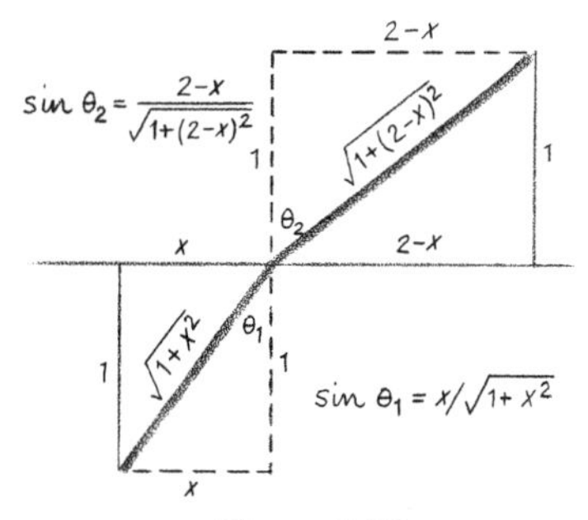

Figure 5.76

The numerator $x/\sqrt{1+x^2}$ is equal to the sine of the angle between the ray in the water and the perpendicular to the surface of the water; and the denominator is the sine of the angle between the perpendicular and the ray in the air (Figure 5.76). These angles are denoted θ_1 and θ_2, respectively; thus,

$$\frac{v_1}{v_2} = \frac{\sin\theta_1}{\sin\theta_2}. \tag{2}$$

This equation is known as Snell's Law. It states that the ratio of the velocities in the two media is equal to the ratio of the sines of the angles between the rays and the perpendicular to the surface. The first angle is the angle of **incidence** and the second is the angle of **refraction.**

Equation (2) is generally true, even if the distances between A and B, or the distances from the surface, are changed. ▲

5.11 EXERCISES

1 A rectangular box with an open top is to be made from a piece of sheet-metal 1 foot wide and 2 feet long by cutting a square from each corner and bending up the sides.

(a) Let x be the length of the edge of a cut corner. Express the volume of the box as a function of x, and state the domain of the function. (See Example 2, Section 4.1.)

(b) What is the volume of the largest box that can be made in this way? (See Example 1.)

2 A pumping station is to be built on the edge of a river between 2 neighboring towns A and B. The latter are 4 and 5 miles, respectively, from the nearest points C and D on the river; C and D are 12 miles apart. Where should the station be built to make the total length of pipeline a minimum? (See Exercise 16, Section 4.1.)

3 In constructing an office building, the builder estimates that the initial costs (site purchase, foundation, and so on) as 450 times as much as the first floor, the second floor costs twice as much as the first, the third floor costs 3 times as much as the first, and so on. Let x be the number of floors that are planned.

(a) Express the average cost of a floor as a function of x; let c stand for the cost of the first floor. (See Example 4, Section 4.1.)

(b) Assume x to be a continuous variable, taking on not only integer values but all positive real values. Find the value of x which minimizes the average cost.

4 A supermarket expects to sell 10,000 boxes of a certain breakfast cereal in 1 year. The cost of holding 1 box for 1 year is 50¢ (cost of space, loss due to damage or theft, and so on). The fixed cost of placing an order for a shipment is \$1; and the cost of shipment of an individual box is 1¢. How many boxes should be ordered in each shipment to minimize the total cost of inventory? (See Example 5, Section 4.1.)

5 A 40-foot rope forms the perimeter of a rectangle. Find the dimensions of the rectangle that maximize the area. (See Example 3.)

6 A rectangular plot is designed so that a long straight wall is to form 1 side. 500 feet of fencing is to be used to enclose the remaining 3 sides. Find the greatest area that can be so enclosed. (See Example 3.)

7 An 8-foot fence is located 1 foot from the base of the wall of a building. An extension ladder is to be placed to lean against the wall in such a way that the ladder rests on the fence. Find the length of the shortest ladder that can be used. (See Exercise 2, Section 4.1.)

8 A man in a rowboat is 1 mile from the nearest point A on a straight shoreline. He wishes to go as quickly as possible to B, on the shoreline 6 miles from A. He plans to row on a straight line to a point on the shore between A and B, and walk the remaining distance to B. He can walk 5 mph and row 3 mph. Using Snell's Law, (2), determine how far from A the boat should be landed. Then confirm this result using the method of minimization given in this section. What if B is 10 miles from A? 20 miles?

9 Two planes are traveling at the same altitude, and their routes cross at a point P. One plane is flying south at 400 mph, and the other west at 600 mph. At a particular moment the first is 30 miles north of P, and the second is 40 miles east. How close will the planes ever come? (See Exercise 11, Section 4.1.)

10 The top and the bottom of a box are square. The box is designed to hold a volume of 100 cubic inches. The top and the bottom are made from the same material, which is twice as expensive as the material used for the 4 walls. Find the dimensions of the box to minimize the cost. (See Example 3.)

11 A closed right circular cylinder is to contain a volume V. Determine the ratio of the radius of the base circle to the height that minimizes the surface area. (See Example 3.)

12 A printer has the following price schedule for the production of a 250-page book:

(i) The first 10,000 copies cost \$4 each.

(ii) For every group of 100 books exceeding 10,000, there is a reduction of 2¢ each on all the books; that is, the first 100 cost \$3.98 each, the second 100 cost \$3.96 each, and so on.

- (a) Let x be the number of hundreds of books exceeding 10,000. Express the revenue of the printer as a function of x.
- (b) What value of x maximizes the revenue of the printer? (See Example 4, Section 4.1.)

13 A technical publisher estimates that the cost of selling a certain book may be broken down as follows:

(a) There is a fixed overhead cost of \$1000.

(b) The cost of selling the first 500 books is 50¢ per book.

(c) For every group of 100 books over 500, the cost of selling an individual book rises by 2¢: it costs 52¢ to sell books numbered 501 to

600, 54¢ for books 601 to 700, and so on. How many hundreds of books should the publisher plan to sell to minimize the average cost of selling a book? (See Example 4, Section 4.1.)

14 The profit of an economic unit (factory, utility, office) is its revenue minus its cost. These are functions of the output:

$$R(x) = \text{Revenue at level } x \text{ of output,}$$
$$C(x) = \text{Cost at level } x \text{ of output.}$$

(See Example 2, Section 5.3.) According to the theory of mathematical economics, the profit $R(x) - C(x)$ is a maximum when the marginal revenue is equal to the marginal cost:

$$\left(\frac{d}{dx}\right) R(x) = \left(\frac{d}{dx}\right) C(x).$$

Using the method of this section explain the basis for this "theorem" of economic theory. What assumption is sufficient for its validity?

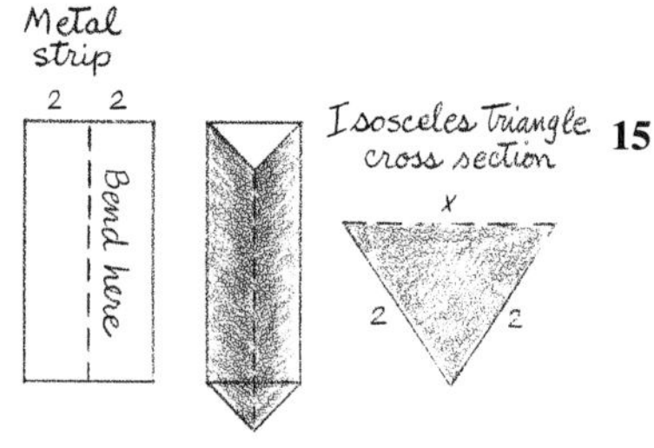

Figure 5.77

15 A biology student is constructing a small water trough for his mice. The sides of the trough are made from a rectangular strip of metal by bending it down the middle. The strip is 4 inches wide. How wide should the top of the trough be made to maximize the volume? (Figure 5.77.) [*Hint:* Maximize the cross-sectional area, which is an isosceles triangle with 2 sides of length 2. Let x be the width of the other side of the triangle. Express the area in terms of x, and maximize it.]

16 The morning after a dinner party the hostess finds a cigarette hole near the corner of her tablecloth. It is located 2 and 3 inches from the 2 edges of the tablecloth. She decides to cut off the corner on a straight line through the hole (and then sew a hem). Find the minimum area that can be cut off in this way (Figure 5.78).

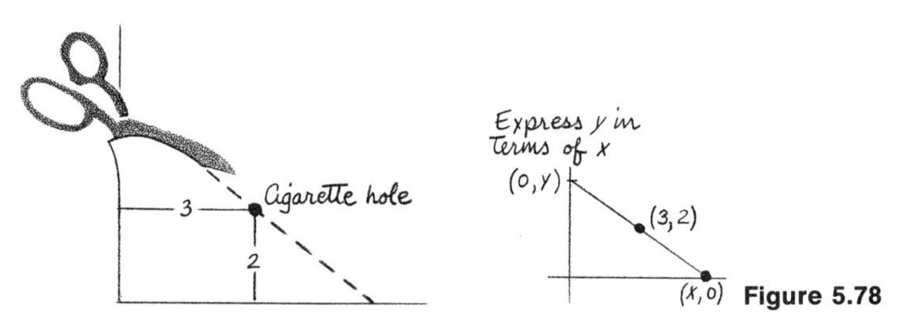

Figure 5.78

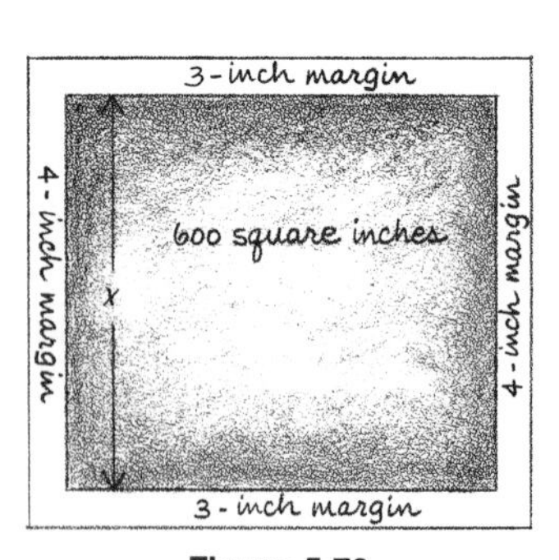

Figure 5.79

17 A newspaper layout editor has to place a rectangular advertisement containing 600 square inches of print, and with top and bottom margins of 3 inches, and side margins of 4 inches.

(a) Let x be 1 side of the rectangle formed by the print. Express the total area of the advertisement as a function of x (Figure 5.79).

(b) Find the minimum area.

18 According to the regulations of the United States Postal Service, the girth plus the length of a fourth-class parcel must not exceed 100 inches (Figure 5.80). Find the maximum permissible volume for a package with 2 square sides.

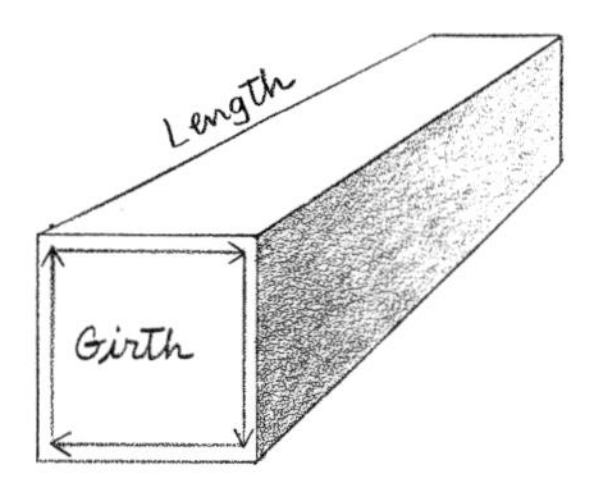

Figure 5.80

19 An 18-foot ladder leans on an 8-foot fence, forming an angle of x radians with the ground. Find the angle at which the horizontal protrusion of the ladder beyond the fence is a maximum (Figure 5.81).

20 A smuggler wants to fit a small cylindrical vial inside a hollow rubber ball with a 3-inch diameter. Let x be the angle (in radians) shown in Figure 5.82.

(a) Express the height of the cylinder in terms of x.

(b) Express the radius of the base circle in terms of x.

(c) Using (a) and (b) express the volume as a function of x.

(d) Find the volume of the largest cylinder that can be made in this way.

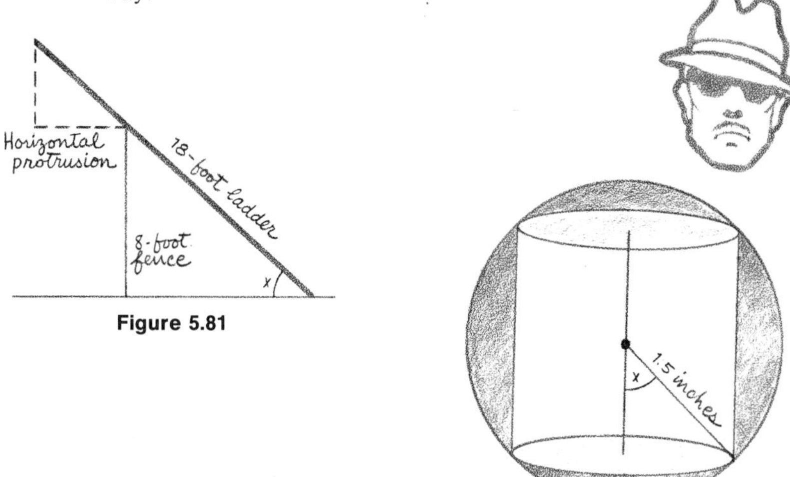

Figure 5.81

Figure 5.82

21 In Section 1.7, we showed by a geometric argument that light is reflected from a surface according to the Law of Reflection: The angle of incidence is equal to the angle of reflection. This was based on the principle that light takes the shortest path: The angles of incidence and reflection are equal when the path is shortest. This result can also be deduced by the method of this section.

(a) Let x be the lateral distance from the source A to the point P on the reflector (Figure 5.83). Express the total length of the path from the source to the receiver as a function of x. Take the lateral distance from A to B as 1, and let the perpendicular distances of A and B to the reflector be a and b.

(b) Show that the function is a minimum when the **cosines** of the angles of incidence and reflection are equal.

(c) What is the relation of this result to the cowboy-and-horse and pipeline problems, Example 2 and Exercise 2, respectively?

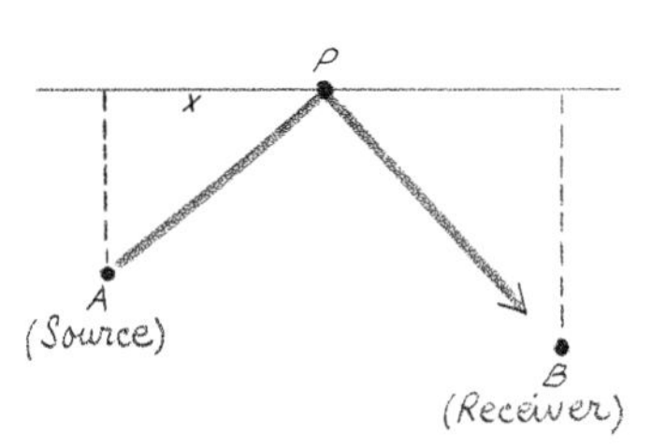

Figure 5.83

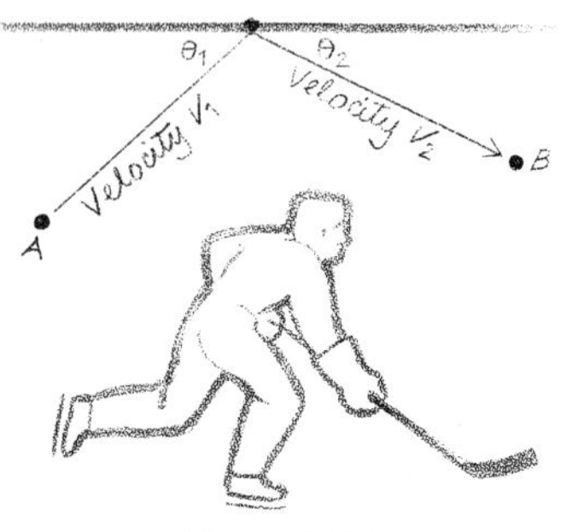

Figure 5.84

22 Here is an extension of the Law of Reflection to hockey pucks bouncing off the side boards. A hockey player at A passes the puck to a point B by bouncing it against the wall at P. The velocity of the puck leaving the stick of the player is v_1 ft/sec. After hitting the wall, the puck's speed is reduced to v_2 ft/sec. Let θ_1 and θ_2 be the angles of incidence and reflection, respectively (Figure 5.84). Show that the total time for the pass from A to B is a minimum when

$$\frac{\cos\theta_1}{\cos\theta_2} = \frac{v_1}{v_2}.$$

Compare this with Snell's Law, Equation (2). As in Exercise 21, let a and b be the perpendicular distances to the side board, and let 1 be the lateral distance from A to B.

5.12 Newton's Method for the Solution of Equations

In this section we show how the derivative may be used in obtaining approximate solutions for equations of a general kind. Such equations might arise in locating critical points, where we set the first derivative of a function equal to 0. The method can be used in many other situations where a complicated equation has to be solved.

Let $f(x)$ be a differentiable function. Consider the equation

$$f(x) = 0. \tag{1}$$

Suppose we want its solution, that is, the value x for which it is satisfied. Assume that we have an initial guess x_0 for the true solution. **Newton's method** is based on a formula for improving the guess, and getting a **first approximation** x_1 to the true solution:

$$x_1 = x_0 - \frac{f(x_0)}{f'(x_0)}. \tag{2}$$

A **second approximation** x_2 is obtained from x_1 by applying the same formula but with x_1 and x_2 in place of x_0 and x_1, respectively:

$$x_2 = x_1 - \frac{f(x_1)}{f'(x_1)}.$$

By the same procedure, we can get a third, fourth, . . . , nth approximation. Each approximation x_n is derived from its predecessor by the equation of the type (1). Under certain conditions on the function f, the nth approximation converges to the true solution as n tends to ∞:

$$\lim_{n\to\infty} x_n = x.$$

A simple condition which is sufficient for the convergence to the true solution is the following: The true solution x and the initial guess x_0 should belong to an interval in the domain where f has 1 of the 4 shapes in Figure 5.85. In terms of f and its derivatives this condition means that f' and f'' are not equal to 0, but each maintains its same sign throughout the interval.

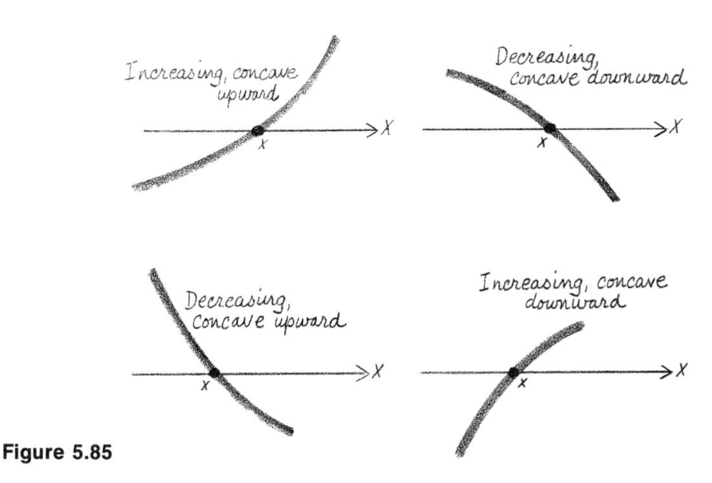

Figure 5.85

Formula (2) is based on formula (2) of Section 5.6, which is an approximate, not a strict equation. If x_1 is the solution of (1) and x_0 is a point near x_1, then we have (approximately)

$$f(x_1) - f(x_0) = f'(x_0)(x_1 - x_0). \tag{3}$$

(Here we have used x_0 and x_1 in place of x and x^*, respectively.) By Equation (1), the approximate relation (3) reduces to

$$-f(x_0) = f'(x_0)(x_1 - x_0). \tag{4}$$

When f' is not equal to 0, then (4) is equivalent to (2).

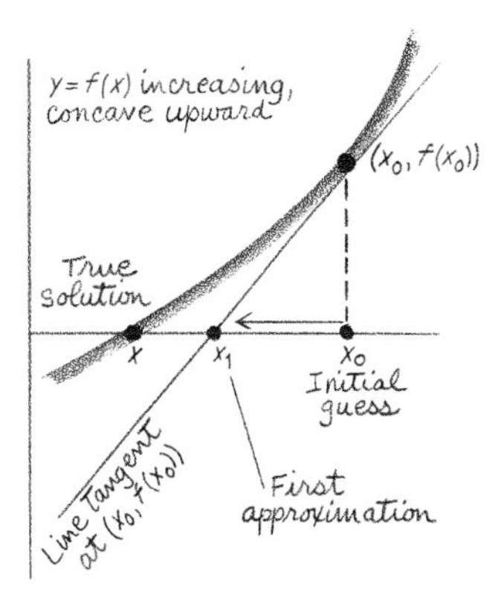

Figure 5.86

Using a diagram we will show how formula (2) yields better and better approximations to the true solution. A typical function f appears in Figure 5.86. The slope of the line tangent to the curve through the point $(x_0, f(x_0))$ is, by definition, $f'(x_0)$. The equation of this line is

$$y - f(x_0) = f'(x_0)(x - x_0)$$

(by the point-slope formula, Section 1.2). Let (x_1, y_1) be the point on the line where it cuts the x-axis; then $y_1 = 0$, and so x_1 satisfies Equation (4), which is equivalent to (2). In the picture it appears that x_1 is closer to the solution x than is the guess x_0, and that the successive

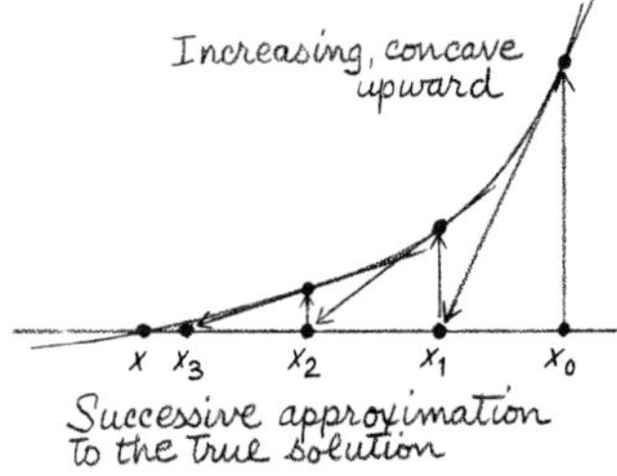

Figure 5.87

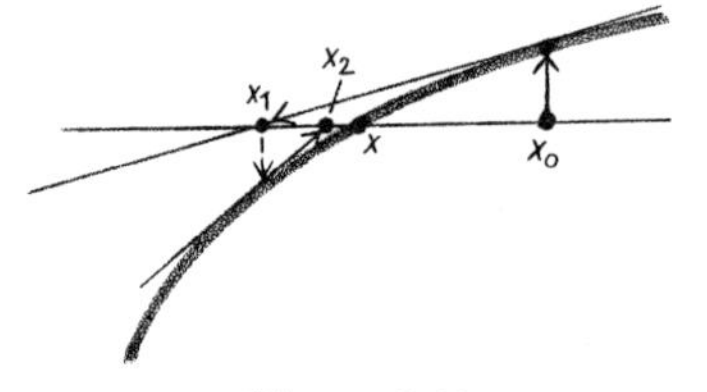

Figure 5.88

approximations get closer and closer (Figure 5.87). Another case, that of an increasing function concave downward, is shown in Figure 5.88.

The steps in Newton's method are:

1. By examining several values of the function f near the solution x, choose x_0 as close as possible to it.
2. Check that f' and f'' maintain their signs (and are not equal to 0) throughout an interval containing x and x_0.
3. Apply formula (2). If the solution is sufficiently accurate for the purpose desired, then stop. If not, apply the formula again to obtain a second approximation. And so on.

The following procedure is often used to get the solution to a desired number of decimal places: Calculate the successive approximations to the given number of places, and stop as soon as two successive approximations agree.

EXAMPLE 1

Let us approximate $\sqrt[3]{5}$ as closely as possible with just one application of (2). This root is the solution of the equation $x^3 - 5 = 0$. Here $f(x)$ is the function $x^3 - 5$; and so $f'(x) = 3x^2$ and $f''(x) = 6x$. The solution is necessarily positive because the (real) cube root of a positive number is itself positive. The first two derivatives do not change sign over the positive axis: both are positive. Therefore, any positive x_0 may be used as the initial guess. However, we try to choose the initial guess as close as possible to the true solution in order that the first approximation x_1 be as close as possible to the solution: If x_0 is close to x, then x_1 will be even closer. Since $1^3 = 1$ and $2^3 = 8$, we try several numbers between 1 and 2:

$$(1.5)^3 = 3.375, \qquad (1.7)^3 = 4.913,$$
$$(1.6)^3 = 4.096, \qquad (1.8)^3 = 5.832.$$

See Table I. We expect the solution to be just slightly more than 1.7; hence, we choose $x_0 = 1.7$. Apply formula (2) with $f(x) = x^3 - 5$ and $f'(x) = 3x^2$:

$$x_1 = 1.7 - \frac{(1.7)^3 - 5}{3(1.7)^2} = 1.7 + 0.0100 = 1.7100.$$

This differs by only 0.00002 from the value given in Table I. ▲

EXAMPLE 2

Consider the cubic equation $x^3 + 2x - 5 = 0$. First we show that it has only one real solution. It may be written as $x^3 = 5 - 2x$.

The graph of the function on the left, $y = x^3$, rises from $-\infty$ to ∞ as x runs over the line from $-\infty$ to ∞. The graph of the (linear) function on the left, $y = 5 - 2x$, similarly falls from ∞ to $-\infty$. Hence the graphs intersect at exactly one point. Next we estimate the location of the solution by testing integer values of x and evaluating $f(x) = x^3 + 2x - 5$:

x	-2	-1	0	1	2
$f(x)$	-17	-8	-5	-1	7

Since $f(x)$ changes from negative to positive in the interval from $x = 1$ to $x = 2$, the solution must be here. Using Table I, we test several more values to 1 decimal place:

x	1.1	1.2	1.3	1.4	1.5
x^3	1.331	1.728	2.197	2.744	3.375
$2x$	2.2	2.4	2.6	2.8	3.0
$f(x)$	-1.469	-0.872	-0.203	0.544	1.375

It follows that the solution is between 1.3 and 1.4. To simplify the computations with formula (2), let us choose x_0 to be a rational number between 1.3 and 1.4; here we shall choose $x_0 = \frac{4}{3} = 1.333. . . .$

In order to apply (2), we check the conditions on f' and f'':

$$f'(x) = 3x^2 + 2, \qquad \textbf{which is positive for all } x;$$
$$f''(x) = 6x, \qquad \textbf{which is positive for positive } x;$$

In particular f'' is positive in the interval from 1.3 to 1.4.

Now apply (2) with $x_0 = \frac{4}{3}$, $f(x) = x^3 + 2x - 5$, and $f'(x) = 3x^2 + 2$:

$$x_1 = \tfrac{4}{3} - \frac{(\frac{4}{3})^3 + 2(\frac{4}{3}) - 5}{3(\frac{4}{3})^2 + 2} = \tfrac{4}{3} - \frac{4^3 + 8(3)^2 - 5(3)^3}{3^2 4^2 + 2(3)^3}$$
$$= \tfrac{4}{3} - \frac{1}{198} = 1.333. . . -0.0050505. . .$$
$$= 1.328.$$

This is correct to the given number of places; it can be verified by applying (2) again. ▲

5.12 EXERCISES

Find all solutions to 2 decimal places.

1 Find the roots by Newton's method; check in Table I:
(a) $\sqrt{4.3}$ (b) $\sqrt{110}$ (c) $\sqrt[3]{8.2}$ (d) $\sqrt[3]{9}$

2 Find the roots by Newton's method; check in Table I:
(a) $\sqrt{5.7}$ (b) $\sqrt{71}$ (c) $\sqrt[3]{5.9}$ (d) $\sqrt[3]{47}$

3 Evaluate by Newton's method.
(a) $(17)^{1/4}$ [*Hint:* $2^4 = 16$.]
(b) $(4)^{2/3}$ [*Hint:* First get the solution of $x^3 = 4^2$.]
(c) $(20)^{1/4}$ [*Hint:* First solve $x^2 = \sqrt{20}$.]
(d) $(19)^{-1/3}$

4 Evaluate by Newton's method.
(a) $(10)^{4/3}$ [*Hint:* $10(10)^{1/3}$.]
(b) $(8)^{1/4}$ [*Hint:* See 3*c*.]
(c) $(25)^{-2/3}$ [*Hint:* $(0.2)(0.2)^{1/3}$.]
(d) $(10)^{-3/2}$

5 Solve by Newton's method:
(a) $x^3 + 2x - 4 = 0$ (b) $x^3 - 3x^2 + 3 = 0$ (3 solutions)

6 Solve by Newton's method:
(a) $x^3 + x - 1 = 0$ (b) $x^4 + x - 3 = 0$ (2 solutions)

7 Find the solution of $x^3 - x^2 + x + 1 = 0$ between -1 and 0.

8 Find the (unique) solution of the equation $2x^3 - x^2 + x - 3 = 0$.

9 Find the solution of $x^3 - 3x^2 - 2x + 5 = 0$ between -1.5 and -1.0.

10 Find the solution of $x^3 + 2x^2 - 7x + 1 = 0$ between 0 and 0.5.

11 Using Table II, find the positive solution of the equation

$$e^{-x} + \tfrac{1}{2}x - 1 = 0.$$

[*Hint:* Compare e^{-x} and $1 - \frac{1}{2}x$.]

12 Using Table IV, solve $\cos x = x$.

13 Using Table IV, solve $x - 0.3 \sin x - 1 = 0$.

14 Find the solution of $x = 5 \log x$ which is between 1.2 and 1.3; use Table III.

15 Find the solution of $e^x + x - 2 = 0$; use Table II.

16 Find the positive solution of $x^2 - 2 \cos x = 0$; use Table IV.

CHAPTER

6

Integral Calculus

6.1 The Area Under the Graph of a Nonnegative Function

Let $f(x)$ be a nonnegative function (one assuming only nonnegative values) with a domain that is a finite closed interval I. The graph is always on or above the x-axis, and forms a region over I between the graph and the axis (Figure 6.1). We are going to discuss the measurement of the **area** of this region. Although the reader might have an intuitive idea of what we call **area** and its measurement in square units, he probably knows how to measure area in only certain cases: squares, rectangles, triangles, and circles. If $f(x)$ is a constant function, then its graph is a horizontal line; thus, the area under it is a rectangle, and its size in square units is equal to its height times its width (Figure 6.2). For an arbitrary linear function the graph is a trapezoid or a triangle, and the area is the sum of the areas of a rectangle and a triangle (Figure 6.3). In general, the **area of a region** is a nonnegative number associated with the region and measures its "size."

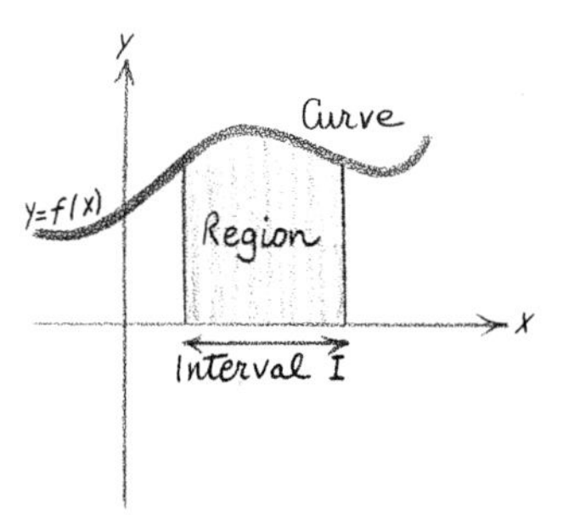

Figure 6.1

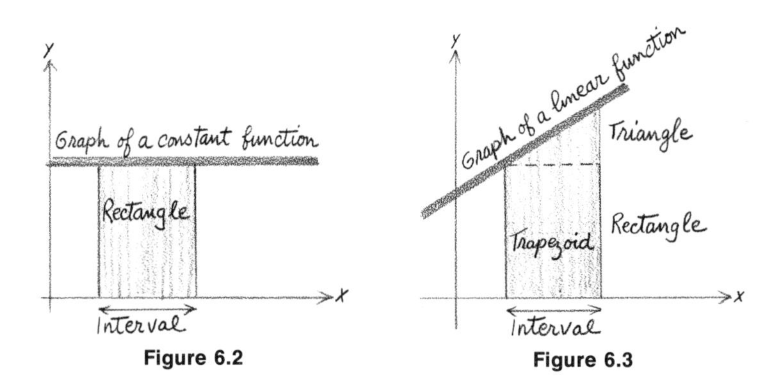

Figure 6.2

Figure 6.3

There are only a few functions for which the area under the graph can be defined and directly computed by methods of elementary geometry. Even algebraically simple functions fail to have obvious area computations. Consider the function $f(x) = x^2$. If we take the domain I as the closed interval from 0 to 2, then the area is that of the shaded region in Figure 6.4; however, there is no apparent way of computing its value in square units.

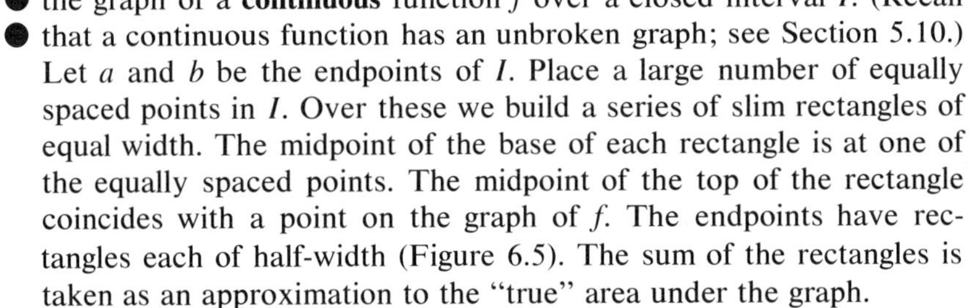

Now we shall define and show how to compute the area under the graph of a **continuous** function f over a closed interval I. (Recall that a continuous function has an unbroken graph; see Section 5.10.) Let a and b be the endpoints of I. Place a large number of equally spaced points in I. Over these we build a series of slim rectangles of equal width. The midpoint of the base of each rectangle is at one of the equally spaced points. The midpoint of the top of the rectangle coincides with a point on the graph of f. The endpoints have rectangles each of half-width (Figure 6.5). The sum of the rectangles is taken as an approximation to the "true" area under the graph.

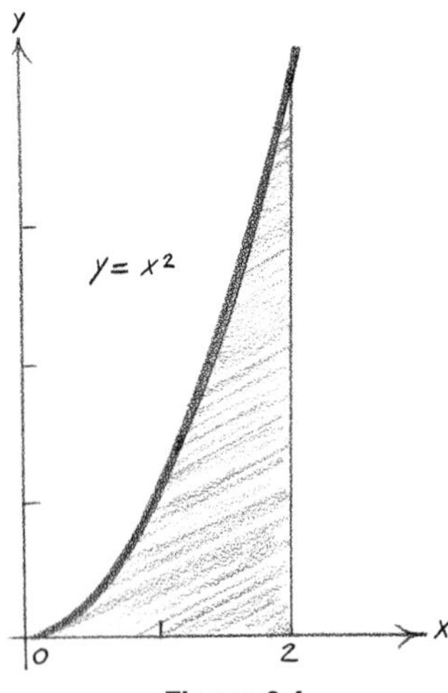

Figure 6.4

What is the "true" area? Suppose we repeat this process, increasing the number of equally spaced points. The rectangles become slimmer, and the discrepancy between the tops of the rectangles and the graph of the function becomes smaller (Figure 6.6). Then the approximation to the "true" area is better. It seems plausible that by successive repetition of this process, we could get the true area to any desired degree of accuracy. It is so. One of the basic theorems of integral calculus states that, as the number of subdivisions of I increases indefinitely, the sequence of the sums of the areas of the rectangles converges to a **limit.** The area under the curve is **defined** as this limit. The proof of this important theorem is contained in more advanced books on integral calculus, and is based on mathematics probably not accessible to the reader of this book; however, our geometric argument involving Figures 6.5 and 6.6 should convince him at this stage.

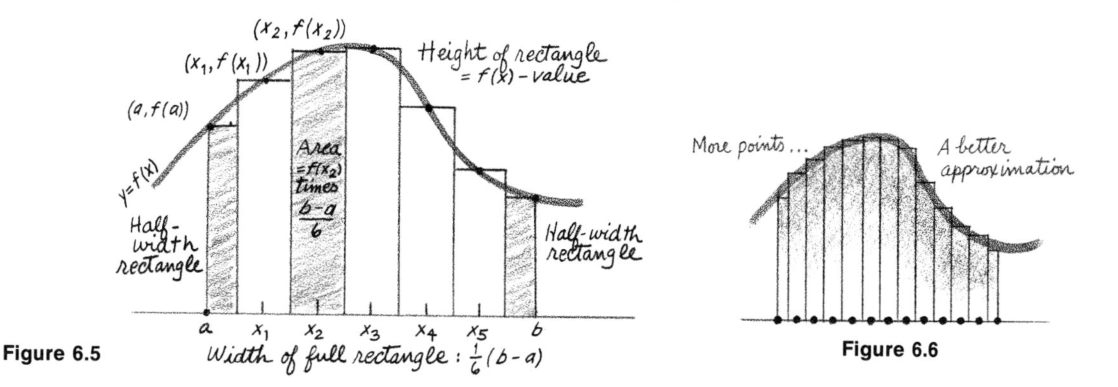

Figure 6.5

Figure 6.6

In evaluating the sum of the areas of the rectangles under a curve, it is not necessary to draw the graph and the rectangles. There is a simple formula, the **trapezoidal rule,** which results from the description above, and which can be used for a direct evaluation of the area. Suppose that the interval is divided into n equal parts by means of $n-1$ equally spaced points $x_1, \ldots, x_{n-1}$, as in Figure 6.5. The length of the base of a full rectangle is $(b-a)/n$. The first and the last rectangles are of half-width, $(b-a)/2n$. We compute the areas of the rectangles from left to right. The area of the first rectangle is the length of the base times the height,

$$\frac{(b-a)f(a)}{2n}.$$

The second has the area

$$\frac{(b-a)f(x_1)}{n},$$

the third $(b-a)f(x_2)/n, \ldots,$ and the next to the last $(b-a)f(x_{n-1})/n$. The last has the area $(b-a)f(b)/2n$. It follows that the sum of the areas is

$$\frac{b-a}{n}\left(\tfrac{1}{2}f(a)+f(x_1)+\cdots+f(x_{n-1})+\tfrac{1}{2}f(b)\right). \qquad (1)$$

This is the formula for the trapezoidal rule.

The number of points, $n-1$, used in formula (1) depends on several factors: how complicated the function is and what accuracy is desired. The more points we use, the more accurate our estimate of the area; however, the labor of calculation increases. In all the examples and exercises, the number of points will be specified.

EXAMPLE 1

Use formula (1) to estimate the area under the graph of $f(x) = 1 + x + x^2$ over the interval from 0.5 to 1.0. We shall use $n = 5$, so that $n - 1$, or 4, equally spaced points are placed between 0.5 and 1.0:

$$x_1 = 0.6, \quad x_2 = 0.7, \quad x_3 = 0.8, \quad x_4 = 0.9.$$

Evaluate the function at these points and the endpoints:

x	0.5	0.6	0.7	0.8	0.9	1.0
$f(x)$	1.75	1.96	2.19	2.44	2.71	3.00

Apply formula (1) with $a = 0.5$, $b = 1.0$, and $n = 5$:

$$\frac{0.5}{5}(0.88 + 1.96 + 2.19 + 2.44 + 2.71 + 1.50) = 1.168.$$

Later we shall verify by the exact method of integral calculus that the true area under the curve is $1\frac{1}{6} = 1/1666\ .\ .\ .$, which is very close to our estimate. ▲

The tables of powers, roots, reciprocals, exponentials, logarithms, and trigonometric functions can be used in formula (1) in calculating areas under corresponding functions.

EXAMPLE 2

Estimate the area under the graph of $f(x) = \sin x$ from $x = 0.0$ to $x = 1.0$ using (1) with $n = 10$. The values of the sine function are obtained from Table IV:

x	0.0	0.1	0.2	0.3	0.4	0.5
$f(x)$	0	0.09983	0.19867	0.29552	0.38942	0.47943

x	0.6	0.7	0.8	0.9	1.0
$f(x)$	0.56464	0.64422	0.71736	0.78333	0.84147

Apply formula (1) with $n = 10$, $a = 0$, and $b = 1$:

$$(0.1)(0 + 0.09983 + \cdot\cdot\cdot + 0.78333 + 0.42074) = 0.45932.$$

By the method of integral calculus to be studied later, the area is shown to be 0.45970 (to 5 places), which is very close to our estimate. ▲

At this point the reader probably has two good questions:

1. What is the method of integral calculus for the computation of area?
2. Do areas have practical use?

The method of calculus will be introduced in Sections 6.3 and 6.4. An application of the area concept to the description of statistical data is given in Section 6.2.

6.1 EXERCISES

In the following exercises the reader is asked to compare areas of certain regions with the approximations based on formula (1).

1 The area under the graph of the function $f(x) = 2x$ over the interval from $x = 0$ to $x = 1$ is a triangle. Compute the area (a) by the elementary formula for the area of a triangle, and (b) by using formula (1) with $n = 10$.

2 Find the area under the graph of $f(x) = 1 + 2x$ from $x = 0.5$ to $x = 1.5$ (a) by elementary geometry, and (b) by formula (1) with $n = 10$.

3 Put $f(x) = x^2$. In integral calculus we shall show that the area under the graph of this function over an interval from a to b is given by the formula $(b^3 - a^3)/3$. Find the area under the curve from $x = 0$ to $x = 1$ (a) by using this formula, and (b) by formula (1) with $n = 10$.

4 It is shown in integral calculus that the area under the graph of $f(x) = 1/x$, $x > 0$, from $x = a$ to $x = b$ is given by the formula $\log b - \log a$. Find the area under the curve from $x = 1$ to $x = 2$ (a) by this formula, and (b) by formula (1) with $n = 10$, and the values of the reciprocals in Table I.

5 The area under the graph of the exponential function $f(x) = e^x$ over the interval from $x = a$ to $x = b$ is given in calculus by the simple formula $e^b - e^a$. Find the area under the curve from $x = 0$ to $x = 1$ (a) by this formula, and (b) by formula (1) with $n = 10$.

6 For the negative exponential $f(x) = e^{-x}$, the formula in Exercise 5 is changed to $e^{-a} - e^{-b}$. Repeat the calculations of Exercise 5 in this case.

7 The area under the graph of the sine function over an interval from a to b contained in the interval from 0 to π is given by a formula involving the cosine: $\cos a - \cos b$. Find the area under the curve from $x = 0$ to $x = 1$ (a) by this formula, and (b) by formula (1) with $n = 10$.

8 The area under the cosine is given in terms of the sine: If $-\pi/2 \leq a < b \leq \pi/2$, then the area under the cosine from a to b is $\sin b - \sin a$. Find the area under the cosine function from $x = -0.5$ to $x = 0.5$ (a) by this formula, and (b) by formula (1) with $n = 10$.

6.2 Histograms and Population Densities

A **population** is a group of persons or objects. A **numerical characteristic** is a measurement associated with each member of the population; for example, the birth weight of each baby in a group of babies. Data on numerical characteristics arise in surveys of populations and in scientific experiments. One of the useful tools in the description of such data is the **histogram,** which is an area graph in the xy-plane.

A histogram is constructed from a given set of data in the following way:

1. The set of measurements obtained from the population is grouped by **class intervals** along the x-axis. For example, the weights of babies might be grouped in "500-gram intervals" such as "At least 1.0 (thousands of grams) but less than 1.5," "At least 1.5 but less than 2.0," and so on. These are **closed classes** because they have finite upper and lower endpoints. There is often an **open class** at one or the other end of the scale; for example, the class "At least 4.5."
2. Next we compute the **class proportions,** which are the proportions of the population in the various classes.
3. Over each closed class interval we draw a horizontal segment whose length is equal to that of the class interval, and whose height above the x-axis is equal to the ratio

$$\frac{\text{Class proportion}}{\text{Length of the class interval}}.$$

It follows that the area of the rectangle under this segment (and over the class interval) is equal to the class proportion.

The collection of line segments over the closed class intervals forms the histogram. The total area under the histogram is equal to the sum of the proportions of the population in the various closed classes; thus, the total area is less than or equal to 1 accordingly as there is or is not at least one open class.

In many examples the closed classes are of equal length; then the height of the histogram is directly proportional to the corresponding class proportion.

EXAMPLE 1

Consider a population consisting of babies; and the numerical characteristic is the **birth weight** in grams. In the report, "Evaluation of Obstetric and Related Data Recorded on Vital Records and Hospital Records: District of Columbia, 1952," by Oppenheimer *et al.,* U.S. Department of Health, Education and Welfare, the proportions of babies whose births were registered in Washington, D.C. in 1952 were arranged according to weight. Here the classes are meant to include the left but not the right endpoint; however, this has no effect on the construction of the histogram.

Weight (thousands of grams)	Percentage of babies
0 –1.0	0.6
1 –1.5	0.9
1.5–2.0	2.0
2.0–2.5	6.6
2.5–3.0	21.5
3.0–3.5	39.7
3.5–4.0	22.9
4.0–4.5	5.1
4.5 and over	0.7

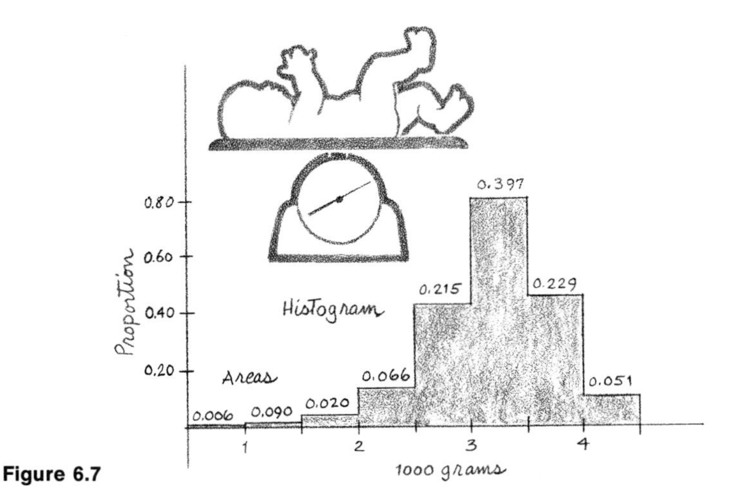

Figure 6.7

The histogram appears in Figure 6.7. The area under the histogram and over a class interval is equal to the proportion of the population in that interval; therefore the height of the histogram is the proportion divided by the length of the interval. It follows that the height over the interval from 0 to 1.0 is 0.6, the same as the proportion. The height over the interval from 1.0 to 1.5 is 1.8, which is twice the proportion 0.9. The last interval, 4.5 and over, is an **open** class because it has no endpoint on the right; the histogram is not drawn over this interval. ▲

A nonnegative function $f(x)$ which "fits" the given histogram is called a **density function** for the population: there often exists a function such that the areas under the various parts of the function are equal, or at least approximately equal to the corresponding areas under the histogram. Statisticians often know, on the basis of past experience, that certain data follow the graph of a known density function f; they can compute the various proportions of the population by means of areas under the graph of f. There is a large stock of such functions (various shapes) from which the statistician can choose to fit a given set of data. The common characteristics of all density functions are:

1. They are all nonnegative; and
2. The area under the entire graph is equal to 1.

The first is a consequence of the fact that proportions are, by their nature, nonnegative; and the second by the fact that the sum of the proportions of the entire population is, by definition, equal to 1.

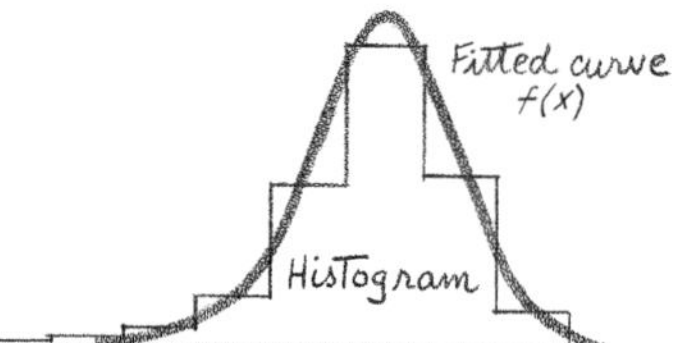

Figure 6.8

An example of a density function $f(x)$ for the histogram in Figure 6.7 is drawn in Figure 6.8; note that the areas of the histogram and the function are approximately equal. Later, we shall furnish formulas for some densities, and show how to find the areas under them. The density for a particular set of data is not uniquely determined by the data: several densities may fit. A particularly well-known density is the **normal** density; it has a bell-shaped curve. It fits many kinds of data arising in applications. It will be discussed in detail in the sections on probability and statistics (for example, Section 8.9).

6.2 EXERCISES

1 The following table contains neonatal deaths of babies whose births were registered in Washington, D.C., in 1952, arranged according to birth weights; the source is the same as that for the data in Example 1:

Weight (thousands of grams)	Percentage of babies
0 –1.0	27.4
1.0–1.5	20.6
1.5–2.0	15.7
2.0–2.5	11.4
2.5–3.0	10.5
3.0–3.5	8.3
3.5–4.0	4.3
4.0–4.5	1.5
4.5 and over	0.3

Draw the histogram and then fit a smooth density.

2 The following table contains the ages of American grooms at the times of their first marriages, from Marriage Statistics Analysis, 1963, U.S. Department of Health, Education and Welfare, 1968:

Age in years		
At least	But not	Percentage of all grooms
0	18	1.6
18	20	14.7
20	25	56.3
25	30	16.7
30	35	5.4
35	45	3.6
45	55	1.0
55	65	0.5
65 and over		0.2

Draw the histogram and then fit a smooth density. The intervals are 0 to 18, 18 to 20, 20 to 25, and so on.

3 The following information is from the report of the Committee on Recommendations to Medical and Dental Schools of Washington Square College, New York University, concerning admissions of seniors to medical and dental schools in 1971. Applicants are classified according to grade-point average at the time of application.

Average	Total applicants	Total acceptances
2.00–2.49	13	6
2.50–2.99	17	10
3.00–3.49	31	26
3.50–4.00	22	19

Draw separate histograms for the proportions of applicants and acceptances and fit densities to each.

6.3 Integration of Basic Functions

The function $f(x) = \frac{1}{2}x^2$ has the derivative $f'(x) = x$. The latter is obtained by the process of differentiation. The reverse process, that of going from the derivative back to the original function, is called **integration.** (The term **antidifferentiation** is also used.) Thus, $\frac{1}{2}x^2$ is obtained from x by integration. The former is called an **integral** of the latter and is obtained by **integrating** the latter function.

But x seems to have another integral: $\frac{1}{2}x^2 + 1$. And more: $\frac{1}{2}x^2 - 3$, $\frac{1}{2}x^2 + 1.3$, and so on. These are all integrals of x because when they are differentiated they are transformed into x. All the integrals of x have a common form: $\frac{1}{2}x^2 + C$, where C is some constant number. For each number C there is an integral, and we refer to the function

$\frac{1}{2}x^2 + C$ as **the integral** of x. It is also called the **indefinite integral** of x because the number C is not specified. This function is not a single function but represents a **class** of functions, each differing from others in the class by some constant.

In general, if $f_1(x)$ and $f_2(x)$ are two integrals of a given function $g(x)$, then there is a fixed number C such that $f_1(x) = f_2(x) + C$ for all x in the domain. The reason is as follows. The difference of the two functions has the derivative 0:

$$\left(\frac{d}{dx}\right)(f_1(x) - f_2(x)) = f_1'(x) - f_2'(x) = g(x) - g(x)$$
$$= 0.$$

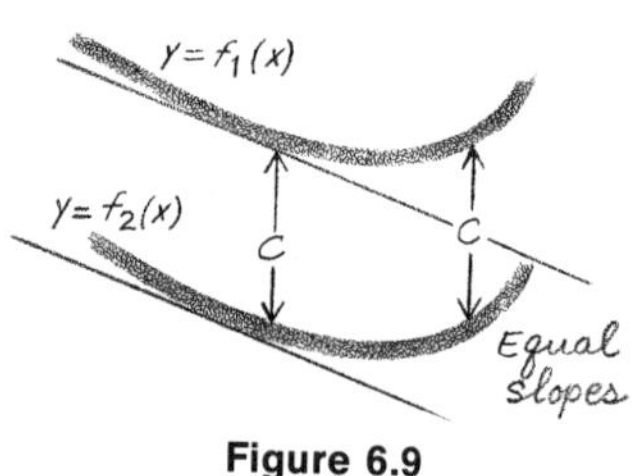

Figure 6.9

Therefore, the slope of the tangent line to the graph of $y = f_1(x) - f_2(x)$ is 0 at every point. Thus, $f_1 - f_2$ is a constant function: $f_1(x) - f_2(x) = C$ for some fixed number C; hence, $f_1(x) = f_2(x) + C$. The graphs of f_1 and f_2 are similar except that their heights differ by the fixed amount C (Figure 6.9). The slopes of the tangent lines are identical at every point.

We integrate elementary functions by **hindsight:** Since we know that $\frac{1}{2}x^2$ has the derivative x, we can conclude that $\frac{1}{2}x^2 + C$ is the integral of x. Similarly, $\frac{1}{3}x^3 + C$ is the integral of x^2, $x + C$ is the integral of the constant function 1. In general, for any exponent p not equal to -1, the integral of x^p is $x^{p-1}/(p + 1) + C$. In other words, we raise the exponent by 1, and then divide by the exponent. In order to integrate, we must know how to differentiate well. We construct this table of integrals, using formula (5) of Section 5.1, formulas (1) to (6) of Section 5.2, and formulas (2) and (3) of Section 5.4:

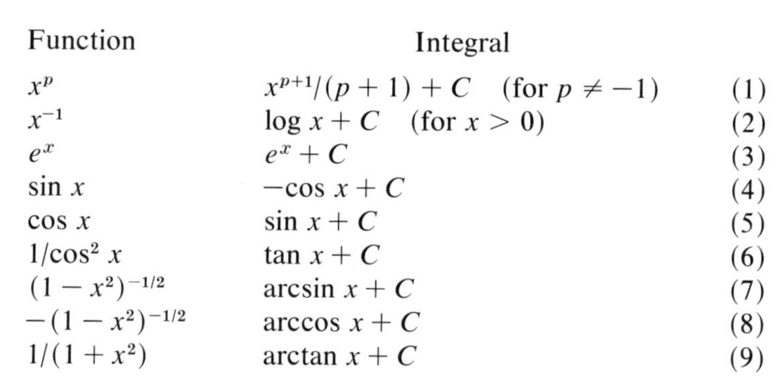

Function	Integral	
x^p	$x^{p+1}/(p+1) + C$ (for $p \neq -1$)	(1)
x^{-1}	$\log x + C$ (for $x > 0$)	(2)
e^x	$e^x + C$	(3)
$\sin x$	$-\cos x + C$	(4)
$\cos x$	$\sin x + C$	(5)
$1/\cos^2 x$	$\tan x + C$	(6)
$(1 - x^2)^{-1/2}$	$\arcsin x + C$	(7)
$-(1 - x^2)^{-1/2}$	$\arccos x + C$	(8)
$1/(1 + x^2)$	$\arctan x + C$	(9)

This is hardly a complete theory of integration. Many functions in calculus are built from these elementary functions by addition, multiplication, composition, and so on. The derivatives of the more complex functions were found in terms of those of the elementary ones by means of the rules in Sections 5.4 and 5.5: the rule for the derivative of a sum, product, quotient, and composition. By reversing these rules, we can get some general rules for the process of integration.

In differential calculus, we call f' the derivative of f; hence, f is an integral of f'. But now we shall consider the function to be integrated as a function in its own right, not just as the derivative of some other function. Now we let $f(x)$ represent the function which is to be integrated.

Here are the rules for integration which are obtained from the corresponding rules for differentiation:

Rule 1. The integral of a constant times a function is equal to the constant times the integral of the function:

$$\textbf{Integral of } A\, f(x) = A \textbf{ times integral of } f(x).$$

EXAMPLE 1

The integral of $2x^3$ is 2 times the integral of x^3: $2(\frac{1}{4}x^4 + C)$, or $\frac{1}{2}x^4 + 2C$. It is customary to omit the coefficient 2 of C because C is an arbitrary constant, and therefore, so is $2C$. So we write the integral as $\frac{1}{2}x^4 + C$. ▲

Rule 2. The integral of the sum of two (or more) functions is equal to the sum of the corresponding integrals:

$$\textbf{Integral of } f(x) + g(x) = \textbf{Integral of } f(x) + \textbf{Integral of } g(x).$$

EXAMPLE 2

The integral of $x^2 + \sin x$ is $(x^3/3 + C_1) + (-\cos x + C_2)$, where the C's are arbitrary constants. It follows that $C_1 + C_2$ is also arbitrary; hence, we write the integral as $x^3/3 - \cos x + C$, where $C = C_1 + C_2$. ▲

EXAMPLE 3

The integral of a polynomial is obtained by raising each exponent by 1, dividing each term by the corresponding exponent, and then adding a constant. For example, the integral of $2 - x - x^2 - 3x^3$ is $2x - \frac{1}{2}x^2 - \frac{1}{3}x^3 - \frac{3}{4}x^4 + C$. ▲

Rules 1 and 2 follow from the corresponding rules in Section 5.4 for the derivatives of a multiple and a sum. The rule of integration associated with the rule for the derivative of a product will be studied in Section 6.5. We point out that it is **not** true that the integral of a product is the product of the integrals. There is no general rule associated with the rule for the differentiation of a quotient. But there is a very useful method of integration associated with the chain rule (Section 5.5). This is often called the method of **substitution.**

If the function to be integrated is of the form

$$u'(x)f'(u(x)), \tag{10}$$

then its integral is the composite function

$$f(u(x)) + C. \tag{11}$$

EXAMPLE 4

Find the integral of $(1 + 2x)^3$. We proceed by trial-and-error. First we try the candidate

$$\tfrac{1}{4}(1 + 2x)^4.$$

We have raised the exponent by 1, and divided by the exponent. We check by differentiation to see if the integral is correct:

$$\left(\frac{d}{dx}\right) \tfrac{1}{4}(1 + 2x)^4 = (1 + 2x)^3 \cdot \left(\frac{d}{dx}\right) (2x)$$
$$= 2(1 + 2x)^3.$$

(The chain rule has been used.) Our answer misses by a factor of 2, which is a constant. If we divide the original candidate by that factor, then we get the right integral:

$$\left(\frac{d}{dx}\right) [\tfrac{1}{2} \cdot \tfrac{1}{4}(1 + 2x)^4] = (1 + 2x)^3$$

It follows that the integral is $\frac{1}{8}(1 + 2x)^4 + C$.

This integration can also be done more formally by referring to formula (10). The function to be integrated is of the form

$$\frac{1}{8}\frac{d}{dx}(1 + 2x)4(1 + 2x)^3.$$

By formula (11), the integral of the coefficient of $\frac{1}{8}$ is $(1 + 2x)^4$; hence, by Rule 1, the integral of the function is $\frac{1}{8}$ of the latter. ▲

EXAMPLE 5

Integrate $x\sqrt{1 + x^2}$, which is $x(1 + x^2)^{1/2}$. The factor x which is outside the parentheses is a multiple of the derivative of the polynomial which is inside, namely, $1 + x^2$. This suggests that our function was obtained by differentiating $(1+x^2)^{3/2}$ by the chain rule. Let this be our first candidate. By the chain rule,

$$\frac{d}{dx}(1+x^2)^{3/2}=\frac{3}{2}(1+x^2)^{1/2}\frac{d}{dx}(1+x^2)=3x(1+x^2)^{1/2}.$$

This differs from our original function by a factor of 3. To get the integral we seek, we "fix" the candidate by dividing by 3 and obtain

$$\tfrac{1}{3}(1+x^2)^{3/2}+C.$$

It can now be checked by differentiation that this is the integral. ▲

EXAMPLE 6

We shall integrate the product ($\sin x \cos x$) by two different methods. The cosine is the derivative of the sine; hence, by (10) and (11), a candidate for the integral is $\sin^2 x$. By the chain rule, its derivative is

$$\left(\frac{d}{dx}\right)\sin^2 x=2(\cos x)(\sin x).$$

This differs from our function by a factor of 2; hence, the integral we seek is $\frac{1}{2}$ times this:

$$\tfrac{1}{2}\sin^2 x+C.$$

The second way to find the integral is to start with the candidate $\cos^2 x$. Its derivative is

$$\left(\frac{d}{dx}\right)\cos^2 x=2\cos x\,(-\sin x)\qquad\text{or}\qquad -2(\cos x)(\sin x).$$

This misses by a factor of -2; hence, the integral is obtained by multiplying the candidate by $-\frac{1}{2}$:

$$-\tfrac{1}{2}\cos^2 x+C.$$

There appear to be two different answers:

$$\tfrac{1}{2}\sin^2 x+C\qquad\text{and}\qquad -\tfrac{1}{2}\cos^2 x+C.$$

These are really the same because they differ by a constant; indeed, the difference between them is, by the identity, $\sin^2 x+\cos^2 x=1$, equal to $\frac{1}{2}$. ▲

EXAMPLE 7

Consider the quotient of two functions in which the numerator is the derivative of the denominator. Then the integral is the **logarithm** of the denominator. In

$$f(x) = \frac{2x}{1+x^2}$$

the numerator $2x$ is the derivative of the denominator $1+x^2$; hence, the integral is

$$\log(1+x^2) + C.$$

To check this apply the chain rule formula for the logarithm, formula (2) of Section 5.5:

$$\frac{d}{dx}[\log(1+x^2)+C] = \frac{(d/dx)(1+x^2)}{1+x^2} = \frac{2x}{1+x^2}. \quad \blacktriangle$$

EXAMPLE 8

Let us determine the integral of the function xe^{x^2}. The coefficient is a multiple of the derivative $2x$ of the exponent. The chain rule suggests that our function is a multiple of the derivative of e^{x^2}:

$$\frac{d}{dx} e^{x^2} = \frac{d}{dx}(x^2)e^{x^2} = 2xe^{x^2}$$

[see formula (2), Section 5.5]. This differs from our function by a factor of 2; hence, we divide the candidate by 2 to get the desired integral

$$\tfrac{1}{2}e^{x^2} + C. \quad \blacktriangle$$

EXAMPLE 9

Consider a given function $f(x)$, and the function $f(x+b)$ obtained from it by adding b to the independent variable. Then the integral of the latter is the same as that of the former except that b is added to x. For example, the integral of x^3 is $\frac{1}{4}x^4$, and so the integral of $(x+2)^3$ is $\frac{1}{4}(x+2)^4$. This can be verified by the chain rule. ▲

EXAMPLE 10

Consider a given function $f(x)$, and the function $f(bx)$ obtained from it by multiplying x by b, where $b \neq 0$. Then the integral of the latter is obtained from that of the former by two operations: multiplying the **integral** by $1/c$ and then multiplying the **variable** x

by c. For example, the integral of $\cos x$ is $\sin x$, and so the integral of $\cos 2x$ is $\frac{1}{2}\sin 2x + C$. ▲

6.3 EXERCISES

Find the integrals of the functions:

1 (a) x^4 (b) x^{-2} (c) $1 + x^{-1/2}$ (d) $x^2 - 4x + 13$ (e) $3x^4 - 5x^3 - 4x^2 - 2$ (f) $e^x - 1 - x$ (g) $x + 3/x$ (h) $x^2 + \sin x$

2 (a) x^5 (b) x^{-3} (c) $x + x^{3/5}$ (d) $3x^2 - 6x + 7$ (e) $4x^3 - 12x^2 - 4x + 12$ (f) $\sin x - \cos x$ (g) $4/x - \dfrac{1}{\sqrt{1-x^2}}$ (h) $4e^x + 8 \sin x$

3 (a) $\sqrt{1-x}$ (b) $(2x+3)^5$ (c) $x(3x^2+5)^{3/2}$ (d) $(4x^2 - 2x + 3)^4(4x - 1)$ (e) $\dfrac{x^2}{\sqrt{x^3+2}}$ (f) $\dfrac{4x+2}{x^2+x-7}$ (g) $\dfrac{x+1}{x^2+2x+5}$ (h) $\dfrac{1/(2\sqrt{x})}{1+\sqrt{x}}$

4 (a) $(1+x)^{4/3}$ (b) $(4x-1)^{-3}$ (c) $2x(8x^2+3)^{-1/2}$ (d) $(x^2+x+1)^{-3}(2x+1)$ (e) $\dfrac{x^2+x}{(x^3+\frac{3}{2}x^2+1)^4}$ (f) $\dfrac{x}{x^2+5}$ (g) $\dfrac{2x-3}{x^2-3x+1}$ (h) $\dfrac{-7/x}{3+\log x}$

5 (a) $e^x/(e^x+1)^2$ (b) $3xe^{-(1/2)x^2}$ (c) e^{-2x} (d) $e^{\sqrt{x}}/\sqrt{x}$ (e) $\sin(3x-2)$ (f) $\cos x \sin^3 x$ (g) $e^{\cos x} \sin x$ (h) $e^x e^{e^x}$

6 (a) $\dfrac{e^{-2x}}{1+e^{-2x}}$ (b) $\dfrac{e^{1/x}}{x^2}$ (c) $(x+1)e^{x^2+2x+1}$ (d) $\dfrac{e^x+e^{-x}}{e^x-e^{-x}}$ (e) $\cos(x+8)$ (f) $\tan x$ [*Hint:* $\sin x/\cos x$] (g) $\sin x \cos^4 x$ (h) $x \cos x^2$

7 Find the integral of $(4-x^2)^{-1/2}$. [*Hint:* Write this as $\frac{1}{2}(1-(x/2)^2)^{-1/2}$ and then use the formula for the derivative of the arcsine.]

8 Generalize the result of Exercise 7: Find the integral of $(a^2-x^2)^{-1/2}$, where a is an arbitrary number such that $a^2 > x^2$.

6.4 Finding Areas by Integration

In Section 6.1 area was discussed, and an approximate method of computation was given. The technique of integration was developed in Section 6.3. By integration, we shall see how areas can be computed exactly. The method is justified by the following statement, which is a special case of the Fundamental Theorem of Calculus:

The area under the graph of a nonnegative continuous function $f(x)$ over the closed interval from a to b is equal to the difference between the values of the integral of f at b and at a, respectively.

In other words, the area is found by integrating f, evaluating the integral at b and at a, and then computing the difference.

Before explaining the reason for this important statement, we give several examples of its use.

EXAMPLE 1

Consider the function $f(x)$ identically equal to the constant 1, and find the area under it over the interval from 0.5 to 2.0. Here, $f(x) = 1$, and $a = 0.5$ and $b = 2.0$. The integral of f is $x + C$, where C is an unspecified constant. The integral at $x = 2.0$ and at $x = 0.5$ has the values $2 + C$ and $0.5 + C$, respectively. The difference is $(2 + C) - (0.5 + C) = 1.5$. This is the same as the area computed by the elementary formula for the area of the rectangle; indeed, the area under the curve is a rectangle of height 1 and base 1.5. ▲

EXAMPLE 2

Let $f(x)$ be the linear function $f(x) = 1 + x$, and let the interval be from 0 to 2. The integral is $x + \frac{1}{2}x^2 + C$. By the fundamental theorem stated above, the area under the curve is the difference between

$$2 + \tfrac{1}{2}(2)^2 + C \qquad \text{and} \qquad 0 + \tfrac{1}{2}(0)^2 + C,$$

or 4. This is exactly the same as the area obtained as the sum of the areas of a rectangle and a triangle (Figure 6.10). ▲

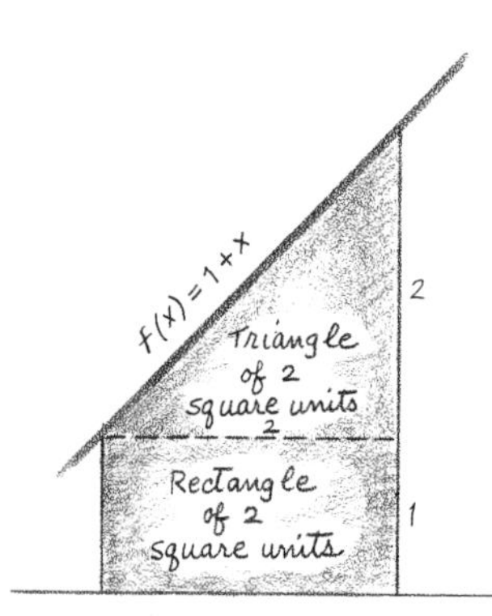

Figure 6.10

EXAMPLE 3

Let $f(x)$ be $1 + x + x^2$ and the interval from 0.5 to 1.0. The in- integral is $x + \frac{1}{2}x^2 + \frac{1}{3}x^3 + C$. The area under the curve is the difference

$$1 + \tfrac{1}{2} + \tfrac{1}{3} + C - [\tfrac{1}{2} + \tfrac{1}{2}(\tfrac{1}{2})^2 + (\tfrac{1}{3})(\tfrac{1}{2})^3 + C] = \tfrac{7}{6} = 1.1666. \ . \ .$$

Compare this to the area obtained by the trapezoidal rule (Section 6.1). ▲

We present a brief geometric explanation of the Fundamental Theorem of Calculus (in the special form given above). For simplicity, let us suppose that a is positive, and that $f(x)$ is nonnegative not only over the interval from a to b but also from 0 to b. With f we associate an area function $A(x)$. This is an "area cumulator": for each positive x, $A(x)$ is defined as the area under the graph of f and over the interval from 0 to x. The area under the graph between a and b is equal to the difference $A(b) - A(a)$; indeed, it is equal to the area up to b minus that up to a. We will show that **$f(x)$ is the derivative of $A(x)$,** which is equivalent to **$A(x)$ is an integral of $f(x)$.** Form the difference quotient

$$\frac{A(x^*) - A(x)}{x^* - x}.$$

By definition, the numerator is the area under the graph of f and over the interval from x to x^*; and the denominator is the length of that interval. If the interval is of small length, then the area in the numerator is approximately that of a slim rectangle of height $f(x)$ (the height of the graph) and width $x^* - x$: $A(x^*) - A(x)$ is approximately equal to $f(x)(x^* - x)$, or, equivalently, the difference quotient of A is approximately equal to f. We conclude that f is the derivative of A (Figure 6.11).

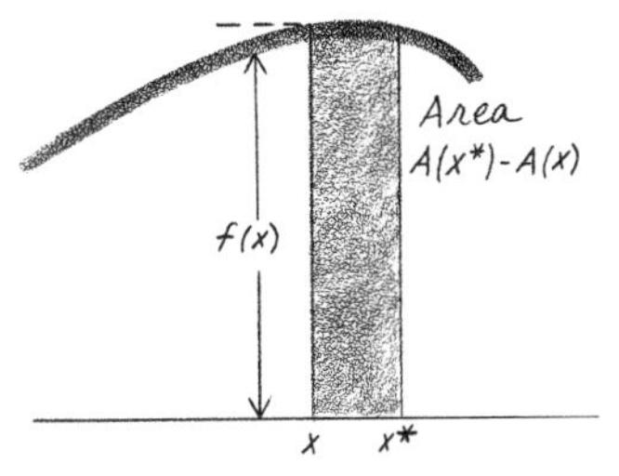

Figure 6.11

A special symbol is used for the difference of the values of the integral of f at b and a:

$$\int_a^b f(x)\,dx. \tag{1}$$

The elongated S on the left stands for the word "sum"; and "dx" on the right for "the differential of x." The number a is the **lower limit of integration** and b is the **upper limit.** This notation is based on the definition of area as the limit of the sum of the areas of many slim rectangles, summed over the interval: f represents the heights of the rectangles and dx the bases (Figure 6.12). The symbol (1) is read as "the integral of $f(x)$ from a to b." The pair of letters dx has no algebraic meaning but is part of the **symbol** for the area.

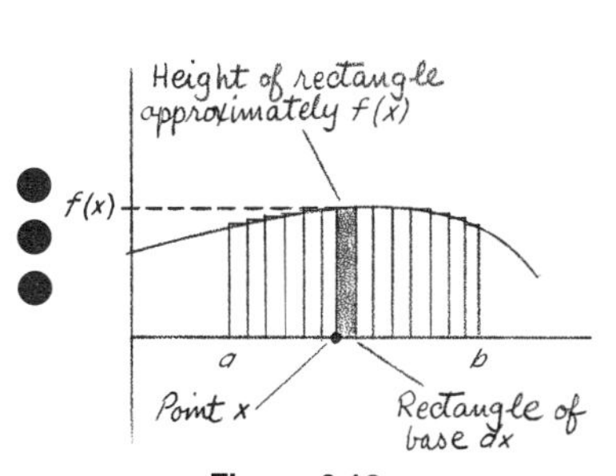

Figure 6.12

In contrast to the "integral of f" (or the **indefinite integral**), the quantity (1) is called the **definite integral** because it has a definite numerical value. It is conventional to use the same symbol as (1) for the indefinite integral but without the limits of integration:

$$\int f(x)\,dx;$$

for example,

$$\int 2x\,dx = x^2 + C.$$

In calculating a definite integral it is convenient to use the following notation. Suppose $f(x)$ has the integral $F(x) + C$; then, by definition,

$$\int_a^b f(x)\,dx = F(b) - F(a).$$

In performing the calculation we write the right-hand side as

$$F(x)\Big|_a^b.$$

The function $f(x)$ in (1) is called the **integrand.** When it is nonnegative, the definite integral may be interpreted as area. When f is not necessarily nonnegative, the definite integral is defined and computed in exactly the same way as before; however, it loses its interpretation as area.

Here are several illustrations of the computation of definite integrals.

EXAMPLE 4

Evaluate

$$\int_1^2 \left(x - \frac{1}{x}\right) dx.$$

The integrand is $x - 1/x$ and its integral is

$$F(x) + C = \tfrac{1}{2}x^2 - \log x + C.$$

Then the definite integral is

$$F(x)\Big|_1^2 = \tfrac{1}{2}x^2 - \log x\Big|_1^2 = \tfrac{1}{2}(2)^2 - \log 2 - \tfrac{1}{2}(1)^2 + \log 1$$
$$= \tfrac{3}{2} - \log 2.$$

Note that the term C, the "constant of integration," cancels itself in the computation of the definite integral. ▲

EXAMPLE 5

Evaluate

$$\int_0^1 x\sqrt{1 + x^2}\;dx.$$

Here $f(x) = x\sqrt{1+x^2}$. By the result of Example 5, Section 6.3, the integral is $(\frac{1}{3})(1+x^2)^{3/2} + C$; hence, the definite integral is

$$\left(\tfrac{1}{3}\right)(1+x^2)^{3/2}\Big|_0^1 = \left(\tfrac{1}{3}\right)(1+1^2)^{3/2} - \left(\tfrac{1}{3}\right)(1+0^2)^{3/2}$$

$$= \left(\tfrac{1}{3}\right)(2^{3/2} - 1). \ \blacktriangle$$

EXAMPLE 6

Find the area under the graph of $f(x) = e^{-x}$ over the interval from $x = 1$ to $x = 3$. The derivative of e^{-x} is, by the chain rule, $-e^{-x}$; therefore, the integral of $-e^{-x}$ is $e^{-x} + C$. It follows from Rule 1 for integration (with $A = -1$) that the integral of e^{-x} is $-e^{-x} + C$. The area is the definite integral

$$\int_1^3 e^{-x}\,dx = -e^{-x}\Big|_1^3 = -e^{-3} - e^{-1} = e^{-1} - e^{-3},$$

which, from Table II, is $0.36788 - 0.049787 = 0.31709$. ▲

EXAMPLE 7

Find the area of the region bounded by the x-axis and the parabola $y = 4x - x^2$. The parabola opens downward. It crosses the x-axis at the points where $y = 0$; that is, $4x - x^2 = 0$. The solutions of this equation are $x = 0$ and $x = 4$. The area under the parabola and over the x-axis extends from $x = 0$ to $x = 4$ (Figure 6.13). It is equal to the definite integral

$$\int_0^4 (4x - x^2)\,dx = 2x^2 - \frac{x^3}{3}\Big|_0^4 = 32 - \frac{64}{3} = \frac{32}{3}. \ \blacktriangle$$

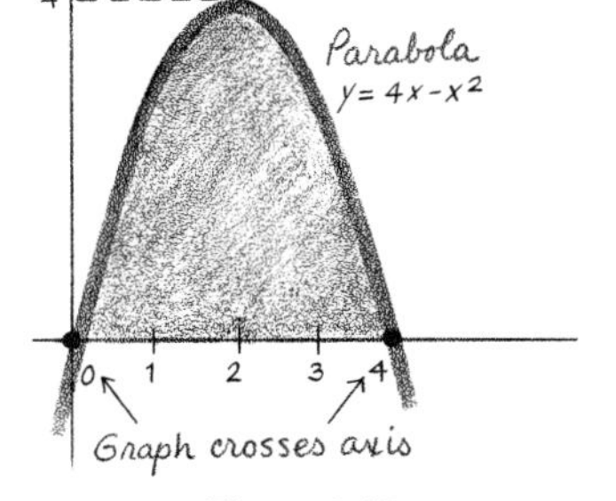

Figure 6.13

EXAMPLE 8

Find the area under the graph of $f(x) = \sin x$ from $x = 0$ to $x = 1$:

$$\int_0^1 \sin x\,dx = -\cos x\Big|_0^1 = (-\cos 1) - (-\cos 0)$$

$$= 1 - \cos 1 = 0.45970.$$

(See Section 6.1, Example 2.) ▲

In the following example, we compute areas under a density function fitted to a histogram based on income data.

EXAMPLE 9

The distribution of income of American families in 1944 is described in the table below, taken from Current Population

Reports, Bureau of the Census, U.S. Department of Commerce, Ser. P-60, No. 33:

Income ($1000's)	0–1	1–2	2–3	3–4	4–5	5–6	6–10	10 and over
Percentage of families	23.2	22.1	20.7	16.1	7.9	4.1	4.2	1.4

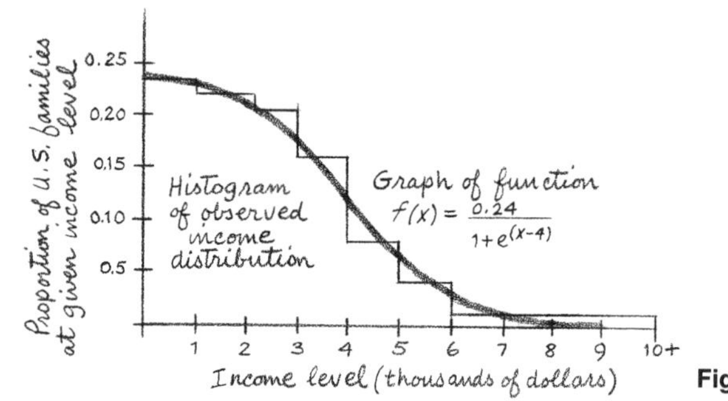

Figure 6.14

The histogram appears in Figure 6.14. Observe that the height of the histogram over the 6–10 bracket has been reduced from 4.2 by a factor of 4 because the range covers 4 units on the x-axis. The open class above 10 has been omitted. The function

$$f(x) = \frac{0.24}{1 + e^{(x-4)}}$$

has been drawn on the same set of axes; it serves as a density function. We shall compute some areas under the density and compare them to the proportions in the table.

In order to integrate $f(x)$, we first have to get it in a form where we can apply the method of substitution involving formulas (10) and (11) of Section 6.3. Divide the numerator and denominator of f by $e^{(x-4)}$:

$$f(x) = \frac{0.24e^{-(x-4)}}{1 + e^{-(x-4)}}.$$

The numerator is -0.24 times the derivative of the denominator; therefore, the integral is -0.24 times the logarithm of the denominator (Section 6.3, Example 7); therefore, the formula for the definite integral from a to b is

$$-0.24 \log(1 + e^{-(x-4)})\Big|_a^b = -0.24[\log(1 + e^{-(b-4)}) - \log(1 + e^{-(a-4)})]$$

$$= -0.24 \log \frac{1 + e^{-(b-4)}}{1 + e^{-(a-4)}}.$$

Substitute $a = 0$ and $b = 1$, then $a = 1$, $b = 2$, . . . to find the various areas:

Interval	Area under f	Proportion in table
0–1	0.233	0.232
1–2	0.221	0.221
2–3	0.195	0.207
3–4	0.149	0.161
4–5	0.091	0.079
5–6	0.044	0.041
6–10	0.031	0.042

Note that every area under the theoretical density function is within 1.2 percentage points of the corresponding observed percentage in the income table.

How was the density found? This requires some experience with data and graphs of functions, not yet available to the reader; therefore, he will not be expected to discover the right density for a given set of data. In every exercise involving a density, it will be given to the reader, who will be asked to compute the areas under it. Our purpose in introducing densities is to show how the topic of area calculation is used in analyzing data. The density concept is also a fundamental part of probability and statistics.

One of the famous laws of mathematical economics is Pareto's Law for the distribution of income. (Pareto was an economist of the late nineteenth and early twentieth centuries.) It states that the proportion of the population whose income is x is inversely proportional to a power of x:

$$\text{Proportion at income level } x = \frac{\text{Constant}}{x^p}.$$

Our data do not fit this formula (Exercise 9). Broad statements such as Pareto's Law are very often not applicable to many kinds of data arising in practice. ▲

6.4 EXERCISES

In Exercises 1 to 4 evaluate the definite integrals. When appropriate the answer may be left in terms of e, log *or radical.*

1 (a) $\int_{-3}^{5} 2x\,dx$

(b) $\int_{-10}^{10} 7\,dx$

(c) $\int_{1}^{4} (x^2 - 2x + 7)\,dx$

(d) $\int_{0}^{1} (5x^2 + 1)^5 x\,dx$

(e) $\int_{1}^{4} \frac{x + 1}{\sqrt{x}}\,dx$

[*Hint:* Express integrand as a sum.]

(f) $\int_0^1 \frac{1}{(2-x)^3}\,dx$

(g) $\int_0^4 \frac{x}{\sqrt{x^2+9}}\,dx$

(h) $\int_0^2 \frac{(2x+3)}{\sqrt{x^2+3x+1}}\,dx$

2 (a) $\int_{-2}^2 x^4\,dx$

(b) $\int_1^4 (x^2-x)\,dx$

(c) $\int_1^4 x(\sqrt{x}-1)\,dx$

(d) $\int_4^7 \sqrt{x-3}\,dx$

(e) $\int_0^1 (x^2+2x+1)^{1/2}(x+1)\,dx$

(f) $\int_0^1 \frac{x^2+x}{(x^3+\frac{3}{2}x^2+2)^4}\,dx$

(g) $\int_{\sqrt{6}}^3 \frac{x}{\sqrt{x^2-5}}\,dx$

(h) $\int_0^3 (3x-1)^{4/3}\,dx$

3 (a) $\int_1^2 e^{-x}\,dx$

(b) $\int_0^1 xe^{-x^2}\,dx$

(c) $\int_1^2 x^{-3}e^{-4/x^2}\,dx$

(d) $\int_0^1 \frac{1}{3-x}\,dx$

(e) $\int_0^{\log 3} \frac{e^x}{1+e^x}\,dx$

(f) $\int_{\pi/4}^{\pi/2} \frac{\cos x}{\sin^2 x}\,dx$

(g) $\int_0^{\pi/2} \sin 2x\,dx$

(h) $\int_0^{\pi/4} \tan x\,dx$

(*See* Exercise 6f, Section 6.3.)

4 (a) $\int_2^5 e^{2x}\,dx$

(b) $\int_4^9 \left(\frac{1}{\sqrt{x}}\right) e^{\sqrt{x}}\,dx$

(c) $\int_1^3 \frac{e^{-x}}{(1+e^{-x})^2}\,dx$

(d) $\int_{10}^{12} \frac{x}{x^2-5}\,dx$

(e) $\int_1^3 \frac{1}{1+e^x}\,dx$

(See Example 9.)

(f) $\int_0^{\pi/4} (\sin x)^3 \cos x\,dx$

(g) $\int_{\pi/4}^{\pi/2} \frac{\cos x - \sin x}{\sin x + \cos x}\,dx$

(h) $\int_{\pi/4}^{\pi} \cos 5x\,dx$

5 Find the area of the region bounded by
(a) the curve $y = 3 - 2x - x^2$ and the x-axis;
(b) the curve $y = x^2 + 1$, the lines $x = -1$ and $x = 1$, and the x-axis;
(c) the curve $y = 1/x$, the lines $x = 1$, $x = 3$, and the x-axis;
(d) the curve $y = \sin x$ and the x-axis, between 0 and π.

6 Find the area of the region bounded by
(a) the curve $y = x - x^2$ and the x-axis;
(b) the curve $y = \sqrt{x}$, the line $x = 1$, and the x-axis;
(c) the curve $y = x/(x^2 + 1)$, the line $x = \sqrt{8}$, and the x-axis;
(d) the curve $y = \cos x \sin^2 x$, the line $x = \pi/2$, and the two axes.

7 Find the areas under the graph of the function

$$f(x)=\frac{(2.4)e^{-2.4(x-3.25)}}{[1+e^{-2.4(x-3.25)}]^2}$$

over these intervals:

(a) 1.5 to 2.0
(b) 2.0 to 2.5
(c) 2.5 to 3.0
(d) 3.0 to 3.5
(e) 3.5 to 4.0
(f) 4.0 to 4.5

Compare these to the corresponding areas under the histogram of the baby weights in Section 6.2, Example 1.

8 Find the areas under the graph of

$$f(x)=xe^{-x}$$

over these intervals:

(a) 0 to 1.0
(b) 1.0 to 1.5
(c) 1.5 to 2.0
(d) 2.0 to 2.5
(e) 2.5 to 3.0
(f) 3.0 to 3.5
(g) 3.5 to 4.0
(h) 4.0 to 4.5

Compare these to the areas under the histogram for weights in neonatal deaths, in Section 6.2, Exercise 1. Note that $xe^{-x}=(d/dx)(1-(x+1)e^{-x})$.

9 Why is it impossible for Pareto's function to fit the income data in Figure 6.14? [*Hint:* Compare the concavity of the histogram and the function.]

10 An ant is clearing a straight tunnel through loose soil. We shall show that under certain simple conditions the time it takes to clear a tunnel is proportional to the **square** of the length of the tunnel. We assume that the digging is negligible, but the main work is in carrying the bits of soil from the inside to the mouth of the tunnel. Suppose that the time it takes to carry out a load of soil is proportional to the volume of the load **times** the length of the tunnel already dug. The soil is homogeneous so that the volume of soil in a piece of the tunnel is proportional to the length of the piece (Figure 6.15). Let $A(x)$ be the amount of time spent carrying out soil for a tunnel of length x.

(a) Show that $A(x^*)-A(x)$ is approximately proportional to x^*-x times x if x^* is just a bit larger than x.
(b) Conclude from (a) that dA/dx is a constant times x. [*Hint:* See the proof of the fundamental theorem.]
(c) Conclude that $A(x)=$ constant times x^2.

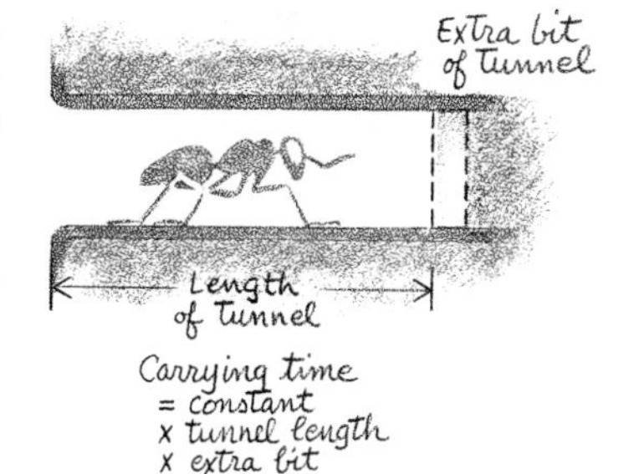

Figure 6.15

6.5 Further Applications of the Definite Integral

The definite integral is the measure of an area when the integrand is nonnegative. What if it assumes negative values, or both positive and negative values?

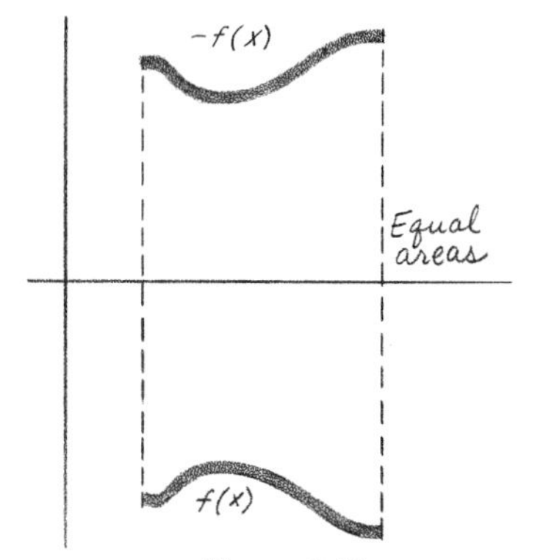

Figure 6.16

If $f(x)$ is negative on the interval from a to b, then $-f(x)$ is the reflected function and is positive: the curve $y=-f(x)$ is obtained from $y=f(x)$ by reflection in the x-axis. Since the area is unchanged by reflection, the area **under** the graph of $-f(x)$ is the same as the area **over** the graph of $f(x)$ (Figure 6.16) and is equal to $\int_a^b -f(x)dx$. By the first rule for integration (Section 6.3), this is equal to the negative of the integral of f,

$$-\int_a^b f(x)\, dx;$$

thus, **the definite integral of a negative function is the negative of the area between the curve and the axis.**

If f and g are two functions such that $f(x)$ is at least equal to $g(x)$ for all x in the interval from a to b, then **the area between the two curves is the integral of the difference of the two functions:**

$$\int_a^b (f(x)-g(x))dx. \tag{1}$$

To see this, suppose first that both f and g are nonnegative. Then the area between the curves is equal to the difference between the larger and the smaller areas (Figure 6.17),

$$\int_a^b f(x)dx - \int_a^b g(x)dx.$$

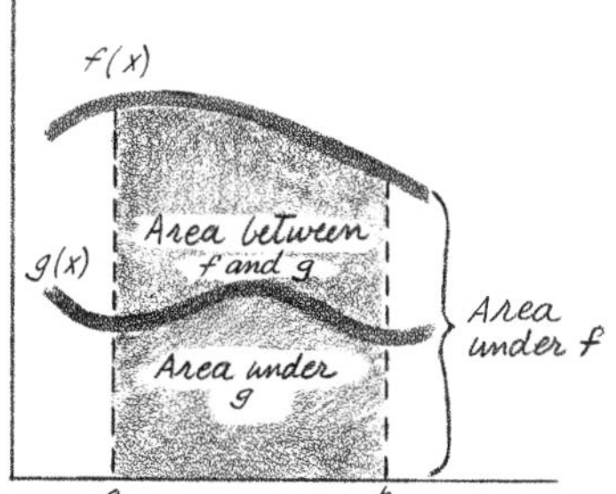

Figure 6.17

By the rules of integration, this is equal to the integral (1). The extension of the reasoning to functions assuming both positive and negative values is indicated in Exercise 11.

EXAMPLE 1

Find the area between the line $y=2x$ and the parabola $y=x^2$ (Figure 6.18). First we find the points of intersection of the curves. Equate the right-hand sides of the equations of the curves: $x^2 = 2x$. The solutions of this equation are $x = 0$ and $x = 2$. Take $f(x) = 2x$ and $g(x) = x^2$; then f is larger than g on the interval considered, and so the area between the curves is given by the integral (1) with $a = 0$, $b = 2$, and $f(x) - g(x) = 2x - x^2$:

$$\int_0^2 (2x - x^2)\,dx = x^2 - \tfrac{1}{3}x^3 \Big|_0^2 = 4 - \tfrac{8}{3} = \tfrac{4}{3}. \ \blacktriangle$$

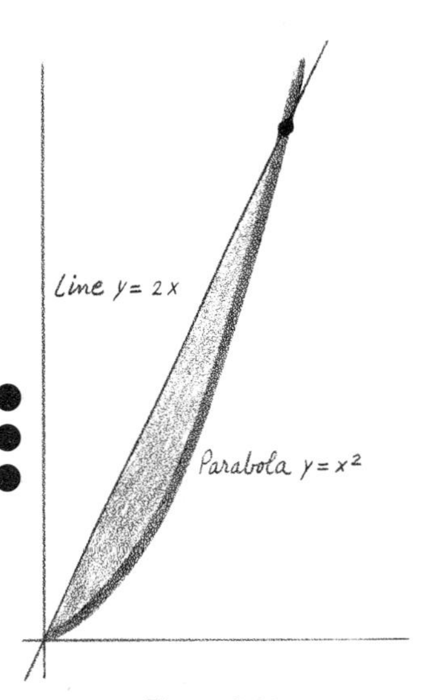

Figure 6.18

Area has an **additive** property: the area of a region consisting of several nonoverlapping subregions is equal to the sum of the areas of the subregions. This is reflected in the following property of the integral:

$$\int_a^b f(x)\,dx = \int_a^c f(x)\,dx + \int_c^b f(x)\,dx. \tag{2}$$

If f is nonnegative and $a < c < b$, then this equation states simply that the area under the graph of f over the interval from a to b is equal to the sum of the areas over the subintervals from a to c and from c to b, respectively.

Two curves may intersect several times. In this case the region between the curves consists of subregions separated by the points of intersection. The total area is then defined as the sum of the areas of the subregions (Figure 6.19).

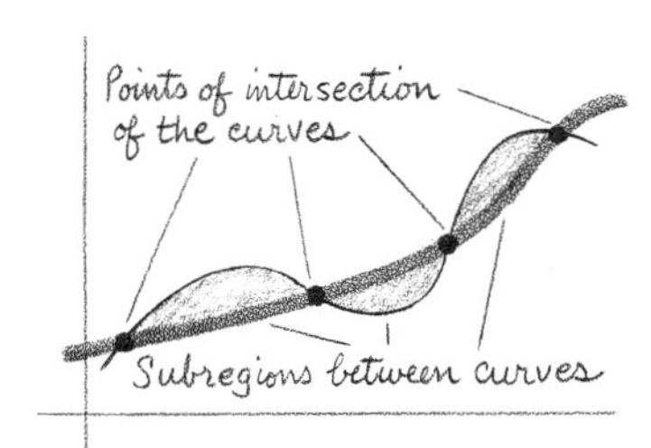

Figure 6.19

EXAMPLE 2

Find the area between the line $y = x + 1$ and the curve $y = x^3 - x^2 - x + 1$. The points of intersection of the curves are obtained by solving

$$x^3 - x^2 - x + 1 = x + 1 \qquad \text{or} \qquad x^3 - x^2 - 2x = 0.$$

The latter is solved by factoring:

$$x^3 - x^2 - 2x = x(x^2 - x - 2) = x(x - 2)(x + 1) = 0;$$

the solutions are $x = -1$, 0, and 2. We compute the area between the curves separately for the interval from -1 to 0 and for the interval from 0 to 2.

We apply formula (1). In the calculation it is not necessary to know which of the functions is the larger. If f in (1) is really **less** than or equal to g, then the integral is the **negative** of the area between the curves,

$$-\int_a^b (g(x) - f(x))\,dx.$$

The area itself is then obtained by changing the sign to positive.

Let us put $f(x) = x^3 - x^2 - x + 1$, and $g(x) = x + 1$; then $f(x) - g(x) = x^3 - x^2 - 2x$. The area from -1 to 0 is computed:

$$\int_{-1}^{0} (x^3 - x^2 - 2x)\,dx = \tfrac{1}{4}x^4 - \tfrac{1}{3}x^3 - x^2 \Big|_{-1}^{0} = \tfrac{5}{12}.$$

This is positive; hence, this is the area. Next we compute

$$\int_{0}^{2} (x^3 - x^2 - 2x)\,dx = \tfrac{1}{4}x^4 - \tfrac{1}{3}x^3 - x^2 \Big|_{0}^{2} = -\tfrac{8}{3};$$

then the area is obtained by changing the sign: $\frac{8}{3}$. The whole area between the curves is the sum of the two areas: $5/12 + 8/3 = 37/12$. ▲

The definite integral has applications to motion problems. Let $f(x)$ be the distance of a moving object from its starting position after x time units of motion. Then $f'(x)$ is the velocity after x time units, and $f''(x)$ the acceleration (Section 5.8). It follows from the relation between the integral and the derivative that:

(i) the distance is the integral of the velocity, and

(ii) the velocity is the integral of the acceleration.

The change in the distance between the time points $x = a$ and $x = b$ is the definite integral of $f'(x)$ from a to b; and the change in the velocity is the definite integral of $f''(x)$.

EXAMPLE 3

The driver of a car stops in an emergency by gradual application of the brake. The deceleration (negative of the acceleration) increases as the car is brought to a stop (Figure 6.20). If a car is traveling at a given speed, how much distance should the driver allow to bring it to a sudden emergency stop? The Automobile Association of America recommends the following distance allowances:

Speed of car (mph)	10	20	30	40	50	60	70
Stopping distance (ft)	10	45	78	125	188	272	381

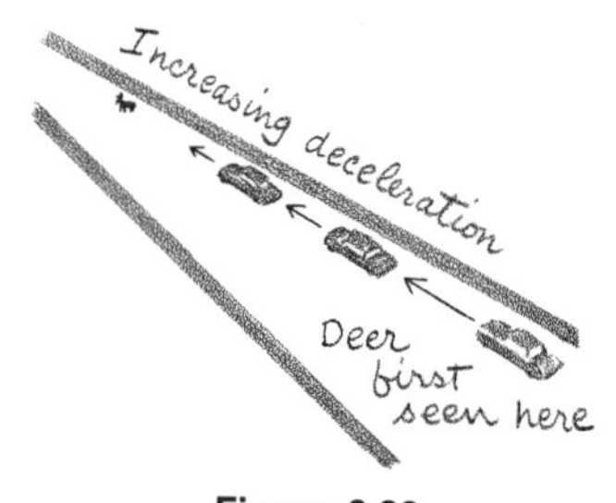

Figure 6.20

Is there any formula or principle underlying this schedule? It appears to be based in part on the assumption that the deceleration of a car moving at moderate speed (20–50 mph) is approximately $12x$ ft/sec per second, x seconds after the application of the brake; in other words, the acceleration is $-12x$ ft/sec^2. Let us test this assumption in the case of a speed of 40 mph. We shall compare the stopping distance based on this assumption about the acceleration to the distance recommended by the AAA, namely, 125 feet.

The speed of 40 mph is equivalent to 59 ft/sec; it is obtained by multiplying the former by a factor of 1.47. The stopping distance is obtained by:

(i) getting the velocity function by integration of the acceleration function; then,

(ii) deducing the **stopping time** by setting the velocity function equal to 0 and solving the resulting equation for the time; and then,

(iii) getting the distance function by integrating the velocity function and then evaluating it at the stopping time.

After t seconds, the loss of velocity due to the deceleration is the definite integral of the deceleration function from 0 to t:

$$\int_0^t (12x)\,dx = 6x^2 \Big|_0^t = 6t^2.$$

Thus, the velocity of the car after x seconds is obtained by taking $t = x$ and subtracting the loss of velocity,

$$59 - 6x^2.$$

The distance traveled in t seconds is the definite integral of the velocity function:

$$\int_0^t (59 - 6x^2)\,dx = 59x - 2x^3 \Big|_0^t = 59t - 2t^3.$$

The car will stop when the velocity is 0, that is, at the time t when

$$59 - 6t^2 = 0 \qquad \text{or} \qquad t = \sqrt{59/6}\,.$$

At this time the distance traveled by the car is given by the distance function with this value of t:

$$59t - 2t^3 = 59\sqrt{\frac{59}{6}} - 2\left(\frac{59}{6}\right)^{3/2} = \frac{59}{6}\sqrt{59 \cdot 6},$$

which is approximately 122 feet. The average distance on the AAA table above is 125 feet.

The calculations for the speeds of 30 and 50 mph are left to Exercises 9 and 10. ▲

6.5 EXERCISES

In Exercises 1 and 2 find the area bounded by the given curves.

1 (a) $y = x^2 - 2, \quad y = -x$
(b) $y = x^2, \quad y = x^3$
(c) $y = x^2, \quad y^2 = x$
(d) $y = x^2 - 4, \quad y = 5$
(e) $y = x^3 - x^2, \quad y = 6x$
(f) $y = x^3 - 2x^2 - 3x, \quad y = 5x$
(g) $y = x^3 + 2x^2 - 7x + 1, \quad y = x + 1$
(h) $y = 12x - x^3, \quad y = x^2$

2 (a) $y = 2 - x^2, \quad y = -x$
(b) $y = x, \quad y = x^{1/3}$
(c) $y = x, \quad y = x^3$
(d) $y = x^4 - 3x^2, \quad y = 4$
(e) $y = \frac{1}{4}x^2, \quad y = (x^2 + 4)/8$
(f) $y = x^3 + 2x^2 - 3x, \quad y = 12x$
(g) $y = x^3 - 9x, \quad y = x^2 + 3x$
(h) $y = 9 - x^2, \quad y = x + 3$

3 Find the area bounded by the curve $y = \sin(x\pi/2)$ and the line $y = x$.

4 Find the area between the curves $y = \sin x$ and $y = \cos x$ over the interval from $x = -\pi$ to $x = \pi$.

5 An object is moved from rest by a force whose acceleration is $5x$ ft/sec^2 after x seconds. (a) Find the formula for the velocity and distance of the object after x seconds. (b) What if the object is initially moving at 10 ft/sec? (See Example 3.)

6 An object is moved from rest with a force whose acceleration function is e^{-x}. Find the formula for the velocity after x seconds. Explain why the motion is almost uniform after a long time period.

7 The author has estimated on the basis of experiments that the acceleration of a baseball in the wind is approximately $\frac{1}{5}$ the wind velocity. Find the displacement of a ball in a 20-mph wind, aloft for 3 seconds. (See Example 3.)

8 A salmon is swimming up a stream flowing at 10 mph. The salmon is tiring; its still-water velocity upstream x hours after entering the mouth of the stream is $50/\sqrt{x}$. After how many hours will the salmon be slowed to a stop? How far upstream will it have traveled? After how many hours will it be washed back to the mouth of the stream?

9 A motorist is traveling 30 mph and brings the car to an emergency stop. Calculate the stopping distance under the assumption of Example 3. Compare it to the AAA recommendation.

10 Repeat Exercise 9 for 50 mph.

11 An object is started from rest by an acceleration of $\sin x$ ft/sec^2. Derive the distance function.

12 Suppose that (1) represents the area between the curves for nonnegative functions f and g. Why is it then also true for functions which are not necessarily nonnegative? [*Hint:* The area is unchanged if a common constant is added to both functions. Also use the fact that a continuous function has a maximum and a minimum on a closed finite interval (Section 5.10).]

6.6 Integration by Parts

The rule for the derivative of the product of two functions has a matching integration rule, called **integration by parts.** Recall the derivative rule:

$$\frac{d}{dx}[f(x)g(x)] = \left(\frac{d}{dx}f(x)\right)g(x) + f(x)\left(\frac{d}{dx}g(x)\right).$$

Integrate on both sides; then it follows from the definition of the integral that

$$f(x)g(x) + C - \int f'(x)g(x)\,dx + \int f(x)g'(x)\,dx.$$

Transpose terms:

$$\int f(x)g'(x)\,dx = f(x)g(x) + C - \int f'(x)g(x)\,dx. \tag{1}$$

For the definite integral from a to b, the constant C cancels and we get

$$\int_a^b f(x)g'(x)\,dx = f(x)g(x)\Big|_a^b - \int_a^b f'(x)g(x)\,dx. \tag{2}$$

Formulas (1) and (2) are used for integrating functions that are products. To use these, we have to properly identify the factors. Suppose that the integral cannot be directly found for the product, but that it is possible to integrate an associated product: that of the **derivative of one factor** times the **integral of the other factor.** Formulas (1) and (2) are just what are needed to integrate the product: $f'(x)$ and $g(x)$ on the right side are the derivative and integral of $f(x)$ and $g'(x)$, respectively, on the left-hand side (Figure 6.21).

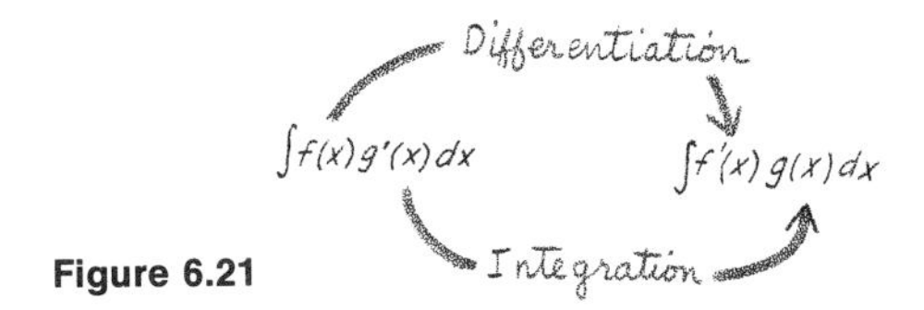

Figure 6.21

In a typical example involving integration by parts, our first task is to properly factor the integrand into a product, and to correctly identify f' and g. There often are several ways to factor the integrand. If one way does not lead to a form that can be integrated by integration by parts, another may be tried. In the following examples we show how to properly identify f' and g after the factors have been chosen.

EXAMPLE 1

Find $\int x \sin x \, dx$. Set up a table of derivatives and integrals for the two factors x and $\sin x$:

	x	$\sin x$
Derivative	1	$\cos x$
Integral	$\frac{1}{2}x^2$	$-\cos x$

Can we integrate

(i) the product of the derivative of x and the integral of $\sin x$, namely, $-\cos x$?

(ii) the product of the integral of x and the derivative of $\sin x$, namely, $\frac{1}{2}x^2 \cos x$?

We can do the first, but not the second. Write the product $x \sin x$ as $f(x)g'(x)$, where $f(x) = x$ and $g(x) = -\cos x$. Apply formula (1):

$$\begin{aligned}\int x \sin x \, dx &= -x \cos x + C - \int \left(\frac{d}{dx}\right) x \, (-\cos x) \, dx \\ &= -x \cos x + C + \int \cos x \, dx \\ &= -x \cos x + \sin x + C.\end{aligned}$$

We can check this by differentiation:

$$\left(\frac{d}{dx}\right) [-x \cos x + \sin x + C] = x \sin x. \ \blacktriangle$$

EXAMPLE 2

Find $\int xe^x \, dx$. The table is:

	x	e^x
Derivative	1	e^x
Integral	$\frac{1}{2}x^2$	e^x

Our choice is between the product of the derivative of x and the integral of e^x, namely, e^x, and the product of the integral of x and the derivative of e^x, namely, $\frac{1}{2}x^2e^x$. We choose the first: put $f(x) = x$, $g(x) = e^x$, and apply formula (1):

$$\int xe^x \, dx = xe^x + C - \int e^x \, dx = xe^x - e^x + C.$$

Check this by differentiation:

$$\left(\frac{d}{dx}\right)(xe^x - e^x + C) = xe^x + e^x - e^x = xe^x. \blacktriangle$$

EXAMPLE 3

Find $\int \log x\, dx$. We already have a formula for the derivative of $\log x$ but not its integral. We express $\log x$ as an artificial product, $1 \cdot \log x$. The table of derivatives and integrals is:

	1	$\log x$
Derivative	0	$1/x$
Integral	x	?

There is no choice but $f(x) = \log x$ and $g(x) = x$:

$$\int 1 \cdot \log x\, dx = x \log x + C - \int 1\, dx = x \log x - x + C.$$

Check by differentiation: $\left(\frac{d}{dx}\right)(x \log x - x + C) = \log x$. ▲

EXAMPLE 4

Find $\int \arcsin x\, dx$. Again we introduce the factor 1, as in Example 3. Our table is

	1	$\arcsin x$
Derivative	0	$(1 - x^2)^{-1/2}$
Integral	x	?

We choose $f(x) = \arcsin x$ and $g(x) = x$:

$$\begin{aligned}\int 1 \cdot \arcsin x\, dx &= x \arcsin x + C - \int x(1 - x^2)^{-1/2} dx \\ &= x \arcsin x + (1 - x^2)^{1/2} + C. \blacktriangle\end{aligned}$$

The method of integration by parts can be used in successive steps.

EXAMPLE 5

Find the area under the graph of the function $f(x) = x^2e^{-x}$ over the interval from 1 to 2,

$$\int_1^2 x^2e^{-x}\, dx.$$

Our table is:

	x^2	e^{-x}
Derivative	$2x$	$-e^{-x}$
Integral	$\frac{1}{3}x^3$	$-e^{-x}$

Choose $f(x) = x^2$ and $g(x) = -e^{-x}$, and apply formula (2):

$$\int_1^2 x^2 e^{-x}\,dx = x^2(-e^{-x})\Big|_1^2 - \int_1^2 2x(-e^{-x})\,dx$$

$$= -4e^{-2} + e^{-1} + 2\int_1^2 xe^{-x}\,dx.$$

We evaluate the last integral by the method of integration by parts, as in Example 2; it is equal to $2e^{-1} - 3e^{-2}$. Substitute in the previous expression:

$$\int_1^2 x^2 e^{-x}\,dx = -4e^{-2} + e^{-1} + 4e^{-1} - 6e^{-2} = 5e^{-1} - 10e^{-2},$$

which, by Table II, is approximately $5(0.36788) - 10(0.13534) = 0.4860$. ▲

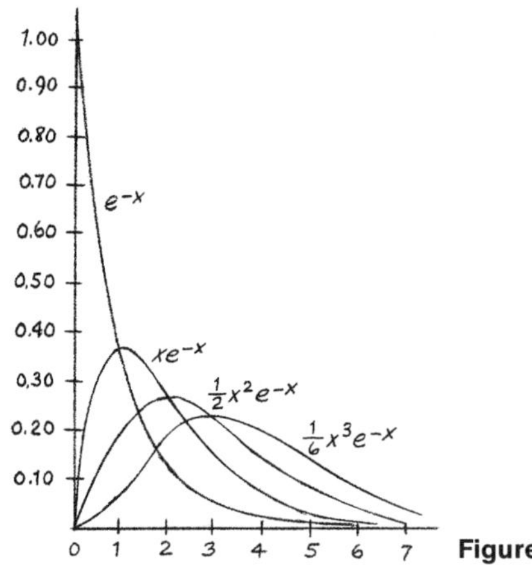

Figure 6.22

The functions

$$ce^{-cx}, \quad c^2xe^{-cx}, \quad \tfrac{1}{2}c^3x^2e^{-cx}, \quad \tfrac{1}{6}c^4x^3e^{-cx}, \ldots,$$

where c is a positive constant, are called **gamma density functions.** The number c is called a **scale parameter.** One such density was used in Section 6.4, Exercise 8 to fit a histogram of baby weights. The graphs of several gamma densities with $c = 1$ appear in Figure 6.22.

The method of integration by parts is used to compute areas under the graph. Suppose we want to integrate $x^m e^{-x}$. To apply formula (1), differentiate x^m and integrate e^{-x}: this yields the product $-mx^{m-1}e^{-x}$. Apply formula (1) again; it yields $m(m-1)x^{m-2}e^{-x}$. After m steps, the power of x is reduced to $x^0 = 1$, so that the other factor, e^{-x}, can then be integrated.

EXAMPLE 6

The following data is from Divorce Statistics Analysis, United States, 1963, U.S. Department of Health, Education and Welfare: it gives the distribution of divorces in the state of Missouri according to the age of the wife at the time of the marriage:

Age	Percentage	Age	Percentage
0–19	9.2	45–49	6.6
20–24	20.8	50–54	5.4
25–29	17.0	55–59	2.3
30–34	13.8	60–64	1.2
35–39	11.1	65 and over	1.3
40–44	11.1		

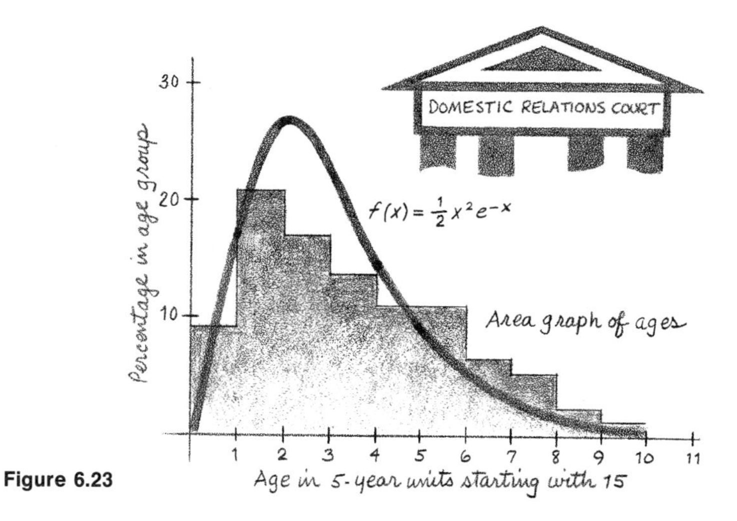

Figure 6.23

The histogram is in Figure 6.23. The units are 5-year age intervals. Note that the age category includes all those whose age is at least equal to the lower limit of the category, but not to the lower limit of the next category; for example, the bracket 20–24 includes all those past their 20th but not their 25th birthday.

Since it is very unlikely for girls to be married below the age of 15, the first age group is considered to be from 15 to 20, not 0 to 20.

The gamma density $f(x) = \frac{1}{2}x^2e^{-x}$ is drawn over the histogram. Let us compare the areas under the curve with the proportions given in the table. First of all, in Example 5 above, we found that

$$\int_1^2 x^2e^{-x}\,dx = 0.4860;$$

hence, the area under the curve is $\frac{1}{2}(0.4860) = 0.2430$. The corresponding proportion for the age group 20–24 is 0.208. The other areas are found by deducing the (indefinite) integral of the density and then applying it to the definite integral over the various intervals (Exercise 5). ▲

The method of integration by parts is suited to the integration of powers of trigonometric functions.

EXAMPLE 7

Find $\int \sin^2 x\,dx$. Write the integrand as the product $\sin x \cdot \sin x$ and apply formula (1) with $f(x) = \sin x$ and $g(x) = -\cos x$:

$$\int \sin^2 x\,dx = -\cos x \sin x + C + \int \cos^2 x\,dx.$$

Write $\cos^2 x$ as $1 - \sin^2 x$, and integrate:

$$\int \cos^2 x\,dx = \int (1 - \sin^2 x)\,dx = \int 1\,dx - \int \sin^2 x\,dx$$
$$= x - \int \sin^2 x\,dx + C.$$

Substitute in the previous equation:

$$\int \sin^2 x\,dx = -\cos x \sin x + x - \int \sin^2 x\,dx + C.$$

Transpose the integral from the right to the left side:

$$2\int \sin^2 x\,dx = -\cos x \sin x + x + C.$$

Solve the equation for the integral:

$$\int \sin^2 x\,dx = -\tfrac{1}{2}\cos x \sin x + \tfrac{1}{2}x + C. \quad ▲$$

EXAMPLE 8

The integral of a power of a sine (or cosine) can be reduced to an integral of power 2 units smaller. By repetition, it can then be reduced to the integral of the first or second power. Take $\int \sin^6 x\, dx$. Apply formula (1) with $f(x) = \sin^5 x$ and $g(x) = -\cos x$:

$$\int \sin^6 x\, dx = -\cos x \sin^5 x + \int 5 \cos^2 x \sin^4 x\, dx + C.$$

Substitute $1 - \sin^2 x$ for $\cos^2 x$ in the last integral. The equation becomes

$$\int \sin^6 x\, dx = -\cos x \sin^5 x + 5 \int \sin^4 x\, dx - \int 5 \sin^6 x\, dx + C.$$

Transpose $-5 \int \sin^6 x\, dx$ and solve the resulting equation:

$$\int \sin^6 x\, dx = -\frac{1}{6} \cos x \sin^5 x + \frac{5}{6} \int \sin^4 x\, dx.$$

This is a **reduction formula;** it gives the integral of the sixth power in terms of the integral of the fourth. By the same procedure, the latter is then reduced to the integral of the square. It is then evaluated by the result of Example 7. ▲

6.6 EXERCISES

1 Integrate by parts:

(a) $\int x \cos x\, dx$

(b) $\int \frac{x}{\sqrt{1+x}}\, dx$

[*Hint:* $f(x) = x$.]

(c) $\int \arccos x\, dx$

[*Hint:* Example 4.]

(d) $\int \frac{x^3}{\sqrt{1-x^2}}\, dx$

[*Hint:* $f(x) = x^2$.]

(e) $\int \sqrt{x} \log x\, dx$

(f) $\int x^2 e^{-(1/2)x^2}\, dx$

[*Hint:* $f(x) = x$.]

2 Integrate by parts:

(a) $\int x^2 \log x\, dx$

(b) $\int \cos^2 x\, dx$

[*Hint:* Example 7.]

(c) $\int x\sqrt{1+x}\, dx$

(d) $\int x^3\sqrt{1+x^2}\, dx$ [*Hint:* $f(x) = x^2$.]

(e) $\int \arctan x\, dx$

(f) $\int x^2(1-x^2)^{-3/2}\, dx$

[*Hint:* $f(x) = x$.]

3 Find by repeated integration by parts:

(a) $\int x^3 \sin x\, dx$

(b) $\int e^x \cos x\, dx$

[*Hint:* Solve for the integral as in Example 7.]

(c) $\int x^2\, e^{-x}\, dx$

(d) $\int (\log x)^2\, dx$

[*Hint:* Example 3.]

(e) $\int x^2 \arcsin x\, dx$

(f) $\int x^5(1 - x^2)^{-1/2}\, dx$

(g) $\int \sin^4 x\, dx$

[*Hint:* Example 8.]

(h) $\int \cos^5 x\, dx$

4 Find by repeated integration by parts:

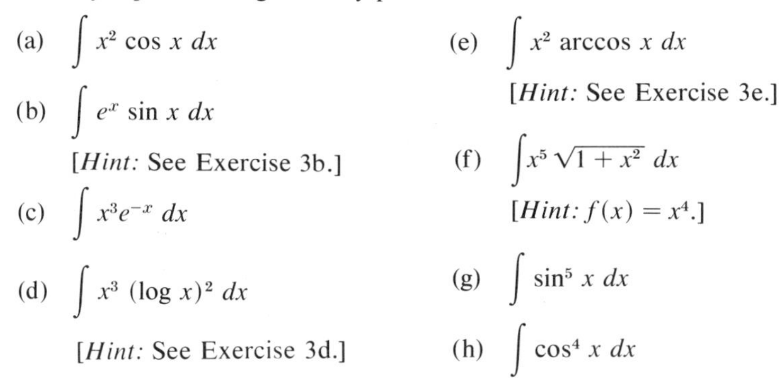

(a) $\int x^2 \cos x\, dx$

(b) $\int e^x \sin x\, dx$

[*Hint:* See Exercise 3b.]

(c) $\int x^3 e^{-x}\, dx$

(d) $\int x^3\, (\log x)^2\, dx$

[*Hint:* See Exercise 3d.]

(e) $\int x^2 \arccos x\, dx$

[*Hint:* See Exercise 3e.]

(f) $\int x^5 \sqrt{1 + x^2}\, dx$

[*Hint:* $f(x) = x^4$.]

(g) $\int \sin^5 x\, dx$

(h) $\int \cos^4 x\, dx$

5 Find the areas under the curve $f(x) = \frac{1}{2}x^2e^{-x}$ over the intervals from 0 to 1, 1 to 2, . . . , 9 to 10. Use the result of Exercise 3(c). Compare the areas to the proportions in the table in Example 6.

6 Here is the distribution of income for American males in the age group 25–34 in 1958, taken from Current Population Reports, Bureau of the Census, U.S. Department of Commerce, Ser. P-60, No. 33:

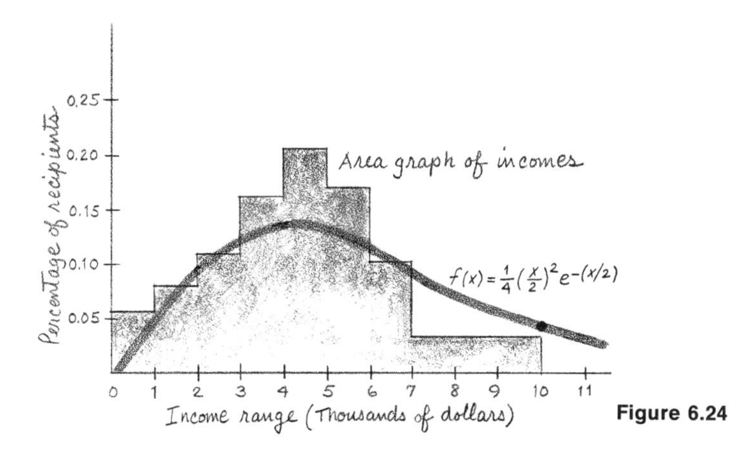

Figure 6.24

Income ($1000's)	0–1	1–2	2–3	3–4	4–5	5–6	6–7	7–10	10+
Percentage	5.4	8.0	11.0	16.1	20.5	17.0	10.1	9.1	3.0

The histogram appears in Figure 6.24. The gamma density $f(x) = \frac{1}{4}(x/2)^2e^{-(x/2)}$ is fitted to it. Find the areas under the density over the intervals (a) from 1 to 2, (b) from 2 to 3, (c) from 6 to 7; and compare these to the areas under the histogram. [*Hint:* Integrate by parts as in Exercise 3c.]

6.7 Improper Integrals and Distributions

The calculation of area is based on the Fundamental Theorem of Calculus (Section 6.4). Its assumption includes the condition that the interval which is the domain of the continuous function is **closed** and **finite.** This ensures that the function has a maximum and a minimum (Section 5.10), so that the region under the graph has an area which is a finite number (Figure 6.25).

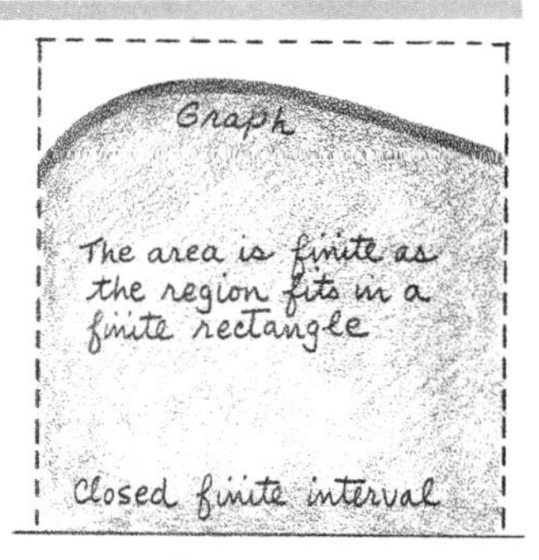

Figure 6.25

If the interval is either not closed or not finite, then it is still possible to define "area" but this must be done in a special way. We consider these two possibilities:

(i) If the interval is not closed, then the continuous function may tend to infinity at one or the other endpoint. The range of the function is then infinite, and so the "area" under the graph may be finite or infinite (Figure 6.26).

(ii) If the interval itself is infinite in length (for example, half-line or the whole axis), then the "area" may be finite or infinite (Figure 6.27).

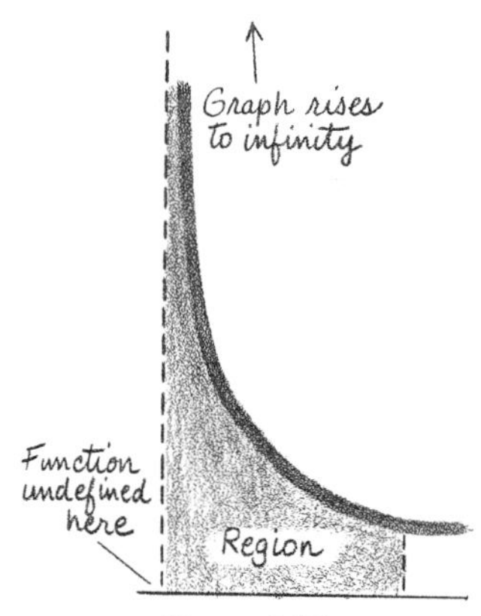

Figure 6.26

Areas of such regions are defined in the following way. We take a closed finite interval within the original domain of the function and compute the subarea under the graph over this interval (by means of the fundamental theorem). Then we let this interval "swell" to include the whole domain. The corresponding limit of the computed subareas is called the **area under the graph of the function over the domain.** This area is finite or infinite accordingly as the computed subareas converge to a finite limit or to ∞ (Figure 6.28).

One point requires a comment: Why do the swelling subareas necessarily tend to a finite or infinite limit? This follows from a basic assumption about the real number system, the **completeness axiom.** According to this axiom, an increasing sequence of real numbers (one whose terms are in increasing order of magnitude) either tends to a **finite** limit, or if not, to ∞ (Section 5.3). This seems to be a reasonable assumption: It means simply that if a "car" with only forward gears but no reverse gears moves to the right along the x-axis, then it will

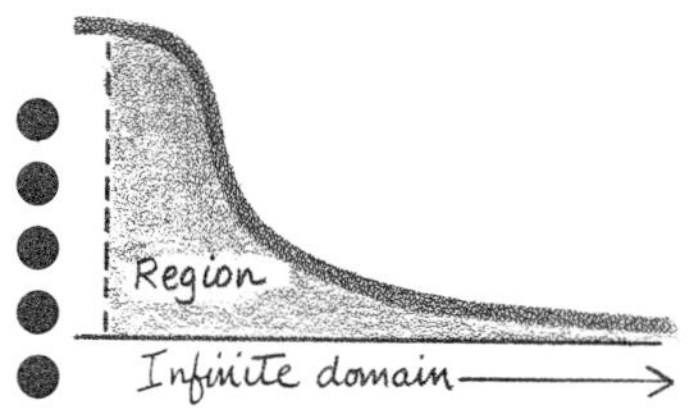

Figure 6.27

either move closer and closer to **some** point on the axis (the limit) or else go on to ∞ (Figure 6.29). (You might consider this geometrically "obvious.") This property guarantees the **existence** of the limit of the subareas: they become larger and larger as the subintervals increase, and so must tend to a limit, finite or infinite.

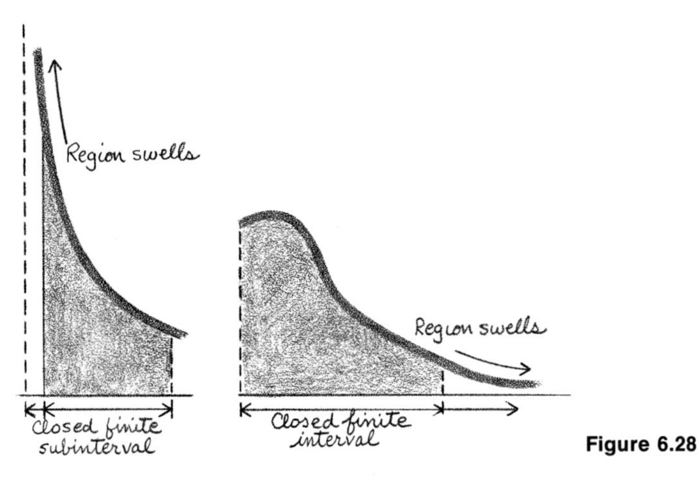

Figure 6.28

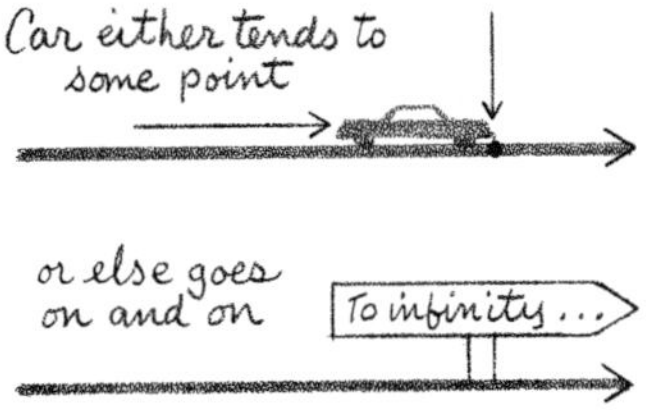

Figure 6.29

Let f be a nonnegative continuous function whose domain is the interval from a to b, excluding a and including b. The area is, as before, denoted

$$\int_a^b f(x)\,dx.$$

This integral is called **improper** because $f(x)$ is undefined at $x = a$. The integral is defined as the limit

$$\lim_{t \to a} \int_t^b f(x)\,dx$$

where t is a point just to the right of a and the integral is taken over the closed interval from t to b. Similarly, if the right endpoint b is not included in the domain, then the improper integral is defined as the limit of the integral over the closed interval with right endpoint at t, with t tending to b.

EXAMPLE 1

The function $f(x) = \tan x$ is continuous at each point of the interval from 0 to $\pi/2$, except that it is not defined at the latter point. [Some evidence for the continuity may be found in its "unbroken" graph (Section 4.6).] Let us attempt to proceed formally with the definite integral

$$\int_0^{\pi/2} \tan x \, dx.$$

The integrand may be written as $\sin x/\cos x$; therefore, the (indefinite) integral is $-\log(\cos x) + C$, and

$$\int_0^{\pi/2} \tan x \, dx = -\log(\cos x) \Big|_0^{\pi/2} = -\log\left(\cos \frac{\pi}{2}\right) + \log(\cos 0)$$

$$= \log 0 + \log 1.$$

Now $\log 1 = 0$, but $\log 0$ is undefined (Section 4.4). So the blind use of the definite integral has led us to a meaningless expression for the area.

As stated above, the integral of the tan function is an **improper** integral because the function is undefined at $\pi/2$. It is defined as the limit of the integral from 0 to t, where $t < \pi/2$, and where t tends to $\pi/2$. We have

$$\int_0^t \tan x \, dx = -\log(\cos t).$$

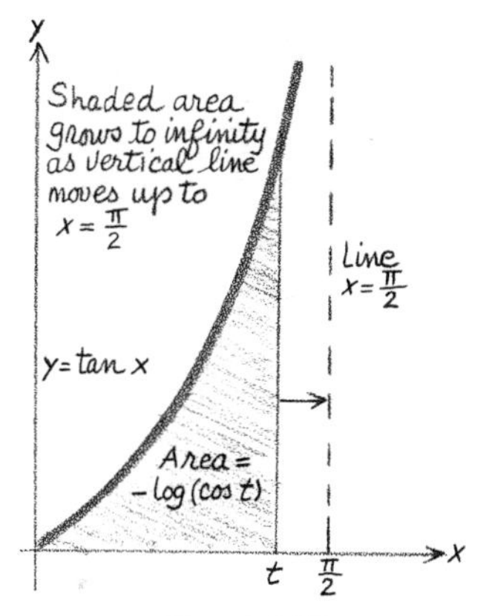

Figure 6.30

As t gets larger and approaches $\pi/2$, the cosine approaches 0, and the negative logarithm gets larger and larger; for example,

t	1	1.30	1.50	1.55	1.57
$\cos t$	0.54030	0.26750	0.07074	0.02079	0.00080
$-\log \cos t$	0.6162	1.3168	2.6493	3.8728	7.1310

It appears (and it can be rigorously shown) that

$$\lim_{t \to \pi/2} \int_0^t \tan x \, dx = \infty.$$

Therefore, the area under the curve is infinite in measure (Figure 6.30). The improper integral is called **divergent.** ▲

EXAMPLE 2

Next we consider the area under the graph of $f(x) = -\log x$ over the interval from 0 to 1 (Figure 6.31). Formally, this is the definite integral $\int_0^1 -\log x \, dx$. By integration by parts (see Section 6.6, Example 3),

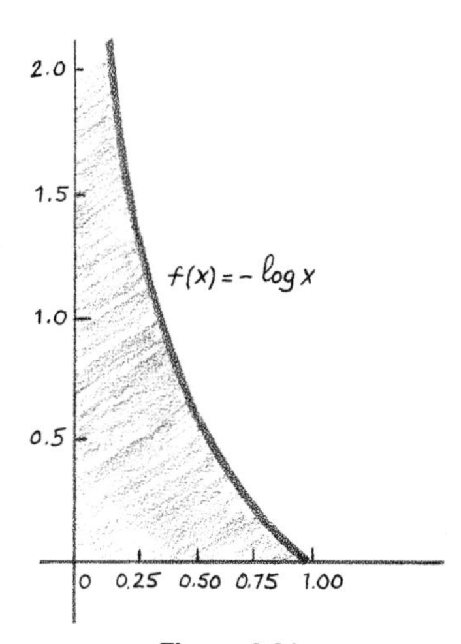

Figure 6.31

$$\int_0^1 -\log x \, dx = -x \log x \Big|_0^1 + x \Big|_0^1$$
$$= -1 \cdot \log 1 + 0 \cdot \log 0 + 1.$$

Now $\log 1 = 0$, but $\log 0$ has no meaning so that neither does the product $0 \cdot \log 0$. To properly evaluate the integral we use the same procedure as for the tan function. Let t be a positive number smaller than 1. The definite integral from t to 1 is

$$\int_t^1 -\log x \, dx = -x \log x \Big|_t^1 + x \Big|_t^1 = t \log t + 1 - t.$$

Let t approach 0. The second term in the last expression, $1 - t$, becomes 1. The first term, $t \log t$, is harder to follow: t tends to 0, but $\log t$ tends to $-\infty$. But the product is dominated by the first factor t; for example,

t	10^{-1}	10^{-5}	10^{-10}	10^{-100}
$t \log t$	$-2.3(10)^{-1}$	$-1.2(10)^{-4}$	$-2.3(10)^{-9}$	$-2.3(10)^{-98}$

So $t \log t$ apparently approaches 0 as t tends to 0:

$$\lim_{t \to 0} t \log t = 0.$$

This can be rigorously proved by the tools of calculus; however, at this point, the table above should convince the reader.

It follows that the area under the curve from t to 1 swells to 1 as t tends to 0; hence, we write

$$\int_0^1 -\log x \, dx = 1.$$

Such an improper integral is called **convergent:** it has a finite value. ▲

● The definition of the improper integral can be extended to all continuous functions, not necessarily positive, on a finite but not closed domain:

If f is defined and continuous on the open interval from a to b, and also at the right endpoint b, but not at a, then the improper integral of f over the interval is defined as

$$\int_a^b f(x)\,dx = \lim_{t \to a} \int_t^b f(x)\,dx,$$

provided the limit exists. It is called **convergent** if the limit is finite. If the integral from t to b tends to ∞ as t tends to a, then the integral is

divergent. Similarly, the integral is defined when f is undefined or discontinuous at b:

$$\int_a^b f(x)\,dx = \lim_{t \to b} \int_a^t f(x)\,dx.$$

The second type of improper integral arises when the domain is infinite; for example, when the domain is a half-line or the whole real axis.

EXAMPLE 3

Find the area under the graph of the exponential function $f(x) = e^{-x}$ over the positive axis. We calculate the area up to an arbitrary positive point t on the x-axis; then we let t tend to infinity and determine the limit of the area. This is similar to the "swelling" method used in Examples 1 and 2. The area from 0 to t is

$$\int_0^t e^{-x}\,dx = -e^{-x}\Big|_0^t = 1 - e^{-t}.$$

When t tends to 0, the second term on the right, e^{-t}, tends to 0; therefore, the area tends to 1:

$$\lim_{t \to \infty} \int_0^t e^{-x}\,dx = 1.$$

We write this limit as the integral with ∞ as the upper limit of integration:

$$\int_0^\infty e^{-x}\,dx = 1;$$

hence, the area under the curve is 1. This is a convergent improper integral. ▲

We summarize the definition and evaluation of the improper integral for an infinite domain: The integral of $f(x)$ over the positive axis is defined as

$$\int_0^\infty f(x)\,dx = \lim_{t \to \infty} \int_0^t f(x)\,dx.$$

The integral over the negative axis is the limit of the integral from t to 0 as t tends to $-\infty$. The integral over the whole line,

$$\int_{-\infty}^\infty f(x)\,dx,$$

is the sum of the limits over the positive and the negative axes, respectively. The integrals are convergent or divergent accordingly as the limits are finite or infinite.

EXAMPLE 4

Is the integral

$$\int_0^\infty e^{-x^2}\,dx$$

convergent or divergent? We examine the integral from 0 to t, where $t > 1$, and then let $t \to \infty$. The region under the graph of e^{-x^2} consists of two parts: a bounded part, from 0 to 1, and an unbounded part, from 1 to t, which grows with t (Figure 6.32). The area of the first region is evidently finite. The area of the second region increases with t. We will show that it actually tends to a finite limit, so that the integral is convergent. The graph of e^{-x^2} is below that of e^{-x} on the half-line $x > 1$. The area under the latter was shown, in Example 3, to be finite; therefore, the area under e^{-x^2} is necessarily smaller because the region is "smaller" (Figure 6.32). It follows that the integral cannot diverge; therefore, it must converge. In other words, since the area is either finite or infinite, it must in this case be finite. ▲

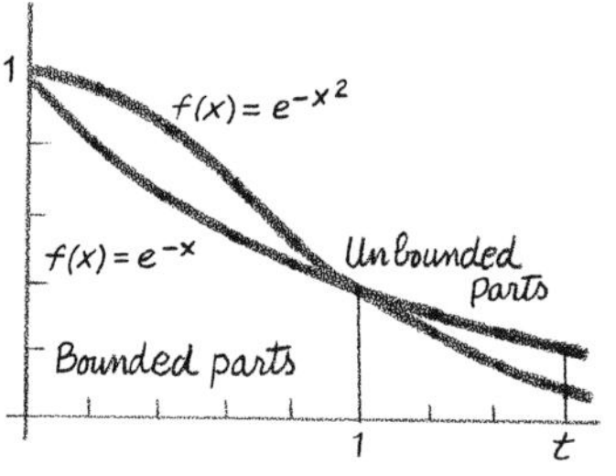

Figure 6.32

EXAMPLE 5

The integral

$$\int_{-\infty}^\infty e^{-x^2}\,dx$$

is the sum of the improper integrals

$$\int_0^\infty e^{-x^2}\,dx \quad \text{and} \quad \int_{-\infty}^0 e^{-x^2}\,dx.$$

The integrand is a **symmetric** function: $e^{-(-x)^2} = e^{-x^2}$; hence, the two integrals are equal (Figure 6.33). They are convergent according to the result of Example 4. ▲

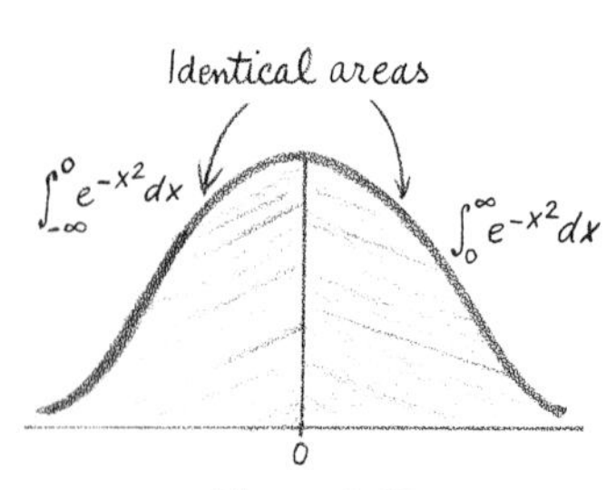

Figure 6.33

EXAMPLE 6

The relation

$$\lim_{x\to\infty} x^p e^{-x} = 0, \qquad \text{for } p > 0 \tag{1}$$

is often used in evaluating improper integrals. The numerical illustration of this fact is left to Exercise 8.

Let us determine whether the improper integral

$$\int_0^\infty x^2 e^{-(1/2)x^2}\, dx$$

is convergent. We first calculate the integral from 0 to t, and then let t become infinite. Apply the formula for integration by parts with $f(x) = x$ and $g(x) = -e^{-(1/2)x^2}$.

$$\int_0^t x^2 e^{-(1/2)x^2}\, dx = -xe^{-(1/2)x^2}\Big|_0^t + \int_0^t e^{-(1/2)x^2}\, dx.$$

Let t tend to infinity. By (1), the first term on the right-hand side tends to 0 and the second term to $\int_0^\infty e^{-(1/2)x^2}\, dx$; hence,

$$\int_0^\infty x^2 e^{-(1/2)x^2}\, dx = \lim_{t\to\infty} \int_0^t x^2 e^{-(1/2)x^2}\, dx = \int_0^\infty e^{-(1/2)x^2}\, dx.$$

The latter integral was shown in Example 4 to be convergent; hence, the former is also convergent. ▲

Let $f(x)$ be a nonnegative function such that the improper integral over the whole line (total area under the curve) is convergent and equal to 1:

$$\int_{-\infty}^\infty f(x)\, dx = 1.$$

Such a function is a density (see Section 6.2). For arbitrary x, let $F(x)$ be the improper integral over the half-line up to x:

$$F(x) = \int_{-\infty}^x f(t)\, dt.$$

In other words, $F(x)$ is the area under the density to the left of the point x. If $f(x)$ has been fitted to a histogram, then $F(x)$ is approximately the proportion of the population whose numerical characteristic is less than or equal to x. Note the variable of integration t in the integral of f; it is used in place of x to distinguish it from the variable x in the upper limit of integration. F is called a **distribution function.** ●

The difference $F(b) - F(a)$ represents the area under the curve to the left of b **minus** the area to the left of a; hence, if a is less than b, then the difference is the area under the density **between** a and b:

$$F(b) - F(a) = \int_a^b f(x)\,dx.$$

It measures the proportion of the population whose value is between a and b (Figure 6.34).

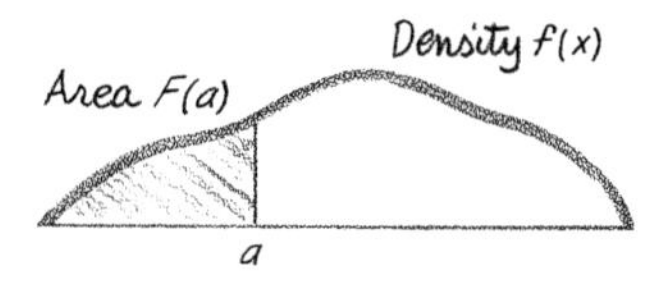

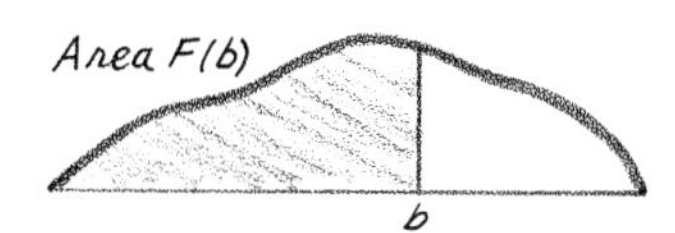

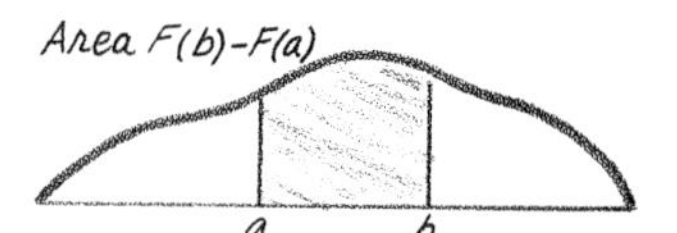

Figure 6.34

A distribution function F has the following three characteristic properties:

i. **$F(x)$ is nondecreasing as x increases: If $x < x'$, then $F(x) \le F(x')$.** More area is picked up as the point x moves to the right.

ii. **$F(x)$ tends to 1 as x tends to $+\infty$: $\lim_{x\to\infty} F(x) = 1$.** As x gets larger and larger the area under the curve swells to 1.

iii. **$F(x)$ tends to 0 as x tends to $-\infty$: $\lim_{x\to-\infty} F(x) = 0$.** The area to the left of x becomes very small as x moves far to the left.

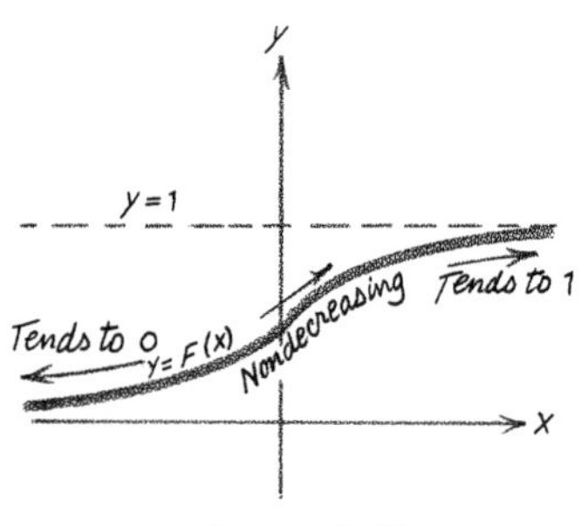

Figure 6.35

The graph of a typical distribution function appears in Figure 6.35.

When the numerical characteristic of a population assumes just nonnegative values, then the density is often chosen so that it is positive only for nonnegative x. The density is equal to 0 for negative x; hence, $F(x) = 0$ for negative x and

$$\int_{-\infty}^{\infty} f(x)\,dx = \int_0^{\infty} f(x)\,dx = 1$$

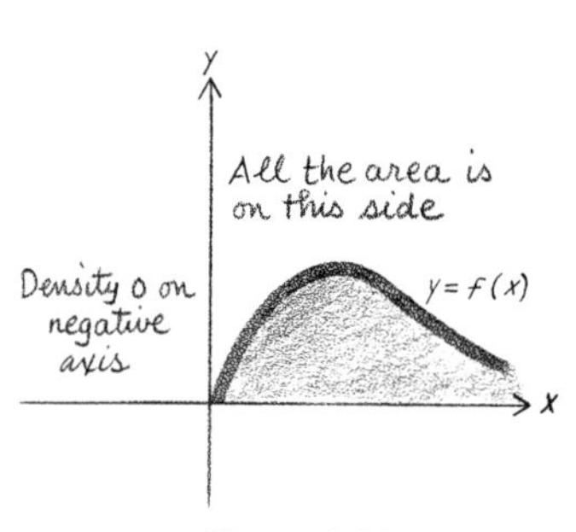

Figure 6.36

(Figure 6.36).

Numerical characteristics usually have **bounded** domains, formed by finite intervals. For example, the weight of a baby at the time of birth is certainly greater than 0, and certainly less than, shall we say, 300 pounds. The histogram of weights of babies should have no area outside this finite interval. The densities that are fitted to histograms by statisticians are often positive over the whole axis or over the positive part of the axis; however, the area in the "tails" is negligible.

EXAMPLE 7

The gamma density $\frac{1}{2}x^2e^{-x}$ was fitted to the histogram of the ages of the divorced women at the times of their marriages (Figure 6.23). The density is defined as equal to 0 for negative x; thus,

$$F(x) = 0, \qquad \text{for negative } x.$$

For $x > 0$, the area under the density to the left of x is equal to the area under $\frac{1}{2}x^2e^{-x}$ between 0 and x:

$$F(x) = \int_0^x \tfrac{1}{2}t^2e^{-t}\,dt.$$

(Note that the variable of integration is changed to distinguish it from the x in the upper limit of integration.) By integration by parts we find that the integral is equal to

$$F(x) = 1 - e^{-x}(1 + x + \tfrac{1}{2}x^2).$$

This increases with x; indeed, the derivative is equal to the density $\frac{1}{2}x^2e^{-x}$, which is positive. It tends to 1 as $x \to \infty$ by the relation (1). The distribution is equal to 0 for negative x; hence, it trivially tends to 0 as $x \to \infty$. It follows that F has the three characteristic properties of a distribution function.

Since the **total** area under the density is 1, the area to the **right** of the abscissa point x is 1 **minus** the area to the left:

$$\text{Area to the right of } x = 1 - F(x).$$

In our case it is equal to $e^{-x}(1 + x + \frac{1}{2}x^2)$. This is very small if x is sufficiently large; for example, when $x = 10$, it is $e^{-10}(1 + 10 + 50) = 0.00002724$. In a practical problem such an area can be ignored (Figure 6.37). ▲

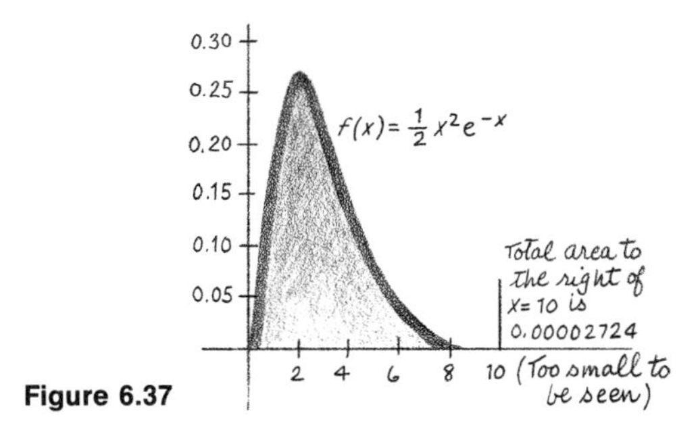

Figure 6.37

Distributions will be studied in more detail in the chapters on probability and statistics.

6.7 EXERCISES

In Exercises 1 and 2, determine which integrals are convergent, and evaluate them.

1 (a) $\int_0^1 1/x\, dx$

(b) $\int_1^\infty 1/x^2\, dx$

(c) $\int_0^\infty e^{-3x}\, dx$

(d) $\int_0^\infty x^{-1/2}\, dx$

(e) $\int_0^\infty \frac{1}{1+e^x}\, dx$

(f) $\int_0^\infty \sin^2 x\, dx$

(g) $\int_{-\infty}^\infty \frac{1}{1+x^2}\, dx$

(h) $\int_0^\infty e^{-x} \cos x\, dx$

2 (a) $\int_1^\infty 1/x\, dx$

(b) $\int_0^1 x^{-3/2}\, dx$

(c) $\int_0^\infty xe^{-4x}\, dx$

(d) $\int_1^3 (x-1)^{-1/2}\, dx$

(e) $\int_{-\infty}^\infty \frac{e^{-x}}{(1+e^{-x})^2}\, dx$

(f) $\int_{-\infty}^0 e^{-x} \sin x\, dx$

(g) $\int_{-\infty}^\infty x^3 e^{-(1/2)x^2}\, dx$

(h) $\int_0^1 \frac{x}{\sqrt{1-x^2}}\, dx$

3 For which values of p is each integral convergent or divergent?

(a) $\int_0^1 x^p\, dx$

(b) $\int_1^\infty x^p\, dx$

(c) $\int_0^1 x^p \log x\, dx$

(d) $\int_1^\infty x^p \log x\, dx$

4 For which values of k is each integral convergent or divergent?

(a) $\int_0^\infty e^{kx}\, dx$

(b) $\int_0^\infty xe^{kx}\, dx$

(c) $\int_0^\infty x^p e^{kx}\, dx$, where $p > 0$

5 By comparison with the integrals in Exercises 3 and 4, determine which of these integrals are convergent:

(a) $\int_0^\infty \frac{1}{1+x^{3/2}}\, dx$

(b) $\int_{-\infty}^\infty \frac{x}{1+x^2}\, dx$

(c) $\int_0^\infty \frac{\log x}{1+x^3}\, dx$

(d) $\int_0^\infty \sqrt{x^2+1}\, e^{-x}\, dx$

(e) $\int_0^\infty \frac{x \log x}{1+x^{5/2}}\, dx$

(f) $\int_{-\infty}^\infty x^4 e^{-x^2}\, dx$

6 The year 1952 was a moderately good one for the average American worker. The rate of unemployment was approximately 3.1 percent of the labor force. The duration of unemployment for the unemployed workers is described in the following table (*Source:* U.S. Dept of Labor, Bureau of Labor Statistics):

No. of weeks unemployed:	4 or less	5–10	11–14	15–26	26+
Proportion of unemployed	0.613	0.202	0.065	0.077	0.043

The histogram is in Figure 6.38. The unit on the horizontal axis is a 4-week period. The category "5–10 weeks" is understood to be "fewer than 10 weeks but more than 4." The other categories are interpreted in the same way. The density function e^{-x} is fitted to the histogram. Find the distribution function and show that it has the three characteristic properties. Compute the areas under the density over the intervals (a) from 0 to 1; (b) from 1 to 2.5; (c) from 2.5 to 3.5; (d) from 3.5 to 6.5. Compare these to the proportions in the table. (See Example 7.)

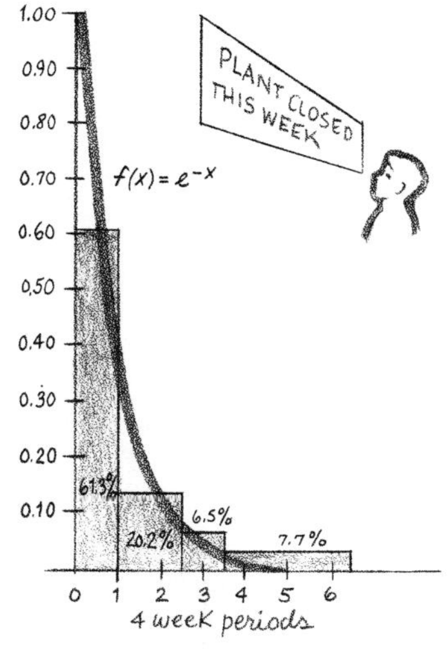

Figure 6.38

7 Statistics indicate that the likelihood of divorce decreases with the duration of marriage. In the following table, the divorces decreed in the state of Idaho in 1963 are distributed according to the duration of the marriage (*Source:* Divorce Statistics Analysis, U.S., 1963, National Center for Health Statistics, Ser. 21, No. 13, 1967, U.S. Department of Health, Education and Welfare):

Years of marriage	Percentage of divorces	Years of marriage	Percentage of divorces
1 or less	10.5	8–9	2.9
1–2	13.9	9–10	3.3
2–3	9.8	10–14	11.5
3–4	8.9	15–19	8.5
4–5	7.2	20–24	4.2
5–6	5.2	25–29	2.6
6–7	4.5	30+	2.0
7–8	5.1		

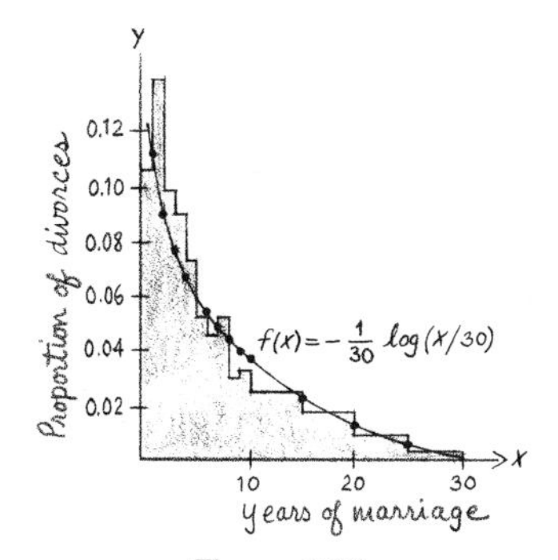

Figure 6.39

The histogram is in Figure 6.39, and the density function $f(x) = (-1/30)\log(x/30)(0 < x \leq 30)$, is fitted to it. The density is defined as 0 outside the interval (0,30). Find the distribution function $F(x)$ and show that it has the three characteristic properties. Compute the areas under the density over the intervals (a) from 0 to 5; (b) from 5 to 10; (c) from 10 to 15; (d) from 15 to 20. Compare these to the corresponding proportions in the table.

(Note that the 1-year categories include only the first year mentioned; for example, the category 2–3 means all marriages that ended after the second anniversary but before the third. The category 10–14 means after the tenth but before the fifteenth anniversary. See Example 7.)

8 Compute the value of $x^p e^{-x}$ for the following values of x and p: $p = 1, 2$ and $x = 1, 5, 10, 15, 20$. Use Table II, and the fact $e^{15} = (e^{7.5})^2$.

9 Let $f(x)$ be a density function. Using the properties of the improper integral show that for every real c, the function $f(x - c)$ is also a density function. How are the graphs of the two functions related?

10 Let $f(x)$ be a density function and c a positive real number. Show that $cf(cx)$ is also a density. How are the graphs of the two functions related?

11 Let $F(x)$ be the distribution function corresponding to the density $f(x)$. Find the distribution corresponding to the density $f(x - c)$, defined in Exercise 9.

12 Repeat Exercise 11 for the distribution of $cf(cx)$.

6.8 The Mean of a Distribution

Imagine two weights placed along a straight thin rod. The **center of gravity** of the weights is the point on the rod where the two weights balance each other. According to the laws of mechanics, the center of gravity is at a point between the weights, where the ratio of the **distances** to the weights is the inverse of the **weight ratio.** This is explained in terms of the **balance scale** in Figure 6.40.

There is a more formal way of describing the center of gravity. Suppose that the rod is marked like the x-axis. Let x and x' be the points where the weights are placed, and m and m' the corresponding weights (in appropriate units). Then the center of gravity is at the point

$$\bar{x} = \frac{xm + x'm'}{m + m'}.$$

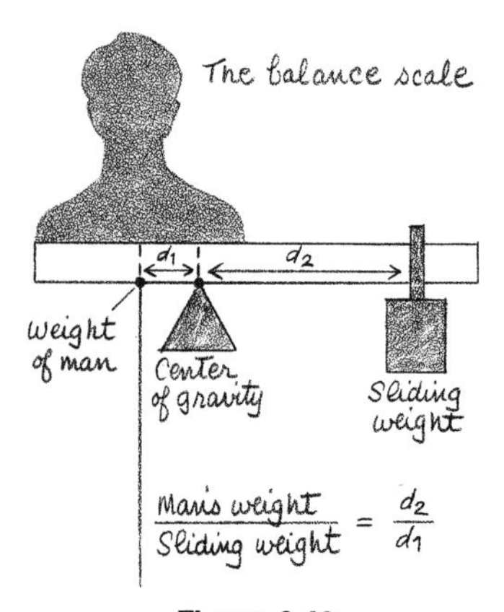

Figure 6.40

The numerator is the sum of the products of the abscissa times the corresponding weight; the denominator is the sum of the weights. The center of gravity may also be written as a weighted sum of the abscissas:

$$\bar{x} = x\left(\frac{m}{m + m'}\right) + x'\left(\frac{m'}{m + m'}\right);$$

the abscissas are weighted by the corresponding proportions.

More generally, suppose that several weights $m_1, m_2, \ldots, m_n$ are placed at the points $x_1, x_2, \ldots, x_n$ on the x-axis; then the center of gravity is defined as

$$\bar{x} = \frac{x_1 m_1 + \cdots + x_n m_n}{m_1 + \cdots + m_n}. \tag{1}$$

EXAMPLE 1

Weights of magnitudes 2, 6, 4, and 5 are placed at the points -1, 1, 4, and 7, respectively. Then, by formula (1), the center of gravity is at

$$\bar{x} = \frac{2(-1) + 6(1) + 4(4) + 5(7)}{2 + 6 + 4 + 5} = \frac{50}{12} = 4\tfrac{1}{6}$$

(Figure 6.41). ▲

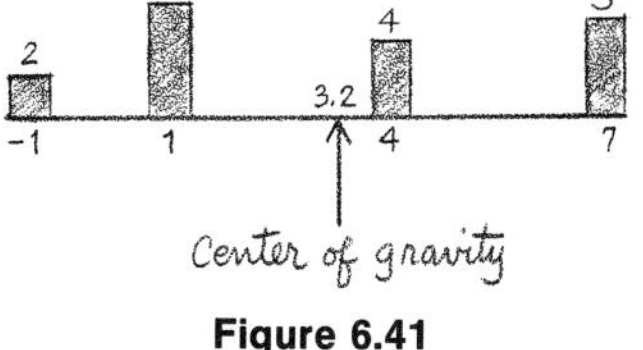

Figure 6.41

The concept of center of gravity has an important interpretation in the context of statistics: If each member of a population is considered to be a "weight" on the x-axis, placed at a point whose abscissa is equal to his numerical characteristic, then the center of gravity of the population is identical with the **average** number in the population.

EXAMPLE 2

Suppose that 10 students in a class received the following grades in a brief mathematics quiz; each examination is scored from 0 to 10:

9, 8, 9, 10, 7, 9, 8, 6, 8, 4.

The grades are arranged in the following table:

Grade	0	1	2	3	4	5	6	7	8	9	10
No. of students	—	—	—	—	1	—	1	1	3	3	1

The average grade is the sum divided by 10: $(9 + 8 + \cdots + 4)/10 = 7.8$. This can be computed in a different way, as a center of gravity. If each grade weighs 1 unit, then the grade distribution is like a distribution of weights on a numbered rod: there is 1 weight unit at 4, 6, 7, and 10, respectively, and 3 at 8 and 9, respectively. By formula (1), the center of gravity is

$$\bar{x} = \frac{4(1) + 6(1) + 7(1) + 8(3) + 9(3) + 10(1)}{1 + 1 + 1 + 3 + 3 + 1}$$

$$= 7.8.$$

This coincides with the average computed earlier. ▲

If $f(x)$ is the density function of a numerical characteristic of a population, then the **mean** of the density is defined as the improper integral

$$\bar{x} = \int_{-\infty}^{\infty} x f(x)\, dx. \qquad (2)$$

This is defined only if the two improper integrals

$$\int_0^\infty xf(x)\,dx \qquad \text{and} \qquad \int_{-\infty}^0 xf(x)\,dx$$

are convergent. The mean $\bar{x}$ is the **theoretical** average value of the numerical characteristic in the population. Imagine that the x-axis is subdivided into many small intervals. The proportion of the population whose number is in a small interval from x to x^* is the area under the graph of $f(x)$ over the interval from x to x^*. It is approximately a rectangle of height $f(x)$ and width $x^* - x$. The **average** of the population can be obtained by weighting x by the population proportion, approximately $f(x)(x^* - x)$, and summing over all the intervals (Figure 6.42). This is the same process as the integration of $f(x)$ except that $f(x)$ is replaced here by the product $xf(x)$. It follows that the sum of the weighted x-values is the integral $\bar{x}$ in (2).

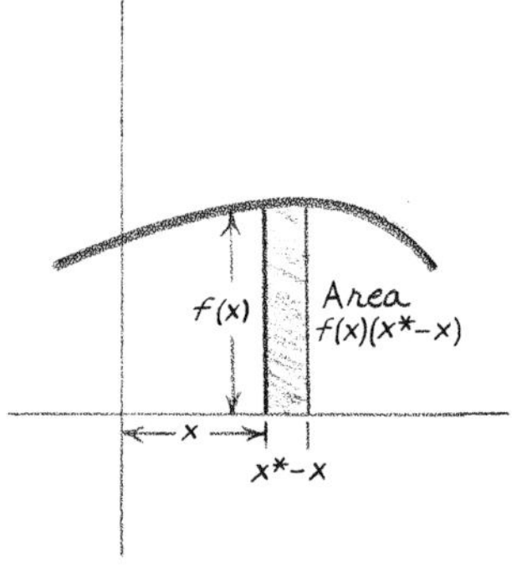

Figure 6.42

If the density is equal to 0 over a portion of the axis, then, in computing the integral (2), that portion may be left out of the domain of integration. For example, if the density is equal to 0 for negative values of x, then $\bar{x}$ is given by the integral

$$\int_0^\infty xf(x)\,dx. \tag{3}$$

EXAMPLE 3

Find the mean of the density $f(x)$ defined as

$$f(x) = 0, \qquad \text{for } x < 0$$
$$= 3e^{-3x}, \qquad \text{for } x \geq 0.$$

By formula (3) it is equal to

$$\bar{x} = \int_0^\infty x(3e^{-3x})\,dx.$$

Integrate by parts:

$$\int_0^\infty x(3e^{-3x})\,dx = x(-e^{-3x})\Big|_0^\infty + \int_0^\infty e^{-3x}\,dx.$$

The first term on the right-hand side is equal to 0 because, by formula (1) of Section 6.7,

$$\lim_{x\to\infty} xe^{-x} = 0.$$

The second term is equal to

$$\left(\tfrac{1}{3}\right)(-e^{-3x})\Big|_0^\infty = \tfrac{1}{3};$$

hence, the mean of the density is $\frac{1}{3}$ (Figure 6.43). ▲

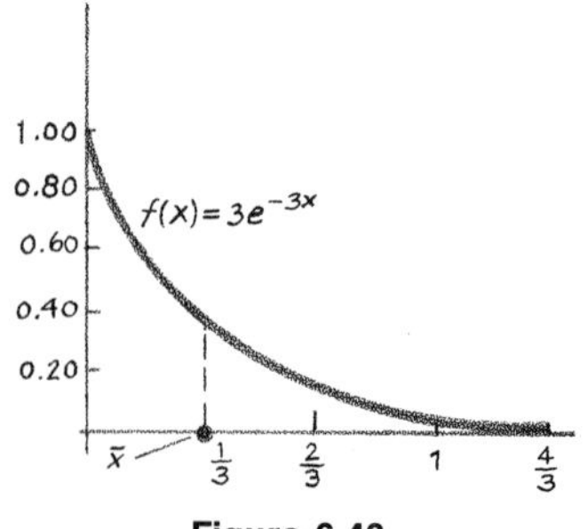

Figure 6.43

EXAMPLE 4

Find the mean of the density function defined as

$$f(x) = \frac{3x^2}{26}, \qquad \text{for } x \text{ between 1 and 3,}$$
$$= 0, \qquad \text{elsewhere.}$$

The density is equal to 0 outside the interval from 1 to 3; hence, $\bar{x}$ is equal to

$$\int_1^3 f(x)\,dx = \left(\frac{3}{26}\right)\int_1^3 x \cdot x^2\,dx = \left(\frac{3}{26}\right)\int_1^3 x^3\,dx$$
$$= \frac{3^5 - 3}{4(26)} = \frac{30}{13} = 2.3 \qquad \text{(approximately)}$$

(Figure 6.44). ▲

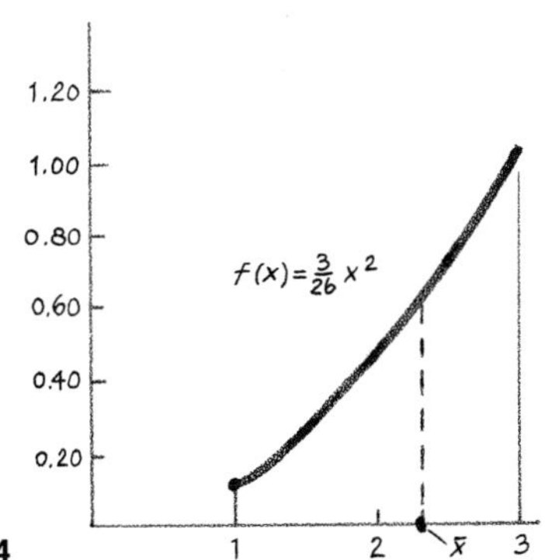

Figure 6.44

6.8 EXERCISES

In Exercises 1 and 2 find the means of the given densities.

1 (a) $f(x) = \frac{1}{2}e^{-x/2}$, for $x \geq 0$
$= 0$, for $x < 0$

(b) $f(x) = 2x^{-3}$, for $x \geq 1$
$= 0$, for $x < 1$

(c) $f(x) = \cos x$, for $0 \leq x \leq \pi/2$
$= 0$, elsewhere

(d) $f(x) = (2/\pi)(1 - x^2)^{-1/2}$, for $0 \leq x \leq 1$
$= 0$, elsewhere

(e) $f(x) = -\log x$, for $0 < x \leq 1$
$= 0$, elsewhere

(f) $f(x) = x \sin x$, for $0 \leq x \leq \pi/2$
$= 0$, elsewhere

(g) $f(x) = (2/\pi)\cos^2 x$, for $0 \leq x \leq \pi$
$= 0$, elsewhere

(h) $f(x) = 4(1 - x)^3$, for $0 \leq x \leq 1$
$= 0$, elsewhere

2 (a) $f(x) = \frac{1}{2}x^2 e^{-x}$, for $x \geq 0$
$= 0$, for $x < 0$

(b) $f(x) = 3x^{-4}$, for $x \geq 1$
$= 0$, for $x < 1$

(c) $f(x) = \frac{1}{2}\sin x$, for $0 \leq x \leq \pi$
$= 0$, elsewhere

(d) $f(x) = (4/\pi)(1 - x^2)^{1/2}$, for $0 \leq x \leq 1$
$= 0$, elsewhere

(e) $f(x) = -4x \log x$, for $0 < x \leq 1$
$= 0$, elsewhere

(f) $f(x) = (1/\sqrt{2\pi})e^{-(1/2)x^2}$

(g) $f(x) = (2/\pi)\sin^2 x$, for $0 \leq x \leq \pi$
$= 0$, elsewhere

(h) $f(x) = (1/66)(3x^2 - x)$, for $2 \leq x \leq 4$
$= 0$, elsewhere

3 What is the average age of the divorced women for the population given by the density in Section 6.6, Example 6? (Use the age units.)

4 Find the average duration of unemployment for the unemployed American worker in 1952; use the density in Section 6.7, Exercise 6. (Note: $x = 4$ weeks.)

5 Using the density in Section 6.6, Exercise 6, find the average income of American males in the age group 25–34 in 1958.

6 Find the average duration of marriage for divorced couples in the state of Idaho when the density is given by the function in Section 6.7, Exercise 7.

7 The **conditional mean** of a population relative to an interval from a to b is

$$\bar{x}(a,b) = \frac{\int_a^b xf(x)\,dx}{\int_a^b f(x)\,dx}.$$

This is the population average for those in the interval from a to b. (See Figure 6.45.) Find the conditional means for the densities relative to the given intervals:

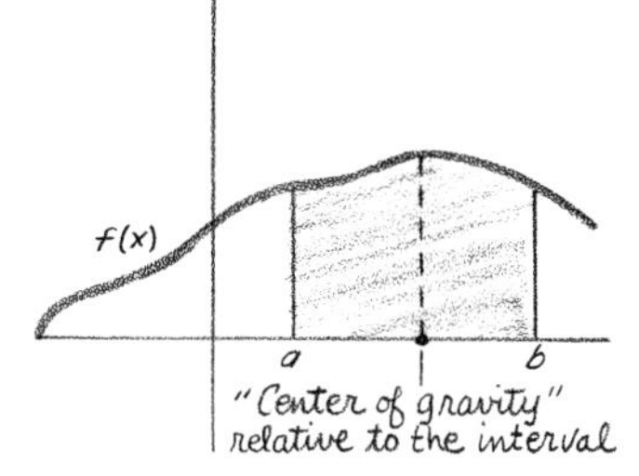

Figure 6.45

	Interval	Density
(a)	0 to 1	e^{-x}
(b)	3 to 4	xe^{-x}
(c)	$\pi/4$ to $\pi/2$	$\cos x$
(d)	5 to 10	$3x^{-4}$

8 We refer to Exercise 4: Find the average duration of unemployment for those unemployed more than 4 weeks. (Here $b = \infty$.)

9 We refer to Exercise 5: Find the average income for those in the range \$1000–\$6000.

10 We refer to Exercise 6: Find the average duration of marriage for divorced couples whose marriage lasted at least 7 years.

6.9 Volumes by Integration

We have already used certain formulas for volumes: those of balls, cylinders, and cones. By means of integral calculus we can justify and derive **these** formulas, and also many others.

The volumes that can be calculated by the methods of this section are **volumes of revolution;** they are generated by revolving a plane area around an axis in three-dimensional space. For example, a cylinder is the volume swept out by the revolution of a rectangle about an axis (Figure 6.46). Similarly, the ball is obtained from a semicircle (Figure 6.47); and the cone from a triangle (Figure 6.48). Other interesting volumes of revolution are the **torus** (doughnut, Figure 6.49) and the hollow cylinder (Figure 6.50).

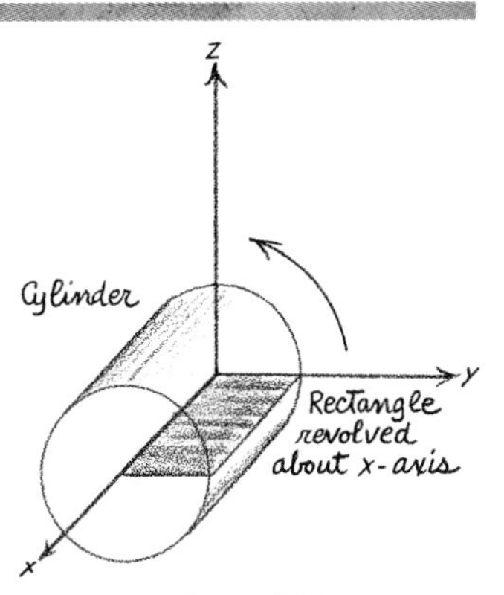

Figure 6.46

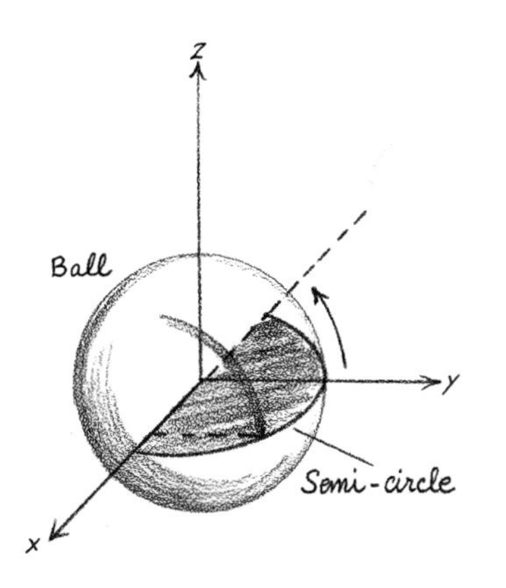

Figure 6.47

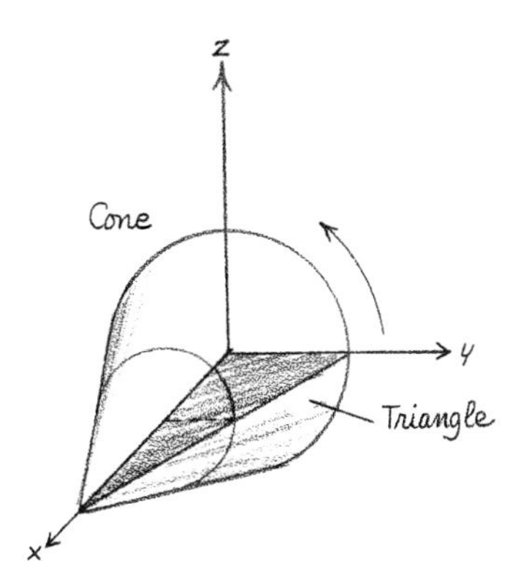

Figure 6.48

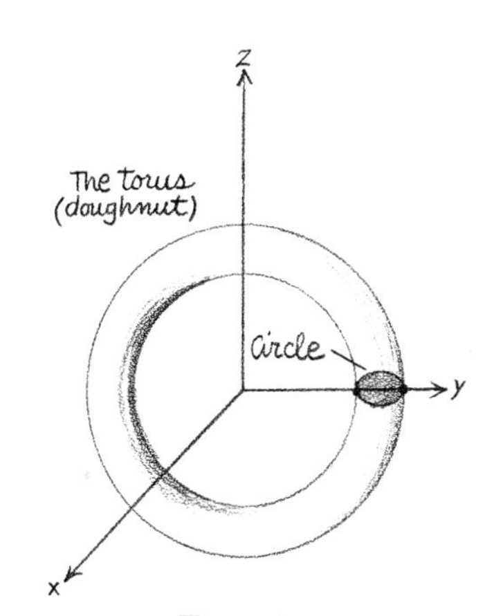

Figure 6.49

There are two simple formulas for the volume generated by the revolution of a plane area. Let the area be that under the graph of a nonnegative function $f(x)$ over an interval from a to b. **If the revolution is about the x-axis, then the volume of revolution is, in cubic units, equal to the integral**

$$V = \pi \int_a^b [f(x)]^2 dx. \tag{1}$$

If the revolution is about the y-axis, the volume is

$$V = 2\pi \int_a^b x f(x) dx. \tag{2}$$

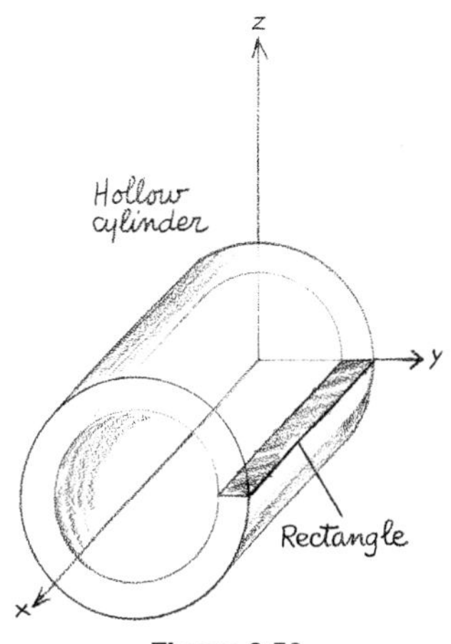

Figure 6.50

The first formula is called the **disk** formula and the second is the **shell** formula. These names are descriptive of their derivations. The first formula is based on the fact that the volume of the solid is approximately the sum of the volumes of many thin disks. When the portion of the area under the curve and over a small interval on the x-axis is revolved about that axis, then the volume swept out is like a thin disk or coin (Figure 6.51). The volume of the disk is the product of the cross-sectional area and the thickness. The cross-sectional area is the area of a circle of radius $f(x)$, namely, $\pi[f(x)]^2$; the thickness of the disk is $x^* - x$; therefore the volume is $\pi[f(x)]^2(x^* - x)$. We sum the volumes of all the disks. This is just like the summation of the areas of rectangles to get the total area under a curve (Section 6.4); the difference is that the height $f(x)$ of the rectangle is replaced here by the cross-sectional area of the disk, $\pi[f(x)]^2$. It follows that the sum of the volumes of the disks is an approximation to the definite integral (1).

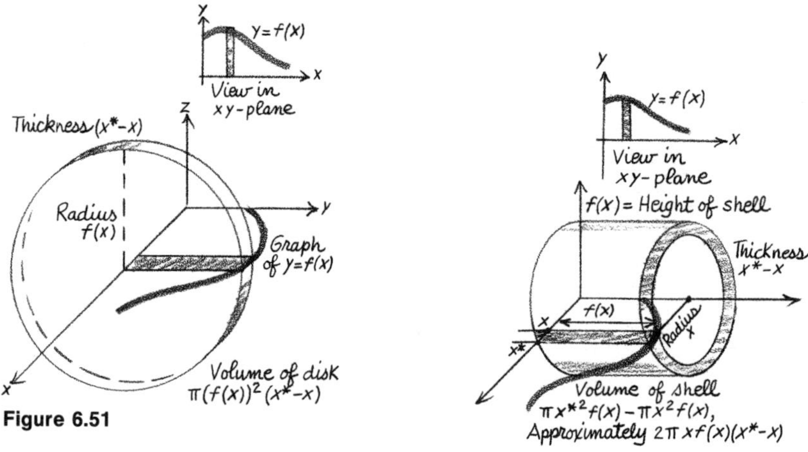

Figure 6.51

Figure 6.52

The second formula is based on the fact that the volume can be represented as the sum of the volumes of many thin cylindrical hollow shells. Such a shell is generated when the portion of the area under the curve and over a small interval on the x-axis is revolved about the y-axis (Figure 6.52). The volume of the shell is the difference between the volumes of two concentric cylinders of radii x and x^*, respectively, and heights $f(x)$,

$$\pi x^{*2} f(x) - \pi x^2 f(x) = \pi(x^{*2} - x^2) f(x);$$

by factorization, this is equal to

$$\pi(x^* + x)(x^* - x) f(x).$$

If x^* is very close to x, then the sum $x^* + x$ is nearly equal to $2x$; hence, the volume of the shell is approximately $2\pi x f(x)(x^* - x)$. Summing over all the intervals, we obtain the definite integral (2).

EXAMPLE 1

Let us derive the formula for the volume of a ball of radius r. First we shall use the disk formula (1). The ball can be obtained by revolving a semicircle about the x-axis [Figure 6.53(a)]. The equation of the circle of radius r is $x^2 + y^2 = r^2$; therefore, the area under the semicircle is the same as the area under the graph of the function $y = \sqrt{r^2 - x^2}$. The volume of half the ball is obtained by revolving the area over the interval from $x = 0$ to $x = r$ about the x-axis. Here $f(x)$ is $\sqrt{r^2 - x^2}$. By formula (1),

$$V = \pi \int_0^r (\sqrt{r^2 - x^2})^2 dx = \pi \int_0^r (r^2 - x^2)\, dx = \pi \left(r^2 x - \frac{x^3}{3} \right) \Bigg|_0^r$$
$$= \left(\tfrac{2}{3}\right)\pi r^3.$$

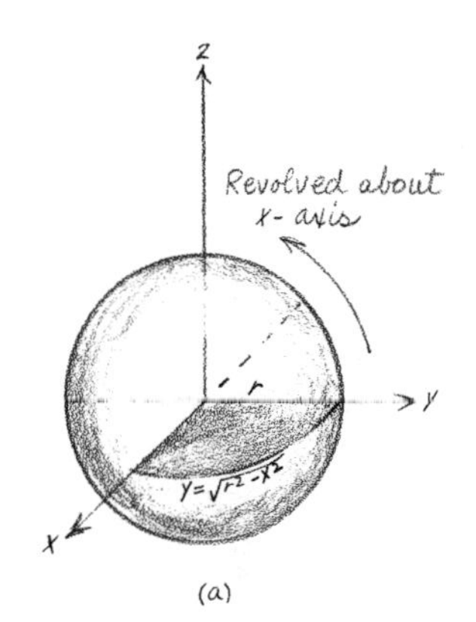

(a)

The volume of the whole ball is obtained by doubling this: we get the well-known formula $V = \left(\frac{4}{3}\right)\pi r^3$.

Let us compute the same volume by the shell method. The top half of the ball is obtained by revolving the quarter circle under the curve $y = \sqrt{r^2 - x^2}$ about the y-axis [Figure 6.53(b)]. Apply formula (2) with $f(x) = \sqrt{r^2 - x^2}$:

$$V = 2\pi \int_0^r x\sqrt{r^2 - x^2}\, dx = -2\pi\left(\tfrac{1}{3}\right)(r^2 - x^2)^{3/2} \Bigg|_0^r$$
$$= \left(\tfrac{2}{3}\right)\pi r^3.$$

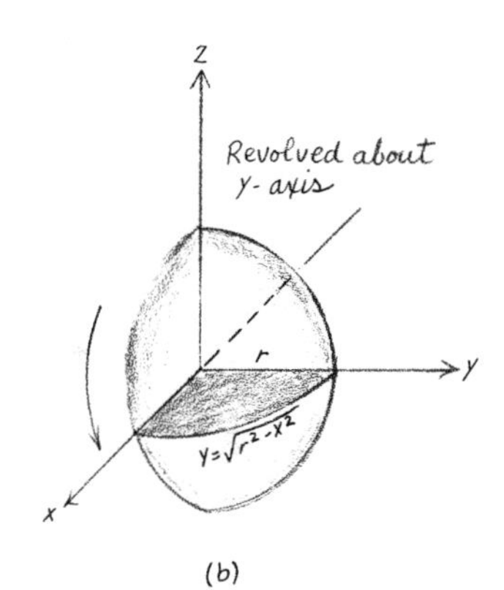

(b)

Figure 6.53

Double this to obtain $V = \left(\frac{4}{3}\right)\pi r^3$. ▲

EXAMPLE 2

An important volume associated with the **normal density function** in the theory of probability and statistics is that obtained by revolving the portion of the area under the density to the right of the origin about the y-axis. Let $f(x)$ be the function $e^{-(1/2)x^2}$. When the area under it is revolved about the y-axis, the volume generated is like a circular mound (Figure 6.54). The volume formula (2) involves a convergent improper integral:

$$V = 2\pi \int_0^\infty x e^{-(1/2)x^2}\, dx = 2\pi\left(-e^{-(1/2)x^2}\right) \Bigg|_0^\infty = 2\pi. \ ▲$$

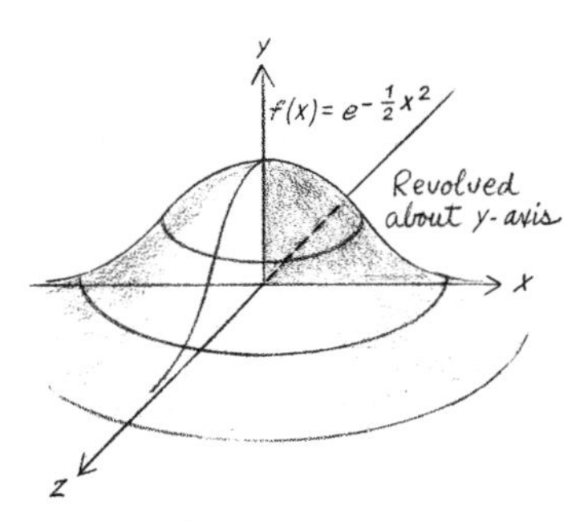

Figure 6.54

6.9 EXERCISES

In Exercises 1 and 2 find the volumes obtained by revolving the given area about the x*-axis.*

1 (a) Area under $f(x) = x^2$, interval from 0 to 1.
(b) Area under $f(x) = 3x$, interval from 0 to 1.
(c) Area under $f(x) = \sin x$, interval from $\pi/4$ to $\pi/2$.
(d) Area under $f(x) = (1 - \frac{1}{4}x^2)^{1/2}$, interval from 0 to 1.

2 (a) Area under $f(x) = \sqrt{x}$, interval from 1 to 2.
(b) Area under $f(x) = x^2 + x$, interval from 0 to 1.
(c) Area under $f(x) = xe^{-x}$, interval from 0 to 1.
(d) Area under $f(x) = \cos^2 x$, interval from 0 to $\pi/2$.

In Exercises 3 and 4 find the volumes obtained by revolving the areas about the y*-axis; the given intervals are on the* x*-axis.*

3 (a) Area under $f(x) = 2$, interval from 0 to 2.
(b) Area under $f(x) = \sqrt{16 - x^2}$, interval from 0 to $\sqrt{7}$.
(c) Area under $f(x) = \sin x$, interval from 2π to $\frac{5}{2}\pi$.
(d) Area under $f(x) = e^x$, interval from 0 to 1.

4 (a) Area under $f(x) = 3(1 - x)$, interval from 0 to 1.
(b) Area under $f(x) = x\sqrt{1 + x^3}$, interval from 0 to 2.
(c) Area under $f(x) = \log x$, interval from 1 to 4.
(d) Area under $f(x) = xe^{-x}$, half-line $x \geq 1$.

5 Find the volume obtained by revolving the area bounded by the y-axis, the parabola $4y = x^2$, and the line $y = 2$ about the y-axis. [*Hint:* Consider the volume as the **difference** of two volumes generated by areas under the graphs of nonnegative functions.]

6 The volume of the right circular cone of base radius r and height h is $V = \frac{1}{3}\pi r^2 h$. Derive this formula by using (1) and then using (2). [*Hint:* See Figure 6.48.]

7 Find the volume of the solid generated by revolving the upper half of the ellipse $x^2/9 + y^2/4 = 1$ about the x-axis.

8 Consider the ellipse $x^2/a^2 + y^2/b^2 = 1$. Find the formula for the volume generated by revolving the right half of the ellipse about the y-axis, and the formula for the volume generated by revolving the top half of the ellipse about the x-axis.

6.10 Linear and Related Differential Equations

In Chapter 4 we derived the forms of many kinds of functions which arise in scientific problems. There we used only the **static** aspects of the relation between the independent and dependent variables. In Section 5.3 we introduced the notion of the **dynamic** aspect of a relation between variables. It can be used to deduce the formulas for functions in many areas where the static approach would not work (Figure 6.55). The dynamic approach involves the method of **differential equations.** Let $y = f(x)$ be a function of x, and write $y' = dy/dx$, the derivative of f. A **first-order** differential equation is one involving x, y, and y'. The information about a function is often given in terms of such an equation. The explicit functional formula $y = f(x)$ can be obtained by **solving** the equation.

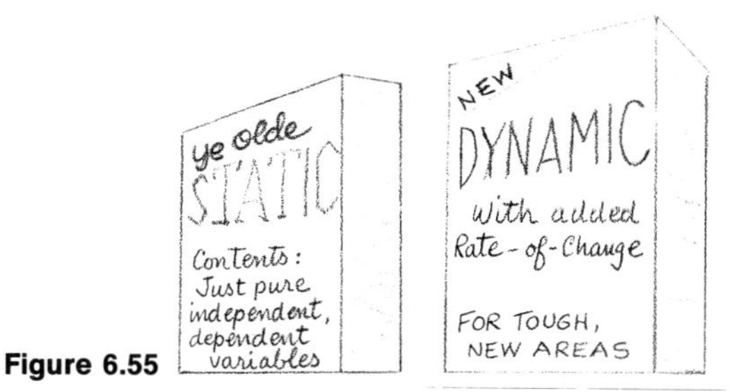

Figure 6.55

EXAMPLE 1

In Section 4.3 we showed that the size of a quantity growing like continuously compounded interest is given by the exponential function $f(x) = Ce^{kx}$; here C is the initial size and k the nominal growth rate. Let us now derive the same function by the method of differential equations. The characteristic feature of continuous, constant-rate growth is that the **rate** of change in the size of the population, the rate of change in y, is directly proportional to the instantaneous size y:

$$\frac{dy}{dx} = ky \qquad \text{or} \qquad y' = ky, \tag{1}$$

where k is a constant. This is a differential equation for the unknown function y. Suppose $y \neq 0$. Let us divide both sides of the equation by y: $y'/y = k$. By the chain rule, the left-hand side of this equation is equal to the derivative of the logarithm of y; hence,

$$\left(\frac{d}{dx}\right) \log y = k.$$

Integrate each member: $\log y = kx + C$. Since the logarithm has the base e, the equation is equivalent to $y = e^{kx+C}$, or $y = e^{kx}e^{C}$. Since C is an arbitrary constant, e^C is an arbitrary positive constant; hence, we can replace e^C by the symbol C:

$$y = Ce^{kx}. \tag{2}$$

In this way we have derived the formula for exponential growth from the differential equation (1).

If k is the nominal rate of **decay** in a process, then the differential equation (1) is replaced by $y' = -ky$, and (2) by $y = Ce^{-kx}$. ▲

Unlike some other books on elementary calculus, our text does not develop techniques for solving various kinds of differential equations. Our aim is to identify the solutions of one kind of differential equation and its variations: the **first-order linear differential equation.** This equation is sufficiently general to be useful in a large class of applied problems. It is a more general version of (1):

$$y' = p(x)y + q(x), \tag{3}$$

where p and q are given functions of x, and y is the unknown function. Equation (3) states that y' is not simply proportional to y; rather, it is a linear function of y with coefficients depending on x. In other words, the rate of change of y at "time x" is a linear function of y with "time-dependent" coefficients. The significance of the coefficients will be explained in Section 6.11. In the very special case (1) the coefficients p and q are identically equal to k and 0, respectively.

What function y satisfies (3)? We do not go into equation-solving techniques, but just state the **solution:**

$$y = e^{\int p(x)dx} \left\{ \int e^{-\int p(x)dx} \, q(x)\,dx + C \right\}. \tag{4}$$

As in Section 6.3, $\int p(x)dx$ stands for the (indefinite) integral of $p(x)$. Let us verify that the function y really satisfies Equation (3). By the definition of the integral,

$$\left(\frac{d}{dx}\right) \int p(x)\,dx = p(x),$$

$$\left(\frac{d}{dx}\right) \left\{ \int e^{-\int p(x)dx} \, q(x)\,dx + C \right\} = e^{-\int p(x)dx} \, q(x);$$

hence, by the chain rule,

$$\left(\frac{d}{dx}\right) e^{\int p(x)dx} = p(x)\,e^{\int p(x)dx}.$$

We use these relations and the product rule to differentiate the right-hand side of (4):

$$y' = e^{\int p(x)dx} \left(\frac{d}{dx}\right) \left\{\int e^{-\int p(x)dx} q(x)\,dx + C\right\}$$

$$+ \left\{\left(\frac{d}{dx}\right) e^{\int p(x)dx}\right\} \left\{\int e^{-\int p(x)dx} q(x)\,dx + C\right\}$$

$$= q(x) + p(x)y;$$

hence, Equation (3) is satisfied. The function (4), with the arbitrary constant C, is the **only** one satisfying (3). The proof is outlined in Exercise 8.

EXAMPLE 2

Let us find the solution of

$$y' = xy + x^3, \tag{5}$$

such that $y = 1$ when $x = 0$. Here $p(x) = x$ and $q(x) = x^3$. We have

$$\int p(x)\,dx = \tfrac{1}{2}x^2,$$

where the constant of integration, C, is left out until the calculation is completed. To find the integral in the brace in (4), we integrate by parts:

$$\int e^{-(1/2)x^2} x^3\,dx = \int x^2(xe^{-(1/2)x^2})\,dx = x^2(-e^{-(1/2)x^2}) + 2\int xe^{-(1/2)x^2}\,dx$$

$$= -x^2e^{-(1/2)x^2} - 2e^{-(1/2)x^2} + C.$$

Therefore, the solution is of the form

$$y = e^{(1/2)x^2}\{-x^2e^{-(1/2)x^2} - 2e^{-(1/2)x^2} + C\} = -x^2 - 2 + Ce^{(1/2)x^2}.$$

We are given the **initial condition** that $y = 1$ when $x = 0$; we substitute these in the equation above, and solve for C: $C = 3$. It follows that the solution is

$$y = -x^2 - 2 + 3e^{(1/2)x^2}.$$

It is directly shown that this function y actually satisfies (5): $y' = -2x + 3xe^{(1/2)x^2}$, and so

$$xy + x^3 = -x^3 - 2x + 3xe^{(1/2)x^2} + x^3 = y'. \ \blacktriangle$$

There is another type of differential equation that is not of the form (3) but that can be transformed into it. This differential equation arises in problems of **bounded growth,** such as populations or epidemics (Section 5.8, Example 9). It is of the form

$$y' = Ay(B - y), \tag{6}$$

where A and B are constants, not depending on x. The equation is solved by changing the function y to the reciprocal $z = 1/y$, showing that z satisfies a first-order linear differential equation, solving the latter, and then changing from z back to y.

EXAMPLE 3

Find the solution of

$$y' = 3y(4 - y) \tag{7}$$

subject to the initial condition $y = 1$ when $x = 0$. Put $z = 1/y$; then $y = 1/z$, and, by the chain rule, $y' = -z'/z^2$. Substitute the function z for y in (7):

$$\frac{-z'}{z^2} = \left(\frac{3}{z}\right)\left(4 - \frac{1}{z}\right)$$

or

$$z' = -12z + 3.$$

This is of the form (3) with $p(x) = -12$ and $q(x) = 3$; hence, by (4), the solution is

$$z = e^{-12x}\left\{3\int e^{12x}\,dx + C\right\} = \frac{1 + 4Ce^{-12x}}{4}.$$

Since C is arbitrary, so is $4C$; hence, z can be expressed as

$$z = \frac{1 + Ce^{-12x}}{4}.$$

It follows that

$$y = \frac{1}{z} = \frac{4}{1 + Ce^{-12x}}.$$

This is a **logistic function.** To determine C, put $x = 0$ and $y = 1$; then $C = 3$. The solution of (7) with the given initial condition is

$$y = \frac{4}{1 + 3e^{-12x}}. \quad \blacktriangle$$

Another equation similar to (6) is

$$y' = A(B - y)(D - y), \tag{8}$$

where the capital letters represent constants. This equation arises in a growth process involving the fusion of two components. It is reduced to a linear differential equation in a manner similar to (6).

EXAMPLE 4

Find the solution of

$$y' = 4(1 - y)(2 - y) \tag{9}$$

subject to the initial condition $y = \frac{1}{2}$ when $x = 0$. Here we take the function z as the reciprocal of one of the factors, $(1 - y)$ or $(2 - y)$. Suppose we put

$$z = \frac{1}{2 - y};$$

then

$$y = 2 - \frac{1}{z} \quad \text{and} \quad y' = \frac{z'}{z^2}.$$

Substitute the function z for y in (9):

$$\frac{z'}{z^2} = 4\left(\frac{1}{z} - 1\right)\left(\frac{1}{z}\right) \quad \text{or} \quad z' = -4z + 4.$$

This is of the form (3) with $p(x) = -4$ and $q(x) = 4$; hence, by Equation (4), the solution for z is

$$z = e^{-4x}\left\{\int 4e^{4x}\,dx + C\right\} = 1 + Ce^{-4x}.$$

Before proceeding to substitute for y, let us first determine C. By the definition of z, we have $z = \frac{2}{3}$ when $y = \frac{1}{2}$, that is, when $x = 0$; therefore, from the general form of z with $x = 0$, it follows that

$$\tfrac{2}{3} = 1 + C \quad \text{or} \quad C = -\tfrac{1}{3}.$$

The explicit form of z is then

$$z = 1 - \tfrac{1}{3}e^{-4x}.$$

Now substitute for z in terms of y:

$$y = 2 - \frac{1}{z} = \frac{1 - \frac{2}{3}e^{-4x}}{1 - \frac{1}{3}e^{-4x}}. \ \blacktriangle$$

We summarize the results of this section on the solution of the given differential equations:

(i) The first-order linear differential equation (3) has the solution (4).

(ii) Equation (6) is transformed into linear differential form by substituting $z = 1/y$, solving for z, and then for y.

(iii) Equation (8) is solved like (6) after substituting $z = 1/(B - y)$ or $z = 1/(D - y)$.

6.10 EXERCISES

1 Find the solution of the equation subject to the initial condition:

(a) $y' = y + 1, \quad y = 2$ when $x = 0$.

(b) $y' = y + x, \quad y = 1$ when $x = 0$.

2 Find the solution of the equation subject to the initial condition:

(a) $y' = 2y - 5, \quad y = 4$ when $x = 0$.

(b) $y' = y + \sin x, \quad y = 1$ when $x = 0$.

In Exercises 3 and 4, find the general form of the solution of the equation. The relation $e^{\log A} = A$ is to be used several times.

3 (a) $y' = -y/x + (x^2 + 1)/x$

(b) $y' = (2/x)y + 1$

(c) $y' = y \cos x + \cos x$

(d) $y' = -2xy + 2xe^{-x^2}$

(e) $y' = xy + x$

(f) $y' = -y/\tan x + \cos x$

4 (a) $y' = -3y/x + 2x$

(b) $y' = y + e^{-x}$

(c) $y' = -y/x + 3x^2 - 5$

(d) $y' = y + x^2e^x$

(e) $y' = y/x + \log x$

(f) $y' = y(2x - 1)x^{-2} + 1$

In Exercises 3 and 4, find the general form of the solution of the equation. The relation $e^{\log A} = A$ is to be used several times.

5 (a) $y' = 10y(4 - y), \quad y = 2$ when $x = 0$.

(b) $y' = y(2 - y), \quad y = \frac{1}{2}$ when $x = 0$.

(c) $y' = \frac{1}{2}(y - 1)(y - 2), \quad y = \frac{1}{3}$ when $x = 0$.

(d) $y' = 3(10 - y)(12 - y), \quad y = 5$ when $x = 0$.

6 (a) $y' = 5y(10 - y), \quad y = 2$ when $x = 0$.

(b) $y' = \frac{1}{2}y(3 - y), \quad y = 1$ when $x = 0$.

(c) $y' = 4(8 - y)(10 - y), \quad y = 2$ when $x = 0$.

(d) $y' = 7(3 - y)(4 - y), \quad y = 1$ when $x = 0$.

7 Prove the uniqueness of solution (4) of Equation (3) in the particular case when the function $q(x)$ is identically equal to 0. [*Hint:* Use the method of Example 1.]

8 Here is the outline of the proof of the uniqueness of the general solution (4):

(a) Suppose that y_1 and y_2 are solutions of (3). Put $z = y_1 - y_2$. Show that z satisfies $z' = p(x)z$.

(b) Deduce the form of z from the result of Exercise 7.

(c) What do we conclude from (b) about the relation between the forms of y_1 and y_2? Explain why the solution (4) is considered to be unique.

9 Let y satisfy (6). By means of the substitution $z = 1/y$, show that the formula for the general solution of (6) is

$$y = \frac{B}{1 + BCe^{-ABx}},$$

where C is an arbitrary constant. Show that this function can be put in the form of the logistic function, as defined in Section 5.8, Example 9.

10 Let y satisfy (8). By means of the substitution $z = 1/(B - y)$, show that the formula for the general solution of the equation is

$$y = \frac{D + BCe^{A(D-B)x}}{1 + Ce^{A(D-B)x}},$$

where C is an arbitrary constant.

6.11 Applications of Differential Equations

In this section we show how to set up a differential equation for an unknown function y on the basis of information about the function and its derivative. We present applications from various fields.

EXAMPLE 1

The population of tigers in a developing area of the world is declining because the habitat is being taken over by man's projects (Figure 6.56). The tigers' birth rate is declining, and they are being exterminated at an increasing rate by hunger and hunters. Suppose that the initial population is 20,000, and that the annual birth rate after x years is given by the formula

$$\frac{2x}{1 + x^2}$$

Figure 6.56

per individual tiger. This decreases for $x \geq 1$. This formula signifies that the increase in the population due to births is, after x years, equal to $2x/(1 + x^2)$ times the population size at the time. Suppose, further, that the tigers die at an annual rate of $8x$ thousand per year during the year x: 8000, 16,000, and 24,000 in the first, second, third years, ... The differential equation describing the rate of change in the population at time x is

$$y' = \left(\frac{2x}{1 + x^2}\right) y - 8000x; \tag{1}$$

and the initial condition is $y = 20{,}000$ when $x = 0$. The first term on the right-hand side of Equation (1) represents the increase due to births, and the second term the decrease due to deaths. This equation is a first-order linear differential equation with

$$p(x) = \frac{2x}{1+x^2} \qquad \text{and} \qquad q(x) = -8000x$$

[Section 6.10, Equation (3)]. We substitute in the formula for the solution of the equation. By integration,

$$\int p(x)\,dx = \int \frac{2x}{1+x^2}\,dx = \log(1+x^2),$$

$$e^{\int p(x)dx} = e^{\log(1+x^2)} = 1 + x^2$$

$$\int e^{-\int p(x)dx}\, q(x)\,dx = \int \frac{-8000x}{1+x^2}\,dx$$

$$= -4000 \log(1+x^2).$$

The solution takes the form

$$y = (1+x^2)\{-4000 \log(1+x^2) + C\}.$$

When $x = 0$, the right-hand side of the equation is equal to C; hence, according to the given initial condition, $C = 20{,}000$; thus, the complete solution is

$$y = (1+x^2)\{20{,}000 - 4000 \log(1+x^2)\}. \tag{2}$$

This expresses the size of the population x years later.

How long will it take until the population is completely eliminated? Set $y = 0$ in Equation (2) and solve for x:

$$(1+x^2)\{20{,}000 - 4000 \log(1+x^2)\} = 0,$$

or

$$5 = \log(1+x^2) \qquad \text{or} \qquad e^5 = 1 + x^2$$
$$\text{or} \qquad x^2 = e^5 - 1 = 147;$$

hence, $x = \sqrt{147}$, which is just slightly larger than 12. Under the given conditions the tiger will be extinct in the area in 12 years. ▲

Our next application of differential equations is to the theory of epidemics.

EXAMPLE 2

A dormitory houses many college students. One contracts a cold. The others are susceptible to it. This student spreads the cold to others when he coughs or sneezes in their presence. Suppose that this student does so in the presence of 2 other dormitory residents every day. As each person becomes infected, he also transmits the infection to 2 others every day (Figure 6.57). Some of the receivers may already have the cold; in that case, a new cold does not arise. Every susceptible person is bound to catch the cold sooner or later according to the following model. Let y be the proportion of infectives at time x (measured in days); then $1-y$ is the proportion of susceptibles. If the susceptibles and infectives mix with each other without regard to their states of health, then the proportion of new cases in one day is equal to the product of the current proportion of infectives (y), the current proportion of susceptibles ($1-y$), and the number of contacts per infective (2). Therefore, we get the following differential equation for y':

$$y' = 2y(1-y). \tag{3}$$

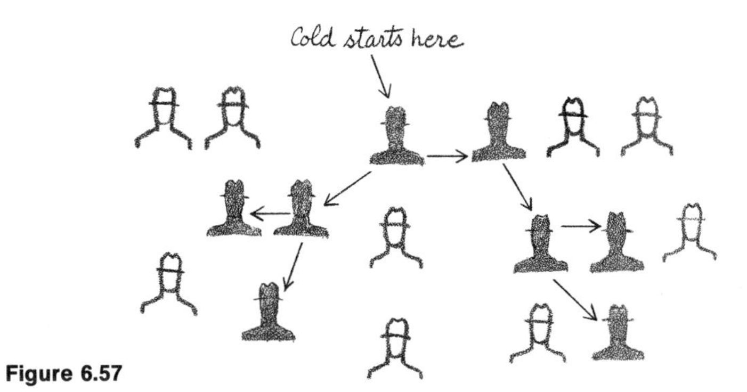

Figure 6.57

This is of the type of Equation (6) of Section 6.10. It is solved by means of the substitution $z = 1/y$. The equation in z becomes

$$\frac{-z'}{z^2} = \left(\frac{2}{z}\right)\left(1 - \frac{1}{z}\right), \quad \text{or} \quad z' = -2z + 2.$$

The latter is a first order linear equation with $p(x) = -2$ and $q(x) = 2$; hence, the solution is

$$z = e^{-2x}\{e^{2x} + C\} = 1 + Ce^{-2x}.$$

We conclude that the proportion of infectives at time x is given by the formula

$$y = \frac{1}{z} = \frac{1}{1 + Ce^{-2x}}. \tag{4}$$

Recall that the graph of this function is above the x-axis for all values of x, from $-\infty$ to ∞ (Figure 5.52). This implies that there must always be some infectives in the population. On the other hand, the graph is always below the line $y = 1$: thus, there are always some susceptibles. But, in fact, y tends to 0 as x becomes large and negative, and to 1 as x becomes large and positive: There are few cases if we go far back in time and few susceptibles if we go far forward in time.

The constant C is determined by additional information given as an initial condition. Suppose, at a particular time, which, for convenience, we call $x = 0$, the proportion of infectives is $\frac{1}{4}$. This is the initial condition

$$y = \tfrac{1}{4} \text{ when } x = 0. \tag{5}$$

When this pair of x and y values is put in (4), the value of C is 3; therefore, the proportion of infectives has the specific form

$$y = \frac{1}{1 + 3e^{-2x}}.$$

From this we can compute the proportion of infectives at various times; for example,

Number of days after $x = 0$	1	2	3	4
Proportion of infectives	.71	.95	.99	1.00.

Within 4 days almost every student is infected.

The differential equation of the form $y' = Ay(B - y)$ is not appropriate in the case of a disease which is severe enough to prevent the mingling of the infectives and the susceptibles; for example, when the disease causes incapacitation through fever or enforced quarantine. ▲

Equation (8) of Section 6.10,

$$y' = A(B - y)(D - y),$$

arises in the formation of "compounds" from "elements" of two kinds: salt molecules from sodium and chlorine molecules, "transactions" from "buyers" and "sellers," engaged couples from boys and girls, and so on.

EXAMPLE 3

In a certain office there are 20 eligible young men and 25 eligible young women. All hope to marry someone in the group. The monthly rate of engagements in the office after x months is directly proportional to the product of the numbers of eligible men and women, respectively, present at time x. Let us deduce the formula for the number of engagements that have taken place up to time x (Figure 6.58).

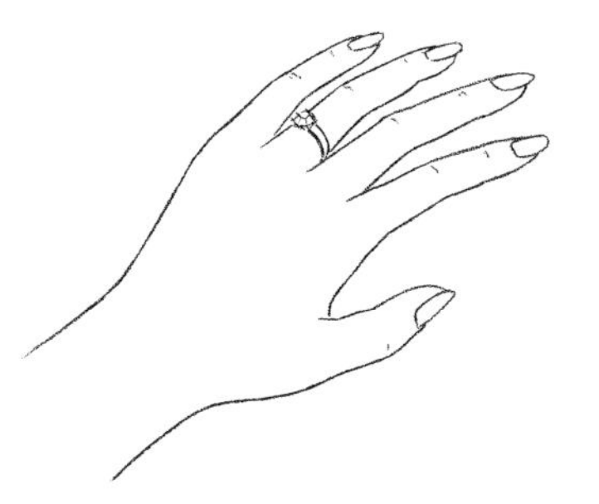

Figure 6.58

Let y be the number of engagements that have already taken place; then the number of eligibles remaining is $20 - y$ men and $25 - y$ women. By assumption, the monthly rate is proportional to the product of these:

$$y' = A(20 - y)(25 - y), \tag{6}$$

where A is a positive constant to be determined. This is of the type mentioned before the start of the example. We solve it by the change of variable $z = 1/(25 - y)$; Equation (6) becomes

$$\frac{z'}{z^2} = A\left(\frac{1}{z} - 5\right)\left(\frac{1}{z}\right) \qquad \text{or} \qquad z' = -5Az + A.$$

This is first-order linear, with $p(x) = -5A$ and $q(x) = A$; hence, the solution is

$$z = e^{-5Ax}\left\{A \int e^{5Ax}\, dx + C\right\} = \tfrac{1}{5} + Ce^{-5Ax}.$$

Since there are no engagements at time $x = 0$, the initial condition is $y = 0$ when $x = 0$. In terms of z this is: $z = 1/25$ when $x = 0$; therefore, the constant C is determined by putting $x = 0$ and $z = 1/25$ in the general solution, and we obtain

$$\frac{1}{25} = \frac{1}{5} + C \qquad \text{or} \qquad C = \frac{-4}{25}.$$

The explicit form of z is then

$$z = \frac{5 - 4e^{-5Ax}}{25};$$

hence,

$$y = 25 - \frac{1}{z} = \frac{100 - 100e^{-5Ax}}{5 - 4e^{-5Ax}}. \tag{7}$$

If A is given, then y can be computed directly from (7). On the other hand, if it is unknown, it can be found if we know one pair of (x,y)-values. For example, suppose that we have observed that 5 engagements have taken place after the first month ($x = 1$ and $y = 5$). Substitute these in (7) and solve for $5A$:

$$5 = \frac{100 - 100e^{-5A}}{5 - 4e^{-5A}} \qquad \text{or} \qquad e^{5A} = \frac{16}{15},$$

so that $5A = \log 16 - \log 15 = 0.0645$. The final form of y is then

$$y = \frac{100 - 100e^{-0.0645x}}{5 - 4e^{-0.0645x}} \quad \text{or} \quad y = \frac{100(1 - (15/16)^x)}{5 - 4(15/16)^x}. \tag{8}$$

Here is the cumulative engagement record for the first 12 months; it is obtained by successive substitution of x-values in (8) and the rounding of y to the nearest integer:

Month	Total number of engagements	Month	Total number of engagements
1	5	7	15
2	8	8	15
3	10	9	16
4	12	10	16
5	13	11	17
6	14	12	17

Note that the rate of new engagements goes down as the pool of eligibles decreases. The first 5 engagements took place in 1 month; however, the last 5 were over an 8-month period. ▲

6.11 EXERCISES

1 The birth rate in a certain country is increasing with time and the death rate decreasing. Suppose that the birth rate x years after January 1, 1973 is $20 + 0.1x$ per thousand, and the death rate $10 - 0.04x$ per thousand (per year). Find the formula for the size of the population (in thousands) after x years if the initial size is 4 million. (See Example 1.)

2 A disease is affecting the trees in a national forest. It first strikes at time $x = 0$, when there are N trees. The number of trees killed per year x years later is $N/(x + 3)$. (The rate declines over the years because the live trees are cared for and the diseased ones removed or pruned.) Every year $N/10$ new trees are planted. Find the formula for the number of living trees as a function of x. When is the number a minimum? What proportion of the original forest does this number represent? (See Example 1.)

3 A company has 1000 employees, paid the same salary s. Automation is introduced. Workers are released from their positions at the rate of $2x$ per year x years after the start of automation: 2 in the first year, 4 in the second, 6 in the third, etc. The salary of an individual worker still employed after x years is $se^{0.05x}$ (continuous growth at 5 percent per year). Find the formula for the total payroll of the company after x years. After how many years is the payroll a maximum? (See Example 1.)

4 Rumors spread like a contagious disease. In a town of 1000, a single individual tells a bit of gossip to two others in one hour. Each of the latter tells two others during the next hour. In general, every person who knows the gossip bit tells two others hourly, around the clock (Figure 6.59). If 100 persons know the story at time $x = 0$, how long will it take for 900 residents to know it? (See Example 2.)

Figure 6.59

5 In Section 5.8, Example 9, it was shown that the population of the United States grew from 1790 to 1910 according to the logistic function

$$f(x) = \frac{197{,}273{,}000}{1 + e^{-0.3134(x-12.4)}},$$

where x is the number of decades after 1790. What was the projected limit? When should $\frac{3}{4}$ capacity have been reached? (See Example 2.)

6 Suppose that the population of an expanding town has been growing according to the logistic curve, and that $\frac{1}{4}$ of the capacity has been used by 1964, and $\frac{1}{2}$ by 1970. When will 90 percent of the capacity be taken? (See Example 2.)

7 A suburban town has a population capacity of 10,000. The increase in the proportion of used capacity is equal to $\frac{1}{10}$ times the product of the proportions of the used and unused capacities. If the population is 5000 on January 1, 1970, approximately when will the population reach 8000? 9000? (See Example 2.)

8 A certain neighborhood has enough garbage cans to support a population of 1000 cats. Two hundred cats were counted on April 1, 1973. The cats increased their numbers at a rate proportional to the product of the proportions of the used and the unused cat-capacity. On April 1, 1974, the count was 300. How long will it take for the population to reach 500? (See Example 3.)

9 In the presence of a catalyst, the atoms of 2 elements unite in equal proportions to form a compound. The rate of formation of the compound is assumed to be proportional to the product of the quantities of the 2 elements still present. Suppose that a reaction begins with 3 and 2 units of

the respective elements of the 2 kinds. If half the atoms of the second element are consumed by the end of 1 minute, find the formula for the amount consumed after t minutes, where t is an arbitrary positive time. (See Example 3.)

10 Suppose that, in the chemical reaction in Exercise 9, 2 atoms of the first kind combine with 1 of the second kind. Find the differential equation for the number of atoms of the second kind consumed after x minutes. Then answer the same question as in Exercise 9. (See Example 3.)

11 At a college dance there are 125 men and 150 women. As soon as the band starts to play, the rate at which dancing couples are formed is proportional to the product of the numbers of men and women still without partners. Suppose that 20 couples are formed in the first 10 seconds. (a) Find the formula for the number of couples formed in the first t seconds. (b) How long will it take for 75 women to be on the dance floor? (See Example 3.)

12 A new housing development has 300 homes for sale. There are 1200 potential customers. The rate at which the houses are bought is proportional to the product of (i) the number of houses currently still for sale, and (ii) the number of customers currently remaining in the market. Suppose that 50 homes are sold in the first month after they are put on sale. Find the formula for the number sold after x months. (See Example 3.)

6.12 Power Series

This section is about certain sums with infinitely many terms. These are called **infinite series.** It is like a train with infinitely many cars (Figure 6.60). If the cars were all of the same length, then the train would extend to infinity. However, if the cars were of successively smaller sizes, then it is conceivable that the whole train might have a **finite** length, even though it has infinitely many cars.

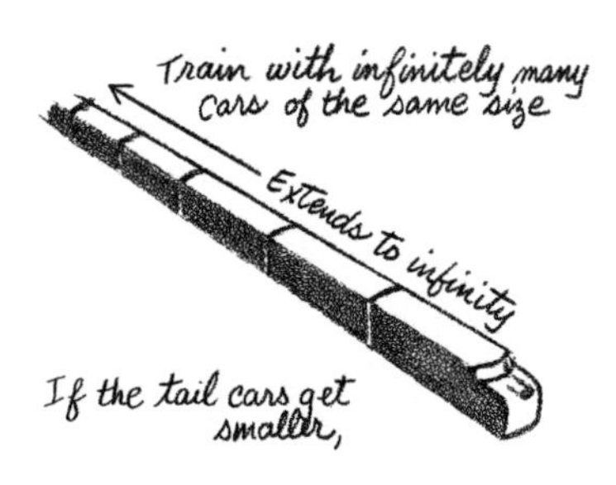

it might have finite length

Figure 6.60

A power series is a sum of infinitely many terms,

$$a_0 + a_1x + a_2x^2 + a_3x^3 + \cdots + a_nx^n + \cdots . \tag{1}$$

The a's are the **coefficients** of the series, and x is a variable. It is a "polynomial with infinitely high powers." The series (1) is denoted

$$\Sigma a_nx^n; \tag{2}$$

the greek letter **sigma** (Σ) stands for summation, like the symbol $\int$ for the integral. The subscript n is the **index of summation:** it runs over the set of integers $0, 1, 2, \ldots$.

EXAMPLE 1

A simple series is the geometric series

$$\Sigma x^n = 1 + x + x^2 + \cdot \cdot \cdot + x^n + \cdot \cdot \cdot . \tag{3}$$

The sum of the first n terms is

$$S_n = 1 + x + x^2 + \cdot \cdot \cdot + x^{n-1}. \tag{4}$$

The series (3) is obtained from the sum (4) by letting n tend to infinity. The behavior of the partial sum S_n depends on the value of x:

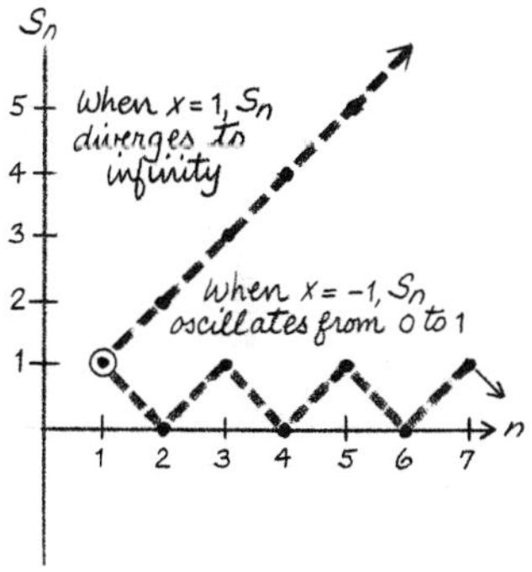

(i) If $x = 1$, then $S_n = 1 + 1 + \cdot \cdot \cdot + 1$ (n terms), and so $S_n = n$; for example, $S_2 = 2$, $S_{14} = 14$, ... S_n grows with n: it becomes infinite as n does. Such a series is said to **diverge** to infinity; and we write $\Sigma x^n = \infty$. If $x = -1$, then S_n shifts back and forth between 0 and 1: $S_1 = S_3 = S_5 = \cdot \cdot \cdot = 1$, and $S_2 = S_4 = S_6 = \cdot \cdot \cdot = 0$. The series **oscillates,** and is an **oscillatory** divergent series (Figure 6.61).

Figure 6.61

Suppose $x \neq \pm 1$. Write S_n in the form

$$S_n = \frac{1 - x^n}{1 - x} = \frac{1}{1 - x} - \frac{x^n}{1 - x}. \tag{5}$$

[The reader can verify by elementary algebra that this expression is equivalent to the original form (1).]

(ii) If x is strictly between -1 and $+1$ (for example, $x = 0.5$), then the second term on the right-hand side of (5) tends to 0 as n becomes infinite:

$$\lim_{n \to \infty} \frac{x^n}{1 - x} = 0.$$

Thus, the difference of the two terms approaches $1/(1 - x)$ (Figure 6.62):

$$\lim_{n \to \infty} S_n = \frac{1}{1 - x}.$$

In this case we say that the series **converges** and we write

$$\Sigma x^n = \frac{1}{1 - x}. \tag{6}$$

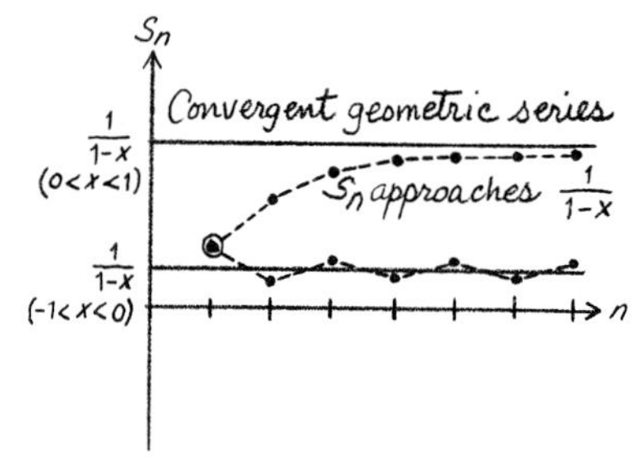

Figure 6.62

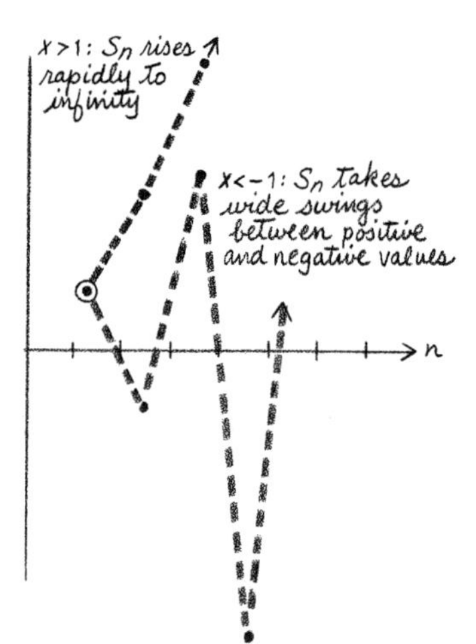

Figure 6.63

(iii) If $x > 1$, then the second term on the right-hand side of (5) is $x^n/(x-1)$, with a preceding positive sign. The series diverges to infinity because x exceeds 1 and the denominator is positive. If $x < -1$, then the second term on the right-hand side of (5) grows in absolute value, but its sign changes back and forth: S_n fluctuates between large positive and large negative values. For example, if $x = -2$, then the successive values of S_n are

$$1, -1, 3, -5, 11, -21, 43, \ldots .$$

This is an oscillatory divergent series (Figure 6.63). ▲

EXAMPLE 2

The geometric series (3) occurs in many parts of mathematics. Here we present an application to economic theory. The economists of the 1930s, particularly the Keynesians, were concerned with reviving dormant business. They taught that the economy would grow if consumers and businessmen would spend more and save less. They were often accused of making the private virtue of thrift into a public sin. One of the economic theories was that of the "multiplier effect": If a businessman pays \$1 to his worker, it would create 2, 3, or even 4 times as much spending money in the economy. The reason is as follows: Suppose that each person saves, on the average, $\frac{1}{3}$ of every dollar received. The worker saves $\frac{1}{3}$ of that and spends $\frac{2}{3}$ at the grocer. The grocer saves $\frac{1}{3}$ of that and spends the remaining $\frac{2}{3}$ at the wholesale market. The wholesaler gives $\frac{2}{3}$ of this to the trucking company, who pays $\frac{2}{3}$ to the driver, who pays $\frac{2}{3}$ to the . . . , and so on (Figure 6.64). The total amount of spending money created is, according to this derivation, an infinite series whose terms are the amounts spent by the individuals. Formula (6) gives the value of the series:

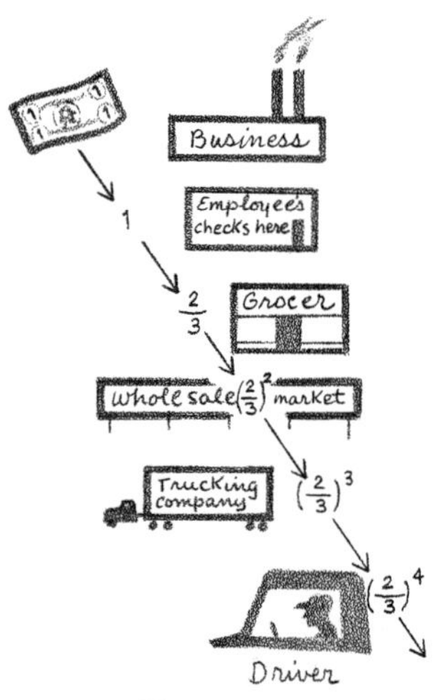

Figure 6.54

$$\Sigma(\tfrac{2}{3})^n = \frac{1}{1-\frac{2}{3}} = 3.$$

Three dollars for one! A more general formulation is: If each individual spends a proportion x of what he receives, then the total amount of spending money generated is given by (6), and is equal to the **multiplier** $1/(1-x)$. If people have a high "propensity to spend," then x is close to 1, and the multiplier is large. ▲

The geometric series is typical of the general power series (1). Let S_n be the sum of the first n terms:

$$S_n = a_0 + a_1x + \cdot\cdot\cdot + a_{n-1}x^{n-1}.$$

The series is said to **converge** for a particular value x if the corresponding sequence of partial sums, S_n, converges to a finite limit as n tends to infinity (see Section 5.3). The series may or may not converge; this depends on the value x. The series (1) certainly converges for the value $x = 0$ because, in this case, $S_n = a_0$ for all n. The series may converge for every x. Or for every $x \neq 0$, it may fail to converge. Finally, it may converge for some x and not for others. In the last case, it can be shown that there exists a positive number A such that the series converges for all x in the open interval from $-A$ to A, but fails to converge at every x outside this interval. The latter is called the **interval of convergence** of the series (Figure 6.65). The series may or may not converge at either endpoint. In the geometric series the interval of convergence is from -1 to 1.

Power series represent functions. They are useful in the computation of tables of functions; for example, the tables of trigonometric, logarithmic, and exponential functions. The first several terms of a convergent power series often carry most of the weight and the remaining terms are negligible. Power series will arise in our study of discrete probability distributions (Section 9.2).

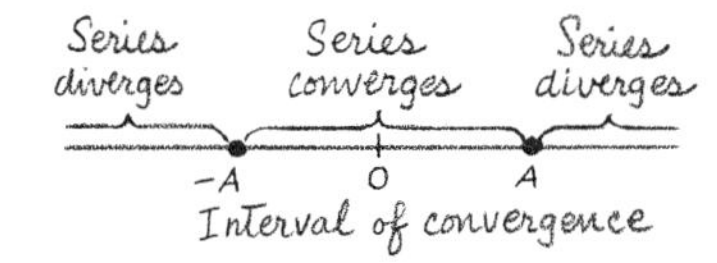

Figure 6.65

If we are given a function $f(x)$, how do we determine the coefficients (a_n) such that the series (2) is equivalent to the function? In other words, given f, how do we determine the a's so that

$$f(x) = a_0 + a_1x + a_2x^2 + \cdot\cdot\cdot ? \quad (7)$$

Maclaurin's formula states that

$$a_0 = f(0), \qquad a_1 = f'(0), \qquad a_2 = \frac{1}{2}f''(0),$$

$$a_3 = \frac{1}{3\cdot 2}f^{(3)}(0), \ldots, \qquad a_n = \frac{1}{n(n-1)(n-2)\cdot\cdot\cdot 1}f^{(n)}(0), \quad (8)$$

where, as usual, $f(0), f'(0), \ldots$ are the values of the function and its derivatives at $x = 0$. Then the series (7) becomes

$$\sum \frac{f^{(n)}(0)}{n(n-1) \cdots 1} x^n,$$

or, more conventionally and conveniently denoted,

$$\sum \frac{f^{(n)}(0)}{n!} x^n, \qquad \text{where } n! = 1 \cdot 2 \cdots (n-1)n. \tag{9}$$

In this form the series is called the **Maclaurin** series for f. We do not give the conditions under which the series converges; these are outlined in Exercises 17 and 18. In each example and exercise a given series should be assumed to converge for the given values of x. We trust that the reader will be satisfied at this point with a simple but formalized proof of the statement that: If f has derivatives of all orders in an interval containing the origin, and if it can be represented as a series (2), then the coefficients of the series are necessarily those given by Maclaurin's formula (8). For the "proof," set $x = 0$ in (7); then $f(0) = a_0$; thus, the first coefficient is the right one. Next differentiate on each side of (7). The left-hand side is $f'(x)$. If we formally differentiate term-by-term on the right-hand side, we obtain $a_1 + 2a_2x + 3a_3x^2 + \cdots$. Now set $x = 0$: $f'(0) = a_1$, and so the second coefficient a_2 in (8) is correct. Differentiate again; then $f''(x) = 2a_2 + 6a_3x + 12a_4x^2 + \cdots$; thus, $f''(0) = 2a_2$, and so the third coefficient is correct. To get each coefficient, we differentiate the series term-by-term, and then set $x = 0$. This proof is not meant to be completely rigorous; the missing point is the justification of the taking of the derivative of the sum term-by-term. An important fact about power series is that they may be differentiated term-by-term at every x inside the interval of convergence.

EXAMPLE 3

The series formula (6) is consistent with (8); here $f(x) = (1-x)^{-1}$. Then $f'(x) = (1-x)^{-2}, f''(x) = 2(1-x)^{-3}, \ldots, f^{(n)}(x) = n!(1-x)^{-(n+1)}$; hence, $f^{(n)}(0)/n! = 1$ for all n. ▲

EXAMPLE 4

The following series can be obtained from (8):

$$\frac{1}{1+x} = \Sigma(-x)^n \qquad (\text{which is } 1 - x + x^2 - x^3 + \cdots). \tag{10}$$

This can also be obtained directly from (6) by replacing x by $-x$. The series converges for $-1 < x < 1$. ▲

EXAMPLE 5

The following series can be obtained by replacing x by x^2 in (6):

$$\frac{1}{1-x^2} = 1 + x^2 + x^4 + \cdot \cdot \cdot = \Sigma x^{2n}. \tag{11}$$

The interval of convergence is from -1 to 1. ▲

EXAMPLE 6

The following series converges for $-1 < x < 1$:

$$\frac{1}{(1-x)^2} - 1 + 2x + 3x^2 + 4x^3 + \cdot \cdot \cdot = \Sigma n x^{n-1}. \tag{12}$$

This can be obtained either from (8) or by termwise differentiation of each side of (6). ▲

EXAMPLE 7

The series for the logarithm of $1 - x$ is

$$\log(1-x) = -x - \tfrac{1}{2}x^2 - \tfrac{1}{3}x^3 - \cdot \cdot \cdot = -\sum \frac{x^{n+1}}{n+1}; \tag{13}$$

it converges for $-1 < x < 1$. It can be obtained from (8) or by termwise integration of the series (6) in the integrand in the formula below:

$$\int_0^t \frac{1}{1-x}\,dx = -\log(1-t).$$

When x is replaced by $-x$, the series (13) becomes

$$\log(1+x) = x - \tfrac{1}{2}x^2 + \tfrac{1}{3}x^3 - \cdot \cdot \cdot \tag{14}$$

Let us demonstrate how (13) is also obtained from (8). We have $f(x) = \log(1-x)$; hence,

$$f'(x) = -(1-x)^{-1}, \qquad f''(x) = -(1-x)^{-2},$$
$$f^{(3)}(x) = -2(1-x)^{-3}, \; . \; . \; . \; ,$$

and so

$$f(0) = 0, \qquad f'(0) = -1, \qquad f''(0) = -1,$$
$$f^{(3)}(0) = -2, \; . \; . \; . \; . \; ▲$$

Now we present the series for the exponential and trigonometric functions:

$$e^x = 1 + x + \tfrac{1}{2}x^2 + \frac{1}{3!}x^3 + \cdots + \frac{1}{n!}x^n + \cdots \tag{15}$$

$$= \sum \frac{x^n}{n!}.$$

The series for e^{-x} is obtained by replacing x by $-x$:

$$e^{-x} = 1 - x + \tfrac{1}{2}x^2 - \frac{1}{3!}x^3 + \cdots. \tag{16}$$

The series for the trigonometric functions are

$$\sin x = x - \frac{x^3}{3!} + \frac{x^5}{5!} - \cdots, \tag{17}$$

$$\cos x = 1 - \tfrac{1}{2}x^2 + \frac{x^4}{4!} - \frac{x^6}{6!} + \cdots, \tag{18}$$

$$\tan x = x + \frac{x^3}{3} + \frac{2}{15}x^5 + \cdots. \tag{19}$$

Power series can be used for the numerical evaluation of definite integrals in which the integrand cannot be directly integrated. This can be done whenever the domain of integration is contained within the interval of convergence of the series, and does not include the endpoints of the interval.

EXAMPLE 8

Approximate the definite integral

$$\int_0^1 e^{-x^2}\,dx.$$

Expand the integrand in a series, using formula (16) with x^2 in place of x:

$$e^{-x^2} = 1 - x^2 + \tfrac{1}{2}x^4 - \frac{1}{6}x^6 + \frac{1}{24}x^8 - \cdots.$$

Integrate term-by-term from 0 to 1:

$$\int_0^1 e^{-x^2}\,dx = \int_0^1 \left(1 - x^2 + \tfrac{1}{2}x^4 - \frac{1}{6}x^6 + \frac{1}{24}x^8 - \cdots\right)dx$$

$$= x - \frac{1}{3}x^3 + \frac{1}{10}x^5 - \frac{1}{42}x^7 + \frac{1}{216}x^9 \Big|_0^1 - \cdots$$

$$= 1 - \frac{1}{3} + \frac{1}{10} - \frac{1}{42} + \frac{1}{216} - \cdots$$

$$= 0.748 \qquad \text{(approximately)}. \blacktriangle$$

If two functions have convergent series, then the sum, difference, product, and quotient (provided the denominator is not 0) of the functions has a series that converges and that can be computed from the series of the two functions.

EXAMPLE 9

Find the series for $f(x) = e^{x^2} \sin x$. If we try to use Maclaurin's formula, we find that the higher derivatives are quite complicated; for example, the first two derivatives are

$$f'(x) = e^{x^2} \cos x + 2xe^{x^2} \sin x,$$

and

$$f''(x) = e^{x^2} \sin x + 4xe^{x^2} \cos x + 4x^2 e^{x^2} \sin x.$$

A simpler way is to compute the product of the individual series:

$$e^{x^2} = 1 + x^2 + \frac{1}{2} x^4 + \frac{1}{3!} x^6 + \frac{1}{4!} x^8 + \cdots,$$

$$\sin x = x - \frac{1}{3!} x^3 + \frac{1}{5!} x^5 - \frac{1}{7!} x^7 + \cdots.$$

Construct a multiplication table for the two series:

	1	x^2	$\frac{1}{2}x^4$	$\frac{x^6}{3!}$	$\frac{x^8}{4!}$	$\cdots$
x	x	x^3	$\frac{1}{2}x^5$	$\frac{x^7}{3!}$	$\frac{x^9}{4!}$	$\cdots$
$\frac{-x^3}{3!}$	$\frac{-x^3}{3!}$	$\frac{-x^5}{3!}$	$\frac{-x^7}{2 \cdot 3!}$	$\frac{-x^9}{(3!)^2}$	$\frac{-x^{11}}{3!4!}$	$\cdots$
$\frac{x^5}{5!}$	$\frac{x^5}{5!}$	$\frac{x^7}{5!}$	$\frac{x^9}{2 \cdot 5!}$	$\frac{x^{11}}{5!3!}$	$\frac{x^{13}}{5!4!}$	$\cdots$
$\frac{-x^7}{7!}$	$\frac{-x^7}{7!}$	$\frac{-x^9}{7!}$	$\frac{-x^{11}}{2 \cdot 7!}$	$\frac{-x^{13}}{7!3!}$	$\frac{-x^{15}}{7!4!}$	$\cdots$
$\cdots$	$\cdots$	$\cdots$	$\cdots$	$\cdots$	$\cdots$	$\cdot$

Now arrange the terms in the product according to ascending powers of x. The powers are $x, x^3, x^5, \ldots$. Terms of the same power lie along the diagonal columns of the table above:

$$x$$
$$x^3 - x^3/3! = (5/6)x^3$$
$$\tfrac{1}{2}x^5 - x^5/3! + x^5/5! = (41/120)x^5$$
$$x^7/3! - x^7/(2 \cdot 3!) + x^7/5! - x^7/7! = (461/5040)x^7$$

The sums above are the first 4 terms of the series of the product. Any number of terms can be obtained by extending the table. ▲

Let x_0 be a point in the domain of a function $f(x)$. Suppose that f has derivatives of all orders in an interval containing x_0; then under conditions analogous to those needed for the validity of the Maclaurin series, $f(x)$ has a representation in powers of $(x - x_0)$:

$$f(x) = \sum \frac{f^{(n)}(x_0)}{n!}(x - x_0)^n, \tag{20}$$

where the derivatives are evaluated at x_0. This is called the **Taylor series** for $f(x)$ at x_0 (see Exercise 19).

6.12 EXERCISES

In Exercises 1 and 2 find the first 4 terms of the power series of the function.

1 (a) $(1 - 2x)^{-1}$
(b) $\sin(x + \pi/6)$
(c) $(1 + x)^{1/2}$
(d) e^{x+2}
(e) $(1 - x)^{3/2}$

2 (a) $\log(1 - \tfrac{1}{4}x)$
(b) $\cos(x + \pi/4)$
(c) $(1 - x)^{-1/2}$
(d) $\arcsin x$
(e) $(1 + x)^{-5/2}$

3 Find the series for $\arctan x$ by term-by-term integration of the series for $1/(1 + x^2)$. [*Hint:* Replace x in (10) by x^2.]

4 Find the series for $(1 - x)^{-3}$ by differentiating both sides of (12).

5 Show that the differentiation in Example 6 is justified: Differentiate each side of the identity,

$$1 + x + x^2 + \cdot \cdot \cdot + x^{n-1} = \frac{1}{1 - x} - \frac{x^n}{1 - x},$$

and then let n tend to infinity.

6 Show that the series in 2(d) can be obtained from that in 2(c) by integration after replacing x by x^2.

7 Verify the series (15) and (16) using formula (8).

8 Verify the series (17), (18), and (19) using formula (8).

9 Show that the series (18) for the cosine may be obtained by differentiating the series (17) for the sine.

10 Use the first 4 terms of the series (16) with $1/x$ in place of x to evaluate the integral

$$\int_{2}^{2.5} e^{-1/x}\,dx.$$

11 Use the first 3 terms of the series (14) with x^2 in place of x to evaluate the integral

$$\int_{0.2}^{0.5} \log(1+x^2)\,dx.$$

12 Find the first 3 terms of the series of the composite function $\log(\cos x)$. [*Hint:* Note that its derivative is $-\tan x$.]

13 By the method of Example 9 find the first 3 terms of the series for these functions:

(a) $\sin^2 x$ (c) $\dfrac{x^2}{1-x+x^2-x^3}$ (Use long division.)

(b) $e^{-x}\cos x$ (d) $\dfrac{1}{1+e^{-x}}$ (Use (16) and long division.)

14 Same as Exercise 13:

(a) $\cos^2 x$ (c) $\dfrac{\sin x}{1-x^2}$

(b) $\sin x \log(1+x)$ (d) $\dfrac{e^x}{1+x+x^2}$

15 The series for a composite function can be obtained from those of the original pair of functions. For example, $e^{\sin x}$ may be written as

$$\begin{aligned} 1+\sin x+\frac{\sin^2 x}{2!}+\frac{\sin^3 x}{3!}+\cdots \\ = 1+\left(x-\frac{x^3}{3!}+\cdots\right)+\frac{1}{2}\left(x-\frac{x^3}{3!}+\cdots\right)^2 \\ +\left(\frac{1}{3!}\right)\left(x-\frac{x^3}{3!}+\cdots\right)^3+\cdots. \end{aligned}$$

The powers are then arranged in increasing order. Find the series for this function up to terms with powers 3.

16 By the same method as for Exercise 15, find the series for

(a) $e^{(1-\cos x)}$ (c) $(1+\sin x)^{-1/2}$

(b) $\log(1+\sin x)$ (d) $\cos(e^x-1)$

up to terms with powers 4.

17 In this exercise we sketch the proof of **Taylor's formula with integral remainder;** it is a preliminary to the theorem stating the conditions under which the power series really converges to the given function that it represents.

(a) Verify the relation

$$f(x) - f(0) = \int_0^x f'(t)\,dt.$$

(b) Show that

$$\int_0^x f'(t)\,dt = xf'(0) - \int_0^x (t-x)f''(t)\,dt.$$

[*Hint:* Integrate by parts.]

(c) Conclude from (a) and (b) that

$$f(x) = f(0) + xf'(0) - \int_0^x (t-x)f''(t)\,dt.$$

(d) By integration by parts, show that

$$\int_0^x (t-x)f''(t)\,dt = \tfrac{1}{2}x^2 f''(0) - \tfrac{1}{2}\int_0^x (t-x)^2 f^{(3)}(t)\,dt.$$

(e) Conclude from (c) and (d) that

$$f(x) = f(0) + xf'(0) + \tfrac{1}{2}x^2 f''(0) - \tfrac{1}{2}\int_0^x (t-x)^2 f^{(3)}(t)\,dt.$$

(f) Continuing the process of successive integration by parts show that $f(x)$ is equal to the sum of the first n terms of the series (9) **plus** the remainder

$$R_n = \frac{1}{(n-1)!}\int_0^x (t-x)^{n-1} f^{(n)}(t)\,dt.$$

18 Show that the series (9) converges to $f(x)$ if and only if the remainder term, R_n, defined in Exercise 17, converges to 0 as n tends to infinity.

19 In the text we gave a formal proof that the coefficients in (8) are the right ones for the function $f(x)$. Using the same type of proof, show that the coefficients in the Taylor series (20) are also the right ones when the function is represented in powers of $x - x_0$. [*Hint:* Differentiate both sides of (20) and then set $x = x_0$.]

CHAPTER

7

Functions of Two Variables

7.1 Partial Derivatives

In Section 4.1 we defined the concept of function as the relation between a dependent and an independent variable. The concept can be extended to functions of two or more independent variables (and a single dependent variable). We shall consider just the case of two independent variables; the theory for more than two is similar but more complicated in technique and notation.

The letters x and y will be used for independent variables and z for the dependent variable. A functional relation is expressed by the equation

$$z = f(x,y). \tag{1}$$

The graph of such a function is the set of points (x,y,z) in three-dimensional space satisfying Equation (1): it is a **surface.** The domain is contained in the xy-plane. The value z corresponding to a given pair (x,y) is the height (positive or negative) of the point (x,y,z) over the point $(x,y,0)$ in the plane.

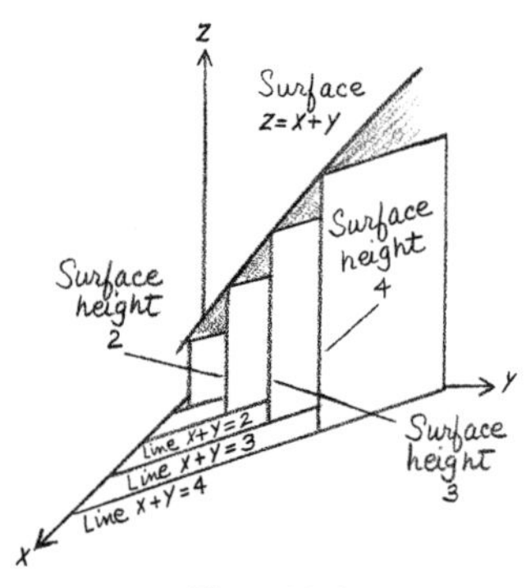

Figure 7.1

EXAMPLE 1

Consider the linear function $f(x,y) = x + y$. The value of the function at a point in the domain is the sum of the coordinates; for example, $f(3,2) = 3 + 2 = 5$. The surface appears in Figure 7.1. It is 5 units above the xy-plane at the point (3,2) in the domain. ▲

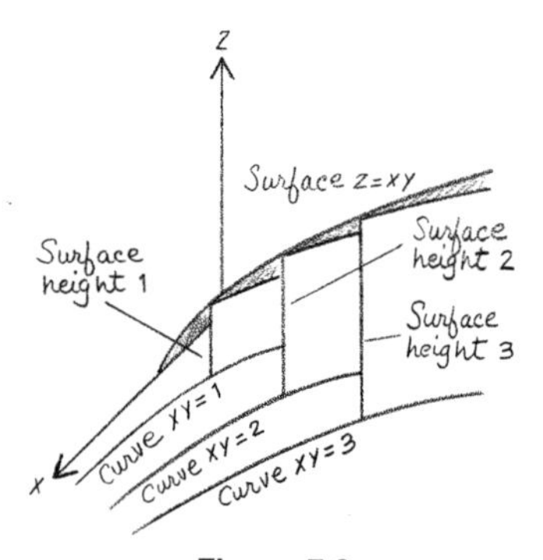

Figure 7.2

EXAMPLE 2

Consider the function $f(x,y) = xy$. The height of the surface above a point in the xy-plane is the product of the coordinates. The portion of the surface in the first octant appears in Figure 7.2. Here we will not discuss the details of surface sketching; however, the reader may confirm the given picture by using the methods of Section 2.3. There the surfaces were sketched by considering their intersections with various planes. ▲

EXAMPLE 3

Surfaces studied in solid analytic geometry may be considered to represent functions of two variables:

(i) the elliptic paraboloid

$$z = \frac{1}{2c}\left(\frac{x^2}{a^2} + \frac{y^2}{b^2}\right),$$

and

(ii) the hyperbolic paraboloid

$$z = \frac{1}{2c}\left(\frac{x^2}{a^2} - \frac{y^2}{b^2}\right).$$

These surfaces were analyzed in Section 2.3. ▲

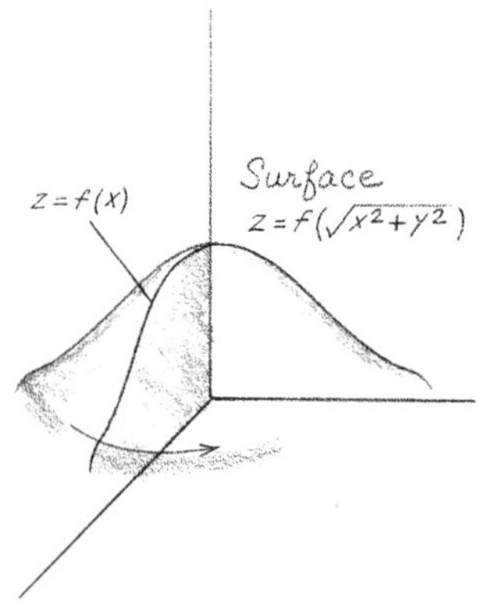

Figure 7.3

EXAMPLE 4

Let $f(x)$ be a function of a single variable with domain on the nonnegative axis. If the curve is drawn in the xz-plane [the graph of $z = f(x)$] and is revolved about the z-axis, then the surface which is generated represents the function

$$z = f\left(\sqrt{x^2 + y^2}\right).$$

Such a function is called **radial** because its values depend on (x,y) only through its distance $\sqrt{x^2 + y^2}$ from the origin: the function value is the same along circles of given radii about the origin (Figure 7.3). The same surface and function are obtained by revolving the curve $z = f(y)$ about the z-axis. ▲

If the variable y is held fixed, then $f(x,y)$ is a function of x alone. Its graph appears in a plane perpendicular to the y-axis, and is formed by the intersection of the function surface and the plane. For example, if the variable y is held fixed, equal to b, then the corresponding function of x, $f(x,b)$, has a graph in the plane $y = b$, formed by its intersection with the surface (Figure 7.4). If $f(x,b)$ is differentiable with respect to x, then the derivative is called the **partial derivative of f with respect to x.** For each fixed value of y, the partial derivative with respect to x is a function of x alone; however, this function may be a **different function** for the various y-values; therefore, the partial derivative depends on both x and y; that is, it is a function of x and y. It is denoted

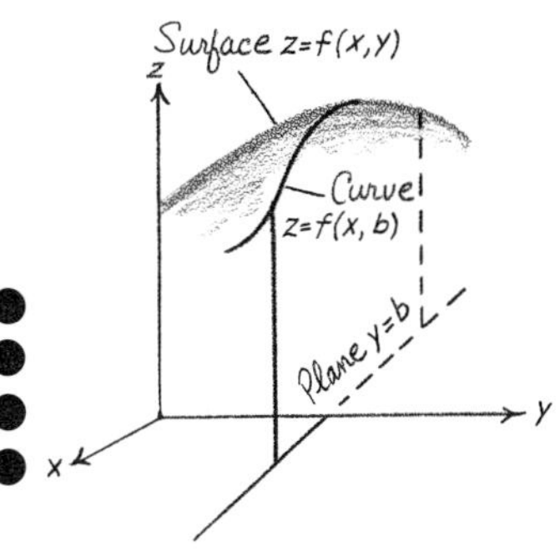

Figure 7.4

$$\frac{\partial f(x,y)}{\partial x}, \qquad \text{or, more briefly,} \qquad \frac{\partial f}{\partial x}.$$

The geometric interpretation of the partial derivative is taken from the case of a single variable. The partial derivative is the slope of the line formed by the intersection of the surface (1) and a plane perpendicular to the y-axis (Figure 7.5).

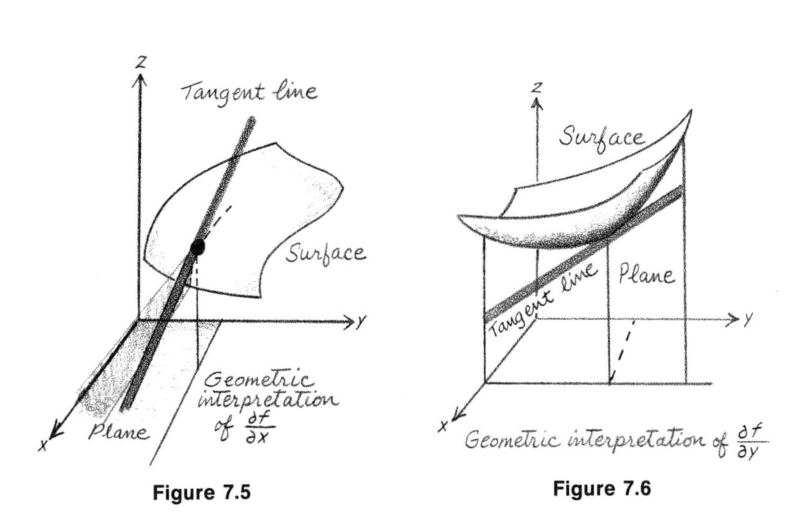

Figure 7.5

Figure 7.6

Similarly, we define the partial derivative with respect to y. For fixed x, $f(x,y)$ is a function of y alone. If it is differentiable with respect to y, then the **partial derivative with respect to y** is denoted

$$\frac{\partial f(x,y)}{\partial y} \qquad \text{or} \qquad \frac{\partial f}{\partial y}.$$

The geometric interpretation of $\partial f/\partial y$ is similar, except that the roles of x and y are interchanged (Figure 7.6).

Higher-order partial derivatives are also defined. The second partial derivative with respect to x is the partial derivative of the partial derivative, and is denoted

$$\frac{\partial}{\partial x}\left(\frac{\partial f(x,y)}{\partial x}\right) \quad \text{or} \quad \frac{\partial^2 f(x,y)}{\partial x^2}, \quad \text{or simply} \quad \frac{\partial^2 f}{\partial x^2}.$$

Similarly, the second partial derivative with respect to y is denoted

$$\frac{\partial}{\partial y}\left(\frac{\partial f(x,y)}{\partial y}\right) \quad \text{or} \quad \frac{\partial^2 f(x,y)}{\partial y^2}, \quad \text{or simply} \quad \frac{\partial^2 f}{\partial y^2}.$$

EXAMPLE 5

Find the partial derivatives of the third-degree polynomial $f(x,y) = x^3 + x^2y - xy^2 + 2y^3$:

$$\frac{\partial f}{\partial x} = 3x^2 + 2xy - y^2, \qquad \frac{\partial^2 f}{\partial x^2} = 6x + 2y,$$

$$\frac{\partial f}{\partial y} = x^2 - 2xy + 6y^2, \qquad \frac{\partial^2 f}{\partial y^2} = -2x + 12y.$$

The higher partial derivatives are computed in the same way. ▲

An entirely new kind of higher derivative arises in the theory of functions of two variables: the **mixed** or **cross** partial derivative. It is obtained by successive differentiation with respect to different variables. If we differentiate $\partial f/\partial x$ with respect to y, we obtain the mixed derivative

$$\frac{\partial}{\partial y}\left(\frac{\partial f}{\partial x}\right) = \frac{\partial^2 f}{\partial y \partial x}.$$

Similarly, if we differentiate first with respect to y and then x, we obtain

$$\frac{\partial}{\partial x}\left(\frac{\partial f}{\partial y}\right) = \frac{\partial^2 f}{\partial x \partial y}.$$

For many functions in calculus the two mixed partial derivatives, obtained by differentiating in both orders, are equal:

$$\frac{\partial^2 f}{\partial x \partial y} = \frac{\partial^2 f}{\partial y \partial x}. \tag{2}$$

A theorem of calculus states that the relation (2) holds if at least one of the above mixed partial derivatives is a **continuous** function of (x,y). Continuity of functions of two variables is discussed in Section 7.3.

EXAMPLE 6

Verify Equation (2) in the particular case of the function $f(x,y) = x^3 + x^2y - xy^2 + 2y^3$, given in Example 5:

$$\frac{\partial}{\partial y}\left(\frac{\partial f}{\partial x}\right) = \frac{\partial}{\partial y}(3x^2 + 2xy - y^2) = 2x - 2y,$$
$$\frac{\partial}{\partial x}\left(\frac{\partial f}{\partial y}\right) = \frac{\partial}{\partial x}(x^2 - 2xy + 6y^2) = 2x - 2y. \blacktriangle$$

EXAMPLE 7

Equation (2) holds if $f(x,y)$ is a product of a function of x and a function of y: If $f(x,y) = f_1(x)f_2(y)$, then

$$\frac{\partial^2 f}{\partial x \partial y} = \frac{\partial f_1}{\partial x}\frac{\partial f_2}{\partial y} = \frac{\partial^2 f}{\partial y \partial x}.$$

Equation (2) is true also if f is the sum of functions which are products of functions of x and y, respectively; indeed, the sum is differentiated term-by-term. ▲

Functions of two or more variables arise frequently in applications, even more than functions of a single variable. The scientist finds that a variable usually depends on two or more factors, not a single one.

EXAMPLE 8

Stay in school and live longer. This seems to be the message of the following data expressing the age-specific death rate as a function of two independent variables: age and amount of formal education. In a fixed educational achievement group, the death rate naturally rises with age. At a fixed age level, the death rate declines as the amount of education increases. The following data are from Socioeconomic Characteristics of Deceased Persons, U.S. 1962–1963 Deaths, National Center for Health Statistics, Ser. 22, No. 9, U.S. Department of Health, Education and Welfare:

Deaths per 1000 Population, 25 Years and Older

	Education		
Age group	Elementary or none	High school	College and more
25–44	3.6	2.1	1.6
45–54	9.8	6.5	5.9
55–64	20.0	15.0	13.6
65+	66.8	53.8	55.1

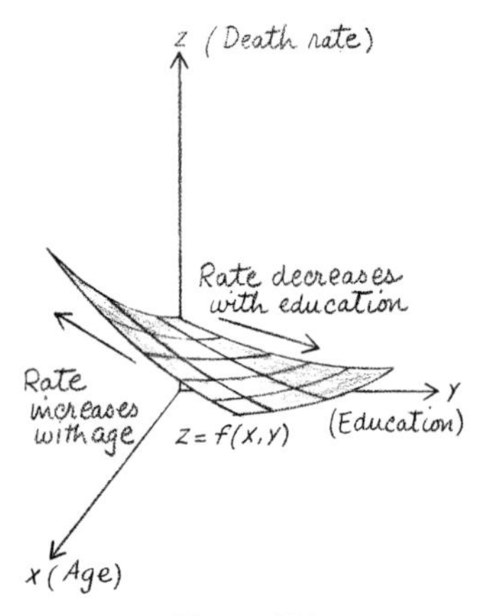

Figure 7.7

Let x represent age, y the number of academic years of formal education, and z the age-specific death rate. The table above furnishes some information about the variation of z with x and y. Here we have converted the factor **education** into a variable **years of formal education.** The characteristic features of the surface of this function are sketched in Figure 7.7. ▲

EXAMPLE 9

The death rate for bronchitis among American males doubled (and more) in the period from 1950 to 1964. Here are the age-specific death rates per 100,000 population, taken from Mortality from Diseases Associated with Smoking, U.S. 1950–1964, National Center for Health Statistics, Ser. 20, No. 4, U.S. Department of Health, Education and Welfare, 1966:

					Age Group			
Year	**All**	**25–34**	**35–44**	**45–54**	**55–64**	**65–74**	**75–84**	**85+**
1950	1.6	0.1	0.2	0.9	3.5	7.0	13.9	51.5
1951	1.6	0.1	0.3	0.9	3.5	7.5	14.4	38.7
1952	1.6	0.1	0.3	1.1	3.3	7.4	14.4	28.9
1953	1.8	0.1	0.2	1.0	4.0	8.1	15.0	32.5
1954	1.6	0.1	0.1	0.7	3.7	7.2	15.3	21.4
1955	1.6	0.1	0.2	0.7	3.9	7.9	12.5	28.9
1956	1.8	0.1	0.2	0.8	4.4	7.8	15.6	29.7
1957	2.0	0.0	0.3	1.1	4.7	10.4	15.5	30.9
1958	2.2	0.1	0.3	1.2	5.7	11.8	17.7	30.8
1959	2.3	0.1	0.3	1.4	5.5	13.0	17.8	34.6
1960	2.6	0.1	0.3	1.8	6.0	15.7	22.5	34.8
1961	2.5	0.1	0.4	1.6	6.0	14.9	22.0	31.1
1962	2.9	0.1	0.3	1.8	7.0	18.0	26.0	41.6
1963	3.5	0.1	0.4	2.1	8.6	23.2	32.6	50.4
1964	3.6	0.1	0.5	2.2	8.6	23.2	33.2	53.2

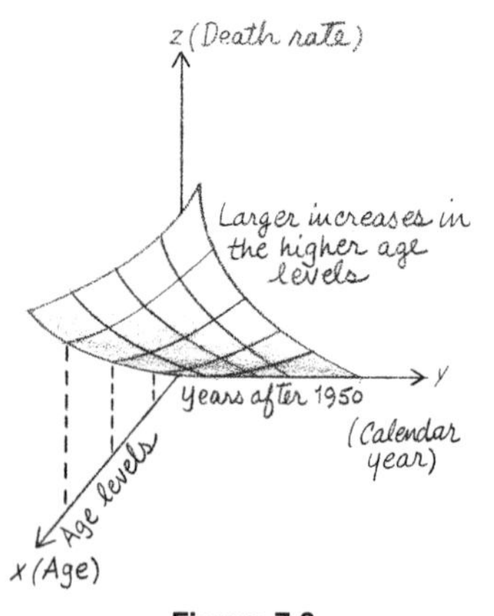

Figure 7.8

Note that there was little change in the death rate for males below the age of 35 over the 15-year period. The greatest increase took place in the groups over 55. The annual increase in the death rate roughly increases with the age level. This has been attributed by some authorities to increased smoking. In this example the independent variables are time (in calendar years) and age. A surface with the properties of this function is sketched in Figure 7.8. ▲

7.1 EXERCISES

In Exercises 1 and 2 find the first partial derivatives with respect to each variable.

1 (a) $f(x,y) = x^2 + 2y^2 + 5x$
(b) $f(x,y) = x^2y^3 + 6x + 5y$
(c) $f(x,y) = e^{xy}$
(d) $f(x,y) = \cos(x + y)$
(e) $f(x,y) = \log(\sqrt{x^2 + y^2})$
(f) $f(x,y) = (x^2 + y^2)^{-1/2}$
(g) $f(x,y) = \arccos \sqrt{x^2 + y^2}$
(h) $f(x,y) = \sin \dfrac{x}{\sqrt{x^2 + y^2}}$

2 (a) $f(x,y) = x^2y + 7x + 3y + 10$
(b) $f(x,y) = x^2y^5 - 5x^3y - 2xy + 11$
(c) $f(x,y) = x/y$
(d) $f(x,y) = \sin xy$
(e) $f(x,y) = \sqrt{x^2 + y^2}$
(f) $f(x,y) = e^{\sqrt{x^2+y^2}}$
(g) $f(x,y) = \arcsin \sqrt{x^2 + y^2}$
(h) $f(x,y) = \dfrac{xy}{\sqrt{x^2 + y^2}}$

In Exercises 3 and 4 find $\partial^2 f/\partial x^2$, $\partial^2 f/\partial y^2$, $\partial^2 f/\partial x \partial y$, *and* $\partial^2 f/\partial y \partial x$ *for the given function* $f(x,y)$:

3 (a) $x^4y + 3x^2y^3 - x + 2$
(b) $3x^2y^4 - 18xy^2 + 5x$
(c) e^{-xy}
(d) $\cos xy$
(e) $\sqrt{x^2 + y^2}$
(f) $\dfrac{x + y}{x - y}$
(g) $\sin(x + y)$
(h) e^{x^2+xy}

4 (a) $(x + y)^4$
(b) $2x^3y^2 - 17x + 3y^2$
(c) $x/(x + y)$
(d) $e^{x^2+y^2}$
(e) $\log \dfrac{x}{\sqrt{x^2 + y^2}}$
(f) $\sqrt{x^2 - y^2}$
(g) $\sin xy$
(h) $e^{x/y}$

7.2 Tangent Planes and Differentials

In Section 2.4 we defined the tangent plane for several surfaces in solid analytic geometry. Now we show how the partial derivatives of a function of two variables are used to define the tangent plane for the surface which represents the function. Let us begin with the intuitive notion of tangent plane. Imagine a smooth, curved surface like that of the shell of an egg; and a straight plane like a pane of glass. Put a small pencil mark on the surface of the egg, and another on the glass. Now bring the egg and the glass together so that they touch exactly at the two marks. In that position, the plane is said to be tangent to the surface at the given point (Figure 7.9).

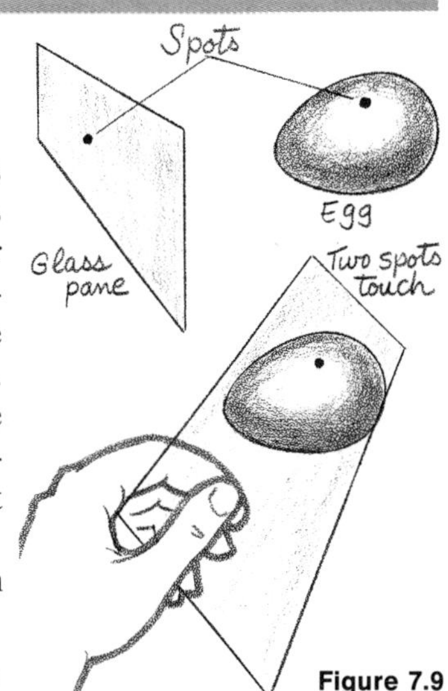

Figure 7.9

Let us recall the general formula for the equation of the plane in three-dimensional space [Section 2.2, Equation (1)]:

$$A(x - x_0) + B(y - y_0) + C(z - z_0) = 0. \quad (1)$$

There is a formula for the equation of the plane tangent to the surface of the function $z = f(x,y)$ at a point (a,b,c). Put

$$\left(\frac{\partial f}{\partial x}\right)_{(a,b)} = \text{Value of } \frac{\partial f}{\partial x} \text{ at } x = a, \quad y = b,$$

$$\left(\frac{\partial f}{\partial y}\right)_{(a,b)} = \text{Value of } \frac{\partial f}{\partial y} \text{ at } x = a, \quad y = b.$$

If the plane tangent to the surface at (a,b,c) is not parallel to the z-axis, then it has the equation

$$z - c = \left(\frac{\partial f}{\partial x}\right)_{(a,b)} (x - a) + \left(\frac{\partial f}{\partial y}\right)_{(a,b)} (y - b). \tag{2}$$

The proof is as follows. An arbitrary plane through (a,b,c) has the equation (1) with $a = x_0$, $b = y_0$, and $c = z_0$. If the plane is not parallel to the z-axis, then the constant C above is not equal to 0; therefore, the equation may be divided through by C, and rewritten as

$$z - c = \left(\frac{-A}{C}\right) (x - a) - \left(\frac{B}{C}\right) (y - b). \tag{3}$$

When this plane intersects the plane $x = a$, it forms the line

$$z - c = -\left(\frac{B}{C}\right) (y - b). \tag{4}$$

If the plane (3) is tangent to the surface, then the line (4) must have the slope $(\partial f/\partial y)_{(a,b)}$ (Section 7.1); therefore,

$$\left(\frac{\partial f}{\partial y}\right)_{(a,b)} = \frac{-B}{C}.$$

By a similar argument with the plane $y = b$, it follows that

$$\left(\frac{\partial f}{\partial x}\right)_{(a,b)} = \frac{-A}{C}.$$

Equation (2) now follows from Equation (3) (Figure 7.10).

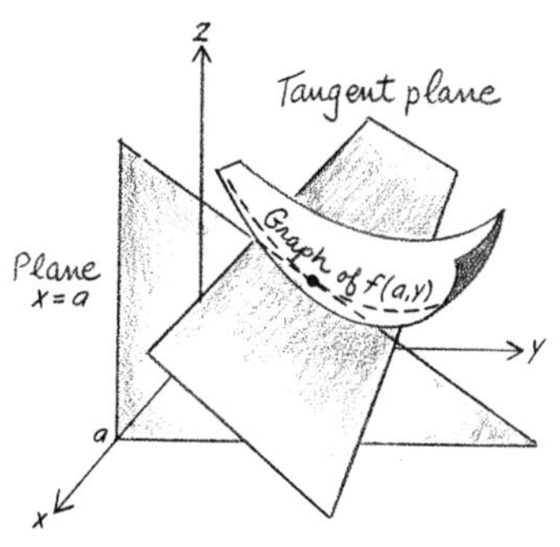

Figure 7.10

EXAMPLE 1

The function

$$f(x,y) = \sqrt{1 - x^2 - y^2}$$

has a surface identical with the half-sphere $x^2 + y^2 + z^2 = 1$ above the xy-plane. (To see this, just solve the equation for z.) The partial derivatives are

$$\frac{\partial f}{\partial x} = \frac{-x}{\sqrt{1 - x^2 - y^2}} \quad \text{and} \quad \frac{\partial f}{\partial y} = \frac{-y}{\sqrt{1 - x^2 - y^2}}.$$

If (a,b,c) is a point on the hemisphere, then c is positive, and

$$\left(\frac{\partial f}{\partial x}\right)_{(a,b)} = \frac{-a}{\sqrt{1 - a^2 - b^2}} = -\frac{a}{c},$$

$$\left(\frac{\partial f}{\partial y}\right)_{(a,b)} = \frac{-b}{\sqrt{1 - a^2 - b^2}} = -\frac{b}{c}.$$

Therefore, by (2), the equation of the tangent plane is

$$z - c = -\left(\frac{a}{c}\right)(x - a) - \left(\frac{b}{c}\right)(y - b)$$

or

$$a(x - a) + b(y - b) + c(z - c) = 0.$$

This is identical with the form of the tangent plane given in Equation (1) of Section 2.4, where $(a,b,c) = (x_0,y_0,z_0)$. ▲

EXAMPLE 2

Find the equation of the plane tangent to the surface of the function

$$f(x,y) = \frac{x}{x + y}$$

at the point $(2,1,\frac{2}{3})$. The partial derivatives are

$$\left(\frac{\partial f}{\partial x}\right)_{(2,1)} = \left(\frac{y}{(x + y)^2}\right)_{(2,1)} = \tfrac{1}{9},$$

$$\left(\frac{\partial f}{\partial y}\right)_{(2,1)} = \left(\frac{-x}{(x + y)^2}\right)_{(2,1)} = -\tfrac{2}{9};$$

thus, the equation of the tangent plane is

$$z - \tfrac{2}{3} = (\tfrac{1}{9})(x - 2) - (\tfrac{2}{9})(y - 1). \ ▲$$

Figure 7.11

The tangent plane serves the function of two variables as the tangent line does for the function of one variable. The surface of a function with partial derivatives is locally indistinguishable from the tangent plane. If one cuts out a very small piece from the surface, it is nearly flat (Figure 7.11). It looks just like the corresponding portions of the tangent planes that touch the piece. The significance of this is that **in analyzing a function in a small neighborhood of a given point we may, as an approximation, replace the formula $z = f(x,y)$ of the function by the formula for the tangent plane.** This is similar to replacing a small piece of the graph of a function of one variable by the tangent line (Section 5.3).

This principle is used in generalizing the method of differentials (Section 5.6) to functions of two variables. Let (a,b,c) be a fixed point on the surface of the function f, and (x^*,y^*,z^*) a second point on the surface, close to the former, so that $c = f(a,b)$ and $z^* = f(x^*,y^*)$. If f were a linear function, then we would have

$$z^* - c = \left(\frac{\partial f}{\partial x}\right)_{(a,b)} (x^* - a) + \left(\frac{\partial f}{\partial y}\right)_{(a,b)} (y^* - b). \tag{5}$$

If f is not linear but has **continuous** (see Section 7.3) first-order partial derivatives around the point (a,b) in the domain, then the right-hand side is approximately equal to the left-hand side for **all** (x^*,y^*) near (a,b). (The continuity of the partial derivatives signifies that the surface is "smooth," so that it is "locally approximable" by a plane, the surface of a linear function.)

The reasoning for (5) is an extension of that for the local linear approximation for a differentiable function of a single variable (Section 5.3). For fixed y^*, we approximate $f(x^*,y^*)$ by a linear function of x^*:

$$f(x^*,y^*) = f(a,y^*) + \left(\frac{\partial f}{\partial x}\right)_{(a,y^*)} (x^* - a) \qquad \text{(approximately)}.$$

Now approximate $f(a,y^*)$:

$$f(a,y^*) = f(a,b) + \left(\frac{\partial f}{\partial y}\right)_{(a,b)} (y^* - b) \qquad \text{(approximately)}.$$

Substitute the latter in the former:

$$f(x^*,y^*) = f(a,b) = \left(\frac{\partial f}{\partial x}\right)_{(a,b)} (x^* - a) + \left(\frac{\partial f}{\partial y}\right)_{(a,b)} (y^* - b) \qquad \text{(approximately)}.$$

This is equivalent to (5).

Equation (5) is the fundamental differential formula for a function of two variables. It states that

$$\textbf{Change in } z = \frac{\partial f}{\partial x} \cdot \textbf{Change in } x + \frac{\partial f}{\partial y} \cdot \textbf{Change in } y. \tag{6}$$

In applications the point (a,b,c) is denoted (x,y,z), and formula (5) is written

$$z^* - z = \frac{\partial f}{\partial x}(x^* - x) + \frac{\partial f}{\partial y}(y^* - y). \tag{7}$$

EXAMPLE 3

Evaluate $\sqrt{x^2 + y^2}$ for $x = 2.98$ and $y = 4.01$. Here $z = f(x,y) = \sqrt{x^2 + y^2}$. The value of z at $x = 3.00$ and $y = 4.00$ is $\sqrt{9 + 16} = 5$. We estimate the difference between this and the desired value of z by means of Formula (7).

The partial derivatives of f at (3,4) are

$$\left(\frac{\partial f}{\partial x}\right)_{(3,4)} = \left(\frac{x}{\sqrt{x^2 + y^2}}\right)_{(3,4)} = \frac{3}{5},$$

$$\left(\frac{\partial f}{\partial y}\right)_{(3,4)} = \left(\frac{y}{\sqrt{x^2 + y^2}}\right)_{(3,4)} = \frac{4}{5}.$$

Apply formula (7) with $(x,y,z) = (3,4,5)$, $x^* = 2.98$, and $y^* = 4.01$, and solve for z^*:

$$\begin{aligned} z^* - 5 &= (\tfrac{3}{5})(2.98 - 3.00) + (\tfrac{4}{5})(4.01 - 4.00) \\ &= 0.008 - 0.012 = -0.004, \end{aligned}$$

and so $z^* = 5 - 0.004 = 4.996$. ▲

EXAMPLE 4

Recall that the volume of a cylinder of radius r and height h is $V = \pi r^2 h$. If the radius is increased by 0.5 percent and the height by 1.0 percent, what is the approximate percentage change in the volume?

By formula (6),

$$\text{Change in } V = \frac{\partial V}{\partial r} \cdot \text{Change in } r + \frac{\partial V}{\partial h} \cdot \text{Change in } h,$$

or

$$\text{Change in } V = 2\pi rh \cdot \text{Change in } r + \pi r^2 \cdot \text{Change in } h.$$

Divide the left-hand side of this equation by V, and the right-hand side by its equivalent, $\pi r^2 h$:

$$\frac{\text{Change in } V}{V} = 2\,\frac{\text{Change in } r}{r} + \frac{\text{Change in } h}{h}.$$

This means that the percentage change in V is equal to twice the percentage change in r plus the percentage change in h. Therefore, the percentage change in V is

$$2(0.5)\% + 1.0\% = 2.0\%. \ \blacktriangle$$

From the "plane approximation" to the surface of a function, there also follows a new chain rule, for functions of two variables. Let $z = f(x,y)$ be a function of x and y. If the latter are themselves functions of a third variable t, we write x as $x(t)$ and y as $y(t)$. If these are substituted in f, then z becomes a function of the single variable t: $z = f(x(t),y(t))$. Let dx/dt, dy/dt, and dz/dt be the derivatives with respect to t, and $\partial z/\partial x$ and $\partial z/\partial y$ the partials of z with respect to x and y. **The chain rule formula is**

$$\frac{dz}{dt} = \frac{\partial z}{\partial x}\frac{dx}{dt} + \frac{\partial z}{\partial y}\frac{dy}{dt}. \tag{8}$$

This is easily verified in the case of a linear function of two variables (Exercise 16). We can explain the formula for a general function $z = f(x,y)$ on the basis of the local linear approximation, as for formula (5).

EXAMPLE 5

Put

$$z = (t^2 + 1)^2(2 - t)^2$$

and find dz/dt. We write this as a simple function of x and y, and the latter as functions of t. Suppose we put $x = t^2 + 1$, and $y = 2 - t$; then we have

$$z = x^2y^2.$$

We need the following partial derivatives in order to apply (8):

$$\frac{\partial z}{\partial x} = 2xy^2, \qquad \frac{\partial z}{\partial y} = 2x^2y, \qquad \frac{dx}{dt} = 2t, \qquad \frac{dy}{dt} = -1.$$

Then

$$\frac{dz}{dt} = 2xy^2(2t) + 2x^2y(-1) = 2xy(2yt - x).$$

Now substitute the given forms of x and y:

$$\frac{dz}{dt} = 2(t^2 + 1)(2 - t)(-3t^2 + 4t - 1). \ \blacktriangle$$

7.2 EXERCISES

In Exercises 1 and 2 find the equation of the plane tangent to the surface at the given point. (See Example 2.)

1

	Surface	Point
(a)	$z = 3 - x^2/9 - y^2/16$	$x = 2,\quad y = 2,\quad z = 83/36$
(b)	$z = xy$	$x = 1,\quad y = 1,\quad z = 1$
(c)	$z = x \log y$	$x = 1,\quad y = 1,\quad z = 0$
(d)	$z = \cos^2 xy$	$x = \pi/2,\quad y = 1/2,\quad z = 1/2$
(e)	$z = y \sin(x + y)$	$x = \pi/2,\quad y = 0,\quad z = 0$
(f)	$z = \sin xy \cos xy$	$x = \pi/4,\quad y = 1,\quad z = 1/2$

2

	Surface	Point
(a)	$z = y^2/9 - x^2/16$	$x = 15,\quad y = 75/4,\ z = 225/9$
(b)	$z = x^2 + y^2$	$x = 1,\quad y = 1,\quad z = 2$
(c)	$z = \sqrt{4 - x^2 - y^2}$	$x = 1,\quad y = 1,\quad z = \sqrt{2}$
(d)	$z = (1 + x)/\cos xy$	$x = 0,\quad y = 0,\quad z = 1$
(e)	$z = e^{x^2+y^2}$	$x = 1,\quad y = -1,\quad z = e^2$
(f)	$z = e^{-x} \sin y$	$x = 0,\quad y = \pi/4,\quad z = \sqrt{2}/2$

In Exercises 3 and 4 find the approximate values of the functions at the indicated points. (See Example 3.)

3

	Function	Point
(a)	$z = \sqrt{x^2 + y^2}$	$x = 3.03,\quad y = 4.02$
(b)	$z = xy^2 + x^3y^4$	$x = 0.01,\quad y = 0.99$
(c)	$z = \log x \log y$	$x = 2.02,\quad y = 3.98$
(d)	$z = \sin xy$	$x = \pi/4 + 0.01,\quad y = 0.99$
(e)	$z = \arcsin\ (x + y)$	$x = 0.1,\quad y = -0.2$
(f)	$z = e^{x+y}$	$x = \log 2 + 0.02,\quad y = -0.01$

4

	Function	Point
(a)	$z = 2x^2y^3$	$x = 1.98,\quad y = 2.02$
(b)	$z = x/(x^2 + y^2)$	$x = 3.98,\quad y = 2.01$
(c)	$z = (\log x)/(1 + y^2)$	$x = e + 0.02,\quad y = -0.01$
(d)	$z = \sin x \cos y$	$x = \pi/4 - 0.03,\quad y = \pi/4 - 0.01$
(e)	$z = y^2 \cos x$	$x = \pi/2 + 0.02,\quad y = 1.01$
(f)	$z = \tan xy$	$x = 0.01,\quad y = 0.99$

In Exercises 5 and 6 find the derivative of the given function with respect to t by means of the chain rule. (See Example 5.)

5 (a) $(3t + 1)^2(2 - 3t)^2$

(b) $(t + 1)^4(t^3 + 3)^2$

(c) $\dfrac{t^2 - 1}{t + 3}$

(d) $\dfrac{\sin t}{1 + \cos^2 t}$

6 (a) $(t+2)^4(3t-4)^3$
(b) $(t^2+1)(t+1)^5$
(c) $\dfrac{t^4-7}{t^2+1}$
(d) $\sin^2 t \cos^4 t$

Exercises 7–15 are solved by using Equation (7).

7 The height of a cone is increased from 10 to 10.02 inches and the radius is decreased from 2 to 1.97 inches. Find the change in volume. (Note $V = \frac{1}{3}\pi r^2 h$.)

8 The inner dimensions of a closed rectangular box with square base are 10, 20, and 20 inches, respectively. The walls are 0.2 inches thick, and weigh 5 ounces per cubic inch. Find the approximate weight of the box.

9 Prove the following result about the "propagation of errors": The relative error in the product of two approximate numbers is (approximately) equal to the sum of the relative errors in each. For the proof, use the method in Example 4.

10 An object floats in water if it is less dense than water. A floating homogeneous object has mass M_1 above the water and mass M_2 below it (Figure 7.12). The ratio of the densities is then shown to be given by

$$D = \frac{M_2}{M_1 + M_2}.$$

In this way the density relative to water can be measured by placing the object in water and observing M_1 and M_2. Suppose that M_1 and M_2 are measured as 3.0 and 4.2 pounds, respectively, with possible errors of 0.05. Find the largest possible error in the ratio D, and the largest possible relative error.

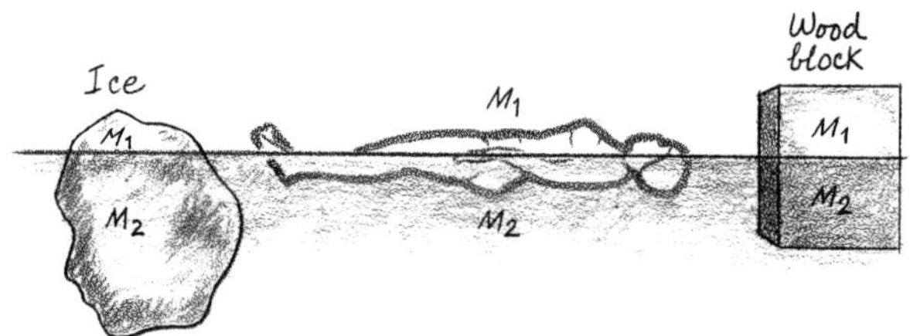

Figure 7.12

11 The output of a firm varies directly as a product of (i) a positive power p of the quantity of labor employed, and (ii) a positive power q of the quantity of capital invested:

$$\text{Ouptut} = \text{Constant } x^p y^q, \qquad \text{where } x = \text{labor}, \quad y = \text{capital}.$$

Find the formula relating the percentage change in output to the percentage changes in labor and capital. [*Hint:* See Example 4.]

12 Telephone calls arrive at the information desk of a large bus terminal at an average rate of m per minute. The calls are completed at an average rate of n per minute. It is shown in queuing theory that if m is less than n, then, under appropriate conditions, the average number of callers waiting for service at a given time is m/n. If the number of calls increases by 3 percent and the number of calls completed by 4 percent, what is the percentage change in the average number of callers waiting for service? In general, if the calls increase by p percent and the completions by q percent, what is the percentage change in the numbers waiting? (See Example 4.)

13 If every individual and corporation in the economy spends a proportion x of every dollar received, then, by the multiplier principle (Section 6.12, Example 2), the amount of money generated by an infusion of y dollars is $y/(1-x)$. If y is increased from 1 to 1.03 million, and x decreased from 0.60 to 0.58, what is the change in the amount of money generated?

14 We refer to the inventory problem in Section 4.1, Example 5. Put

$y =$ Cost of holding 1 unit per year,
$s =$ No. of units sold over the year,
$x =$ Fixed cost of a single shipment.

Then the optimal number of items to order in a single shipment (to minimize the cost of inventory) is given by the formula

$$N = \sqrt{\frac{2sx}{y}}.$$

If x and y increase by 1 and 3 percent, respectively, what is the percentage change in N?

15 Patients arrive at the emergency room of a hospital at a daily average rate of x per hour over the period from 8 A.M. to 10 P.M. Under certain assumptions of randomness, it is shown in the theory of probability that the average proportion of days for which the **first** emergency patient arrives at least y hours after 8 A.M. is given by the formula $A = e^{-xy}$.
(a) If x changes from 4 to 4.05 and y from 0.50 to 0.48, what is the corresponding change in A?
(b) Derive the following formula:
Relative change in $A = -xy$(relative change in x + relative change in y)

16 Verify the chain rule, formula (8), in the particular case where x and y are linear functions of t,

$$x = At + B \qquad \text{and} \qquad y = Ct + D,$$

and where the equation of the function is the equation of a plane,

$$z - c = H(x-a) + G(y-b),$$

where H and G are constants.

17 Find the formula for the direction cosines of the normal to the plane (2) (see Section 2.2).

7.3 Maxima and Minima of Functions of Two Variables

The first and second derivatives were used to determine the extremes of a function of a single variable (Section 5.10). The partial derivatives are now used to locate the extremes of a function of two variables.

First we discuss the typical domains of functions of two variables. The domain is a set in the plane, possibly the whole plane. If the domain has a boundary, it is assumed to be smooth (Figure 7.13). The domain is **closed** if it contains the boundary; it is **bounded** if it can be fitted into some circle of finite radius about the origin (Figure 7.14). The boundary of the domain corresponds to the endpoints of the interval in the single variable case; and a **closed bounded domain** corresponds to a closed finite interval.

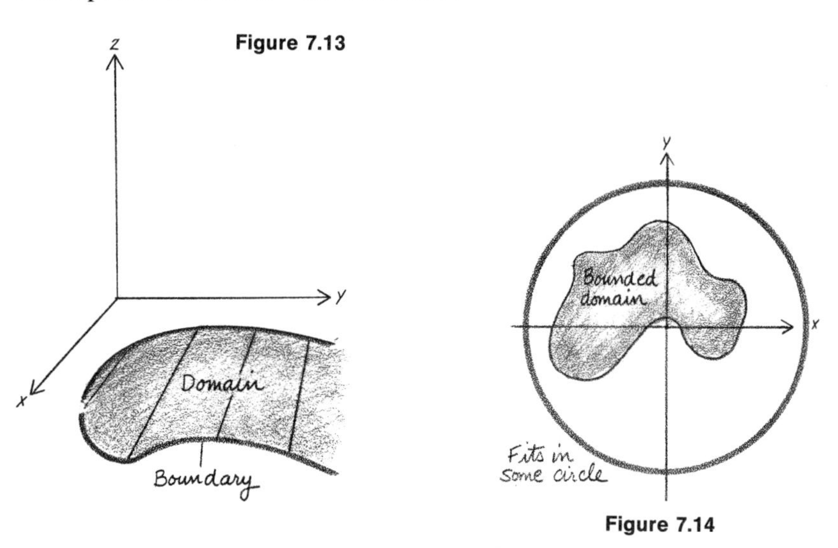

Figure 7.13

Figure 7.14

Let (x,y) be a fixed point in the domain of a function f, and (x^*,y^*) an arbitrary point in the domain, distinct from the first. We say that f is **continuous at the point** (x,y) if $f(x^*,y^*)$ tends to $f(x,y)$ when (x^*,y^*) tends to (x,y), that is, when the distance between the two points tends to 0. This is denoted

$$\lim_{(x^*,y^*)\to(x,y)} f(x^*,y^*) = f(x,y).$$

As for continuity of a function of a single variable (Section 5.10) this means that two values in the range are "close" whenever the corresponding values in the domain are close. A function is **continuous on its domain** when it is continuous at every one of its points (Figure

7.15). Such a function has an unbroken surface, like an untorn sheet. The Extreme Value Theorem also holds for functions of two variables: A continuous function on a closed bounded domain assumes a maximum and a minimum value at some points in the domain (Figure 7.16).

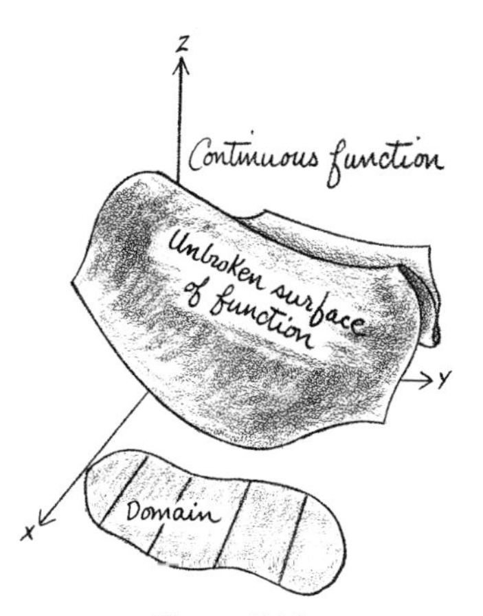

Figure 7.15

The maximum of a function on a domain is usually called the **absolute maximum,** and the minimum the **absolute minimum.** This is to distinguish these from the **relative** maximum and minimum, defined below. If a function with a smooth surface has a maximum or minimum over a domain, these can turn up only at particular kinds of points in the domain: at

1. a **cap,** also called a **relative maximum,**
2. a **cup,** also called a **relative minimum,**
3. a **boundary point;** this corresponds to the endpoint of an interval in the one-variable case,
4. a **peak,** or rough edge. This is usually a point where at least one of the partial derivatives is discontinuous or does not exist. It corresponds to a point of nondifferentiability in the one-variable case (Section 5.10). In all our examples, the partial derivatives will be continuous whenever they are defined, so that the student will have only to check the **existence** of the first partial derivatives.

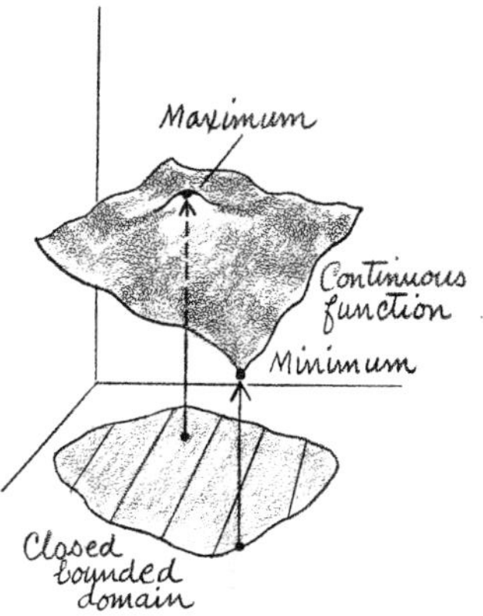

Figure 7.16

A typical surface appears in Figure 7.17; the various kinds of points are located.

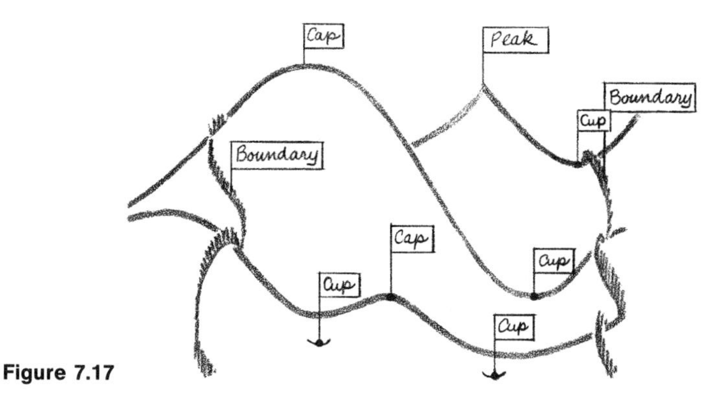

Figure 7.17

To determine the maximum and minimum of a function, we examine its values at the points above. The following criterion is used to locate a cap or cup:

Let $f(x,y)$ have continuous first- and second-order partial derivatives throughout a given set in the domain; this set is assumed to have a "smooth" boundary. Suppose the following conditions hold a given point in the interior (not the boundary) of the set:

(i) $$\frac{\partial f}{\partial x} = 0 \quad \textbf{and} \quad \frac{\partial f}{\partial y} = 0;$$

(ii) the product $$\frac{\partial^2 f}{\partial x^2}\frac{\partial^2 f}{\partial y^2} \quad \textbf{exceeds} \quad \left(\frac{\partial^2 f}{\partial x \partial y}\right)^2.$$

Then there is a

$$\textbf{cap at the point if } \frac{\partial^2 f}{\partial x^2} \textbf{ is negative,}$$

and a

$$\textbf{cup at the point if } \frac{\partial^2 f}{\partial x^2} \textbf{ is positive.}$$

[We remark that the hypotheses of this proposition guarantee that the mixed second-order partial derivative in (ii) is also equal to $\partial^2 f/\partial y\, \partial x$.]

Here is the geometric basis of the criterion. A point where the equations in (i) hold is called a **critical point.** Here the equation of the tangent plane is $z = c$ [Section 7.2, Equation (2)], and the plane is horizontal, that is, perpendicular to the z-axis. Since the tangent plane is horizontal at every cap and cup (Figure 7.18), we may, in seeking such points, restrict our search to critical points. The other conditions of the theorem serve to identify the caps and cups among the critical points. Suppose you slice the surface with a plane perpendicular to the xy-plane. If the surface is cut through a point on the top of a cap, then the cross-sectional curve is concave downward, with a maximum at the cap-point. This is true for every angle at which the surface is cut through the cap-point (Figure 7.18). If the surface is similarly cut at the bottom of a cup, then the cross-sectional curve is concave upward. It can be shown that these geometric conditions are implied by the algebraic conditions in the above criterion (Exercises 17, 18).

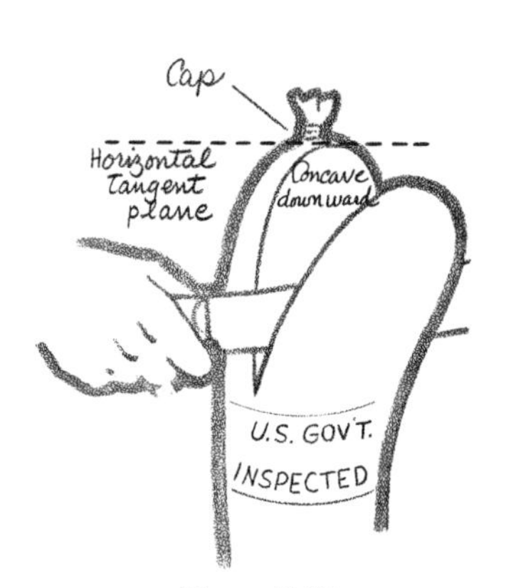

Figure 7.18

If the domain of a function is the entire plane, then there are no boundary points, and therefore, no maxima or minima at points of the type (3). If the partial derivatives exist and are continuous throughout the domain, then there are no "peak" points of the type (4).

EXAMPLE 1

Determine the caps or cups of the function

$$f(x,y) = 9 - (x-1)^2 - (y+2)^2.$$

The first partial derivatives are

$$\frac{\partial f}{\partial x} = -2x + 2, \qquad \frac{\partial f}{\partial y} = -2(y + 2).$$

The critical points are obtained by setting these equal to 0 and solving for x and y: $x = 1$, $y = -2$.

Next we verify condition (ii). The mixed partial derivative is equal to 0:

$$\frac{\partial^2 f}{\partial y \partial x} = \left(\frac{\partial}{\partial y}\right)(-2x + 2) = 0.$$

The second partial derivatives are

$$\frac{\partial^2 f}{\partial x^2} = -2, \qquad \frac{\partial^2 f}{\partial y^2} = -2.$$

The product of these is 4, and certainly exceeds the mixed derivative, which has the value 0.

The second partial derivative with respect to x is negative; hence, by the criterion above, there is a cap at this critical point. ▲

To determine the extremes of a function on a domain, it suffices to consider its values at critical, boundary, and "no-derivative" points.

EXAMPLE 2

Let us determine the extrema of the function

$$f(x,y) = xy - \frac{1}{3}x^3 - \frac{1}{3}y^3$$

over the whole plane. To find the critical points we compute the first partial derivatives:

$$\frac{\partial f}{\partial x} = y - x^2, \qquad \frac{\partial f}{\partial y} = x - y^2.$$

Set each partial equal to 0, and solve the resulting system of equations,

$$y - x^2 = 0, \qquad x - y^2 = 0.$$

Solve for x in the second equation ($x = y^2$) and substitute in the first equation:

$$y - (y^2)^2 = 0, \qquad \text{or} \qquad y(1 - y^3) = 0.$$

The solutions are $y = 0$ and $y = 1$. The corresponding x-values are $x = 0$ and $x = 1$, respectively. It follows that the critical points are (0,0) and (1,1).

Next we check to see if condition (ii) holds at these points. The second-order derivatives are

$$\frac{\partial^2 f}{\partial x^2} = -2x \qquad \frac{\partial^2 f}{\partial y^2} = -2y \qquad \frac{\partial^2 f}{\partial x\, \partial y} = 1.$$

At the point (0,0) we find

$$\frac{\partial^2 f}{\partial x^2}\frac{\partial^2 f}{\partial y^2} - \left(\frac{\partial^2 f}{\partial x\, \partial y}\right)^2 = 0 - 1 = -1,$$

so that (ii) does not hold; hence, (0,0) is neither a cap or cup point. At the point (1,1), we find

$$\frac{\partial^2 f}{\partial x^2}\frac{\partial^2 f}{\partial y^2} - \left(\frac{\partial^2 f}{\partial x\, \partial y}\right)^2 = 4 - 1 = 3,$$

so that (ii) is satisfied. Finally we note that

$$\frac{\partial^2 f}{\partial x^2} = -2$$

at (1,1) so that there is a cap.

We assert that this is the point where the maximum of the function—the absolute maximum—is located. The only other possible points are boundary points and points where the function surface has a rough edge. Now the domain of the function is the whole plane, so that it has no boundary. The surface also has no rough edges because the partials exist everywhere and are continuous. Since there are no boundary points or rough points, the absolute maximum is necessarily at the only cap point, (1,1).

The value of the function at the cap is $f(1,1) = 1/3$. ▲

EXAMPLE 3

The function $\sqrt{x^2 + y^2}$ has an obvious minimum at the origin, where it is equal to 0. There are two ways to verify this by means of our criterion. First we note that the function is positive; hence the function is minimized if and only if the square of the function is minimized. Thus it suffices to minimize $x^2 + y^2$. This is easily done, as in Example 2. Let us also try to verify this by the use of our criterion, without first squaring f. The first partials are

$$\frac{\partial f}{\partial x} = \frac{x}{\sqrt{x^2 + y^2}} \qquad \text{and} \qquad \frac{\partial f}{\partial y} = \frac{y}{\sqrt{x^2 + y^2}}.$$

These fail to exist at the origin because there the numerator and denominator are both equal to 0; hence, there is a rough point on the surface at the origin (Figure 7.19). There are no critical points; if we set both partials equal to 0, the only solution of the set of equations is $x = y = 0$. However, the derivatives are not defined at this point. It follows that the only extreme occurs at the origin; it is a minimum because the function is positive everywhere else. ▲

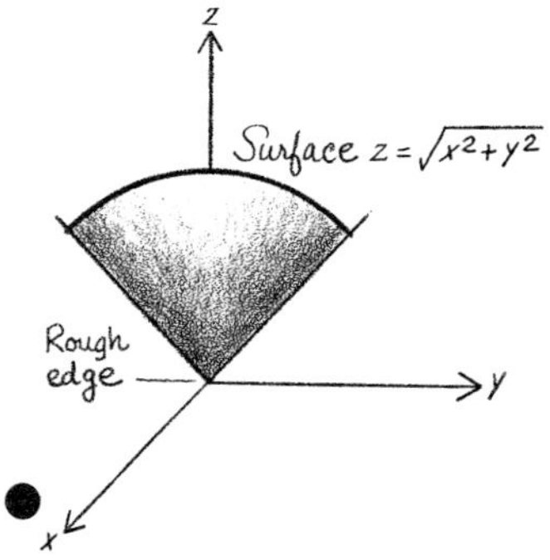

Figure 7.19

Next we introduce a maximum problem with boundary points. ●

EXAMPLE 4

Find the maximum value of the product of 3 nonnegative numbers whose sum is 1. In other words, determine the maximum of xyz subject to $x + y + z = 1$, and x, y, and z all nonnegative. Also, find the values of the variables for which the maximum is attained.

First we reduce this to a problem of minimizing a function of **two** variables. Solve the linear equation $x + y + z = 1$ for z, and substitute in the product xyz, to form a function of x and y:

$$f(x,y) = xy(1 - x - y).$$

Since the variables x and y are nonnegative, and their sum does not exceed 1, the domain of the function is the triangle in the plane bounded by the coordinate axes and the line $x + y = 1$ (Figure 7.20).

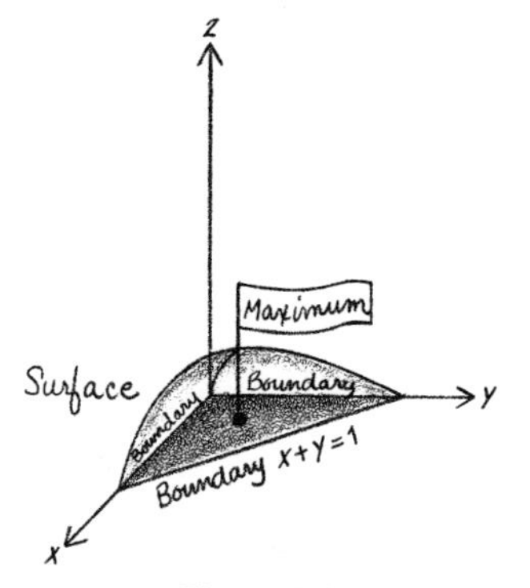

Figure 7.20

The function has the value 0 along the boundary: If a point is on the x-axis, then its coordinate y is 0, and so is $xy(1 - x - y)$. Similarly, it is seen to be equal to 0 along the other two lines forming the boundary.

Let us look for caps inside the triangle. Set the first partial equal to 0, and solve for x and y:

$$\frac{\partial f}{\partial x} = y(1 - 2x - y) = 0,$$

$$\frac{\partial f}{\partial y} = x(1 - x - 2y) = 0.$$

If either x or y is equal to 0, then we get a boundary point; hence, we look only at solutions with both x and y positive, by solving the system

$$1 - 2x - y = 0, \qquad 1 - x - 2y = 0.$$

The solution is $x = y = \frac{1}{3}$.

We check condition (ii) of our criterion.

$$\frac{\partial^2 f}{\partial x^2} = \frac{\partial^2 f}{\partial y^2} = -\frac{2}{3}, \qquad \frac{\partial^2 f}{\partial y\, \partial x} = -\frac{1}{3}.$$

Since $\partial^2 f/\partial x^2$ is negative, there is a cap at $x = \frac{1}{3}$, $y = \frac{1}{3}$. This must be the point where the maximum occurs; indeed, $f(x,y)$ has the value 1/27 at this point, the value 0 on the boundary, and has no rough points because the partial derivatives exist throughout the domain. ▲

EXAMPLE 5

A closed rectangular box is to be constructed with a specified surface area A. What should be the dimensions of the box to maximize the volume?

Let x, y, and z stand for the length, width, and height, respectively. The total area of the 6 faces of the box is $2xy + 2xz + 2zy$ (Figure 7.21). It is specified that the area is equal to A:

$$2xy + 2xz + 2zy = A.$$

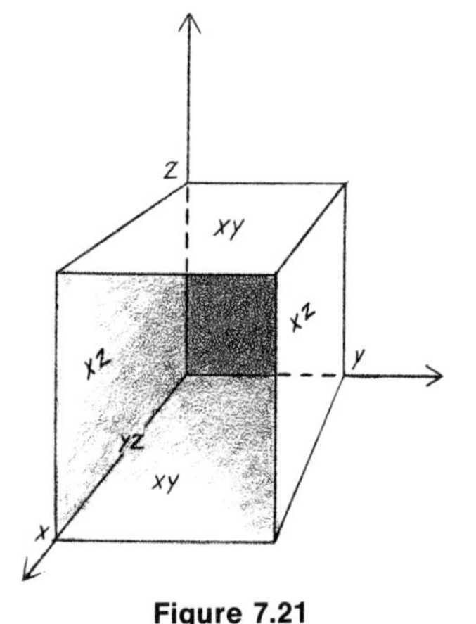

Figure 7.21

The volume of the box is xyz. The volume is minimized when the **square** of the volume is minimized. Now the square of the volume may be written as

$$(xyz)^2 = xy \cdot xz \cdot yz = \frac{2xy}{A} \cdot \frac{2xz}{A} \cdot \frac{2zy}{A} \cdot \frac{A^3}{8}.$$

The linear equation for the area may be written as

$$\frac{2xy}{A} + \frac{2xz}{A} + \frac{2yz}{A} = 1.$$

It follows from the result of Example 4 that the square of the volume is minimized when

$$\frac{2xy}{A} = \frac{2xz}{A} = \frac{2yz}{A} = \frac{1}{3},$$

or, equivalently, $x = y = z = \sqrt{A/6}$; thus, the box should be a cube. The volume of the cube is $xyz = A^{2/3}/6^{2/3}$. ▲

A function does not always have a cap or cup at a critical point. If part (i) of the fundamental critical point criterion holds, then there is a **saddlepoint** if

$$\left(\frac{\partial^2 f}{\partial x\, \partial y}\right)^2 \quad \text{exceeds} \quad \frac{\partial^2 f}{\partial x^2} \cdot \frac{\partial^2 f}{\partial y^2},$$

and a **degenerate point** if these two quantities are equal.

EXAMPLE 6

The function $f(x,y) = xy$ has the first partial derivatives

$$\frac{\partial f}{\partial x} = y, \qquad \frac{\partial f}{\partial y} = x.$$

These are equal to 0 at the origin. We have

$$\frac{\partial^2 f}{\partial x^2} = \frac{\partial^2 f}{\partial y^2} = 0, \qquad \frac{\partial^2 f}{\partial x\, \partial y} = \frac{\partial^2 f}{\partial y\, \partial x} = 1.$$

It follows that there is a saddlepoint at the origin (Figure 7.22). (See Section 2.3.) ▲

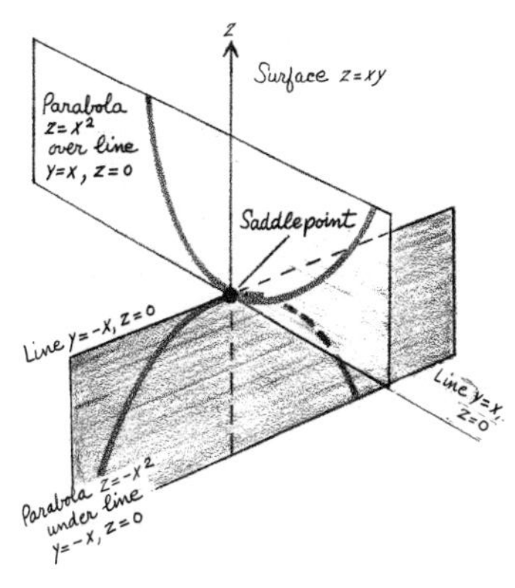

Figure 7.22

EXAMPLE 7

The function of $f(x,y) = x^3 - 3xy^2$ has the first partial derivatives

$$\partial f/\partial x = 3x^2 - 3y^2, \qquad \partial f/\partial y = -6xy.$$

These are both equal to 0 at the origin. The second order derivatives are

$$\frac{\partial^2 f}{\partial x^2} = 6x, \qquad \frac{\partial^2 f}{\partial y^2} = -6x, \qquad \frac{\partial^2 f}{\partial x \partial y} = \frac{\partial^2 f}{\partial y \partial x} = -6y.$$

These are all equal to 0 at the origin; hence, there is a degenerate point, neither a cap nor a cup (Figure 7.23 a,b). ▲

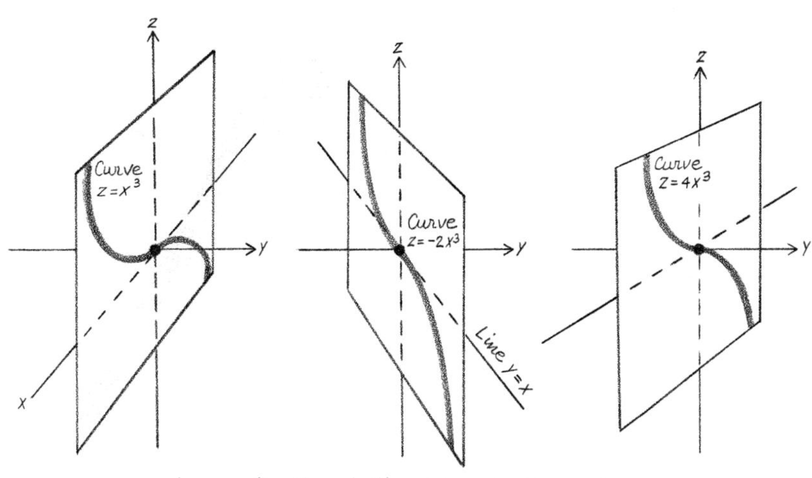

Three routes through the origin over the surface $z = x^3 - 3xy^2$

(a)

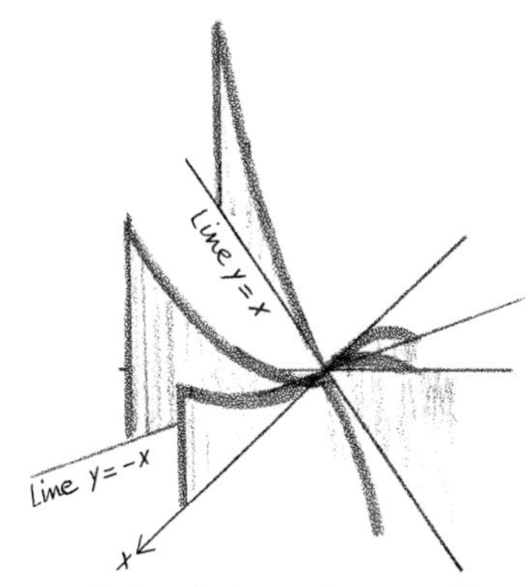

Combined view of three routes

(b)

Figure 7.23

7.3 EXERCISES

In Exercises 1 and 2 locate all caps, cups, saddlepoints, and degenerate points. The given function is $f(x,y)$.

1 (a) $2x^2 + 2xy + 14x - 2y^2 + 22y - 8$
(b) $4x^2 + y^2$
(c) $3x^2 - 2xy - 2x + 3y^2 + 1$
(d) $2x - 6y - x^2 - y^2 - 9$
(e) $x^3 + y^3 - 2xy$
(f) $6x^3y^2 - x^4y^2 - x^3y^3$
(g) $x^3 - 3x^2 + 3x + y^3 - 2xy + 2y - 1$
(h) $x^2y - x^3y + x^2y^2$
(i) $x^4 + y^4 + 32x - 4y + 52$
(j) $x^2y^2 - 5x^2 - 5y^2$
(k) ye^{-x^2}
(l) $\cos(x + y)$

2 (a) $x^2 - 6x + y^2 + 4y + 5$
(b) $xy - y^2 - x^2$
(c) $2x^3 + 16y^3 - 9xy$
(d) $4xy - 2x^2y - xy^2$
(e) $25x + 36y + 49/x + 64/y$
(f) $x^3 + y^3 + 3xy - 3y^2 + 3y - 1 - 3x$
(g) $x^4 + y^5 - 32x - 5y$
(h) $(\frac{1}{3})x^3y^2 - (\frac{1}{2})x^4y^2 + (\frac{1}{3})x^3y^3$
(i) $x^3y - \frac{1}{2}x^2y^2 - x^2y$
(j) $(x - 1)y - (x - 1)^3 - y^3$
(k) $\sin(x + y)$
(l) $\log(1 + x^2 + y^2)$

In Exercises 3 and 4 find the extrema (absolute maxima and minima) of the functions over the given domains:

3 (a) $f = x - y$ on the rectangle bounded by the lines $x = 4$, $y = 2$, and the axes.
(b) $f = x^2 + y^2$ on the region bounded by the ellipse $x^2/9 + y^2/4 = 1$. (To find the extremes on the boundary, solve for y in the equation of the ellipse, substitute in f, and express the latter as a function of x alone on the boundary.)
(c) $f = x^2 - 2y^2$ on the square bounded by the lines $x = 0$, $x = 3$, $y = -1$, $y = 2$.

4 (a) $f = x^2 + y$ on the disk bounded by the circle of radius 1, centered at (0,1). (See Exercise 3b.).
(b) $f = x^2 - x + y^2 - \frac{1}{2}y + \frac{5}{16}$ on the triangle with the vertices (0,0), (0,1), (1,1).
(c) $f = x^{2/3} + y^{2/3}$ on the square bounded by the lines $x = \pm 1$, $y = \pm 1$.

In Exercises 5 and 6 find the point in the plane nearest to the given point P: minimize the square of the distance from P to the point (x,y,z) *satisfying the plane equation.*

5 (a) plane $x + y + z = 1$, point $(0,0,0)$
(b) plane $x + y + z = 1$, point $(1,1,0)$
(c) plane $x - 2y - z = 4$, point $(-1,0,1)$

6 (a) plane $x + y - z = -1$, point $(0,0,0)$
(b) plane $x - 3y - z = -2$, point $(0,3,2)$
(c) plane $2x - y + 2z = 16$, point $(0,0,0)$

7 Find the maximum of the product of 3 nonnegative numbers x, y, and z subject to the condition that $2x + y + z = 1$. (See Example 4.)

8 Generalize the result of Exercise 7: Maximize xyz subject to the condition $ax + by + z = c$, where x, y, and z are nonnegative variables, and a, b, and c are nonnegative constants. (See Example 4.)

9 A large rectangular box is to have an aluminum bottom, cardboard sides, and a plastic top. The cost of aluminum is 1¢ per square foot, cardboard 0.3¢, and plastic 0.6¢. The volume is prescribed to be a certain number V. What dimensions (length, width, height) will minimize the cost of the box? (Let x, y, and z be the dimensions. Express the costs of the walls of the box as functions of these variables; then substitute for z from the volume equation, and minimize the cost, as a function of x and y.)

10 A rectangular bin is designed to have a capacity of 4000 cubic feet and with an open top. What dimensions will minimize the surface area?

11 U.S. Postal Service regulations state that a package sent by parcel post shall have the sum of the length and girth not exceeding 100 inches. Find the dimensions of the rectangular box of maximum volume that can be sent in this way (Figure 7.24).

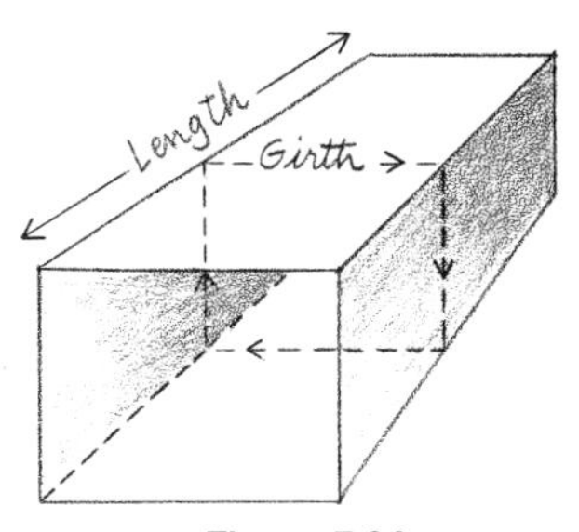

Figure 7.24

12 Generalize the result of Exercise 9: Minimize $Axy + Bxz + Cyz$ subject to the condition $xyz = V$, where A, B, and C are positive constants, and the variables are positive.

13 Three ants are at points in a plane below the surface of the soil; these have the coordinates (2,1), (−1,1), and (1,−2), respectively. They plan to bore straight tunnels to a common point (x,y). The time it takes to clear a tunnel is proportional to the square of the length of the tunnel (see Section 6.4, Exercise 10). At what point should they join the tunnels to minimize the total number of "ant-hours" of digging? (Figure 7.25.)

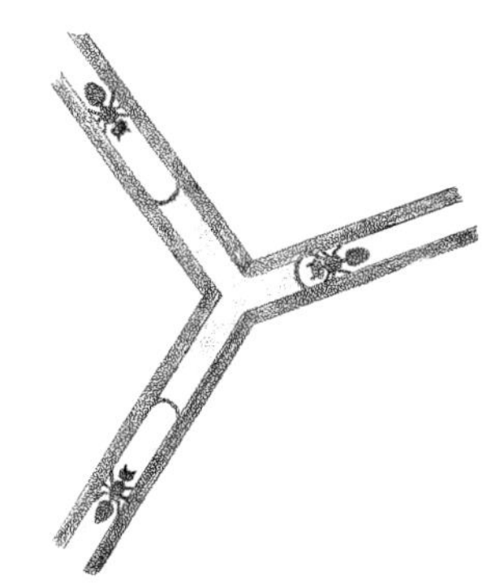
Figure 7.25

14 In Exercise 13, suppose that the ant at (2,1) can dig twice as fast as the other 2. In this case where should they join the tunnels?

15 Generalize the result of Exercise 13: Let there be m ants at given points $(x_1,y_1), \ldots, (x_m,y_m)$. Find the formulas for the coordinates of the optimal meeting point.

16 Generalize the result of Exercise 15: Let there be different speeds for the various ants. The ant at (x_1,y_1) can dig a_1 units of tunnel in 1 time unit, the ant at (x_2,y_2) a_2 units, . . . , and the ant at (x_m,y_m) a_m units. Where should the tunnels meet?

17 Let $f(x,y)$ be a function with first- and second-order partial derivatives and equal mixed derivatives. For a fixed number c, form the composite function $g(x) = f(x,xc)$, which is a function of the variable x alone. Show that its derivative as a function of x is equal to a combination of the partial derivatives of f with respect to x and y:

$$\frac{dg}{dx} = \frac{\partial f}{\partial x} + c\,\frac{\partial f}{\partial y};$$

and that the following relation holds for its second derivative:

$$\frac{d^2g}{dx^2} = \frac{\partial^2 f}{\partial x^2} + 2c\,\frac{\partial^2 f}{\partial x\,\partial y} + c^2\,\frac{\partial^2 f}{\partial y^2}.$$

[*Hint:* Apply the chain rule for composite functions.]

18 We outline the proof of the criterion for a cap; the proof for the cup is similar. For simplicity, we choose the point in the domain as the origin.

(a) Show that if

$$\left(\frac{\partial f}{\partial x}\right)_{(0,0)} + \left(c\,\frac{\partial f}{\partial y}\right)_{(0,0)} = 0,$$

then the curve which is the intersection of the surface of $z = f(x,y)$ with the plane $y = xc$ through the origin has a critical point at the origin. [*Hint:* Use the first result of Exercise 17.]

(b) Show that if

$$\frac{\partial^2 f}{\partial x^2} + 2c\,\frac{\partial^2 f}{\partial x\,\partial y} + c^2\,\frac{\partial^2 f}{\partial y^2} < 0$$

at the origin, then the curve in (a) has a maximum at the origin. [*Hint:* Use the second result of Exercise 17.]

(c) If the hypotheses of (a) and (b) hold for every real number c, then there is a cap at the origin. Why?

(d) Show that the conditions of the cap criterion imply the hypotheses of (a) and (b) above. (For the proof of (b), recall the fact that if the quadratic form in (b) is positive for **every** real number c, then the quadratic equation obtained by setting the form equal to 0 has no real solutions; finally, use the quadratic formula to identify the conditions on the coefficients.)

7.4 Extremal Problems with Constraints and Lagrange's Method

In this section we consider the problem of determining the extreme values of a function $f(x,y)$ of two variables when there is a restriction on the independence of x and y. Such a restriction is given in the form of an equation

$$g(x,y) = 0, \tag{1}$$

where g is a function of two variables. The most direct way of finding the extremes of $f(x,y)$ when its variables satisfy (1) is to solve the equation for one variable in terms of the other, express f as a function of a single variable, and then find its extremes by the method for one variable. However, this is not always practically possible because Equation (1) might be difficult to solve.

Here is a summary of **Lagrange's method** of solving such an extremal problem. (J. L. Lagrange was a French mathematician of the eighteenth and nineteenth centuries.) Instead of considering the function $f(x,y)$, we may consider $f(x,y) - \lambda g(x,y)$, where λ is an arbitrary real number; indeed, the second term is equal to 0, by Equation (1). Then we locate the critical points by setting the partial derivatives of this function equal to 0, and solving these, together with (1), for the unknowns x, y, and λ. The resulting set of equations is known as the set of **Lagrange equations:**

$$\frac{\partial f}{\partial x} = \lambda \frac{\partial g}{\partial x}, \qquad \frac{\partial f}{\partial y} = \lambda \frac{\partial g}{\partial y}, \qquad g(x,y) = 0. \tag{2}$$

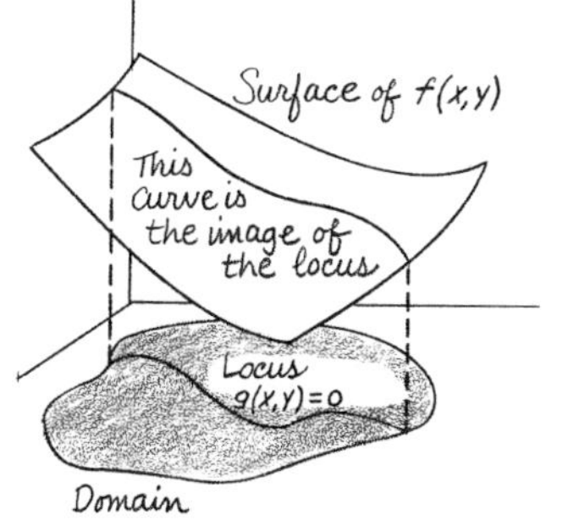

Figure 7.26

The geometric interpretation of the extremal problem with a constraint is portrayed in Figure 7.26. The function is maximized over the locus of points in the domain satisfying Equation (1).

EXAMPLE 1

Find the maximum and the minimum of the linear function $f(x,y) = x + 2y$ over the region bounded by the ellipse

$$\frac{x^2}{9} + \frac{y^2}{4} = 1.$$

The first partial derivatives of f are

$$\frac{\partial f}{\partial x} = 1 \qquad \text{and} \qquad \frac{\partial f}{\partial y} = 2.$$

There are no critical points, or points where the derivatives are undefined; thus, the extreme values of the function are necessarily assumed on the boundary, the ellipse.

Put

$$g(x,y) = \frac{x^2}{9} + \frac{y^2}{4} - 1 = 0,$$

and find the extremes of f subject to (1). The Lagrange equations (2) are

$$\text{(i)} \qquad 1 = \frac{2\lambda x}{9},$$

$$\text{(ii)} \qquad 2 = \frac{\lambda y}{2},$$

$$\text{(iii)} \qquad \frac{x^2}{9} + \frac{y^2}{4} = 1.$$

Divide the first equation by the second and solve for x in terms of y: $x = 9y/8$; then substitute in Equation (iii) and solve for y:

$$\frac{9y^2}{8^2} + \frac{y^2}{4} = 1 \qquad \text{or} \qquad y = \pm\frac{8}{5}.$$

The corresponding values of x are then found by substituting in Equation (iii): $x = \pm 9/5$. It follows that the extremes on the boundary are located at some of the four points $(\pm 9/5, \pm 8/5)$. We compute the values of the function at these points and identify the largest and the smallest:

$$f\left(\frac{9}{5}, \frac{8}{5}\right) = 5, \qquad f\left(\frac{-9}{5}, \frac{-8}{5}\right) = -5,$$

$$f\left(\frac{9}{5}, \frac{-8}{5}\right) = \frac{-7}{5}, \qquad f\left(\frac{-9}{5}, \frac{8}{5}\right) = \frac{7}{5}.$$

The maximum is 5, and it occurs at $(9/5, 8/5)$; and the minimum is -5, at $(-9/5, -8/5)$. ▲

EXAMPLE 2

Find the point in the disk bounded by the circle $x^2 + y^2 = 1$ which is nearest to the point (4,3). This point is the one for which the distance function $\sqrt{(x-4)^2 + (y-3)^2}$ is a minimum; or, equivalently, for which the squared distance

$$f(x,y) = (x-4)^2 + (y-3)^2$$

is a minimum. The partial derivatives are

$$\frac{\partial f}{\partial x} = 2(x-4), \qquad \frac{\partial f}{\partial y} = 2(y-3).$$

The only critical point is (4,3), but it is outside the disk. It follows that there are no cups inside the disk, and also no points where the derivatives are undefined; thus, the minimum is necessarily attained on the circle bounding the disk. Our problem is now to minimize f on the circle, or, equivalently, to minimize f under the condition $g(x,y) = x^2 + y^2 - 1 = 0$. The Lagrange equations are

$$2(x-4) = 2\lambda x,$$
$$2(y-3) = 2\lambda y,$$
$$x^2 + y^2 = 1.$$

Solve for x and y in the first and second equations, respectively: $x = 4/(1-\lambda)$, $y = 3/(1-\lambda)$. Then substitute in the third equation:

$$\left(\frac{4}{1-\lambda}\right)^2 + \left(\frac{3}{1-\lambda}\right)^2 = 1 \quad \text{or} \quad (1-\lambda)^2 = 25.$$

Solving this for λ, we get $\lambda = 6$ and $\lambda = -4$. Now substitute for λ in the equations for x and y:

$$x = \pm\tfrac{4}{5}, \qquad y = \pm\tfrac{3}{5}.$$

The four candidates for the minimum are $(\pm\frac{4}{5},\pm\frac{3}{5})$. The one for which $f(x,y)$ is smallest is $(\frac{4}{5},\frac{3}{5})$; and the minimum value is 4 (Figure 7.27). ▲

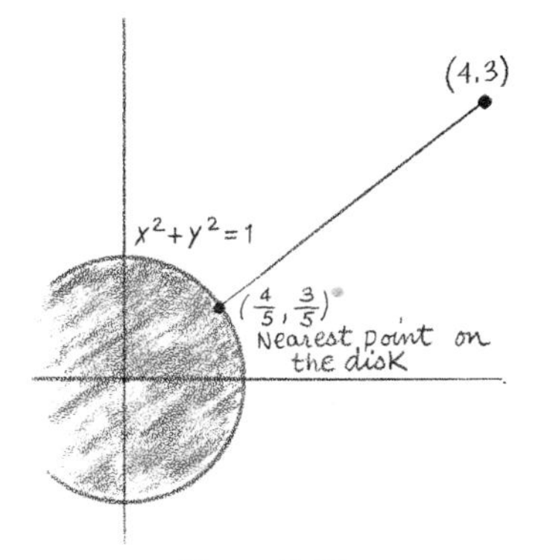

Figure 7.27

EXAMPLE 3

A bone surgeon has to operate on the arm of a patient to treat a subsurface infection. He is to cut 2 small wedges in the infected area. In this problem we consider how the wedges should be cut to minimize the loss of flesh in cutting.

We refer to Figure 7.28, where the cross section of the arm is shown. The points on the surface where the wedges are to start are given. They are to meet at a common point on the surface of the bone. The cross section of each wedge is an isosceles triangle in which the base is $\frac{1}{5}$ the height. The infected area is presumed to have a circular cross section.

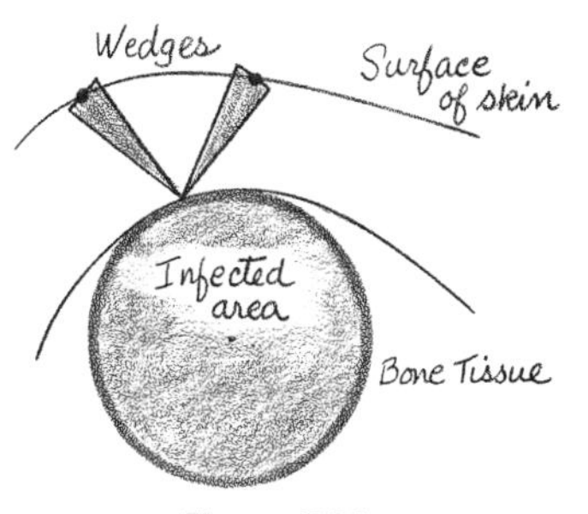

Figure 7.28

We impose a coordinate system. The origin is taken as the center of the circle of the infection, which has radius 1. Let the coordinates of the points from which the cuts are made be denoted (a,b) and (c,d), respectively. The altitudes of the triangles are equal to the distances from the latter points to the point (x,y)

on the circle. The base of the triangle is $\frac{1}{5}$ of the altitude (Figure 7.29). It follows that the sum of the areas of the triangles is

$$(\tfrac{1}{10})[(x-a)^2+(y-b)^2]+(\tfrac{1}{10})[(x-c)^2+(y-d)^2].$$

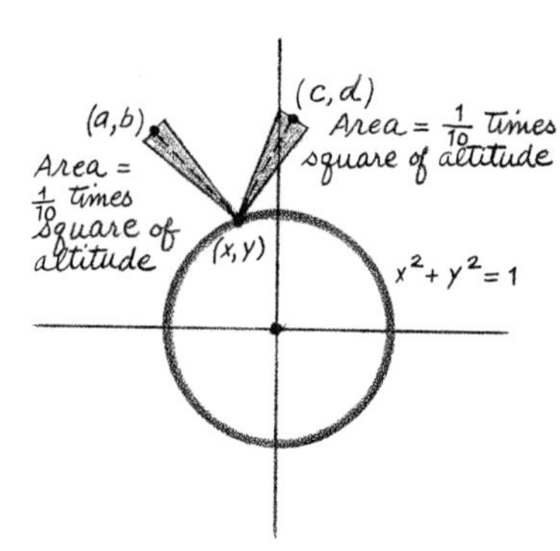

Figure 7.29

We have to minimize this, subject to the condition that (x,y) is on the circle; or, equivalently, minimize

$$f(x,y)=(x-a)^2+(x-c)^2+(y-b)^2+(y-d)^2$$

subject to

$$g(x,y)=x^2+y^2-1=0.$$

The Lagrange equations take the form

$$\begin{aligned}(x-a)+(x-c)&=\lambda x,\\ (y-b)+(y-d)&=\lambda y,\\ x^2+y^2&=1.\end{aligned}$$

Solve the first equation for x and the second for y, in terms of λ:

$$x=\frac{a+c}{2-\lambda},\qquad y=\frac{b+d}{2-\lambda}.$$

Substitute in the third equation, and solve for λ:

$$(a+c)^2+(b+d)^2=(2-\lambda)^2,$$

or

$$\lambda=2\pm\sqrt{(a+c)^2+(b+d)^2}.$$

Solving for x and y, we obtain

$$x=\pm\frac{a+c}{\sqrt{(a+c)^2+(b+d)^2}},\qquad y=\pm\frac{b+d}{\sqrt{(a+c)^2+(b+d)^2}}.$$

There are four possible points where the minimum may occur; the actual minimizing point depends on the particular values of the given numbers a, b, c, and d.

As a numerical illustration, suppose that the wedges are to be cut from the points (1,2) and (0,3). Here $a = 1$, $b = 2$, $c = 0$, and $d = 3$, and

$$x = \frac{\pm 1}{\sqrt{26}}, \qquad y = \frac{\pm 5}{\sqrt{26}}.$$

It is geometrically evident, and directly verified, that the point minimizing f is the one with positive coordinates, $(1/\sqrt{26}, 5/\sqrt{26})$. ▲

7.4 EXERCISES

In Exercises 1 and 2 find the maximum and minimum of the given function over the indicated domain. If one or the other does not exist, then state so.

1 (a) $f(x,y) = 2x - x^2 + 4y + y^2$ on the line $x + 2y = 2$
(b) $f(x,y) = xy$ on the circle $x^2 + y^2 = 4$ [*Hint:* $\lambda = \pm\frac{1}{2}$]
(c) $f(x,y) = (x - 3)^2 + (y - 4)^2$ on the disk bounded by $x^2 + y^2 = 1$
(d) $f(x,y) = xy$ on the region bounded by the triangle with vertices (0,0), (0,1), (1,0)
(e) $f(x,y) = (x^2 + y^2)^{-1/2}$ on the half-space below the line $x + 2y = -3$ [*Hint:* Maximize $x^2 + y^2$]

2 (a) $f(x,y) = xy + y^2$ on the line $2x + y = 1$
(b) $f(x,y) = y^2 - x^2$ on the disk bounded by $x^2 + y^2 = 1$ [*Hint:* $\lambda = \pm 1$]
(c) $f(x,y) = x - y$ on the ellipse $x^2/4 + y^2/25 = 1$
(d) $f(x,y) = 2x^2 + y^2$ on the region bounded by the triangle with vertices (0,0), (1,0), (0,3)
(e) $f(x,y) = (x - 4)^2 + y^2$ on the circle $x^2 + (y - 3)^2 = 1$

3 Suppose that the cost of boring a tunnel in the side of a mountain is proportional to the square of the length of the tunnel. Two tunnels are dug from points on the side of the mountain and are to meet at a point on a circular core inside the mountain. The tunnels are dug from the points (4,5) and (2,3). The circle has the equation $x^2 + y^2 = 1$. To what point along the circle should the tunnels be dug to minimize the sum of the costs of the 2 tunnels?

4 Find the point in the region bounded by the coordinate axes and the line $2x + y = 3$ that is closest to the point (2,1).

5 Find the points on the circle $x^2 + y^2 = 36$ that are closest and farthest, respectively, from (2,−3).

6 This is an example of an economic allocation problem solved by means of the Lagrange equations. Suppose that the output of a firm is proportional to the product of (a) the square root of the number of supervisors, and (b) the number of regular employees. For simplicity, suppose that all supervisors receive the same salary, and all regular employees the same salary. Furthermore, we suppose that the ratio of a supervisor's salary to an employee's salary is 5 to 4. If the company has a fixed amount of money for the payroll, what should be the ratio of supervisors to employees for maximum output?

7 Generalize the result of Exercise 6. Suppose the output is proportional to $x^p y^q$, where x and y are two factors of production, and p and q positive numbers. Let r be the ratio of the unit cost of x to that of y. Find the formula for the ratio of units of x to y that should be applied to maximize the output. Assume as before that there is a fixed amount available to pay for the cost of the two factors.

8 Extend Example 3 above (bone surgery) to the case of 3 wedges cut from points (x_1,y_1), (x_2,y_2), and (x_3,y_3).

9 Suppose in Example 3 that the base of the wedge from (a,b) is $\frac{1}{5}$ of the altitude, but the base of the other wedge is $\frac{3}{5}$ the altitude. How is the area minimized in this case?

10 Let $f(x,y)$ be an arbitrary linear function: $f(x,y) = ax + by$, where a and b are constants. Why are the maximum and minimum values, when they exist, assumed on the boundary of the domain? If a line segment is contained in the domain, why are the extremes of the function on the segment assumed at the endpoints of the segment? These results can be used to prove the validity of the method of extreme points in the solution of linear programming problems (Section 3.9).

7.5 Double Integrals and Bivariate Densities

Recall that integration is the reverse of differentiation for functions of a single variable. Double integration of functions of two variables is the operation reversing second-order mixed partial differentiation.

Let $F(x,y)$ be a function of x and y, and $f(x,y)$ its mixed partial derivative:

$$f(x,y) = \frac{\partial^2 F(x,y)}{\partial x\, \partial y} = \frac{\partial^2 F(x,y)}{\partial y\, \partial x}.$$

Then $F(x,y)$ is the **double integral** of f. As in the one-variable case, the integral is not unique: it has two **additive functions** (in the place of an additive constant). $F(x,y)$ is not the only function whose derivative is f; another is

$$F(x,y) + G_1(x) + G_2(y),$$

where G_1 and G_2 are arbitrary differentiable functions of x and y, respectively. The reader can check this by differentiation: G_1 is eliminated by the differentiation with respect to y, and G_2 by that with respect to x. These arbitrary functions will have no role in our computations because we shall be concerned only with **definite** integrals. Just as the arbitrary constant C is ignored in the calculation of the single definite integral, so are G_1 and G_2 ignored in the calculation of the double definite integral.

This definition of "double integral" is not as general as the usual definition; ours is commonly called the **iterated integral.** At this elementary level, the double integral is always computed by reduction to an iterated integral. For this reason we have chosen to avoid the real distinction between the two concepts.

EXAMPLE 1

Find the double integral of $f(x,y) = x + x^2y$, disregarding the arbitrary functions. Integrate first with respect to x, holding y fixed:

$$\int (x + x^2y)\,dx = \tfrac{1}{2}x^2 + \tfrac{1}{3}x^3y.$$

Now hold x fixed, and integrate with respect to y:

$$\int (\tfrac{1}{2}x^2 + \tfrac{1}{3}x^3y)\,dy = \tfrac{1}{2}x^2y + \tfrac{1}{6}x^3y^2.$$

This can be checked by partial differentiation. The same integral is obtained by integration first with respect to y and then with respect to x. ▲

EXAMPLE 2

The double integral of $f(x,y) = \sin(x + y)$ is $-\sin(x + y)$: If we integrate first with respect to x, holding y fixed, then

$$\int \sin(x + y)\,dx = -\cos(x + y);$$

and then integrating over y, we obtain

$$-\int \cos(x + y)\,dy = -\sin(x + y).$$

The same result is obtained when the order of integration is reversed. ▲

Let us show how the definite integral is calculated, and why the arbitrary functions G have no part. Fix x and consider $f(x,y)$ as a function of y alone. Its definite integral over an interval from c to d is

$$\int_c^d f(x,y)\,dy.$$

Consider the integral to be now a function of x, and integrate over an interval from a to b:

$$\int_a^b \left(\int_c^d f(x,y)\,dy\right) dx.$$

This is written without parentheses as

$$\int_a^b \int_c^d f(x,y)\,dy\;dx. \tag{1}$$

If we integrate first with respect to x and then with respect to y, we obtain

$$\int_c^d \left(\int_a^b f(x,y)\,dx\right) dy,$$

which we write as

$$\int_c^d \int_a^b f(x,y)\,dx\;dy. \tag{2}$$

The two double integrals (1) and (2) are equal for almost all functions f arising in practical problems. In examples and exercises, they should be assumed equal.

EXAMPLE 3

Evaluate

$$\int_0^2 \int_1^3 (x^2y + xy^2)\,dy\;dx.$$

Integrate first over y:

$$\int_1^3 (x^2y + xy^2)\,dy = \left(\tfrac{1}{2}x^2y^2 + \tfrac{1}{3}xy^3\right)\Big|_1^3 = 4x^2 + \frac{26x}{3}.$$

Now integrate over x:

$$\int_0^2 \left(4x^2 + \frac{26x}{3}\right) dx = \tfrac{1}{3}(4x^3 + 13x^2)\Big|_0^2 = 28.$$

The same value is obtained by integrating first over x and then over y. Note that the arbitrary G-functions have no role in the computation. ▲

While definite integrals of nonnegative functions of a single variable are considered areas, the double definite integrals of nonnegatives functions of two variables are **volumes.** If $f(x,y)$ is nonnegative, then the definite integral (1) represents the volume under the surface $z=f(x,y)$ and over the rectangle in the xy-plane bounded by the lines $x=a$, $x=b$, $y=c$, and $y=d$ (Figure 7.30). We give the geometric basis of this assertion.

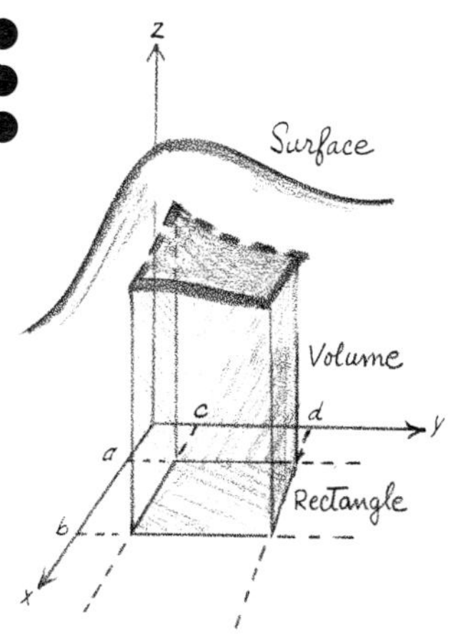

Figure 7.30

Imagine the rectangle cut into many narrow strips parallel to the y-axis. The volume under the entire surface is the sum of the volumes over the strips. Each of the latter volumes is like a thin slab, and its measurement is approximately equal to the product of the lateral area of one side of the slab and the thickness of the slab. The lateral area of a slab running along the line from (x,c) to (x,d) is the definite integral

$$\int_c^d f(x,y)\,dy;$$

indeed, this is the formula for the area under the curve $z=f(x,y)$ where x is fixed. The thickness of the slab is the width of the strip, which is equal to the length Δx of the interval on the x-axis (Figure 7.31). It follows that the volume of the slab is approximately equal to the product of the lateral area and the thickness:

$$\int_c^d f(x,y)\,dy \cdot \Delta x. \tag{3}$$

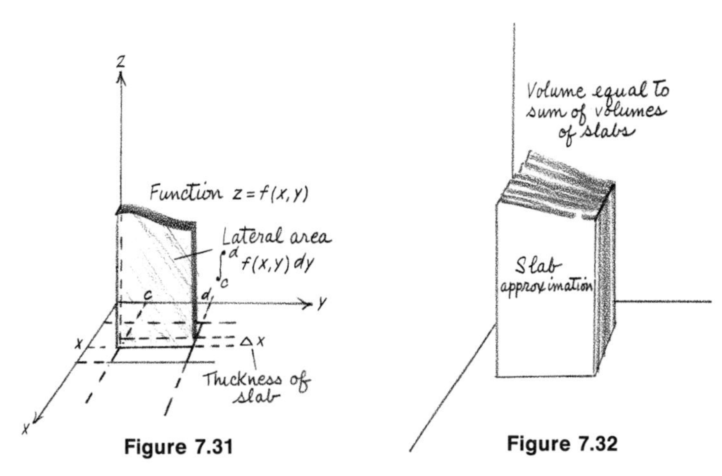

Figure 7.31

Figure 7.32

The total volume is the sum of the slab volumes (Figure 7.32). This sum is approximately equal to the integral of the lateral area "function" from a to b:

$$\int_a^b \left(\int_c^d f(x,y)\,dy\right) dx.$$

The reasoning is similar to that for defining the area under a curve as an integral (Section 6.3). The thickness Δx of the slab plays the role of the length of the base of the rectangle, and the lateral area,

$$\int_c^d f(x,y)dy,$$

plays the role of the function with the given graph.

This procedure for measuring the volume is similar to what you would do to measure the volume of a loaf of bread. The bread may be cut into slices. The volume of each slice is approximately the product of the lateral area and the thickness. The volume of the whole loaf is then obtained as the sum of the volumes of the individual slices (Figure 7.33).

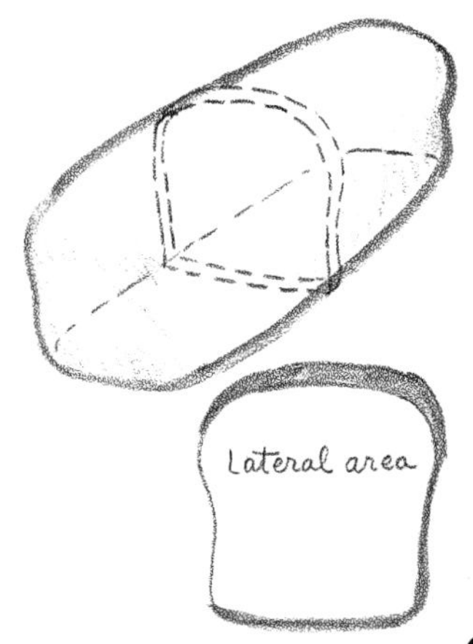

Figure 7.33

A **bivariate** numerical characteristic of a population is a **pair** of numbers associated with each member of the population. A **bivariate density function** is often used to describe the distribution of the bivariate characteristic in the population; this is a generalization of the density function of one variable (Section 6.2). A function $f(x,y)$ is a bivariate density if it satisfies the following two conditions:

(i) It is nonnegative.

(ii) The double improper integral

$$\int_{-\infty}^{\infty}\int_{-\infty}^{\infty} f(x,y)\,dy\,dx$$

converges and is equal to 1. The double improper integral is considered to be the improper integrals over x and y, respectively.

EXAMPLE 4

Consider the function

$$\begin{aligned} f(x,y) &= 0, && \text{for all } x \le 0 \text{ or } y \le 0,\\ &= xe^{-x(1+y)}, && \text{for both } x > 0 \text{ and } y > 0. \end{aligned}$$

This is positive only in the first quadrant of the plane; therefore, the double improper integral reduces to that over the first quadrant,

$$\int_0^{\infty}\int_0^{\infty} xe^{-x(1+y)}dy\,dx.$$

Let us verify that this is a bivariate density. It is certainly nonnegative. Let us evaluate the double integral by integration first over y and then over x:

$$\int_0^\infty xe^{-x}e^{-xy}dy = e^{-x}\int_0^\infty xe^{-xy}\,dy = e^{-x}(-e^{-xy})\Big|_0^\infty = e^{-x},$$

$$\int_0^\infty e^{-x}\,dx = 1.$$

We get the same result by integrating first with respect to x and then with respect to y. By integration by parts with respect to x, with y fixed, we obtain

$$\int xe^{-x(1+y)}\,dx = -\frac{x}{1+y}\,e^{-x(1+y)} - \frac{1}{(1+y)^2}\,e^{-x(1+y)} + C.$$

It follows that

$$\int_0^\infty xe^{-x(1+y)}\,dx = \frac{1}{(1+y)^2}.$$

Finally, by integration over y, we obtain

$$\int_0^\infty \frac{1}{(1+y)^2}\,dy = 1. \blacktriangle$$

The distribution of a bivariate characteristic can be portrayed by means of a **two-dimensional histogram,** or volume graph. Let x and y be the components of the bivariate characteristic. The volume graph is a collection of flat rectangular surfaces over the plane. The volume under a surface over a portion of the plane is equal to the proportion of the population whose pairs (x,y) are in that part of the plane. A bivariate density function $f(x,y)$ is considered to describe the bivariate characteristic if the volumes under its surface $z=f(x,y)$ are approximately equal to the corresponding volumes under the volume graph. In other words, the integral (1) is approximately equal to the proportion of the population whose x-characteristic is between a and b inclusive, and whose y-characteristic is between c and d inclusive.

EXAMPLE 5

An interdisciplinary area of current interest is the study of sociological factors in medical care. A U.S. Government investigation was concerned with the relation between poverty and prenatal care of infants. Legitimate live births in the United States in 1963 were classified according to two characteristics: the family income and the weight of the newborn. It is presumed that the weight of the newborn is an indicator of prenatal care: a substantial weight indicates that the fetus was well nourished during pregnancy. In the table below the percentage distribution of births according to the two factors for mothers in the age group 20–24 is shown; this has been calculated from the data in Variations in Birth Weight, Legitimate Live Births, U.S., 1963, National Center for Health Statistics, Ser. 22, No. 8, U.S. Department of Health, Education and Welfare, Table 6.

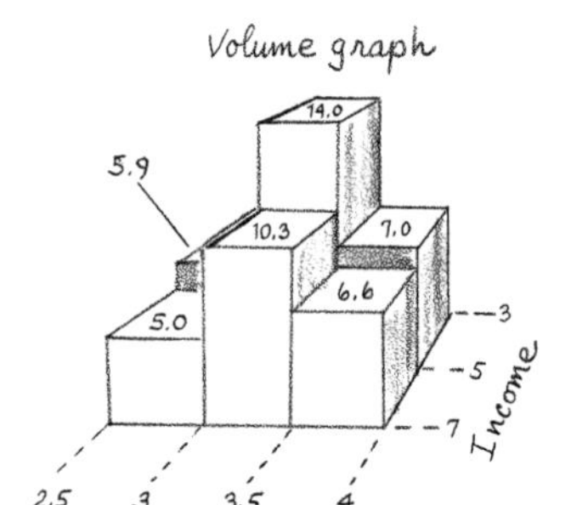

Figure 7.34

Percentages of Births

Income ($1000's)	Birth weight (thousands of grams)				
	below 2.5	**2.5–3**	**3–3.5**	**3.5–4**	**4+**
Below 3	2.4	4.5	9.4	5.5	0.9
3–5	2.4	5.9	14.0	7.0	2.5
5–7	1.7	5.0	10.3	6.6	1.9
7+	1.0	4.0	7.4	4.8	1.3
Unknown:	1.8				

The volume graph appears in Figure 7.34. Income is measured along the x-axis, and weight along the y-axis. The open-ended classes are not included.

There are many possible functions $f(x,y)$ that can serve as the density for these data. We present one such function:

$$f(x,y) = 0, \qquad \text{for } x \leq 0,$$

$$= \frac{x}{16} e^{-x^2/32} \cdot \frac{3.4e^{-3.4(y-3.25)}}{[1 + e^{-3.4(y-3.25)}]^2}, \qquad \text{for } x > 0. \qquad (4)$$

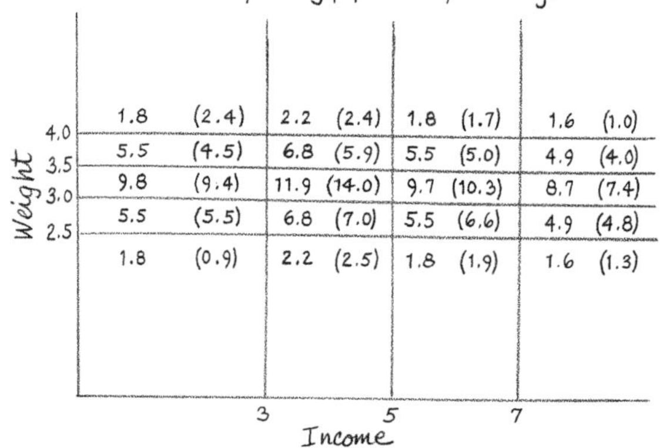

Figure 7.35

It is left to the reader to verify that this function has the two characteristic properties of a bivariate density (Exercise 3). The volumes under the various portions of this surface are very close to the corresponding population proportions. These are compared in the chart in Figure 7.35; the population percentages are given there in parentheses.

Let us show how these volumes are computed from formula (4). The double integral is relatively easy to compute because the integrand is a product of functions of x and y. As an example, we now calculate the volume under f for the income bracket 5–7 and the weight bracket 3.5–4,

$$\int_5^7 \int_{3.5}^4 f(x,y)\,dy\,dx.$$

Integrate first over y. The second factor in (4) is the derivative of

$$\frac{1}{1+e^{-3.4(y-3.25)}};$$

hence,

$$\int_{3.5}^4 f(x,y)\,dy = \left(\frac{x}{16}\right) e^{-x^2/32} \left.\frac{1}{1+e^{-3.4(y-3.25)}}\right|_{3.5}^4$$
$$= \left(\frac{x}{16}\right) e^{-x^2/32}(0.2269).$$

The product of the first 2 factors in the last expression is the derivative of

$$-e^{-x^2/32};$$

hence, integrating over x, we obtain

$$(0.2269)\left(-e^{-x^2/32}\right)\Big|_5^7 = (0.2269)(24.19) = 0.0549$$
$$\approx 0.055.$$

The other volumes are computed in a similar way. ▲

If $f(x,y)$ is a bivariate density, then the improper integral

$$\int_{-\infty}^{\infty} f(x,y)\,dy = f_1(x) \tag{5}$$

is a function of x. f_1 is also a density—of a single variable. It is non-negative and its integral is 1:

$$\int_{-\infty}^{\infty} f_1(x)\,dx = \int_{-\infty}^{\infty}\int_{-\infty}^{\infty} f(x,y)\,dy\,dx = 1.$$

It is called the **marginal** density of the x-component of the bivariate characteristic. Similarly, the integral

$$\int_{-\infty}^{\infty} f(x,y)\,dx = f_2(y) \tag{6}$$

is the marginal density of the y-component. f_1 and f_2 are densities for the marginal totals of the population percentages.

EXAMPLE 6

(Continuation of Example 5.) If we sum across the rows of the two-way table of birth percentages, we get the marginal income totals:

Income	Below 3	3–5	5–7	7+
Percent	22.7	31.8	25.5	18.5

The marginal density of x, obtained from the bivariate density (4), is

$$f_1(x) = \int_{-\infty}^{\infty} f(x,y)\,dy = 0, \qquad \text{for } x \leq 0,$$

$$= \left(\frac{x}{16}\right) e^{-x^2/32}, \qquad \text{for } x > 0.$$

This density fits the income distribution above. The areas under the curve corresponding to the income brackets are found by integration to be

Income	Below 3	3–5	5–7	7+
Area · 100	24.42	29.74	24.19	21.65

These are compared with the marginal totals in Figure 7.36.

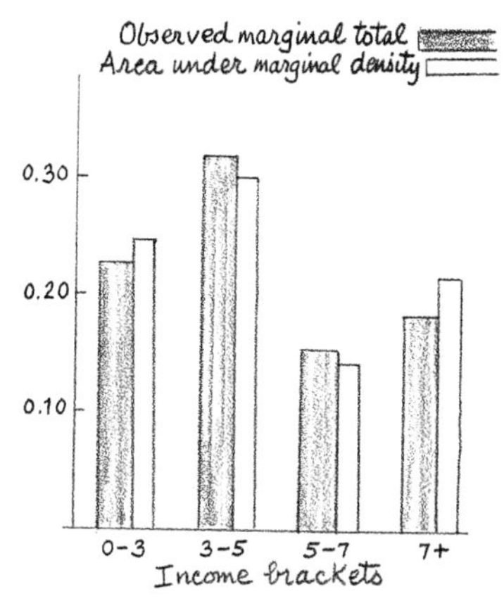

Figure 7.36

The marginal density of y is

$$f_2(y) = \int_{-\infty}^{\infty} f(x,y)\,dx = \frac{3.4e^{-3.4(y-3.25)}}{[1 + e^{-3.4(y-3.25)}]^2}.$$

The areas under this density fit the corresponding marginal weight totals (see Figure 7.37):

Weight	Below 2.5	2.5–3	3–3.5	3.5–4	4+
Marginal total	7.5	19.4	41.1	23.9	6.6
Area under density	7.25	22.69	40.13	22.69	7.24

▲

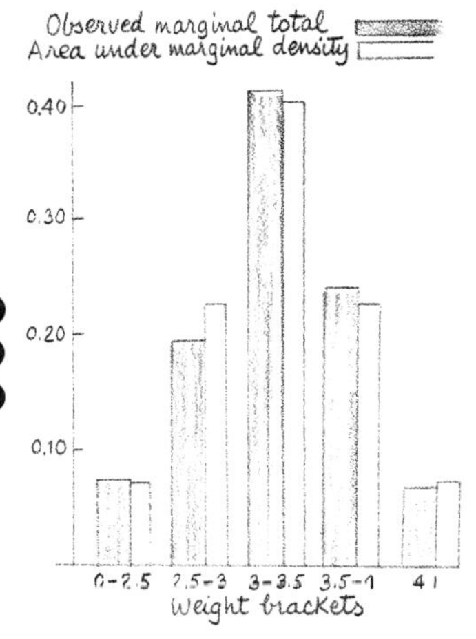

Figure 7.37

The two components of the bivariate characteristic are called **independent** if the bivariate density is equal to the product of the marginal densities:

$$f(x,y) = f_1(x)f_2(y). \tag{7}$$

In this case the integral (1) factors into the product of single integrals:

$$\int_a^b \int_c^d f_1(x)f_2(y)\,dy\,dx = \int_a^b f_1(x)\left(\int_c^d f_2(y)\,dy\right) dx,$$

because x is fixed with respect to integration over y; and the last expression is equal to

$$\int_a^b f_1(x)\,dx \cdot \int_c^d f_2(y)\,dy,$$

because the second integral is constant. This result signifies that the proportion of the population such that $a \leq x \leq b$ **and** $c \leq y \leq d$ is equal to the **product** of the proportions for which $a \leq x \leq b$ and for which $c \leq y \leq d$, respectively. This means that the two components are without influence on each other.

EXAMPLE 7

(Continuation of Example 6.) The density chosen for the data in Example 5 does have the property (7). This implies that the family income is without influence on the birth weight of the newborn: A baby born in a poor family arrives with the same nutrition as in a prosperous family. This conclusion might have to be accepted with qualifications. First of all, the data include only live births; there might be a higher proportion of neonatal deaths among the poor (or even among the rich). Secondly, the study includes only **legitimate** births; and there might be strong association between poverty and illegitimacy, which would not be found in these data.

We do not suggest that this is a proof that there is no relation between the two factors, birth weight and family income. The particular density chosen strongly suggests this. There are refined statistical tests of independence which can be used in place of this crude method; however, these will not be considered here.

Let us demonstrate the numerical significance of the asserted independence of family income and birth weight. The table below is a multiplication table. The second column contains the marginal income totals from Example 6. The second row contains the marginal weight totals. Each entry in the table is a product of the marginal percentages in the corresponding row and column. The actual population percentages from Example 5 are given in parentheses. Note how close the products are to the population percentages in most of the cases.

Birth Weight		**Below 2.5**	**2.5–3**	**3–3.5**	**3.5–4**	**4+**
Income	**Marginal totals**	**7.5**	**19.4**	**41.1**	**23.9**	**6.6**
Less than 3	22.7	1.7 (2.4)	4.4 (4.5)	9.3 (9.4)	5.4 (5.5)	1.5 (0.9)
3–5	31.8	2.4 (2.4)	6.2 (5.9)	13.0 (14.0)	7.6 (7.0)	2.1 (2.5)
5–7	25.5	1.9 (1.7)	4.9 (5.0)	10.5 (10.3)	6.1 (6.6)	1.7 (1.9)
7+	18.5	1.4 (1.0)	3.6 (4.0)	7.6 (7.4)	4.4 (4.8)	1.2 (1.3)

This implies that the distribution of weights is approximately the same **within** the various income levels. For example, the proportion of births in the 3–3.5 weight range relative to all those in the income bracket below 3 is approximately the same as the corresponding proportion of those in the 5–7 income bracket:

$$\frac{9.4}{22.7} \approx 41.4 \qquad \text{(approximately)}$$

for the first, and

$$\frac{10.3}{25.5} \approx 40.4 \qquad \text{(approximately)}$$

for the second. ▲

Property (7), signifying independence, is not typical of bivariate densities. The density does not factor when there is some relation between the components. The best known and most used bivariate density is the **bivariate normal density.** It is the two-dimensional version of the normal density mentioned in Section 6.9, Example 2. We shall give here only a geometric description of the surface and a numerical example. For a more complete description of the bivariate normal density and its applications, the reader should consult one of the many introductory texts on statistics.

EXAMPLE 8

A man's blood pressure is measured by a **sphygmomanometer,** and is given in terms of the height of a column of mercury (in millimeters). The **systolic** pressure is that during the contraction of the heart muscle; and the **diastolic** pressure is that during its relaxation. A strong relation is, naturally, expected between the two measured pressures for a single individual. The two kinds of measurements for men and women of various ages were recorded in a U.S. Government report, Blood Pressure of Adults by Age and Sex, United States, 1960–1962, National Center for Health Statistics, Ser. 11, No. 4, U.S. Department of Health, Education and Welfare. In Figure 7.38 we give the distribution of 10,281 men of ages 25–34 according to systolic and diastolic pressure. The actual numbers, not the proportions, are given. Note that the systolic pressure distribution shifts to the right as the diastolic pressure increases; similarly, the diastolic pressure distribution shifts upward as the systolic pressure increases. This signifies a **statistical correlation** between the two pressures.

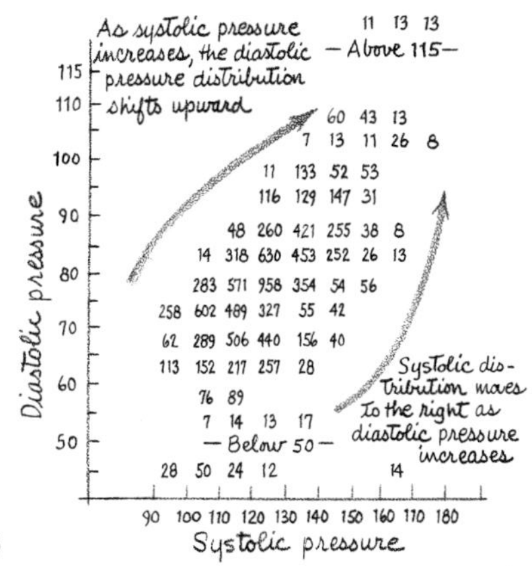

Figure 7.38

The bivariate normal density is a surface like that of a smooth mound with a single summit. All the cross-sectional curves are bell-shaped normal densities. The lines of equal altitude are ellipses with the major axis along the diagonal line in the case of "positive correlation," and along the perpendicular to the diagonal in the case of "negative correlation." The blood pressures seem to fit the normal density with positive correlation: The volumes under the surface are approximately equal to the proportions of the population in the various blood pressure categories (Figure 7.39). ▲

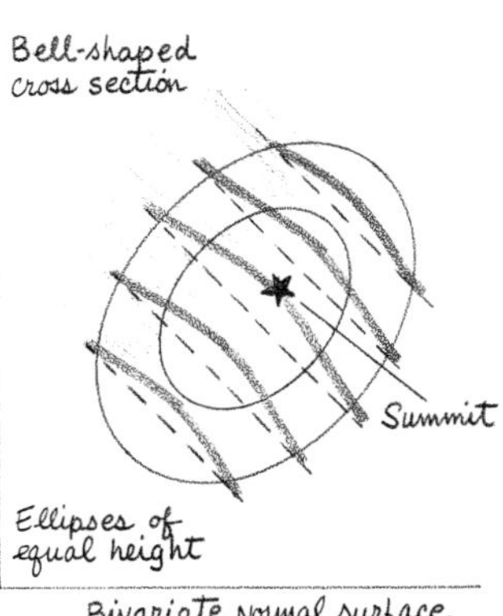

Figure 7.39

We conclude this section with an application of double integration to the evaluation of the integral of the normal density.

EXAMPLE 9

The standard one-variate normal density function has the formula

$$\phi(x) = \frac{1}{\sqrt{2\pi}} e^{-(1/2)x^2}$$

Let us show that it is really a density: not only is it nonnegative, but **the area under the curve is equal to 1.** Form the double integral

$$\int_{-\infty}^{\infty} \int_{-\infty}^{\infty} \phi(x)\phi(y)\,dy\,dx.$$

As in Example 5, the double integral factors into a product of single integrals, which, in this case, are identical:

$$\int_{-\infty}^{\infty} \phi(x)\,dx \cdot \int_{-\infty}^{\infty} \phi(y)\,dy = \left(\int_{-\infty}^{\infty} \phi(x)\,dx\right)^2.$$

On the other hand, the double integral may be written as

$$\int_{-\infty}^{\infty} \int_{-\infty}^{\infty} \left(\frac{1}{2\pi}\right) e^{-(1/2)(x^2+y^2)}\,dy\,dx.$$

The integrand is a function of two variables obtained by revolving the graph of the function

$$\left(\frac{1}{2\pi}\right) e^{-(1/2)x^2}$$

about the z-axis. By the result of Example 2 of Section 6.9, the volume of this solid is equal to 1. It follows that

$$\left(\int_{-\infty}^{\infty} \phi(x)\,dx\right)^2 = 1 \qquad \text{or} \qquad \int_{-\infty}^{\infty} \phi(x)\,dx = 1.$$

Since the area under the curve is invariant under shifts, it also follows that for any fixed number c,

$$\int_{-\infty}^{\infty} \phi(x-c)\,dx = 1. \tag{8}$$

▲

7.5 EXERCISES

In Exercises 1 and 2 compute the value of the given double integral.

1 (a) $\int_0^3 \int_0^2 xy^2\,dy\,dx$ (b) $\int_{-1}^1 \int_1^2 (x^2+y^2)\,dy\,dx$

(c) $\int_0^1 \int_1^2 \left(\frac{x}{y}\right) dy\, dx$

(d) $\int_0^2 \int_2^3 (x+y)^{-4} dy\, dx$

(e) $\int_0^{\sqrt{5}} \int_0^2 \frac{xy}{\sqrt{x^2+y^2}}\, dx\, dy$

(f) $\int_0^\infty \int_0^\infty e^{-(x^2+y^2)} xy\, dy\, dx$

(g) $\int_{-\pi/2}^{\pi/2} \int_0^{\pi/2} \sin(x+y)\cos(x+y) dy\, dx$
(See Sections 6.3, 6.6)

(h) $\int_0^1 \int_0^1 \frac{e^{-(x+y)}}{[1+e^{-(x+y)}]^2}\, dy\, dx$
(See Sections 6.3, 6.4)

2 (a) $\int_0^1 \int_1^2 (x^2y^4 - x^3y^2) dy\, dx$

(b) $\int_0^1 \int_1^2 (x+y)^2 dx\, dy$

(c) $\int_0^2 \int_1^4 x\, dy\, dx$

(d) $\int_0^3 \int_0^4 xy\sqrt{x^2+y^2}\, dy\, dx$

(e) $\int_1^2 \int_0^1 ye^{xy}\, dx\, dy$

(f) $\int_0^{\pi/2} \int_0^{\pi/2} \cos(x+y) dy\, dx$

(g) $\int_0^1 \int_0^1 \frac{1}{x+y}\, dy\, dx$
(See Section 6.6)

(h) $\int_1^2 \int_0^1 (x+y)^{-2} dy\, dx$

3 Show that the function (4) possesses the two defining properties of a bivariate density.

4 Verify that the volumes under the density (4) are those given in Figure 7.35, for these rectangles:
(a) $3 \le x \le 5,\quad 3 \le y \le 3.5$ (b) $0 \le x \le 3,\quad 2.5 \le y \le 3$

5 Verify that the functions are bivariate densities:
(a) $f(x,y) = 0,$ for either $x \le 0$ or $y \le 0$,
$= 4xe^{-2x(1+y)},$ for both $x > 0$ and $y > 0$. (See Example 4.)
(b) $f(x,y) = 0,$ for either $x \le 0$ or $y \le 0$,
$= x^2e^{-x(1+y)},$ for both $x > 0$ and $y > 0$. (See Example 4.)
(c) $f(x,y) = \frac{1}{\pi^2} \cdot \frac{1}{1+x^2} \cdot \frac{1}{1+(x+y)^2}$ [Hint: Recall $\frac{d}{dx}$ arctan x.]
(d) $f(x,y) = \frac{1}{2\pi}\, e^{-(1/2)x^2} e^{-(1/2)(y-x)^2}$ [*Hint:* (Equation (8).]

6 Using the method of Example 9 evaluate the double integral

$$\int_0^\infty \int_0^\infty e^{-\sqrt{x^2+y^2}}\, dy\, dx.$$

7 Let a population have a bivariate density of the form given in Example 4. Find the proportion of the population for which $0 \le x \le 1$, $1 \le y \le 2$. Compare this to the product of the corresponding marginal proportions. Are the two characteristics independent?

8 A population has a bivariate density

$$f(x,y) = \frac{e^{-x-y}}{(1+e^{-x})^2(1+e^{-y})^2}.$$

Find the proportion of the population for which $0 \le x \le 2, -1 \le y \le 1$. Are the components independent?

9 The volume of a cylinder can be computed by double integration. The value is the same as that obtained, in Section 6.9, by the formula for a volume of revolution. Half a cylinder of height h and radius r is the volume under the surface $z = \sqrt{r^2 - x^2}$ and over the rectangle $-r \le x \le r$, $0 \le y \le h$ (Figure 7.40).

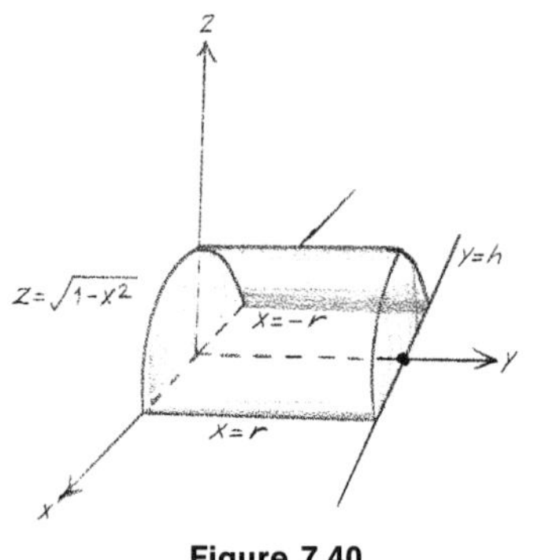

Figure 7.40

(a) Show by a geometric argument and the Fundamental Theorem of Calculus that the definite integral

$$\int_{-r}^{r} \sqrt{r^2 - x^2}\, dx$$

is equal to $\pi r^2/2$.

(b) Show that the volume of the half-cylinder is

$$\int_0^h \int_{-r}^{r} \sqrt{r^2 - x^2}\, dx\, dy.$$

Evaluate this. Then deduce the formula for the whole cylinder.

● **10** Let $f(x,y)$ be a bivariate density. For every pair (x,y) put

$$F(x,y) = \int_{-\infty}^{x} \int_{-\infty}^{y} f(u,v)\, dv\, du.$$

F is the bivariate distribution function.

(a) Show that if $a < b$, then

$$F(b,y) - F(a,y) = \int_a^b \int_{-\infty}^{y} f(u,v)\, dv\, du.$$

(b) Extend the result in (a); show that if $a < b$ and $c < d$, then

$$F(b,d) - F(a,d) - F(b,c) + F(a,c) = \int_a^b \int_c^d f(u,v)\, dv\, du.$$

(c) Define

$$F(x,\infty) = \int_{-\infty}^{x} \int_{-\infty}^{\infty} f(u,v)\, dv\, du.$$

●● This is the **marginal distribution function of** x. Show that it is the integral of the marginal density:

$$F(x,\infty) = \int_{-\infty}^{x} f_1(u)\, du.$$

●● **11** Let $f(x,y)$ be a bivariate density with marginals $f_1(x)$ and $f_2(y)$. The **correlation coefficient** is defined as the ratio

$$\frac{\int_{-\infty}^{\infty}\int_{-\infty}^{\infty} xyf(x,y)\,dy\,dx - \int_{-\infty}^{\infty} xf_1(x)\,dx \cdot \int_{-\infty}^{\infty} yf_2(y)\,dy}{\left\{\int_{-\infty}^{\infty} x^2 f_1(x)\,dx - \left(\int_{-\infty}^{\infty} xf_1(x)\,dx\right)^2\right\}^{1/2}\left\{\int_{-\infty}^{\infty} y^2 f_2(y)\,dy - \left(\int_{-\infty}^{\infty} yf_2(y)\,dy\right)^2\right\}^{1/2}},$$

provided the integrals in the denominator converge. Compute the correlation for these bivariate densities:

(a) $f(x,y) = 0$, for $x \le 0$ or $y \le 0$
$= e^{-x-y}$, for $x > 0$ and $y > 0$.

(b) $f(x,y) = 3xy(x+y)$, for $0 \le x \le 1$, $0 \le y \le 1$
$= 0$ elsewhere

12 Using the definition of the correlation coefficient in Exercise 11, show that it is equal to 0 whenever the components are independent.

13 The converse to the statement in Exercise 11 is not generally true: the correlation may be 0 but the components do not have to be independent. Here is a counterexample. Let $f(x)$ be a density function of a single variable, equal to 0 on the negative axis. Let $f(x,y)$ be the bivariate density obtained from it by revolving the curve about the z-axis and dividing by $2\pi\sqrt{x^2+y^2}$:

$$f(x,y) = \frac{f(\sqrt{x^2+y^2})}{2\pi\sqrt{x^2+y^2}}.$$

(a) Show, by a geometric argument, that the center of gravity of either marginal distribution is equal to 0; and then conclude that the second term in the numerator of the correlation coefficient is equal to 0.

(b) Show that the first term in the numerator of the correlation coefficient is also equal to 0. [*Hint:* Consider the integrals over the four quadrants of the plane, and show that their sum is equal to 0.]

(c) Why does this example provide a counterexample to the converse to Exercise 11? [*Hint:* Consider a simple case such as $f(x) = x/2$, for $0 \le x \le 1$, and 0 elsewhere.]

7.6 Double Integrals and More General Volumes

In Section 7.5, the computed volumes were over rectangles in the plane. Now we shall use the double integral to compute volumes over more general areas.

Let A stand for an area in the plane. Suppose that its lower and upper boundaries are the graphs of two functions, $y = g(x)$ and $y = h(x)$, respectively. The left-hand and right-hand boundaries are either lines parallel to the y-axis (Figure 7.41); or one or both are points formed by the intersection of the upper and lower boundaries [Figures 7.42(a), (b)].

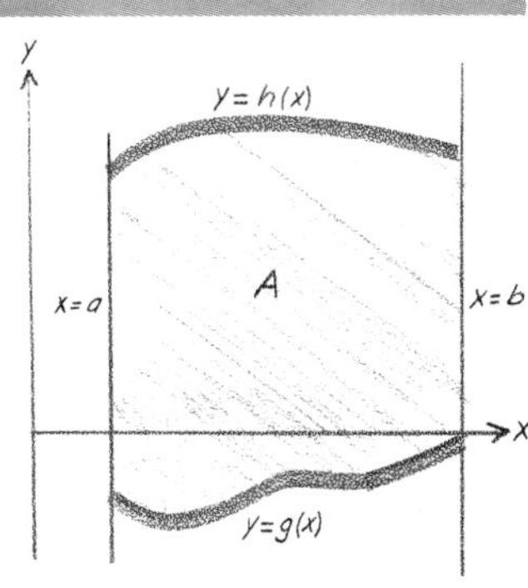

Figure 7.41

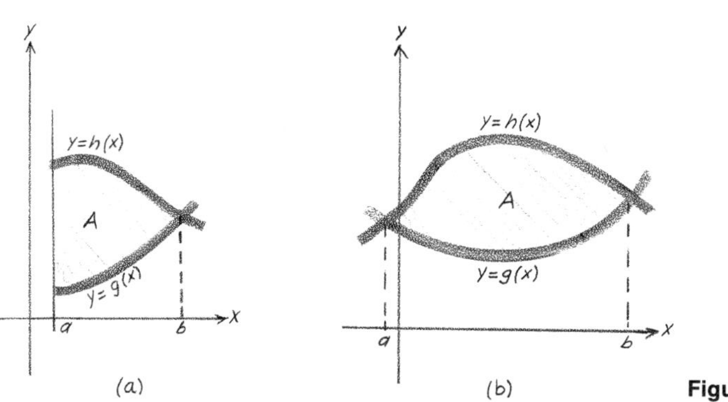

Figure 7.42

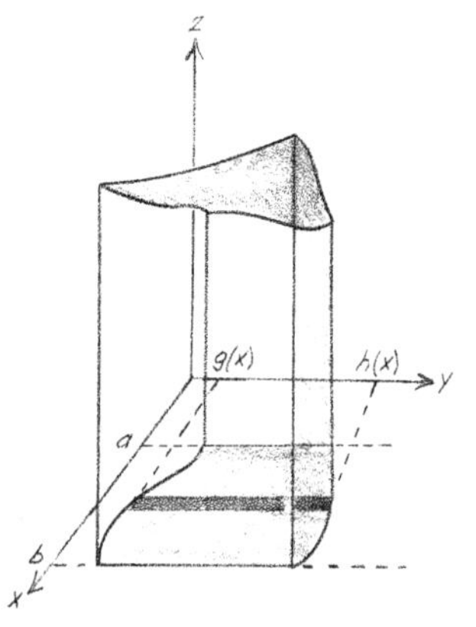

Figure 7.43

The volume under the surface of a nonnegative function $z = f(x,y)$ over such an area in the plane is equal to the double integral

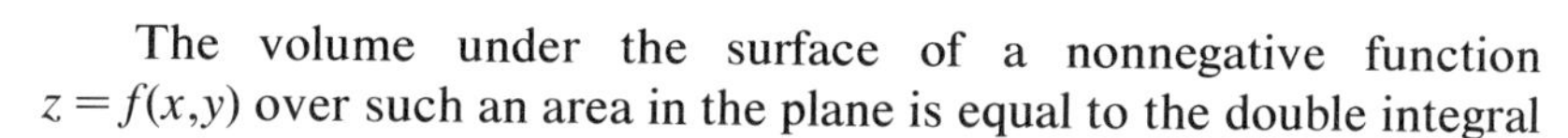

$$\int_a^b \left(\int_{g(x)}^{h(x)} f(x,y)\,dy \right) dx. \tag{1}$$

The inner integral over y is taken from $g(x)$ to $h(x)$, which are functions of x; the **order** of integration cannot be changed. The proof that (1) gives the volume follows by a simple extension of the proof for the rectangle in Section 7.5: The "bread slices" run from $g(x)$ to $h(x)$ instead of from c to d (Figure 7.43).

EXAMPLE 1

Find the volume under the surface $z = xy$ over the quarter disk of radius 1 in the first quadrant of the xy-plane; the disk is centered at (0,0). The upper boundary for the area is the curve $y = \sqrt{1 - x^2}$; the lower boundary is the x-axis; the left-hand boundary is the y-axis; and, finally, the right-hand boundary is a point on the line $x = 1$ (Figure 7.44). It follows from the general formula (1) that the double integral for the volume is

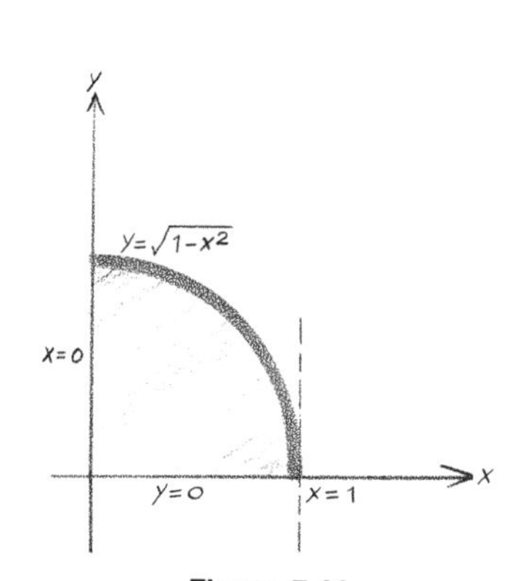

Figure 7.44

$$\int_0^1 \int_0^{\sqrt{1-x^2}} xy\,dy\,dx.$$

Integrate first with respect to y:

$$\int_0^{\sqrt{1-x^2}} y\,dy = \tfrac{1}{2}y^2 \Big|_0^{\sqrt{1-x^2}} = \tfrac{1}{2}(1 - x^2).$$

Then multiply by x, and integrate with respect to x:

$$\int_0^1 x \cdot \tfrac{1}{2}(1 - x^2)\,dx = \tfrac{1}{4}x^2 - \tfrac{1}{8}x^4 \Big|_0^1 = \tfrac{1}{8}. \ \blacktriangle$$

EXAMPLE 2

Find the volume under the plane $4x + 3y + 2z = 6$, over the portion of the xy-plane in the first quadrant. The latter area is a triangle bounded by the axes and the line $4x + 3y = 6$, which is formed by the intersection of the plane $4x + 3y + 2z = 6$ with the xy-plane (Figure 7.45). The height of the plane above the xy-plane is obtained by solving the plane equation for z: $z = 3 - 2x - 3y/2$. The upper boundary of the area is the line $4x + 3y = 6$; we solve this for y, and write its equation as $y = 2 - 4x/3$. The left- and right-hand boundaries are along the lines $x = 0$ and $x = \frac{3}{2}$. It follows that the double integral for the volume is

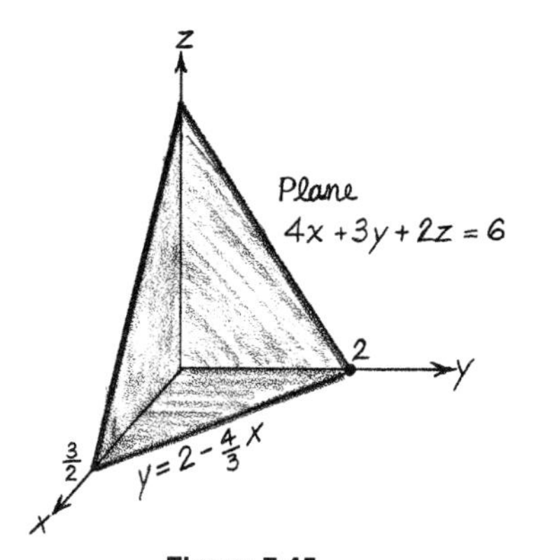

Figure 7.45

$$\int_0^{3/2} \int_0^{2-4x/3} \left(3 - 2x - \frac{3y}{2}\right) dy\, dx.$$

Integrate first with respect to y:

$$\int_0^{2-4x/3} \left(3 - 2x - \frac{3y}{2}\right) dy = 3 - 4x + \tfrac{4}{3}x^2.$$

Now integrate over x from 0 to $\frac{3}{2}$:

$$\int_0^{3/2} (3 - 4x + \tfrac{4}{3}x^2)\, dx = \tfrac{3}{2}. \blacktriangle$$

EXAMPLE 3

Find the volume under the surface $z = e^{-x-y}$ over the triangle bounded by the coordinate axes and the line $x + y = a$ (Figure 7.46). It is equal to the double integral

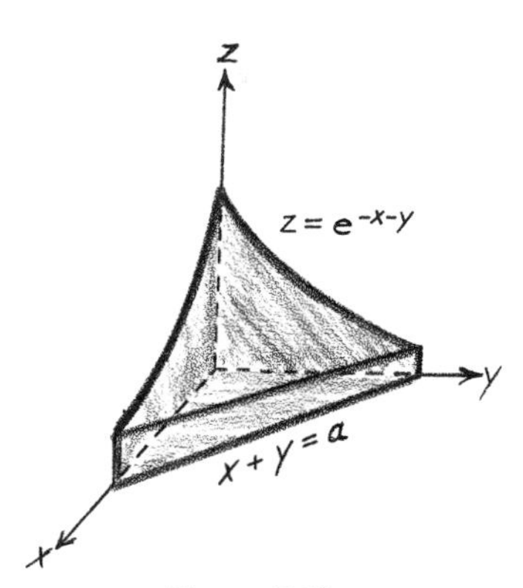

Figure 7.46

$$\int_0^a \int_0^{a-x} e^{-x-y} dy\, dx.$$

Integrate over y:

$$\int_0^{a-x} e^{-x-y} dy = e^{-x} \int_0^{a-x} e^{-y}\, dy = e^{-x}(1 - e^{-a+x})$$
$$= e^{-x} - e^{-a}.$$

Now integrate over x:

$$\int_0^a (e^{-x} - e^{-a})\, dx = 1 - e^{-a} - ae^{-a}. \blacktriangle$$

7.6 EXERCISES

1 Evaluate:

(a) $\displaystyle\int_0^1 \int_{-x^2}^{x^2} (x + y)\, dy\, dx$ (b) $\displaystyle\int_0^1 \int_0^x (x + y)^5 dy\, dx$

(c) $\int_0^1 \int_0^x \frac{1}{x+y}\, dy\, dx$

(d) $\int_0^1 \int_0^x xy\sqrt{x^2+y^2}\, dy\, dx$

(e) $\int_2^3 \int_0^{1/x} xy\, dy\, dx$

(f) $\int_1^2 \int_0^x \frac{y}{x^2+y^2}\, dy\, dx$

(g) $\int_0^\infty \int_0^x xye^{-(x^2+y^2)}\, dy\, dx$

(h) $\int_0^{\pi/2} \int_0^x (\sin x + \sin y)\, dy\, dx$

2 Evaluate:

(a) $\int_1^2 \int_0^{x^2+1} xy^2\, dy\, dx$

(b) $\int_0^2 \int_0^x (x^2+y^2)\, dy\, dx$

(c) $\int_0^{1/2} \int_0^{\sqrt{1-x^2}} (xy^3+y)\, dy\, dx$

(d) $\int_0^1 \int_0^x \frac{xy}{\sqrt{x^2+y^2}}\, dy\, dx$

(e) $\int_0^2 \int_0^x xe^{-y}\, dy\, dx$

(f) $\int_1^4 \int_1^{x^2} \left(\frac{x}{y}\right) dy\, dx$

(g) $\int_0^{\pi/2} \int_{-x}^x \cos(x+y)\, dy\, dx$

(h) $\int_0^1 \int_0^x y(1-y^2)^{-3/2}\, dy\, dx$

[*Hint:* Use $\frac{d}{dx}$ arcsin x]

3 Find the volume under the plane $x + 5y + 3z = 15$ and over the first quadrant portion of the xy-plane.

4 Find the volume under the plane $x + 4y + 9z = 36$ and over the first quadrant portion of the xy-plane.

5 Interpret the double integral

$$\int_0^1 \int_0^{\sqrt{1-x^2}} e^{-(1/2)(x^2+y^2)}\, dy\, dx$$

as a quarter of the volume of a solid of revolution. Then use the appropriate formula of Section 6.9 to evaluate it.

6 Verify the formula for the volume of a cylinder of radius r and height h by integrating the constant function $f(x,y) = h$ over the disk of radius r about the origin (see Exercise 9, Section 7.5).

7 Find the volume under the plane $3x + 4y + z = 12$ over the portion of the disk of radius 1 centered at the origin in the first quadrant. [*Hint:* Find $\int_0^1 \sqrt{1-x^2}\, dx$ by Figure 7.44]

8 Find the volume under the plane $z = 2y$ and above the region bounded by $x = 3$, $y = 0$, and the circle $x^2 + y^2 = 36$.

9 Find the volume under the plane $z = \frac{1}{2}x + y$ and above the region in the first quadrant bounded by a portion of the ellipse $x^2/9 + y^2/4 = 1$.

10 Find $\frac{1}{8}$ the volume of a ball of radius r by means of the double integral of $\sqrt{r^2 - x^2 - y^2}$ over the part of the disk of radius r, centered at the origin in the first quadrant. [*Hint:* Apply the result of Exercise 9a, Section 7.5, with $\sqrt{r^2 - y^2}$ in place of r.]

CHAPTER

8 Discrete Probability

8.1 The Basic Counting Principle

As a necessary preliminary to the study of probability, we introduce the elements of the branch of mathematics known as **combinatorial analysis.** It provides ways of counting the numbers of certain arrangements or operations, and it will be used in the calculation of probabilities. We state the basic counting principle:

If one operation can be done in m different ways, and following it, another operation can be done in n different ways, then the two operations can be done together in the stated order in $m \cdot n$ ways. In other words, the number of ways of doing both in the given order is equal to the product of the numbers of ways of doing each individually.

EXAMPLE 1

A person is planning a trip from New York to San Jose, California. He plans to fly from New York to San Francisco, and then take the bus to San Jose. There are 4 different airlines with suitable flight schedules, and 2 different bus lines. Then the number of different ways that he can complete the trip is 8; this is the product of the number of ways of taking a plane (4) and the number of ways of taking a bus (2) (Figure 8.1). ▲

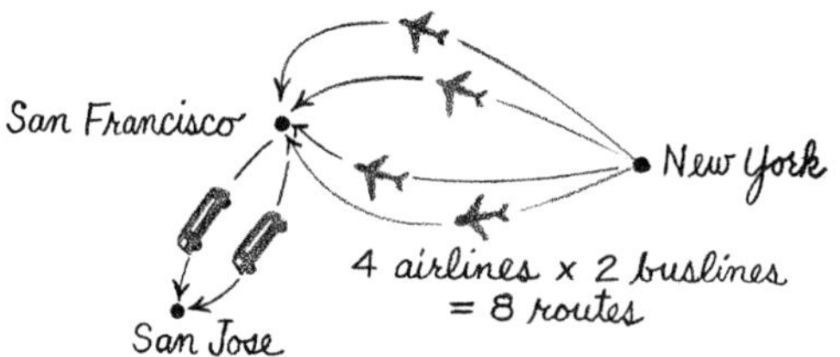

Figure 8.1

The proof of the counting principle is based on a chart similar to the multiplication table: an m-by-n matrix. Let each row of the matrix correspond to one way of doing the first operation, and each column correspond to a way of doing the second operation. Then each entry of the matrix corresponds to one way of doing both operations in the given order; for example, the entry in the second row and third column corresponds to doing the first operation in the second way, and the second operation in the third way. Since an m-by-n matrix has $m \cdot n$ entries, it follows that there are $m \cdot n$ ways of doing both operations.

The basic counting principle can be extended to more than two successive operations:

If one operation can be done in n_1 ways, and after that, a second operation in n_2 ways, and after that, a third in n_3 ways, and so on, then the number of different ways of performing all the operations in the given order is equal to the product $n_1 \cdot n_2 \cdot n_3 \cdot \cdot \cdot$.

EXAMPLE 2

Suppose that the traveler in Example 1 plans to go by bus from San Jose to Los Angeles and that there are 2 bus lines convenient for him. In how many different ways can he complete the entire trip from New York to Los Angeles? By the extended counting principle, the number of ways is the product of the numbers of ways for traveling between the successive cities, namely, $4 \cdot 2 \cdot 2 = 16$ (Figure 8.2). ▲

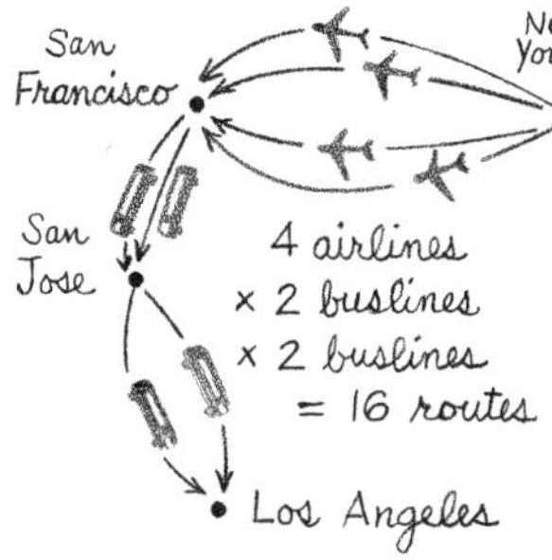

Figure 8.2

EXAMPLE 3

The outcome of a toss of a coin is either **heads** or **tails;** thus, each toss can be done in exactly 2 ways. When the coin is tossed twice, there are 2 "operations," each with 2 possible "ways" of being done. By the counting principle, the 2 tosses have $2 \cdot 2 = 4$ possible outcomes (Figure 8.3). If the coin is tossed 3 times, then the number of possible outcomes is $2 \cdot 2 \cdot 2 = 8$ (Figure 8.4). In general, the number of outcomes of n tosses is $2 \cdot 2 \cdot 2 \cdots 2$ (n factors) $= 2^n$. ▲

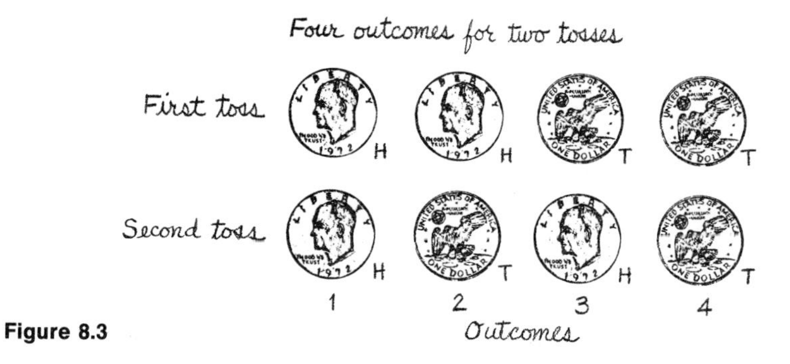

Figure 8.3

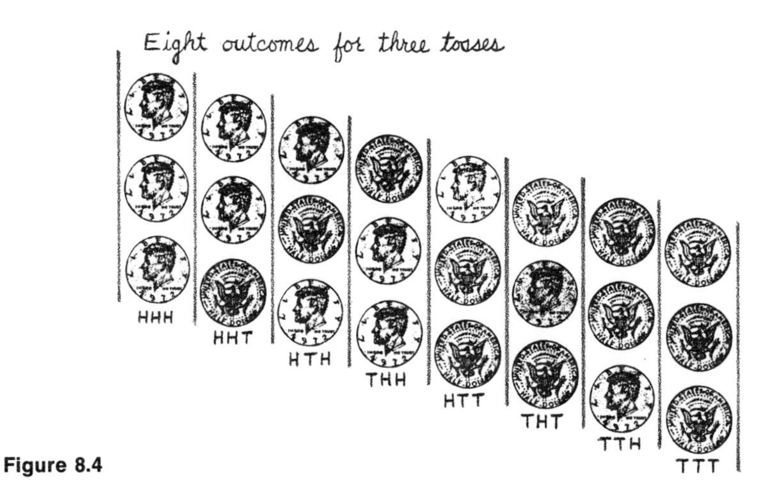

Figure 8.4

In applying the counting principle, we often have to calculate a product of successive integers such as $1 \cdot 2 \cdot 3$, or $4 \cdot 5 \cdot 6$. It is useful in these situations to have the factorial expression, $n!$, defined as the product of the first n integers, $1 \cdot 2 \cdots n$. (This was used in Section 6.12.) for example,

$$1! = 1, \quad 2! = 1 \cdot 2 = 2, \quad 3! = 1 \cdot 2 \cdot 3 = 6, \quad 4! = 1 \cdot 2 \cdot 3 \cdot 4 = 24,$$

and so on. We see from the list above that each factorial is obtained from its predecessor by simply multiplying by the number of the former factorial; for example,

$$3! = 2! \cdot 3, \qquad 4! = 3! \cdot 4, \qquad 5! = 4! \cdot 5,$$

and so on. In general,

$$n! = n \cdot (n-1)!. \tag{1}$$

The relation (1) is valid for integers n at least equal to 2. For convenience, we define the factorial of 0 as

$$0! = 1.$$

With this definition, the relation (1) becomes valid also for $n = 1$.

A **permutation** of a finite set of objects is an arrangement of the set in a particular order. Suppose that we have 3 poles and 3 different flags. The number of different ways in which the various flags can be put on the different poles is the number of permutations of the set of flags (Figure 8.5). There are 6 such permutations. This follows from the counting principle: The setting of the 3 flags is a result of 3 successive operations. The first flag can be hung in 3 different ways, the second in 2 ways, and the third in 1 way; hence, the number of ways of hanging all the flags is $3 \cdot 2 \cdot 1 = 6$. This reasoning is extended to the number of permutations of a set of n objects:

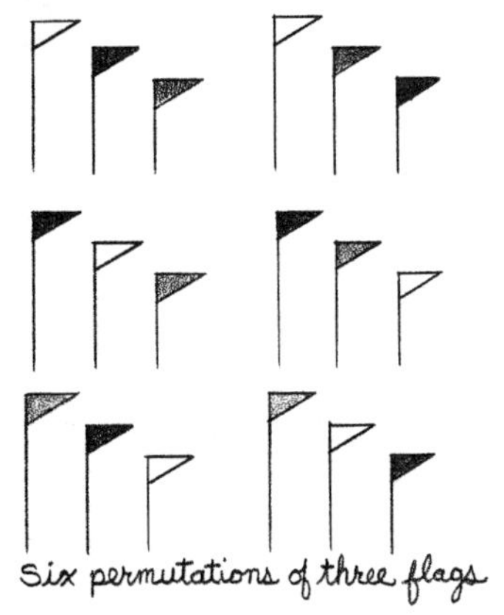

Figure 8.5

There are $n!$ permutations of a set of n objects.

EXAMPLE 4

If 10 horses are in a race, then the number of ways in which the horses can finish first, second, . . . tenth is $10! = 3{,}628{,}800$ (Figure 8.6). ▲

Figure 8.6

Next we consider **the number of permutations of a set of n objects taken r at a time.** Suppose, in Example 4, that we are interested only in the horses that finish first, second, or third. In other words, we distinguish orderings of the horses that differ on at least one of the first 3 places. In how many different ways can horses finish first, second, and third? This number is called the number of permutations of 10 horses taken 3 at a time. This number is much smaller than the total number of permutations of the horses. There are 10 ways to choose the horse that places first, and after that, 9 ways to choose the horse that finishes second, and finally, 8 ways to choose the horse that finishes third. By the counting principle, the number of ways in which the horses can finish first, second, and third is the product $10 \cdot 9 \cdot 8 = 720$.

There is a general formula for the number of permutations of n objects taken r at a time, where $r \leq n$; this number is denoted $P(n,r)$:

$$P(n,r) = n(n-1)(n-2) \cdots (n-r+1) \qquad (r \text{ factors}). \quad (2)$$

In words, the number of such permutations is the product of r integers, starting with n and going down by 1 on each factor. Another way of writing $P(n,r)$ is

$$P(n,r) = \frac{n!}{(n-r)!}. \quad (3)$$

The proof of the general formula (2) is, like that of the particular case of the first 3 horses to finish, a direct consequence of the counting principle.

EXAMPLE 5

If there are 12 horses in a race, then the number of ways the horses can finish first, second, third, and fourth is

$$P(12,4) = 12 \cdot 11 \cdot 10 \cdot 9 = 11{,}880. \blacktriangle$$

In the particular case where the number of objects taken is equal to the whole number in the set, then $r = n$, and $P(n,n) = n!$.

8.1 EXERCISES

1 It takes 4 operations to manufacture a particular product. A plant has 4 machines for the first operation, 1 machine for the second, 6 for the third, and 2 for the fourth. In how many different ways can the product be made?

2 The manager of a baseball team has to select players for various positions. There are 2 first basemen, 3 second basemen, 4 shortstops, and 1 third baseman. In how many different ways can he choose men for the 4 positions?

3 A traveler driving from New York to Minneapolis through Chicago has a choice of 4 highways from New York to Chicago, and 3 from Chicago to Minneapolis. Find the number of different routes for the complete trip. If he returns to New York through Chicago, in how many ways can he make the entire round trip?

4 Find the total number of possible outcomes for 6 tosses of a coin.

5 When a die is tossed, there are 6 possible outcomes. How many outcomes are possible when 2 dice are tossed? What if a pair of dice is tossed twice?

6 A restaurant offers 2 different entrees, 4 main dishes, 3 desserts, and 3 beverages. How many different menus are possible?

7 A clothing store has shirts of either cotton or dacron, in 4 colors, and 3 sizes. How many different kinds of shirts are there?

8 From a group of 7 Americans, 5 Britons, and 3 Frenchmen, how many different 3-man committees can be formed so that there is 1 representative from each country on the committee?

9 A student registering for his courses has a choice of 2 mathematics sections, 3 biology sections, 3 English sections, and 1 history section. How many different programs are possible?

10 A nominating committee suggests 2 names for President, 3 for Vice President, 4 for Secretary, and 2 for Treasurer. How many different election outcomes are possible?

11 Compute:
(a) $P(6,3)$
(b) $P(31,4)$
(c) $P(7,7)$
(d) $P(15,5)$
(e) $P(7,4)$

12 Compute:
(a) $P(19,3)$
(b) $P(9,6)$
(c) $P(8,8)$
(d) $P(201,2)$
(e) $P(600,1)$

13 Six candidates run for the same office. In how many different orders can they finish (a) first and second; (b) first, second, and third; (c) first through sixth?

14 From a team of 9 starting players in how many ways can the manager choose a batting order? In how many ways can he choose the third, fourth, and fifth batters in that order?

15 Suppose that every automobile license plate is to have 5 digits. How many different plates can be made in such a way that an individual digit does not occur more than once on the same plate?

16 How many different 2-digit numbers can be formed with the digits 3, 4, 5, 6, 7 if the number is not to contain 2 equal digits? What if it may have equal digits?

17 A stockbroker has a list of stocks suggested for his customers. The list is to contain 5 stocks, ranked according to their potential. These are obtained from a master list of 50 stocks. In how many ways can the broker present an ordered list of 5 stocks?

18 In how many ways can a nominating committee select a President, Vice President, Secretary, and Treasurer from a single panel of 12 suggested candidates?

19 In how many different ways can 200 contestants win first, second, and third prizes?

20 In how many different ways can a party of 6 persons, which includes a man and his wife, be seated around a circular table with 6 chairs in such a way that the man and his wife are separated by at least one other person? In this case an "arrangement" of the sitters is identified by the system of left and right neighbors of each person; we do not distinguish orderings that are obtained one from the other by simply rotating the table.

8.2 Combinations

A **combination** is a subset of a finite set of objects. If the set has n objects, and the subset has $r \leq n$, then the subset is called **a combination of n objects taken r at a time.**

The combination differs from the permutation in that the former is defined without regard to order.

EXAMPLE 1

We illustrate the difference between the combination and the permutation. Suppose there is an election for 2 seats on a council, and that there are 3 candidates: A, B, and C. The 2 candidates with the largest numbers of votes are awarded the seats. In how many different ways can the candidates finish? There are two ways to view the outcomes, as permutations and as combinations. If we consider the ranks of the 2 winners in terms of relative numbers of votes, that is, we order the winners, then each outcome is a permutation of 3 candidates taken 2 at a time. There are $P(3,2)$ permutations representing the orders of the 2 winners:

$$AB \quad BA \quad AC \quad CA \quad BC \quad CB.$$

If we are interested only in which candidates won but not in their order, then we consider the outcomes as combinations of 3 candidates taken 2 at a time. There are just 3 combinations: AB, AC, and BC. Even though AB and BA are different permutations, they still represent the same combination. ▲

We state a general formula for the number of combinations of n objects taken r at a time. This number is denoted $C(n,r)$, and it is given by the formula

$$C(n,r) = \frac{P(n,r)}{r!}. \qquad (1)$$

In words: **The number of such combinations is equal to the corresponding number of permutations divided by $r!$.**

Figure 8.7

The reason for the formula is as follows: Every combination of n objects, taken r at a time, can be ordered in $r!$ different ways; hence, from every such combination, we can build $r!$ different permutations by simply ordering the members (Figure 8.7). Therefore, the number of permutations is $r!$ times the number of combinations: $P(n,r) = r!C(n,r)$. The latter is exactly the same as Equation (1).

EXAMPLE 2

A committee of 4 members is to be selected from a panel of 12. Every selection of 4 is a combination of 12 objects taken 4 at a time; there are $C(12,4)$ such combinations. On the other hand, if the committee members are given ranks, then the selection of the committee is in terms not only of membership but also in terms of order of membership. Therefore, each selection is a permutation; there are $P(12,4)$ such permutations. The members of a committee of 4 can be ordered in 4! ways; hence, the number of permutations is equal to 4! times the number of combinations; and so,

$$C(12,4) = \frac{P(12,4)}{4!} = \frac{12 \cdot 11 \cdot 10 \cdot 9}{4 \cdot 3 \cdot 2 \cdot 1} = 495. \blacktriangle$$

The combination numbers $C(n,r)$ have a **symmetry property:**

$$C(n,r) = C(n,n-r). \tag{2}$$

In words: **The numbers of combinations of r objects out of n is equal to the number of combinations of the complementary number $n - r$ out of n.** For example, there are as many combinations of 12 objects taken 4 at a time as there are of 12 objects taken 8 at a time. The reason is that the selection of 4 objects to form a combination is equivalent to the selection of the remaining 8 to be excluded from the combination (Figure 8.8).

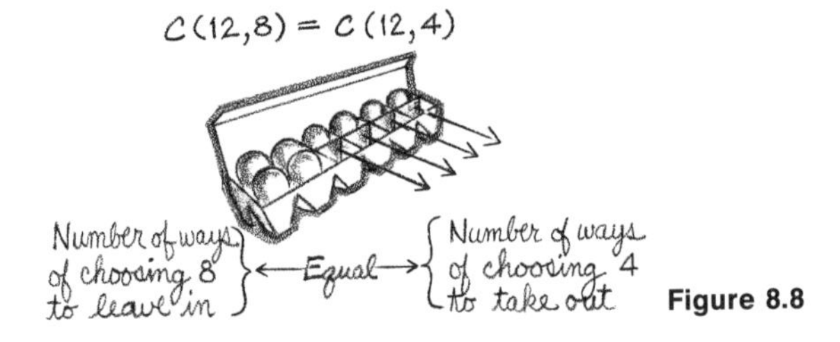

Figure 8.8

For convenience, we define the combination for $r = 0$:

$$C(n,0) = C(n,n) = 1; \tag{3}$$

this is consistent with $0! = 1$.

In some problems involving combinations, we cannot use formula (1) alone, but must combine it with the counting principle.

EXAMPLE 3

Let us again consider the problem of selecting a committee of 4 from a panel of 12; however, this time the panel is said to consist of 7 men and 5 women. In how many ways can we form a committee consisting of 2 men and 2 women? We first analyze the numbers of ways of selecting the men and the women separately; then we combine these by the counting principle. Since the panel has 7 men, there are $C(7,2) = 21$ ways of selecting 2 men for the committee. Similarly, there are $C(5,2) = 10$ ways of selecting the 2 women. Since the committee is formed by the selection of the men and then of the women (or in the reverse order), it follows that the committee can be chosen in $C(7,2) \cdot C(5,2) = 210$ ways (Figure 8.9). ▲

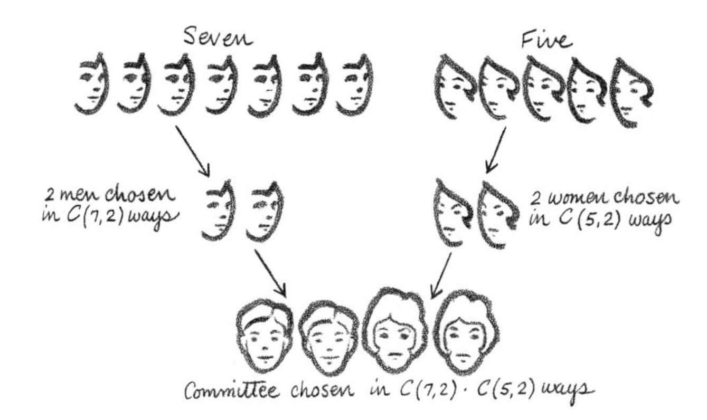

8.2 EXERCISES

1 Evaluate:

(a) $C(6,2)$
(b) $C(30,3)$
(c) $C(12,12)$
(d) $C(12,3)$
(e) $C(12,9)$

2 Evaluate:

(a) $C(21,4)$
(b) $C(8,6)$
(c) $C(11,11)$
(d) $C(110,3)$
(e) $C(200,199)$

3 How many subcommittees of 3 congressmen can be chosen from a committee of 13?

4 An instructor lets the class choose 10 questions out of 12 on an examination. How many different combinations of questions may be answered?

5 A young baseball fan has 30 pictures of players but space for only 5 to hang on his wall. In how many ways can he choose the 5?

6 In a 10-horse race in how many different ways can the horses qualify for 1 of the first 3 places, without regard to order among the first 3?

7 How many basketball teams of 5 men, without regard to position, can be formed from a group of 9?

8 Ten longshoremen line up for work but only 4 are needed. In how many ways can the 4 be chosen?

9 A pack of playing cards has 52 cards. A **hand** consists of 5. How many different hands are possible?

10 A psychologist wishes to test the reaction of 12 persons to a certain sound. If there are 16 persons available for the experiment, in how many different ways can he choose the group?

11 A panel consists of 6 men and 5 women. In how many ways can a committee of 3 men and 2 women be selected?

12 A student has to answer 8 questions out of 10 on an examination. (a) How many different combinations can he choose? (b) How many if the first 3 questions must be answered? (c) How many if at least 4 of the first 5 must be answered?

13 A large class of 100 students is to be divided into 3 sections, of sizes 15, 25, and 60. Write a formula [in terms of $C(n,r)$] for the number of ways in which this can be done.

14 A committee has 7 members including a chairman. How many subcommittees of 4 can be formed if the chairman has to be one of its members? What if he is not to be one of its members?

15 There are 12 chairs in a row. (a) In how many different ways can 2 persons be seated? (b) In how many ways can they be seated in adjacent chairs? (c) In how many ways can they be seated so that there is at least one vacant chair separating them?

16 Five judges are to be appointed from a panel of 7 Democrats and 6 Republicans. In how many ways can 2 Democrats and 3 Republicans be selected?

8.3 The Binomial Theorem

The binomial theorem is an important formula expressing an integer power of a binomial, $(A + B)^n$, as a sum of terms containing powers of A and B. By ordinary multiplication we find:

$$\begin{aligned}
(A + B)^1 &= A + B, \\
(A + B)^2 &= A^2 + 2AB + B^2, \\
(A + B)^3 &= A^3 + 3A^2B + 3AB^2 + B^3, \\
(A + B)^4 &= A^4 + 4A^3B + 6A^2B^2 + 4AB^3 + B^4.
\end{aligned}$$

We see that the power $(A+B)^n$ for $n=1, 2, 3, 4$ can be obtained in this way:

The first term is A^n.

The exponent of A decreases by 1 in each successive term, and the exponent of B increases by 1. The sum of the powers of A and B is always n.

The coefficients of the successive terms are the combination numbers $C(n,0)$, $C(n,1)$, $C(n,2)$, ... For this reason, the numbers $C(n,r)$ are also called **binomial coefficients.** As an example consider the expansion with $n=4$. The powers in the successive terms are A^4, A^3B, A^2B^2, AB^3, and B^4. The numerical coefficients are

$$C(4,0)=1, \quad C(4,1)=4, \quad C(4,2)=6, \quad C(4,3)=4, \quad C(4,4)=1.$$

What is the reason for this relation between the binomial expansion and the combination numbers $C(n,r)$? The expansion is obtained by a process of multiplication and collection of terms with common powers. Let us find $(A+B)^4$ by direct multiplication, but with the collection of terms postponed to the last step:

$$\begin{aligned}(A+B)^4 &= (A+B)(A+B)(A+B)(A+B)\\ &= (AA+BA+AB+BB)(A+B)(A+B)\\ &= (AAA+BAA+ABA+BBA+AAB\\ &\quad +BAB+ABB+BBB)(A+B)\\ &= (AAAA+BAAA+ABAA+BBAA+\cdot\cdot\cdot).\end{aligned}$$

In each of the 16 terms there is 1 letter, either A or B, from the first factor $(A+B)$, 1 from the second factor $(A+B)$, 1 from the third, and 1 from the fourth. The number of terms containing 1 factor B and 3 factors A is the same as the number of combinations of 4 objects, here 4 positions, taken 1 at a time; hence, the coefficient of A^3B is $C(4,1)=4$. Similarly, the number of terms with 2 A-factors and 2 B-factors is equal to the number of combinations of 4 taken 2 at a time; therefore, the coefficient of A^2B^2 is $C(4,2)=6$.

In general, the binomial expansion of the power n has $n+1$ terms, and is given by the formula

$$\begin{aligned}(A+B)^n = C(n,0)A^n &+ C(n,1)A^{n-1}B + C(n,2)A^{n-2}B^2\\ &+\cdot\cdot\cdot\ C(n,n-1)AB^{n-1}+C(n,n)B^n. \quad (1)\end{aligned}$$

The $(r+1)$st term is of index r, and is denoted

$$C(n,r)A^{n-r}B^r. \quad (2)$$

The binomial coefficients are **symmetric:** they form the same sequence whether read in ascending or descending order of index; for example, the coefficients for $n=3$ are 1, 3, 3, 1 and for $n=4$ are 1, 4, 6, 4, 1.

This is the result of the symmetry of the combination numbers [Section 8.2, Equation (2)]. The bar graphs of the binomial coefficients for $n = 4, 6, 8, 10$ appear in Figure 8.10, together with a list of the numbers for $n = 1$ through $n = 10$.

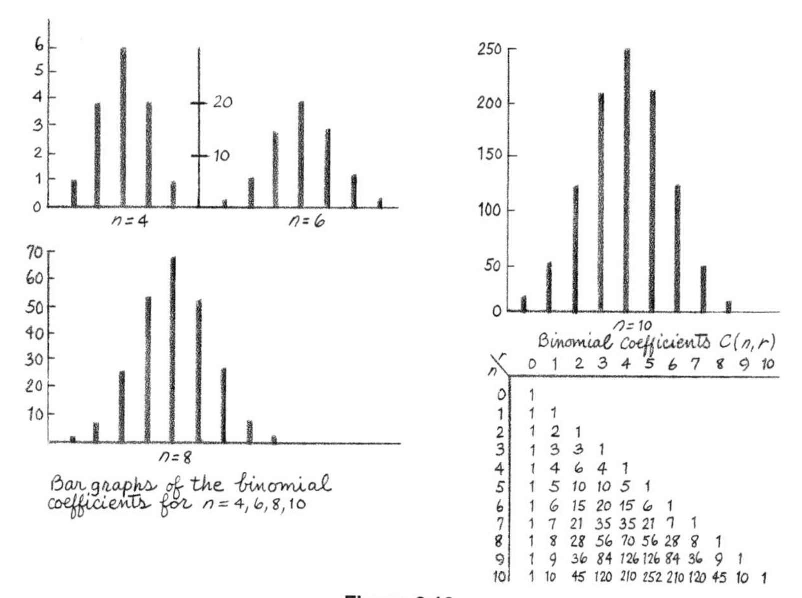

Figure 8.10

EXAMPLE 1

Expand $(A + 2B^2)^5$ and simplify each term. Here $2B^2$ takes the place of B in formula (1). The binomial coefficients for $n = 5$ are

$$1, 5, 10, 10, 5, 1;$$

thus,

$$\begin{aligned}(A + 2B^2)^5 &= A^5 + 5A^4(2B^2) + 10A^3(2B^2)^2 + 10A^2(2B^2)^3 \\ &\quad + 5A(2B^2)^4 + (2B^2)^5 \\ &= A^5 + 10A^4B^2 + 40A^3B^4 + 80A^2B^6 + 80AB^8 \\ &\quad + 32B^{10}. \blacktriangle\end{aligned}$$

When the sign in the binomial is changed to (−), the expansion of $(A - B)^n$ is obtained very simply from that of $(A + B)^n$: write $A - B$ as $A + (-B)$ and apply formula (1) with $-B$ in place of B. The terms with even powers of $-B$ will be the same as in the expansion of $(A + B)^n$; however, the terms with odd powers will differ in sign; for example,

$$(A - B)^3 = A^3 - 3A^2B + 3AB^2 - B^3.$$

EXAMPLE 2

Compute $(0.99)^5$ to 2 decimal places. Write this as a power of a binomial and apply the formula:

$$\begin{aligned}(0.99)^5 &= (1-0.01)^5 = 1-5(0.01)+10(0.01)^2\\ &\quad -10(0.01)^3+5(0.01)^4-(0.01)^5\\ &= 1-0.05+0.001-\cdots\\ &= 0.95 \text{ (to 2 places)}.\end{aligned}$$

The remaining terms are insignificant to the first 2 places. This method of calculation is useful in compound interest and decay problems. ▲

8.3 EXERCISES

In Exercises 1 and 2, expand by the binomial theorem.

1 (a) $(a+x)^6$
(b) $(x-2y)^4$
(c) $(ax-b)^5$
(d) $(x^2+x)^7$
(e) $(x/2+2)^5$

2 (a) $(2x-a)^7$
(b) $(3a+2b)^5$
(c) $(x-\frac{1}{2})^6$
(d) $(2x/3-3/2)^4$
(e) $(x^2-3)^5$

In Exercises 3 and 4 find and simplify the first 4 terms of the expansion.

3 (a) $(y-3x)^{10}$
(b) $(2x^2+\frac{1}{2})^7$
(c) $(n-2)^9$
(d) $(2a+1/a)^8$
(e) $(x/3-y/2)^5$

4 (a) $(b+\frac{1}{2})^{11}$
(b) $(1-t/3)^9$
(c) $(x-\frac{2}{3})^6$
(d) $(1+y/x)^{12}$
(e) $(1+5y/2x)^5$

In Exercises 5 and 6, find the indicated term of the expansion in terms of C(n,r).

5 (a) $(x-2)^{15}$, fifth term
(b) $(x+y/2)^{16}$, thirteenth term
(c) $(2n-3)^9$, fourth term

6 (a) $(t/2+2)^{12}$, seventh term
(b) $(x^3-4)^{10}$, third term
(c) $(1-y/2)^{30}$, fifth term

7 Compute to 4 decimal places:
(a) $(1.01)^6$
(b) $(0.99)^6$
(c) $(1.001)^4$
(d) $(0.98)^7$

8 Compute to 4 decimal places:
(a) $(1.003)^5$
(b) $(0.998)^6$
(c) $(1.008)^5$
(d) $(0.995)^4$

9 Recall that the sum of the principle and interest on c dollars compounded m times per year at an annual rate of $100k$ percent is given by the formula

$$c\left(1+\frac{k}{m}\right)^m$$

(Section 4.3). Find the value of \$100 compounded 4 times per year at the annual rate of 6 percent. Also find the value for 8 compoundings per year at the same interest rate.

10 A piece of machinery depreciates by 2 percent of its value every year. If its original price is \$220, find its value after 6 years. [*Hint:* This is a problem in compounded decay.]

8.4 An Application of the Binomial Theorem: The Distribution of the Density of Atmospheric Particles

The table of binomial coefficients is called **Pascal's triangle.** (Pascal was a mathematician of the seventeenth century, one of the founders of the theory of probability.) It is usually drawn in the form of an isosceles triangle:

$$\begin{array}{ccccccccccc} & & & & & 1 & & & & & \\ & & & & 1 & & 1 & & & & \\ & & & 1 & & 2 & & 1 & & & \\ & & 1 & & 3 & & 3 & & 1 & & \\ & 1 & & 4 & & 6 & & 4 & & 1 & \\ 1 & & 5 & & 10 & & 10 & & 5 & & 1 \\ & & & & & \cdots & & & & & \end{array}$$

This array has the following property: The entry in any row is the sum of the entries immediately above it in the preceding row; for example, the entry 10 in the sixth row is the sum of 4 and 6, which are above it in the fifth row. This property of the binomial coefficients is expressed by the equation

$$C(n,r) = C(n-1,r-1) + C(n-1,r). \qquad (1)$$

This formula may be proved by substituting the algebraic expressions which the symbols represent, and then reducing one side to the other (Exercise 1).

We mention another property of the binomial coefficients: **Their sum is equal to 2^n for every power n;** that is,

$$C(n,0) + C(n,1) + C(n,2) + \cdots + C(n,n) = 2^n. \qquad (2)$$

This is verified by substituting $A = B = 1$ in the formula for the binomial expansion of $(A + B)^n$.

These coefficients have a physical interpretation in terms of sand flowing through a rectangular network. Sand is poured into a pipe at the top (Figure 8.11). At the bottom of this segment of pipe is a joint where 2 pipes open, 1 to the left and 1 to the right. Half the sand goes into each. At the bottom of each of these are 2 more openings; and half the sand again goes into each. And so on. The amounts of sand deposited at the joints in a given row are proportional to the binomial coefficients (Figure 8.12).

Figure 8.11

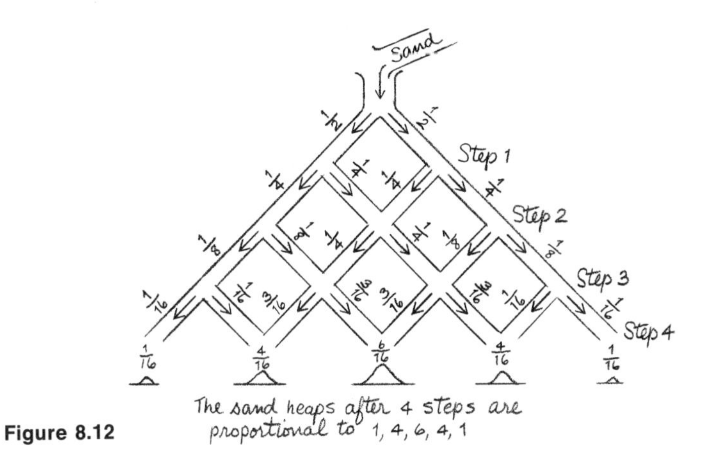

Figure 8.12

The relation between the binomial coefficients and the sand network illustrates the mathematical properties of certain **diffusion** processes, such as the dispersion of effluents from a smokestack into the atmosphere. The effluent consists of a large amount of many small particles, like sand grains. As they emerge from the stack in the presence of a steady wind, they are carried parallel to the ground in the direction of the wind. They also have an up-and-down movement, perpendicular to the ground. A typical "plume" is shown in Figure 8.13. It has been observed that the density of the effluent (particles per volume air) at a fixed distance from the stack varies with the height above the ground, and that the distribution of this density is roughly bell-shaped. This can be mathematically explained by comparing the vertical movement of the effluent to the movement of the sand through the grid. We may imagine the motion of the effluent particles as consisting of small steps, taken at fixed, closely spaced time points. The particles all start at the same height, at the mouth of the stack. At the first time point, half move up 1 unit, and half down 1

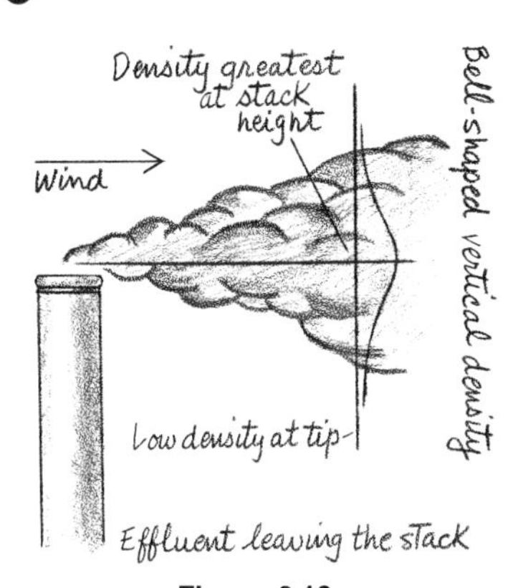

Figure 8.13

unit. At the second time point, the same process is repeated for each of the two groups, and so on (Figure 8.14). In this way the vertical behavior of the particles is exactly like that of the sand flowing through the network. After n steps, the particles are distributed in proportions equal to the binomial coefficients. As one can see from the bar graphs of the coefficients, Figure 8.10, the distribution becomes approximately bell-shaped as the number of steps increases. The "limiting" distribution turns out to be the **normal distribution.** We have already discussed some properties of this distribution in Chapter 6, in connection with population densities. It will be discussed in more detail later in this chapter.

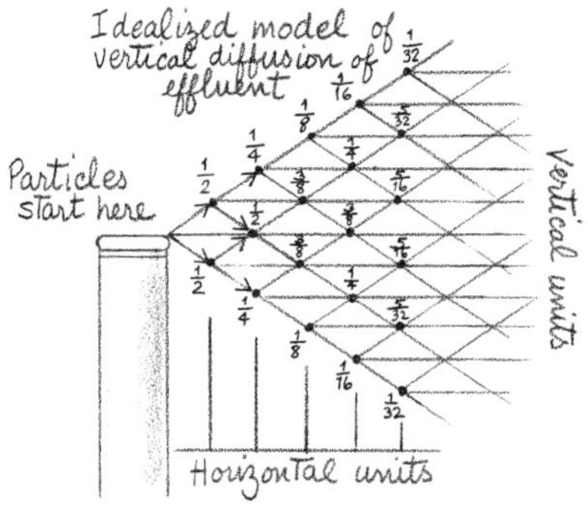

Figure 8.14

8.4 EXERCISES

1 We outline the proof of Equation (1):

(a) Show, directly from the definition, that

$$C(n,r) = \frac{n!}{(n-r)!\, r!}.$$

(b) Show that the right-hand side of Equation (1) is equal to

$$\frac{(n-1)!}{(n-r)!\,(r-1)!} + \frac{(n-1)!}{(n-1-r)!\, r!}.$$

(c) By means of the formula $n! = n\,(n-1)!$, show that the expression in (c) is equivalent to the left-hand side of Equation (1).

2 Every day a gambler bets \$1 on the outcome of each of n tosses of a fair coin. His fortune after each day's series of bets is (in dollars) equal to the difference between the numbers of individual bets won and lost; it may be positive or negative. Suppose on the average that he wins half the individual bets and loses half. After the first bet his fortune is either 1 or -1; after 2 bets, it is -2, 0, or 2; and after 3 bets it is -3, -1, 1, or 3; and so on. Over the long run, what is the proportion of days on which his fortune is -2, 0, or 2, respectively, after the second bet? Indicate the connection between this and the sand grid. What is the relation between the binomial coefficients and the distribution of his fortune after n tosses according to proportions of days? For example, what is the relation between the binomial coefficients for $n = 6$ and the proportion of days on which his fortune is, after 6 tosses, equal to a specified number, such as 2?

8.5 Probability for Equally Likely Outcomes

Probability theory is the study of the mathematics of **random experiments.** Such an experiment is an act or operation with a set of possible outcomes; the **actual** outcome cannot be predicted with certainty before the act. For example, when a coin is tossed, it may show either heads or tails when it lands, so that there are 2 possible outcomes; and the actual outcome on any particular toss cannot be predicted with certainty. The same is true of the toss of a die, and its 6 outcomes. In contrast to the random experiment, a nonrandom experiment has a predictable outcome, only 1 possible outcome; for example, the outcome of the experiment of pouring water on a lighted cigarette is predictable: the fire will be extinguished.

We are interested in the theory of probability because many problems in the sciences may be analyzed from the point of view of random experiments. The **probability** of a particular outcome of a random experiment is a numerical measure of the likelihood that it will be the one selected as the actual outcome of the experiment. Suppose a coin is tossed. Which is the more likely outcome, heads or tails? If the coin is not bent, loaded, or shaved, then we feel that the 2 outcomes are equally likely; we assign each the same numerical probability (Figure 8.15). It is the convention that the sum of the probabilities of the outcomes of the experiment shall be equal to 1. For this reason, the outcomes heads and tails are each assigned the probability $\frac{1}{2}$. Coin-tossing experiments confirm our feeling that heads and tails are "equally likely." If the coin is tossed 2, 4, or 6 times, it is of course not true that half the tosses result in heads and half in tails.

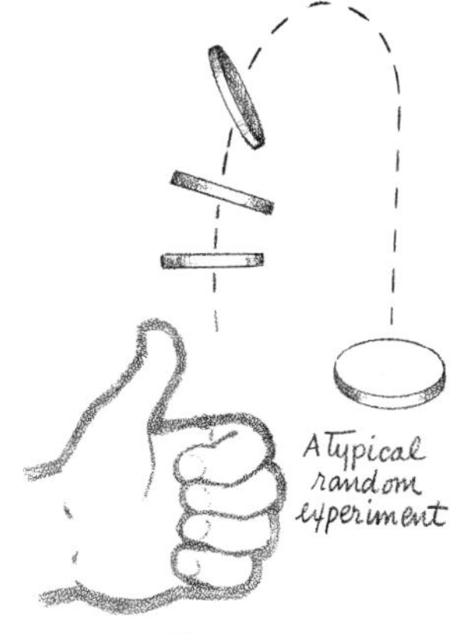

Figure 8.15

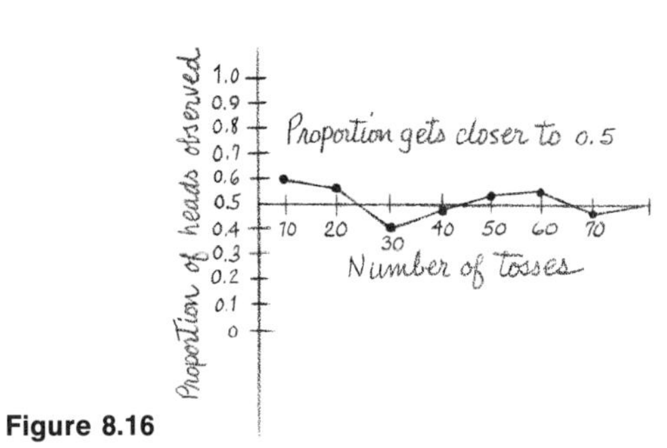

Figure 8.16

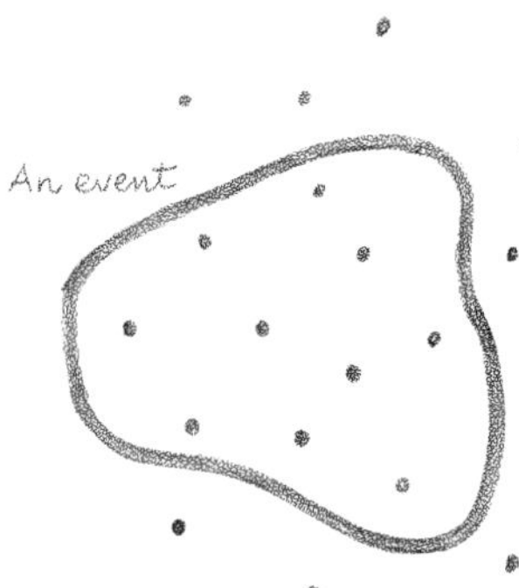

Figure 8.17

However, if the coin is tossed a large number of times, then the **proportion** of heads tends to be close to $\frac{1}{2}$. The record of a long series of tosses is portrayed in Figure 8.16.

When a die is tossed, there are 6 possible outcomes, represented by the digits 1, . . . , 6. If the die is "honest," then each of these outcomes is considered equally likely, and is assigned probability $\frac{1}{6}$. If an honest die is repeatedly tossed, then in the long run the proportions of 1's, 2's, . . . and 6's tend to be close to $\frac{1}{6}$.

Suppose a bowl contains n tickets, numbered 1 through n and that one is drawn at "random." The number appearing on the ticket actually drawn is the result of the random experiment. If the drawing is fair, then each ticket is equally likely, and is assigned the probability $1/n$.

We summarize: **In a random experiment with N equally likely outcomes, each outcome is assigned the probability $1/N$.**

An **event** is a set of outcomes. In an experiment with equally likely outcomes, the probability of an event is defined as the ratio

$$\frac{\text{Number of outcomes forming the event}}{\text{Total number of outcomes of the experiment}}. \tag{1}$$

This is the same as the sum of the probabilities of the outcomes forming the event (Figure 8.17).

EXAMPLE 1

An honest die is tossed once. Consider the event, "An even number appears." This is formed by the 3 outcomes: 2, 4, and 6. Therefore, the probability is $\frac{3}{6}$, or $\frac{1}{2}$. ▲

Now we discuss more complex experiments derived from the coin, die, and ticket-in-the-bowl experiments. These are multiple experiments, corresponding to several tosses and several drawings.

EXAMPLE 2

A fair coin is tossed twice. According to the counting principle there are 4 outcomes; these are

H H, H T, T H, and T T.

Here H H means that heads occurs on both tosses; H T means heads on the first and tails on the second; and so on. We regard each of these 4 outcomes as equally likely, and assign each the same probability $\frac{1}{4}$. Consider the event that exactly 1 head appears in the 2 tosses. It consists of the 2 outcomes H T and T H; hence, by formula (1), the probability of this event is $\frac{1}{2}$.

If the coin is tossed 3 times, there are 8 outcomes, which we regard as equally likely:

H H H	H T H	T T H	T H T
H H T	T H H	H T T	T T T.

Each has probability $\frac{1}{8}$.

If the coin is tossed n times, then, by the counting principle, there are $2 \cdot 2 \cdot \cdot \cdot 2 = 2^n$ possible outcomes; these are of the form

H or T on the first toss, . . . , H or T on the nth toss.

Each outcome has the same probability $1/2^n$. ▲

EXAMPLE 3

A die is tossed twice. Since there are 6 outcomes on each of the 2 tosses, it follows from the counting principle that there are 36 outcomes for the 2 tosses. These are listed below as pairs of integers from 1 through 6. The first and second members of each pair correspond to the scores on the first and second tosses of the die, respectively. These are taken to be equally likely, so that each has probability 1/36.

1 1	1 2	1 3	1 4	1 5	1 6
2 1	2 2	2 3	2 4	2 5	2 6
3 1	3 2	3 3	3 4	3 5	3 6
4 1	4 2	4 3	4 4	4 5	4 6
5 1	5 2	5 3	5 4	5 5	5 6
6 1	6 2	6 3	6 4	6 5	6 6

▲

EXAMPLE 4

A bowl contains n tickets, labeled with the integers 1 through n. Suppose that r are drawn at random. The set of outcomes of this random experiment consists of all permutations of the n tickets taken r at a time. There are $P(n,r)$ such permutations. Each is considered equally likely, and is assigned the probability $1/P(n,r)$. For example, if a bowl contains 4 labeled tickets, and 2 are drawn at random, then there are $P(4,2) = 12$ outcomes, listed below in terms of the first and second tickets drawn:

Outcomes

First	1	1	1	2	2	2	3	3	3	4	4	4
Second	2	3	4	1	3	4	1	2	4	1	2	3

Consider the event, "1 on the first ticket drawn." It consists of 3 outcomes; hence, it has the probability $\frac{3}{12}$, or $\frac{1}{4}$. Similarly, the event, "1 on the second ticket drawn" consists of 3 outcomes, and has the same probability. ▲

Now we show how to compute the probabilities of certain events which naturally arise in the random experiments described in the examples above.

EXAMPLE 5

A fair coin is tossed 5 times. Find the probability of the event "Exactly 3 heads (and 2 tails) appear." By formula (1), this is equal to the ratio

$$\frac{\text{No. of outcomes with exactly 3 heads}}{\text{No. of outcomes for 5 tosses}}.$$

There are $2^5 = 32$ outcomes for the 5 tosses (see Example 2). The number of outcomes with exactly 3 heads is $C(5,3) = 10$. The reason is that the outcomes with 3 heads are formed like combinations of 5 taken 3 at a time. The 5 tosses are like the set of 5 objects. The number of outcomes with 3 heads is like the number of ways of choosing 3 of the 5 for the outcome heads (Figure 8.18). The probability of the event is 10/32 or 5/16.

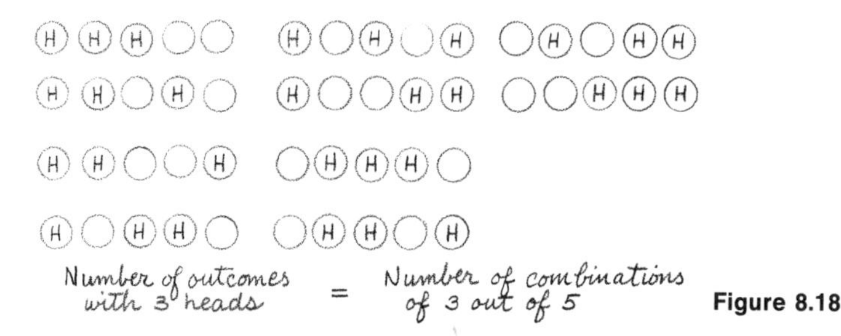

Figure 8.18

By an extension of this reasoning, we get a general formula for the probability of exactly r heads in n tosses:

$$\text{Probability of } r \text{ heads} = \frac{C(n,r)}{2^n}, \qquad \text{for } r = 0, 1, \ldots, n. \qquad (2)$$

The event "At least 4 heads" consists of the outcomes with exactly 4 heads and those with exactly 5 heads. There are $C(5,4)$ and $C(5,5)$ of these, respectively. It follows that the probability is the **sum** of these, divided by 32, or $6/32 = 3/16$. ▲

EXAMPLE 6

Dice throwers are interested in the sum of the scores appearing on a pair of dice, or equivalently, on 2 throws of a single die. The outcomes are listed in Example 3 above. Let us find the probability that the sum of the scores is equal to 5. The event "The sum of the scores is 5" consists of the outcomes

4 1   3 2   2 3   and   1 4

Therefore, the probability is 4/36, or 1/9. The event "The sum of the scores is 8" consists of the outcomes

6 2   5 3   4 4   3 5   and   2 6;

hence, the probability is 5/36 (Figure 8.19). ▲

Figure 8.19 Outcomes forming the event "Sum of scores = 8"

EXAMPLE 7

One hundred tickets are placed in a lottery bowl. Three are going to be drawn at random. Suppose that you have purchased 10 of the tickets in the bowl. Find the probability that yours is the first ticket drawn, and that another of yours also appears on at least 1 of the second or third draws. The number of outcomes of the random experiment is $P(100,3) = 100 \cdot 99 \cdot 98 = 970{,}200$. There are 3 different kinds of permutations forming the event under consideration:

(i) Your tickets are drawn on each of the 3 draws. The number is the same as the number of permutations of 10 objects, taken 3 at a time; hence, there are $P(10,3) = 720$ such outcomes.

(ii) Your tickets are drawn first and second but not third. The number of such outcomes is the product of $P(10,2)$, the number of ways of drawing yours on the first 2, and

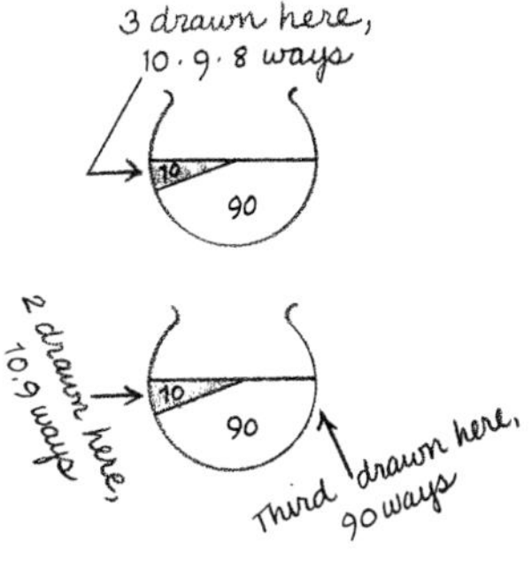

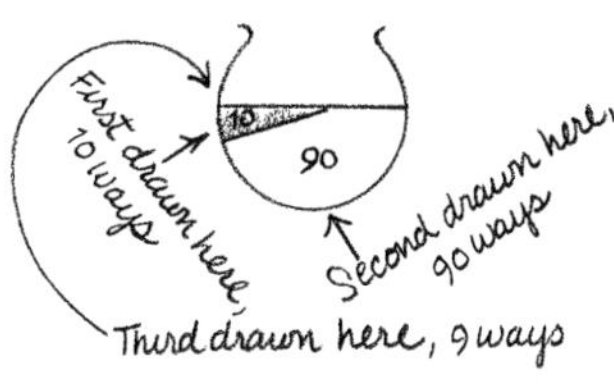

Three different kinds of permutations in the event

Figure 8.20

$P(90,1)$, the number of ways of drawing none of yours, but instead, 1 of the remaining 90, on the third. It follows that there are 8100 such outcomes.

(iii) Your tickets are drawn first and third, but not second. The reasoning is the same as in (ii), and the number is also 8100 (Figure 8.20).

It follows that there are $720 + 8100 + 8100 = 16{,}920$ outcomes in the event described; hence, the probability is the ratio 16,920/970,200, which is approximately 0.174. ▲

8.5 EXERCISES

In Exercises 1 and 2 find the probability of the event associated with the tosses of the fair coin.

1 (a) 1 head, 3 tosses
(b) At least 3 heads, 3 tosses
(c) At least 1 head, 4 tosses
(d) At most 2 heads, 4 tosses
(e) At most 3 heads, 5 tosses
(f) 4 heads, 12 tosses
(g) 3 heads, 6 tosses
(h) 5 heads, 10 tosses

2 (a) 3 heads, 3 tosses
(b) 2 heads, 4 tosses
(c) At least 2 heads, 4 tosses
(d) At most 2 heads, 5 tosses
(e) 7 heads, 10 tosses
(f) 3 heads, 7 tosses
(g) 4 heads, 8 tosses
(h) At most 4 heads, 9 tosses

3 Find the probabilities of these scores on the toss of a pair of dice: 2, 3, 7, 4.

4 It is often assumed that the sex of a child born in a family is like the outcome of the toss of a fair coin, with probability $\frac{1}{2}$ each for a boy and a girl. (The probability for a boy is apparently a bit larger than $\frac{1}{2}$.) Let us assume that the size of the family is independent of the actual sexes of the children already born; for example, the will to have another child is neither weakened nor strengthened by a large proportion of either boys or girls already in the family. Which is more likely, at least 1 boy among 3 siblings, or at least 2 among 6? (See Example 5.)

5 A student is unprepared for a 10-question true-false examination. He answers each question according to the outcome of the toss of a fair coin, "true" for heads and "false" for tails. If the answers to all questions are "false," what is the probability of his scoring at least 60 percent correct? What if the answers are all true? Does it make any difference whether the answers to the questions are "true" or "false"? (See Example 5.)

6 An ice cream vending machine has 2 service outlets. Each successive customer chooses 1 of the 2 at random, like the outcome of the toss of a coin. If there are 6 bars left in each outlet, what is the probability that 1 of these outlets will have been emptied by the time 8 customers have purchased bars? (See Example 5.)

7 The outcomes of the successive games in the World Series are often compared to the outcomes of the tosses of a coin, provided the teams are about evenly matched. (Recall that the first team to win 4 is the winner, and the series stops after the fourth game won.) Find the probability that
(a) a given team wins the first 4
(b) either team wins the first 4
(c) a given team wins 4 out of the first 5
(d) either team wins 4 out of the first 6
(e) the series lasts 7 games (See Example 2 with $n = 7$.)

8 A gambler tosses a coin and stops as soon as the first head appears. (He stops after the first toss if it appears first.) He will also not toss more than 8 times. Describe the outcomes in the event, "The first head appears on the sixth toss," and find its probability. (See Example 2 with $n = 8$.)

9 A pair of dice is tossed. Find the probability that the sum of the scores is an even number. (See Example 3.)

10 A single die is tossed 3 times. (a) By the counting principle, determine the number of outcomes. (b) Describe the set of outcomes. (c) Which outcomes form the event "The sum of the scores is equal to 4"? (d) Find its probability. (See Example 3.)

11 One hundred numbered tickets are placed in a bowl and 2 are drawn in order at random. Find the probability that (a) both numbers drawn are even; (b) the first is odd; (c) the first is odd and the second is even. (See Example 4.)

12 Three tickets are drawn at random from a bowl containing 50 numbered tickets. (a) Find the probability that the first one drawn will have a number from 1 through 5. (b) Find the probability that either the second or third or both will have a number from 6 through 10. (c) Find the probability that the first will have a number from 1 through 5 **and** the second or third or both will show a number from 6 through 10. (See Example 7.)

13 Two tickets are drawn from a bowl of 10. Find the probability that the number on the first ticket drawn is greater than the number on the second. (See Example 7.)

14 Three tickets are drawn from a bowl of 5. Find the probability that the numbers on the tickets drawn appear in increasing order of magnitude. (See Example 7.)

15 (Draft lottery.) 365 tickets are drawn one at a time at random. Find the probability that No. 37 is
(a) not among the first 5
(b) not among the first 10
(See Example 7.)

16 Two friends in the lottery in Exercise 15 have the numbers 37 and 148, respectively. Find the probability that neither of their numbers is among the first 5 drawn; and the probability that exactly one of them will have his number drawn among the first 4. (See Example 7.)

17 The **median** of a set of distinct numbers is that number in the set such that equally many numbers in the set are greater and smaller than it; for example, the median of the set

$$1, \quad 5, \quad 4, \quad 13, \quad 2$$

is 4 because there are 2 larger and 2 smaller than it. (We shall define the median here only for a set with an odd number of members.) A bowl contains 10 numbered tickets, and 3 are drawn at random. Find the probability that No. 6 is drawn and is equal to the median of the numbers drawn.

18 In Exercise 17, find the probability that No. 4 is drawn and is the median.

8.6 Unordered Random Samples

In drawing tickets from a bowl, as in the case of the draft lottery, the order in which the tickets are drawn is most important. However, in other problems of drawing objects at random from a "bowl," the order of appearance is unimportant; it is the identity of the objects drawn that is important. In this case the analysis of the events and the computation of the probabilities is simpler. Let the bowl contain n numbered tickets, and let r be drawn at random, where $r \leq n$. Such a set of tickets is an **unordered random sample,** or simply an **unordered sample.** If we disregard the order in which the tickets are drawn, then the sample represents a combination of n objects taken r at a time. Since all permutations are equally likely under random sampling, it follows that all combinations are also equally likely; indeed, each combination has the same number of permutations (Section 8.2). In other words, if we consider the outcomes of the drawings as combinations (that is, we disregard the order of the drawing), then every combination has the same probability; the latter is equal to $1/C(n,r)$.

EXAMPLE 1

Four numbered tickets are in a bowl, and 2 are drawn at random. There are 6 combinations as outcomes:

$$1\ 2, \quad 1\ 3, \quad 1\ 4, \quad 2\ 3, \quad 2\ 4, \quad 3\ 4.$$

Twelve permutations are generated by these, 2 from each combination. Since each permutation has the same probability, $\frac{1}{12}$, each combination has probability $\frac{1}{6}$. ▲

In discussing unordered samples, we refer to the experiment of drawing colored balls from an urn instead of numbered tickets from a bowl. The basic example is that of an urn with balls of 2 colors; for example, white and red. The balls are indistinguishable except for color. Put

$$\begin{aligned} n &= \text{No. of balls in the urn,} \\ w &= \text{No. of white balls in the urn,} \\ r &= \text{No. of balls drawn in the sample,} \\ k &= \text{No. of white balls drawn in the sample.} \end{aligned}$$

It follows that

$$\begin{aligned} n - w &= \text{No. of red balls in the urn,} \\ r - k &= \text{No. of red balls drawn in the sample.} \end{aligned}$$

By definition of the probability for equally likely outcomes, **the probability that the sample contains exactly k white (and, therefore, $r - k$ red) balls is**

$$\frac{\text{No. of combinations with } k \text{ white (and } r - k \text{ red)}}{\text{No. of combinations of } n \text{ balls taken } r \text{ at a time}}. \qquad (1)$$

(This is valid as long as k does not exceed w or r.) We shall show that this probability is given by

$$\frac{C(w,k)C(n-w,r-k)}{C(n,r)}. \qquad (2)$$

The reason for (2) is as follows. A combination with k white and $r - k$ red balls is formed by choosing k white out of the total of w in the urn, and then a combination of $r - k$ red from the $n - w$ in the urn. The first combination is formed in $C(w,k)$ ways, and the second in $C(n - w,\ r - k)$ ways. By the counting principle, both combinations are formed in a number of ways equal to the product. By definition, the denominator in (2) is equal to that in (1). It follows that the ratios are equal (Figure 8.21).

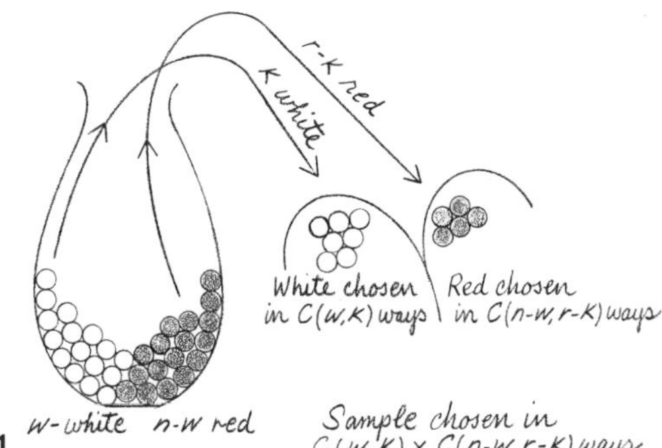

Figure 8.21

The system of probabilities (2) for $k = 0, 1, \ldots$ is called the **hypergeometric distribution.**

EXAMPLE 2

A retail store inspector is testing the quality of containers of cheese. There are 20 containers of cheese on the shelf, and 3 are stale. The inspector takes 2 at random, and checks them. We think of the containers on the shelf as balls in an urn, and the containers of stale cheese as the white balls, and the others as the red. What is the probability that neither of the inspected containers is stale? According to (2), the probability is

$$\frac{C(3,0) \cdot C(17,2)}{C(20,2)} \quad \text{or} \quad \frac{17 \cdot 16}{20 \cdot 19} = 0.716.$$

The probability that exactly 1 container has stale cheese is

$$\frac{C(3,1)C(17,1)}{C(20,2)} = \frac{102}{380} = 0.268.$$

The probability that both are stale is

$$\frac{C(3,2)C(17,0)}{C(20,2)} = \frac{6}{380} = 0.016. \ \blacktriangle$$

The model of sampling from an urn with balls of 2 colors can be generalized to more than 2 colors. Suppose there are n_1 balls of color 1, n_2 balls of color 2, . . . , and n_m balls of color m. Then **the probability that the sample has k_1 balls of color 1, k_2 of color 2, . . . , and k_m of color m is**

$$\frac{C(n_1,k_1) \ . \ . \ . \ \cdot C(n_m,k_m)}{C(n,r)}; \tag{3}$$

here, the numbers of balls of various colors in the urn have the sum n, and the numbers in the sample have the sum r. The proof of (3) is similar to that of (2): The numbers of ways of picking the balls of various colors is equal to the product of the individual combinations.

EXAMPLE 3

An apartment house has 10 families with annual incomes below \$9000, 10 families with income from \$9000 to \$15,000, and 5 families with income above \$15,000. A voting pollster selects 4 families at random for the purpose of finding out their voting preferences. Find the probability that the sample contains at least 1 from each of the 3 income groups. There are 3 ways in which the sample includes all income groups: 2 from the lower group, and 1 from each of the 2 others; 2 from the middle group and 1 from each of the others; and 2 from the upper group and 1 from each of the others. There are

$$C(10,2) \cdot C(10,1) \cdot C(5,1)$$

combinations corresponding to the first of the 3; and the same number for the second. The number of combinations of the third kind is

$$C(10,1) \cdot C(10,1) \cdot C(5,2).$$

The probability that each income group is represented is the sum of the numbers of these combinations, divided by the total number of combinations, $C(25,4)$:

$$\frac{45}{253} + \frac{45}{253} + \frac{20}{253}, \qquad \text{or approximately } 0.43. \ \blacktriangle$$

8.6 EXERCISES

In Exercises 1 and 2 find the probability of getting k *white balls in a sample of* r *when the urn contains* w *white balls out of* n.

1 (a) $n = 4$, $w = 2$, $r = 2$, $k = 0, 1, 2$
(b) $n = 5$, $w = 3$, $r = 2$, $k = 0, 1, 2$
(c) $n = 5$, $w = 2$, $r = 2$, $k = 0, 1, 2$
(d) $n = 6$, $w = 3$, $r = 3$, $k = 0, 1, 2, 3$

2 (a) $n = 4$, $w = 3$, $r = 2$, $k = 1, 2$
(b) $n = 5$, $w = 3$, $r = 3$, $k = 1, 2, 3$
(c) $n = 6$, $w = 3$, $r = 2$, $k = 0, 1, 2$
(d) $n = 6$, $w = 4$, $r = 3$, $k = 1, 2, 3$

3 An urn has 3 red, 3 white, and 4 blue balls. A sample of 3 is drawn at random. Find the probability of getting
(a) 3 white balls
(b) balls of the same color
(c) balls of at least 2 different colors
(d) balls of 3 different colors

4 An urn has 4 red, 5 white, and 6 blue balls. A sample of 4 is drawn at random. Find the probability of getting
(a) 2 red and 1 of each of white and blue balls
(b) all white balls
(c) balls of 3 different colors
(d) balls of at most 2 different colors

5 Fifty mice have been infected and 40 of these have contracted a disease. A sample of 3 is drawn at random and tested. Find the probability that at least 2 of the mice will have the disease. (See Example 2.)

6 A tank of experimental fruit flies contains 3 different strains: 6 of type I, 6 of type II, and 8 of type III. A sample of 4 is chosen at random. Find the probability that at least 1 of each of the strains appears in the sample. (See Example 3.)

7 A batch of 20 bulbs contains 4 defectives. A sample of 4 is drawn at random, and each bulb in the sample is tested. Find the probability that the sample contains

(a) at least 1 defective
(b) at least 2 defectives (See Example 2.)

8 A class consists of 15 boys and 10 girls. Three are chosen at random for a psychological experiment. Find the probability that the sample consists of

(a) 3 girls
(b) 3 boys
(c) at least 1 of each
(d) a majority of boys
(e) a majority of girls (See Example 2.)

9 Of the workers in an office, 15 are supporters of the Democratic party and 10 are supporters of the Republican party. A sample of 5 workers is chosen at random. What is the probability of a majority of Democrats in the sample? (See Example 2.)

10 An economist draws a sample of 3 families from a list of 25 families with the following income distribution:

Income (\$1000's)	Below 5	5–10	10+
Number	15	5	5

Find the probability that

(a) the lower group is represented in the sample
(b) at least 2 income groups, including the lower, are represented
(c) at least 2 income groups are represented
(d) all 3 groups are represented (See Example 3.)

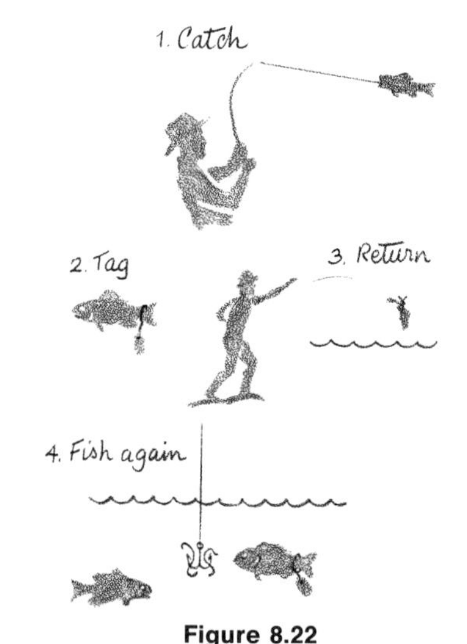

Figure 8.22

11 The following method has been used to estimate the number of fish in a lake. A sample of M fish is taken from the lake, tagged by means of some mark, and returned to the water. After a while, another sample is drawn. Let p be the proportion of tagged fish in the second sample; then the estimate of the population size is M/p. This method can be analyzed from the point of view of probability. Suppose a box has n red balls where $n \geq 4$. Four balls are drawn out, painted white, and returned to the box. Then a sample of 4 is drawn at random. Find the probability that the sample contains exactly 2 white balls if

(a) $n = 8$ (b) $n = 6$ (c) $n = 12$

Under which of these values of n are you most likely to find **exactly** 2 white balls in the sample? (Figure 8.22.) Here $M = 4$ and $p = \frac{1}{2}$.

8.7 Bernoulli Trials

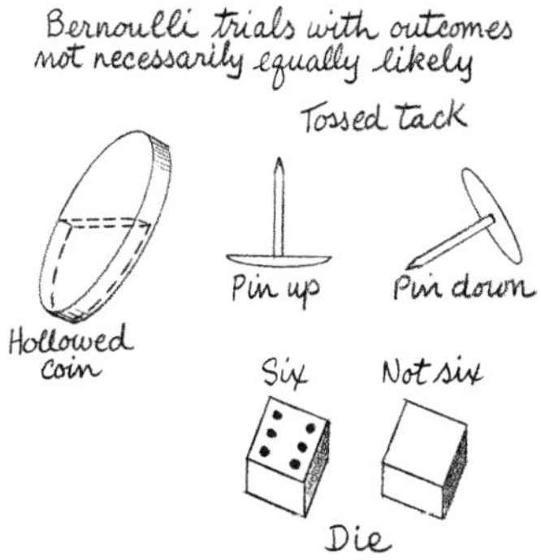

If a coin is shaved or bent, then the 2 outcomes, heads and tails, might no longer be equally likely. If a tack is thrown at random on a table, then it lands either on its head or its side, pin up or pin down. It seems that a common tack is more likely to fall pin down. You can check this by throwing a tack several times. If a die is tossed, and you bet on "6," then there are, as far as you are concerned, only 2 outcomes. **6** or **not 6.** If the die is fair, then these outcomes have the (unequal) probabilities $\frac{1}{6}$ and $\frac{5}{6}$, respectively. These are several examples of random experiments with 2 outcomes that are **not necessarily** equally likely; such an experiment is called a **Bernoulli trial** (Figure 8.23). [There were two mathematicians with this name who worked on the theory of probability: James Bernoulli (1654–1705), and Daniel Bernoulli (1700–1782). Bernoulli trials are apparently named after the former.] The particular case in which the outcomes are equally likely is also a Bernoulli trial.

Figure 8.23

The 2 outcomes of a Bernoulli trial are referred to as "success" and "failure." In coin-tossing, heads is called success and tails failure. The probabilities for these are not necessarily $\frac{1}{2}$. We denote by a number p, where p is between 0 and 1 inclusive, the probability of success. If p is greater than $\frac{1}{2}$, then success is more probable than failure; if it is less than $\frac{1}{2}$, then success is less probable than failure. If p is close to 1, then success is highly probable. The case $p = 1$ is an extreme case; success occurs with "probability 1," which means that it is certain to occur. In this case the random experiment is not really random, because there is only one possible outcome, success. If p is close to 0, then success is very improbable; if it is equal to 0, then success is impossible, and the experiment has only 1 outcome.

Since, by convention, the probabilities of success and failure should have the sum 1, we define the probability of failure as $q = 1 - p$. In what follows, we shall always suppose that p is strictly between 0 and 1 (between and not equal to 0 or 1), so that the experiment is truly a random one.

If the coin is tossed twice, then, as in the case of the balanced coin, there are 4 possible outcomes; however, the probabilities are not necessarily equal. We assign the probabilities in this way:

Outcome	H H	H T	T H	T T
Probability	p^2	pq	qp	q^2

The reason for this assignment is discussed in Section 8.8. Let us accept it for the moment as given. The rule used here is that the probability of an outcome is a product of the two factors p and q: a factor p for each success, and q for each failure. These probabilities are positive because p and q are positive. The sum of the 4 probabilities is equal to 1 because $p + q = 1$; indeed, by the binomial theorem, we have

$$p^2 + pq + qp + q^2 = (p + q)^2 = 1^2 = 1.$$

As in the case of the balanced coin, the number of heads appearing in 2 tosses may be 0, 1, or 2; however, the probabilities are no longer equal to $\frac{1}{4}$, $\frac{1}{2}$, and $\frac{1}{4}$, respectively. The event "No heads in 2 tosses" consists of the outcome T T; hence, the probability is q^2. The event "Exactly 1 head in 2 tosses" consists of H T and T H. Their probabilities are pq and qp, respectively; hence, the probability of the event is $pq + qp = 2pq$. Similarly, the probability of 2 heads in 2 tosses is p^2. The system of probabilities

No. of heads	0	1	2
Probabilities	q^2	$2pq$	p^2

is called the **probability distribution** of the number of heads in 2 tosses; equivalently, it is the distribution of the **number of successes** in 2 Bernoulli trials.

Suppose a coin is tossed 3 times. There are 8 possible outcomes. The probabilities are assigned in the following way:

Outcome	Probability	Outcome	Probability
H H H	p^3	T T T	q^3
H H T	p^2q	T T H	q^2p
H T H	p^2q	T H T	q^2p
T H H	p^2q	H T T	q^2p

Each probability is a product of 3 factors p and q, with a factor p for each H, and a factor q for each T. The sum of the probabilities is

$$p^3 + 3p^2q + 3pq^2 + q^3,$$

which, by the binomial theorem, is equal to $(p + q)^3 = 1$. The probability distribution of the number of heads is

No. of heads	0	1	2	3
Probability	q^3	$3q^2p$	$3qp^2$	p^3

For example, the event "2 heads in 3 tosses" consists of the 3 outcomes

H H T, H T H, and T H H,

which have the common probability p^2q; therefore, the probability of 2 heads is equal to the sum of the 3 probabilities, $3p^2q$. The other probabilities are computed in a similar way (Figure 8.24).

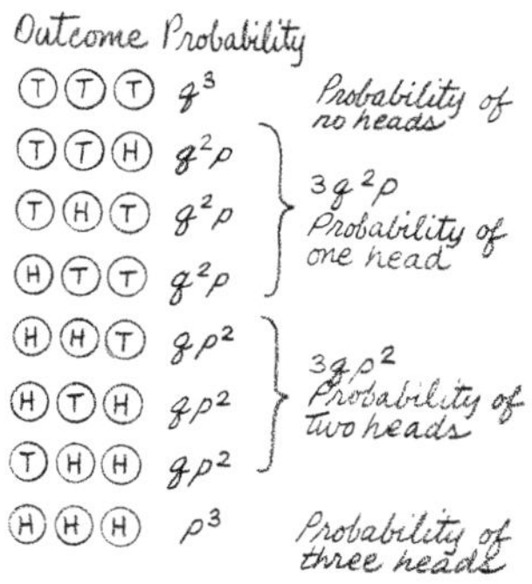

We generalize this to n tosses of a coin, or n Bernoulli trials. The system of outcomes is the system of all multiplets of n letters that can be formed from the 2 letters H and T. The letters indicate the outcomes; for example, the outcome corresponding to the multiplet H T H . . . T has heads on the first and third tosses, and tails on the second and last tosses. There are 2^n outcomes (Section 8.5, Example 2). The probability assigned to an outcome is a product of n factors p and q, with a p-factor for each head and a q-factor for each tail. The probability of exactly k heads in n tosses is

$$C(n,k)p^kq^{n-k}, \qquad k = 0, 1, \ldots, n. \tag{1}$$

The reason is as follows: The event "k heads in n tosses" consists of all outcomes with exactly k heads and $n - k$ tails. **Each** such outcome has probability p^kq^{n-k}. There are $C(n,k)$ such outcomes; indeed, each arrangement of k heads among n tosses is a combination of n objects taken k at a time (Section 8.5, Example 5). Therefore, the sum of the probabilities of all outcomes with exactly k heads is equal to the common probability p^kq^{n-k} multiplied by the number $C(n,k)$ of outcomes; this is the reason for (1).

The system of probabilities (1) is called the **binomial distribution** because its terms are those obtained by expanding $(p + q)^n$ according to the binomial theorem (Section 8.3). The sum of the terms is $(p + q)^n = 1$. It is the distribution of the number of successes in n Bernoulli trials, where the probability of success on each trial is p. Table VI (Appendix) contains the values of the terms of the distribution for various n and p. The values are given explicitly for $p \leq 0.50$. The values for $p > 0.50$ are obtained from these by symmetry: The probability of k successes in n trials for a given p is equal to the probability of $n - k$ successes in n trials for the corresponding value $1 - p$. For example, the probabilities for $p = 0.6, 0.7, \ldots$ are obtained by reversing the probabilities for 0.4, 0.3, respectively: the probability of 2 successes in 6 trials with $p = 0.7$ is equal to the probability of 4 successes with $p = 0.3$, namely, 0.0595.

Outcome	Probability	
TTT	q^3	Probability of no heads
TTH	q^2p	
THT	q^2p	$3q^2p$ Probability of one head
HTT	q^2p	
HHT	qp^2	
HTH	qp^2	$3qp^2$ Probability of two heads
THH	qp^2	
HHH	p^3	Probability of three heads

Figure 8.24

EXAMPLE 1

A coin is tossed 6 times. Find the probability of obtaining exactly 2 heads. According to formula (1) with $n = 6$ and $k = 2$, the probability is

$$C(6,2)p^2q^4 \qquad \text{or} \qquad 15p^2q^4. \ \blacktriangle$$

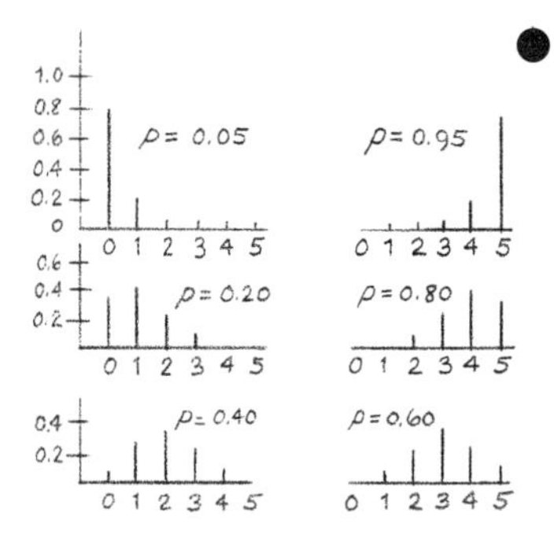

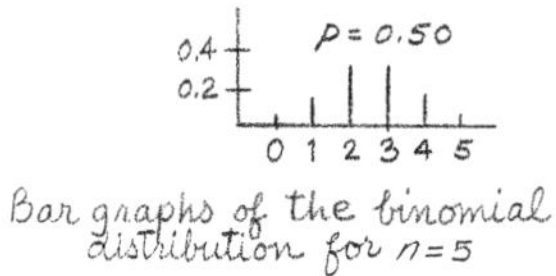

Figure 8.25

In the particular case $p = q = \frac{1}{2}$, the binomial distribution (1) reduces to the system (2) of Section 8.5.

Bar graphs of various binomial distributions with $n = 5$ appear in Figure 8.25. Note that the distribution is symmetric about the point 2.5 when $p = \frac{1}{2}$. When p is less than $\frac{1}{2}$, the distribution is "piled" on integers smaller than 2.5; and when p is greater than $\frac{1}{2}$, it is piled on larger integers. The integer carrying the highest probability is the **most probable value,** or **modal** value. It can be shown that this is the integer which is greater than $5p - q$ but not more than $5p + p$. For example, when $p = \frac{1}{3}$ and $q = \frac{2}{3}$, the most probable value is 2 because

$$5p + p = 2 \qquad \text{and} \qquad 5p - q = 1.$$

In general, the most probable value is that integer which is greater than $np - q$ but not more than $np + p$. In some cases there are 2 consecutive most probable values; for example, the values 0 and 1 for $n = 4$, $p = 0.2$, which have probability 0.4096. Since p and q are between 0 and 1, the most probable value is within 1 unit of np. This is the "central" value of the general binomial distribution, and is called the **expected value.** It coincides with the center of gravity of the probability distribution; this will be discussed in more detail in Section 8.13.

While formula (1) gives the probability for an exact number k of heads, it can also be used to calculate the probability that the number of heads is one of a set of integers from among 0 through n. The probability that the number of heads belongs to a given set of integers is the sum of the probabilities (1) over all k in that set; for example, the probability that there are at least 2 heads in 4 tosses is the sum of the probabilities (1) for $n = 4$ over the integers $k = 2, 3, 4$; that is,

$$C(4,2)p^2q^2 + C(4,3)p^3q + C(4,4)p^4.$$

Another way of calculating this sum is based on the fact that the sum of all the terms in the distribution is equal to 1: we may subtract the complementary terms of the distribution from 1 to get the same sum, namely,

$$1 - C(4,0)q^4 - C(4,1)pq^3.$$

This is a more convenient method when the set of integers is relatively large.

EXAMPLE 2

The sexes of the successive children born in a given family are compared to the outcomes of successive Bernoulli trials (see Section 8.5, Exercise 4). With apologies to the ladies, we continue with the arbitrary but conventional practice of identifying the birth of a male as a "success," and that of a female as a "failure." In the earlier exercise, we noted that it is often assumed that $p = \frac{1}{2}$, but in reality, p is a bit more than $\frac{1}{2}$, namely 0.51. Let us compare the distributions of the number of males in a family of 4 siblings under $p = 0.5$ and $p = 0.51$:

No. of males	0	1	2	3	4
Probability for $p = 0.5$	$\frac{1}{16}$	$\frac{1}{4}$	$\frac{3}{8}$	$\frac{1}{4}$	$\frac{1}{16}$
Probability for $p = 0.51$	0.06	0.24	0.37	0.26	0.07

So the difference between the distributions is small if the number of siblings is not too large.

Let us compute the probability that the number of males is at most 3, under the hypothesis $p = 0.51$. It is the sum of the probabilities for $k = 0$, 1, 2, and 3. A simple way to get this probability is to subtract the complementary probability for $k = 4$, namely, 0.07, from 1; we obtain $1 - 0.07 = 0.93$.

The probability of at least 2 males is the sum of the probabilities for $k = 2, 3, 4$, namely,

$$0.37 + 0.26 + 0.07 = 0.70.$$

This can also be obtained by subtracting the sum of the complementary probabilities from 1:

$$1 - 0.06 - 0.24 = 0.70. \quad \blacktriangle$$

EXAMPLE 3

Suppose you are told that a hollowed wooden nickel is biased, that is, the probability of heads is not $\frac{1}{2}$. Furthermore, you are told that p is either 0.3 or 0.7, but not which. How shall you determine which value of p is the correct one? A reasonable procedure is to toss the coin several times and observe the numbers of heads and tails. If the number of heads is much greater, then you decide that $p = 0.7$; if tails is much greater, then choose $p = 0.3$; finally, if the numbers are close, you would make no decision without further experimentation.

This procedure is an application of the general statistical principle of **maximum likelihood,** due to R. A. Fisher, one of the founders of the modern theory of statistics in the early twentieth century. We decide in favor of that value of p which is **more likely** to lead to the result actually observed. Suppose the coin is tossed 5 times; then the number of heads observed is from 0 through 5. From Table VI we get the distributions of the number of heads under the two hypotheses:

		No. of Heads					
		0	1	2	3	4	5
Probability for	$p = 0.7$	0.0024	0.0284	0.1323	0.3087	0.3602	0.1681
	$p = 0.3$	0.1681	0.3602	0.3087	0.1323	0.0284	0.0024

Consider the event that 5 heads are observed; its probabilities are 0.1681 under the hypothesis $p = 0.7$ and 0.0024 under $p = 0.3$. Thus, this event is 84 times more likely to happen under $p = 0.7$ than under $p = 0.3$. For this reason we feel that we should decide in favor of $p = 0.7$ if 5 heads are actually observed. By symmetry, we decide in favor of $p = 0.3$ if 5 tails are observed.

The probabilities for 4 heads are 0.3602 under the hypothesis $p = 0.7$, and 0.0284 under $p = 0.3$; the ratio of these probabilities is almost 13. If 4 heads are observed, it is still reasonable to select $p = 0.7$.

If 3 heads are observed, you may have doubts about deciding without further investigation. In this case the ratio of probabilities is 0.3087/0.1323, or 2.34. If you had to make a decision, then it is reasonable to choose $p = 0.7$ because it is the more likely one. ▲

8.7 EXERCISES

In Exercises 1 and 2 compute the probabilities (1) for the given values of n, k, *and* p; *then compare these to the values given in Table VI.*

1 (a) $n = 2, \quad k = 1, \quad p = 0.1$
(b) $n = 4, \quad k = 3, \quad p = 0.1$
(c) $n = 3, \quad k = 2, \quad p = 0.1$
(d) $n = 2, \quad k = 2, \quad p = 0.2$
(e) $n = 3, \quad k = 3, \quad p = 0.8$
(f) $n = 2, \quad k = 1, \quad p = 0.7$
(g) $n = 3, \quad k = 3, \quad p = 0.3$
(h) $n = 4, \quad k = 3, \quad p = 0.7$
(i) $n = 6, \quad k = 1, \quad p = 0.5$
(j) $n = 7, \quad k = 3, \quad p = 0.6$

2 (a) $n = 3, \quad k = 0, \quad p = 0.1$
(b) $n = 4, \quad k = 3, \quad p = 0.9$
(c) $n = 2, \quad k = 0, \quad p = 0.2$
(d) $n = 3, \quad k = 1, \quad p = 0.8$
(e) $n = 4, \quad k = 0, \quad p = 0.2$
(f) $n = 3, \quad k = 2, \quad p = 0.7$
(g) $n = 3, \quad k = 1, \quad p = 0.3$
(h) $n = 6, \quad k = 2, \quad p = 0.5$
(i) $n = 7, \quad k = 2, \quad p = 0.5$
(j) $n = 7, \quad k = 4, \quad p = 0.6$

In Exercises 3 and 4 find the probabilities that the number of heads is at least equal to the specified number. Use Table VI for the computation.

3 (a) at least 1 head in 2 tosses, $p = 0.1$
(b) at least 2 heads in 4 tosses, $p = 0.9$

(c) at least 2 heads in 3 tosses, $p = 0.2$
(d) at least 1 head in 4 tosses, $p = 0.4$
(e) at least 3 heads in 7 tosses, $p = 0.5$

4 (a) at least 2 heads in 3 tosses, $p = 0.6$
(b) at least 3 heads in 4 tosses, $p = 0.1$
(c) at least 2 heads in 5 tosses, $p = 0.7$
(d) at least 1 head in 4 tosses, $p = 0.3$
(e) at least 5 heads in 6 tosses, $p = 0.5$

5 Find the probability of getting at least 3 six's in 5 tosses of a die.

6 On a multiple choice examination a student finds that he does not know the answers to 4 of the questions. There are 4 choices as answers for each of the questions. He chooses each answer at random, selecting the correct one with probability $p = \frac{1}{4}$. Find the probability that he answers at least 2 correctly.

7 A candy-bar machine has 5 outlets. Each successive customer chooses an outlet at random. If each outlet contains 8 bars, what is the probability that the outlet on the extreme right will be empty by the time 10 customers have bought bars? (Figure 8.26.)

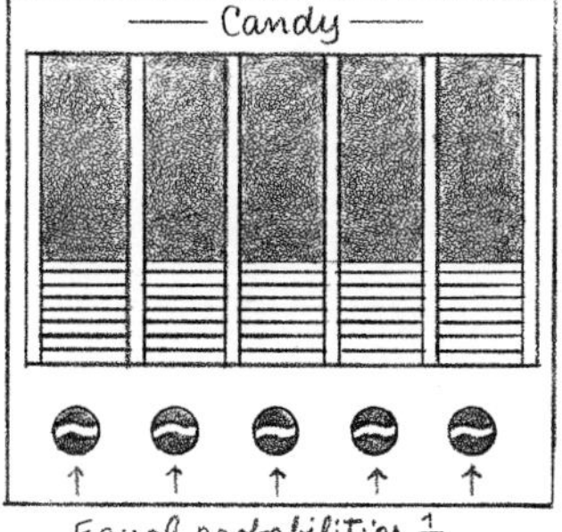

Figure 8.26

8 An employment agent sends resumés to several employers on behalf of a client. The probability that an employer responds to the letter is $\frac{1}{5}$. If letters are sent to 8 employers, what is the probability of at least 3 responses?

9 When a patient is treated for a disease he either recovers or not. The outcome is like that of a Bernoulli trial: success for recovery and failure for nonrecovery. Let p be the probability that a patient recovers. When a group of n patients has the disease, which we suppose to be not communicable, then the number of recoveries is like the number of successes in n Bernoulli trials with probability p of success on each trial. Suppose that an investigator wants to determine the rate of recovery under a certain treatment. The natural rate of recovery in the absence of treatment is given as 50 percent; and the rate of recovery with the proposed treatment is supposed to be 80 percent. Eight patients are treated. What is the most probable number of recoveries if the treatment is worthless, that is, $p = 0.5$? What is the most probable number when the treatment is as effective as claimed, that is, $p = 0.8$? Consider the event, "At least 6 patients recover." What is the probability when $p = 0.5$? When $p = 0.8$?

10 The probability that a certain baseball player will get a hit on his turn to bat is 0.3. Find the probability that he gets at least 2 hits in 5 turns.

11 The probability that a student will awake in time to get to class in the morning is 0.9. Find the probability that he is late at most once in 7 days.

12 The probability that an automobile will have to be returned to the manufacturer for mechanical adjustments is 0.2. If 8 automobiles are sold 1 day in a particular store, what is the probability that at least 1 will have to be returned?

13 The principle of maximum likelihood applies to several values of p (see Example 3). Suppose that the probability that the graduate of a particular high school completes college is p; and that the number in a group of n from that school is like the number of successes in n Bernoulli trials with probability p. It is assumed that p is one of the numbers 0.1, 0.2, . . . , 0.9, but it is not known which it is. A group of 10 enters college, and 7 complete their studies. Under which value of p is the probability of this event the highest? (This gives the "maximum likelihood estimate" of p.)

14 We refer to the sand grid in Section 8.4. Suppose at each joint the proportions of sand going to the right and to the left are p and q, respectively. Show that the piles at each level are proportional to the terms of the binomial distribution (1); consider the particular cases $n = 1, 2, 3, 4$.

8.8 Sampling with Replacement and Bernoulli Trials

The procedure described in Section 8.5 (Example 4) for drawing tickets from a bowl is known as **sampling without replacement.** This is to distinguish it from another procedure, called **sampling with replacement.** In the latter, the first ticket is drawn at random and put back in the bowl. Then a second ticket is drawn at random and returned. And so on. Every ticket has the same probability of being drawn on each drawing, and may be drawn several times. If r tickets are drawn at random in this way, then there are n^r possible outcomes; indeed, there are n ways in which each ticket can be drawn, and so by the counting principle, there are $n \cdot n \ldots \cdot n = n^r$ ways of drawing all the tickets. Each outcome is an ordered arrangement of r tickets with possible repetitions.

In sampling without replacement, the number of outcomes is $P(n,r)$. We shall show that **if n is large relative to r, then sampling with and without replacement are practically the same.** The intuitive explanation is that it is very unlikely for a ticket to be chosen twice if the number of draws is small relative to the number of tickets in the bowl. The mathematical justification is as follows. There are more outcomes possible when the sampling is with replacement because repetitions may occur; however, there are "relatively few more." When the sampling is without replacement there are $P(n,r)$ possible outcomes; with replacement, n^r; and the ratio of the two is

$$\frac{P(n,r)}{n^r}. \tag{1}$$

This is very close to 1 if n is large relative to r; for example, when $r = 2$,

$$\frac{P(n,2)}{n^2} = \frac{n(n-1)}{n^2} = 1 - \frac{1}{n},$$

which is close to 1 for large n. [In general, for each fixed r,

$$\lim_{n\to\infty} \frac{P(n,r)}{n^r} = 1$$

(see Section 5.3).]

Here is an important application. An urn contains n balls, w white, and $n-w$ red. r balls are drawn at random in order **with replacement.** Let p be the proportion of white balls in the urn, $p = w/n$. Consider a particular outcome of r drawings, resulting in k white and $r-k$ red balls. The k draws resulting in white balls can be made in w^k ways; the $r-k$ draws resulting in red balls can be made in $(n-w)^{r-k}$ ways; and there are n^r outcomes in all. It follows that **the probability of such an outcome is the ratio**

$$\frac{w^k(n-w)^{r-k}}{n^r} = \frac{w^k(n-w)^{r-k}}{n^k \cdot n^{r-k}} = p^k(1-p)^{r-k}.$$

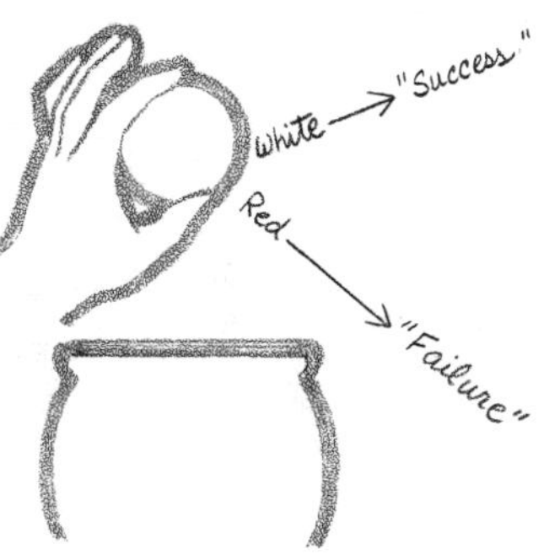

Figure 8.27

This is the same as the probability of an outcome with k successes and $r-k$ failures in r Bernoulli trials. If we consider the drawing of a white ball as success and a red as failure, then the distribution of the number of whites is the same as the distribution of the number of successes, namely, it is the binomial distribution with r in place of n (Figure 8.27). Incidentally, this also explains the assignment of the probabilities for Bernoulli trials in Section 8.7.

If n is large relative to r, then, as shown above, sampling with and without replacement are nearly the same; therefore, the binomial distribution may be used in place of the hypergeometric distribution [formula (2), Section 8.6].

EXAMPLE 1

An urn contains 4 white and 6 red balls. If 2 are drawn at random without replacement, then, by formula (2), Section 8.6, the probabilities of 0, 1, and 2 whites are

$$\frac{C(4,0)C(6,2)}{C(10,2)} = \frac{1}{3}, \qquad \frac{C(4,1)C(6,1)}{C(10,2)} = \frac{8}{15} = 0.533,$$

$$\frac{C(4,2)C(6,0)}{C(10,2)} = \frac{2}{15} = 0.133,$$

respectively. If the drawings are with replacement, then, by the binomial distribution with $n = 2$ and $p = \frac{2}{5}$, the probabilities are 0.360, 0.480, and 0.160, respectively. Thus, there is a moderate difference between sampling with and without replacement for 10 balls. Next, suppose that the urn holds 100 balls of which 40 are white. In sampling without replacement, the probabilities of drawing 0, 1, and 2 white balls are 0.358, 0.485, and 0.158. The probabilities for sampling with replacement are the same as before: 0.360, 0.480, and 0.160. Thus, the differences between the probabilities in sampling with and without replacement are very small when $n = 100$. ▲

EXAMPLE 2

Suppose that a wholesale consumer orders a large lot of identical manufactured items from a producer; for example, a photography supplies dealer orders a lot of flashbulbs. The consumer wants to know the quality of the lot: each bulb is either defective (will not light properly) or nondefective. Let p stand for the fraction of defective items in the lot; it is unknown. Although some defectives are expected in a large lot because the manufacturing process is not perfect, the consumer does not want a lot with too many defectives. He can be absolutely sure of the quality of the lot only if he tests all the bulbs; however, such complete testing would use up all the bulbs. Hence, he must compromise his certainty of the quality, and test at most a small fraction. After testing he decides whether to buy the lot or not: to **accept** or **reject.**

Consider this testing procedure: n bulbs are drawn at random without replacement and tested. The lot is accepted or rejected accordingly as the number of defectives found in the sample is less than or equal to a predetermined number c, or greater than c. For example, the test may involve a sample of 4 bulbs, and the lot is accepted if not more than $c = 1$ is found to be defective. Such a procedure is a **sampling plan** based on the pair (n,c). A good sampling plan has a small probability of accepting a lot with a large proportion p of defectives, and a large probability of accepting a lot with a small p.

Let us derive the probability of accepting the lot when the proportion of defectives is an arbitrary number p. The defectives in the lot are like the white balls in the urn; those in the sample are like the white drawn at random without replacement. When the sample is small relative to the size of the lot, we may use the binomial distribution as the distribution of the number of defectives in the sample. The probability of accepting the lot is then computed as a sum of terms of the binomial distribution. For example, if 4 bulbs are tested, and the lot is to be accepted if not more than 1 is defective, then the probability of acceptance is equal to the sum of the terms of index 0 and 1:

$$q^4 + 4pq \qquad \text{or} \qquad (1-p)^4 + 4p(1-p).$$

This expression is denoted $L(p)$; it is a function of p for values of p from 0 through 1. The values are found as sums of entries of Table VI (Appendix):

p	$L(p)$
0.0	1.0000
0.1	0.9477
0.2	0.8192
0.3	0.6517
0.4	0.4752
0.5	0.3125

p	$L(p)$
0.6	0.1782
0.7	0.0837
0.8	0.0282
0.9	0.0037
1.0	0.0000

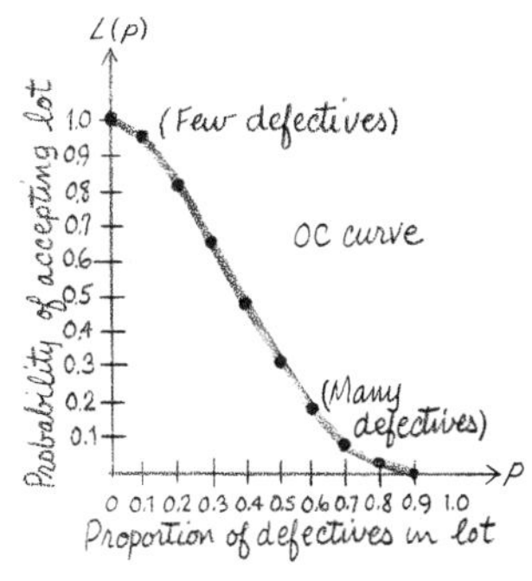

Figure 8.28

The graph of $L(p)$ appears in Figure 8.28. Note that the probability of acceptance decreases as the proportion of defectives increases from 0 to 1. The probability of acceptance is close to 1 when p is close to 1. The function $L(p)$ is the **operating characteristic** of the sampling plan, and its graph is called the "OC curve." ▲

EXAMPLE 3

The electorate of a certain city consists of supporters of 2 parties, which we shall call the Blue and the Green parties. Let p be the proportion favoring the Blue party. Before the election a poll taker wants to estimate the proportion p. He takes a sample of n voters at random without replacement, and observes the number of Blue party supporters in the sample. The electorate is the urn and the voters are the colored balls; and the sample is small relative to the electorate; thus, the number of Blues in the sample has the binomial distribution (approximately). For example, suppose that $p = 0.6$, and that a sample of 8 voters is taken; then the probability that the sample contains k Blue party supporters is

$$C(8,k)\,(0.6)^k(0.4)^{8-k}, \qquad \text{for } k = 0, 1, \ldots, 8.$$

These can be found in Table VI (Appendix).

One of the most remarkable features of modern election-night "projections" is that the elections are accurately predicted on the basis of an insignificant proportion of the total number of votes. Later we shall show what size samples are required for proper projections; this will be discussed in the section on the normal approximation to the binomial (Section 8.10). ▲

As the last topic in this section, we consider Bernoulli trials with **variable probabilities.** Suppose there are n urns, each containing white and red balls. Let $p_1, \ldots, p_n$ be the proportions of white balls, and $q_1, \ldots, q_n$ the complementary proportions of red in the respective urns. One ball is sampled at random from each of the urns. The outcomes are sets of n balls, white and red. The probability of an outcome is a product of n p-factors and q-factors, a p-factor for each white ball and a q-factor for each red. This is like tossing n biased coins with probabilities $p_1, \ldots, p_n$, respectively. The proof for the assignment of the probabilities to the outcomes is very much like the proof above for sampling with replacement from a single urn.

As an example, consider 2 urns with proportions p_1 and p_2 of white balls, and proportions q_1 and q_2 of red. The 4 outcomes and their probabilities are

white, white	p_1p_2	white, red	p_1q_2
red, white	q_1p_2	red, red	q_1q_2

8.8 EXERCISES

In Exercises 1 and 2 compare the distributions of the number of white balls drawn in sampling with and without replacement.

1 (a) 3 white, 2 red; sample of 2
(b) 5 white, 5 red; sample of 3
(c) 50 white, 50 red; sample of 2

2 (a) 2 white, 4 red; sample of 2
(b) 4 white, 6 red; sample of 3
(c) 40 white, 60 red; sample of 2

3 Five bulbs are selected at random from a large lot. Find the probability that at most 1 is defective for $p = 0.1, 0.2, 0.3$, and 0.4.

4 In a large population, 60 percent of the labor force has an annual income of less than \$9000. In a sample of 8 workers, find the probability that at least 5 earn less than that amount.

5 If a certain city has a registration of 75 percent Democrats and 25 percent Republicans, what is the probability that a sample of 7 voters will contain at least 5 Democrats?

6 President Johnson received approximately 60 percent of the popular vote in Nevada in the election of 1964. (The exact percentage was 58.5 percent.) In a sample of 10 voters, what would be the probability of finding at least 5 who voted for him?

7 Ten percent of a cattle herd has a disease. If 6 are selected at random, what is the probability that at least 3 have the disease?

8 Approximately 120,000 bachelor and professional degrees in the field of education were awarded in the United States in the years 1965–1969. Approximately $\frac{3}{4}$ of these were awarded to women. In a random sample of 9 such degree holders, what is the probability of getting from 6 through 8 women (inclusive)?

9 According to a U.S. Government report, 9.5 percent of American adults had definite hypertensive disease in the period 1960–1962 (Heart Disease in Adults, U.S. Department of Health, Education and Welfare, Ser. 11, No. 6). Find the probability of at least 2 hypertensives in a random sample of 10 adults; assume $p = 0.1$ instead of 0.095.

10 Ten percent of women ever married and in the age group from 35 to 39 have never had children. Find the probability that in a sample of 5 women from this group at most 2 had a child.

11 If 40 percent of the students at a certain college favor a particular national foreign policy, and 7 students are chosen at random, what is the probability that at most 4 favor the policy?

12 Twenty percent of patients given a certain medication will have side effects. Find the probability that among 12 patients, the number developing side effects is between 2 and 4 inclusive.

13 One-third of the television sets produced by a manufacturer have to be recalled from the dealers for mechanical corrections. If a dealer has 5 sets of this kind, what is the probability that at least 2 have to be returned?

14 Three tickets are drawn at random with replacement from a bowl containing 10. In how many ways can these 3 be drawn? In how many ways can 3 distinct tickets be drawn? Use this to find the probability that the 3 tickets drawn at random with replacement are distinct.

15 Generalize the result of Exercise 14: Show that if r tickets are drawn with replacement, then the probability that they are all distinct is given by the ratio (1). ●●●

16 Each of 6 persons at a party is asked to choose by himself a number from 1 through 10 at random. Find the probability that they will choose 6 distinct numbers.

17 In this exercise we have the famous **birthday problem:** In a group of n persons, what is the probability that at least 2 have the same birthday? We imagine that each person chooses his birthday by himself at random from among the 365 days of the year; therefore, the n birthdays are chosen at random with replacement from among 365 "tickets." Find the probability for $n = 5$. [*Hint:* Use the method of the previous 3 exercises. First find the probability of distinct birthdays.] ●●●

18 Here is an outline of a direct verification that the hypergeometric distribution is very close to the binomial when n is large relative to r; we outline the proof for $r = 2$. The probabilities of drawing 0, 1, and 2 white balls are ●●●●

$$\frac{C(w,0)C(n-w,2)}{C(n,2)}, \qquad \frac{C(w,1)C(n-w,1)}{C(n,2)}, \qquad \frac{C(w,2)C(n-w,0)}{C(n,2)},$$

respectively. Now replace w and $n - w$ by np and $n(1-p) = nq$, and let n be very large. Using the definition of the combination number, show that the 3 probabilities above are very close to the corresponding terms of the binomial distribution.

8.9 Normal Approximation to the Binomial Distribution: The Case $p = \frac{1}{2}$

In many practical and theoretical problems involving the binomial distribution, the number of trials n is very large. The computation of probabilities of certain events is then difficult because the number of terms in the distribution is large and the individual terms are small. For example, if a fair coin is tossed 500 times, and we wish to calculate the probability that at least 300 heads occur, then we have to sum the last 201 terms of the binomial distribution for $p = \frac{1}{2}$ and $n = 500$. Most of these terms are of insignificant size, and, as a matter of fact, are equal to 0 for the first several decimal places; however, there are so many of them that we have to be sure that their sum is also insignificant before throwing out the summands. Such problems are solved by using the **normal approximation** to the binomial distribution.

The proof of this approximation depends on much more mathematics than is learned in a term or even a year of calculus; hence, we omit the proof, but instead give an intuitive, numerical explanation with the use of diagrams. We discuss only the case $p = \frac{1}{2}$; however, the approximation is extended in the next section to the case of general p.

According to formula (2) of Section 8.6, the term of index k of the binomial distribution for $p = \frac{1}{2}$ is

$$C(n,k)\left(\tfrac{1}{2}\right)^k\left(\tfrac{1}{2}\right)^{n-k} \qquad \text{or} \qquad C(n,k)2^{-n}$$

[see Section 8.5, formula (2)]. Thus for each n, the terms are proportional to the binomial coefficients for $k = 0, 1, \ldots, n$, and their sum is 1. The bar graph of the probabilities is therefore the same as the bar graph of the binomial coefficients except that the scale on the vertical axis is changed by a factor 2^{-n}. The bar graph of the probabilities for $n = 10$ appears in Figure 8.29(a); that for the binomial coefficients is in Figure 8.10. In Figure 8.29(b), the bar graph of the probabilities is converted to a **histogram,** an area graph; the bars are replaced by rectangles of unit width, centered at the integers, and whose heights are equal to the corresponding bar-lengths. The areas of these rectangles are equal to the corresponding probabilities.

Next we fit a **density** to the histogram (see Section 6.2). (It is not necessary that the reader now be familiar with the material in that section.) First we replace the rectangles by trapezoids of correspondingly equal areas [Figure 8.29(c)]. Finally, we replace the upper boundary of the system of trapezoids by a smooth curve [Figure

8.29(d)] in such a way that the areas of the rectangles are practically the same as the corresponding areas under the smooth curve.

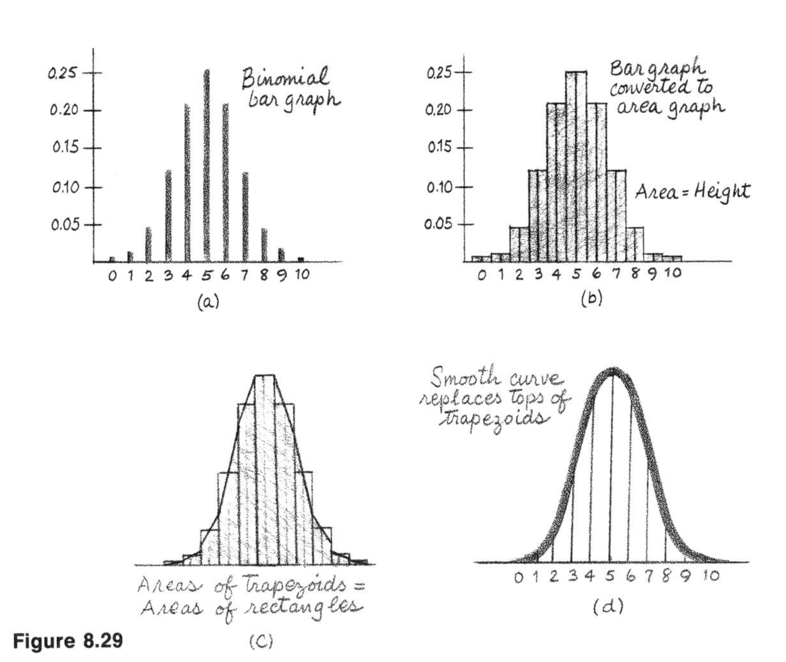

Figure 8.29

It follows from the construction above that the probability $C(n,k)2^{-n}$ of the binomial distribution is approximately equal to the area under the curve in Figure 8.29(d) over the interval from $k - \frac{1}{2}$ to $k + \frac{1}{2}$; for example, for $n = 10$ and $k = 4$, the probability is the area under the curve from 3.5 to 4.5. It also follows that the sum of several terms of the binomial is approximately equal to the sum of the corresponding areas under the curve.

Let us now identify this smooth curve and the areas under it. Note that it is symmetric about the point $x = 5$, and falls down rapidly to the left and to the right of that point [Figure 8.29(d)]. It is bell-shaped. We expect the total area under the curve to be equal to 1 because it is approximately equal to the sum of the terms of the binomial distribution; indeed, it **is** equal to 1. This curve is a member of the family of **normal** curves. All members of the family are related in the following way: Each member can be obtained from any other by either or both of two processes:

(i) shifting the curve horizontally to the right or left;

(ii) stretching the curve horizontally, and squeezing it vertically by the same factor; or stretching vertically and squeezing horizon-

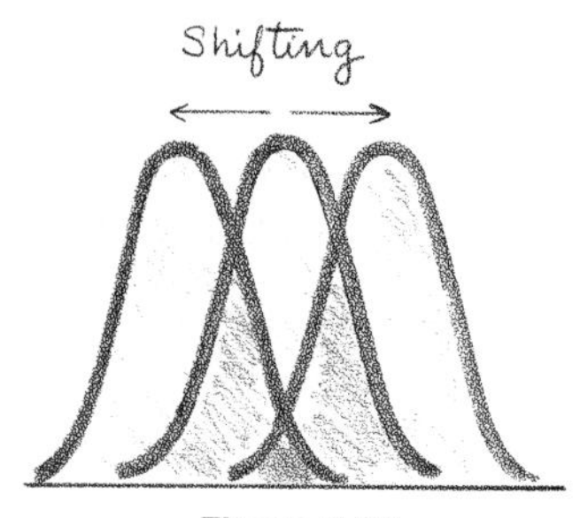

Figure 8.30

tally. For example, stretching one way by a factor of 2, and then cutting the other way by reducing it to $\frac{1}{2}$ the original dimension.

In Figure 8.30 we show how the curve in Figure 8.29(d) is shifted to the right or left; it is clear that the shift does not change the total area under the curve. The stretch-squeeze operation on the same curve is demonstrated in Figure 8.31. This operation also does not change the total area under the curve. The reason is that the operation of stretching one dimension and squeezing the other dimension by a compensating factor does not alter the area of a rectangle; hence, since the area under the curve is the sum of the areas of rectangles, it is also not affected by the operation. It follows that the area under every normal curve is the same, and is equal to 1.

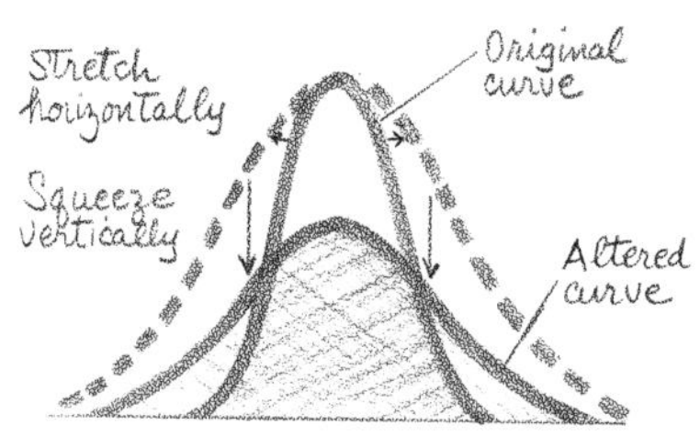

Figure 8.31

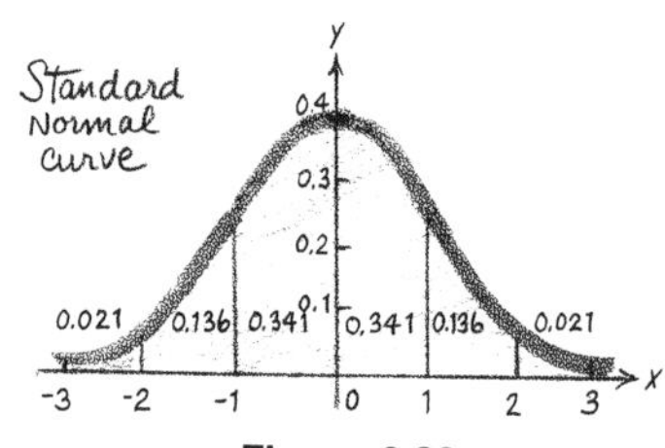

Figure 8.32

A particular member of the normal family is identified as the **standard normal curve.** Its graph appears in Figure 8.32. It is symmetric about the origin. The areas under the main portions of the curve are also indicated in Figure 8.32. Note that the units on the vertical scale are larger than those on the horizontal; otherwise, the curve would be drawn too flat.

[For those who have already studied Section 4.3, we remark that the curve is the graph of the function

$$\phi(x) = \frac{1}{\sqrt{2\pi}} e^{-(1/2)x^2} \tag{1}$$

The symmetry of the curve is expressed by the relation

$$\phi(x) = \phi(-x).$$

The values of this function for nonnegative x are in Table VII of the Appendix; the other values are obtained from the symmetry property. In Section 7.5, Example 9, we showed that the total area under the curve is equal to 1.]

For each number x, let $\Phi(x)$ be the area under the standard normal curve to the left of the point with abscissa x. As a function of

x, $\Phi(x)$ is called the **standard normal distribution function.** It has all the properties of a population distribution (Section 6.7). It increases with x, tends to 0 for large negative values of x, and tends to 1 for large positive values (Figure 8.33).

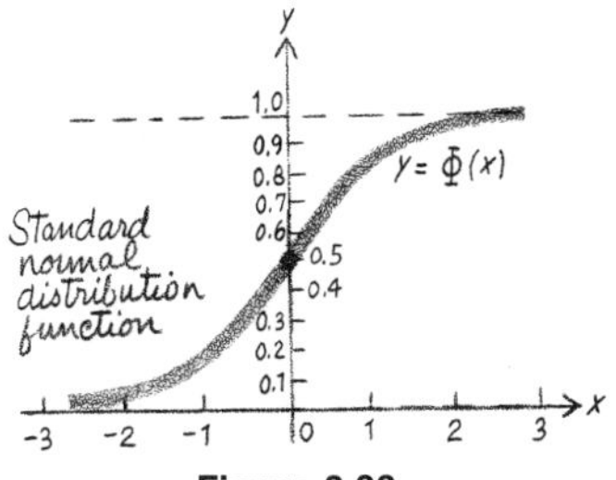

Figure 8.33

[For the reader who has already studied Chapter 6: According to the theory of integral calculus, the area under the curve $y = \phi(x)$ between the points a and b on the x-axis is the definite integral

$$\int_a^b \phi(x)\,dx.$$

It happens that there is no analytic method for finding the integral; hence, it has to be done by an approximate numerical method such as the trapezoidal rule (Exercise 14). By the Fundamental Theorem of Calculus it follows that the distribution function $\Phi(x)$ is related to the density function $\phi(x)$ by the equation

$$\Phi(x) = \int_{-\infty}^{x} \phi(t)\,dt;$$

and so from the relation between the integral and the derivative, we get

$$\phi(x) = \frac{d}{dx}\Phi(x).]$$

Since the area under the curve is equal to 1, it follows that $1 - \Phi(x)$ represents the area to the **right** of the point x; hence, by the symmetry of the curve it also follows that the area to the **right** of a point is equal to the area to the left of the negative of the point:

$$\Phi(-x) = 1 - \Phi(x) \tag{2}$$

(Figure 8.34).

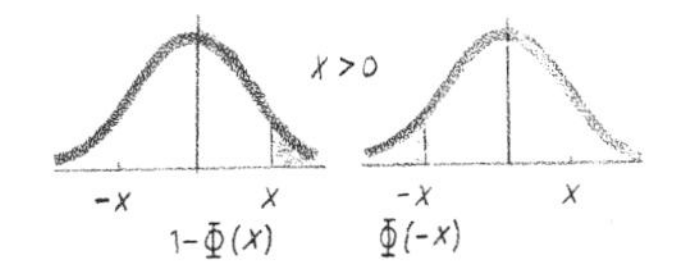

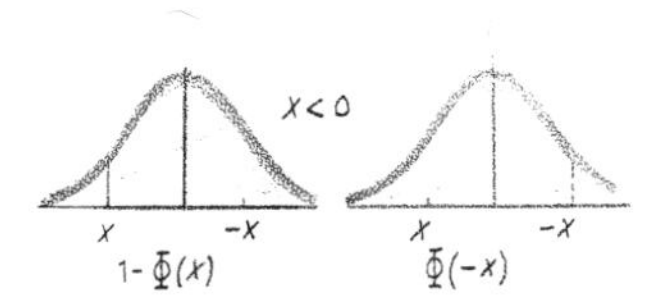

Figure 8.34

For each number z greater than or equal to 0, Table VIII (Appendix) furnishes the value of the difference $\Phi(z) - \Phi(0)$, the area under the curve from 0 to the point with abscissa z. From these values we can find the values of $\Phi(z)$ and of all differences $\Phi(b) - \Phi(a)$. In addition to the relation (2), the following observations are useful in computation of areas:

(i) $\Phi(0) = 1 - \Phi(0) = \frac{1}{2}$. This follows from the fact that half the area is to the right of the origin and half is to the left [Figure 8.35(a)].

(ii) $\Phi(z) = \frac{1}{2} + (\Phi(z) - \Phi(0))$. For positive z the area to the left of z is obtained by adding $\frac{1}{2}$, which is the area to the left of 0, to the tabled value [Figure 8.35(b)].

(iii) If $0 < a < b$, then the difference $\Phi(b) - \Phi(a)$ is obtained from the table as

$$\Phi(b) - \Phi(0) - [\Phi(a) - \Phi(0)]$$

[Figure 8.35(c)].

(iv) If $a < 0 < b$, then the difference is obtained from the table as

$$\Phi(b) - \Phi(0) + [\Phi(-a) - \Phi(0)]$$

[Figure 8.35(d)].

(v) If $a < b < 0$, then the difference is obtained from the table as $\Phi(-a) - \Phi(-b)$, as in (iii) above.

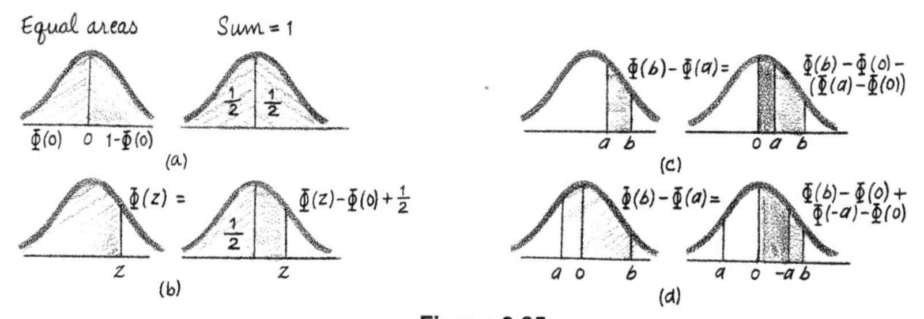

Figure 8.35

EXAMPLE 1

$1 - \Phi(1.71)$ is written as the difference between the entire area to the right of 0, and the area from 0 to 1.71:

$$1 - \Phi(1.71) = 0.5000 - 0.4564 = 0.0436.$$

This is also equal to the area left of -1.71, which is $\Phi(-1.71)$. The area to the left of 1.71 is, by (ii) above, $0.5000 + 0.4564 = 0.9564$. ▲

EXAMPLE 2

If $a = 0.94$ and $b = 2.25$, then, by (iii) above, the area from a to b is

$$\Phi(2.25) - \Phi(0) - [\Phi(0.94) - \Phi(0)] = 0.4878 - 0.3264$$
$$= 0.1614.$$

If a is negative and b positive, for example, $a = -0.94$ and $b = 2.25$, then $\Phi(2.25) - \Phi(-0.94)$ is, by (iv) above, $0.4878 + 0.3264 = 0.8142$. ▲

Now we state the special case, for $p = \frac{1}{2}$, of the normal approximation theorem for the binomial distribution; it is known as the **DeMoivre-Laplace theorem,** after A. DeMoivre (1667–1754) and P. S. Laplace (1749–1827).

The probability of k successes in n Bernoulli trials with $p = \frac{1}{2}$ is approximately equal to the area under the standard normal curve from

$$\frac{k - \frac{1}{2}n - \frac{1}{2}}{\frac{1}{2}\sqrt{n}} \quad \text{to} \quad \frac{k - \frac{1}{2}n + \frac{1}{2}}{\frac{1}{2}\sqrt{n}} \tag{3}$$

for large values of n; in mathematical terms, $C(n,k)2^{-n}$ is approximately

$$\Phi\left(\frac{k - \frac{1}{2}n + \frac{1}{2}}{\frac{1}{2}\sqrt{n}}\right) - \Phi\left(\frac{k - \frac{1}{2}n - \frac{1}{2}}{\frac{1}{2}\sqrt{n}}\right). \tag{4}$$

More generally for any two integers $a < b$, the sum of the probabilities $C(n,k)2^{-n}$ for $k = a, \ldots, b$ is approximately equal to

$$\Phi\left(\frac{b - \frac{1}{2}n + \frac{1}{2}}{\frac{1}{2}\sqrt{n}}\right) - \Phi\left(\frac{a - \frac{1}{2}n - \frac{1}{2}}{\frac{1}{2}\sqrt{n}}\right). \tag{5}$$

The accuracy of the approximation increases with n.

EXAMPLE 3

Let us apply the approximation in the case $n = 10$. The pair of numbers in (3) is

$$\frac{2(k - 5.5)}{3.16228}, \qquad \frac{2(k - 4.5)}{3.16228}.$$

In the table below we compare the differences (4) with the corresponding probabilities of the binomial:

k	**Normal approximation (4)**	$C(10,k)2^{10}$
0	0.0020	0.0010
1	0.0114	0.0098
2	0.0435	0.0439
3	0.1140	0.1172
4	0.2034	0.2051
5	0.2510	0.2461
6	0.2034	0.2051
7	0.1140	0.1172
8	0.0435	0.0439
9	0.0114	0.0098
10	0.0020	0.0010

Observe that the distribution and the approximation are symmetric about $\frac{1}{2}n = 5$, so that it suffices to give just half the set of values. The approximation is good even for smaller n; for example, when $n = 4$, we have

k	Normal approximation	$C(4,k)2^1$
0,4	0.067	0.0625
1,3	0.242	0.2500
2	0.383	0.3750

For $n = 16$, the approximation differs from the binomial by at most 0.001 in any term. ▲

EXAMPLE 4

A balanced coin is tossed 25 times. Find the probability of getting at most 10 heads. This is equal to the sum of the terms of the binomial distribution with $n = 25$ and $p = \frac{1}{2}$ for $k = 0, 1, \ldots, 10$. According to the approximation theorem, the probability is approximately given by (5) with $a = 0$, $b = 10$, and $n = 25$: this is $\Phi(-0.8) - \Phi(-9.2)$, which by the symmetry of the distribution, is $\Phi(9.2) - \Phi(0.8)$. Now the area under the normal curve outside the interval from -4 to 4 is, by Table VIII, smaller than 0.0001; thus, to the given number of decimal places, we have

$$\Phi(9.2) = 1.0000.$$

It follows that

$$\Phi(9.2) - \Phi(0.8) = 1 - \Phi(0.8) = 0.5000 - 0.2897 = 0.2103;$$

the latter is obtained from the note (ii) above on the use of Table VIII. ▲

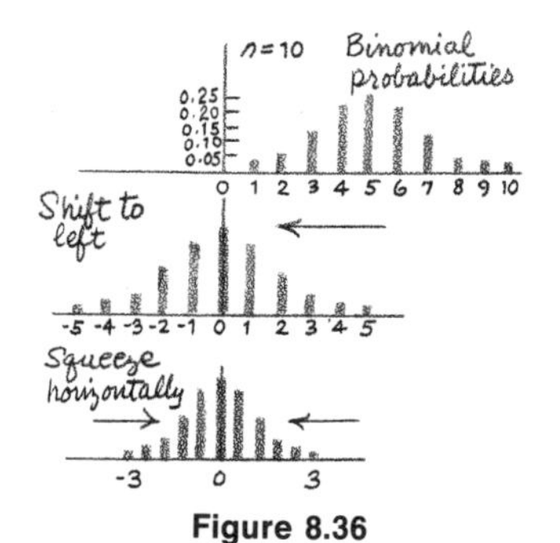

Figure 8.36

In conclusion, we briefly explain the appearance of the term $\frac{1}{2}n$ in the numerators in (3) and the factor $\frac{1}{2}\sqrt{n}$ in the denominator. Recall that np is the "expected value" of the number of successes, and is within 1 unit of the most probable value (Section 8.7, Example 1). In the particular case $p = \frac{1}{2}$, the expected value is $\frac{1}{2}n$. Subtraction of this term from k in (3) shifts the distribution to the left so that it becomes centered at 0. The second operation, division by a multiple of $\sqrt{n}$, reduces the horizontal scale and pulls the distribution toward the center at 0. These two operations give the binomial distribution the image of the standard normal (Figure 8.36). Note that the squeezing operation keeps most of the probability between -3 and $+3$; and all of the area under the normal curve, except for 0.001, is contained between these two points.

In the next section we shall extend the approximation to the binomial distribution with p not necessarily equal to $\frac{1}{2}$; and we shall consider several applications.

8.9 EXERCISES

In Exercises 1 and 2 find the area under the standard normal curve between the given values.

1 (a) from $z = 0$ to $z = 2.12$
(b) from $z = -0.65$ to $z = 0$
(c) left of $z = 1.44$
(d) left of $z = -1.25$
(e) right of $z = 0.94$
(f) right of $z = -0.37$
(g) from $z = -0.84$ to $z = -2.03$
(h) from $z = -0.29$ to $z = 1.37$

2 (a) from $z = 0$ to $z = 1.87$
(b) from $z = -0.14$ to $z = 0$
(c) left of $z = 1.35$
(d) left of $z = -1.80$
(e) right of $z = 2.12$
(f) right of $z = -1.07$
(g) from $z = 0.54$ to $z = 2.00$
(h) from $z = -1.34$ to $z = 0.98$

In Exercises 3 and 4 find the value of z *(by means of Table VIII) so that the area involving* z *is the one given.*

3 (a) Area from 0 to z is 0.3729
(b) Area left of z is 0.9357
(c) Area right of z is 0.0455
(d) Area right of z is 0.8159
(e) Area left of z is 0.1660
(f) Area between $-z$ and z is 0.0320

4 (a) Area from 0 to z is 0.2704
(b) Area left of z is 0.8531
(c) Area right of z is 0.1867
(d) Area right of z is 0.8438
(e) Area left of z is 0.0054
(f) Area between $-z$ and z is 0.8812

In Exercises 5 and 6 compare the term of the binomial distribution with the normal approximation.

5 (a) $C(3,1)2^{-3}$
(b) $C(8,3)2^{-8}$
(c) $C(5,2)2^{-5}$
(d) $C(7,4)2^{-7}$
(e) $C(9,6)2^{-9}$

6 (a) $C(8,4)2^{-8}$
(b) $C(7,3)2^{-7}$
(c) $C(9,2)2^{-9}$
(d) $C(12,5)2^{-12}$
(e) $C(11,6)2^{-11}$

7 A coin is tossed 9 times. Using the normal approximation find the probability of (a) at least 7 heads; (b) at most 6 heads; (c) more than 3 but fewer than 7 heads.

8 A coin is tossed 36 times. Using the normal approximation, find the probability of (a) at least 20 heads; (b) at most 18 heads; (c) more than 15 but fewer than 23 heads.

9 Two competing newspaper vendors are in a railroad station. Each morning 1000 commuters pass and buy papers. The vendors offer identical papers and service so that there is no reason for a customer to prefer one or the other vendor. We suppose that each customer chooses a vendor at random; the choices of the customers are like the outcomes of 1000 tosses of a fair coin, with heads for one vendor and tails for the other. Using the normal approximation, find the probability that a particular vendor sells more than (a) 540 papers on a morning; (b) 550 papers.

10 In Exercise 9, at least how many newspapers should a vendor have in stock on a particular morning so that the probability that the demand exceeds the supply be at most 0.02?

11 The proportion of heads observed in a given number n of tosses is the ratio **No. of heads/n.** (a) If a coin is tossed 100 times, what is the approximate probability that the proportion of heads is strictly between 0.45 and 0.55? (b) What if it tossed 1000 times? (c) 5000 times?

12 Repeat Exercise 11 for the proportion limits 0.49 and 0.51.

13 Let $y = \phi(x)$ be the equation of the standard normal curve.

(a) Find the equation of the curve obtained by shifting it m units to the left.

(b) Find the equation of the curve obtained by stretching it vertically by s units, and reducing it horizontally by a factor of s.

(c) Combine the two operations above: Find the equation obtained by first shifting to the left by m units and then stretching and squeezing by s.

14 **(For the reader who has studied Section 6.1.)** The areas in Table VIII can be obtained from the values of the function $\phi(x)$ in Table VII by approximate methods. Find these areas using the trapezoidal rule; then compare them to those given in Table VIII: (a) area from 0 to 0.10; (b) area from 1.00 to 1.10; (c) area from 2.00 to 2.10.

8.10 Normal Approximation to the Binomial: The General Case

The normal approximation is valid for the binomial distribution not only for $p = \frac{1}{2}$ but also for any p between 0 and 1. The approximation is not as accurate as in the case $p = \frac{1}{2}$ when n is small; however, the accuracy increases with n. In Figure 8.37 the rectangular area graphs of the binomial are shown for $p = \frac{1}{5}$ and $n = 5$, 10, 20, and 40; and a normal curve is fitted to each. Note how the approximation improves as n increases. The general form of the approximation theorem is the **DeMoivre-Laplace theorem** (Section 8.9), stated as follows:

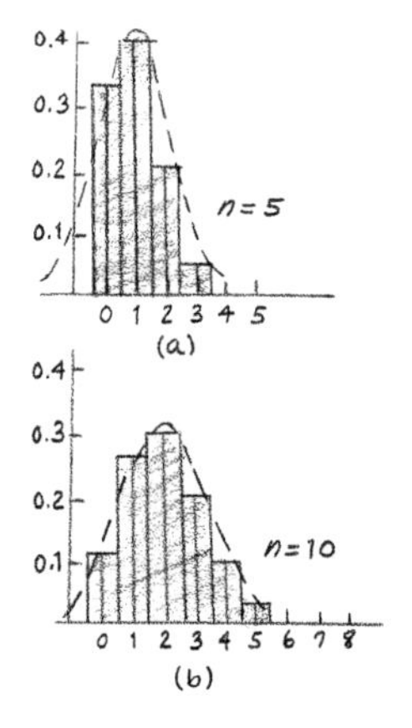

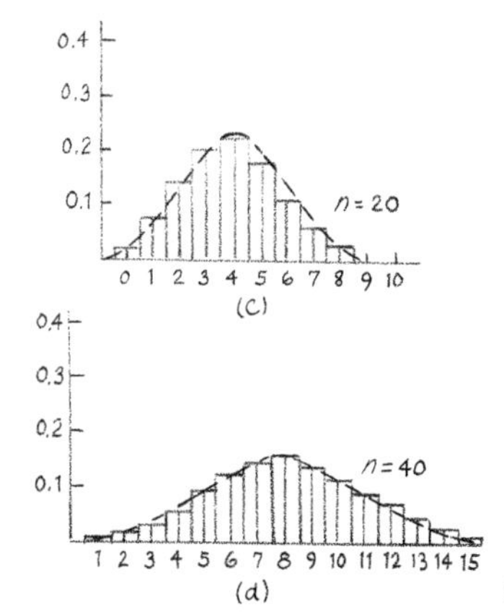

Figure 8.37

The probability that the number of successes in n Bernoulli trials with probability p of success is between the integers a and b inclusive ($a < b$) is approximately equal to the area under the standard normal curve from

$$\frac{a - np - \frac{1}{2}}{\sqrt{npq}} \quad \text{to} \quad \frac{b - np + \frac{1}{2}}{\sqrt{npq}}. \tag{1}$$

The accuracy of the approximation increases with n.

(For those who have read about **limits** in Section 5.3, we can state this in the form of a limit theorem; see Exercise 20.)

The reason for the subtraction of np is similar to that for the subtraction of $\frac{1}{2}n$ in the case $p = \frac{1}{2}$: np is the **expected value,** the central value of the binomial distribution, so that its subtraction shifts the center of the distribution to the origin. Division by $\sqrt{npq}$ pulls the terms of the distribution together, so that practically all the probability is concentrated on values from -3 to $+3$.

EXAMPLE 1

Let us consider the normal approximation in the case $n = 225$ and $p = 0.2$: 225 Bernoulli trials with probability 0.2 of success on each. The **expected** number of successes is $np = 45$. Let us find the probability that the observed number of successes is between 40 and 50, inclusive. According to the theorem, with $q = 0.8$, $a = 40$, and $b = 50$, the probability is approximately the area under the standard normal curve from

$$\frac{40 - 45 - \frac{1}{2}}{\sqrt{(4/25)225}} \quad \text{to} \quad \frac{50 - 45 + \frac{1}{2}}{\sqrt{(4/25)225}},$$

or

$$\Phi\left(\frac{5.5}{6}\right) - \Phi\left(-\frac{5.5}{6}\right) = 2\Phi(0.92) - 1 = 0.6424. \ \blacktriangle$$

EXAMPLE 2

Psychologists have used the following statistical experiment to prove the existence of extrasensory perception, called ESP (Figure 8.38). (This is the supposed ability of human beings to communicate by unknown methods not involving the 5 senses.) An experimenter and a subject play a card game with a deck of 25 cards, consisting of 5 kinds of cards, 5 of each kind. The experimenter first shuffles the cards and then draws them one at a time. The subject does not see the cards. Before each drawing he guesses the kind of card that is to be drawn. The experimenter

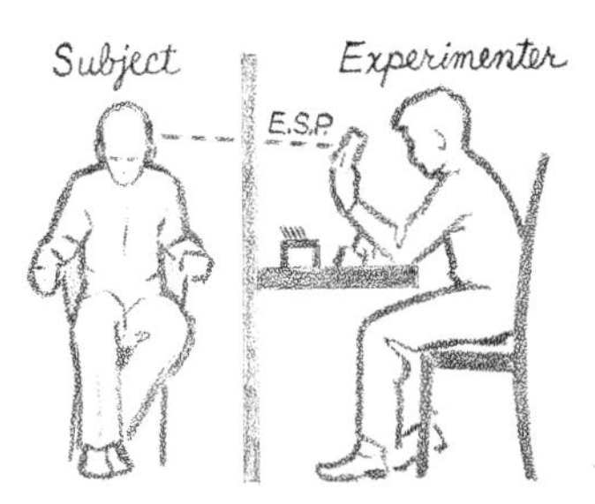

Figure 8.38

records the number of correct guesses from among the 25. If a subject has no power of ESP, then his guesses are just random: he has probability $\frac{1}{5}$ of being correct on each guess. It follows that the correct guesses are like successes in 25 Bernoulli trials with probability $p = \frac{1}{5}$. Under the alternative that the subject does have powers of ESP, the probability of his correctly guessing an individual card is greater than $\frac{1}{5}$; however, he is not expected to correctly guess every card. Thus, ESP is defined as the ability to correctly guess with probability higher than $\frac{1}{5}$; it is a **statistical** definition.

Under the assumption of random guessing, the expected number of correct guesses is $np = 25/5 = 5$. It is "unlikely" that the subject will guess "significantly" more than 5. If he does, then we reject the assumption of random guessing in favor of the alternative that the subject's probability of correct guessing is larger than $\frac{1}{5}$. Now let us be more specific about the terms "unlikely" and "significantly." For example, is it likely that the subject will correctly guess at least 6 more than the expected number, that is, at least 11 cards? Let us compute the probability under $p = \frac{1}{5}$ by means of the normal approximation: it is the area under the standard normal curve from

$$\frac{11 - 5 - \frac{1}{2}}{2} \quad \text{to} \quad \frac{25 - 5 + \frac{1}{2}}{2},$$

or

$$\Phi(10.25) - \Phi(2.75) = 1.0000 - 0.9970 = 0.0030.$$

From this we see that under the assumption of random guessing there are only 3 chances in 1000 that the subject will correctly guess at least 11. The probability is higher when p is greater than $\frac{1}{5}$; for example, when $p = \frac{1}{2}$, the probability is, by the normal approximation, about 0.7881. The psychologist has a good argument for claiming that p is greater than $\frac{1}{5}$, so that his subject does have ESP. ▲

EXAMPLE 3

Suppose that fever in a certain childhood disease is expected to last at most 4 days in 60 percent of cases. In other words, the probability that a particular child has the fever at most 4 days is 0.6. In another disease with similar symptoms, the probability of no fever after 4 days is 0.8 (Figure 8.39). Suppose that one of these 2 illnesses affects 120 children in a particular school. The school doctor wants to determine which of the 2 illnesses is present. In the first disease the expected number of normal temperatures after 4 days of infection is $np = 120(0.6) = 72$; and in the second case, it is $120(0.8) = 96$. The doctor used the following rule for deciding which of the 2 diseases was present: If 84 or more of the children had fever lasting at most 4 days, then he would decide that the second disease was present; otherwise, he would decide in favor of the first disease. Find the probability that he correctly identifies the disease when (a) the first disease is really present, and (b) the second is really present.

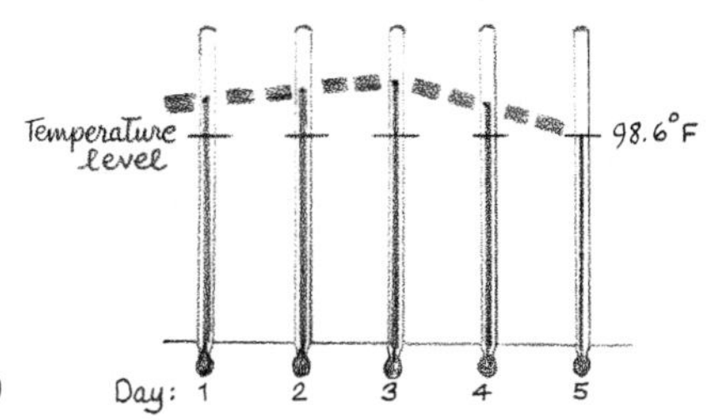

Figure 8.39

The probability for (a) is the probability of at most 83 successes in 120 Bernoulli trials with $p = 0.6$. By the normal approximation, it is the area under the standard normal curve from

$$\frac{0 - 72 - \frac{1}{2}}{\sqrt{120(0.6)(0.4)}} \quad \text{to} \quad \frac{83 - 72 + \frac{1}{2}}{\sqrt{120(0.6)(0.4)}},$$

or

$$\Phi\left(\frac{\sqrt{5}(11.5)}{12}\right) - \Phi\left(-\frac{\sqrt{5}(72.5)}{12}\right) = \Phi(2.142) - \Phi(-13.51)$$
$$= 0.9838.$$

The probability for (b) is the probability of at least 84 successes in 120 Bernoulli trials with $p = 0.8$. By the normal approximation it is the area under the standard normal curve from

$$\frac{84 - 96 - \frac{1}{2}}{(2/5)\sqrt{120}} \quad \text{to} \quad \frac{120 - 96 - \frac{1}{2}}{(2/5)\sqrt{120}}.$$

By calculations similar to those above, this area is found to be

$$\Phi(6.03) - \Phi(-2.85), \qquad \text{or approximately } 0.9978$$

We see that in the case of the first disease, the chances of his being right are more than 98 in 100, while in the case of the second, more than 99 in 100. ▲

In a typical kind of statistical problem the probability p is unknown, and we wish to estimate it. The conventional estimate of p is the observed proportion of successes in n trials; this estimate is denoted $\hat{p}$:

$$\hat{p} = \frac{\text{No. of successes}}{n}.$$

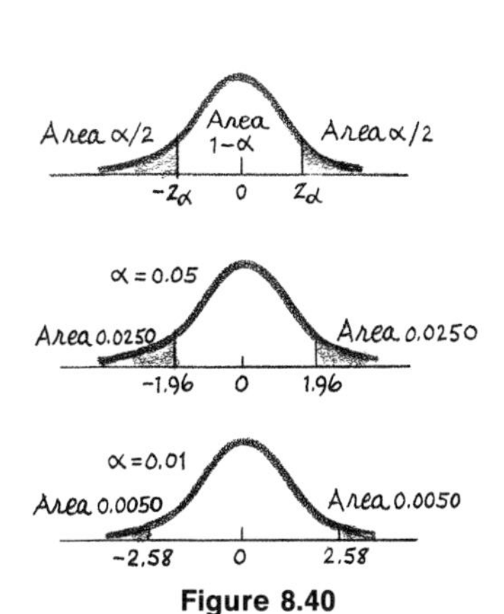

Figure 8.40

A natural question is: How far is this estimate likely to be from the true value of p? The normal approximation theorem furnishes a formula for the **probable deviation** between the estimate and the unknown value p.

For every number α, $0 < \alpha < 1$, let z_α represent the point on the horizontal axis of the graph of the standard normal curve such that the sum of the areas to the right of z_α and to the left of $-z_\alpha$ is equal to α (Figure 8.40). For example, if $\alpha = 0.05$, then $z_\alpha = 1.96$. (This notation differs from that in other books, where this is denoted $z_{\alpha/2}$. Our notation is typographically simpler.)

For given α, the expression

$$\frac{z_\alpha}{2\sqrt{n}} \tag{2}$$

is called the **probable deviation** of p for probability $1 - \alpha$. **The probability that $\hat{p}$ differs from p by at most $z_\alpha/2\sqrt{n}$ is at least $1 - \alpha$ if n is large.** This statement follows directly from the normal approximation theorem; the proof is outlined in Exercises 18 and 19. The probable deviation is a measure of the maximum error we make in using $\hat{p}$ as an estimate of p; and it is subject to the given probability $1 - \alpha$. Note that the probable deviation varies inversely as the square root of the number of trials; in particular, the probable deviation tends to 0 as n increases. This means that the estimate tends to become more and more accurate as the number of trials increases (Figure 8.41).

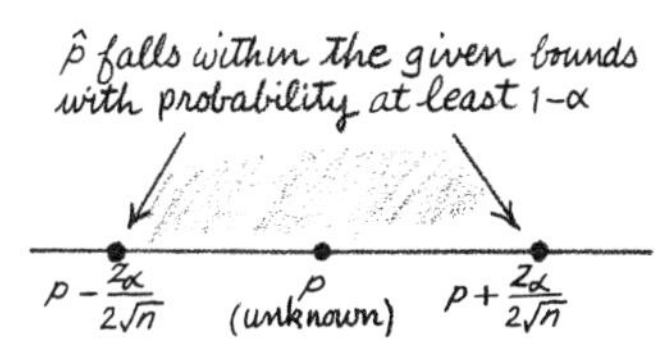

Figure 8.41

EXAMPLE 4

A coin with unknown p is tossed 100 times. Find the **probable deviation for probability** 0.95. Here $1 - \alpha = 0.95$ so that $\alpha = 0.05$, and $z_{0.05} = 1.96$. By formula (2), the probable deviation is 0.098. This signifies that the estimate $\hat{p}$ differs from p by at most 0.098 with probability at least 0.95; in other words, the chances are at least 95 in 100 that our estimate is correct to within 0.098 units. ▲

EXAMPLE 5

A coin with unknown probability p is tossed n times, and $\hat{p}$ is computed as the estimate of p. Suppose for a given probability $1-\alpha$ we want the probable deviation to be a specified number d. How large must the number of tosses n be so that we obtain the desired probable deviation? This number n is obtained by setting d equal to expression (2) and solving for n:

$$n = \frac{z_\alpha^2}{4d^2}. \tag{3}$$

For example, a poll taker wants to estimate the proportion p of the electorate favoring the Blue party (Section 8.8, Example 3). Suppose that he wants his estimate to be within 0.03 units of the true p, and with probability at least equal to 0.98; in other words, the desired probable deviation is 0.03 and the probability is 0.98. Here $\alpha = 0.02$, and so $z_{0.02} = 2.33$. Then n is computed by means of (3):

$$n = \frac{(2.33)^2}{4(0.03)^2} = 1508.$$

We conclude that he requires a sample of at least 1508 voters. ▲

8.10 EXERCISES

1 In 200 Bernoulli trials with $p = 0.3$ find the approximate probability that the number of successes is between
(a) 66 and 72
(b) 54 and 70
(c) 40 and 50
(d) 55 and 200
(e) 0 and 54
(f) 10 and 70
inclusive.

2 In 500 Bernoulli trials with $p = 0.6$ find the probability that the number of successes is between
(a) 310 and 320
(b) 280 and 290
(c) 325 and 390
(d) 290 and 315
(e) 0 and 280
(f) 320 and 500
inclusive.

3 A pair of dice is repeatedly tossed. Let p be the probability that the sum of the scores on a single toss is either 7 or 11.
(a) Show that $p = 2/9$. (See Section 8.5, Example 3.)
(b) Find the probability that "7 or 11" appears on at least 35 out of 100 tosses. (See Example 3.)

4 When a pair of dice is tossed, we say that a double occurs whenever the same faces appear on both dice.
(a) Find the probability p of a double on a toss.
(b) In 600 tosses of a pair of dice, what is the probability that the number of doubles is between 90 and 110 inclusive? (See Example 3.)

5 A student is completely unprepared for a multiple choice examination consisting of 100 questions, with 4 choices on each question. Find the probability that he will correctly guess at least 32 answers if he chooses 1 out of the 4 at random on each question. (See Example 3.)

6 A vending machine has 4 identical service outlets. Each customer chooses 1 of the 4 at random. If 200 customers purchase items from the machine, find the probability that they will select at most 65 items from a particular outlet. (See Example 3.)

7 In an ESP experiment what is the probability, under the assumption of random guessing, of correctly calling at least 125 out of 500? 250 out of 1000? Also compute these probabilities for $p = \frac{1}{4}$. (See Example 2.)

8 A skeptic is willing to accept the existence of ESP if a subject is able to correctly guess a sufficiently large number N of cards out of 10,000. The number N is chosen so that the probability of correctly guessing at least N cards is 0.001 under the assumption of random guessing. Find the value of N. (See Example 2.)

9 A market research organization wants to estimate the proportion p of residents of a certain area having 2 or more television sets. A sample of 120 residents is taken. Let $\hat{p}$ be the proportion in the sample. Find the probable deviation of $\hat{p}$ from p for (a) probability 0.95; (b) probability 0.99. (See Example 4.)

10 A bridge traffic controller wants to determine the proportion of vehicles heading from the bridge to a certain highway during the morning rush hours. Cars and trucks are stopped at random, and their destinations are noted. Let $\hat{p}$ be the proportion of a sample of n vehicles heading for the highway. If the probable deviation is to be 0.05 and the probability $1 - \alpha = 0.98$, what is the size n of a sufficiently large sample? (See Example 5.)

11 On the basis of a sample of 700 voters, a poll finds a certain proportion $\hat{p}$ in favor of candidate A. Find the probable deviation for probability 0.96. (See Example 4.)

12 A polling organization wants to "call" an election within 3 percentage points, and be correct with probability at least 0.90. How large a sample do they need? What if they raise the probability of being correct to at least 0.98? (See Example 5.)

13 Sixty percent of patients recover from a certain intestinal virus within 2 days. In a sample of 150 patients, find the probability that at least 75 and at most 100 recover within the given time. (See Example 3.)

14 Every seed sown in a plot of ground has probability $\frac{1}{5}$ of germinating and growing into a plant. If 600 seeds are sown, what is the probability of getting at least 100 plants? At least 160? (See Example 3.)

15 Every seed of a certain plant is assumed to have probability $p = \frac{1}{5}$ of growing into a plant. If the planter wants to be assured of at least 100 plants with probability at least 0.9772, how many seeds have to be sown? (See Example 2.) To simplify the computation, drop the term $\frac{1}{2}$ in Equation (1); then solve a quadratic equation for $\sqrt{n}$.

16 Suppose that 10 percent of all horses in a certain area eventually contract a specific disease and die. A group of 250 horses is given a certain inoculation; of these, only 10 eventually get the disease. Is it plausible that the treatment is really worthless, and that the small number of illnesses is due to chance alone? [*Hint:* Find the probability of at most 10 successes in 250 trials with $p = 0.1$.] (See Example 3.)

17 Suppose the probability p that a patient recovers from a cold within 3 days is 0.6. It is proposed that Vitamin C is effective in curing colds. A group of 100 cold sufferers is given the vitamin. The effectiveness of the treatment is decided in the following way: If at least 70 of the patients recover within the given time, then the vitamin is concluded to be effective; otherwise, it is considered to be of no value.

(a) What is the probability of calling the vitamin effective when it is really worthless, that is, $p = 0.6$?

(b) Find the probability that the test results in a decision against the vitamin even though it really raises p to 0.8. (See Example 3.)

18 If p is a number between 0 and 1, inclusive, then the product $p(1-p) = pq$ is at most equal to $\frac{1}{4}$; this maximum value is assumed when $p = q = \frac{1}{2}$.

(a) Give a proof of this inequality based on these two facts: $(p+q)^2 = 1$ and $(p-q)^2$ is never negative. [*Hint:* Deduce the properties of the cross-product term $2pq$.]

(b) For those who have analytic geometry, demonstrate the inequality by drawing the locus of points (x,y) satisfying $y = x(1-x)$.

(c) For those who have already studied differential calculus, prove the inequality by maximizing $p(1-p)$ as a function of p on the interval from 0 to 1.

19 *Prove:* The probability that $\hat{p}$ differs from p by at most (2) is at least $1-\alpha$, if n is large. For the proof note that $\hat{p} =$ No. of successes/n, apply the normal approximation with

$$a = np - \frac{z_\alpha \sqrt{n}}{2} \qquad \text{and} \qquad b = np + \frac{z_\alpha \sqrt{n}}{2},$$

and then use the fact from Exercise 18 that $\sqrt{pq} \leq \frac{1}{2}$.

20 The normal approximation theorem can be stated in the form of a theorem about a limit (see Section 5.3).

(a) Show that if n is large, then the terms $\pm\frac{1}{2}$ in (1) may be dropped without significant change in the ratios.

(b) Define

$$A = \frac{a - np}{\sqrt{npq}}, \qquad B = \frac{b - np}{\sqrt{npq}};$$

and show that the probability that the number of successes is between a and b inclusive is approximately equal to the area under the standard normal curve between A and B.

(c) Using the results of (a) and (b), state the normal approximation as a **limit,** with a and b depending on n.

8.11 The Poisson Approximation to the Binomial Distribution

In many applications involving Bernoulli trials, the number of trials n is very large but the probability p of success is small; and the expected value np is of moderate size. We denote the expected value by $\lambda = np$. Under these conditions, the probabilities in the binomial distribution can be approximated by the **Poisson distribution** (after S. D. Poisson, 1781–1840). Although the binomial depends on the two parameters n and p, the Poisson depends on the single parameter λ. Table IX (Appendix) contains the values of this distribution for various λ. Graphs of the distribution for $\lambda = 1$, 4, and 6 appear in Figure 8.42.

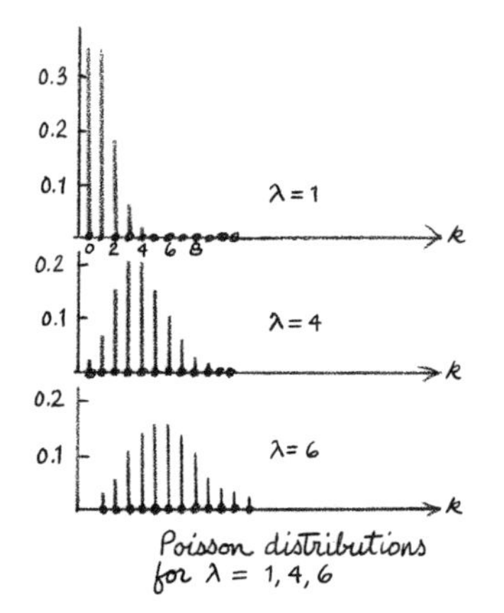

Figure 8.42

For the benefit of those who have learned the definition of the number e in Section 4.3 we derive the formula for the Poisson distribution. Let us first estimate the probability of no successes in n trials. By the definition of the binomial and of the number λ, the probability is equal to

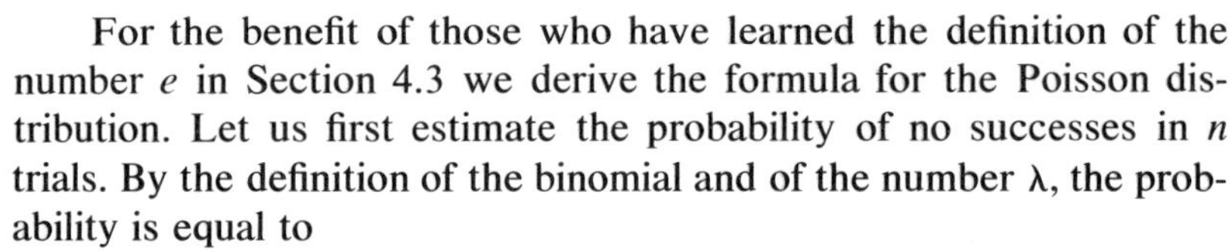

$$q^n = (1-p)^n = \left(1 - \frac{\lambda}{n}\right)^n.$$

As n tends to infinity, with λ constant, this converges to $e^{-\lambda}$. Similarly, the probability of exactly 1 success is

$$np(1-p)^{n-1} = n\frac{\lambda}{n}\left(1 - \frac{\lambda}{n}\right)^{n-1};$$

this converges to $\lambda e^{-\lambda}$. The probability of 2 successes is

$$C(n,2)p^2(1-p)^{n-2} = \frac{n(n-1)}{2}\left(\frac{\lambda}{n}\right)^2\left(1 - \frac{\lambda}{n}\right)^{n-2},$$

and this converges to $\frac{1}{2}\lambda^2 e^{-\lambda}$. By similar reasoning, we find that the probability of k successes, for fixed k, tends to

$$\frac{\lambda^k}{k!}e^{-\lambda}, \qquad \text{for } k = 0, 1, \ldots. \tag{1}$$

This is the formula for the Poisson distribution.

Note that the index k in (1) runs through all the integers that are not negative. For any k, no matter how large, the probability of k successes tends to (1) as n tends to infinity. It appears that the binomial distribution, which has a finite number n of terms, is being approximated by a distribution with an infinite number of terms. However, in actual numerical calculations this does not cause any problems: all the terms (1) from some point have a sum which is negligibly small. For example, when $\lambda = 1$ all the terms for $k = 8$ and more have a sum not exceeding 0.0001; thus, the distribution (1) is essentially restricted to values of k from 0 through 7. For this reason the sum of the tabulated probabilities is usually 0.9999, not 1.0000.

In the table below we compare 3 binomial distributions with $np = 1$ to the Poisson distribution with $\lambda = 1$:

No. of successes	Binomial probability			Poisson
k	$n = 10$ $p = 0.1$	$n = 20$ $p = 0.5$	$n = 100$ $p = 0.01$	$\lambda = 1$
0	0.349	0.358	0.366	0.3679
1	0.387	0.377	0.370	0.3679
2	0.194	0.189	0.185	0.1839
3	0.057	0.060	0.061	0.0613
4	0.011	0.013	0.015	0.0153
5	0.001	0.002	0.003	0.0031
6	. . .	. . .	. . .	0.0005
7	. . .	. . .	. . .	. . .

Note that the approximation improves as n increases.

There are many applications of the Poisson distribution; their primary common characteristic is that they involve randomly occurring "rare" events, such as accidents, rare diseases, rare accomplishments, opportunities, and so on.

EXAMPLE 1

A residential fire insurance company has a large number n of policy holders. The probability that a particular home is destroyed by fire during a given year is a very small number p; such an event is very unlikely. We can think of the numbers of homes so destroyed during 1 year as the number of successes in n Bernoulli trials with probability p; each home corresponds to a trial and each destruction to a "success" (Figure 8.43). (Although we do not ordinarily consider a burned home as a "success," we do it here only for the purpose of identifying the probability distribution as the binomial.) To apply our distribution, we shall consider only fires that result in essentially complete loss of the home, not just partial damage. However, the more realistic model involving partial loss can be treated by more complicated distributions considered in advanced books on probability. These have been successfully used in actuarial work. This area of mathematics is sometimes called **collective risk** theory; it has been developed mainly by Scandinavian mathematicians.

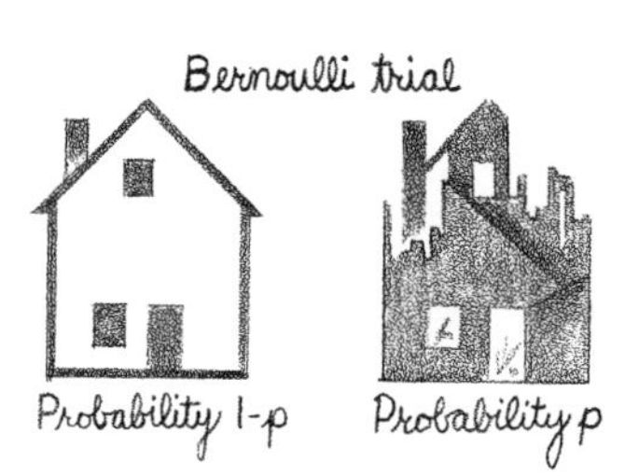

Figure 8.43

Suppose there are $n = 10{,}000$ policies, and that in 1 year each home has probability $p = 0.0005$ of burning; here $\lambda = np = (10{,}000)(0.0005) = 5$. The probabilities that exactly 0,

1, . . . , 15 houses burn are given by the entries of the column $\lambda = 5$ of Table IX. The sum of the probabilities for $k \geq 16$ is negligible. If each house is insured for exactly \$20,000, and if the company seeks to cover all possible fire losses with absolute certainty, then it would have to charge a premium of \$20,000 to each policy holder; however, no one would buy such insurance. It is very unlikely that all the houses will be destroyed in 1 year. So the company reduces the premium and bears a small probability of not being able to pay the claims. Since the probability of more than 15 fires during the year is not greater than 0.0001, the company might be willing to disregard the possibility of more than 15 fires, since it has less than 1 chance in 10,000. In this case the maximum total liabilities (for 15 fires) is \$20,000 (15) = \$300,000. If each of the 10,000 policy holders paid an equal share of this amount, the individual premiums would be \$30. Even so there is a large probability that the company will have a profit at the end of the year: If x is the number of homes destroyed, then the company earns \$20,000($15 - x$). For example, with probability 0.0005 it earns \$20,000 (14 homes lost), with probability 0.0013 it earns \$40,000 (13 lost), and so on.

If the company is willing to take a larger risk, it can charge lower premiums. For example, suppose it is willing to bear a probability 0.002 of not being able to meet all claims. In this case it can count on a maximum of 12 claims; indeed, the sum of the probabilities for 13, 14, 15, . . . claims is $0.0013 + 0.0005 + 0.0002 = 0.0020$. The loss of 12 homes results in a liability of \$240,000. If each policy holder pays an equal share of it, the premium is \$24. ▲

EXAMPLE 2

Certain diseases affect a very small proportion p of the population. These are usually not communicable, so that the presence of the disease in one person does not affect its presence in another. An example of this kind of disease is muscular dystrophy. Cholera is rare in the United States; however, it is very communicable so that if it occurred somewhere it would likely afflict a large group in the area. We do not consider a disease of the latter type. If the population consists of n persons, then the presence of the disease in a person is considered **success;** its probability is p. Absence of the disease is **failure,** and its probability is q. The n persons are like n Bernoulli trials: just as the outcomes of the tosses of a coin do not influence each other, so the occurrences of the disease in the various individuals do not influence each other. The number of cases is the number of successes in n trials. If n is large, p small, and $\lambda = np$ moderate, then the Poisson approximation is appropriate. Suppose that the disease is known to occur in 1 person per thousand in a population; then the probabilities of exactly 0, 1, . . . cases in a town of 3000 are given by the entries of Table IX in the column $\lambda = 3$. The probability of at least 2 cases is the sum of the probabilities for $k = 2$, 3, ... A simple way of computing this sum is to subtract the complementary probabilities for $k = 0$ and 1 from 1:

$$1 - 0.0498 - 0.1494 = 0.8008. \quad \blacktriangle$$

EXAMPLE 3

The Poisson distribution applies to events that occur "at random" in time or space. Consider the arrival of calls at a telephone exchange during a 1-hour period on a business day; they seem to arrive at random time points. It has long been known that the number of calls arriving in a time period of fixed length has a Poisson distribution. Let λ be the average number of calls in 1 unit of time (1 hour, $\frac{1}{2}$ hour, and so on); then the number to arrive in such a time period has the Poisson distribution with parameter λ. λ is called the **intensity** of the traffic of calls. The number to arrive in T time units has a Poisson distribution, but with λT in place of λ. The theoretical reason for the Poisson distribution is as follows. Imagine that the time interval from 0 to T is subdivided into many small intervals of equal length; let n be the number of such intervals. If λ is the average number of calls per unit interval, then λT is the average for an interval of T units, and $\lambda T/n$ the average for an interval of length T/n. Now it is very unlikely that more than one call will arrive in a very small interval. The probability of one call is $\lambda T/n$, and the probability of no calls is approximately $1 - \lambda T/n$; this yields an average of $\lambda T/n$ (Figure 8.44). The calls come from independent sources, so that the arrivals of the calls are without mutual influence; furthermore, the intensity is assumed to be constant over the given time interval. Each time interval of length T/n corresponds to a Bernoulli trial: a success (call) occurs with probability $\lambda T/n$, and a failure (no call) with probability $1 - \lambda T/n$. There are n such intervals in the time period from 0 to T; hence, the number of calls has the binomial distribution with $\lambda T/n$ in place of p. If n is large (and we assume this to be so), then, according to the Poisson approximation, the distribution is the Poisson with λT in place of λ.

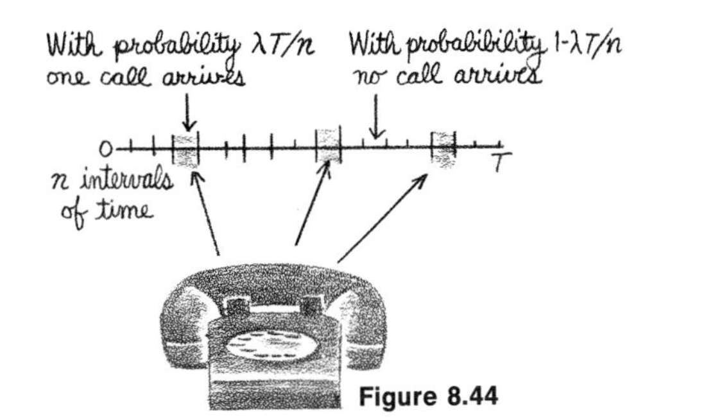

Figure 8.44

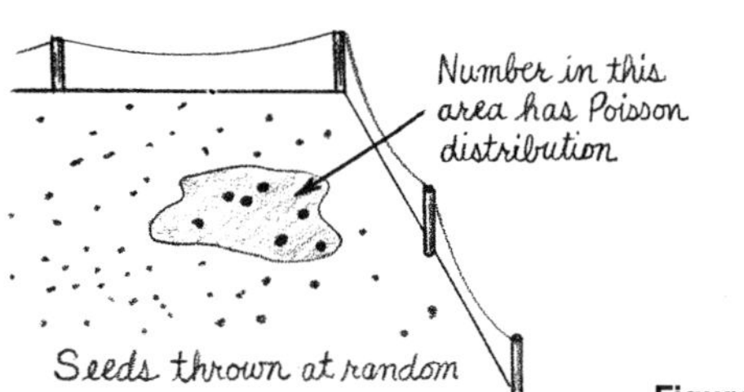

Figure 8.45

The Poisson distribution also applies to distributions in two and three dimensions. If seeds are scattered at random over a piece of ground, then the number falling in a relatively small area of specified size has a Poisson distribution; here the parameter λ is the average number of seeds per unit area. The reason for this is that each seed has a very small probability of falling in the given area (success) but there is a large number of seeds (trials) (Figure 8.45). The same reasoning applies to the number of fish in a given portion of a lake, the number of pollen grains in a portion of the atmosphere, and so on. ▲

EXAMPLE 4

One of the rare events of the major league baseball season is the "no-hitter." In each game there is a very small probability p that the pitcher will achieve the no-hitter: each game is a Bernoulli trial and **success** means a **no-hitter.** Then the number of no-hitters during a complete season has a binomial distribution with large n and small p, so that it can be approximated by the Poisson. In reality, p varies from game to game according to the teams involved; however, it appears that this deviation from the theoretical assumption of a uniform probability p is not serious. It seems that the distribution of the number of no-hitters per season over the 60-year period from 1901 to 1960 is close to the Poisson. (During these years each of 16 teams played 154 games per season. In 1961 the numbers of games and teams increased.) In the table below we give the numbers of years in which $k = 0$, $1, \ldots$ no-hitters occurred. These numbers are roughly porportional to the corresponding terms of the Poisson distribution with $\lambda = 1.5$; the last column shows the probabilities multiplied by 60.

They represent the "expected number" of no-hitters under the assumption of the Poisson distribution (Figure 8.46).

Figure 8.46

No. of no-hitters k	No. of years	Poisson probability times 60
0	12	13.4
1	20	20.1
2	12	15.1
3	8	7.5
4	5	2.8
5	0	0.8
6	2	0.2
7	1	0.0
8 and more	0	0.0

▲

EXAMPLE 5

A classical example of observations fitting the Poisson distribution is that of the numbers of deaths in the old Prussian army caused by kicks of horses; it is due to Bortkiewicz (1898). He found the number of corps, out of a total of 200 corps, in which k such deaths occurred, for $k = 0, 1, \ldots$. Then he compared these numbers to the terms of the Poisson distribution with $\lambda = 0.61$, and multiplied the latter by 200. The results are in the table below:

No. of deaths	No. of corps	Poisson probability times 200
0	109	108.7
1	65	66.3
2	22	20.2
3	3	4.1
4	1	0.6
5 or more	0	0.0

▲

8.11 EXERCISES

In Exercises 1 and 2 use the Poisson approximation to find the probability of the given number of successes in a large number of trials.

1 (a) 4 successes when $\lambda = 3$
(b) 5 successes when $\lambda = 7$
(c) 6 successes when $\lambda = 2$
(d) at least 3 successes when $\lambda = 1$
(e) at least 2 successes when $\lambda = 6$
(f) at most 10 successes when $\lambda = 5$
(g) at most 8 successes when $\lambda = 4$
(h) between 2 and 11 inclusive when $\lambda = 5$

2 (a) 8 successes when $\lambda = 5$
(b) 9 successes when $\lambda = 6$
(c) 2 successes when $\lambda = 3$
(d) at least 1 when $\lambda = 4$
(e) at least 9 when $\lambda = 5$
(f) at most 2 when $\lambda = 6$
(g) at most 7 when $\lambda = 2$
(h) between 1 and 15, inclusive when $\lambda = 6$

3 An insurance company pays exactly $50 to the parent of a child injured on the way to school or in it. The company finds that the injury rate is 1 per 200 students. If 2000 students are covered by this insurance, and if the company wants to be able to meet all claims with probability at least (0.9928, what is the maximum number of claims it should plan to get? Find the premium that each parent should pay to meet this number of claims. (See Example 1.)

4 In 1967 the death rate for travelers on scheduled American passenger trains was about 1 per billion passenger miles. Among passengers traveling approximately 50 miles, the rate is therefore 1 per 20 million trips. If a company insures passengers on 100 million trips during 1 year, and each passenger is insured for $10,000 for accidental death, what is the maximum number of deaths it should consider if it wishes to meet all claims with probability at least 0.9998? What is a "fair" premium for each trip? (See Example 1.)

5 The annual rate of accidental death in American homes is about 1 per 10,000 population. Find the probability of at least 5 accidental deaths in a city of 60,000. (See Example 2.)

6 The marriage rate in the United States in 1965 was about 93 marriages per 10,000 population. What is the probability of finding at least 1 newlywed in a group of 225 women? (Assume that "10,000 population" implies "5000 women.") Use the nearest value of λ in Table IX. (See Example 2.)

7 The incidence of polio during the period 1949–1954 was about 25 per 100,000 population. (Polio was a relatively rare disease.) In a city of 20,000 what is the probability of at most 10 cases? (See Example 2.)

8 We refer to the method of industrial acceptance sampling in Section 8.8, Example 2. When the number n of sampled items is large and p is small, we can use the Poisson approximation to the binomial to compute the operating characteristic of the sampling plan. Suppose 100 items are sampled from a large lot, and it is rejected if more than 2 items are found to be defective. Find the probability of accepting the lot ($L(p)$) for $p = 0.01, 0.02, 0.03, 0.04$, and 0.05.

9 In 1950 the rate of tetanus infection in the United States was about 1 per 30,000 population. What was the probability of 2 or more cases in a city of 180,000? (See Example 2.)

10 Calls arrive at a switchboard at a railroad terminal at a rate of 2 per minute. Find (a) the probability of at least 15 calls in a period of 4 minutes; (b) the probability of at most 5 calls in the same time. (See Example 3.)

11 In 1935 the Supreme Court of the United States ruled that much of President Franklin Roosevelt's New Deal was unconstitutional. The President yearned for the resignation of the conservative justices and their replacement with those more favorable to his program. A well-known statistician, W. A. Wallis, reported in the Journal of the American Statistical Association, Vol. 31 (1936), pp. 376–380, that the annual number of Court vacancies, by death or resignation, for the 96 years from 1837 to 1932, followed a Poisson distribution. The annual numbers of vacancies were distributed as follows:

No. of vacancies k	0	1	2	3	More than 3
No. of years with k vacancies	59	27	9	1	0

Compare this distribution of vacancies with that under the Poisson distribution with $\lambda = 0.5$; multiply the latter probabilities by 96. (See Example 4.)

12 A lake contains 1 fish for every 1000 gallons of water. Find the probability of not finding a single fish in 7000 gallons. (See Example 3.)

13 Small nuts are mixed in the syrup used to make chocolate candy bars. If the average ounce of syrup contains 2 nuts, and if a bar is made from 5 ounces of syrup, what is the probability that a bar will contain fewer than 5 nuts? (See Example 3.)

14 Smugglers enter the port of a certain city at a rate of 1 per 200 incoming travelers. Find the probability that an incoming boat with 600 persons has at least 2 smugglers. (See Example 2.)

15 Engine failures affect the buses of a large transit company at the average rate of 2 per day. Each engine failure requires the service of a mechanic for 1 day. The comapny wants to have enough mechanics to service the buses that fail. But it does not want too many because they have to be paid even when they do not work. Suppose it wants sufficiently many so that the probability that a defective bus will have a mechanic to repair it is at least 0.95. What is the minimum number of mechanics it should hire? (See Example 1.)

16 In 1960 the death rate for motor vehicle accidents was about 21.2 per 100,000 population. If you had to insure 20,000 persons, and wanted to meet all claims with probability at least 0.99, for how many such accidents would you allow? To simplify the problem use the rate of 20 per 100,000. (See Example 1.)

17 The intensity of bird calls from a certain tree in a park is about 4 every 10 seconds. Suppose you listen for 15 seconds. Find the smallest integer n such that the probability of at least n calls during the period is at most 0.1. (See Example 3.)

18 For those who have studied Section 6.12: Prove that the sum of the terms of the Poisson distribution is equal to 1; that is, the series of terms (1) has the sum 1.

8.12 Sample Space, Probability, and Conditional Probability

We have already considered several specific models of random experiments: Bernoulli trials, die tossing, and sampling from urns and bowls, with and without replacement. These have the following common characteristics:

(i) The set of outcomes is finite, and each outcome is assigned a probability, a number between 0 and 1, inclusive.

(ii) The sum of the probabilities is 1.

(iii) An event is a set of outcomes, and the probability of an event is the sum of the probabilities of the outcomes belonging to the event.

The set of all outcomes is the **sample space,** and is denoted by the letter S. Events are denoted by the letters $A, B, C, \ldots$. The probability of an event A is written $P(A)$; thus, $0 \le P(A) \le 1$. By (ii) above it follows that $P(S) = 1$. The **impossible event,** denoted I, is the hypothetical event consisting of no outcomes; its probability is 0. The impossible event plays the role of "zero" for events.

From the given events we can form others by **binary operations.** (The reader who has already studied elementary set theory will recognize these as set operations.) These are illustrated through **Venn diagrams.** The **union of 2 events** A and B, denoted $A \cup B$, is the set of outcomes belonging to A or B or both (Figure 8.47). The **intersection** of A and B, denoted $A \cap B$, consists of the outcomes belonging to both (Figure 8.48). The **complement** of A, denoted ${}^cA$, is the set of outcomes not belonging to A (Figure 8.49). Two events are **disjoint,** or **mutually exclusive,** if they have no common outcomes, that is, their intersection is the impossible event (Figure 8.50). If A and B are disjoint events, then

$$P(A \cup B) = P(A) + P(B);$$

indeed, the sum of the probabilities in either event is the sum of the probabilities of those in each (Figure 8.50). More generally,

$$P(A \cup B) = P(A) + P(B) - P(A \cap B); \tag{1}$$

in fact, the last term $P(A \cap B)$ is subtracted to compensate for the "double counting" in $P(A)$ and $P(B)$ (Figure 8.51).

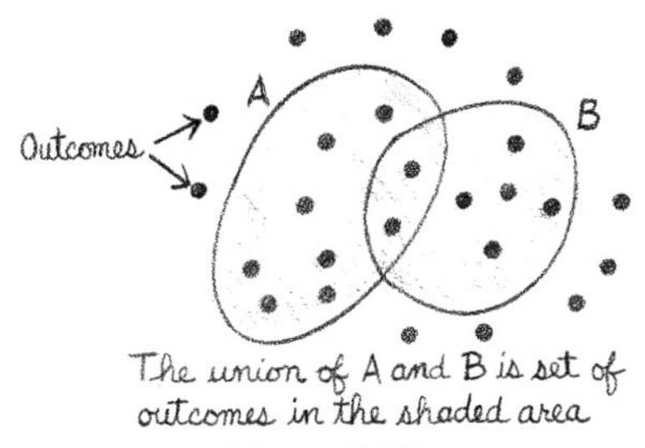

Figure 8.47

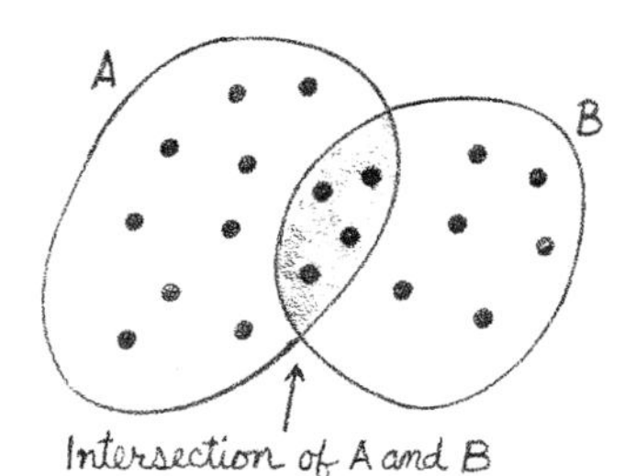

Figure 8.48

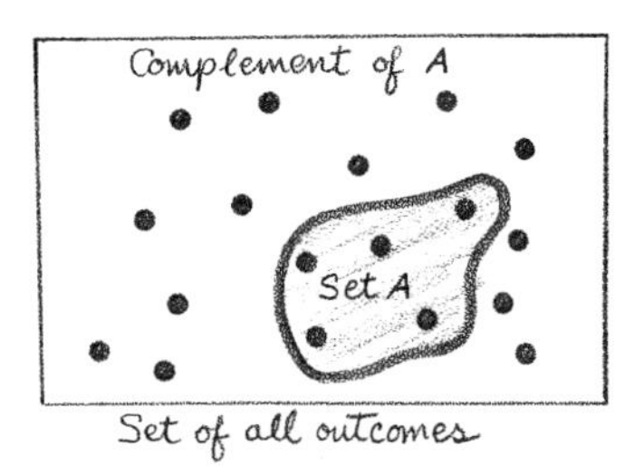

Figure 8.49

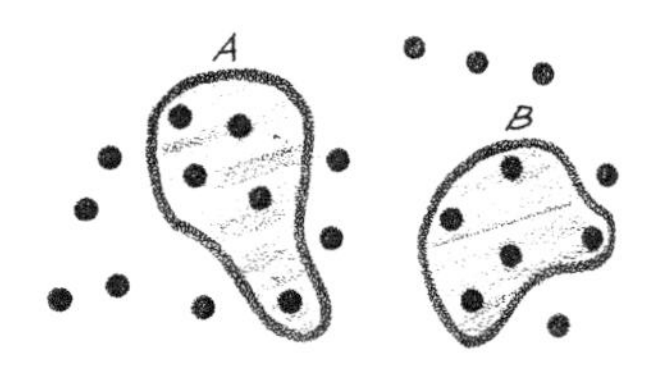

Figure 8.50

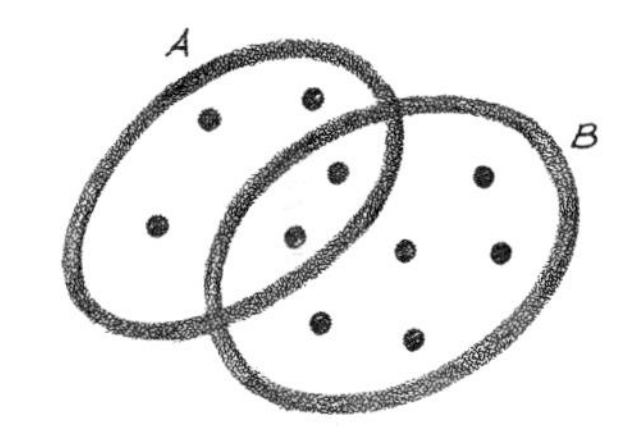

Figure 8.51

The probability of the complement is 1 minus the probability of the event:

$$P({}^cA) = 1 - P(A).$$

EXAMPLE 1

A fair coin is tossed 3 times. The event A = "Heads on the first toss" consists of the 4 outcomes HHH, HHT, HTH, HTT. It has probability $\frac{1}{2}$. The event B = "Heads on the second toss" consists of HHT, HHH, THT, THH. It also has probability $\frac{1}{2}$.

The intersection $A \cap B$ consists of the 2 outcomes in both, HHH and HHT. The union $A \cup B$ consists of the 6 outcomes HHH, HHT, HTH, HTT, THT, THH. Its probability is $\frac{3}{4}$; this is consistent with Equation (1). ▲

Let A and B be events such that the latter has positive probability; then we define the **conditional probability of A given B.** as

$$P(A \mid B) = \frac{P(A \cap B)}{P(B)}.$$

It represents the probability of A relative to the event B: it reflects the likelihood of A in the presence of B.

EXAMPLE 2

The prevalence of definite coronary heart disease in American adults in 1960–1962 is given in the table below (in thousands) by age and sex. The source is: Coronary Heart Disease in Adults, United States, 1960–1962, National Center for Health Statistics, Ser. 11, No. 10, U.S. Department of Health, Education and Welfare.

Coronary Heart Disease (Adults)

Total 18–79 years	Both sexes	Men	Women
	3125	1945	1180
18–24 years	...	...	...
25–34 years	60	42	19
35–44 years	177	120	57
45–54 years	517	352	165
55–64 years	1111	726	384
65–74 years	1064	575	489
75–79 years	195	130	64

Suppose a patient is drawn at random from the population. Let A be the event that the patient is male, and B the event that the patient is 65 or older. Then

$$P(A) = \frac{\text{No. of men}}{\text{No. of patients}} = \frac{1945}{3125} = 0.6224;$$

$$P(B) = \frac{\text{No. of patients 65 and over}}{\text{No. of patients}} = \frac{1259}{3125} = 0.4029;$$

$$P(A \cap B) = \frac{\text{No. of male patients 65 and over}}{\text{No. of patients}} = \frac{705}{3125} = 0.2256;$$

$$P(A \cup B) = P(A) + P(B) - P(A \cap B) = 0.7997.$$

The conditional probabilities are

$$P(B \mid A) = \frac{\text{No. of male patients 65 and over}}{\text{No. of male patients}} = \frac{705}{1945} = 0.3625;$$

$$P(A \mid B) = \frac{\text{No. of male patients 65 and over}}{\text{No. of patients 65 and over}} = \frac{705}{1259} = 0.5600.$$

Observe that the conditional probabilities of A and B are smaller than the corresponding (unconditional) probabilities. This signifies that the proportion of men among all patients 65 and older is smaller than the proportion of men among all the patients; and that the proportion of those above 65 among all men is smaller than the proportion of those above 65 among all patients. This might signify that female coronary patients live longer than the males. ▲

The operations of union and intersection can be performed for more than 2 events. If $A, B, C, \ldots$ are events, then their union is denoted $A \cup B \cup C \cup \ldots$, and consists of all outcomes belonging to **at least** 1 of them. The intersection, denoted $A \cap B \cap C \cap \ldots$, consists of the outcomes belonging to all of the events. Several events are **mutually exclusive** if every 2 of them are disjoint (Figure 8.52). If $A, B, C, \ldots$ are such events, then

$$P(A \cup B \cup C \ldots) = P(A) + P(B) + P(C) + \cdots \quad (2)$$

More generally, for an arbitrary event E,

$$P(E \cap (A \cup B \cup C \ldots)) = P(E \cap A) + P(E \cap B) + P(E \cap C) + \cdots \quad (3)$$

Figure 8.53).

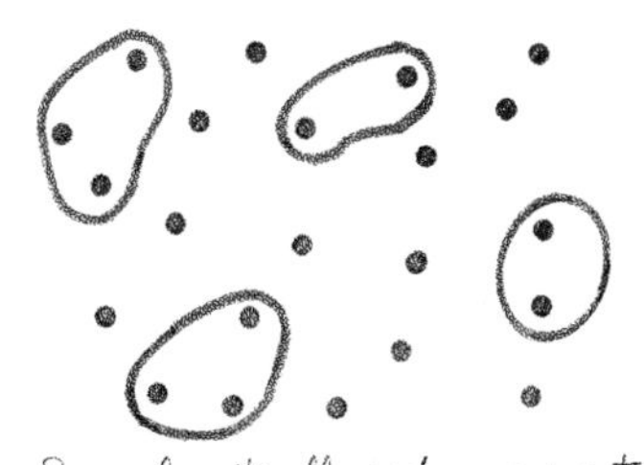

Figure 8.52

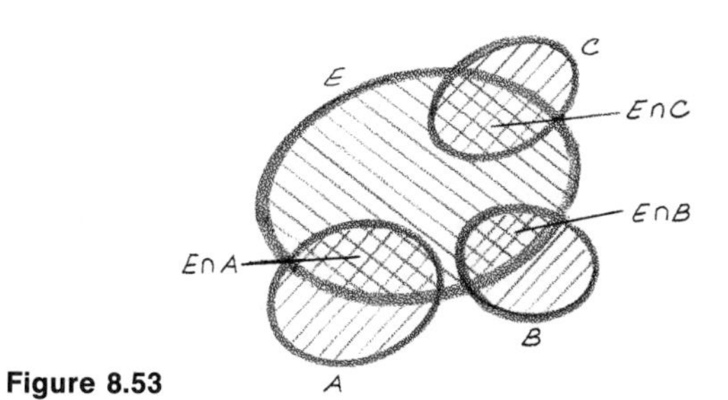

Figure 8.53

Suppose there are 2 urns, labeled I and II; these contain n_1 and n_2 balls, respectively, of which w_1 and w_2 are white and the others red. One urn is selected by a random experiment: urn I is selected with probability p and urn II with probability $q = 1 - p$. These probabilities are called **a priori** probabilities. Suppose that a ball is taken at random from the selected urn, but that we do not know which urn it is, I or II. The color of the ball is observed. What is the probability that it came from urn I or II? This is the **a posteriori** probability. This experiment is **compound;** it has 4 outcomes corresponding to the urn selected and the color of the ball drawn from it:

urn I, white ball	urn I, red ball
urn II, white ball	urn II, red ball.

The probabilities are assigned as follows: The probability of an outcome is the product of the a priori probability of selecting the given urn and the probability of selecting the ball of the given color from that urn. The outcomes and their probabilities are

		BALL	
		White	Red
URN	I	pw_1/n_1	$\dfrac{p(n_1 - w_1)}{n_1}$
	II	qw_2/n_2	$\dfrac{q(n_2 - w_2)}{n_2}$

The justification is similar to that for Bernoulli trials (Section 8.8).

Let us compute the a posteriori probability of urn I given that a white ball is drawn. By definition, this is the ratio

$$\frac{\text{Probability that urn I is selected and a white ball is drawn}}{\text{Probability that a white ball is drawn}}.$$

The numerator is pw_1/n_1; and the denominator is the sum $pw_1/n_1 + qw_2/n_2$; thus, the a posteriori probability is

$$\frac{pw_1/n_1}{pw_1/n_1 + qw_2/n_2}. \tag{4}$$

More generally, if there are s urns and balls of t colors, then, by the counting principle, there are st outcomes, corresponding to the urn drawn and the ball taken from it. The probability of an outcome is the product of

(i) the a priori probability of selecting the urn, and

(ii) the probability of drawing a ball of the given color from the selected urn.

Let $A_1, A_2, \ldots$ be events corresponding to the selection of urns I, II, . . . , and B an event corresponding to a ball of a given color; then $P(A_1), P(A_2), \ldots$ are the a priori probabilities, and $P(B \mid A_1)$, $P(B \mid A_2), \ldots$ are the proportions of balls of the given color in the various urns. There is a well-known formula for the a posteriori probability of a particular urn given the color of the ball drawn. The probability of urn k is

$$P(A_k \mid B) = \frac{P(B \mid A_k)P(A_k)}{P(B \mid A_1)P(A_1) + P(B \mid A_2)P(A_2) + \cdots}, \tag{5}$$

for $k = 1, 2, \ldots$. This is **Bayes' formula** (after Thomas Bayes, eighteenth century). It is called the law of the **probabilities of causes.** The proof is based directly on the definition of conditional probability and the "total probability formula" (3); it is outlined in Exercise 12 below.

EXAMPLE 3

Consider 3 urns with these compositions:

	Urn I	Urn II	Urn III
Red	3	2	1
White	1	2	1
Blue	1	1	3

The a priori probabilities of the urns are $\frac{1}{4}$, $\frac{1}{2}$, and $\frac{1}{4}$, respectively. Let us compute the a posteriori probability of urn I, given a white ball drawn. By Bayes' formula (5), this is

$$\frac{(\frac{1}{5})(\frac{1}{4})}{(\frac{1}{5})(\frac{1}{4}) + (\frac{2}{5})(\frac{1}{2}) + (\frac{1}{5})(\frac{1}{4})},$$

or $\frac{1}{6}$. Similarly, the a posteriori probability of urn II is

$$\frac{(\frac{2}{5})(\frac{1}{2})}{(\frac{1}{5})(\frac{1}{4}) + (\frac{2}{5})(\frac{1}{2}) + (\frac{1}{5})(\frac{1}{2})},$$

or $\frac{2}{3}$. Finally, the probability for urn III is $\frac{1}{6}$. ▲

EXAMPLE 4

A factory has 3 machines, A, B, and C, which produce 40, 40, and 20 percent of the items, respectively. The proportions of defective items produced are 0.02, 0.01, and 0.03, respectively. An item is drawn at random and found to be defective. Find the probability that it was produced by the first machine.

Let A_1, A_2, and A_3 be the events corresponding to the 3 machines, and B the event corresponding to the production of a defective. The a priori probabilities of the events A are 0.4, 0.4, and 0.2. The probabilities of B given the various events A are

$$P(B \mid A_1) = 0.02, \qquad P(B \mid A_2) = 0.01, \qquad P(B \mid A_3) = 0.03.$$

It follows from Bayes' formula that

$$P(A_1 \mid B) = \frac{(0.02)(0.4)}{(0.02)(0.4) + (0.01)(0.4) + (0.03)(0.2)} = \tfrac{4}{9}.$$

This is slightly higher than the a priori probability of machine A. This signifies that the proportion of items produced by A is higher among defective items than among all items. The a posteriori probabilities for machines B and C are $\frac{2}{9}$ and $\frac{1}{3}$, respectively. ▲

8.12 EXERCISES

1 A coin is tossed 3 times. Let A be the event "At least 1 head appears," B the event "At least 1 tail," and C the event "More heads than tails."
- (a) Enumerate the outcomes in each of these.
- (b) Enumerate the outcomes in $A \cup B$, $A \cup C$, $B \cup C$, and $A \cup B \cup C$.
- (c) Enumerate the outcomes in $A \cap B$, $A \cap C$, $B \cap C$, and $A \cap B \cap C$.
- (d) Enumerate the outcomes in the complements of A, B, and C.

2 A coin is tossed 4 times. Let A be the event "Heads and tails on the first and second tosses, respectively," B the event "Tails and heads on the third and fourth, respectively," and C the event "Tails on the second and third tosses." Enumerate the outcomes in these events, their unions, intersections, and complements. (See Example 1.)

3 A pair of dice is tossed. List the outcomes in the events A: "The sum of the scores is at least 5" and, B: "The sum of the scores is an even number"; and the outcomes in their union and intersection. (See Example 3, Section 8.5.)

4 Two tickets are drawn at random from a bowl containing tickets labeled 1, 2, 3, and 4. Let A be the event "Two odd-numbered tickets are drawn" and B the event "The sum of the numbers on the tickets drawn is at least equal to 5." Show that these are disjoint. Compute the probabilities of the events and their union.

5 The age and sex distribution of Americans suffering from definite **angina pectoris** in 1960–1962 is given (in thousands) in the table below; the source is the same as for Example 2.

Age	Both sexes	Male	Female
18–24	. . .	. . .	. . .
25–34	13	13	. . .
35–44	14	. . .	14
45–54	273	141	133
55–64	600	380	221
65–74	542	261	281
75–79	104	40	64
Total	1548	835	713

A patient is drawn at random. Let A_1 and A_2 be the events that the patient is male and female, respectively. Let $B_1, B_2, \ldots, B_6$ be the events corresponding to the age groups 25–34, 35–44, . . . , 75–79, respectively. Compute the probabilities of the events A and the events B; the conditional probabilities of A_1, given $B_3, \ldots, B_6$, respectively; and the conditional probabilities of $B_3, \ldots, B_6$ given A_1. (See Example 2.)

6 The educational achievements of American adults (age 25 and over) by sex in 1962 are summarized in the table below. (*Source:* U.S. Bureau of the Census, Current Population Reports, Ser. P-20, No. 121)

(In thousands)

	Total	**Elementary or none**	**High school**	**College plus**
Male	48,283	18,140	20,331	9812
Female	52,381	18,124	25,897	8360
Total	100,664	36,264	46,228	18,172

A person is drawn at random. Let A_1 and A_2 be the events "Male" and "Female," and B_1, B_2, and B_3 the events corresponding to the three educational achievement groups, respectively. Compute the probabilities of these events, and the conditional probabilities of A-events given B-events, and the conditional probabilities of B-events given A-events. (See Example 2.)

7 The theory of conditional probability is useful in diagnostic tests. Suppose that a serious disease may have a preliminary symptom; however, not all those with the symptom have the disease and not all those with the disease have the symptom. It is found that among all those having the disease 40 percent had the symptom; and among all those without the disease, 5 percent had the symptom. The proportion of the population having the disease is 0.01. Find the conditional probability that a person has the disease, given that he has the symptom. [*Hint:* Let A_1 and A_2 be the events corresponding to the presence and absence of the disease; and B_1 and B_2 the events corresponding to the presence and absence of the symptom. Find the a posteriori probabilities of the A-events given the B-events.] (See Example 4.)

8 A city is divided into high, low, and medium crime rate areas. The probabilities that a home in an area is burglarized at least once during the year, and the proportions of homes in the areas are:

	High crime	**Medium crime**	**Low crime**
Probability of burglary	0.40	0.20	0.05
Proportion of homes in area	0.30	0.40	0.30

If the home of a burglary victim is chosen at random from a list of homes that have been burglarized at least once, what is the probability that the home is in a (a) high crime; (b) medium crime; (c) low crime area? (See Example 4.)

9 A banker classifies his debtors as high, medium, and low risks. Their probabilities of defaulting are 0.05, 0.01, and 0.001, respectively. The proportions of debtors in the various categories are 0.2, 0.4, and 0.4, respectively. What is the probability that a defaulter is of the high risk type? Medium risk? Low risk? (See Example 4.)

10 An urn contains n tickets, numbered $1, \ldots, n$. These are drawn out at random without replacement. Let A be the event "Ticket #1 is drawn last," and B_k the event "Ticket #1 is not among the first k drawn," for $k < n$. Find the formulas for

(a) $P(A)$
(b) $P(B_k)$
(c) $P(A \cap B_k)$
(d) $P(A \mid B_k)$.

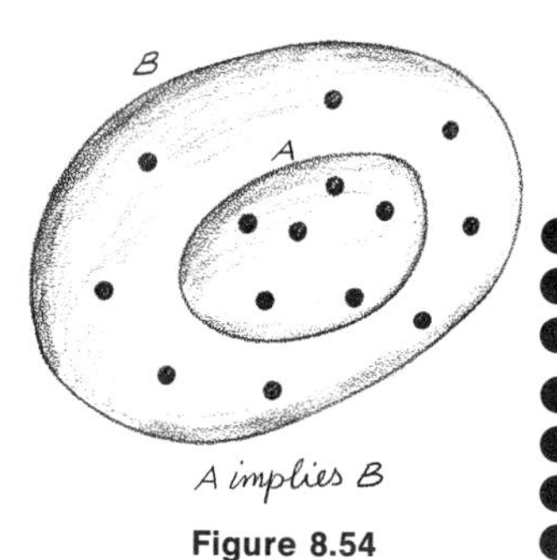

Figure 8.54

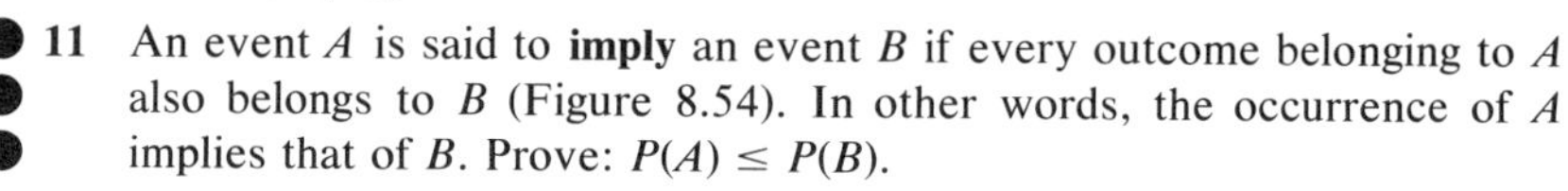

11 An event A is said to **imply** an event B if every outcome belonging to A also belongs to B (Figure 8.54). In other words, the occurrence of A implies that of B. Prove: $P(A) \leq P(B)$.

12 Here is an outline of the proof of Bayes' formula.

(a) Show that the numerator in (5) is equal to $P(B \cap A_k)$.
(b) Show that the terms in the denominator are equal to $P(B \cap A_1)$, $P(B \cap A_2), \ldots$.
(c) Show that the sum in the denominator is equal to $P(B)$. [*Hint:* Apply formula (3), and note that the A's are disjoint.]

8.13 Discrete Random Variables

A random variable is a number associated with the actual outcome of a random experiment. With each possible outcome of the experiment we associate a particular number. The experiment is carried out, and the **observed** value of the random variable is the number associated with the outcome that actually occurred. For example, if a gambler bets \$1 that the outcome of the toss of a coin will be heads, then his net gain after the toss is a random variable which assumes the value \$1 for the outcome heads, and $-\$1$ for the outcome tails.

A random variable may be considered to be a **function** whose domain is the sample space of the random experiment, and whose range is a set of real numbers. (These terms are defined in Sections 4.1 and 8.12.) This differs from the functions studied in calculus, where the domain is also a set of numbers, or pairs of numbers in the plane. The term **random variable** is based on the fact that the observed value of the function (here called a variable) is determined by chance factors.

A **discrete** random variable has either finitely many values, or else an infinite number of values which can be arranged in a sequence (see Example 4 below). Let X stand for such a random variable, and x (lower case) for a particular value of X. The event associated with the value x, denoted $\{X = x\}$, is the set of outcomes for which X assumes the value x (Figure 8.55). The probability of this event is denoted $P(X = x)$; it is read "The probability that (capital) X is equal to (little) x." The table of $P(X = x)$ for the various values of x is called the **probability distribution** of X. Since distinct values of x are associated with disjoint events, the sum of the terms of a probability distribution is 1.

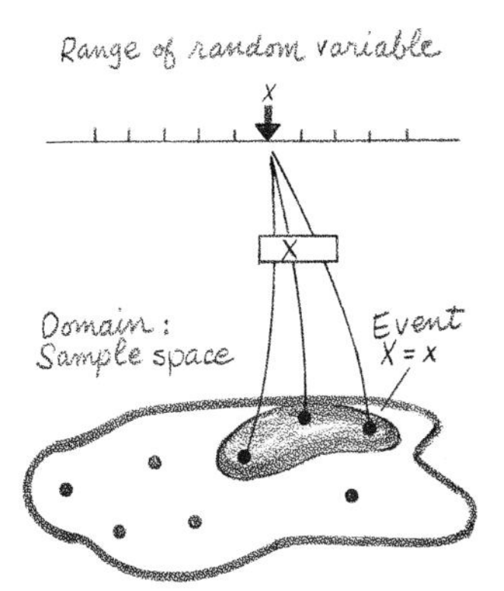

Figure 8.55

EXAMPLE 1

A coin with probability p of heads is tossed 3 times. Let X be the random variable representing the number of heads that appear. The possible values of X are 0, 1, 2, and 3. The events associated with each of these values and their probabilities are:

Value	Event	Outcomes	Probability
0	$X = 0$	TTT	q^3
1	$X = 1$	HTT, THT, TTH	$3pq^2$
2	$X = 2$	HHT, HTH, THH	$3p^2q$
3	$X = 3$	HHH	p^3

EXAMPLE 2

A fair coin is tossed 4 times. Let X be the number of times in the course of the tossing that the numbers of heads and tails are equal. For example, in the outcome HTHH there are equal numbers of heads and tails (1 and 1) after the second toss and no others; hence, $X = 1$. We can represent the outcomes by records of the successive numbers of heads minus the corresponding numbers of tails. Each record has a graph which starts at 0 and moves 1 unit up or down accordingly as heads or tails appears (Figure 8.56). Since the coin is balanced, the 16 outcomes are equally likely, each with probability $\frac{1}{16}$. The event $X = 0$ is associated with graphs which stay either above the horizontal axis or below it for all tosses; there are no **ties.** The event $X = 1$ consists of graphs which touch the axis exactly once, either at the second or fourth toss. The event $X = 2$ consists of graphs which touch the axis at both the second and the fourth tosses (Figure 8.57). The probabilities of these events are $\frac{3}{8}$, $\frac{3}{8}$, and $\frac{1}{4}$, respectively; therefore, the probability distribution of X is

x	0	1	2
$P(X = x)$	$\frac{3}{8}$	$\frac{3}{8}$	$\frac{1}{4}$ ▲

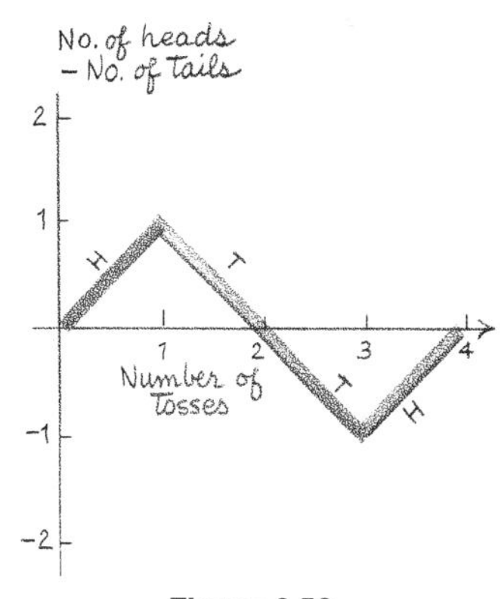

Figure 8.56

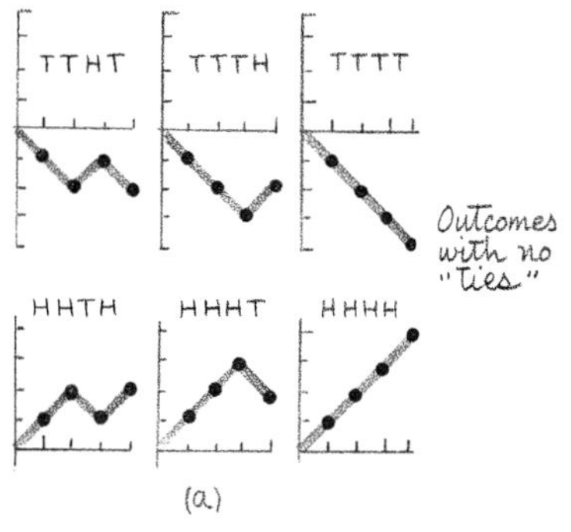

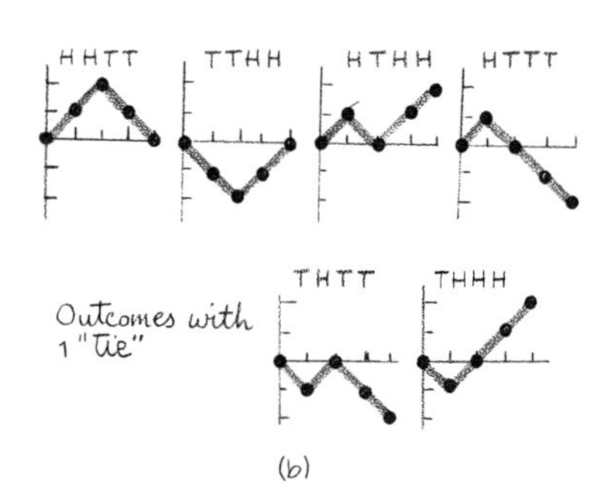

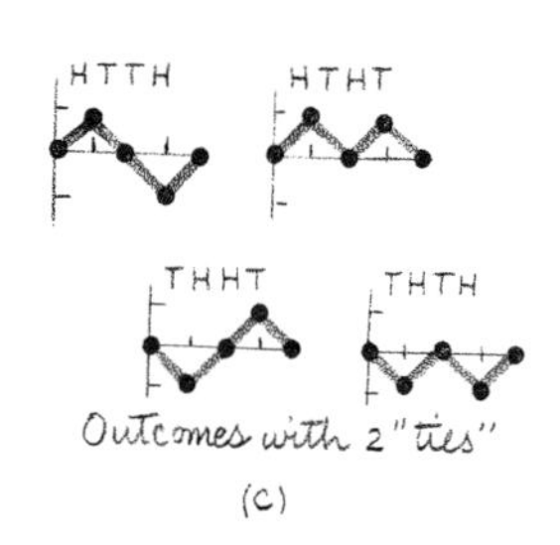

Figure 8.57

Outcomes

Radio buyer	Camera buyer	Clock buyer	X
			3
			1
			1
			1
			0
			0

Figure 8.58

EXAMPLE 3

A radio, camera, and clock have been ordered from an appliance store by 3 different customers. The shipping clerk has the names of the customers, but forgot to record the item which each customer is to receive. He decides to send them at random to the 3 customers. Let X be the number of customers (0, 1, or 3) who receive the appliance they ordered. Find the probability distribution of X.

There are 6 outcomes for the random experiment, namely, the 6 permutations of the 3 appliances. One of these, the correct set of orders, results in 3 satisfied customers. There are 3 permutations where just 1 customer is satisfied, and 2 where none are (Figure 8.58). Thus, the distribution of X is

x	0	1	3
$P(X = x)$	$\frac{1}{3}$	$\frac{1}{2}$	$\frac{1}{6}$ ▲

EXAMPLE 4

Here is an example of a random variable assuming an infinite number of possible values, the integers 0, 1, 2, ... A coin with probability p of heads is tossed until, for the first time, heads appears. Let X be the number of tails which turned up before the first heads. It may assume any one of the values 0, 1, 2, ... The value assumed may even be arbitrarily large. For example, if the first 10 million tosses resulted in tails, then X is at least equal to that number. While the probability is very small, q^{-10^7}, it is still positive, so that there is a chance that it may assume that value. While there are some technical problems in a rigorous definition of the sample space for this open-ended sequence of tosses, we shall avoid these problems and use the intuitive notion of the first head to calculate the distribution of X. The random variable takes on the value 0 when heads appear on the very first toss. Since the probability of heads on the first toss is p, we have $P(X=0)=p$. X assumes the value 1 if tails appears on the first toss and heads on the second; hence, $P(X=1)=pq$. It assumes the value 2 if tails appear on the first 2 tosses, and heads on the third: $P(X=2)=pq^2$. More generally, X is equal to k if the first k tosses result in tails and the $(k+1)$st in heads:

$$P(X=k)=pq^k, \qquad k=0, 1, 2, \ldots . \qquad (1)$$

This is known as the **geometric distribution.** The sum of the terms of the distribution is equal to 1; this can be verified by means of the formula for the sum of the terms of the infinite geometric series (Section 6.12). ▲

The **expected value** of a random variable X is the weighted sum of the values of the random variable, where each value is weighted by the corresponding probability. If $x_1, x_2, \ldots$ are the values of X, and $P(X=x_1), P(X=x_2), \ldots$ the corresponding probabilities, then the expected value of X, denoted $E(X)$, is defined as

$$E(X)=x_1P(X=x_1)+x_2P(X=x_2)+\cdots . \qquad (2)$$

This sum may be interpreted as the center of gravity of the masses $P(X=x)$ placed at the points x (Section 6.8). It is certainly well defined when X assumes finitely many values x; however, the case of infinitely many x, as in Example 4 above, will be discussed in Section 9.2. $E(X)$ represents the "average value" of X.

EXAMPLE 5

Let X be the number of dots appearing when a die is tossed. It assumes integer values 1 through 6. If the die is balanced, then each of the values has the same probability $\frac{1}{6}$. Then

$$E(X)=\frac{1}{6}(1+2+\cdots+6)=3.5.$$

This is the center of gravity of equal masses placed at the 6 integer points. Suppose now that the die is loaded so that additional weight is put on the face with 1 dot, and so that the probabilities are

Face	1	2	3	4	5	6
Probability	$\frac{1}{4}$	$\frac{5}{32}$	$\frac{5}{32}$	$\frac{5}{32}$	$\frac{5}{32}$	$\frac{1}{8}$

In this case the expected value is

$$E(X) = 1(1/4) + 2(5/32) + 3(5/32) + 4(5/32) + 5(5/32) + \\ + 6(1/8) = 3.1875.$$

The center of gravity is smaller than in the first case because there is an additional mass on the smaller values (Figure 8.59). ▲

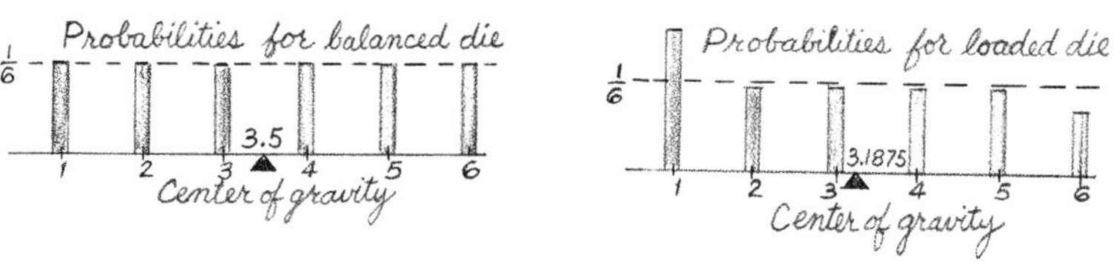

Figure 8.59

EXAMPLE 6

Let X be the number of successes in 1 Bernoulli trial; it has the probability distribution

x	0	1
$P(X=x)$	q	p

Then $E(X) = 0 \cdot q + 1 \cdot p = p$. In 2 trials the number of successes has the binomial distribution with $n = 2$; therefore,

$$E(X) = 0 \cdot P(X=0) + 1 \cdot P(X=1) + 2 \cdot P(X=2) \\ = 2pq + 2p^2 = 2p(p+q) = 2p$$

(because $p + q = 1$). It can be shown that if X is the number of successes in n trials, then

$$E(X) = np. \tag{3}$$

This was mentioned in Section 8.7. The proof of (3) for general n will be carried out in Section 9.2. ▲

8.13 EXERCISES

1 A fair coin is tossed 3 times. Let X represent the number of heads or tails, whichever is **larger.** (It assumes the values 2 and 3.) Find the probability distribution of X and its expected value.

2 A fair coin is tossed 3 times. Let X be the length of the longest run of successive heads; for example, the longest run of heads for the outcome HTH is 1 and for the outcome THH is 2. Find the distribution and the expected value of X.

3 A fair coin is tossed 4 times. Let X be the difference between the number of heads on the first 2 tosses and the number on the second 2 tosses. Find the distribution of X and its expected value.

4 A balanced die is tossed twice.
(a) Find the expected value of the sum of the 2 scores.
(b) Find the expected value of the difference of the scores.

5 (We refer to Section 8.5, Exercise 7.) We think of the successive outcomes of the games of the World Series as the outcomes of the tosses of a fair coin. The Series is decided as soon as 1 of the teams wins 4 games. Let X be the number of games played until the decision. Find the distribution and expected value of X.

6 A man bets $1 that a coin tossed once turns up heads. If he wins, he quits. If he loses, he bets $2 on a second toss. If he wins he quits. Otherwise, he bets $4 on a third toss, and then stops. Let X be the net amount (positive or negative) that he takes with him when his plan is complete. Find the distribution of X and its expected value when the probability p of heads is (a) $\frac{1}{2}$; (b) $\frac{1}{4}$; (c) $\frac{3}{4}$.

7 Suppose that in some monolithic state the rulers decided to remove babies from the care of their natural mothers and distribute them at random to adoptive parents. If 4 babies were taken from their mothers and returned to them in random order, 1 to a mother, find the distribution and expected value of the number returned to their natural mothers. (See Example 3.)

8 A man has 5 keys on a chain. Exactly 1 of these fits the lock that he is trying to open. He has 2 procedures for finding the right key. Under the first, he arranges the keys in random order, tries them 1 by 1 until he finds the right one, and then stops. Under the second procedure, he selects the keys at random **with replacement;** thus, it is possible that the wrong key might be tried several times before the right one is found. Let X be the number of times that a key is unsuccessfully tried. Find the distribution of X under each procedure.

9 An urn contains 5 white and 4 red balls. Two are drawn at random. Let X be the number of white balls in the sample. Find $E(X)$.

10 A nurse at a children's clinic is studying the relation between the health of children and their siblings. She asks the mothers of children brought to the clinic about the health of the siblings. Some of the children might not have siblings. Suppose that there are 10 children in the clinic on one morning, and that 4 do not have siblings. The nurse selects 4 of the children present at random. Let X be the number having at least 1 sibling. Find $E(X)$.

11 A fair coin is tossed 4 times. Let X be the maximum difference between the cumulative numbers of heads and tails during the tosses; for example, the maximum difference for the outcome THHT is 1. Find the distribution of X. (The initial difference is 0.)

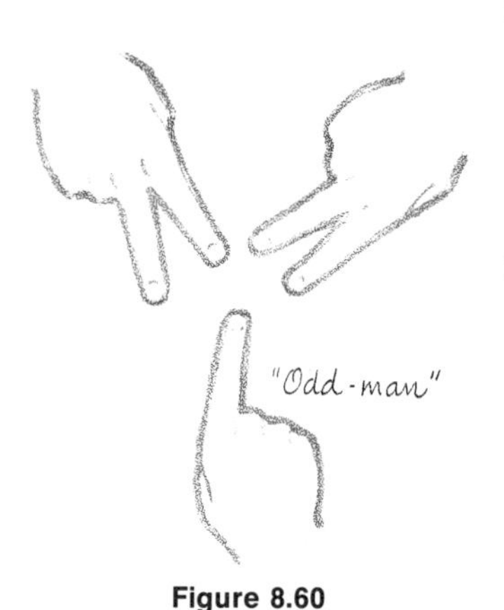

Figure 8.60

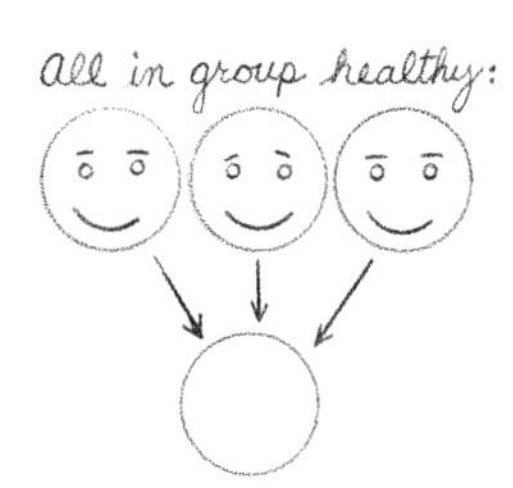

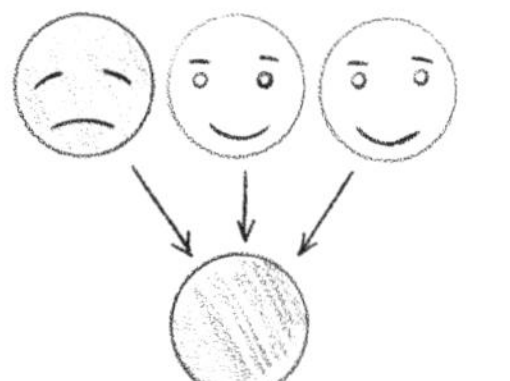

Figure 8.61

12 A coin with probability p of heads is tossed until either 3 heads or 3 tails appear for the first time. Let X be the required number of tosses. Find the distribution of X when (a) $p = \frac{1}{2}$ (b) $p = \frac{1}{3}$ (c) $p = \frac{2}{3}$.

13 A salmon is swimming up a strongly flowing river. There are 3 points where the rocks are hazardous and the current very forceful. The probability that he is successful in an attempt to pass such a point is $p = 0.4$. If he fails, he tries again. If he experiences 4 failures along the series of obstacles, he is so exhausted that he gives up. Let X be the number of attempts made; successful or not. Find its distribution.

14 After a half-mile on a hot sunny road 3 hikers discover that they have forgotten their food, and want 1 of the 3 to go back and get it. They decide who should go by the game of "odd man out." At each play each person puts out 1 or 2 fingers at random (Figure 8.60). The person who is the only one to show 1 finger or the only one to show 2 fingers is the **odd man**, and is assigned to go get the food. If all 3 fingers are the same, the play is repeated sufficiently many times until the odd man is determined. Let X be the number of plays made before the odd man is chosen. Find its distribution. (See Example 4.)

15 (Group testing.) During World War II large numbers of men were called for military service and had to be tested for certain diseases by means of blood tests. In order to reduce the number of necessary laboratory analyses, the method of group testing was devised. The blood samples of groups of k men are pooled and a single analysis made on the pooled samples. If the test is negative (no disease agents are present), then none of the k men in the group have the disease, and only a single laboratory analysis is required. If the test on the pooled sample has a positive result, then any one of the k men in the pool might have the disease, and so each of the individual blood samples have to be tested individually; this results in $k + 1$ laboratory analyses (Figure 8.61). Let p be the prevalence of the disease in the general population, the proportion of the population having the disease. Each man represents a Bernoulli trial: **success** signifies that he has the disease and **failure** that he does not; and p is the probability of success. Let X be the number of analyses required for a group of k men. (a) Find the distribution and expected value of X. (b) Find the formula for the difference between the number of analyses for k samples without group testing (k) and the expected number under group testing. (c) Evaluate this difference for $k = 5$ and $p = 0.01$.

16 Local telephone calls seem to end at random. The duration appears to be determined in the following way. At the end of each small time unit there is a "Bernoulli trial": the call is terminated with probability p and continued with probability q. In this way the length of the conversation is proportional to the number of failures to occur before the first success; hence, it has the geometric distribution. Suppose at each 20-second interval there is probability $p = \frac{1}{6}$ that the conversation ends, and probability $\frac{5}{6}$ that it continues. Find the probability that the length of the call is between 1 and 2 minutes, inclusive. (See Example 4.)

CHAPTER

9

Continuous Probability and Statistics

9.1 Continuous Random Variables and Their Distributions

A continuous random variable is, like a discrete one (Section 8.13), a number associated with the outcome of a random experiment. But it differs from the discrete in that it assumes one of a continuum of values, not just one of a discrete set (finite or an infinite sequence). The difference between discrete and continuous mathematical variables was briefly discussed in Section 4.1. Now we give precise definitions of the two kinds of random variables. The theory of continuous random variables depends much on integral and differential calculus.

Consider a population with a numerical characteristic. For each x let $F(x)$ be the proportion of the population for which the numerical characteristic is less than or equal to x; then $F(x)$ is the distribution function of the numerical characteristic (Section 6.7). If a member of the population is drawn at random, then the numerical value of the member actually drawn is a **random variable** X. For a given x the

probability that $X \leq x$ is the **proportion of the population** for which $X \leq x$:

$$P(X \leq x) = F(x). \tag{1}$$

For the proof of (1), consider the population as an urn with balls of 2 colors, one corresponding to values less than or equal to x, and the other to values greater than x. The probability of picking a ball of the given color is equal to the corresponding proportion of balls in the urn.

By virtue of Equation (1), $F(x)$ is called the **distribution function of the random variable X.** All properties of population distributions also hold for distributions of random variables: they are non-decreasing, tending to 0 as $x \to -\infty$, and tending to 1 as $x \to \infty$ (Section 6.7). The distribution function of a random variable is uniquely determined by the latter; however, the converse is not true: two random variables, defined for different populations, may have the same distribution function. For example, the number of heads on the tosses of a coin and the number of cured patients may have the same binomial distribution.

EXAMPLE 1

Consider the population of cattle in a given region, of age 6 months. For each calf of this age, the numerical characteristic of interest is the weight gain (in pounds) over the following 3-month period. Let $F(x)$ be the distribution function of the weight gains in the population. A calf is taken at random and its weight gain X is observed. Then X is a random variable with the distribution function $F(x)$ (Figure 9.1). ▲

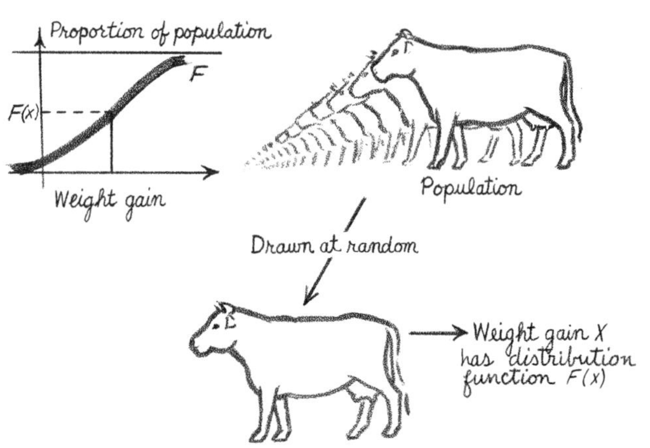

Figure 9.1

EXAMPLE 2

A gun is fired at a circular target. The distance of the point of impact to the center of the target is the outcome of the random experiment of firing the gun; (Figure 9.2). There are two ways of looking at the sample space of the random experiment. The first is to consider the system of outcomes as the set of all points on the face of the target; the random variable is then the function defined as the distance of the point to the center of the target. A second way to visualize the sample space is from the point of view of a sample from a population. Each shot is a member of an infinite population of hypothetical shots. The numerical characteristic of the population is the distance of the point of impact to the center. Any particular shot is a member taken at random from this population. If $F(x)$ is the distribution function of the numerical characteristic, then it is also the distribution function of the random variable X. ▲

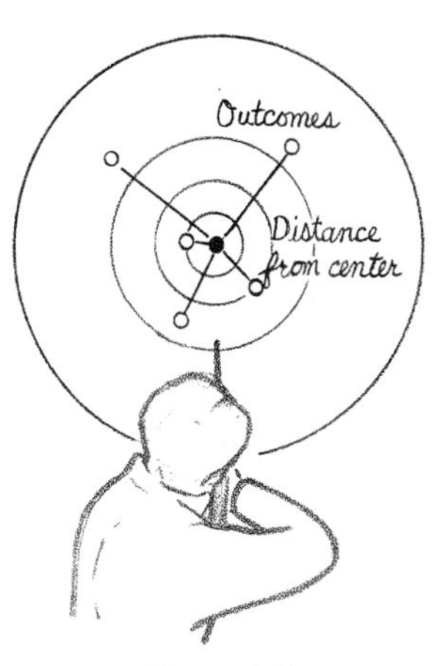

Figure 9.2

There are two primary classes of random variables: **discrete** and **continuous.** The distribution determines the membership of the random variable in either of these classes. X is discrete if its distribution is a **step-function,** increasing in successive jumps, and constant between them. The jumps occur at the values of the random variable. For example, if $F(x)$ is the function in Figure 9.3, with jumps at the 6 x-values, then the random variable assumes just these 6 values, and is discrete.

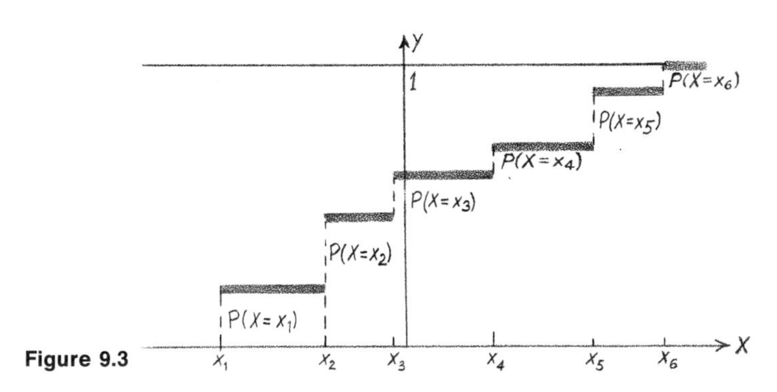

Figure 9.3

EXAMPLE 3

Let X have the binomial distribution with $n = 1$; it assumes the values 1 and 0 with probabilities p and q, respectively. Then

$$\begin{aligned} P(X \le x) &= 0, && \text{if } x < 0 \\ &= q, && \text{if } 0 \le x < 1 \\ &= 1, && \text{if } x \ge 1. \end{aligned}$$

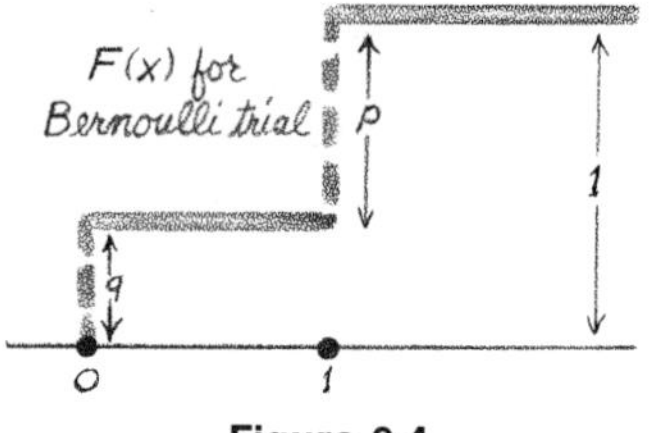
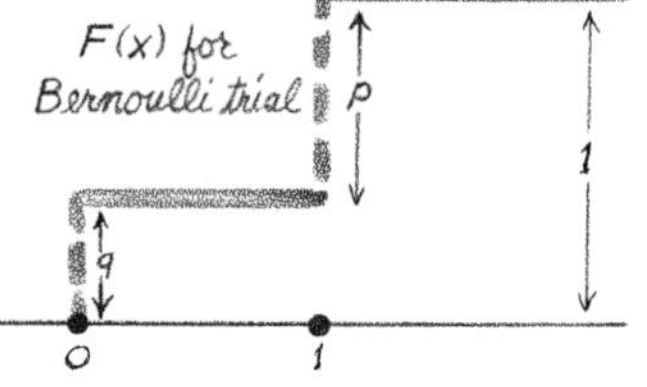

Figure 9.4

There are jumps of p and q at the points 1 and 0, respectively (Figure 9.4). ▲

EXAMPLE 4

The distribution function of the score on a single toss of a die is a step function on the interval from 1 to 6, with a jump of $\frac{1}{6}$ at each integer point (Figure 9.5). ▲

Figure 9.5

A random variable X is **continuous** if $F(x)$ is continuous as a function of x (Section 5.10). We recall that a function is continuous when it has an unbroken graph. Note that the step functions in Figures 9.3, 9.4, and 9.5 are not continuous because they have breaks at the jump-points. Continuity of F means that the difference $F(b) - F(a)$ is "small" whenever the difference $b - a$ is small (Figure 9.6). Now $F(b) - F(a)$ is the proportion of the population whose characteristic is between a and b; and it is the probability that the random variable falls between a and b. The continuity of the distribution implies that the probability that X falls in a very small prescribed interval is small. In particular, the probability that X assumes a **particular** value is 0. This means that if a value x is given before the random experiment, then we are almost "certain" that X will not assume that value. But there may be a positive probability of falling in a **set** of values if the set is sufficiently "thick."

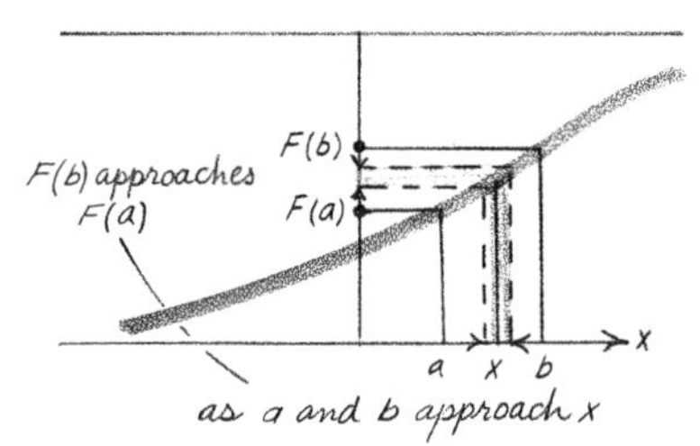

Figure 9.6

EXAMPLE 5

A space capsule returning from flight is to land in a certain set on the surface of the ocean. Several ships are in the region. The actual landing point cannot be predicted, but it is the outcome of a random experiment: the sample space is the set of points on the surface of the ocean in that region. What if the capsule strikes one of the ships as it lands? Space scientists are almost "sure" that it will not happen; indeed, the distribution of the landing point is continuous, so that each point occupied by a boat has probability 0 of being struck. In reality the boat occupies an area of positive area measurement; however, it is insignificant relative to the area of the entire region (Figure 9.7). ▲

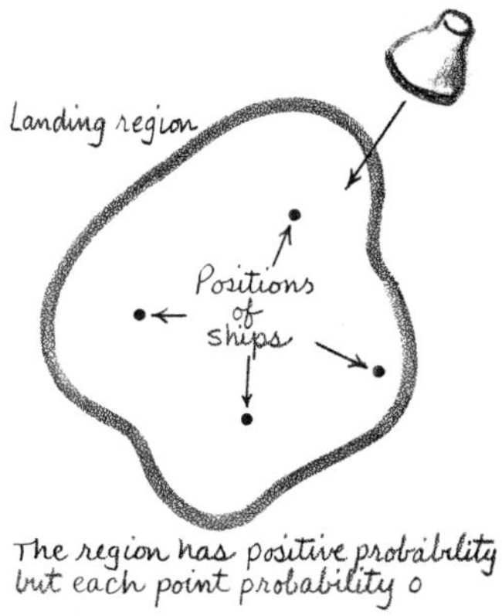

The region has positive probability but each point probability 0

Figure 9.7

If the underlying population has a density function $f(x)$, then the distribution function is continuous. The reason is as follows. The difference $F(b) - F(a)$ is the integral of the density over the interval from a to b (Section 6.7). If the interval is small, then the area is approximately that of a rectangle whose base is the interval; hence, the area is small if the base is small (Figure 9.8).

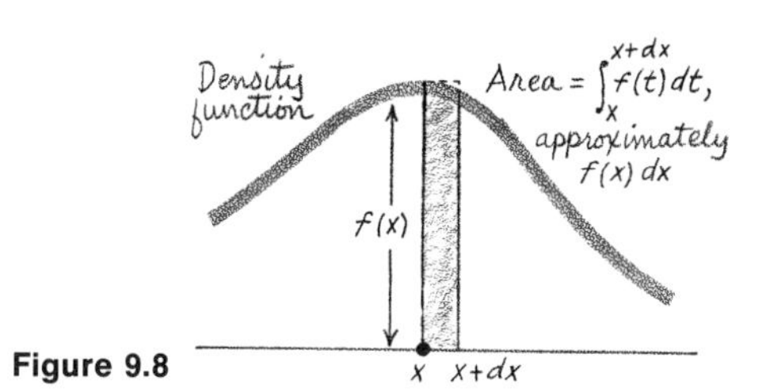

Figure 9.8

Not every distribution is either of the discrete or the continuous type. It may be a mixture of the two, with continuous and discrete pieces. These mixtures occur only infrequently in applications, so that we do not discuss them here.

Now we give the simplest example of a continuous random variable.

EXAMPLE 6

We define the **uniform distribution on the unit interval:**

$$\begin{aligned} F(x) &= 0, \quad && \text{for } x \leq 0 \\ &= x, && \text{for } 0 < x < 1 \\ &= 1, && \text{for } x \geq 1. \end{aligned}$$

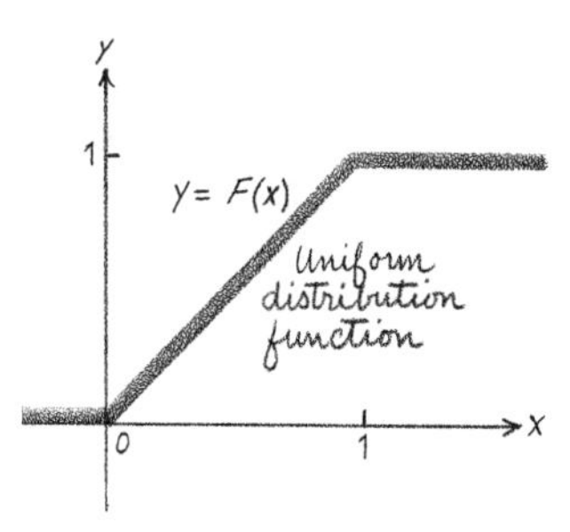

Figure 9.9

(Figure 9.9). It is differentiable at all points except $x = 0$ and $x = 1$; and the density $f(x)$ is equal to 1 on the interval between 0 and 1, and equal to 0 outside the interval. From the point of view of probability it does not matter how we define the density at the particular points $x = 0$ and $x = 1$; indeed, we are interested only in **areas** under the density, and these are not affected by changes in the function at individual points. For simplicity we will therefore define the uniform density as being equal to 1 for x between 0 and 1, and equal to 0 elsewhere (Figure 9.10). This distribution is called uniform because the probability that X falls in a given

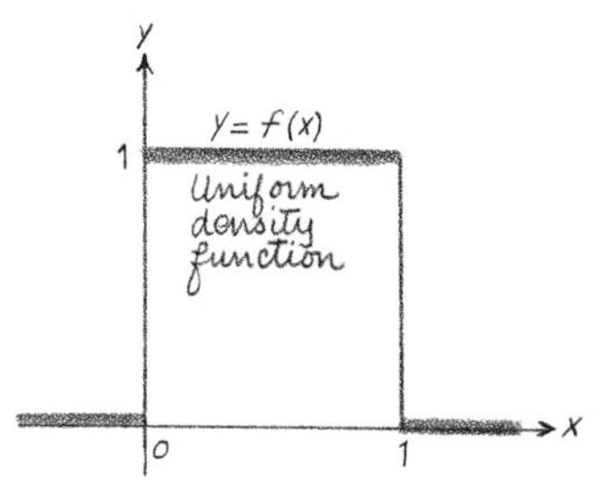

Figure 9.10

subinterval of the interval from 0 to 1 depends only on the length of the subinterval, and not on the location. It arises as the distribution of the abscissa of a point chosen "at random" on the interval from 0 to 1. If a population has a histogram which is constant over this interval, then the uniform density may be fitted to it, and the numerical characteristic is considered to have a uniform distribution.

The **uniform distribution** on an arbitrary interval from a to b has the distribution function

$$F(x) = \begin{cases} 0, & \text{for } x \le a \\ \dfrac{x-a}{b-a}, & \text{for } a < x < b \\ 1, & \text{for } x \ge b; \end{cases}$$

and the density function is

$$f(x) = \begin{cases} \dfrac{1}{b-a}, & \text{for } a < x < b \\ 0, & \text{elsewhere.} \end{cases}$$

The probability that X falls in a subinterval from c to d is the length of this subinterval divided by the length of the interval from a to b: $(d-c)/(b-a)$. ▲

Populations that occur in applications are usually finite; hence, the random variables they generate are necessarily discrete because they can assume at most finitely many values. But they may often be considered as continuous variables: If a smooth density can be fitted to the histogram, then the integral of the density may be taken to be the approximate distribution function of the random variable. The family of **normal** distributions (and densities) appears to fit many kinds of populations. The standard normal density function

$$\phi(x) = \frac{1}{\sqrt{2\pi}} e^{-(1/2)x^2} \tag{2}$$

was defined in Section 8.9, formula (1). Let μ be an arbitrary real number and let σ be an arbitrary positive real number; then the function

$$\frac{1}{\sigma}\phi\left(\frac{x-\mu}{\sigma}\right) = \frac{1}{\sqrt{2\pi}\sigma} e^{-(x-\mu)^2/2\sigma^2} \tag{3}$$

is the normal density with mean μ and standard deviation σ. The graph of this function is obtained from the standard normal by shifting the center from 0 to μ, stretching the horizontal scale by a factor σ, and reducing the vertical by $1/\sigma$. If $\sigma > 1$, then the density is low and flat; and if $\sigma < 1$, it is tall and slim (Figure 9.11). The area under a normal density is 1 (see Section 8.9 and Exercise 17 below). The distribution function corresponding to the density (3), the normal distribution with mean μ and standard deviation σ, is obtained from the standard normal distribution function by simply replacing x by $(x - \mu)/\sigma$: If $\Phi(x)$ is the standard normal distribution function, the normal distribution with mean μ and standard deviation σ is

$$\Phi\left(\frac{x-\mu}{\sigma}\right). \tag{4}$$

The values can be computed from those in Table VIII (Appendix).

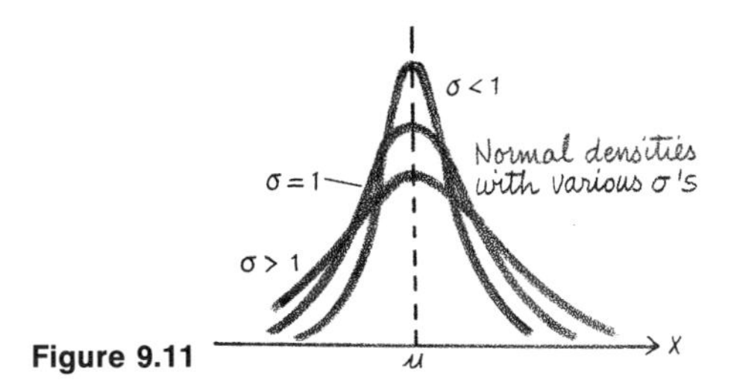

Figure 9.11

There are several other names for the normal distribution: **Gaussian distribution,** after the famous scientist and mathematician, Karl F. Gauss (1777–1855); **Laplacian distribution,** after P. S. Laplace (1749–1827); and **error function** (erf), because of its applications in the analysis of errors in measurement.

The normal distribution was introduced in Section 8.9 as an approximation to the binomial. There are continuous random variables occurring in many kinds of data where the distributions themselves are normal. For this reason the normal distribution has had a central part in the foundations of the modern statistical theory. While stating the importance of this distribution, we add that not all data follow the normal distribution. In any application the validity of the normal distribution should be established, not assumed, by the investigator.

EXAMPLE 7

Normal distributions often arise in data where the numerical characteristic of a member of the population is the result of a large number of "small, independent effects." In particular, this has often been found for certain biological characteristics: heights, weights, intelligence quotients, and others.

In the table below we are given the distribution of the weights of the members of 6 National League baseball teams at the start of the 1970 baseball season; there were 154 players:

Weight range (pounds)	Number of players	Weight range (pounds)	Number of players
150–159	6	200–209	23
160–169	8	210–219	17
170–179	28	220–229	5
180–189	34	230–239	1
190–199	32		

The histogram appears in Figure 9.12, and a normal curve is fitted to it. The approximate values of μ and σ can be obtained by inspecting the graph. The curve is centered near 190; hence, $\mu = 190$. Practically all the area under the curve is contained within 40 pounds of the mean 190. For any normal distribution, the area under the curve is contained almost entirely within 3σ units of the mean (Exercise 18). It follows that $3\sigma = 40$ in our case, so that σ is approximately 13.

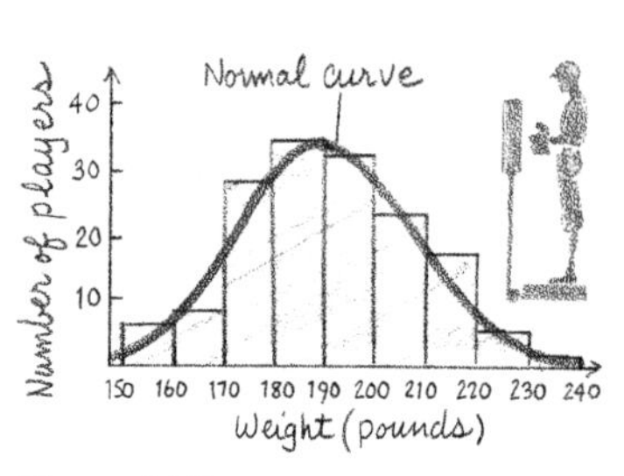

Figure 9.12

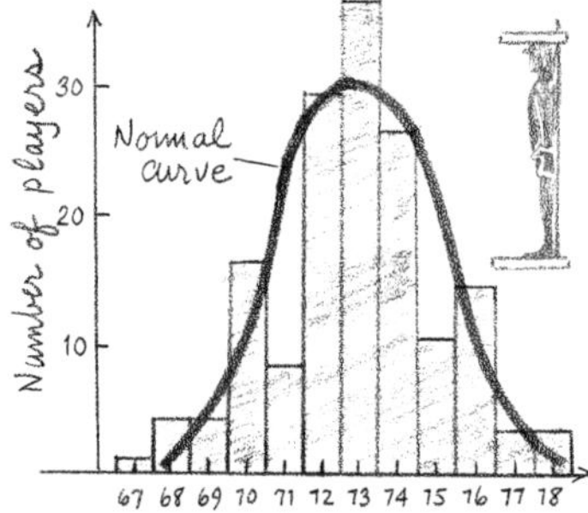

Figure 9.13

In the next table we are given the distribution of the heights (in inches) of the same population of players; the histogram and the fitted normal curve appear in Figure 9.13:

Height	No. of players	Height	No. of players
67	1	73	36
68	4	74	26
69	4	75	10
70	16	76	14
71	8	77	3
72	29	78	3

It appears from the graph that $\mu = 73$ and $\sigma = \frac{5}{3}$. ▲

Now we illustrate the use of Table VIII for the calculation of the probabilities for a general normal distribution.

EXAMPLE 8

Suppose that X has a normal distribution with mean 2.4 and standard deviation 3. Let us find the probability that $1 \le X \le 4$. The distribution function is

$$F(x) = \Phi\left(\frac{x - 2.4}{3}\right).$$

It follows that

$$\begin{aligned} P(1 \le X \le 4) &= F(4) - F(1) = \Phi(0.53) - \Phi(-0.47) \\ &= 0.1808 + 0.2019 = 0.3827. \end{aligned}$$

We see that the probability is obtained from that in the standard normal table by first "standardizing" the variable, subtracting the mean and dividing by the standard deviation. ▲

Now we illustrate a general procedure for the calculation of the probabilities from the density function.

EXAMPLE 9

Let X be a random variable with the density

$$\begin{aligned} f(x) &= \tfrac{4}{15}x^3, && \text{for } 1 < x < 2 \\ &= 0, && \text{elsewhere.} \end{aligned}$$

Find the probability that X falls in the interval from

(a) 1.1 to 1.5; (b) 0.5 to 1.5; (c) 0 to 10; (d) 1.6 to 3.

We note that it does not matter whether the endpoints of the interval are included or not. We first derive the distribution function:

$$\begin{aligned} F(x) &= 0, && \text{for } x \le 1 \\ &= (1/15)(x^4 - 1), && \text{for } 1 < x < 2 \\ &= 1, && \text{for } x \ge 2. \end{aligned}$$

The probabilities are:

(a) $P(1.1 \le X \le 1.5) = F(1.5) - F(1.1) = 3.5984/15 = 0.23989.$

(b) $P(0.5 \le X \le 1.5) = F(1.5) - F(0.5)$
$= 5.0625/15 = 0.33750.$

(c) $P(0 \le X \le 10) = F(10) - F(0) = 1 - 0 = 1.$

(d) $P(1.6 \le X \le 3) = F(3) - F(1.6) = 1 - F(1.6)$
$= 9.4464/15 = 0.62976.$ ▲

9.1 EXERCISES

In Exercises 1 and 2 show that each of the functions is a density function of a random variable. (Show that it is nonnegative and has the integral 1.) Find the distribution function, and compute the probability of the given event.

1 (a) $f(x) = 2x$, for $0 < x < 1$
$= 0$, elsewhere. Find $P(-\frac{1}{2} \le X \le \frac{1}{2})$.

(b) $f(x) = 1/(2\sqrt{x})$, for $0 < x < 1$
$= 0$, elsewhere. Find $P(X > \frac{1}{4})$.

(c) $f(x) = 1 + x$, for $-1 < x \le 0$
$= 1 - x$, for $0 < x \le 1$
$= 0$, elsewhere. Find $P(-\frac{3}{4} \le X \le \frac{2}{3})$.

(d) $f(x) = \dfrac{2}{\pi\sqrt{1 - x^2}}$, for $0 < x < 1$
$= 0$ elsewhere. Find $P(0 < X < \frac{1}{2})$.

(e) $f(x) = \dfrac{e^x}{(1 + e^x)^2}$. Find $P(-1 \le X \le 1)$.

2 (a) $f(x) = 3x^2$, for $0 < x < 1$
$= 0$, elsewhere. Find $P(\frac{1}{4} < X \le \frac{3}{4})$.

(b) $f(x) = (\frac{1}{3})x^{-2/3}$, for $0 < x < 1$
$= 0$ elsewhere. Find $P(x > \frac{1}{8})$.

(c) $f(x) = (\frac{3}{2})(1 + x)^2$, for $-1 < x < 0$
$= (\frac{3}{2})(1 - x)^2$, for $0 \le x < 1$.
$= 0$ elsewhere. Find $P(-\frac{1}{2} < x < \frac{1}{2})$.

(d) $f(x) = \dfrac{1}{\pi(1 + x^2)}$. Find $P(0.20271 < X \le 0.68414)$.

(e) $f(x) = 0$, for $x < 0$
$= \frac{1}{4}xe^{-(1/2)x}$, for $x \ge 0$. Find $P(1 < X \le 3)$.

3 Let X have the density function

$$f(x) = 4x^3, \quad \text{for } 0 < x < 1$$
$$= 0. \quad \text{elsewhere.}$$

Find the number A such that $P(X \ge A) = P(X \le A)$. Also find the number B such that $P(X \ge B) = 0.9984$.

4 Let X have the density function

$$f(x) = 3(1 - x)^2, \quad \text{for } 0 < x < 1$$
$$= 0, \quad \text{elsewhere.}$$

Find the distribution function; and then find the number A such that $P(X \le A) = \frac{7}{8}$.

5 The exponential distribution (see Section 6.7, Exercise 6) arises as a distribution of **waiting times.** It is the continuous version of the geometric distribution (Section 8.13, Example 4).

(a) In n Bernoulli trials, what is the probability of at least 1 success if the probability of success on a single trial is p?

(b) Put $x = np$; and suppose, as in the Poisson approximation, that n increases and p decreases but x is of moderate size. Show that the probability in (a) tends to $1 - e^{-x}$. This is the exponential distribution function.

(c) Let X be the time to the first success. Show, under the conditions in (b), that $P(X \leq n)$ is approximately equal to $1 - e^{-x}$.

6 The general exponential density is of the form

$$f(x) = ce^{-cx}, \quad \text{for } x > 0; \qquad \text{and} = 0, \quad \text{for } x \leq 0.$$

It is used to represent the distribution of the **lifetimes** of nonaging organisms or mechanical components—the waiting time to their deaths (see Exercise 5). Suppose that the lifetime of a bulb following 500 hours of use has an exponential distribution with $c = 0.01$. (The **aging** of the bulb is practically complete, and so it may burn out at any random time.) Find the probability that it lasts

(a) at most 50 hours more
(b) between 100 and 200 more hours
(c) at least 400 more
(d) exactly 73.1 hours

7 The weekly sales of gasoline (in gallons) of a filling station is a random variable X with the density function

$$\begin{aligned} f(x) &= \tfrac{3}{2}(1 - x)^{1/2}, && \text{for } 0 < x < 1 \\ &= 0, && \text{elsewhere.} \end{aligned}$$

Sketch the density. Derive the distribution function and find the number A such that $P(X \geq A) = 27/125$.

8 A grocery store sells a beverage in a bottle which is returnable for a deposit. The proportion of bottles sold daily that are eventually returned is a random variable with the density function

$$\begin{aligned} f(x) &= 12x(1 - x)^2, && 0 < x < 1 \\ &= 0, && \text{elsewhere.} \end{aligned}$$

Find the probability that the proportion of returned bottles is between 0.1 and 0.5.

9 The proportion of students absent from a high school on a normal day is a random variable with a density function of the form

$$\begin{aligned} f(x) &= kx(1 - x^2)^4, && \text{for } 0 < x < 1 \\ &= 0, && \text{elsewhere.} \end{aligned}$$

(a) Find the value of k which is consistent with the properties of f as a density.

(b) Find the probability that more than 50 percent of the students are absent.

10 In a large city the daily consumption of electricity (in millions of kilowatts) is a random variable with the density

$$f(x) = \tfrac{1}{9}xe^{-x/3}, \qquad \text{for } x > 0$$
$$= 0, \qquad \text{for } x \le 0.$$

If the power plants have a capacity of 12 million kilowatts, what is the probability that the demand will exceed the supply on a given day?

11 A 10-sided die has the numbers 0.1, 0.2, . . . , 0.9, 1.0 printed on the respective faces. When the die is tossed each face has the same probability of appearing. Let X be the number on the face that actually turns up. Describe the distribution function of X.

12 Suppose that the multisided die in Exercise 11 has n faces, marked by the fractions

$$\frac{1}{n}, \frac{2}{n}, \ldots, \frac{n-1}{n}, 1.$$

Show that the distribution function of the number appearing on the face of the die tends to the **uniform distribution on the unit interval** (Example 6) as n tends to infinity. In this way we see that the uniform distribution is the limit of a sequence of discrete distributions.

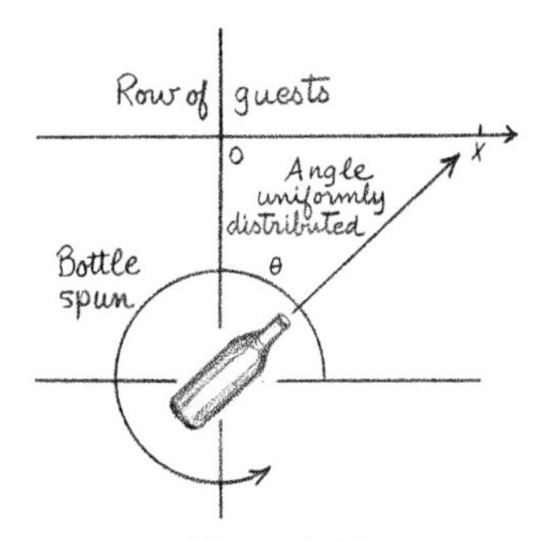

Figure 9.14

13 (Modified game of **Spin-the-Bottle.**) Imagine a party with infinitely many guests, lined up along the x-axis. The host wants to choose one of them at random. A bottle is placed at the point $(0, -1)$ on the y-axis. It is spun and then comes to rest with one end of the bottle pointing to some individual. The angle θ between the y-axis and the axis of the bottle is considered to be uniformly distributed between $-\pi/2$ and $\pi/2$ (Figure 9.14), that is, it has the density

$$f(\theta) = 1/\pi, \qquad \text{for } -\pi/2 < \theta < \pi/2$$
$$= 0, \qquad \text{elsewhere.}$$

Let X be the abscissa of the point on the x-axis to which the end of the bottle points. Find the formula for $P(X \le x)$.

14 The uniform distribution appears in the latter digits of tabulated functions. Consider the last digits of log N in Table III, for $N = 1, 2, \ldots, 100$: 0, 1, 6, ... Consider these as the values of the numerical characteristic of a population. Draw the histogram. Compare it to the uniform density.

15 In a certain trade the hourly wage X of a worker chosen at random has a normal distribution with mean \$3.60 and standard deviation \$0.45. Find the probability that the wage of a randomly selected worker is between \$3.00 and \$3.50. (See Example 8.)

16 The oranges grown in an orchard have weights (in ounces) which have a normal distribution with mean 7 and standard deviation 1.3. Find the probability that the weight of an orange selected at random (a) exceeds 10 ounces; (b) is less than 5 ounces; (c) is between 6 and 8.5 ounces. (See Example 8.)

17 By differentiation with respect to x and application of the chain rule, show that the derivative of the function (4) is the normal density (3). From this conclude that the area under the density is 1.

18 Show that the area under the normal density (3), from $\mu - k\sigma$ to $\mu + k\sigma$ is equal to the area under the **standard** normal density from $-k$ to k.

19 It has long been observed that when a physical quantity is repeatedly measured by an instrument, the numbers recorded are different from one another; and the histogram of the population of observed measurements is approximately a normal curve. This is the experimental foundation of the **law of errors;** and this explains the name **error function** for the normal curve. The mean of the normal distribution is the true constant being measured, and the standard deviation σ is determined by the scale of measurement and the precision of the instrument. Find the probability that the observed measurement is within (a) σ units of the true constant; (b) 3σ units; (c) 3.5σ units. (See Example 8.)

9.2 Expected Value, Variance, and Chebyshev's Inequality

We have defined the **expected value** of a discrete random variable as the weighted sum of its values, the center of gravity of the probability "masses" (Section 8.13). The natural extension of the center of gravity to populations with densities is the **mean** of the population (Section 6.8). For this reason we define the expected value $E(X)$ of a random variable X with a density $f(x)$ as the mean of the population with density $f(x)$:

$$E(X) = \int_{-\infty}^{\infty} xf(x)\,dx. \tag{1}$$

As in Section 6.8, it is defined only when the improper integral is "absolutely convergent." Examples and exercises involving the mean of a density are also given in that section. Here we give an application to the normal distribution.

EXAMPLE 1

In Section 6.8, Example 2, it was shown that the mean of a standard normal density is equal to 0:

$$\int_{-\infty}^{\infty} x\,\phi(x)\,dx = 0.$$

From this we can deduce that the mean of a normal density with "mean μ and standard deviation σ" is μ. The improper integral

$$E(X) = \int_{-\infty}^{\infty} x\frac{1}{\sigma}\,\phi\left(\frac{x-\mu}{\sigma}\right) dx$$

is, by definition (Section 6.7), the limit of

$$\int_{-A}^{B} x \frac{1}{\sigma} \phi\left(\frac{x-\mu}{\sigma}\right) dx$$

as A and B tend to ∞. By the method of substitution of variables (Section 6.3), the integral above is

$$\int_{(-A-\mu)/\sigma}^{(B-\mu)/\sigma} (x\sigma + \mu)\phi(x)\,dx.$$

As A and B become infinite, the latter integral becomes the sum of improper integrals,

$$\sigma \int_{-\infty}^{\infty} x\phi(x)\,dx + \mu \int_{-\infty}^{\infty} \phi(x)\,dx.$$

By the formula above for the mean of $\phi(x)$, the first integral in the sum is 0. The second integral is the area under $\phi(x)$; hence, it is equal to 1. It follows that $E(X) = \mu$. In this way we have verified that the distribution which we have **named** "normal with mean μ and standard deviation σ" really has a mean equal to μ according to the definition (1). ▲

Let $g(x)$ be a function of x. In some applications we consider, instead of the random variable X, the function of the random variable $g(X)$. Then the latter is itself a random variable, because its value is determined by the outcome of the experiment that determines the value of X; it is a composite function on the sample space. [There are some technical conditions that g has to satisfy ("Borel measurability") but these are always satisfied in applications so that we will ignore them.] In order to calculate the expected value of $g(X)$ according to the definition, we would first have to determine the distribution of $g(X)$ and then calculate its mean. But there is a simple formula for the mean of the distribution of the function in terms of that of the distribution of the original random variable X and the function g. When X is a discrete random variable with values $x_1, x_2, \ldots$, the expected value of $g(X)$ is

$$Eg(X) = g(x_1)P(X = x_1) + g(x_2)P(X = x_2) + \cdots; \qquad (2)$$

the values of g are weighted by the corresponding probabilities for X. If X is a continuous random variable with a density function $f(x)$, then the expected value of $g(X)$ is the "integral analog" of the sum (2):

$$Eg(X) = \int_{-\infty}^{\infty} g(x)f(x)\,dx. \qquad (3)$$

The proof of (2) is outlined in Exercise 9.

EXAMPLE 2

Let X be the radius of a mature cucumber of a certain variety, measured in inches. Suppose that X has the density

$$f(x) = 6\left(\frac{x}{2}\right)^2\left(1-\frac{x}{2}\right), \qquad \text{for } 0 < x < 2$$
$$= 0, \qquad \text{elsewhere.}$$

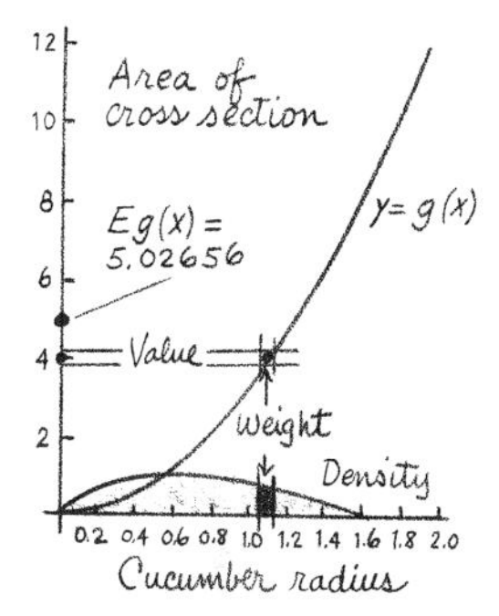

Figure 9.15

Let us find the expected value of the cross-sectional area of a randomly selected cucumber. The cross-sectional area of a cucumber of radius x is $g(x) = \pi x^2$; hence,

$$Eg(X) = \int_0^2 g(x)f(x)dx = \int_0^2 \pi x^2 6\left(\frac{x}{2}\right)^2\left(1-\frac{x}{2}\right) dx$$
$$= \frac{8\pi}{5} = 5.02656.$$

The expected value represents the weighted sum of πx^2, weighted with respect to the density $f(x)$ (Figure 9.15). ▲

$E(X)$ serves as a measure of **location** of the "probable value" of the random variable X because it is a weighted sum of the values. Another important characteristic of a random variable is the **spread** or **dispersion** of its values about the expected value. This is measured by the **variance** of the random variable. It is defined as the expected value of $g(X)$ where $g(x)$ is the particular function

$$g(x) = (x - E(X))^2,$$

which is the square of the difference between the value x and the expected value $E(X)$. It follows that the variance of X, denoted Var(X), is

$$\text{Var}(X) = (x_1 - E(X))^2P(X=x_1) + (x_2 - E(X))^2P(X=x_2) + \cdots, \quad (4)$$

when X is discrete; and is

$$\text{Var}(X) = \int_{-\infty}^{\infty} (x - E(X))^2 f(x)dx \quad (5)$$

when X is continuous and has the density $f(x)$. Note that $E(X)$ is a fixed number, not a variable, in the above integrand.

EXAMPLE 3

Let X be the number of successes in 1 Bernoulli trial with probability p of success; it has the binomial distribution with $n = 1$. Then $E(X) = p$, and

$$\mathrm{Var}(X) = (0-p)^2q + (1-p)^2p = pq(p+q) = pq.$$

For 2 trials, we have $E(X) = 2p$, and

$$\begin{aligned}\mathrm{Var}(X) &= (0-2p)^2q^2 + 2(1-2p)^2pq + (2-2p)^2p^2 \\ &= 2pq.\end{aligned}$$

It can be shown that if X is the number of successes in n trials, then $E(X) = np$ and

$$\mathrm{Var}(X) = npq \tag{6}$$

(Exercises 11 and 12). ▲

EXAMPLE 4

Let X have the uniform density on the interval from 0 to 1. Then $E(X) = \frac{1}{2}$, and

$$\mathrm{Var}(X) = \int_0^1 (x-\tfrac{1}{2})^2 dx = \tfrac{1}{12}. \; ▲$$

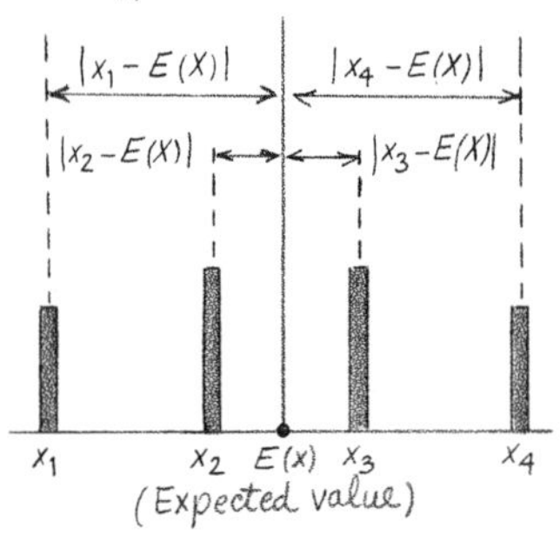

Figure 9.16

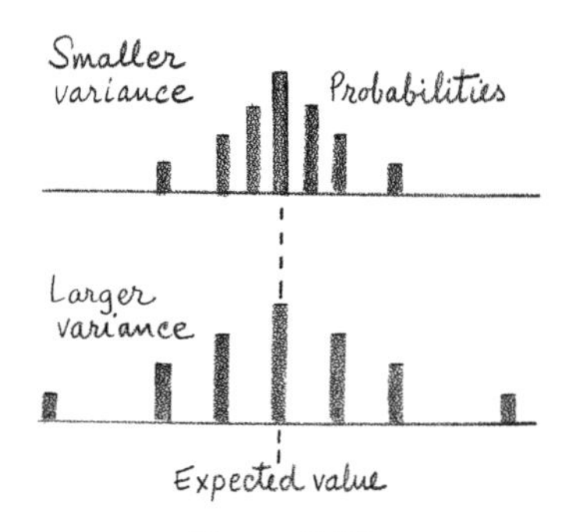

Figure 9.17

The variance, defined by (4) and (5), is the sum of the squares of the deviations of the values of the random variable from its expected value (Figure 9.16), weighted by the corresponding probabilities. A small variance indicates that the distribution is fairly concentrated about the expected value (Figure 9.17). A precise statement of this is the **Chebyshev inequality** (after P. L. Chebyshev, 1821–1894):

The probability that X assumes a value differing from $E(X)$ by at least k units is not more than the ratio of Var(X) to k^2:

$$P(|X - E(X)| \geq k) \leq \frac{\mathrm{Var}(X)}{k^2} \tag{7}$$

(Figure 9.18).

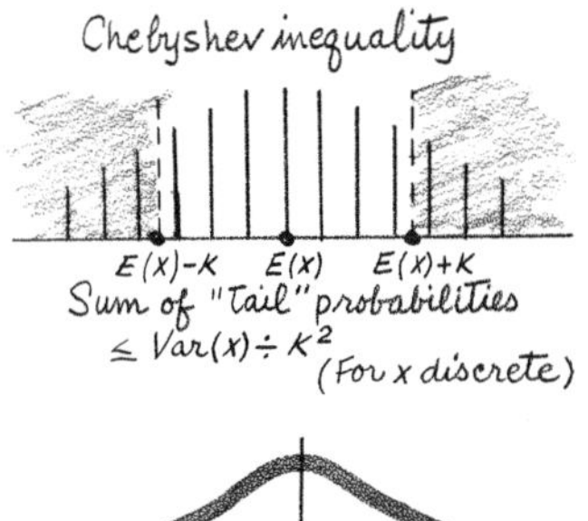

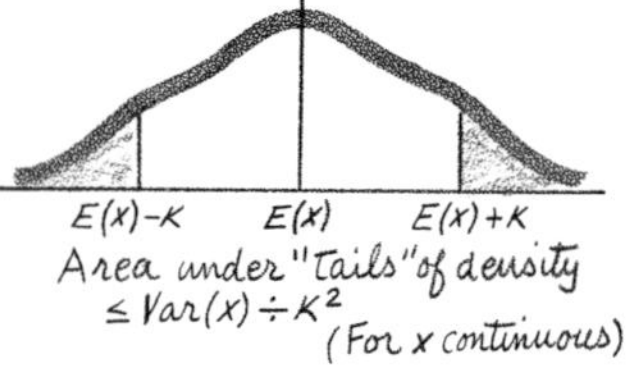

Figure 9.18

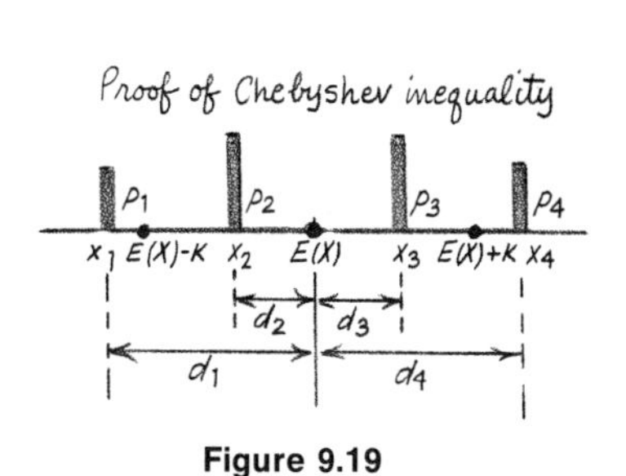

Figure 9.19

1. Variance $= d_1^2 p_1 + d_2^2 p_2 + d_3^2 p_3 + d_4^2 p_4$
2. Exclude terms corresponding to X-values within K-units of $E(x)$: variance $\geq d_1^2 p_1 + d_4^2 p_4$
3. Since $d_1 \geq K$, $d_4 \geq K$: $d_1^2 p_1 + d_4^2 p_4 \geq K^2 p_1 + K^2 p_4 = K^2(p_1 + p_4)$
4. $P(|X - E(x)| \geq K) = p_1 + p_4$
5. Therefore, by above four statements: $\mathrm{Var}(x) \geq K^2 P(|X - E(x)| \geq K)$ or $\boxed{P(|X - E(x)| \geq K) \leq \frac{1}{K^2} \mathrm{Var}(x)}$

The proof of the Chebyshev inequality is outlined in Figure 9.19. It is elementary, depending only on several simple estimates. The power of the inequality is that it applies to any distribution; however, it is not useful in particular numerical examples because the bound is so crude, since it is meant to be universal, for all distributions.

EXAMPLE 5

Let us calculate the variance for the normal density with mean μ and standard deviation σ. If X has the standard normal density, then

$$\mathrm{Var}(X) = \int_{-\infty}^{\infty} x^2 \phi(x)\,dx$$

because $E(X) = 0$. By integration by parts, it can be shown that the integral above is equal to 1 (Exercise 10). The variance for a random variable with a normal density with mean μ and standard deviation σ is

$$\int_{-\infty}^{\infty} (x-\mu)^2 \frac{1}{\sigma} \phi\left(\frac{x-\mu}{\sigma}\right) dx.$$

By the change of the variable method used in Example 1, it follows that the integral above is equal to

$$\sigma^2 \int_{-\infty}^{\infty} x^2 \phi(x)\,dx = \sigma^2.$$

The variance is the square of the standard deviation. (The standard deviation of any distribution, not necessarily normal, is defined as the positive square root of the variance.)

Let us now apply the Chebyshev inequality to the normal distribution, in the particular case $\sigma = 1$.

$$P(|X-\mu| \geq k) \leq \frac{1}{k^2}.$$

The bound on the right-hand side above is computed for several values of k and compared to the actual values of the probability on the left-hand side, obtained from Table VIII:

k	Probability	Chebyshev bound $1/k^2$
0.5	0.6170	4.0
1.0	0.3174	1.0
1.5	0.1336	0.444
2.0	0.0456	0.250
2.5	0.0124	0.160
3.0	0.0026	0.111
3.5	0.0004	0.089

▲

Let X be a random variable, and $g(x)$ a linear function of x: $g(x) = ax + b$, where a is not 0. Here is a general result about the relations between the expected value and variance of X and $g(X)$, respectively:

$$E(aX + b) = aE(X) + b, \tag{8}$$

$$\operatorname{Var}(aX + b) = a^2 \operatorname{Var}(X); \tag{9}$$

in other words, **the expected value of the linear function is the corresponding linear function of the expected value, and the variance of the linear function is the square of the multiplicative factor times the variance.** The proof of (8) is based directly on the linearity property of the sum and the integral; and the proof of (9) depends on the fact that the differences between the values of the random variable and the expected value are unchanged under shifts in the "location" of the distribution (Exercises 16–19).

Here is another way of expressing the variance:

$$\operatorname{Var}(X) = E(X^2) - (E(X))^2; \tag{10}$$

in other words, the variance is the expected value of the square of X **minus** the square of the expected value of X. We give the proof in the case where there is a density function $f(x)$; the proof in the discrete case is similar. Expand the square under the integral sign in (5), and integrate term-by-term:

$$\int_{-\infty}^{\infty} [x^2 - 2xE(X) + (E(X))^2] f(x)\,dx$$
$$= \int_{-\infty}^{\infty} x^2 f(x)\,dx - 2\int_{-\infty}^{\infty} xE(X)f(x)\,dx + \int_{-\infty}^{\infty} (E(X))^2 f(x)\,dx.$$

Since $E(X)$ is a constant, it may be factored from under the sign of integration; therefore, the last expression is equal to

$$\int_{-\infty}^{\infty} x^2 f(x)\,dx - 2E(X)\int_{-\infty}^{\infty} xf(x)\,dx + (E(X))^2 \int_{-\infty}^{\infty} f(x)\,dx,$$

which, by the relations $\int_{-\infty}^{\infty} f(x)\,dx = 1$ and $\int_{-\infty}^{\infty} xf(x)\,dx = E(X)$, is equal to

$$\int_{-\infty}^{\infty} x^2 f(x)\,dx - 2(E(X))^2 + (E(X))^2 = E(X^2) - (E(X))^2.$$

In calculating the variance of a discrete random variable, formula (10) is preferable to (4) because the rounding error in the expected value is not introduced until the last step in the computation.

9.2 EXERCISES

In Exercises 1 and 2 find the expected value and variance of X *for the given discrete distribution.*

1 (a)

x	2	3	9
$P(X=x)$	$\frac{1}{3}$	$\frac{1}{6}$	$\frac{1}{2}$

(b)

x	-3	-1	4	5
$P(X=x)$	0.4	0.1	0.2	0.3

(c)

x	3	7	11	15
$P(X=x)$	$\frac{1}{6}$	$\frac{1}{3}$	$\frac{1}{3}$	$\frac{1}{6}$

2 (a)

x	1	2	4
$P(X=x)$	$\frac{1}{3}$	$\frac{1}{2}$	$\frac{1}{6}$

(b)

x	-7	-5	-3	0
$P(X=x)$	0.2	0.3	0.4	0.1

(c)

x	-1	2	5	8
$P(X=x)$	0.4	0.1	0.2	0.3

In Exercises 3 and 4 find the expected value and variance of random variables with these densities.

3 (a) $f(x) = 3x^2/7$, for $1 < x < 2$; $= 0$, elsewhere

(b) $f(x) = x/4$, for $1 < x < 3$; $= 0$, elsewhere

(c) $f(x) = -x$, for $-1 \le x < 0$
$= x$, for $0 \le x \le 1$
$= 0$, elsewhere

(d) $f(x) = \dfrac{1}{2\sqrt{x}}$, for $0 < x < 1$; $= 0$, elsewhere

(e) $f(x) = 12x^2(1-x)$, for $0 < x < 1$; $= 0$, elsewhere

4 (a) $f(x) = 2x$, for $0 < x < 1$; $= 0$, elsewhere
(b) $f(x) = 3x^2$, for $0 < x < 1$; $= 0$, elsewhere

(c) $f(x) = x, \quad$ for $0 < x < 1$
$\quad = 2 - x, \quad$ for $1 \le x < 2$
$\quad = 0, \quad$ elsewhere

(d) $f(x) = \frac{3}{2}x^{1/2}, \quad$ for $0 < x < 1; \quad = 0,$ elsewhere

(e) $f(x) = 12x(1-x)^2, \quad$ for $0 < x < 1; \quad = 0,$ elsewhere

5 Find the general formulas in terms of a and b for $E(X)$ and $\mathrm{Var}(X)$ for the uniform density on the interval from a to b: $f(x) = 1/(b-a)$, for $a < x < b$; $= 0$, elsewhere.

6 Find the formulas for $E(X)$ and $\mathrm{Var}(X)$ for the gamma density

$$f(x) = \frac{1}{n!}x^n e^{-x}, \quad \text{for } x > 0; \qquad = 0, \quad \text{elsewhere.}$$

[*Hint:* See Section 6.7.]

7 Find $E(X)$ and $\mathrm{Var}(X)$ when X is the
(a) score on a single toss of a die
(b) sum of the scores on the toss of a pair of dice

8 A point P is chosen at random on the upper half of a circle of radius 1 about the origin. (This means that the angle between the radius to P and the x-axis is uniformly distributed on the interval from 0 to π.) Let Y be the y-coordinate of the random point. Compute $E(Y)$ and $\mathrm{Var}(Y)$.

9 Let X be a discrete random variable, assuming the values x_1, $x_2, \ldots$ with probabilities $p_1, p_2, \ldots,$ respectively. For a function $y = g(x)$, let $y_1 = g(x_1)$, $y_2 = g(x_2), \ldots.$
(a) What are the values of the random variable $Y = g(X)$ and their corresponding probabilities?
(b) Why is $E(Y)$ given by the sum (2)?

10 Using the definition of the standard normal density and integration by parts, show that $\int_{-\infty}^{\infty} x^2\phi(x)\,dx = 1$.

11 Write the binomial expansion of $(q+p)^n$ as

$$(q+p)^n = \sum_0^n C(n,k)p^k q^{n-k},$$

where the notation Σ means summation over the indicated indices. Derive the formula

$$np(q+p)^{n-1} = \sum_1^n C(n,k)kp^k q^{n-k}$$

by (i) differentiating both sides of the former equation with respect to p, and (ii) multiplying both sides by p. Deduce from this that $E(X) = np$.

12 Starting with the second equation in Exercise 11, deduce the third equation

$$n(n-1)p^2(q+p)^{n-2} = \sum_2^n C(n,k)k(k-1)p^k q^{n-k}.$$

From this conclude that
(a) $n(n-1)p^2 = E(X^2) - E(X)$, and so
(b) $E(X^2) = n(n-1)p^2 + np$
(c) $\text{Var}(X) = np(1-p) = npq$

13 In n Bernoulli trials with probability p of success, let $\hat{p}$ be the proportion of successes observed,

$$\hat{p} = \frac{\text{No. of successes}}{n}.$$

In Section 8.10, formula (2), it is shown that the probable deviation of $\hat{p}$ from p decreases like the reciprocal of $\sqrt{n}$, and so $\hat{p}$ is a good estimate of p when n is large. We can get a simpler although cruder result of the same type without the normal approximation, using only the expected value and variance of the number of successes, followed by an application of the Chebyshev inequality.
(a) Compute $E(\hat{p})$ and $\text{Var}(\hat{p})$ by means of formulas (8) and (9).
(b) By applying (7) show that $P(|\hat{p} - p| \geq k) \leq pq/nk^2$, for any $k > 0$ (no matter how small). What is the limit of the probability as $n \to \infty$? This result is known as **Bernoulli's law of large numbers.**

14 Let X be the number of failures appearing before the first success in Bernoulli trials with probability p of success; it has the geometric distribution

$$P(X = k) = pq^k, \qquad k = 0, 1, 2, \ldots.$$

Verify that $E(X) = q/p$:
(a) Differentiate both sides of the formula for the geometric series,

$$\frac{1}{1-q} = 1 + q + q^2 + \cdots$$

with respect to q (see Section 6.12).
(b) Then multiply both sides of the resulting equation by pq, and use the relation $p = 1 - q$.

15 Let X have the Poisson distribution with parameter λ. Since the Poisson approximates the binomial with $\lambda = np$, and since the latter is the expected number of successes, we expect that λ is the expected value of X (taken from the Poisson). This can be verified directly from the expression for the Poisson distribution:

$$P(X = k) = \frac{\lambda^k e^{-\lambda}}{k!}, \qquad k = 0, 1, 2, \ldots.$$

(a) Differentiate both sides of the formula for the series of e^λ

$$e^\lambda = \sum \frac{\lambda^k}{k!}$$

with respect to λ (Section 6.12).
(b) Then multiply both sides of the resulting equation by $e^{-\lambda}\lambda$.

16 Let X be a discrete random variable assuming the values x_1, $x_2, \ldots$ with probabilities $p_1, p_2, \ldots$, respectively. Verify formula (8).

17 Let X have the density function $f(x)$. Verify formula (8).

18 Using Equation (8), show that the variance (9) is independent of the quantity b.

19 Let X have the density $f(x)$. Verify formula (9) under the assumption $b = 0$; then use the result of Exercise 18 to conclude that it is true for any b.

20 The standard deviation of a random variable was defined in Example 4 as the square root of the variance. Show that the Chebyshev inequality may be expressed in this way: The probability that X differs from $E(X)$ by at least a constant multiple of the standard deviation is less than or equal to the reciprocal of the constant squared:

$$P(|X - E(X)| \geq k \cdot \textbf{standard deviation}) \leq \frac{1}{k^2}.$$

9.3 Sampling from Populations and Maximum Likelihood Estimation

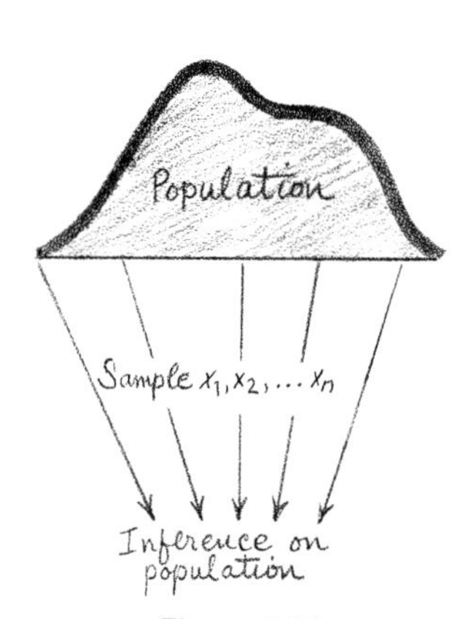

Figure 9.20

Statistics is the science of making decisions about the numerical characteristics of populations. For example, suppose a biologist tests the effect of a nutrient on the growth of cattle. Here the numerical characteristic is the growth, measured in weight or stature gain in appropriate units. The decision is whether or not the nutrient should be used to increase growth. A fundamental method of statistics is the drawing of a relatively small sample from the population, observing the numerical characteristics X_1, $X_2, \ldots$ of the members of the sample, and making the decision on the basis of the latter observations. The decision is based on **inference from the sample to the population.** For example, the biologist will decide to use the nutrient if it improves the growth of the members of the population; such growth might be inferred from the observed growth of the members of the sample (Figure 9.20).

Consider a population with a numerical characteristic with the distribution function $F(x)$. The numerical characteristic X of a member drawn at random is a random variable with the distribution function $F(x)$ (Section 9.1). Populations arising in applications are always large relative to the size of the sample; otherwise, there would be no need to take a **sample:** the decision could be made after examining **every** member of the population if it were small enough. If a sample is drawn at random from a large population, then, for the mathematical analysis, we may assume that the sample had been

drawn with replacement (Section 8.8). Let $X_1, X_2, \ldots, X_n$ be the random variables representing the numerical values of the members of the sample. **For any set of n real numbers $x_1, \ldots, x_n$, the probability that the random variables all satisfy the inequalities $X_1 \leq x_1, \ldots, X_n \leq x_n$ is given by the product of the values of the distribution function at the values $x_1, \ldots, x_n$:**

$$P(X_1 \leq x_1, \ldots, X_n \leq x_n) = F(x_1) \cdots F(x_n). \tag{1}$$

Formula (1) is based on the analogy between the drawing of the sample with replacement and Bernoulli trials with variable probabilities (the end of Section 8.8). Each member of the sample corresponds to a Bernoulli trial. The outcome of the trial, success or failure, is determined by a number x: there is success if the random variable X assumes a value less than or equal to x, and failure if not; and the probability of success is $p = F(x)$. For the given set $x_1, \ldots, x_n$, there is success on the ith trial if $X_i \leq x_i$ and failure otherwise. The probabilities are assigned to the outcomes of the n trials according to the product rule. The probability on the left-hand side of (1) is the probability of success on all n trials; by the product rule, it is equal to the product of the corresponding probabilities on the right-hand side (Figure 9.21).

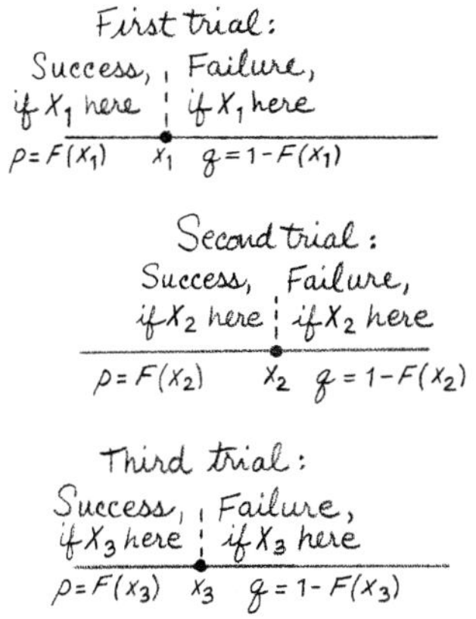

Figure 9.21

The product function on the right-hand side of (1) is called the **joint distribution function** of the sample $X_1, \ldots, X_n$. A set of random variables such that the probability of the joint occurrence of the inequalities $X_1 \leq x_i$ is equal to the product of the corresponding probabilities, for every $x_1, \ldots, x_n$, is called a set of **mutually independent** random variables.

Suppose that the population has a discrete characteristic. Let $x_1, \ldots, x_n$ be a set of n values from the population. Then, by the same reasoning as for (1), **the probability of the joint occurrence of the events**

$$X_1 = x_1, \ldots, X_n = x_n$$

is the product of the corresponding probabilities:

$$P(X_1 = x_1, \ldots, X_n = x_n) = P(X_1 = x_1) \cdots P(X_n = x_n). \tag{2}$$

The function on the right-hand side of this equation is the **likelihood function of the sample.**

If the population has a density function $f(x)$, the product of the density values at $x_1, \ldots, x_n$ is the **likelihood function of the sample:**

$$f(x_1) \cdots f(x_n). \tag{3}$$

Recall that the density function of a continuous random variable measures the "local probability": the probability that the random variable assumes a value in a small interval is approximately equal to the density times the length of the interval (Section 9.1). There is an analogous interpretation for the joint density (3) of the sample. Let us first consider the case $n = 2$. Take a rectangle in the plane, with sides parallel to the axes; for example, the rectangle with corner points (a,c), (a,d), (b,c), and (b,d), where $a < b$ and $c < d$. The probability that the point with the coordinates (X_1,X_2) (a random sample of 2 observations) falls in the rectangle is the probability that

$$a \leq X_1 \leq b \qquad \text{and} \qquad c \leq X_2 \leq d.$$

By the reasoning above, this is the product of the corresponding probabilities,

$$P(a \leq X_1 \leq b) \cdot P(c \leq X_2 \leq d),$$

which, by the definition of the density, is

$$\int_a^b f(x)\,dx \cdot \int_c^d f(y)\,dy;$$

finally, this may be written as the double integral

$$\int_a^b \int_c^d f(x)f(y)\,dy\,dx.$$

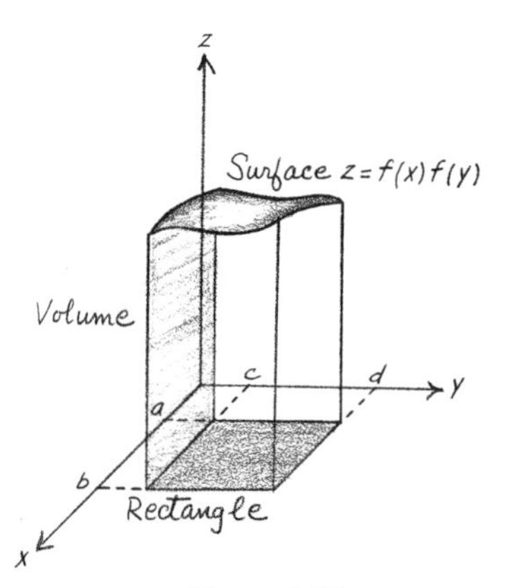

Figure 9.22

This is the volume under the surface $z = f(x)f(y)$ over the rectangle (Figure 9.22). When the rectangle is small, the volume is approximately equal to the area of the rectangle times the height $f(x)f(y)$ of a point on the surface above it: $f(x)f(y)dx\,dy$, where dx and dy are the length and width, respectively, of the rectangle. In the case of a sample of n observations, the probability that the sample values $X_1, \ldots X_n$ fall in intervals from x_1 to $x_1 + dx_1, \ldots,$ from x_n to $x_n + dx_n$, respectively, is, by the same reasoning, equal to the likelihood function (3), times the product of the lengths of the intervals:

$$f(x_1) \cdots f(x_n)\,dx_1 \cdots dx_n.$$

Thus, the likelihood function is proportional to the "local probability" that the sample assumes particular values.

EXAMPLE 1

Consider a "Bernoulli" population, one having the binomial distribution with $n = 1$. A random variable X from this population assumes the values 1 and 0 with probabilities p and q, respectively:

$$P(X = x) = p^x q^{1-x}, \qquad x = 0,\ 1.$$

The likelihood function (2) is obtained by substitution of $x_1, \ldots, x_n$, and the formation of the product:

$$p^{x_1}q^{1-x_1} \cdot \ldots \cdot p^{x_n} q^{1-x_n}.$$

By rearrangement of factors, the latter may be written as

$$p^{(x_1+\cdots+x_n)}q^{n-(x_1+\cdots+x_n)}. \tag{4}$$ ▲

EXAMPLE 2

For a sample of n observations from a normal population with mean μ and standard deviation σ, the likelihood function is

$$\frac{1}{\sigma^n}\,\phi\left(\frac{x_1-\mu}{\sigma}\right)\cdot \ldots \cdot \phi\left(\frac{x_n-\mu}{\sigma}\right),$$

which is

$$\frac{1}{(2\pi)^{n/2}\sigma^n}\, e^{-(1/2\sigma^2)[(x_1-\mu)^2+\cdots+(x_n-\mu)^2]}. \tag{5}$$ ▲

EXAMPLE 3

The likelihood of the sample from a normal population is easy to express as a product because f is positive for all x. The likelihood function is more complicated when the density is equal to 0 somewhere. Consider a sample of 2 observations X_1 and X_2 from a population with the exponential density

$$f(x) = 0, \quad \text{for } x \le 0; \qquad = ce^{-cx}, \quad \text{for } x > 0.$$

The product $f(x)f(y)$ is 0 if either $f(x)$ or $f(y)$ is; therefore, it is equal to 0 when either $x \le 0$, or $y \le 0$, or both. The likelihood function is

$c^2e^{-c(x+y)}$, if both x and y are positive
0, elsewhere (Figure 9.23).

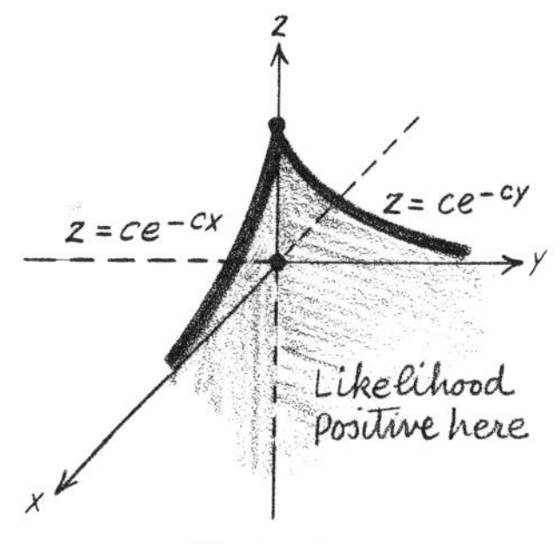

Figure 9.23

More generally, for a sample of n observations, the likelihood function is

$c^ne^{-c(x_1+\cdots+x_n)}$, if all $x_1, \ldots, x_n$ are positive
0, if at least one x_i is less than or equal to 0. ▲

One of the important topics of statistical theory is the **estimation of unknown parameters.** Consider a population with a distribution function of the form $F(x,\theta)$, where F is a known function of the two variables: it is a distribution function in x for each fixed θ, but the value of the latter variable is unknown. θ is an "unknown parameter" of the distribution function. We wish to **estimate** θ on the basis of a sample of n observations. An **estimator** is a function of n variables $h(x_1, \ldots, x_n)$. The **estimation procedure** based on h is to take a sample $X_1, \ldots, X_n$, evaluate the function at the "sample point" $(X_1, \ldots, X_n)$, and use it as the estimate of θ: The estimated value is $h(X_1, \ldots, X_n)$.

How do we choose the estimator h? The classical procedure is the method of **maximum likelihood** (see Section 8.7, Example 3). Since the distribution function depends on the parameter θ, so does the likelihood function. In the discrete case the likelihood function (2) takes the form

$$P(X_1 = x_1;\theta) \cdots P(X_n = x_n;\theta); \tag{6}$$

and, in the density case, the form

$$f(x_1,\theta) \cdots f(x_n,\theta). \tag{7}$$

For given values of the x's, these are functions of the single variable θ. The **maximum likelihood estimator** of θ is that value of θ for which (6) or (7) is maximized. In other words: Given $x_1, \ldots, x_n$, we choose θ to maximize the likelihood (6) or (7). This value is denoted θ. It depends, of course, on the x's; thus, the maximum likelihood estimator is a function of $x_1, \ldots, x_n$, so that it is an estimator in the sense defined earlier.

Having defined the maximum likelihood estimator, we turn to the problem of determining it in particular cases. It can usually be found by the maximization method of Section 5.10. The functions (6) and (7) are products of functions of θ, so that they are difficult to differentiate directly. The maximization is greatly simplified by considering the logarithm of the likelihood function in place of the function itself: Since $\log x$ increases with x, the likelihood function is maximized whenever the logarithm is; hence, it suffices to find the value θ maximizing the **logarithm** of the likelihood.

Since the logarithm of the product is the sum of the logarithms of the factors, the functions (6) and (7) have the logarithms

$$\log P(X_1 = x_1;\theta) + \cdots + \log P(X_n = x_n;\theta) \tag{8}$$

and

$$\log f(x_1,\theta) + \cdots + \log f(x_n,\theta), \tag{9}$$

respectively. To maximize these, we set the derivative (with respect to θ) equal to 0, and solve the resulting equation for θ; and then verify that the second derivative is negative:

Let $\hat{\theta}$ be the solution of

$$\frac{\partial}{\partial\theta}\log P(X_1 = x_1;\theta) + \cdot\cdot\cdot + \frac{\partial}{\partial\theta}\log P(X_n = x_n;\theta) = 0, \quad (10)$$

or

$$\frac{\partial}{\partial\theta}\log f(x_1,\theta) + \cdot\cdot\cdot + \frac{\partial}{\partial\theta}\log f(x_n,\theta) = 0; \quad (11)$$

and then verify that

$$\left(\frac{\partial^2}{\partial\theta^2}\log P(X_1 = x_1;\theta)\right)_{\hat{\theta}} + \cdot\cdot\cdot + \left(\frac{\partial^2}{\partial\theta^2}\log P(X_n = x_n;\theta)\right)_{\hat{\theta}} < 0, \quad (12)$$

or

$$\left(\frac{\partial^2}{\partial\theta^2}\log f(x_1,\theta)\right)_{\hat{\theta}} + \cdot\cdot\cdot + \left(\frac{\partial^2}{\partial\theta^2}\log f(x_n,\theta)\right)_{\hat{\theta}} < 0, \quad (13)$$

respectively. The subscripts $\hat{\theta}$ signify that the second derivatives are evaluated at $\theta = \hat{\theta}$. Equations (10) and (11) are the **likelihood equations.**

EXAMPLE 4

Let us estimate the parameter p of the Bernoulli population with the discrete distribution

$$p^x q^{1-x}, \qquad x = 0,\ 1.$$

(See Example 1.) Here the unknown parameter p is called θ, and so q is related to it by the equation $q = 1 - \theta$. Then the discrete distribution may be expressed as

$$\theta^x(1-\theta)^{1-x}, \qquad x = 0,\ 1;$$

and the logarithm is

$$x\log\theta + (1-x)\log(1-\theta);$$

and its derivative is

$$\frac{x}{\theta} - \frac{1-x}{1-\theta}.$$

The likelihood equation (10) is obtained from the latter expression by substitution of the values $x_1, \ldots, x_n$, and summation over the n terms:

$$\frac{x_1 + \cdot\cdot\cdot + x_n}{\theta} + \frac{n - (x_1 + \cdot\cdot\cdot + x_n)}{1-\theta} = 0,$$

or, equivalently,

$$x_1 + \cdot\cdot\cdot + x_n - n\theta = 0.$$

The solution of this equation is $\hat{\theta} = (x_1 + \cdot\cdot\cdot + x_n)/n$.

Finally we verify Equation (12). By direct differentiation we find

$$\frac{\partial^2}{\partial\theta^2}(x \log \theta + (1-x)\log(1-\theta)) = -\frac{x}{\theta^2} - \frac{(1-x)}{(1-\theta)^2}.$$

The latter is negative for $x=0$ and $x=1$. It remains negative when the x's are substituted and the terms are summed.

We conclude that the **average** of the observations,

$$\frac{X_1 + \cdot \cdot \cdot + X_n}{n},$$

is the maximum likelihood estimator of the unknown parameter p. This **average function** occurs frequently in statistical theory. It is also referred to as the **sample mean,** and is denoted $\bar{X}$ ("X-bar"):

$$\bar{X} = \frac{X_1 + \cdot \cdot \cdot + X_n}{n}. \quad \blacktriangle$$

EXAMPLE 5

Let us find the maximum likelihood estimator of the mean of a normal population. We write the population density as

$$f(x,\theta) = \frac{1}{\sqrt{2\pi}\sigma} e^{-(x-\theta)^2/2\sigma^2}.$$

The mean μ has been replaced by θ because it is the parameter to be estimated; σ is also unknown but, at the moment, its estimate is not needed. We have

$$\log f(x,\theta) = -\log(\sqrt{2\pi}\sigma) - \frac{1}{2\sigma^2}(x-\theta)^2,$$

and

$$\frac{\partial}{\partial\theta}\log f(x,\theta) = \frac{x-\theta}{\sigma^2}.$$

The likelihood equation (11) assumes the form

$$\frac{(x_1-\theta) + \cdot \cdot \cdot + (x_n-\theta)}{\sigma^2} = 0,$$

or, equivalently,

$$(x_1 + \cdot \cdot \cdot + x_n) - n\theta = 0;$$

and the solution is

$$\theta = \frac{x_1 + \cdot \cdot \cdot + x_n}{n},$$

the sample mean. Equation (12) is directly verified:

$$\frac{\partial^2 \log f(x,\theta)}{\partial \theta^2} = -\frac{1}{\sigma^2},$$

and so the left-hand side of (12) is of the form $-n/\sigma^2$, which is negative. We conclude that the sample mean is the maximum likelihood estimator of the mean of a normal population. ▲

The method above is not always satisfactory for finding the maximum likelihood estimator; indeed, the likelihood equation may be difficult to solve, or may have no solution, or may have many solutions. In such cases the maximum likelihood estimator might be found by a direct analysis of the likelihood function (see Exercise 8).

We conclude this section with an application illustrating the relation between sampling from a population and Bernoulli trials (see Figure 9.21).

EXAMPLE 6

Figure 9.24

A building is designed to withstand winds of speeds up to 150 mph. Let X be the maximum annual observed wind gust; it is a random variable with some distribution function $F(x)$. The maximum annual observed wind gusts for the first n years of the life of the building, $X_1, \ldots, X_n$, form a sample from the population of maximum annual gusts (Figure 9.24). The building lasts at least n years if these observations are all less than or equal to 150. By Equation (1), with all x's equal to 150, the probability of this is $F^n(150)$. If $F(150) = 1$, then no gust will ever exceed 150, and so $F^n(150) = 1$ for all n, and the building will surely never fall in the wind. If $F(150)$ is less than 1, then $F^n(150)$ decreases with increasing n, and so the building will "eventually" fall. How long is such a building expected to last? Let X be the number of years before the first wind gust exceeds 150. It has the geometric distribution with $q = F(150)$: Each year is a Bernoulli trial with **success** if a wind gust exceeds 150, and failure otherwise; and X is the number of failures before the first success; therefore, the expected value of X is

$$\frac{q}{p} = \frac{F(150)}{1 - F(150)}$$

(see Section 9.2, Exercise 14). ▲

9.3 EXERCISES

In Exercises 1 and 2 write the expression for the likelihood function for a sample of n *observations from a population with the given density or discrete distribution. Also find the likelihood equation.*

1 (a) Poisson distribution with parameter λ

(b) $f(x) = \frac{1}{2}e^{x}$, for $x < 0$
$= \frac{1}{2}e^{-x}$, for $x \geq 0$ $(\frac{1}{2}e^{-|x|}$, for all $x)$

(c) $f(x) = \dfrac{1}{\pi(1 + x^2)}$

(d) $f(x) = (\frac{1}{6})x^3 e^{-x}$, for $x > 0$; $= 0$, for $x \leq 0$

2 (a) $f(x) = xe^{-(1/2)x^2}$, for $x > 0$; $= 0$, for $x \leq 0$

(b) $f(x) = \dfrac{e^{-x}}{(1 + e^{-x})^2}$

(c) $f(x) = 1/x^2$, for $x > 1$; $= 0$, for $x \leq 1$

(d) $f(x) = 2x$, for $0 < x < 1$; $= 0$, elsewhere

(e) Geometric distribution

3 Find the maximum likelihood estimate of the parameter λ for n observations from a Poisson population. (See Example 4.)

4 Find the maximum likelihood estimate of the parameter p in sampling from a population with the geometric distribution pq^k, $k = 0, 1, \ldots$. (See Example 4.)

5 Find the maximum likelihood estimate of the parameter c for the population with the density

$$f(x) = ce^{-cx}, \quad \text{for } x > 0; \qquad = 0, \quad \text{elsewhere.}$$

(See Example 5.)

6 Show that if μ is given and σ is unknown, then

$$\left(\frac{1}{n}\right)[(x_1 - \mu)^2 + \cdots + (x_n - \mu)^2]$$

is the maximum likelihood estimate of σ^2. (See Example 5.)

7 Suppose that μ and σ^2 are both unknown. Consider the likelihood of a sample of n observations from a normal population as a function of two variables μ and σ^2. Show that the maximum occurs at

$$\mu = \bar{x} \qquad \text{and} \qquad \sigma^2 = \frac{(x_1 - \bar{x})^2 + \cdots + (x_n - \bar{x})^2}{n}.$$

Compare the estimate of σ^2 with that in Exercise 6. (To maximize the function of two variables, use the method of Section 7.3.)

8 Consider a population with the uniform density on the interval from 0 to T:

$$f(x) = \frac{1}{T}, \quad \text{for } 0 < x < T; \qquad = 0, \qquad \text{elsewhere.}$$

(a) Write the likelihood of the sample of n observations.
(b) If T is unknown, what is the maximum likelihood estimate? [*Hint:* All the observations must be smaller than T.]

9 Let Z stand for the largest of n observations from a population with the distribution function $F(x)$. Show that $P(Z \leq x) = F^n(x)$. [*Hint:* Z is less than or equal to x if and only if the same is true for the X's.]

10 Using the result of Example 9, find the expression for the density function of the random variable Z when F has a density function f.

11 Using the results of Exercises 9 and 10, find the distribution function and density of Z in sampling from a uniform population on the unit interval.

12 Repeat Exercise 11 for the exponential density

$$f(x) = e^{-x}, \quad \text{for } x > 0; \qquad = 0, \quad \text{elsewhere.}$$

13 By the arguments in Exercises 9 and 10 find the distribution and density of the **minimum** of n observations in sampling from a population with the distribution function $F(x)$. [*Hint:* Minimum $\leq x$ if and only if there is at least 1 "success".]

14 As in Exercises 11 and 12, find the distribution and density of the **minimum** for the uniform and exponential densities.

15 Suppose, in Example 6, that the distribution function of the maximum annual wind gust is

$$\begin{aligned} F(x) &= 1 - e^{-(x/50)^2}, && \text{for } x > 0 \\ &= 0, && \text{for } x \leq 0. \end{aligned}$$

Find the expected number of years before a gust is 100 mph.

16 The operating lifetime (in months) of a certain component in a piece of machinery has the distribution function

$$\begin{aligned} F(x) &= 0, && \text{for } x \leq 0 \\ &= \left(\frac{x}{10}\right)^2, && \text{for } 0 < x < 10 \\ &= 1, && \text{for } x \geq 10. \end{aligned}$$

If a machine has 4 such new components, find the probability that at least 3 will still be operating before the end of 6 months. (See Example 6.)

17 In an educational experiment to measure the span of attention of children, a teacher presents a lesson to a group of children until several lose interest. Suppose that the attention span (in minutes) of a child in this group is a random variable with the density function

$$f(x) = \tfrac{1}{5}e^{-(x/5)} \qquad \text{for } x > 0$$
$$= 0, \qquad \text{for } x \leq 0.$$

The teacher starts with a group of 3 children, and stops the lesson as soon as 2 have lost interest. Find the probability that the lesson lasts at least 10 minutes. [*Hint:* Find the probability that at least 2 of the 3 observations from this population are at least equal to 10.] (See Example 6.)

18 Suppose that the intelligence quotients of a population of students has a normal distribution with mean 100 and standard deviation 10. If 4 students are selected at random, what is the probability that at most 2 have an intelligence quotient of at least 120? (See Example 6.)

19 Let X_1 and X_2 be independent random variables, that is, satisfying condition (1) for $n = 2$. Show that the **conditional probability** (Section 8.12) that $X_1 \leq x_1$, given $X_2 \leq x_2$, is the same as the (unconditional) probability: $P(X_1 \leq x_1 \mid X_2 \leq x_2) = P(X_1 \leq x_1)$.

9.4 Sums of Independent Random Variables

Many of the topics of probability and statistics involve **sums of independent random variables.** Recall that a set of mutually independent random variables is one satisfying Equation (1) [or Equation (2)] of Section 9.3: the joint probability factors into the product of individual probabilities. We have already discussed such sums in particular cases. The number of successes in n Bernoulli trials is a sum of independent random variables: the summands are the numbers of successes (0 to 1) on the individual trials. The independence of the summands follows from the definition of the probabilities for n Bernoulli trials as the products of p's and q's. Several maximum likelihood estimates (Section 9.3) are expressed in terms of **averages** of the observations: the **sum** divided by n.

Let S_n stand for the sum of $X_1, \ldots, X_n$, a sample of observations from a population:

$$S_n = X_1 + \cdots + X_n, \tag{1}$$

S_n is a real-valued composite function on the sample space; hence, it is a random variable and so has a well-defined distribution function. This section is concerned with this distribution function. Our main results are (i) a general theorem on the expected value and variance of

S_n, and (ii) the method of calculation of the distribution of S_n from that of the X's. Under (i) we have the following two results:

$$E(S_n) = E(X_1) + \cdots + E(X_n), \tag{2}$$

$$\mathrm{Var}(S_n) = \mathrm{Var}(X_1) + \cdots + \mathrm{Var}(X_n); \tag{3}$$

in other words, **the expected value and variance of the sum are equal to the sums of the expected values and variances, respectively.** If m and v are the mean and variance of the population distribution, then (2) and (3) imply

$$E(S_n) = mn, \tag{4}$$

$$\mathrm{Var}(S_n) = vn. \tag{5}$$

To prove these we note first that if $h(x_1, \ldots, x_n)$ is a function of n variables, then $h(X_1, \ldots, X_n)$ is a random variable because it is a composite function on the sample space. If the population distribution is discrete, then

$$Eh(X_1, \ldots, X_n) = \Sigma h(x_1, \ldots, x_n) P(X_1 = x_1) \cdots P(X_n = x_n), \tag{6}$$

where the sum is over all sets of values $(x_1, \ldots, x_n)$ of the random variables $X_1, \ldots, X_n$. If the population has a density function $f(x)$, then

$$Eh(X_1, \ldots, X_n) = \int_{-\infty}^{\infty} \cdots \int_{-\infty}^{\infty} h(x_1, \ldots, x_n) f(x_1) \cdots f(x_n)\, dx_1 \cdots dx_n. \tag{7}$$

[Although we have formally defined the **double** integral, for functions of two variables, in Section 7.5, and not the more general multiple integral (7) of n variables, the extension of the definition to the latter is direct: the reader may interpret (7) as the repeated integral over x_1, x_2, and so on.] The proof of (6) is similar to the proof of the formula for the expected value of a function of a single random variable [Equation (2), Section 9.2]. Formula (7) is the "integral analog" of (6).

Now we turn to the proof of (2). We shall consider just the case of the density function, and only for $n = 2$; this contains the main ideas of the proof of the general case. By formula (7),

$$E(X_1 + X_2) = \iint (x_1 + x_2) f(x_1) f(x_2)\, dx_2\, dx_1.$$

(Here the limits of integration, $\pm\infty$, have been omitted for typographical convenience.) By the linearity of the integrand, the integral above is the sum

$$\int\int x_1 f(x_1)f(x_2)\,dx_2\,dx_1 + \int\int x_2 f(x_1)f(x_2)\,dx_2\,dx_1$$

$$= \int f(x_2)\,dx_2 \cdot \int x_1 f(x_1)\,dx_1 + \int f(x_1)\,dx_1 \cdot \int x_2 f(x_2)\,dx_2$$

$$= 1 \cdot E(X_1) + 1 \cdot E(X_2).$$

In the proof of formula (3) for the variance, we take $h(x_1,x_2) = [x_1 + x_2 - E(X_1) - E(X_2)]^2$. The integrand is squared, and the integration is done term-by-term (Exercise 20).

Formulas (4) and (5) imply the following important result about the average $\bar{X}$ of a sample of n observations from a population. By formula (8), Section 9.2, with $a = 1/n$, $b = 0$, and X standing for S_n,

$$E(\bar{X}) = E\left(\frac{S_n}{n}\right) = \frac{1}{n}\,mn = m; \tag{8}$$

and by formula (9), Section 9.2,

$$\text{Var}(\bar{X}) = \text{Var}\left(\frac{S_n}{n}\right) = \frac{1}{n^2}\,vn = \frac{v}{n}. \tag{9}$$

The first of these, (8) above, states that **the expected value of the average $\bar{X}$ is the same as that of each individual observation,** and the second, (9) above, states that **the variance of $\bar{X}$ is the individual variance of the observations divided by the number of observations.** This means that the average $\bar{X}$ has the same "location" as the original distribution, but the "spread" is much smaller; in other words, **$\bar{X}$ is much more likely to be close to the mean of the distribution.**

Let us apply the Chebyshev inequality [formula (7), Section 9.2] to $\bar{X}$. By formulas (8) and (9) above for the expected value and variance, we obtain

$$P(|\bar{X} - m| \geq k) \leq \frac{v}{nk^2}. \tag{10}$$

For any prescribed $k > 0$, the bound on the right-hand side above tends to 0 as $n \to \infty$:

$$\lim_{n\to\infty} P(|\bar{X} - m| \geq k) = 0.$$

This limit theorem is known as the **Law of Large Numbers: The average of the observations is very likely to be very close to the population mean when the number of observations is large.** (A special case, Bernoulli's Law of Large Numbers, was outlined in Section 9.2, Exercise 13.) (Figure 9.25.)

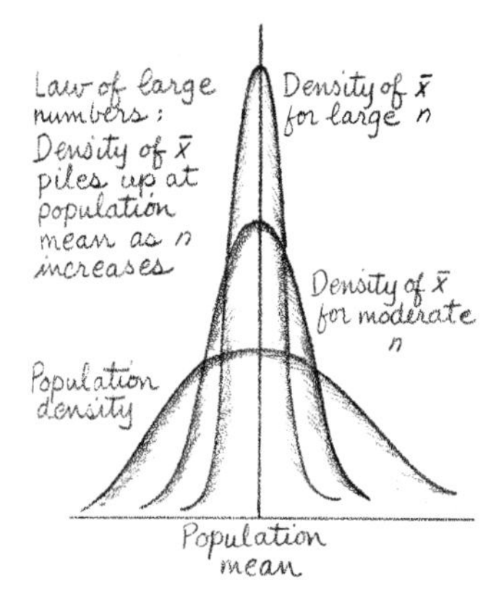

Figure 9.25

EXAMPLE 1

Let us illustrate inequality (10) in the case of a population with $v = 1$, for various k and n. The following table gives those values of the bound for the probability in (10):

	k		
n	1.0	0.1	0.05
100	0.01	1.0	4.0
500	0.002	0.2	0.8
1000	0.001	0.1	0.4
10,000	0.0001	0.01	0.04

For example, the probability that the difference between $\bar{X}$ and the mean exceeds 0.1 in absolute value is at most 0.01 when $n = 10{,}000$. ▲

Now we turn to topic (ii) above, the calculation of the **distribution** of the sum, not just the expected value and variance.

Let X_1 and X_2 be independent and discrete random variables assuming only integer values, and having the common distribution

$$p_k = P(X_1 = k), \qquad k = 0, 1, -1, 2, -2, \ldots .$$

(These are sample observations from a discrete population with integer values.) **For any integer m, the sum $X_1 + X_2$ assumes the value m if and only if the values individually assumed by the X's have the sum m:** The event $\{X_1 + X_2 = m\}$ is the union of the mutually exclusive events

$$\{X_1 = m, X_2 = 0\}, \qquad \{X_1 = m - 1, X_2 = 1\},$$
$$\{X_1 = m + 1, X_2 = -1,\} \qquad \text{and so on.}$$

The probability of this union is the sum of the probabilities of these events [Section 8.12, Equation (2)]. By the independence of the random variables, the probability of the event

$$\{X_1 = m - k, X_2 = k\}$$

is the product of the probabilities for X_1 and X_2, namely, $p_{m-k}p_k$ [Section 9.3, Equation (2)]. It follows that the distribution of the sum is given by the formula

$$P(X_1 + X_2 = m) = p_m p_0 + p_{m-1}p_1 + p_{m+1}p_1 + \cdots + p_{m-k}p_k + \cdots . \qquad (11)$$

EXAMPLE 2

Let X_1 and X_2 be a sample of 2 observations from the discrete population

x	-1	0	1	2
$P(X=x)$	0.2	0.3	0.3	0.2

Here $p_{-1}=0.2$, $p_0=0.3$, $p_1=0.3$, $p_2=0.2$. The sum of the observations assumes values from -2 through 4. The distribution is computed from formula (11):

m	$P(X_1+X_2=m)$
-2	$p_{-1}p_{-1}=0.04$
-1	$p_{-1}p_0+p_0p_{-1}=0.06+0.06=0.12$
0	$p_0p_0+p_{-1}p_1+p_1p_{-1}=0.09+0.06+0.06=0.21$
1	$p_1p_0+p_2p_{-1}+p_0p_1+p_{-1}p_2=0.09+0.04+0.09+0.04=0.26$
2	$p_2p_0+p_1p_1+p_0p_2=0.06+0.09+0.06=0.21$
3	$p_2p_1+p_1p_2=0.06+0.06=0.12$
4	$p_2p_2=0.04$

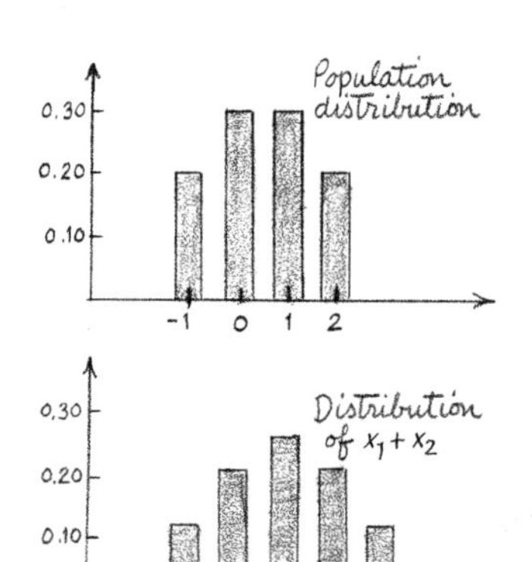

Figure 9.26

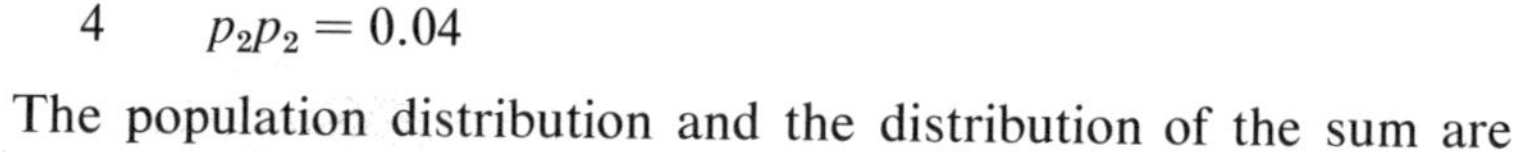
The population distribution and the distribution of the sum are compared in Figure 9.26. ▲

The method for the computation of the distribution of the sum of 2 observations can be extended to 3 and more. By generalizing the reasoning for (11), we find that $P(X_1+X_2+X_3=m)$ is the sum of the products $\{P(X_1+X_2=m-k)P(X_3=k)\}$ over all k. This is similar to (11) except that the sum of the 3 observations is considered to be the sum of the 2 independent variables, X_1+X_2 and X_3.

EXAMPLE 3

Let X_1, X_2, X_3 be a sample of 3 observations from the population with the distribution in Example 2 above. We have:

k	-1	0	1	2
$P(X_3=k)$	0.2	0.3	0.3	0.2

and

k	-2	-1	0	1	2	3	4
$P(X_1+X_2=k)$	0.04	0.12	0.21	0.26	0.21	0.12	0.04

We compute the distribution of $X_1+X_2+X_3$:

$$
\begin{aligned}
P(X_1+X_2+X_3=-3) &= P(X_1+X_2=-2)P(X_3=-1)=0.008\\
P(X_1+X_2+X_3=-2) &= P(X_1+X_2=-1)P(X_3=-1)\\
&\quad + P(X_1+X_2=-2)P(X_3=0)\\
&= 0.024+0.012=0.036;
\end{aligned}
$$

and the complete distribution is found to be

k	Probability	k	Probability
−3	0.008	2	0.207
−2	0.036	3	0.159
−1	0.090	4	0.090
0	0.159	5	0.036
1	0.207	6	0.008

The distributions of the averages of 2 and 3 observations are obtained from those of the corresponding sums by dividing each of the values of the sums by 2 and 3, respectively. These appear in Figure 9.27. Note that the distributions of the averages have the same mean as the population, but are more concentrated about the mean. ▲

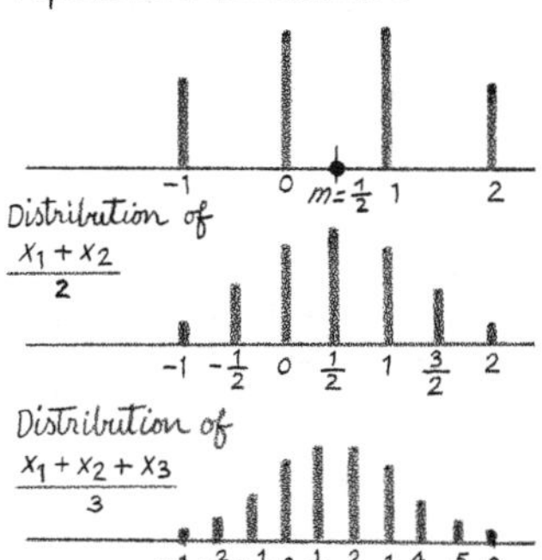

Figure 9.27

Next we consider the case where the population has a density function $f(x)$. **The density of the sum of 2 observations, denoted $h(t)$, is given by the formula**

$$h(t) = \int_{-\infty}^{\infty} f(t-x)f(x)\,dx. \tag{12}$$

This is the integral analog of formula (11): t and x are in place of m and k, respectively, the density function in the place of the discrete probability, and the integral in the place of the sum.

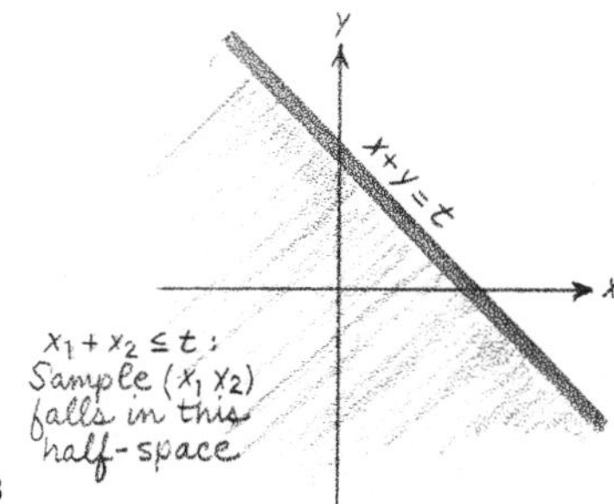

Figure 9.28

To explain (12) recall that the probability that the pair (X_1,X_2) falls in a given rectangle in the plane is the double integral of $f(x)f(y)$ over the rectangle (Section 9.3). As in Section 7.6, this is extended to any set in the plane with a smooth boundary. $X_1 + X_2$ is less than or equal to t when the random point (X_1,X_2) falls in the half-plane $x + y \le t$ (Figure 9.28). It follows that the distribution function $H(t)$ of the sum is the double integral over this half-plane:

$$\int_{-\infty}^{\infty}\int_{-\infty}^{t-x} f(y)f(x)\,dy\,dx,$$

or

$$\int_{-\infty}^{\infty}\left(\int_{-\infty}^{t-x} f(y)\,dy\right) f(x)\,dx = \int_{-\infty}^{\infty} F(t-x)f(x)\,dx.$$

The density (12) is then obtained by differentiation with respect to t (under the integral sign).

The density function of the sum of 3 observations is obtained from (12) by replacing one of the functions f in the integrand by h:

$$\int_{-\infty}^{\infty} f(t-x)h(x)\,dx \qquad \text{or} \qquad \int_{-\infty}^{\infty} h(t-x)f(x)\,dx. \tag{13}$$

This can be extended to sums of more than 3 observations. The reason for (13) is analogous to that in the discrete case.

If the numerical characteristic of the population is nonnegative, then $f(x) = 0$ for $x < 0$. Then the first factor in the integrand in (12) is equal to 0 for $x > t$ and the second factor is 0 for $x < 0$; thus, the integral (12) becomes

$$\begin{aligned} h(t) &= 0, && \text{for } x \leq 0 \\ &= \int_0^t f(t-x)f(x)\,dx, && \text{for } x > 0. \end{aligned} \tag{14}$$

EXAMPLE 4

If the population has the exponential density $f(x) = e^{-x}$ for $x > 0$, and 0 for $x \leq 0$, then the density of the sum of 2 observations is, by (14),

$$\begin{aligned} h(t) &= \int_0^t e^{-(t-x)}\, e^{-x}\,dx && \text{for } t > 0 \\ &= 0, && \text{for } t \leq 0. \end{aligned}$$

The integral above is computed as follows: The factor e^{-t} is a constant because it does not depend on x; hence, it may be factored out of the integrand. The factors e^{x} and e^{-x} have the product 1. Therefore, the integral becomes

$$e^{-t}\int_0^t 1\,dx = te^{-t}, \qquad \text{for } t > 0.$$

By (13) the density function for the sum of 3 observations is

$$\begin{aligned} &\int_0^t e^{-(t-x)}\, xe^{-x}\,dx, && \text{for } t > 0 \\ &0, && \text{for } t \leq 0. \end{aligned}$$

As in the previous paragraph, the integrand may be factored as the product $e^{-t}x$; the first factor may be taken out of the integral; and the remaining portion of the integrand is integrated to get

$$e^{-t}\int_0^t x\,dx = \tfrac{1}{2}t^2e^{-t}, \qquad \text{for } t > 0. \ \blacktriangle$$

EXAMPLE 5

The density of the sum of 2 observations from a standard normal population is, by (12),

$$\left(\frac{1}{2\pi}\right)\int_{-\infty}^{\infty} e^{-(1/2)(t-x)^2-(1/2)x^2}\,dx \qquad \text{or} \qquad \left(\frac{1}{2\pi}\right)e^{-(1/2)t^2}\int_{-\infty}^{\infty} e^{-x^2+tx}\,dx.$$

The latter integral is identified by means of the (already established) fact that the area under the normal density is 1. Complete the square in the exponent in the integrand (see Section 1.3):

$$-x^2 + tx = -(x - \tfrac{1}{2}t)^2 + \tfrac{1}{4}t^2.$$

Then the expression for the density may be written as

$$\frac{1}{2\pi}\,e^{-(1/4)t^2}\int_{-\infty}^{\infty} e^{-(x-t/2)^2}\,dx$$

(because $e^{-(1/4)t^2}$ may be factored out of the integrand). Now the expression above may be written as

$$\frac{1}{\sqrt{2\pi}\;\sqrt{2}}\,e^{-(1/2)(1/2)t^2}\,\frac{\sqrt{2}}{\sqrt{2\pi}}\int_{-\infty}^{\infty} e^{-(1/2)(x-t/2)^2/(1/\sqrt{2})^2}\,dx.$$

Using the ϕ-symbol for the standard normal density (Section 8.9), we may write the latter expression as

$$\left\{\frac{1}{\sqrt{2}}\,\phi\left(\frac{t}{\sqrt{2}}\right)\right\}\left\{\sqrt{2}\int_{-\infty}^{\infty}\phi\left(\frac{x-\frac{1}{2}t}{1/\sqrt{2}}\right)dx\right\}.$$

The first factor is the normal density with mean 0 and standard deviation $\sqrt{2}$; and the second factor is equal to 1 because it represents the area under the normal density with mean $\frac{1}{2}t$ and standard deviation $1/\sqrt{2}$. Therefore, the density of the sum is normal with mean 0 and standard deviation $\sqrt{2}$.

By the method above, and the successive application of (13), it can be shown that the sum S_n of n observations from a normal population with mean μ and standard deviation σ itself has a normal density with mean $n\mu$ and standard deviation $\sqrt{n}\,\sigma$. This is stronger than the results (2) and (3) above because it asserts that the density itself is a normal one. It also follows that the average $\bar{X}$ of n observations has a normal density, but with mean μ and standard deviation $\sigma/\sqrt{n}$ (Figure 9.29). (See Exercise 11.) ▲

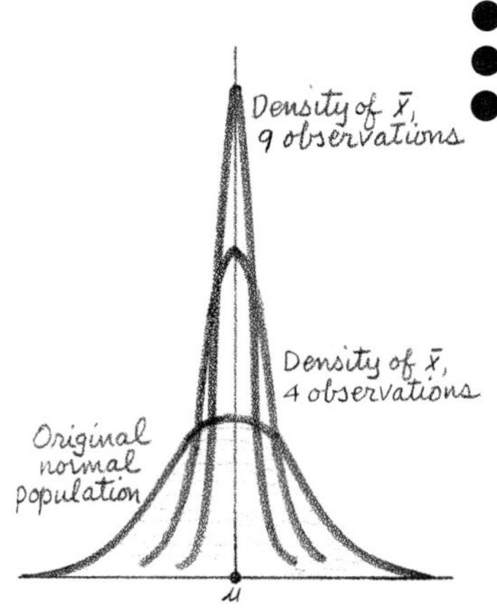

Figure 9.29

Finally we discuss the case where the density is positive only on an interval from a to b, of finite length. Then the integrand in (12) is equal to 0 if x is outside the interval so that (12) becomes

$$h(t) = \int_a^b f(t-x)f(x)dx.$$

The integrand is also equal to 0 if $t - x$ is outside the interval, that is, if $x < t - b$ or $x > t - a$.

EXAMPLE 6

Let $f(x)$ be equal to 0 outside the interval from 0 to 1. If $0 < t < 1$, then

$$h(t) = \int_0^t f(t-x)f(x)dx.$$

If $1 \le t < 2$, then

$$h(t) = \int_{t-1}^1 f(t-x)f(x)dx. \text{ ▲}$$

9.4 EXERCISES

In Exercises 1 and 2 find the distribution of the sum of 2 observations from the given discrete population.

1 (a)

x	-1	1	2	3
$P(X=x)$	0.2	0.3	0.4	0.1

(b)

x	0	1	2	3	4
$P(X=x)$	0.1	0.2	0.3	0.3	0.1

2 (a)

x	0	1	2	3
$P(X=x)$	0.3	0.2	0.4	0.1

(b)

x	-2	-1	0	1	2
$P(X=x)$	0.1	0.2	0.4	0.2	0.1

In Exercises 3 and 4 find $E(X_1 + X_2)$ *and* $Var(X_1 + X_2)$ *in sampling from the given population. Use two methods: Compute the distribution of the sum, and then its expected value and variance; and then derive it from* $E(X)$ *and* $Var(X)$ *by means of formulas (2) and (3).*

3 (a)

x	1	2	3
$P(X=x)$	0.3	0.4	0.3

(b)

x	-1	1	4
$P(X=x)$	0.3	0.2	0.5

4 (a)

x	-1	2	5
$P(X=x)$	0.3	0.3	0.4

(b)

x	3	4	5
$P(X=x)$	0.5	0.2	0.3

In Exercises 5 and 6 find the distribution of the sum of 3 observations from the given population.

5 (a)

x	-1	1
$P(X=x)$	0.5	0.5

(b)

x	-1	0	1
$P(X=x)$	0.2	0.5	0.3

(c)

x	0	1	2
$P(X=x)$	$\frac{1}{3}$	$\frac{1}{3}$	$\frac{1}{3}$

6 (a)

x	0	2
$P(X=x)$	0.4	0.6

(b)

x	-2	0	2
$P(X=x)$	0.3	0.4	0.3

(c)

x	0	2	4
$P(X=x)$	0.4	0.2	0.4

7 Find the density function of the sum of (a) 2, (b) 3 observations in sampling from a uniform population on the unit interval. The graphs appear in Figure 9.30. (See Example 6.)

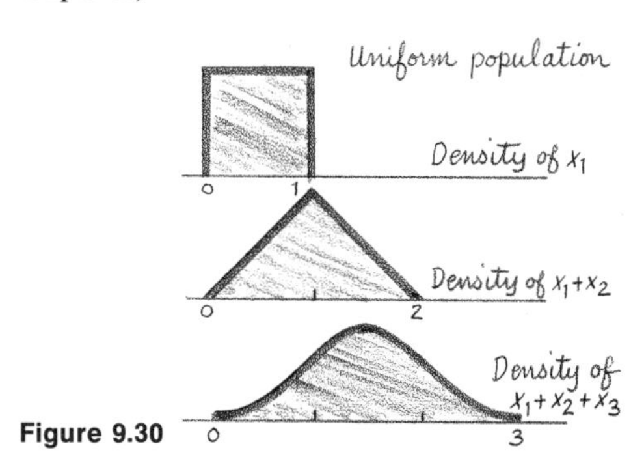

Figure 9.30

8 If the random variable X has the distribution function $F(x)$, then the scaled variable X/c has the distribution function $F(xc)$, for $c > 0$. Prove this directly from the definition. Deduce the density function of X/c. Employing this result and the densities obtained in Exercise 7, derive the densities of the averages of the observations from the uniform population (Figure 9.31).

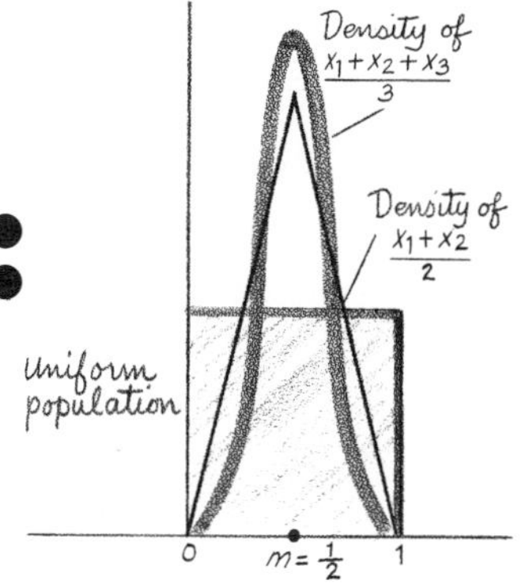

Figure 9.31

In Exercises 9 and 10 find the density of the sum of 2 observations in sampling from the population with the given density.

9 (a) $f(x) = 2x$, for $0 < x < 1$; $= 0$, elsewhere

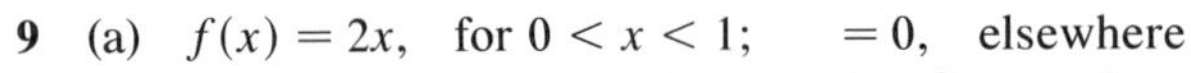

(b) $f(x) = xe^{-x}$, for $x > 0$; $= 0$, for $x \leq 0$

(c) $f(x) =$ normal density with mean 1, standard deviation 3

10 (a) $f(x) = \frac{1}{2}$, for $-1 < x < 1$; $= 0$, elsewhere
(b) $f(x) = \frac{1}{2}x^2e^{-x}$, for $x > 0$; $= 0$, elsewhere
(c) $f(x) =$ normal density with mean -2, standard deviation 2

11 (a) *Show:* If X has a normal distribution with mean μ and standard deviation σ, then the random variable $Y = cX + d$ has the normal distribution with mean $c\mu + d$ and standard deviation $c^2\sigma^2$, where $c \neq 0$. [*Hint:* Compare $P(X \leq x)$ and $P(Y \leq x)$.]
(b) Why does it follow that the average $\bar{X}$ of observations from a normal population has the mean μ and standard deviation $\sigma/\sqrt{n}$? (See Example 5.)
(c) How does the result of this exercise differ from formulas (8) and (9) of Section 9.2?

12 Let X_1, X_2, X_3 be a sample of 3 observations from a binomial population with $n = 1$: the proportion p of the population has the value 1 and the proportion q has the value 0. Using the method of this section, verify that the sums of 2 and 3 observations have the binomial distributions with $n = 2$ and 3, respectively.

13 By means of formulas (2) and (3) derive the expressions for the expected value and variance of a random variable having the binomial distribution (see Section 9.2, Example 3). [*Hint:* Write the random variable as the sum of the numbers of successes (0 or 1) on the n trials.]

14 Compute the expected value and variance of the sum of the scores on the toss of a pair of dice by applying formulas (2) and (3) to the expected value and variance of a single toss. (See Section 9.2, Exercise 7.)

15 Let X_1 and X_2 be observations from a Poisson distribution with parameter λ. (a) Derive the formula for $P(X_1 + X_2 = k)$ for $k = 0, 1, 2, 3$. (b) Can you guess the complete distribution?

16 Lex X_1 and X_2 be observations from a population with the geometric distribution pq^x, $x = 0, 1, \ldots$. Find $P(X_1 + X_2 = m)$ for $m = 0, 1, \ldots, 4$.

17 Let $\bar{X}$ be the average of 12 observations from a normal population with standard deviation 2. Using the Chebyshev inequality, find a bound on the probability that $\bar{X}$ differs from the mean of the population by at least 1.5 units. Then compute the exact probability using the distribution of $\bar{X}$.

18 Let X be a random variable with a standard normal density; and put $Y = X^2$. Then Y is a nonnegative random variable.
(a) Find the density function of Y. [*Hint:* The distribution function can be obtained from the relation $P(Y \leq x) = P(-\sqrt{x} \leq X \leq \sqrt{x})$; then get the density by differentiation.]
(b) Write (but do not evaluate) the integral for the density function of the sum of 2 observations from a population with the same density as Y.

19 Let $h(x_1,x_2)$ be a function of the form $g_1(x_1)g_2(x_2)$. *Show:* If X_1 and X_2 are observations from a population, then $Eh(X_1,X_2) = Eg_1(X_1)\, Eg_2(X_2)$. Give the proof in the case the population has a density function $f(x)$.

20 Let X_1 and X_2 be observations from a population with the density f. By expanding the square and integrating term-by-term, show that

$$E(X_1 + X_2 - E(X_1) - E(X_2))^2 = E(X_1 - E(X_1))^2 + E(X_2 - E(X_2))^2 + 2E[X_1 - E(X_1)][X_2 - E(X_2)].$$

(a) Applying the result of Exercise 19, show that the last term above is equal to 0.

(b) Deduce from the equation above and from the result of (a) that $\text{Var}(X_1 + X_2) = \text{Var}(X_1) + \text{Var}(X_2)$.

21 **(Sample distribution function.)** The following procedure is used to estimate the population distribution function $F(x)$ on the basis of a sample of n observations. For a given x, let $F_n(x)$ be the proportion of the n observations which are less than or equal to x: $F_n(x)$ is a step-function with a jump of magnitude $1/n$ at each value $X_1, \ldots, X_n$ observed on the x-axis (Figure 9.32).

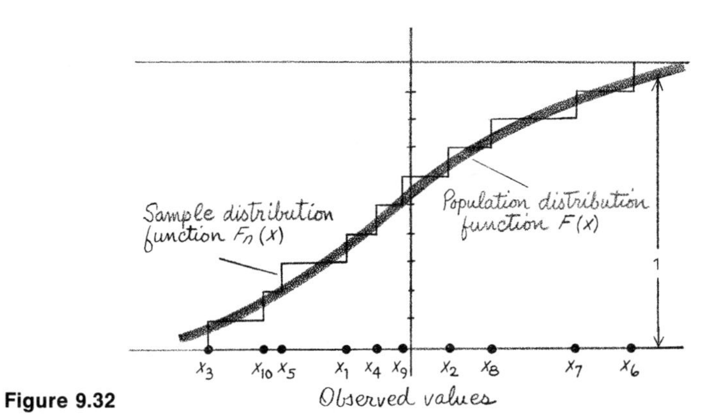

Figure 9.32

(a) Find the expected value and variance of $F_n(x)$ for a fixed x. [*Hint:* For each x, the outcomes of the observations $X_1, \ldots, X_n$ are **successes** or **failures** depending on whether the observation is $\leq x$ or $> x$. The probability of success is $F(x)$, and failure $1 - F(x)$.]

(b) Why is F_n a "good" estimator if n is large?

9.5 Confidence Intervals and Hypothesis Testing

As we have indicated several times, the normal density appears in many populations; therefore, many of the classical statistical procedures were designed for normal populations. Modern statistical theory is not restricted to normal populations, and has developed suitable procedures for general populations; however, in this brief introduction, we will limit ourselves to the special normal case.

At the end of Section 9.4 we observed that if $\bar{X}$ is the average of n observations from a normal population with mean μ and standard deviation σ, then $\bar{X}$ has a normal density with the same mean, but with standard deviation $\sigma/\sqrt{n}$.

The first procedure we consider is that of **confidence intervals.** Suppose that we have a normal population where σ is known but μ is not, and we wish to estimate the latter. The maximum likelihood estimate of μ is $\bar{X}$ (Section 9.3). How **reliable** is this estimate? We may ask: How likely is $\bar{X}$ to be within a given distance of the unknown parameter μ? For a given real number α, $0 < \alpha < 1$, a **confidence interval of coefficient** $1 - \alpha$ is an interval based on the sample which, with probability $1 - \alpha$, contains the unknown parameter value μ. It is constructed in this way. For the given α, let z_α be the point on the x-axis such that the area under the standard normal curve outside the interval from $-z_\alpha$ to z_α is equal to α. (See Section 8.10.) **Then the confidence interval for μ is the**

$$\textbf{Interval from } \bar{X} - \frac{z_\alpha \sigma}{\sqrt{n}} \textbf{ to } \bar{X} + \frac{z_\alpha \sigma}{\sqrt{n}}. \tag{1}$$

This interval contains μ with probability $1 - \alpha$; indeed, the probability that $\bar{X}$ differs in absolute value from μ by $z_\alpha\sigma/\sqrt{n}$ units is, by the definition of z_α and the distribution of $\bar{X}$, equal to

$$P\left(-z_\alpha < \frac{\bar{X} - \mu}{\sigma/\sqrt{n}} < z_\alpha\right) \quad \text{or} \quad 1 - \alpha.$$

Note that the length of the interval, $2z_\alpha\sigma/\sqrt{n}$, is directly proportional to σ and inversely proportional to the square root of n. In particular, it follows that the confidence interval becomes smaller as the size of the sample increases.

The coefficients of the interval in common use are $1 - \alpha = 0.95$, 0.98, and 0.99. The corresponding values z are (Table VIII):

α	0.05	0.02	0.01
z_α	1.96	2.33	2.58

(2)

EXAMPLE 1

Suppose we have a normal population with standard deviation 2, and that the average $\bar{X}$ of a sample of 10 observations is 3.1. Here $\sqrt{n} = 3.16228$. A confidence interval of coefficient 0.95 is, by (1), the interval from

$$3.1 - \frac{1.96(2)}{3.16228} \quad \text{to} \quad 3.1 + \frac{1.96(2)}{3.16228},$$

$$\text{or} \quad 3.1 - 1.2 \quad \text{to} \quad 3.1 + 1.2,$$

or the interval from 1.9 to 4.3.

If the sample size is increased to 25, the confidence interval is smaller: it is the interval from

$$3.1 - \frac{1.96(2)}{5} \qquad \text{to} \qquad 3.1 + \frac{1.96(2)}{5},$$

or from $3.1 - 0.8$ to $3.1 + 0.8$; that is, from 2.3 to 3.9.

It is customary to write the confidence intervals as

$$3.1 \pm 1.2 \qquad \text{and} \qquad 3.1 \pm 0.8,$$

respectively. ▲

If both μ and σ are unknown, then the maximum likelihood estimate of the squared standard deviation, σ^2, is the average sum of squares of the deviations of the sample values from $\bar{X}$:

$$\frac{1}{n}\left[(X_1 - \bar{X})^2 + \cdots + (X_n - \bar{X})^2\right]$$

(Section 9.3, Exercise 7). In practice, the modified estimate

$$s^2 = \frac{1}{n-1}\left[(X_1 - \bar{X})^2 + \cdots + (X_n - \bar{X})^2\right],$$

with division by $n - 1$ in place of n, is used; s is the **sample standard deviation.** The ratio

$$\sqrt{n}\frac{(\bar{X} - \mu)}{\sigma}$$

has the standard normal distribution (see Section 9.4 Exercise 11); but the ratio obtained by replacing σ by s has a distribution known as the **"t-distribution with $n - 1$ degrees of freedom."** Like the normal, it is symmetric and bell-shaped. Here the confidence interval is based on the t-distribution: σ is replaced by its estimate s, and z_α by a corresponding value from the t-density. (There are tables of this distribution in many texts on statistics.) But when n is large, for example, 30 or more, the t-density and the normal are very close to each other; therefore, the confidence interval (1) may be used with the simple modification that the sample estimate s be substituted for the unknown σ:

$$\bar{X} \pm \frac{z_\alpha s}{\sqrt{n}}. \tag{3}$$

The length of this interval is proportional to s, which varies with the values of the sample; however, it can be shown that for large n the distribution of s has a small variance, and so the values of s do not vary much with the sample.

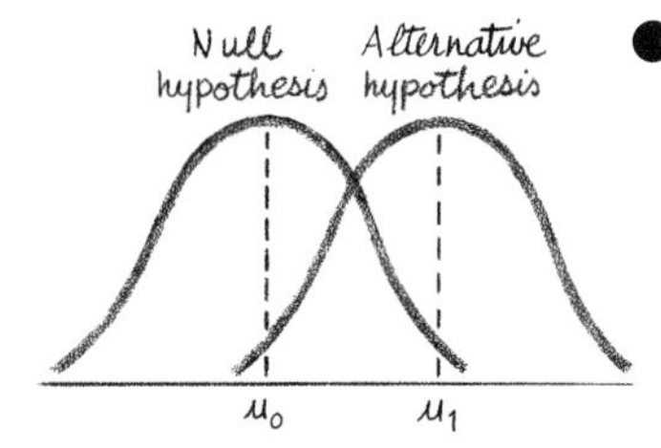

Figure 9.33

Next we consider **hypothesis testing.** Suppose again that we have a normal population with known standard deviation σ, but with two hypothetical mean values, μ_0 and μ_1; however, we do not know which of the two it is. This is a statistical problem of testing the **null hypothesis** μ_0 against the **alternative hypothesis** μ_1. For definiteness, let us suppose that $\mu_0 < \mu_1$ (Figure 9.33). The usual procedure for testing is as follows. Pick a number c between the two hypothetical values of the mean. Take a sample of n observations from the population: If $\bar{X}$ falls below c, we accept the hypothesis μ_0 (reject μ_1); and if $\bar{X}$ falls above c, we accept μ_1 (reject μ_0). The number c is the **critical point** for the test. (We have not mentioned the possibility that $\bar{X} = c$ because the probability is 0 under either hypothesis.) A **Type I Error** is made when the alternative hypothesis is mistakenly accepted, that is, when the null hypothesis is true. A **Type II Error** is the mistaken acceptance of the null hypothesis. The probability of a Type I Error, denoted α, is the probability that $\bar{X}$ exceeds c when the mean is μ_0: since $\bar{X}$ has a normal distribution with mean μ_0 and standard deviation $\sigma/\sqrt{n}$, this probability is

$$\alpha = 1 - \Phi\left(\frac{c - \mu_0}{\sigma/\sqrt{n}}\right). \tag{4}$$

The probability of a Type II Error, denoted β, is the probability that $\bar{X}$ is less than c when μ_1 is the mean:

$$\beta = \Phi\left(\frac{c - \mu_1}{\sigma/\sqrt{n}}\right). \tag{5}$$

The density of $\bar{X}$ has a much smaller variance than that of the underlying population. α is the area to the right of c under the density of $\bar{X}$ for μ_0, and β is the area to the left of c under the density for μ_1 (Figure 9.34).

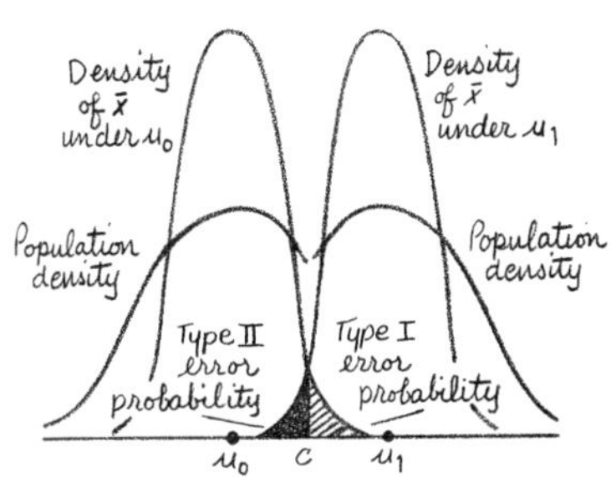

Figure 9.34

EXAMPLE 2

Suppose that the number of boxes of breakfast cereal (in thousands) sold by a manufacturer in a particular area during 1 week is a random variable with a normal density with mean 6 and standard deviation 0.5. The manufacturer wants to increase the mean by putting an offer of stationery on each box (in exchange for cash and boxtops). He hopes that the mean will rise to 7, and leave the standard deviation unchanged at 0.5 (Figure 9.35). After observing 9 weeks of sales, he decides to continue the offer or not accordingly as the mean of the density of sales is 7 or 6, respectively. In other words, he has to test the null hypothesis that the mean is 6 against the alternative 7. His plan is to use 6.6 as the critical point for the test. The null hypothesis is to be accepted if $\bar{X} < 6.6$, and rejected if $\bar{X} > 6.6$. The densities of $\bar{X}$ under the two hypotheses are drawn in Figure 9.36. The Type I Error probability α is, by (4),

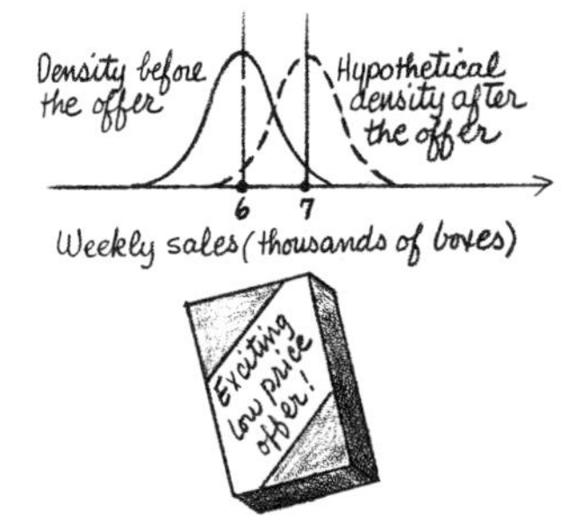

Figure 9.35

$$1 - \Phi\left(\frac{6.6 - 6}{(0.5)/3}\right) = 1 - \Phi(3.6) = 0.0002.$$

Similarly, the Type II Error probability is

$$\Phi\left(\frac{6.6 - 7}{(0.5)/3}\right) = \Phi(-2.4) = 0.0082. \blacktriangle$$

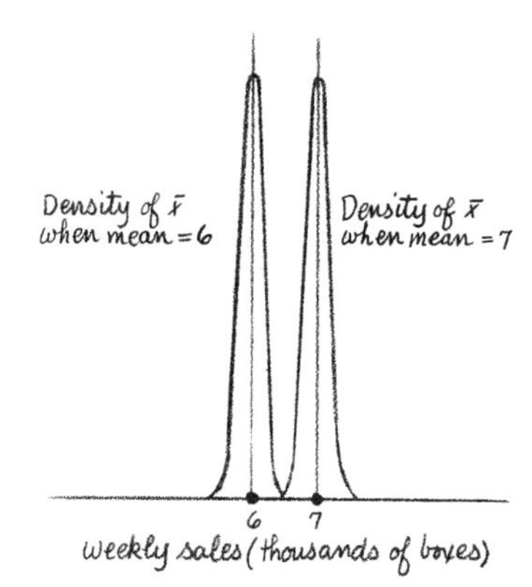

Figure 9.36

In practical problems the alternative hypothesis consists not of a single value μ_1 but of a set of μ-values. The alternative hypothesis is then called **composite.** By contrast, the null hypothesis, consisting of the single value μ_0, is called **simple.** The decision to accept a composite alternative hypothesis is to conclude that the true mean μ is a member of the set. There are three usual types of composite alternative hypotheses against the simple null hypothesis μ_0:

(i) A 1-sided alternative consisting of all μ such that $\mu > \mu_0$ (Figure 9.37). The test procedure is to reject μ_0 (accept the alternative) if $\bar{X}$ exceeds a critical point c.

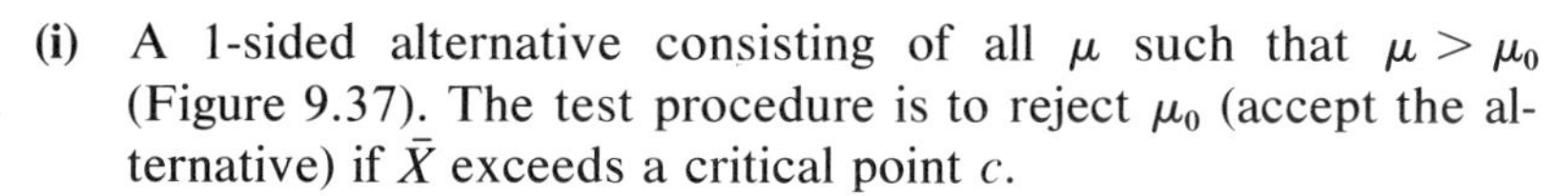

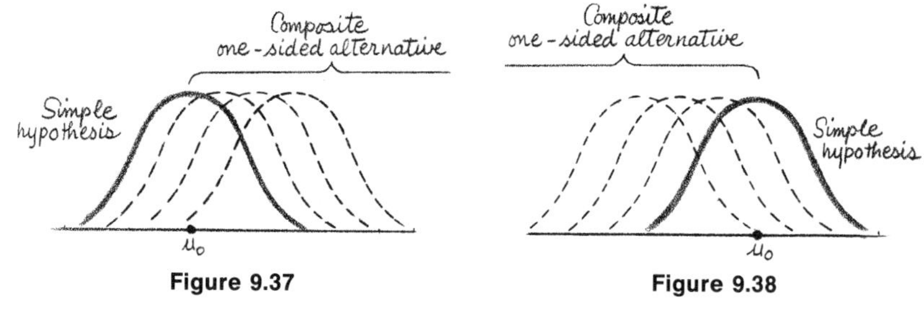

Figure 9.37

Figure 9.38

(ii) A 1-sided composite alternative $\mu < \mu_0$ (Figure 9.38). The procedure is to reject μ_0 if $\bar{X} < c$.

(iii) A 2-sided composite alternative $\mu \neq \mu_0$ (Figure 9.39). The test is to reject μ_0 if the absolute difference $|\bar{X} - \mu_0|$ exceeds a critical value c.

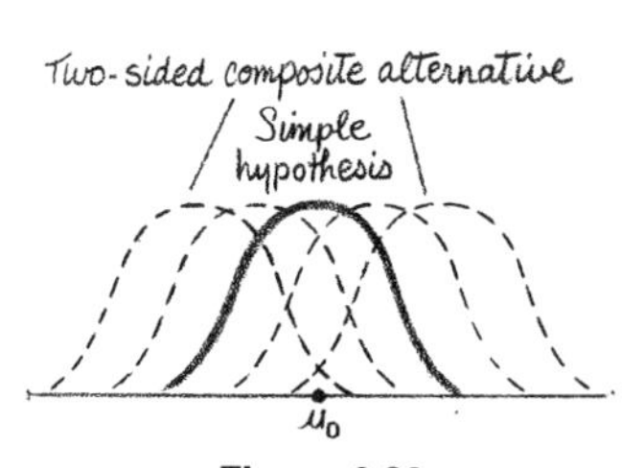

Figure 9.39

The conventional procedure for choosing the number c is as follows: An "acceptable" Type I Error probability is chosen. The usual choice is $\alpha = 0.05$; another is $\alpha = 0.01$. This is called the **level of significance** of the test. For the alternative (i), the hypothesis μ_0 is mistakenly rejected with probability

$$P(\bar{X} > c) = 1 - \Phi\left(\frac{c - \mu_0}{\sigma/\sqrt{n}}\right).$$

By the definition of z_α and the symmetry of the normal density, the area under the curve to the right of $z_{2\alpha}$ is equal to α. If c is chosen to be

$$c = \mu_0 + \frac{z_{2\alpha}\sigma}{\sqrt{n}}, \tag{6}$$

then the probability of mistaken rejection of μ_0 is

$$1 - \Phi(z_{2\alpha}) = \alpha.$$

EXAMPLE 3

Suppose that SAT scores in mathematics have a normal density with mean $\mu_0 = 525$ and standard deviation 75. An educator wants to test whether the students from a certain group of high schools are better on the average than those from all high schools. In terms of a statistical test, he wants to test the simple null hypothesis that the mean of the distribution of this group is 525 against the alternative hypothesis that the mean is larger than 525. (He assumes that the standard deviation is the same in both cases.) He picks $\alpha = 0.05$, and computes the average of 100 observations (test scores in mathematics) from this population. From Table VIII we have $z_{2\alpha} = 1.64$. By (6), the hypothesis $\mu_0 = 525$ is rejected if $\bar{X} > 525 + (1.64)(75)/10 = 537.3$.

For the composite alternative (ii), c is chosen as

$$\mu_0 - \frac{z_{2\alpha}\sigma}{\sqrt{n}}. \tag{7}$$

Finally, for the composite alternative (iii), c is

$$\frac{z_\alpha\sigma}{\sqrt{n}}. \tag{8}$$

Note that the test under (iii) is equivalent to the procedure: accept or reject μ_0 accordingly as the confidence interval of coefficient $1 - \alpha$ contains μ_0 or not. ▲

9.5 EXERCISES

In Exercises 1 and 2 find confidence intervals for the given coefficient, standard deviation, observed $\bar{X}$ *and sample size* n.

		$1-\alpha$	$\bar{X}$	σ	n
1	(a)	0.95	1.2	1	10
	(b)	0.95	0.2	2	16
	(c)	0.98	−1.3	2	20
	(d)	0.99	8.3	4	25
2	(a)	0.95	4.3	1	12
	(b)	0.95	−2.7	3	16
	(c)	0.98	10.0	2	20
	(d)	0.99	−0.3	3	36

In Exercises 3 and 4 find α *and* β *for each test of the given simple null hypothesis against the given alternative.*

		μ_0	μ_1	σ	n	c
3	(a)	0	1	1	10	0.5
	(b)	−1	1	2	16	−0.2
	(c)	3	5	4	25	3.9
	(d)	2	5	3	36	4.0
4	(a)	−1	0	1	10	−0.4
	(b)	0	2	3	16	1.3
	(c)	−2	1	2	25	0.0
	(d)	2	6	4	36	4.3

5 In sampling from a normal population with standard deviation 1, what is the minimum number of observations n sufficient for the confidence interval to be of length at most 0.4 when the coefficient $1-\alpha$ is (a) 0.95; (b) 0.99?

6 The diastolic blood pressure of American women in the age group 25–34 has a normal density with mean approximately 75 and standard deviation approximately 10. (See Section 7.5, Example 8.) An investigator has a hunch that women who have never been pregnant have, on the average, a lower blood pressure than the general population of women in the group. He tests the hypothesis $\mu_0 = 75$ (for women never pregnant) against the alternative $\mu < 75$ at the level of significance 0.01. The sample consists of 100 women. Find the critical point c for the test. What is the probability β if the mean is really 70?

7 In a sample of 100 overdue accounts, a department store finds that the average amount owed is \$20, and that the sample standard deviation is \$5.00. Find a confidence interval of coefficient 0.98 for the average of all overdue accounts. (Assume the density to be normal.)

8 The arrival times of students at a 9:00 A.M. class are measured in minutes deviations from 9 o'clock and have a normal density. For **punctual** students $\mu = 0$ and $\sigma = 2$; for **tardy** students, $\mu > 0$ and $\sigma = 2$. The mathematics professor noticed that a blonde in the first row arrived at the following times during the first 5 lectures of Math 1:

$$8{:}59, \quad 8{:}58, \quad 9{:}01, \quad 9{:}05, \quad 9{:}01.$$

At the level of significance 0.05, is she punctual or tardy?

9 A nutritionist finds that the average number of eggs consumed by 50 randomly selected persons over a 30-day period is $\bar{X} = 28.5$, with a sample standard deviation $s = 2.1$. She assumes that the distribution of eggs consumed in the overall population is normal, and wishes to test the "1-egg-a-day" hypothesis ($\mu_0 = 30$) against the composite alternative $\mu_0 \neq 30$. Is the null hypothesis accepted or rejected at the 0.05 level?

10 Suppose percentage monthly increase in the cost of living in the United States is a random variable from a normal population with a standard deviation of 0.05 percent. The increases (seasonally adjusted) over a 12-month period are considered to be observations from this population. Find the length of the confidence interval for the mean of the population if the confidence coefficient is 0.98.

11 The amount of lead in a certain brand of interior yellow paint is supposed to be about 1 part per hundred. In a sample of 40 cans it is found that the average lead content is 1.1 parts per hundred, and that the sample standard deviation is 0.1. At the level of significance 0.05, is this average sufficiently high to conclude that the mean is really higher than 1 part per hundred?

12 The management of a chain store will operate a branch as long as it maintains an average of 500 customers each day. A particular branch has appeared to be declining. The management decides to use a statistical test on the mean number of customers, and shut the branch if the mean is less than 500. The level of significance is set at 0.01. On 50 randomly selected shopping days, the average number of customers was found to be 496, with a sample standard deviation of 7. Should the store be shut?

13 The monthly number of riders on the transit system of a large city has a normal density with mean 50 (millions) and standard deviation 3. After a fare increase, the average number of riders for the following 6 months was 48.8 million. The standard deviation is assumed to be unchanged. Is this convincing evidence (at the level of significance 0.05) that the number of riders has really decreased as a result of the fare increase?

14 An obscure politician served as Legislator in the State government for many years. The amount of money donated by his supporters to his campaigns for reelection had, over the years, a normal density with mean 25 ($1000's) and standard deviation 3. For the first 4 campaigns after taking a controversial position, the average donations per campaign rose to $30,000. Is he justified, at the level of significance 0.05, in his belief that his financial support has really increased during these 4 campaigns?

15 Verify that the critical point (7) yields a test of significance level α for the alternative $\mu < \mu_0$; and that the same for (8) and the alternative $\mu \neq \mu_0$.

Answers to Odd-Numbered Exercises

SECTION 1.1

1 (a) $2\sqrt{10}$ (b) 4 (c) $\sqrt{13}$ (d) $2\sqrt{5}$ (e) $\sqrt{58}$ (f) $10\sqrt{2}$

3 (a) Two sides equal to $\sqrt{50}$ (b) Two sides equal to $\sqrt{13}$
(c) Two sides equal to $\sqrt{65}$

5 (a) Squares of sides: 116, 29, 145 (b) Squares of sides: 65, 20, 45
(c) Squares of sides: 50, 45, 5

7 (a) $-\frac{2}{3}$ (b) $-\frac{1}{11}$ (c) $\frac{3}{2}$ (d) $\frac{2}{7}$ (e) $-\frac{12}{13}$

9 Slopes of sides:
(a) $-3, 2, -\frac{1}{2}$ (b) $1, -3, \frac{1}{3}$ (c) $2, -\frac{1}{2}, -3$ (d) $\frac{2}{5}, 12, -\frac{5}{2}$
(e) $-\frac{2}{5}, \frac{8}{9}, \frac{5}{2}$

11 (a) Slopes of lines joining the consecutive points: $-\frac{1}{2}, 2, -\frac{1}{2}, 2$
(b) Slopes $2, -4, 2, -4$ (c) Slopes $3, -\frac{1}{3}, 3, -\frac{1}{3}$ (d) Slopes $\frac{1}{4}, -3, \frac{1}{4}, -3$

13 (a) By definition of slope. (b) Apply the Pythagorean Theorem: Since the two smaller triangles are given as right triangles, $D^2 = A^2 + B^2$ and $E^2 = B^2 + C^2$. It follows that the large triangle has a right angle if and only if $(A + C)^2 = D^2 + E^2$.
(c) From (a): $-C/B = -B/A$ if and only if $AC = B^2$. (d) From (b) we see that there is perpendicularity if and only if $A^2 + 2AC + C^2 = A^2 + 2B^2 + C^2$; that is, if and only if $AC = B^2$.

SECTION 1.2

1 (a) $y + 4 = 5(x - 2)$ (b) $y = 3$ (c) $y - 3 = 2(x - 1)$
(d) $y = -\frac{1}{4}(x - 3)$ (e) $y = \frac{2}{3}x$

3 (a) $y = 2$ (b) $x = 5$ (c) $4x - 17y = 0$ (d) $3x - 4y = 12$
(e) $6x - 5y = -29$

5 (a) $m = -4, \quad b = 4$ (b) $m = -3, \quad b = 5$ (c) $m = 2, \quad b = 3$
(d) $m = \frac{1}{2}, \quad b = -\frac{1}{2}$ (e) $m = 7/5, \quad b = 2/5$

7 (a) $(-20,41)$ (b) $(-28/11,7/11)$ (c) $(8/3,4/3)$ (d) $(25/27,2/9)$

9 $y - 3 = -\frac{3}{2}(x + 2)$

11 $(-4,7)$

13 $4\sqrt{2}$

15 $y = 55{,}490 + 0.7(x - 100{,}000)$

SECTION 1.3

1 (a) $x^2 + y^2 = 4$ (b) $(x - 2)^2 + (y + 5)^2 = 9$ (c) $(x + 2)^2 + y^2 = 4$
(d) $(x + 2)^2 + (y + 4)^2 = 16$ (e) $(x - 4)^2 + (y + 3)^2 = 1$

3 (a) center $(-1,2)$, radius 3 (b) center $(2,3)$, radius 2
(c) center $(3,2)$, radius 4 (d) center $(-4,-2)$, radius 3 (e) not circle

5 (a) $x^2 + (y - 3)^2 = 9$ (b) $x^2 + y^2 = 1$ (c) $(x - 2)^2 + (y - 3)^2 = 17$
(d) $(x - 2)^2 + (y - 1)^2 = 13$

7 (a) 2 (b) 1 (c) $3/\sqrt{2}$ (d) 5

9 (a) $y - 4 = \frac{1}{2}(x - 1)$ (b) $y - 1 = x + 3$ (c) $y - 2 = \frac{1}{3}(x - 2)$
(d) $y - 4 = -\frac{5}{8}(x - 3)$

11 (a) (h,k) is equidistant from the two given points
(b) linear equation $6h + 4k = 11$ (c) (h,k) is on the line $x - 3y = 11$
(d) $h = \frac{7}{2}, \quad k = -\frac{5}{2}$ (e) $x^2 - 7x + y^2 + 5y = 14$

SECTION 1.4

3 (a) focus $(0,5)$, directrix $y = -5$ (b) focus $(0,-4)$, directrix $y = 4$
(c) focus $(-1,0)$, directrix $x = 1$ (d) focus $(-8,10)$, directrix $y = 8$
(e) focus $(0,7)$, directrix $y = 9$ (f) focus $(10,4)$, directrix $x = 2$

5 (a) $y^2 = 12x$ (b) $y = -20x^2$ (c) $y + 1 = 8(x + 3)^2$
(d) $y - 1 = (x - 5)^2$ (e) $y - 2 = -(x - 1)^2$

7 (a) focus $(-\frac{1}{3},-\frac{7}{2})$, directrix $y = -\frac{23}{6}$ (b) focus $(0,6)$, directrix $y = 4$
(c) focus $(-\frac{3}{2},-\frac{17}{8})$, directrix $y = -\frac{9}{8}$ (d) focus $(-\frac{3}{2},\frac{3}{2})$, directrix $y = \frac{1}{2}$

9 The focus is 322.5 ft above the low point of the cable. The second point is $11\frac{7}{25}$ ft above the road.

11 51.2 ft

SECTION 1.5

1 (a) center (0,0), semiaxes 11, 7 (b) center (0,0), semiaxes 2, 3
(c) center (0,4), semiaxes 6, 7 (d) center (3,–4), semiaxes 5, 4
(e) center (1,2), semiaxes 1, $\frac{1}{2}$

3 (a) $x^2/16 + y^2/12 = 1$ (b) $x^2/4 + (y-2)^2/8 = 1$ (c) $4x^2/9 + y^2/16 = 1$
(d) $(x-1)^2/9 + (y+1)^2/5 = 1$ (e) $(x-2)^2/5 + (y-1)^2/9 = 1$

5 (a) $\dfrac{2x^2}{3+\sqrt{5}} + \dfrac{2y^2}{1+\sqrt{5}} = 1$ (b) $(x-4)^2/18 + (y+1)^2/9 = 1$
(c) $\dfrac{2x^2}{9+3\sqrt{13}} + \dfrac{2y^2}{11+3\sqrt{13}} = 1$

7 (a) center (0,0), horizontal major axis 8, minor axis 6
(b) center (1,0), horizontal major axis 4, minor axis 2
(c) center (2,–1), vertical major axis 6, minor axis 4
(d) center (–1,4), vertical major axis 10, minor axis 6

9 $x^2/100 + 4y^2/375 = 1$

11 (a) $(\text{distance})^2 = (x-c)^2 + y^2$
(b) $(x-c)^2 + b^2 - b^2x^2/a^2 - (a-c)^2$
$= (x^2 - 2cx + c^2) + (a^2 - c^2) + (a^2 - c^2)x^2/a^2 - a^2 + 2ac - c^2$
$= c^2(x^2/a^2 - 1) - 2c(x-a)$
(c) The only solutions of the equation

$$(x-a)[c^2a^{-2}(x+a) - 2c] = 0$$

are $x = a$ and $x = a(2a/c - 1)$. The second is larger than the first because $a > c$.
(d) The difference between the distances is positive for $x < a$.

13 300

SECTION 1.6

1 (a) center (0,0), horizontal transverse axis, fundamental rectangle with vertices (±4,±4)
(b) center (0,1), horizontal transverse axis, fundamental rectangle with vertices (±2,4), (±2,–2)
(c) center (1,2), vertical transverse axis, fundamental rectangle bounded by $x = 0, 2$; $y = \frac{3}{2}, \frac{5}{2}$
(d) center (6,2), vertical transverse axis, fundamental rectangle bounded by $x = -2, 14$; $y = -12, 8$
(e) center (5,–4), horizontal transverse axis, fundamental rectangle bounded by $x = 17/3, 13/3$; $y = -16/3, -8/3$

3 (a) $x^2/4 - y^2/5 = 1$ (b) $x^2 - 4y^2 = 1$ (c) $(y-1)^2/9 - x^2/4 = 1$
(d) $4(x-4)^2/9 + 4(y-5)^2/7 = 1$ (e) $(y-1)^2/4 - (x+2)^2/5 = 1$
(f) $(y+4)^2/4 + 21(x-1)^2/64 = 1$ (g) $x^2 - 24y^2/25 = 1$
(h) $3y^2/92 - x^2/92 = 1$

5 400/3

7 Put $P = (0,0)$; then $x = 12(1100)$ ft and $y = -5.9(1100)$ ft (approximately).

9 Take P as $(-1,0)$, Q as $(1,0)$; then $x = -0.574$.

11 Square both sides: $(x-c)^2+y^2=4a^2+4a\sqrt{(x+c)^2+y^2}+(x+c)^2+y^2$. Then simplify, bring the radical alone to one side of the equation and square again, and then simplify to obtain $(c^2-a^2)x^2-a^2y^2=a^2(c^2-a^2)$, or $x^2/a^2-y^2/b^2=1$.

SECTION 1.7

1 (a) $\sqrt{2}/4$ (b) $-3\sqrt{7}/7$ (c) $\frac{3}{4}$ (d) $\frac{1}{2}$ (e) $\frac{3}{8}$

3 (a) $\frac{3}{4}$ (b) $-\frac{1}{3}$ (c) $\frac{8}{3}$ (d) $-\sqrt{2}$ (e) $-\frac{3}{5}$

5 $\pm\sqrt{6}/12$

7 slope $\frac{1}{4}$ at $(-\frac{1}{9},-\frac{2}{9})$; undefined at (1,2)

9 slope $\sqrt{2}$ at $(-\sqrt{6},\sqrt{3})$, $(\sqrt{6},-\sqrt{3})$; slope $-\sqrt{2}$ at $(\sqrt{6},\sqrt{3})$, $(-\sqrt{6},-\sqrt{3})$

11 $(\frac{5}{3},\frac{4}{9})$

13 Triangles ADC and ABC are right triangles. They are congruent because $AC=AC$, $DC=BC$.

SECTION 2.1

3 (a) $1/\sqrt{5}$, 0, $2/\sqrt{5}$ (b) $1/\sqrt{14}$, $2/\sqrt{14}$, $3/\sqrt{14}$
(c) $-3/\sqrt{17}$, $-2/\sqrt{17}$, $2/\sqrt{17}$ (d) $-2/\sqrt{29}$, $3/\sqrt{29}$, $-4/\sqrt{29}$

5 (a) $\sqrt{29}$ (b) $\sqrt{14}$ (c) 3 (d) $2\sqrt{6}$ (e) $\sqrt{14}$

7 distances between pairs of points: $\sqrt{14}$, $\sqrt{6}$, $\sqrt{6}$

9 direction numbers 1, 7, -5

11 mutual distances $\sqrt{6}$, $2\sqrt{6}$, $3\sqrt{6}$

13 $(-7,4,9)$, $(1,0,1)$

15 $1(7)+5(-2)+1(3)=0$

17 (a) $7(1)+4(2)+(-5)(3)=0$
(b) $2(-6)+9(-16)+(-12)(-13)=0$

SECTION 2.2

1 (a) $x+y+z=-3$ (b) $y+3+2(z-8)=0$
(c) $-2(x+7)+(y-2)+2(z-9)=0$
(d) $8(x-2)+3(y-5)+2(z-5)=0$ (e) $(x+2)-4(z-1)=0$

3 (a) $9(x-8)-3(y-1)-13(z-5)=0$
(b) $8(x+1)-3(y-4)-8(z-7)=0$
(c) $8(x+1)+12(y+6)+7(z+3)=0$
(d) $4(x-9)+7(y+1)-5(z-3)=0$
(e) $7(x-4)+4(y-6)+4(z-7)=0$

5 (a) $x-1=y/2=z+1$ (b) $x+1=y-1=-z/2$
(c) $(x-1)/2=(y-1)/2=z+2$ (d) $x/4=y/3=z-1$
(e) $(x+1)/3=(y+2)/(-1)=z$

7 (a) $(x-1)/6=(y-5)/7=(z-8)/(-9)$
(b) $(x-4)/6=-(y+8)=(z-2)/6$

(c) $(x+7)/8=(y-4)/(-5)=(z-1)/(-3)$ (d) $x/9=(z-9)/2; \quad y=9$
(e) $(x+5)/2=(y+7)/(-5)=(z+9)/(-8)$

9 $(x-1)/22=(y+2)/13=(z-3)/8$

11 (a) The line through the two points has the directions of the normal to the latter plane.
(b) From (a) we find $A=-B=C$; thus,

$$(x-1)-y+(z-1)=0$$

is the equation of the plane.

13 $(x+1)=(y-2)/2=-(z-1)$; point $(2,8,-2)$

SECTION 2.3

1 (a) circle in xy-plane with center $(0,0)$, radius 4, z-axis direction
(b) ellipse in xy-plane with center $(0,0)$, $a=3$, $b=2$, z-axis direction
(c) parabola in xz-plane, with $a=\frac{1}{2}$, opening along negative x-axis, axis along y-axis
(d) ellipse in yz-plane, center $(-1,1)$, $a=3$, $b=2$, x-axis direction
(e) hyperbola in xz-plane, opening along z-axis, with $a=2$, $b=\sqrt{2}$, center $(0,0)$, y-axis direction

3 (a) circle $y^2+z^2=9$ (b) circle $x^2+z^2=25$ (c) ellipse $x^2+3z^2=\frac{11}{2}$
(d) ellipse $8x^2+9z^2=126$ (e) ellipse $x^2/9+y^2/4=25/16$
(f) hyperbola $y^2/9-z^2/36=\frac{3}{4}$ (g) hyperbola $y^2-x^2=2$
(h) ellipse $x^2/9+z^2/4=9$ (i) parabola $y^2=4(z-1)$ (j) lines $x=\pm y/2$

5 Views: circle from z-axis, ellipse from x- and y-axes. See Figure 2.20.

7 Views: circle from z-axis, hyperbola from x- and y-axes. See Figure 2.21.

9 See Figures 2.25, 2.26, and 2.27.

SECTION 2.4

1 (a) $x-y+\sqrt{2}z=4$ (b) $-x/4+2y/9+\sqrt{11}z/6=1$ (c) $2x+2\sqrt{3}y-z=9$
(d) $x/4-y/5-3z/2=1$ (e) $4x+3y-3z=\frac{7}{2}$ (f) $\sqrt{3}x-y-2z=1$
(g) $y+2z=0$

3 (a) $4/\sqrt{3}$ (b) $5/\sqrt{69}$ (c) $2/(3\sqrt{6})$ (d) $7/\sqrt{10}$ (e) $3/\sqrt{29}$

5 $zz_0-xx_0-yy_0=1$; cup at $(0,0,1)$, cap at $(0,0,-1)$

7 (a) Apply the Pythagorean Theorem to the right triangle formed by 0, P, and P_0.
(b) The distance from 0 to P exceeds 1.

SECTION 3.1

1 (a) $(6 \;\; 4)$ (b) $(6 \;\; 4)$ (c) $(5 \;\; -3 \;\; 0)$ (d) $(5 \;\; -3 \;\; 0)$
(e) $(-1 \;\; -5 \;\; 4)$ (f) $(4 \;\; -2)$ (g) $(-10 \;\; 20 \;\; -25)$ (h) undefined
(i) $(18 \;\; -22 \;\; 14)$

3 (a) $\begin{bmatrix} 4 & 4 \\ 3 & 5 \end{bmatrix}$ (b) $\begin{bmatrix} 5 & 4 \\ 3 & 6 \end{bmatrix}$ (c) $\begin{bmatrix} 5 & 4 \\ 3 & 6 \end{bmatrix}$ (d) $\begin{bmatrix} 2 & 4 \\ 9 & 13 \end{bmatrix}$ (e) $\begin{bmatrix} -4 & -8 \\ -9 & -11 \end{bmatrix}$

(f) $\begin{bmatrix} 1 & 5 \\ -1 & 9 \\ 3 & 7 \end{bmatrix}$ (g) $\begin{bmatrix} 7 & -10 \\ 8 & 3 \\ 6 & -11 \end{bmatrix}$ (h) $\begin{bmatrix} -8 & 5 \\ -7 & -12 \\ -9 & 4 \end{bmatrix}$ (i) $\begin{bmatrix} 13 & -10 \\ 12 & 17 \\ 14 & -9 \end{bmatrix}$

(j) undefined

5 (a) Sum: Line from (0,0) to (7,1) (b) Sum: Line from (0,0) to (−3,3)
(c) Sum: Line from (0,0) to (−1,−5)

7 (a) 4 (b) 22 (c) −16 (d) −18 (e) −18 (f) −10

9 Both are along the line through the origin and (x, y). The first is 3 times as far from the origin as (x, y); and the second is at the mirror image of (x, y) through the origin.

11 Put $X = (x,x')$, $Y = (y,y')$, $Z = (z,z')$; then

$$(X + Y) + Z = (x + y + z,\ x' + y' + z') = X + (Y + Z), \text{ and so on.}$$

13 Put $A = \begin{vmatrix} a_{11} & a_{12} \\ a_{21} & a_{22} \end{vmatrix}$; $B = \begin{vmatrix} b_{11} & b_{12} \\ b_{21} & b_{22} \end{vmatrix}$; $C = \begin{vmatrix} c_{11} & c_{12} \\ c_{21} & c_{22} \end{vmatrix}$. Then use the definition of matrix addition and multiplication by a scalar; and finally use the corresponding algebraic properties of the real numbers.

15 Put $X = (x_1, \ldots, x_n)$; then

$$|cX| = (c^2x_1^2 + \cdots + c^2x_n^2)^{1/2} = |c|(x_1^2 + \cdots + x_n^2)^{1/2} = |c||X|.$$

17 $\cos\theta = x/\text{length}$; $\sin\theta = y/\text{length}$

19 Because $|\cos\phi|$ never exceeds 1 (see Exercise 18), $|\cos\phi| = 1$ if and only if $\phi = 0, \pm\pi, \pm 2\pi, \ldots$

SECTION 3.2

1 (a) not defined (b) $(-7\ \ -18)$ (c) $(30\ \ -17\ \ 10)$
(d) $\begin{bmatrix} -18 & 6 & -12 \\ -19 & 11 & -6 \\ 11 & -6 & 4 \end{bmatrix}$ (e) $\begin{bmatrix} -11 & -4 \\ 18 & 8 \end{bmatrix}$ (f) $\begin{bmatrix} 25 & -8 \\ 6 & 0 \\ 10 & -3 \end{bmatrix}$ (g) undefined

(h) $\begin{bmatrix} 9 & 13 & -11 \\ 11 & 16 & -6 \\ 0 & 4 & -5 \end{bmatrix}$ (i) $\begin{bmatrix} 18 & 9 & 6 \\ 5 & 0 & -1 \\ 15 & -1 & 2 \end{bmatrix}$ (j) $\begin{bmatrix} -14 \\ 19 \end{bmatrix}$ (k) $\begin{bmatrix} 57 \\ 13 \\ 50 \end{bmatrix}$

(l) $[-69\ \ 65\ \ 14]$

3 By direct computation.

5 By direct computation.

7 (a)
$$P^2 = \begin{bmatrix} 0.48 & 0.28 & 0.24 \\ 0.39 & 0.44 & 0.17 \\ 0.39 & 0.28 & 0.33 \end{bmatrix}, \quad P^3 = \begin{bmatrix} 0.444 & 0.312 & 0.244 \\ 0.417 & 0.376 & 0.207 \\ 0.417 & 0.312 & 0.271 \end{bmatrix}$$
(b) $[0.45\ \ 0.36\ \ 0.19]$ $[0.435\ \ 0.344\ \ 0.221]$ $[0.4305\ \ 0.3376\ \ 0.2319]$

9 0.270, 0.369

11 34,000, 34,900, 35,750, 36,650

13 (a)
$$D = \begin{bmatrix} 0 & 1 & 0 & 1 \\ 0 & 0 & 1 & 0 \\ 1 & 0 & 0 & 1 \\ 0 & 1 & 0 & 0 \end{bmatrix}; \quad D^2 = \begin{bmatrix} 0 & 1 & 1 & 0 \\ 1 & 0 & 0 & 1 \\ 0 & 2 & 0 & 1 \\ 0 & 0 & 1 & 0 \end{bmatrix}$$

(b) A: \$300 B: \$300 C: \$300 D: \$200

15

	Hubert	John	George	Ed
(a)	2	1	1	2
(b)	3	2	3	3

SECTION 3.3

1 (a) 10 (b) -2 (c) 11 (d) 0

3 (a) 106 (b) -306 (c) 111 (d) 632

5 (a) 12 (b) 0 (c) 3 (d) 4

7 determinant $= a_{11}a_{22}a_{33} - a_{11}a_{23}a_{32} - a_{12}a_{21}a_{33} + a_{12}a_{23}a_{31} + a_{13}a_{21}a_{32} - a_{13}a_{22}a_{31}$

9 Use the result of Exercise 7.

11 Use the result of Exercise 7.

SECTION 3.4

1 (a) $x = 40/43,\ \ y = 5/43$ (b) $x = 6/11,\ \ y = 14/33$
(c) $x = 25/19,\ \ y = -35/19$ (d) $x = -24/51,\ \ y = 33/51$

3 (a) $x = \frac{11}{2},\ \ y = -5,\ \ z = -\frac{1}{2}$ (b) $x = 2,\ \ y = 1,\ \ z = -1$
(c) $x = 1,\ \ y = -1,\ \ z = 2$ (d) $x = 1,\ \ y = 0,\ \ z = 2$

5 Follow given instructions.

SECTION 3.5

1 (a) $\frac{1}{54}\begin{bmatrix} 9 & 5 \\ -9 & 1 \end{bmatrix}$ (b) $\frac{1}{40}\begin{bmatrix} 0 & -5 \\ 8 & 3 \end{bmatrix}$ (c) $\frac{1}{41}\begin{bmatrix} 1 & 6 \\ 8 & 7 \end{bmatrix}$ (d) $\frac{1}{13}\begin{bmatrix} -1 & -3 \\ 6 & 5 \end{bmatrix}$

3 (a) $\frac{1}{42}\begin{bmatrix} 6 & -21 & 3 \\ 8 & 35 & -3 \\ -8 & 7 & 3 \end{bmatrix}$ (b) $-\frac{1}{17}\begin{bmatrix} -2 & -2 & -3 \\ -10 & 7 & 19 \\ 11 & -6 & -26 \end{bmatrix}$

(c) $\frac{1}{432}\begin{bmatrix} 28 & 42 & -6 \\ 36 & 54 & 54 \\ 58 & -21 & 3 \end{bmatrix}$ (d) $-\frac{1}{442}\begin{bmatrix} -54 & 18 & -4 \\ 42 & -14 & -46 \\ -107 & -38 & 33 \end{bmatrix}$

5 (a) $x = 4,\ \ y = 1$ (b) $x = 2,\ \ y = -1$ (c) $x = 0,\ \ y = 4$
(d) $x = -3,\ \ y = 1$

7 (a) $x = 1,\ \ y = 2,\ \ z = 2$ (b) $x = 1,\ \ y = 0,\ \ z = -1$
(c) $x = 1,\ \ y = -3,\ \ z = -3$ (d) $x = 0,\ \ y = 2,\ \ z = 0$

9 (a) $x = 150,\ \ y = 100$ (b) $x = -\frac{1}{6}b_1 + \frac{2}{3}b_2,\ \ y = \frac{2}{9}b_1 - \frac{5}{9}b_2$

11 liberal arts 400; science 200; engineering 100

13 12 men; 12 women; 48 children

15 Inner product of row i of A and row j of cof A = det A if $i = j$, 0 if $i \neq j$. Row j of cof A = column j of the adjoint of A.

SECTION 3.6

1 (a) $\begin{bmatrix} 1 & \cdots \\ 0 & 1 \end{bmatrix}$ (b) $\begin{bmatrix} 1 & \cdots \\ 0 & 0 \end{bmatrix}$ (c) $\begin{bmatrix} 1 & \cdots & \cdots & \cdots \\ 0 & 1 & \cdots & \cdots \\ 0 & 0 & 0 & 0 \end{bmatrix}$

(d) $\begin{bmatrix} 1 & \cdots & \cdots \\ 0 & 1 & \cdots \\ 0 & 0 & 1 \end{bmatrix}$ (e) $\begin{bmatrix} 1 & \cdots & \cdots \\ 0 & 1 & \cdots \\ 0 & 0 & 1 \\ 0 & 0 & 0 \end{bmatrix}$

3 (a) $\begin{aligned} x + 2y &= 10 \\ y &= 34/7 \end{aligned}$ (b) $\begin{aligned} x - 3y &= 4 \\ y &= -22/19 \end{aligned}$ (c) $\begin{aligned} x + y + 5z &= -2 \\ y + \tfrac{9}{2}z &= -3 \\ z &= -\tfrac{1}{2} \end{aligned}$

(d) $\begin{aligned} x - 2y + 4z &= 0 \\ y + \tfrac{3}{2}z &= 1/3 \\ z &= 3/29 \end{aligned}$ (e) $\begin{aligned} x - 8y + 2z - w &= 2 \\ y - 85z + 95w &= 2 \\ z - (389/347)w &= -9/347 \end{aligned}$

(f) $\begin{aligned} x - 13y &= 15 \\ y &= -35/32 \end{aligned}$

5 $\begin{bmatrix} a & b \\ c & d \end{bmatrix}$ is nonsingular if and only if $ad = bc$. Its rank is less than 2 if and only if there is a number k such that $ak = c$, $bk = d$.

7 Expand in cofactors of the given row.

9 Nonsingularity of a matrix is not affected by elementary row operations.

11 By definition of matrix multiplication.

13 Row operations cannot alter the number of rows with nonzero entries when the matrix is in echelon form.

SECTION 3.7

1 inconsistent

3 $x = 4, \quad y = 2, \quad z = -2$

5 inconsistent

7 inconsistent

9 $x = 2, \quad y = -4, \quad z = \tfrac{1}{2}$

11 $x = 2, \quad y = \tfrac{1}{2}, \quad z = -1, \quad w = 0$

13 $x = z/2, \quad y = -3z/2$

15 $x = 2 - z/2, \quad y = 2 - z/2$

17 $x = -4 - w, \quad y = -w, \quad z = 6 + w$

19 $x = w - 2y, \quad z = 3w, \quad v = 0$

21 No solution: equations are inconsistent.

23 The only solution with nonnegative components is 50 units of carrots and none of the others.

25 $(\tfrac{1}{2}, \tfrac{1}{2})$

27

	healthy	diseased	dead
healthy	0.90	0.08	0.02
diseased	0.05	0.80	0.15
dead	0	0	1

Stationary distribution: All trees dead.

29 $(\frac{1}{8}\ \frac{2}{8}\ \frac{2}{8}\ \frac{2}{8}\ \frac{1}{8})$

31 If $p_{12} + p_{21} > 0$, then the unique stationary distribution is

$$\left(\frac{p_{21}}{p_{12} + p_{21}}, \frac{p_{12}}{p_{12} + p_{21}}\right).$$

If $p_{12} + p_{21} = 0$, then $P = \begin{bmatrix} 10 \\ 01 \end{bmatrix}$, and every distribution is stationary.

SECTION 3.8

1 (a) (0,3) (9,0) (0,0) (b) (0,0) (0,4)
(c) $(1, -\frac{3}{2})$ (1,2) $(-\frac{4}{3},2)$ (d) (3,2) (e) (20,0) (0,15) (2,9)
(f) (1,0) (g) (6,3) (4,5) (1,4)
(h) (12/5,4/5) (−1/2,9/4) (14/24) (i) (1,6) (4,4) (6,2)

3 (a) (4,0,0) (0,4,0) (0,0,4) (0,0,0)
(b) (0,0,0) (6,0,0) (0,2,0) (0,0,6)
(c) (2,0,0) (2,2,0) (0,4,0) (0,0,0) (2,0,2) (0,0,4)
(d) (0,0,0) (0,0,2) (0,6,0) (2,0,0) $(\frac{3}{2},0,\frac{3}{2})$
(e) (0,0,0) (0,0,2) (0,2,0) (1,0,0) $(0,\frac{4}{3},2)$ (3,0,0)

5 (a) (0,0,0,0) $(\frac{4}{3},0,0,\frac{2}{3})$
(2,0,0,0) $(3,0,\frac{1}{2},0)$
(0,0,0,1) $(\frac{8}{3},\frac{2}{3},0,0)$
(0,0,2,0)
(0,2,0,0)
(b) (0,0,0,0) (0,0,0,1) (2,0,0,0) (0,1,0,0)
(0,0,1,0) (2,0,0,1) (4,1,0,0)

SECTION 3.9

1 (a) Extreme points (0,2) (0,0) $(\frac{8}{3},0)$ (2,1); maximum 20, minimum 0
(b) Extreme points (0,0) (0,1) (2,0); maximum 2, minimum 0
(c) Extreme points (0,0) $(0,\frac{5}{2})$ (1,0) (2,1); maximum 5, minimum 0
(d) Extreme points (1,0) (1,1) $(\frac{1}{3},\frac{2}{3})$; maximum 4, minimum 1

3 (a) 16 (b) 17 (c) 15

5 First process once, second process 4 times

7 20 lb potatoes, $\frac{5}{4}$ lb meat

9 50 hr mine I, 15 hr mine II

11 Large boat 6 times

13 Three times method I, 1 time method II

SECTION 3.10

1 (a) $\lambda = 2, 6;$ $\begin{bmatrix} x \\ -x \end{bmatrix} \begin{bmatrix} 3x \\ x \end{bmatrix}$ (b) $\lambda = 5, 1;$ $\begin{bmatrix} x \\ 0 \end{bmatrix} \begin{bmatrix} 0 \\ y \end{bmatrix}$

(c) $\lambda = 2, 3;$ $\begin{bmatrix} 2x \\ x \end{bmatrix} \begin{bmatrix} x \\ x \end{bmatrix}$ (d) $\lambda = 1, 6;$ $\begin{bmatrix} x \\ -x \end{bmatrix} \begin{bmatrix} 4y \\ y \end{bmatrix}$

(e) $\lambda = \frac{1}{2}(9 \pm \sqrt{37});$ $\begin{bmatrix} x \\ (-1 + \sqrt{37})x/6 \end{bmatrix} \begin{bmatrix} (1 - \sqrt{37})y/6 \\ y \end{bmatrix}$

(f) $\lambda = \frac{1}{2}(-1 \pm i\sqrt{31});$ $\begin{bmatrix} x \\ (3 - i\sqrt{31})x/10 \end{bmatrix} \begin{bmatrix} (3 - i\sqrt{31})y/4 \\ y \end{bmatrix}$

3 (a) $\lambda = 2, 3, -3;$ $\begin{bmatrix} 2y \\ y \\ 0 \end{bmatrix} \begin{bmatrix} 0 \\ 0 \\ z \end{bmatrix} \begin{bmatrix} x \\ -2x \\ 0 \end{bmatrix}$ (b) $\lambda = 1, -2;$ $\begin{bmatrix} 0 \\ 0 \\ z \end{bmatrix} \begin{bmatrix} 0 \\ y \\ -y \end{bmatrix}$

(c) $\lambda = 1, 4;$ $\begin{bmatrix} x \\ -x \\ z \\ z \end{bmatrix} \begin{bmatrix} 0 \\ y \\ 0 \\ 0 \end{bmatrix}$

5 The radical in the solution for λ is $\sqrt{(a+c)^2 + 4b^2}$, which is real.

7 Inner product $= ax_1y_1 + b(x_1y_2 + x_2y_1) + cx_2y_2$.

9 Use the definitions of matrix multiplication and addition, and the multiplication of a matrix by a number.

SECTION 3.11

1 (a) $\lambda = 3, -4;$ $\begin{bmatrix} x \\ x \end{bmatrix} \begin{bmatrix} x \\ -\frac{5}{2}x \end{bmatrix}$ (b) $\lambda = 1, 5;$ $\begin{bmatrix} -\sqrt{3}y \\ y \end{bmatrix} \begin{bmatrix} x \\ \sqrt{3}x \end{bmatrix}$

(c) $\lambda = \frac{7}{2}, -\frac{5}{2};$ $\begin{bmatrix} 2y \\ y \end{bmatrix} \begin{bmatrix} -(2/11)y \\ y \end{bmatrix}$ (d) $\lambda = 2, 3, -3;$ $\begin{bmatrix} 0 \\ y \\ 2y \end{bmatrix} \begin{bmatrix} x \\ 0 \\ 0 \end{bmatrix} \begin{bmatrix} 0 \\ -2z \\ z \end{bmatrix}$

3 (a) $Y = -4X_1 - 7X_2$ (b) $Y = \frac{4}{5}X_1 - \frac{1}{5}X_2$ (c) $Y = 3X_2$
(d) $Y = X_1 - X_2 + 2X_3$ (e) $Y = X_1 + X_2 - X_3$ (f) $Y = 26X_1 + 33X_2 - 71X_3$

5 (a) $\begin{bmatrix} 4^k & 2 \cdot 4^{k-1} - 2^{k-1} \\ 0 & 2^k \end{bmatrix}$ (b) $\frac{1}{5}\begin{bmatrix} 4^{k+1} + 9^k & -4^k + 9^k \\ -4^{k+1} + 4 \cdot 9^k & 4^k + 4 \cdot 9^k \end{bmatrix}$

(c) $\frac{1}{2}\begin{bmatrix} 3^k + 1 & 3^k - 1 \\ 3^k - 1 & 3^k + 1 \end{bmatrix}$ (d) $\begin{bmatrix} 1 & 2^{k+2} - 4 & \frac{1}{3}(-33 + 10 \cdot 2^{k+1} + 13(-1)^k) \\ 0 & 2^k & \frac{1}{3}(10 \cdot 2^{k-1} - 5(-1)^k \\ 0 & 0 & (-1)^k \end{bmatrix}$

(e) $\begin{bmatrix} 2^k & 0 & 0 \\ 1 - 2^k & 1 & 2 - 2(-1)^k \\ 0 & 0 & (-1)^k \end{bmatrix}$

7 $\frac{2}{3} - (\frac{5}{12})(0.7)^5$, or approximately 0.60

9 $\begin{bmatrix} 0.2401 & 0.2080 & 0.5519 \\ 0 & 0.6561 & 0.3439 \\ 0 & 0 & 1 \end{bmatrix}$

11 $$\begin{bmatrix} 1 & 0 & 0 & 0 \\ 0.5-0.4(0.8)^{k-1} & \begin{array}{c}0.3(0.6)^{k-1}\\+0.4(0.8)^{k-1}\end{array} & \begin{array}{c}-0.3(0.6)^{k-1}\\+0.4(0.8)^{k-1}\end{array} & 0.5-0.4(0.8)^{k-1} \\ 0.5-0.4(0.8)^{k-1} & \begin{array}{c}-0.3(0.6)^{k-1}\\+0.4(0.8)^{k-1}\end{array} & \begin{array}{c}0.3(0.6)^{k-1}\\+0.4(0.8)^{k-1}\end{array} & 0.5-0.4(0.8)^{k-1} \\ 0 & 0 & 0 & 1 \end{bmatrix}$$

One-half eventually quit.

13 $x_n = (75/2^{n-1})[(3+i)^n + (3-i)^n]$

15 $A^2X = A^2(AX) = A^2(\lambda X) = \lambda A^2X = \lambda \cdot \lambda^2 X = \lambda^3 X$

17 By algebra.

SECTION 4.1

1 (a) horizontal x, vertical $7-x$ (b) $A = x(7-x)$ (c) $0 < x < 7$
(d) area largest at $x = 3\frac{1}{2}$

3 (a) $(4+x^2)^{1/2}$ (b) $6-x$ (c) rowing $\frac{1}{3}(4+x^2)^{1/2}$, walking $\frac{1}{4}(6-x)$
(d) $0 \le x \le 6$
(e)

x	0	1	2	3	4	5	6
T	2.17	2.00	1.94	1.95	1.99	2.04	2.11

5 (a) radius x, area πx^2
(b) density on second disk = density on first disk/x^2; $25/\pi x^2$
(c) Plot $(x, 25/\pi x^2)$ for given x.

7 $(4)10^4/x^2$

9 $x(101-x)$

11 (a) $((3-2t/3)^2 + (4-t)^2)^{1/2}$
(b)

0	1	2	3	4	5
5.00	3.80	2.60	1.41	0.33	1.41

(c) Bus, in 4 min (d) Plot data in (b).

13 $N = 4400/d$

15 (a) $C = 5x/2 + 100 + 1000/x$
(b)

x	5	10	15	20	25	30
C	312.50	225.00	204.17	200.00	202.50	208.33

(c) Plot points in (b).

17 (a) $S = 6x^2$ (b) $V = x^3$ (c) $V = S^{3/2}/6^{3/2}$, $S = 6V^{2/3}$
(d)

V	1	8	27	64
S	6	24	54	96

19 (b) $n+1$
(c) n columns, each having the sum $n+1$: $n(n+1)$
(d) $n(n+1)/2$

SECTION 4.2

1 (a) $y = -\frac{3}{4}x + \frac{8}{3}$ (b) $y - 100 = 3(x-11)$ (c) $y = \frac{4}{5}(x-23)$
(d) $y = 0.7(x-17.5) + 3.75$

3 $y = -0.93x + 68.80$

5 $y = -57x/350 + 121/30$

7 $y = 1.62x + 6.88$ (years -2 -1 0 1 2)

9 $y = 3.8x + 23$ (years -2 -1 0 1 2)

11 $y = 28.1x + 199.4$ (years -2 -1 0 1 2)

13 (a) By definition of vertical deviation.
(b) Expand the squares, and rearrange the terms; and use the fact $\Sigma x = 0$.
(c) Use given hint.
(d) The terms involving m and b are squares, so that they are positive or 0. Their minimum value 0 is obtained for the indicated values of m and b.

SECTION 4.3

1 (a) 1.588, 1.6487 (b) 9.3789, 20.086 (c) 0.31641, 0.36788

3 11, 12.1, 13.31, 14.641 (millions)

5 62,986

7 33%

9 $6e^{0.25t}$; 9.8922, 20.9418, 44.3346 (thousands)

11 60%, 52%, 45%, 38%

13 55.26, 57.52

SECTION 4.4

1 30 years

3 10 days

5 311 years

7 16 years

9 12 months

11 $e^{kt} = e^{\log 2} = 2$

13 If $T = (1/k)\log m$, then $e^{kT} = m$.

15 23 years

SECTION 4.5

1 5 sec; 160 ft/sec

3 3 sec; 144 ft

5 $6\frac{1}{4}$ sec

7 550 ft

9 4 ft/sec/sec

11 (a) $176t/3 - 11t^2$ (b) $176/3 - 22t$ (c) 8/3 sec

13 6 points

15 deceleration 1/36 deg/hr/hr; 2° above normal

SECTION 4.6

1 (a) $\pi/9$ (b) $5\pi/4$ (c) $5\pi/12$ (d) $5\pi/6$ (e) $78\pi/45$

3 (a) 157.5 (b) 202.5 (c) 75 (d) $308\frac{4}{7}$ (e) 36

5 See table.

The three numbers represent sine, cosine, and tangent, respectively:

(a) -0.9397, 0.3420, -2.747 (b) -0.1219, -0.9925, 0.1228
(c) 0.7660, -0.6428, -1.192 (d) -0.6293, -0.7771, 0.8098
(e) 0.9272, -0.3746, -2.475 (f) 0.0872, -0.9962, -0.0875

9 (a) $-\sin 15°$ (b) $\sin 50°$ (c) $-\sin 40°$ (d) $-\sin 30°$ (e) $\cos 27°$
(f) $\cos 40°$ (g) $-\cos 30°$ (h) $-\tan 80°$ (i) $-\tan 30°$ (j) $\tan 80°$

11 (a) $\sin 0$ (b) $\sin \pi/2$ (c) $-\sin \pi/3$ (d) $-\sin \pi/3$ (e) $\cos \pi/4$
(f) $\cos 0$ (g) $-\cos \pi/5$ (h) undefined ($-\tan \pi/2$) (i) $\tan 0$
(j) $\tan \pi/3$

SECTION 4.7

1 (a) $\pi/6$ (b) $-\pi/4$ (c) $5\pi/6$ (d) $\pi/3$ (e) $-\pi/4$ (f) $\pi/3$

3 (a) $\pi/2 - x$ (b) $1/\sqrt{1+x^2}$ (c) $\sqrt{1-x^2}$

5 $\sin^2(\arccos x) + \cos^2(\arccos x) = 1$; therefore, $\sin^2(\arccos x) = 1 - x^2$; therefore, $\sin(\arccos x) = \sqrt{1-x^2}$.

SECTION 4.8

1

	(a)	(b)	(c)	(d)	(e)	(f)
mean	0	0	0	0	1	-2
amplit.	2	4	1	1	1	1
period	2π	2π	$2\pi/3$	$\frac{1}{2}$	π	$2\pi/5$

3 (a) $\frac{1}{2} + \frac{3}{2}\sin(2\pi x + \pi/2)$ (b) $11 + 3\sin(2\pi x/9)$
(c) $\frac{5}{2} + \frac{5}{2}\sin(8\pi x - \pi/2)$ (d) $-3 + 2\sin(x/3 + \pi/2)$

5 (a) The orbit of the moon is like the rim of a wheel moving at constant speed (see Example 2).
(b) 30 days (c) $240\sin(2\pi x + c)$, where c is some constant

7 $61.8 + (17.1)\sin(\pi(x-3)/6)$

x	0	1	2	3	4	5	6	7	8	9	10	11
sine function	44.7	47.0	53.2	61.8	70.4	76.6	78.9	76.6	70.4	61.8	53.2	47.0

9 $9 + 84\sin(\pi(x-3)/6)$

x	0	1	2	3	4	5	6	7	8	9	10	11
sine function	-75	-64	-33	9	51	82	93	82	51	9	-33	-64

11 (a) $4t - t^2$ (b) \$54 (c) \$46 (d) 8 days (e) $50 + 4\sin(\pi t/4)$

13 $0.12 + 0.06\sin(\pi(x+3)/6)$

x	0	1	2	3	4	5	6	7	8	9	10	11
sine function	0.18	0.17	0.15	0.12	0.09	0.07	0.06	0.07	0.09	0.12	0.15	0.17

SECTION 5.1

1 (a) 0 (b) 2 (c) $20x$ (d) $2x+1$ (e) $7-2x$ (f) $3x^2-4x$
(g) $4x^3$ (h) $-28x^6$

3 (a) $-x^{-2}$ (b) $2x^{-2}$ (c) $\frac{1}{4}x^{-3/4}$ (d) $\frac{1}{2}x^{-1/2}+\frac{1}{3}x^{-2/3}$ (e) $1-x^{-2}$
(f) $-3x^{-4}$ (g) $-\frac{1}{2}x^{-3/2}$ (h) $2x+4x^{-3}$

5 (a) $2x$ (b) $-3x^2$ (c) $-x^{-2}$ (d) $-x^{-3}$ (e) $\frac{2}{3}x^{-1/3}$ (f) $-\frac{2}{3}x^{-5/3}$

7 (a) $y=-8x+9$ (b) $y=14x-36$ (c) $y=3x-2$ (d) $y=\frac{3}{4}x-1$
(e) $y=\frac{1}{3}x+\frac{4}{3}$ (f) $y=-\frac{4}{5}x+\frac{9}{5}$

9 Each of the n terms is equal to A^{n-1}, so the sum is nA^{n-1}.

SECTION 5.2

1 (a) derivative $\frac{1}{3}$, approximation 0.33 (b) derivative 1/4.5, approximation 0.22
(c) derivative = approximation = 0.05
(d) derivative 1/36, approximation 0.028

3 (a) 0.99500, 0.995 (b) −0.09983, −0.104 (c) 0.87758, 0.875
(d) −0.47943, −0.484 (e) 0.62161, 0.617 (f) −0.71736, −0.721

5 0.99830, 0.99867, 0.99933, 1.00000, 1.00000; approaches 1

7 $$\frac{\log(x+h)-\log x}{h}=\frac{1}{h}\log\left(\frac{x+h}{h}\right)=\frac{1}{h}\log\left(1+\frac{h}{x}\right)$$
The coefficient of $1/x$ tends to 1 by the case proved in the text.

SECTION 5.3 None.

SECTION 5.4

1 (a) $36x^2-5$ (b) $20x^3+24x$ (c) $3x^2-10x-60x^{-11}$ (d) 0
(e) $\frac{1}{2}x^{-1/2}+\frac{1}{2}x^{-3/2}$ (f) $\frac{1}{3}x^{-2/3}+2x^{-3}$ (g) $\cos x+2\sin x$ (h) $1-1/x$
(i) e^x-1 (j) $-\frac{3}{2}x^{-5/2}+2/x$

3 (a) $\cos x-x\sin x$ (b) $\log x+1$ (c) e^x+xe^x (d) $\cos^2 x-\sin^2 x$
(e) $3x^2\sin x+x^3\cos x$ (f) $2xe^x+(1+x^2)e^x$ (g) $e^x\log x+e^x(1/x)$
(h) $(2/x)\log x$ (i) $(e^x+1)(1+\log x)+(e^x+x)(1/x)$ (j) $3x^2+2x+2$

5 (a) $\dfrac{x^2+4x+10}{(x+2)^2}$ (b) $\dfrac{2-2x^2}{(x^2+3x+1)^2}$ (c) $e^x(x^{-3}-3x^{-4})$
(d) $\dfrac{(1+\log x)\cos x-(1/x)\sin x}{(1+\log x)^2}$ (e) $(1-\log x)/x^2$ (f) $-1/\sin^2 x$
(g) $\dfrac{x/\sqrt{1-x^2}-\arcsin x}{x^2}$ (h) $\dfrac{\pi/2}{\sqrt{1-x^2}(\arccos x)^2}$

7 $2x\cdot x^3+3x^2\cdot x^2$, or $5x^4$

9 (a) Differentiate $x^4+4x^3+6x^2+4x+1$
(b) Derivative $=2(x^2+2x+1)(2x+2)$

11 $f'(x)f(x)+f(x)f'(x)$

13 $\dfrac{\dfrac{1}{g(x)} - \dfrac{1}{g(x^*)}}{x - x^*} = \dfrac{g(x^*) - g(x)}{(x - x^*)g(x)g(x^*)} = -\dfrac{g(x) - g(x^*)}{x - x^*} \dfrac{1}{g(x)g(x^*)}$

The latter converges to $-\dfrac{(d/dx)g(x)}{(g(x))^2}$

15 $f' \dfrac{1}{g} - \dfrac{fg'}{g^2}$, or $\dfrac{f'g - fg'}{g^2}$

17 (a) By definition. (b) $\dfrac{ad - bc}{(cx + d)(cx^* + d)}$

SECTION 5.5

1 (a) $24x(2 + 3x^2)^3$ (b) $80x(4x^2 - 1)^4$ (c) $-3x(4 - 3x^2)^{-1/2}$
(d) $\frac{3}{2}x^2(x^3 + 8)^{-1/2}$ (e) $12(1 - x^2)(3x - x^3 + 1)^3$
(f) $3(2 - x)(3 + 4x - x^2)^{1/2}$ (g) $5x^4/(1 + x)^6$ (h) $\dfrac{2x^2 - 4x + 3}{\sqrt{x^2 - 2x + 2}}$

3 (a) $\dfrac{x(x^2 + 1)^2(53 - 7x^2)}{\sqrt{9 - x^2}}$ (b) $4x(x^2 - 1)\sqrt{x - 1} + \dfrac{(x^2 - 1)^2}{2\sqrt{x - 1}}$
(c) $\dfrac{2x^2 - 2x - 8}{(4 + x^2)^2}$ (d) $\dfrac{x^3 + 6x^2 + 2}{2(1 - x^3)^{3/2}}$ (e) $\dfrac{-4x^2 - 2x - 3}{(x + 1)^2(2x - 3)^3}$
(f) $\dfrac{2x}{\sqrt{x^2 - 1}(x^2 + 1)^{3/2}}$ (g) $(x^2 - 2x + 2)^{-3/2}$ (h) $36x^2(x^3 - 1)^3(2x^3 + 1)^{-5}$

5 (a) $\dfrac{10x}{x^2 + 2}$ (b) $\dfrac{2 - 2x}{2x - x^2}$ (c) $1/\tan x$ (d) $\dfrac{2 \log(x + 1)}{x + 1}$
(e) $1/\sin x \cos x$ (f) $\dfrac{1}{x \log x}$

7 (a) $(2x - 3)e^{x^2 - 3x}$ (b) $-x^{-2}e^{1/x}$ (c) $\dfrac{1}{2\sqrt{x}} e^{\sqrt{x}}$ (d) $(\cos x)e^{\sin x}$
(e) $\dfrac{x}{\sqrt{1 + x^2}} e^{\sqrt{1+x^2}}$ (f) $(\cos x - \sin x)e^{\sin x + \cos x}$

9 (a) $-15 \sin(3x + 2)$ (b) $-4(2x + 1)\sin(2x + 1)^2$ (c) $5 \sin^4 x \cos x$
(d) $-\dfrac{x}{\sqrt{1 - x^2}} \cos\sqrt{1 - x^2}$ (e) $\dfrac{1}{2\sqrt{x}} \cdot \dfrac{1}{\sqrt{1 - x}}$ (f) $\dfrac{1}{x^2 + 1}$

11 $f(u(x)) = c(ax + b) + d$, $f' = ca$

SECTION 5.6

1 (a) $6\frac{1}{12}$ (b) $9\frac{9}{10}$ (c) $3\frac{2}{27}$ (d) $1/3 + 1/972$ (e) $6 - 77/1200$
(f) 1328

3 (a) 0.09486 (b) 0.70739 (c) 0.34587 (d) 0.99 (e) 2.0150
(f) 0.4975 (g) 4.7303 (h) $\pi/6 - 0.02/\sqrt{3}$

5 0.08; 0.015

7 -5.76π; -1.92π

9 0.004

11 15%

13 Lower by 1.62¢ per lb   Elasticity = 1.28

15 8 mi/sec

17 (a) $\dfrac{x^* - x}{x}$ (b) $\dfrac{-f'(x)(x^* - x)}{f(x)}$ (c) Divide (b) by (a).

SECTION 5.7

1 (a) $(dy/dt) = \frac{3}{2}x^{1/2}(dx/dt)$ (b) $z(dz/dt) = 2x(dx/dt) + 3y(dy/dt)$
(c) $dz/dt = x(dy/dt) + y(dx/dt)$ (d) $dy/dt = (\cos x)(dx/dt)$
(e) $\dfrac{dy}{dt} = \dfrac{1 - x^2}{(x^2 + 1)^2}\dfrac{dx}{dt}$ (f) $dy/dt = (1/x)(dx/dt)$ (g) $dz/dt = (dx/dt)/\cos^2 x$
(h) $dy/dt = e^{-x}(1 + e^{-x})^2(dx/dt)$

3 20π mm/week

5 $5/9\pi$ ft/sec

7 3/2 ft/sec

9 (a) $\dfrac{1}{100 + x^2}$ (b) 0.0218 cu ft/sec

11 (a) $R + R \sin x$ (b) R, $\sqrt{3}R$

13 $2180\pi/3$ ft/sec

15 (a) $x = 40 \tan \theta$ (b) $125\pi/24$ ft/sec

17 distance $= (400t^2 - 1024t + 1024)^{1/2}$; -5.6 mph;   12 mph

SECTION 5.8

1 (a) $f'(x) = 3x^2 - 12x + 9$; $f''(x) = 6x - 12$
(b) $f'(x) = 12x^5 + 15x^4 + 10$; $f''(x) = 60x^4 + 60x^3$
(c) $f'(x) = 12x^3 - 36x^2 + 24x$; $f''(x) = 36x^2 - 72x + 24$
(d) $f'(x) = -1/(x + 1)^2$; $f''(x) = 2/(x + 1)^3$
(e) $f'(x) = (2x^2 + 1)(x^2 + 1)^{-1/2}$; $f''(x) = (2x^2 + 3x)(x^2 + 1)^{-3/2}$
(f) $f'(x) = \dfrac{1 - x^2}{(1 + x^2)^2}$; $f''(x) = \dfrac{2x^3 - 6x}{1 + x^2}$

3 (a) $f'(x) = 4 \sin^3 x \cos x$; $f''(x) = 4 \sin^4 x + 12 \sin^2 x \cos^2 x$
(b) $f'(x) = \dfrac{\cos x}{(1 - \sin x)^2}$; $f''(x) = \dfrac{2 \cos^2 x - \sin x}{(1 - \sin x)^3}$
(c) $f'(x) = -\dfrac{\cos x}{\sin^2 x} + \dfrac{\sin x}{\cos^2 x}$
$f''(x) = \dfrac{1 + \cos^2 x}{\sin^3 x} + \dfrac{1 + \sin^2 x}{\cos^3 x}$
(d) $f'(x) = -2 \cos x \sin^2 x + \cos^3 x$
$f''(x) = 2 \sin^3 x - 7 \cos^2 x \sin x$
(e) $f'(x) = \dfrac{x \cos x - \sin x}{x^2}$; $f''(x) = \dfrac{-x^2 \sin x - 2x \cos x + 2 \sin x}{x^3}$

(f) $f'(x) = \dfrac{\sin x}{x^2} - \dfrac{2(1-\cos x)}{x^3}$
$f''(x) = x^{-2}\cos x - 4x^{-3}\sin x + 6x^{-4}(1-\cos x)$
(g) $f'(x) = (1-x^2)^{-1/2}$; $f''(x) = x(1-x^2)^{-3/2}$
(h) $f'(x) = 4x^3\sin^2 x + 2x^4 \sin x \cos x$
$f''(x) = 12x^2\sin^2 x + 16x^3 \sin x \cos x + 2x^4(\cos^2 x - \sin^2 x)$

5 (a) $f'(x) = 1/x$; $f''(x) = -1/x^2$
(b) $f'(x) = x - 2x\log x$; $f''(x) = -1 - 2\log x$
(c) $f'(x) = \dfrac{-2x}{1-x^2}$; $f''(x) = -\dfrac{2(1+x^2)}{1-x^2}$
(d) $f'(x) = -x^{-2}e^{1/x}$ $f''(x) = x^{-4}e^{1/x} + 2x^{-3}e^{1/x}$
(e) $f'(x) = e^{-x} - xe^{-x}$; $f''(x) = xe^{-x} - 2e^{-x}$
(f) $f'(x) = 1/(x\log x)$; $f''(x) = -(1+\log x)(x\log x)^{-2}$
(g) $f'(x) = e^x e^{e^x}$; $f''(x) = e^{2x}e^{e^x} + e^x e^{e^x}$
(h) $f'(x) = xe^{x^2}\log x + (1/2x)e^{x^2}$
$f''(x) = e^{x^2}(2+\log x) + 2x^2e^{x^2}\log x - (1/2x^2)e^{x^2}$

7 $-27/128\pi^2$

9 $x'' = -35/\sqrt{2}$

11 Put $k = \frac{1}{2}\log 2$; then $y'' = 10k^2e^{3k} = 1.3$/hr/hr

SECTION 5.9

1 (a) critical point $(\frac{1}{2},-6\frac{1}{4})$; concave upward
(b) critical point $(\frac{1}{2},-\frac{1}{4})$; concave upward
(c) critical points $(0,0)$, $(2,-4)$; inflection point $(1,-2)$; concave upward for $x > 1$, downward for $x < 1$
(d) no critical points; inflection point $(0,-5)$; concave upward for $x > 0$, downward for $x < 0$
(e) critical points $(0,-1)$, $(-3,-28)$; inflection points $(0,-1)$, $(-2,-17)$; concave upward for $x > 0$, $x < -2$, downward for $-2 < x < 0$
(f) critical points $(-1,0)$, $(\frac{1}{2},-27/16)$; inflection points $(0,-1)$, $(-1,0)$; concave upward for $x > 0$, $x < -1$, downward for $-1 < x < 0$
(g) critical point $(0,1)$; inflection points $(\pm 1/\sqrt{3}, 3/4)$; concave upward for $x < -1/\sqrt{3}$, $x > 1/\sqrt{3}$, downward for $-1/\sqrt{3} < x < 1/\sqrt{3}$
(h) critical point $(2^{-1/3}, 3(2^{-2/3}))$; inflection point $(-1,0)$; concave upward for $x < -1$, $x > 0$, downward for $-1 < x < 0$; break in graph at $x = 0$

3 (a) critical point $(0,1)$; inflection points $(\pm 1, e^{-0.5})$; concave upward for $x < -1$, $x > 1$, downward for $-1 < x < 1$
(b) critical points $(\pm 1, e^{-0.5})$; inflection points $(0,1)$, $(\sqrt{3}, \sqrt{3}e^{-1.5})$, $(-\sqrt{3}, -\sqrt{3}e^{-1.5})$; concave upward for $-\sqrt{3} < x < 0$, $x > \sqrt{3}$, downward for $x < -\sqrt{3}$, $0 < x < \sqrt{3}$
(c) critical point $(e^{-1}, -e^{-1})$; concave upward
(d) critical point $(1,0)$; inflection point $(e,1)$; concave upward for $x < e$, downward for $x > e$

5 (a) critical points $(0,0)$, $(\pi/2,1)$, $(\pi,0)$, $(3\pi/2,1)$, $(2\pi,0)$; inflection points $(\pi/4,1/2)$, $(3\pi/4,1/2)$, $(5\pi/4,1/2)$, $(7\pi/4,1/2)$; concave upward for $0 < x < \pi/4$, $3\pi/4 < x < 5\pi/4$, downward for $\pi/4 < x < 3\pi/4$, $5\pi/4 < x < 7\pi/4$

(b) critical points $(\pi/4, e^{-\pi/4}/\sqrt{2})$, $(5\pi/4, -e^{-5\pi/4}/\sqrt{2})$, $(9\pi/4, e^{-9\pi/4}/\sqrt{2})$, . . . ; inflection points $(\pi/2, 0)$, $(3\pi/2, 0)$, $(5\pi/2, 0)$, . . . ; concave upward for $\pi/2 < x < 3\pi/2$, $5\pi/2 < x < 7\pi/2$, downward for $0 < x < \pi/2$, $3\pi/2 < x < 5\pi/2$, . . .
(c) critical points $(0,3)$, $(\pi,-1)$; inflection points $(\pi/3, 5/4)$, $(5\pi/3, 5/4)$; concave upward for $\pi/3 < x < 5\pi/3$, downward for $0 < x < \pi/3$, $5\pi/3 < x < 2\pi$
(d) critical points $(\pi/4, 1/2)$, $(3\pi/4, -1/2)$, $(5\pi/4, 1/2)$, $(7\pi/4, -1/2)$; inflection points $(0,0)$, $(\pi/2, 0)$, $(\pi, 0)$, $(3\pi/2, 0)$; concave upward for $\pi/2 < x < \pi$, $3\pi/2 < x < 2\pi$, downward for $0 < x < \pi/2$, $\pi < x < 3\pi/2$

SECTION 5.10

1 (a) min at 0, max at 2
(b) min at -1, no max
(c) min at 2, no max

3 (a) no min, max at -3
(b) min at 0, max at -2
(c) min at 2, no max

5 (a) min at ± 2, no max
(b) min at 2, max at 5
(c) min at 2, no max

7 (a) min at -1, max at 0
(b) min at 0, max at 1
(c) min at -1, max at 1

9 (a) no min, max at 0
(b) min at $-\sqrt{8}$, max at $\sqrt{8}$
(c) min at 1, no max

11 (a) min at 0, $\pi/2$; max at $\pi/4$
(b) min at 0, max at $\pi/6$
(c) min at $3\pi/4$; max at $\pi/2$, π

13 (a) no min, max at 0
(b) no min, max at -1
(c) min at 5, max at 0

15 (a) min at 0, $\pm\pi$; max at $\pm\pi/2$
(b) min at 0, $-\pi$; max at $-\pi/2$
(c) min at $5\pi/6$, max at $\pi/2$

17 In the difference quotient in the definition of the derivative, the numerator and denominator are differences of range and domain values, respectively.

19 In Figure 12, let the point of tangency move from endpoint a to endpoint b; follow the direction of the tangent line.

SECTION 5.11

1 (a) $V = x(1-2x)(2-2x)$, $0 \le x \le \frac{1}{2}$ (b) $\sqrt{3}/9$ (approx. 0.19)

3 (a) $450c/x + (c/2)(x+1)$ (b) $x = 30$

5 10 ft by 10 ft

7 $5\sqrt{5}$ ft

9 $10/\sqrt{13}$ mi

11 $r/h = \frac{1}{2}$

13 40 or 41

15 $\sqrt{2}$ in.

17 (a) $(x+6)(600/x+8)$ (b) $240\sqrt{2}+648$ (approx. 987)

19 0.87 radians

21 (a) $\sqrt{x^2+a^2}+\sqrt{(1-x)^2+b^2}$ (b) Derivative is 0 when

$$\frac{x}{\sqrt{x^2+a^2}} = \frac{1-x}{\sqrt{(1-x)^2+b^2}}$$

(c) The three problems involve minimization of the same distances.

SECTION 5.12

1 See Table I.

2 (a) 2.03 (b) 2.52 (c) 2.11 (d) 0.37

5 (a) 1.18 (b) -0.88, 1.35, 2.58

7 -0.53

9 -1.33

11 1.59

13 1.29

15 0.44

SECTION 6.1

1 (a) 1 (b) 1

3 (a) 1/3 (b) 0.335

5 (a) 1.7183 (b) 1.7197

7 (a) 0.45970 (b) 0.45932

SECTION 6.2

1 The height of the histogram over each of the closed class intervals is twice the corresponding percentage, except for the interval from 0 to 1.0, where the height is equal to the percentage.

3 Height of histogram for the respective classes:

(a) 8 10 18.5 13.5

(b) 5 8 21.5 15.5

SECTION 6.3

1 (a) $\frac{1}{5}x^5 + C$ (b) $-x^{-1} + C$ (c) $x + 2x^{1/2} + C$ (d) $\frac{1}{3}x^3 - 2x^2 + 13x + C$
(e) $\frac{3}{5}x^5 - \frac{5}{4}x^4 - \frac{4}{3}x^3 - 2x + C$ (f) $e^x - x - \frac{1}{2}x^2 + C$ (g) $\frac{1}{2}x^2 + 3 \log x + C$
(h) $\frac{1}{3}x^3 - \cos x + C$

3 (a) $-\frac{2}{3}(1-x)^{3/2} + C$ (b) $\frac{1}{12}(2x+3)^6 + C$ (c) $\frac{1}{15}(3x^2+5)^{5/2} + C$
(d) $\frac{1}{10}(4x^2 - 2x + 3)^5 + C$ (e) $\frac{2}{3}(x^3+2)^{1/2} + C$ (f) $2 \log(x^2 + x - 7) + C$
(g) $\frac{1}{2} \log(x^2 + 2x + 5) + C$ (h) $\log(1 + \sqrt{x}) + C$

5 (a) $-(e^x + 1)^{-1} + C$ (b) $-3e^{-(1/2)x^2} + C$ (c) $-\frac{1}{2}e^{-2x} + C$ (d) $2e^{\sqrt{x}} + C$
(e) $-\frac{1}{3}\cos(3x - 2) + C$ (f) $\frac{1}{4}\sin^4 x + C$ (g) $-e^{\cos x} + C$ (h) $e^{e^x} + C$

7 $\arcsin(x/2) + C$

SECTION 6.4

1 (a) 16 (b) 140 (c) 27 (d) 9331/12 (e) $\frac{8}{3}$ (f) $\frac{3}{8}$ (g) 2
(h) $2(\sqrt{11} - 1)$

3 (a) $e^{-1} - e^{-2}$ (b) $\frac{1}{2}(1 - e^{-1})$ (c) $(e^{-1} - e^{-4})/8$ (d) $\log(1.5)$
(e) $\log 2$ (f) $\sqrt{2} - 1$ (g) 1 (h) $\frac{1}{2}\log 2$

5 (a) 32/3 (b) $\frac{8}{3}$ (c) $\log 3$ (d) 2

7 (a) 0.0326 (b) 0.0944 (c) 0.2125 (d) 0.2913 (e) 0.2125
(f) 0.0945

9 The Pareto function has a second derivative of constant sign. However, the concavity of a density fitting the income data changes from negative to positive.

SECTION 6.5

1 (a) 3 (b) $\frac{1}{12}$ (c) $\frac{1}{3}$ (d) 36 (e) 253/12 (f) $49\frac{1}{3}$ (g) $49\frac{1}{3}$
(h) 937/12

3 $(4 - \pi)/2\pi$

5 (a) $5x^2/2$, $5x^3/6$ (b) $5x^2/2 + 10$, $5x^3/6 + 10x$

7 26.4 ft

9 79 ft

11 $-\sin x$

SECTION 6.6

1 (a) $x \sin x + \cos x + C$ (b) $\frac{1}{3}x^3 \log x - x^3/9 + C$
(c) $x \arccos x - (1 - x^2)^{1/2} + C$ (d) $-x^2(1 - x^2)^{1/2} - \frac{2}{3}(1 - x^2)^{3/2} + C$
(e) $\frac{2}{3}x^{3/2} \log x - \frac{4}{9}x^{3/2} + C$ (f) $-x^2 e^{-(1/2)x^2} + 2e^{-(1/2)x^2} + C$

3 (a) $(3x^2 - 6)\sin x + (6x - x^3)\cos x + C$ (b) $\frac{1}{2}e^x(\sin x + \cos x) + C$
(c) $-e^{-x}(x^2 + 2x + 2) + C$ (d) $x(\log x)^2 - 2x \log x + 2x + C$
(e) $(x^3/3)\arcsin x + (x^2/3)(1 - x^2)^{1/2} + \frac{2}{9}(1 - x^2)^{3/2} + C$
(f) $-x^4(1 - x^2)^{1/2} - \frac{4}{3}x^2(1 - x^2)^{3/2} - (8/15)(1 - x^2)^{5/2} + C$
(g) $3x/8 - \frac{1}{4}\cos x \sin^3 x - \frac{3}{8}\cos x \sin x + C$
(h) $(1/5)\sin x \cos^4 x + (4/15)\sin x \cos^2 x + (8/15)\sin x + C$

5 0.0803, 0.2430, 0.2535, 0.1851, 0.1134, 0.0627, 0.0323, 0.0159, 0.0075, 0.0035

SECTION 6.7

1 (a) divergent (b) 1 (c) $\frac{1}{3}$ (d) divergent (e) log 2
(f) divergent (g) π (h) $\frac{1}{2}$

3 (a) convergent: $p > -1$; divergent: $p \leq -1$
(b) convergent: $p < -1$; divergent: $p \geq -1$
(c) convergent: $p > -1$; divergent: $p \leq -1$
(d) convergent: $p < -1$; divergent: $p \geq -1$

5 (a) convergent (b) divergent (c) convergent (d) convergent
(e) convergent (f) convergent

7 $F(x) = 0$ for $x \leq 0$; $= (x/30)(1 - \log(x/30))$ for $0 < x \leq 30$; $= 1$ for $x > 30$

years	area	proportion from table
0–5	0.4653	0.503
5–10	0.2342	0.210
10–15	0.1471	0.115
15–20	0.0904	0.085

9 The function $f(x - c)$ has the same graph as $f(x)$ except that it is displaced c units to the right if $c > 0$, and $-c$ units to the left if $c < 0$; hence, it is also nonnegative, and the area under it is 1.

11 $F(x - c)$

SECTION 6.8

1 (a) 2 (b) 2 (c) $\pi/2 - 1$ (d) $2/\pi$ (e) $\frac{1}{4}$ (f) $\pi - 2$ (g) $\pi/2$
(h) $\frac{1}{5}$

3 30 years

5 \$6000

7 (a) $(e - 2)/(e - 1)$ (approx. 0.418) (b) $(17e - 26)/(4e - 5)$ (approx. 3.44)
(c) $\dfrac{4\pi - \sqrt{2}\pi - 4\sqrt{2}}{8 - 4\sqrt{2}}$ (d) $6\frac{3}{7}$

9 $\dfrac{79e^{2.5} - 624}{13e^{2.5} - 68}$ (approx \$3.75 thousand)

SECTION 6.9

1 (a) $\pi/5$ (b) 3π (c) $(\pi/4)(\pi/2 + 1)$ (d) $11\pi/12$

3 (a) 8π (b) $74\pi/3$ (c) $2\pi(2\pi + 1)$ (d) 2π

5 8π

7 16π

SECTION 6.10

1 (a) $y = 3e^x - 1$ (b) $y = 2e^x - x - 1$

3 (a) $y = x^2/3 + 1 + C/x$ (b) $y = Cx^2 - x$ (c) $y = Ce^{\sin x} - 1$
(d) $y = e^{-x^2}(x^2 + C)$ (e) $y = Ce^{(1/2)x^2} - 1$ (f) $y = \frac{1}{2}\sin x + C/\sin x$

5 (a) $y = \dfrac{4}{1 + e^{-40x}}$ (b) $y = \dfrac{2}{1 + 3e^{-2x}}$ (c) $y = \dfrac{2 - \frac{5}{2}e^{x/2}}{1 - \frac{5}{2}e^{x/2}}$
(d) $y = \dfrac{70 - 60e^{-6x}}{7 - 5e^{-6x}}$

7 If $y' = py$, then $y'/y = p$, and so $\log y = \int p(x)dx + C$, or $y = Ce^{\int p(x)dx}$

9 z satisfies $z' = -ABz + A$. Write y as

$$y = \frac{B}{1 + e^{-AB\left(x - \frac{\log BC}{AB}\right)}}$$

SECTION 6.11

1 $y = 4000e^{10x + 0.07x^2}$

3 $se^{0.05x}(1000 - x^2)$. Maximum after 17.4 years.

5 Limit 197,273,000. Set $0.3134(x - 12.4) = \log 3$, or $x = 15.9$; about the year 1949.

7 10 log 4 years later, or 1984; 10 log 9 years later, or 1992.

9 $y = \dfrac{6(1 - e^{t\log(4/3)})}{2 - 3e^{t\log(4/3)}} = \dfrac{6 - 6(\frac{4}{3})^t}{2 - 3(\frac{4}{3})^t}$

11 (a) $y = \dfrac{150(1 - (105/103)^{t/10})}{1 - (6/5)(105/103)^{t/10}}$
(b) $10\,\dfrac{\log 7 - \log 6}{\log 105 - \log 103}$, or approx. 80 sec

SECTION 6.12

1 (a) $1 + 2x + 4x^2 + 8x^3$ (b) $\frac{1}{2} + (\sqrt{3}/2)x - \frac{1}{4}x^2 - (\sqrt{3}/12)x^3$
(c) $1 + \frac{1}{2}x - \frac{1}{8}x^2 + \frac{1}{16}x^3$ (d) $e^2(1 + x + \frac{1}{2}x^2 + \frac{1}{6}x^3)$ (e) $1 - \frac{3}{2}x + \frac{3}{8}x^2 + (\frac{1}{16})x^3$

3 $\int(1 - x^2 + x^4 - x^6 + \cdots)dx = x - \frac{1}{3}x^3 + (\frac{1}{5})x^5 - (\frac{1}{7})x^7 + \cdots$

5 Derivative of right-hand side:

$$\frac{1}{(1-x)^2} - \frac{(1-x)nx^{n-1} + x^n}{1-x}.$$ The second term tends to 0 for $-1 < x < 1$.

7 For $f(x) = e^x$: $f(0) = f'(0) = \cdots = 1$. For $f(x) = e^{-x}$: $f(0) = f''(0) = \cdots = 1$, $f'(0) = f^{(3)}(0) = \cdots = -1$

9 Differentiate term-by-term.

11 0.0906

13 (a) $x^2 - \frac{1}{3}x^4 + (2/45)x^6$ (b) $1 - x + (\frac{1}{3})x^3$ (c) $x^2 + x^3 + x^6$
(d) $\frac{1}{2} + \frac{1}{4}x + (\frac{3}{16})x^3$

15 $1 + x + \frac{1}{2}x^2 + 0 \cdot x^3$

17 (a) By definition of integral. (b) Write f' as $1 \cdot f'$. (c) By substitution. (d) Integrate $t - x$ with respect to t. (e) By substitution. (f) By repetition.

19 The right-hand side of (20) is $f(x_0) + f'(x_0)(x - x_0) + \frac{1}{2}f''(x_0)(x - x_0)^2 + \cdots$. Now differentiate term-by-term.

SECTION 7.1

1 (a) $\partial f/\partial x = 2x + 5$; $\partial f/\partial y = 4y$
(b) $\partial f/\partial x = 2xy^3 + 6$; $\partial f/\partial y = 3x^2y^2 + 5$
(c) $\partial f/\partial x = ye^{xy}$; $\partial f/\partial y = xe^{xy}$
(d) $\partial f/\partial x = -\sin(x + y)$; $\partial f/\partial y = -\sin(x + y)$
(e) $\partial f/\partial x = x/(x^2 + y^2)$; $\partial f/\partial y = y/(x^2 + y^2)$
(f) $\partial f/\partial x = -x(x^2 + y^2)^{-3/2}$; $\partial f/\partial y = -y(x^2 + y^2)^{-3/2}$
(g) $$\frac{\partial f}{\partial x} = \frac{-x}{(1 - x^2 - y^2)^{1/2}(x^2 + y^2)^{1/2}}$$
$$\frac{\partial f}{\partial y} = \frac{-y}{(1 - x^2 - y^2)^{1/2}(x^2 + y^2)^{1/2}}$$
(h) $$\frac{\partial f}{\partial x} = \frac{y^2}{(x^2 + y^2)^{3/2}} \cos\left(\frac{x}{(x^2 + y^2)^{1/2}}\right)$$
$$\frac{\partial f}{\partial y} = \frac{-xy}{(x^2 + y^2)^{3/2}} \cos\left(\frac{x}{(x^2 + y^2)^{1/2}}\right)$$

3 (a) $\partial^2 f/\partial x^2 = 12x^2y + 6y^3$; $\partial^2 f/\partial y^2 = 18x^2y$
$\partial^2 f/\partial x\,\partial y = \partial^2 f/\partial y\,\partial x = 4x^3 + 18xy^2$
(b) $\partial^2 f/\partial x^2 = 6y^4$; $\partial^2 f/\partial y^2 = 36x^2y^2 - 36x$
$\partial^2 f/\partial x\,\partial y = \partial^2 f/\partial y\,\partial x = 24xy^3 - 36y$
(c) $\partial^2 f/\partial x^2 = y^2e^{-xy}$; $\partial^2 f/\partial y^2 = x^2e^{-xy}$
$\partial^2 f/\partial x\,\partial y = \partial^2 f/\partial y\,\partial x = xye^{-xy} - e^{-xy}$
(d) $\partial^2 f/\partial x^2 = -y^2 \cos xy$; $\partial^2 f/\partial y^2 = -x^2 \cos xy$
$\partial^2 f/\partial x\,\partial y = \partial^2 f/\partial y\,\partial x = -\sin xy - xy \cos xy$
(e) $\partial^2 f/\partial x^2 = y^2(x^2 + y^2)^{-3/2}$; $\partial^2 f/\partial y^2 = x^2(x^2 + y^2)^{-3/2}$
$\partial^2 f/\partial x\,\partial y = \partial^2 f/\partial y\,\partial x = -xy(x^2 + y^2)^{-3/2}$
(f) $\partial^2 f/\partial x^2 = 4y(x - y)^{-3}$; $\partial^2 f/\partial y^2 = 4x(x - y)^{-3}$
$\partial^2 f/\partial x\,\partial y = \partial^2 f/\partial y\,\partial x = -2(x + y)(x - y)^{-3}$
(g) $\partial^2 f/\partial x^2 = \partial^2 f/\partial y^2 = \partial^2 f/\partial x\,\partial y = \partial^2 f/\partial y\,\partial x = -\sin(x + y)$
(h) $\partial^2 f/\partial x^2 = (2x + y)^2e^{x^2+xy} + 2e^{x^2+xy}$; $\partial^2 f/\partial y^2 = (1 + x^2)e^{x^2+xy}$;
$\partial^2 f/\partial x\,\partial y = \partial^2 f/\partial y\,\partial x = (2x^2 + xy + 1)e^{x^2+xy}$

SECTION 7.2

1 (a) $z - 83/36 = -\frac{4}{9}(x - 2) - \frac{1}{4}(y - 2)$ (b) $z - 1 = (x - 1) + (y - 1)$
(c) $z = y - 1$ (d) $z = -\frac{1}{2}(x - \pi/2) - (\pi/2)(y - \frac{1}{2})$ (e) $z = y$ (f) $z = \frac{1}{2}$

3 (a) 0.034 (b) 0.01 (c) 0.1352 (d) $\sqrt{2}\pi(0.00375)$ (e) -0.1
(f) 0.02

5 (a) $6(3t + 1)(2 - 3t)(1 - 6t)$ (b) $2(t + 1)^3(t^3 + 3)(5t^3 + 3t^2 + 6)$
(c) $(t^2 + 6t + 1)(t + 3)^{-2}$ (d) $\dfrac{\cos t}{1 + \cos^2 t}\left[1 + \dfrac{2 \sin^2 t}{1 + \cos^2 t}\right]$

7 $-1.12\pi/3$

9 If $z = xy$, then $dz = y\,dx + x\,dy$.

11 Pct. change in output $= p$ (pct. change in labor) $+\ q$ (pct. change in capital)

13 $137\frac{1}{2}$ thousand

15 (a) $0.055e^{-2}$ (b) Use: $dA = -yA\,dx - xA\,dy$

17 $(\partial f/\partial x)/R,\ (\partial f/\partial y)/R,\ -1/R$, where $R = (1 + (\partial f/\partial x)^2 + (\partial f/\partial y)^2)^{1/2}$

SECTION 7.3

1 (a) saddlepoint $(-5,3)$ (b) cup at $(0,0)$ (c) cup at $(\frac{3}{8},\frac{1}{8})$
(d) cap at $(1,3)$ (e) saddlepoint $(0,0)$, cup at $(\frac{2}{3},\frac{2}{3})$
(f) degenerate points $(x,0)$ for any x, $(0,y)$ for any y, cap $(3,2)$
(g) saddlepoint $(1,0)$, cup at $(\frac{5}{3},\frac{2}{3})$
(h) degenerate at every point $(0,y)$, saddlepoint $(1,0)$, cup at $(\frac{1}{2},-\frac{1}{4})$
(i) cup at $(-2,1)$ (j) cap at $(0,0)$, four saddlepoints $(\pm\sqrt{5},\pm\sqrt{5})$
(k) no critical points (l) Every point (x,y) such that $x + y$ is an integer multiple of π is a degenerate point.

3 (a) $f(4,0) = 4,\ f(0,2) = -2$ (b) $f(3,0) = f(-3,0) = 9,\ f(0,0) = 0$
(c) $f(3,0) = 9,\ f(0,2) = -8$

5 (a) $(\frac{1}{3},\frac{1}{3},\frac{1}{3})$ (b) $(\frac{2}{3},\frac{2}{3},-\frac{1}{3})$ (c) $(-\frac{1}{3},-\frac{4}{5},-\frac{5}{3})$

7 1/54

9 $\frac{1}{2}(3V)^{1/3},\ \frac{1}{2}(3V)^{1/3},\ 4(V/9)^{1/3}$

11 50/3, 50/3, 9/25

13 $(\frac{2}{3},0)$

15 $\left(\dfrac{x_1 + \cdots + x_m}{m}, \dfrac{y_1 + \cdots + y_m}{m}\right)$

17 Follow given hint.

SECTION 7.4

1 (a) $f(-\frac{2}{3},\frac{4}{3}) = 16/3$ max. No min.
(b) $f(\sqrt{2},\sqrt{2}) = f(-\sqrt{2},-\sqrt{2}) = 2$
$f(-\sqrt{2},\sqrt{2}) = f(\sqrt{2},-\sqrt{2}) = -2$
(c) $f(\frac{3}{5},\frac{4}{5}) = 16,\ f(-\frac{3}{5},-\frac{4}{5}) = 36$
(d) $f(\frac{1}{2},\frac{1}{2}) = \frac{1}{4},\ f(0,y) = f(x,0) = 0$ for any x and y
(e) $f(-\frac{3}{5},-\frac{6}{5}) = \sqrt{5}/3$ max. No min.

3 $(\frac{3}{5},\frac{4}{5})$

5 closest $(12/\sqrt{13},-18/\sqrt{13})$, farthest $(-12/\sqrt{13},18/\sqrt{13})$

7 p/rq

9 $x = \pm\dfrac{a + 3c}{((a + 3c)^2 + (b + 3d)^2)^{1/2}},\quad y = \pm\dfrac{b + 3d}{((a + 3c)^2 + (b + 3d)^2)^{1/2}}$

SECTION 7.5

1 (a) 12 (b) 16/3 (c) $\frac{1}{2}\log 2$ (d) 419/21,600
(e) $(19-5\sqrt{5})/3$ (f) $\frac{1}{4}$ (g) 0 (h) $\log \dfrac{2(e+e^{-1})}{(e+1)(1+e^{-1})}$

3 Use the fact that $(d/dx)(1-e^{-x^2/32}) = (x/16)e^{-x^2/32}$, and the corresponding formula for the derivative of $(1+e^{-3.4(y-3.25)})^{-1}$.

5 (a) Use $\int 2xe^{-2xy}\,dy = -e^{-2xy} + C$.
(b) Use (a) above and integrate by parts.
(c) Integral over y is $\arctan(x+y)$.
(d) Use $f(x,y) = \phi(x)\phi(x-y)$

7 Compare $\frac{1}{2}(1-e^{-2}) - \frac{1}{3}(1-e^{-3})$ to $(1-e^{-1})/6$; not independent.

9 (a) The integral represents the area under the semicircle $x^2+y^2=r^2$ and over the x-axis.
(b) By (a) and volume slicing. Volume of half-cylinder $= \pi r^2 h/2$.

11 0

13 (a) By symmetry the "masses" (probabilities) at corresponding points on either side of the x- (or y-) axis are equal.
(b) $$\iint_{\text{quadrant I}} xy\,\frac{f(\sqrt{x^2+y^2})}{\sqrt{x^2+y^2}}\,dx\,dy = -\iint_{\text{quadrant II}} xy\,\frac{f(\sqrt{x^2+y^2})}{\sqrt{x^2+y^2}}\,dx\,dy$$
(c) $(1/2\pi)\,\dfrac{f(\sqrt{x^2+y^2})}{\sqrt{x^2+y^2}}$ is generally not the product of its marginal densities.

SECTION 7.6

1 (a) $\frac{1}{2}$ (b) $\frac{3}{2}$ (c) $\log 2$ (d) $(2\sqrt{2}-1)/15$ (e) $\frac{1}{2}\log\frac{3}{2}$ (f) $\frac{1}{2}\log 2$
(g) $\frac{1}{8}$ (h) $\pi/2$

3 $37\frac{1}{2}$

5 $(\pi/2)(1-e^{-1/2})$

7 $3\pi - 7/3$

9 1

SECTION 8.1

1 48

3 12, 144

5 36, 36^2

7 24

9 18

11 (a) 120 (b) 755,160 (c) 5040 (d) 360,360 (e) 840

13 (a) 30 (b) 120 (c) 720

15 30,240

17 254,251,200

19 7,880,400

SECTION 8.2

1 (a) 15 (b) 4060 (c) 1 (d) 220 (e) 220

3 286

5 142,506

7 126

9 2,598,960

11 200

13 $C(100,15) \cdot C(85,25)$

15 (a) 66; (b) 11; (c) 55

SECTION 8.3

1 (a) $a^6 + 6a^5x + 15a^4x^2 + 20a^3x^3 + 15a^2x^4 + 6ax^5 + x^6$
(b) $x^4 - 8x^3y + 24x^2y^2 - 32xy^3 + 16y^4$
(c) $a^5x^5 - 5a^4x^4b + 10a^3x^3b^2 - 10a^2x^2b^3 + 5axb^4 - b^5$
(d) $x^7(x^7 + 7x^6 + 21x^5 + 35x^4 + 35x^3 + 21x^2 + 7x + 1)$
(e) $\frac{1}{32}x^5 + \frac{5}{8}x^4 + 5x^3 + 20x^2 + 40x + 32$

3 (a) $y^{10} - 30y^9x + 405y^8x^2 - 3240y^7x^3$
(b) $128x^{14} + 224x^{12} + 168x^{10} + 70x^8$
(c) $n^9 - 18n^8 + 144n^7 - 672n^6$
(d) $256a^8 + 1024a^6 + 1792a^4 + 1792a^2$
(e) $(1/243)x^5 - (5/162)x^4y + (5/54)x^3y^2 - (5/36)x^2y^3$

5 (a) $C(15,4)x^{11}2^4$ (b) $C(16,12)x^4(y/2)^{12}$ (c) $-C(9.3)(2n)^6 3^3$

7 (a) 1.0615 (b) 0.9415 (c) 1.0040 (d) 0.8681

9 \$106.14; \$112.64

SECTION 8.4

1 (a) By definition. (b) By (a), with $n - 1$ and $r - 1$ in place of n and r, respectively. (c) Factorization

SECTION 8.5

1 (a) $\frac{3}{8}$ (b) $\frac{1}{8}$ (c) $\frac{15}{16}$ (d) $\frac{11}{16}$ (e) $\frac{13}{16}$ (f) $495/2^{12}$ (g) $\frac{5}{16}$
(h) 63/256

3 1/36, 1/18, 1/6, 1/9

5 193/512 in all cases

7 (a) $\frac{1}{16}$ (b) $\frac{1}{8}$ (c) $\frac{1}{8}$ (d) $\frac{5}{16}$ (e) $\frac{5}{16}$

9 $\frac{1}{2}$

11 (a) 49/198 (b) $\frac{1}{2}$ (c) 50/198

13 $\frac{1}{2}$

15 (a) 72/73 (b) 71/73

17 $\frac{1}{6}$

SECTION 8.6

1 (a) $\frac{1}{6}$, $\frac{4}{6}$, $\frac{1}{6}$ (b) $\frac{1}{10}$, $\frac{6}{10}$, $\frac{3}{10}$ (c) $\frac{3}{10}$, $\frac{6}{10}$, $\frac{1}{10}$ (d) $\frac{1}{20}$, $\frac{9}{20}$, $\frac{9}{20}$, $\frac{1}{20}$

3 (a) 1/120 (b) 1/20 (c) 19/20 (d) 3/10

5 442/490

7 (a) 232/323 (b) 241/969

9 884/1265

11 (a) 18/35 (b) 2/5 (c) 56/165

SECTION 8.7

1 (a) 0.1800 (b) 0.0036 (c) 0.0270 (d) 0.0400 (e) 0.5120
(f) 0.4200 (g) 0.0270 (h) 0.4116 (i) 0.0938

3 (a) 0.1900 (b) 0.9963 (c) 0.1040 (d) 0.8704 (e) 0.7734

5 23/648

7 0.0001

9 4, 7, 0.1445, 0.7969

11 0.0000

13 0.7

SECTION 8.8

1 (a) with: 0.16, 0.48, 0.36; without: 0.10, 0.60, 0.30
(b) with: $\frac{1}{8}$, $\frac{3}{8}$, $\frac{3}{8}$, $\frac{1}{8}$; without: $\frac{1}{12}$, $\frac{5}{12}$, $\frac{5}{12}$, $\frac{1}{12}$
(c) with: $\frac{1}{4}$, $\frac{1}{2}$, $\frac{1}{4}$; without: $\frac{49}{198}$, $\frac{100}{198}$, $\frac{49}{198}$

3 0.9185, 0.7373, 0.5283, 0.3370

5 0.7565

7 0.0159

9 0.2639

11 0.9037

13 131/243

15 $P(n,r)$ = No. of samples with distinct tickets; n^r = No. of samples

17 0.0271

SECTION 8.9

1 (a) 0.4830 (b) 0.2422 (c) 0.9251 (d) 0.1056 (e) 0.1736
(f) 0.6443 (g) 0.1793 (h) 0.5288

3 (a) 1.14 (b) 1.52 (c) 1.69 (d) −0.90 (e) −0.97 (f) 0.04

5 (a) binomial 0.3750, normal 0.3770 (b) binomial 0.2188, normal 0.2186
(c) binomial 0.3125, normal 0.3133 (d) binomial 0.2734, normal 0.2764
(e) binomial 0.1641, normal 0.1596

7 (a) 0.0914 (b) 0.9078 (c) 0.6568

9 (a) 0.0052 (b) 0.0007

11 (a) 0.6318 (b) 0.9982 (c) 1.0000

13 (a) $y = \phi(x + m)$ (b) $y = s\phi(xs)$ (c) $y = s\phi(s(x + m))$

SECTION 8.10

1 (a) 0.1709 (b) 0.7887 (c) 0.0700 (d) 0.8023 (e) 0.1977
(f) 0.9474

3 (a) 8 out of 36 (b) 0.0016

5 0.0668

7 0.0031, 0.0000, 0.5199, 0.5160

9 (a) 0.09 (b) 0.12

11 0.04

13 0.9550

15 598

17 (a) 0.0262 (b) 0.0044

19 The absolute difference between $\hat{p}$ and p is less than or equal to $z_\alpha/2\sqrt{n}$ if and only if $np - z_\alpha\sqrt{n}/2 \leq$ No. of successes $\leq np + z_\alpha\sqrt{n}/2$.

SECTION 8.11

1 (a) 0.1680 (b) 0.1277 (c) 0.0120 (d) 0.0803 (e) 0.9826
(f) 0.9864 (g) 0.9787 (h) 0.9542

3 18 claims, 45¢

5 0.7148

7 0.9864

9 0.9826

11 58, 29, 7, 1, 0

13 0.0293

15 5

17 10

SECTION 8.12

1 (a) A: {HHH HTT THT TTH HHT HTH THH}
B: {TTT THH HTH HHT TTH THT HTT}
C: {HHH HHT HTH THH}

(b) $A \cup B$, $B \cup C$, $A \cup B \cup C$ contain the entire sample space. $A \cup C = A$.
(c) $A \cap B$: All outcomes except HHH, TTT. $A \cap C = C$. $B \cap C$ contains HHT, HTH, THH. $A \cap B \cap C = B \cap C$.
(d) ${}^cA$ contains only TTT. ${}^cB$ contains HHH. ${}^cC$ contains TTT, TTH, THT, HTT.

3 A: All outcomes except the six (1 1), (1 2), (1 3), (2 1), (2 2), (3 1).
B: (1 1), (1 3), (1 5), (2 2), (2 4), . . . , (6 4), (6 6).
$A \cup B$: All outcomes except (1 2) and (2 1).
$A \cap B$: All outcomes in B except (1 1), (1 3), (2 2), (3 1).

5 $P(A_1) = 0.5394$; $P(A_2) = 0.4606$; $P(B_1) = 0.0084$
$P(B_2) = 0.0090$; $P(B_3) = 0.1764$; $P(B_4) = 0.3876$
$P(B_5) = 0.3501$; $P(B_6) = 0.0672$
$P(A_1 \mid B_3) = 0.5165$; $P(A_1 \mid B_4) = 0.6333$; $P(A_1 \mid B_5) = 0.4815$;
$P(A_1 \mid B_6) = 0.3846$; $P(B_3 \mid A_1) = 0.1689$; $P(B_4 \mid A_1) = 0.4551$;
$P(B_5 \mid A_1) = 0.3126$; $P(B_6 \mid A_1) = 0.0479$

7 80/179

9 25/36, 10/36, 1/36

11 Every term in the sum forming $P(A)$ is included in the corresponding sum for $P(B)$.

SECTION 8.13

1
x	2	3	
$P(X=x)$	$\frac{3}{4}$	$\frac{1}{4}$	$E(X) = 2\frac{1}{4}$

3
x	−2	−1	0	1	2	
$P(X=x)$	$\frac{1}{16}$	$\frac{1}{4}$	$\frac{3}{8}$	$\frac{1}{4}$	$\frac{1}{16}$	$E(X) = 0$

5
x	4	5	6	7	
$P(X=x)$	$\frac{1}{8}$	$\frac{1}{4}$	$\frac{5}{16}$	$\frac{5}{16}$	$E(X) = 5\frac{13}{16}$

7
x	0	1	2	4	
$P(X=x)$	$\frac{9}{24}$	$\frac{1}{3}$	$\frac{1}{4}$	$\frac{1}{24}$	$E(X) = 1$

9 10/9

11
x	0	1	2	3	4
$P(X=x)$	$\frac{6}{16}$	$\frac{4}{16}$	$\frac{4}{16}$	$\frac{1}{16}$	$\frac{1}{16}$

13
x	3	4	5	6
$P(X=x)$	0.0640	0.2448	0.3456	0.3456

15 (a) $P(X=1) = (1-p)^k$; $P(X=k+1) = 1-(1-p)^k$
$E(X) = (1-p)^k + (k+1)(1-(1-p)^k)$
(b) $k - E(X) = k(1-p)^k - 1$ (c) 3.76

SECTION 9.1

1 (a) $F(x) = 0$ for $x \le 0$; $= x^2$ for $0 < x < 1$; $= 1$ for $x \ge 1$. probability $= \frac{1}{4}$
(b) $F(x) = 0$ for $x \le 0$; $= \sqrt{x}$ for $0 < x < 1$; $= 1$ for $x \geqslant 1$. probability $= \frac{1}{2}$
(c) $F(x) = 0$ for $x \le -1$; $= \frac{1}{2}(1+x)^2$ for $-1 < x \le 0$; $= 1 - \frac{1}{2}(1-x)^2$ for $0 < x \le 1$; $= 1$ for $x > 1$. probability $= 263/288$
(d) $F(x) = 0$ for $x \le 0$; $= (2/\pi)\arcsin x$ for $0 < x < 1$; $= 1$ for $x \ge 1$. probability $= \frac{1}{3}$
(e) $F(x) = 1/(1 + e^{-x})$ for all x. probability $= (e-1)/(e+1)$

3 $A = 2^{1/4}$; $B = 0.2$

5 (a) $1-(1-p)^n$ (b) $1-(1-x/n)^n \to 1-e^{-x}$
(c) $X \leqslant n$ if and only if there is at least one success in n trials.

7 $F(x) = 0$ for $x \leq 0$; $= 1-(1-x)^{3/2}$ for $0 < x < 1$; $= 1$ for $x \geq 1$. $A = 16/25$

9 (a) $k = 10$ (b) $(\frac{3}{4})^5$

11 jump of $\frac{1}{10}$ at each point 0.1, 0.2, ...

13 $(1/\pi)\arctan x + \frac{1}{2}$

15 0.3211

17 $u = (x-\mu)/\sigma$, $d\Phi/dx = d\Phi/du \cdot du/dx$

19 (a) 0.6826 (b) 0.9974 (c) 0.9996

SECTION 9.2

1 (a) $E(X) = 5\frac{2}{3}$; $\text{Var}(X) = 11\frac{2}{9}$ (b) $E(X) = 1.0$; $\text{Var}(X) = 13.4$
(c) $E(X) = 9$; $\text{Var}(X) = 14\frac{2}{3}$

3 (a) $E(X) = 45/28$; $\text{Var}(X) = 0.07$ (b) $E(X) = 13/6$; $\text{Var}(X) = 11/36$
(c) $E(X) = 0$; $\text{Var}(X) = \frac{1}{2}$ (d) $E(X) = \frac{1}{3}$; $\text{Var}(X) = 4/45$
(e) $E(X) = \frac{3}{5}$; $\text{Var}(X) = 1/25$

5 $E(X) = (a+b)/2$; $\text{Var}(X) = (a-b)^2/12$

7 (a) $E(X) = \frac{7}{2}$; $\text{Var}(X) = 35/12$ (b) $E(X) = 7$; $\text{Var}(X) = 35/6$

9 (a) The values $y_1, y_2, \ldots$ are assumed with probabilities $p_1, p_2, \ldots$, respectively.
(b) Use the result of (a), the definition of $E(Y)$, and the relation $y = g(x)$.

11 $(d/dp)(C(n,0) + C(n,1)pq^{n-1} + C(n,2)p^2q^{n-2} + \cdots) = C(n,1)q^{n-1} + 2C(n,2)pq^{n-2} + \cdots$; note also that $q + p = 1$.

13 (a) $E(\hat{p}) = p$; $\text{Var}(\hat{p}) = pq/n$ (b) Use (a). The limit is equal to 0.

15 (a) $e^\lambda = \Sigma k\lambda^{k-1}/k!$ (b) $\lambda = \Sigma ke^{-\lambda}\lambda^k/k!$

17
$$E(aX+b) = \int_{-\infty}^{\infty} (ax+b)f(x)\,dx = a\int_{-\infty}^{\infty} xf(x)\,dx + b\int_{-\infty}^{\infty} f(x)\,dx$$
$$= aE(X) + b$$

19
$$\text{Var}(aX) = \int_{-\infty}^{\infty} (ax - E(aX))^2 f(x)\,dx = a^2\int_{-\infty}^{\infty} (x-E(X))^2 f(x)\,dx$$
$$= a^2\,\text{Var}(X)$$

SECTION 9.3

1 (a) $\dfrac{e^{-n\lambda}\lambda^{(x_1+\cdots+x_n)}}{x_1!\cdots x_n!}$ (b) $2^{-n}e^{-|x_1|-\cdots-|x_n|}$
(c) $\dfrac{1}{\pi^n(1+x_1^2)\cdots(1+x_n^2)}$
(d) $6^{-n}(x_1\cdots x_n)^3e^{-x_1-\cdots-x_n}$, if all $x_i > 0$
0, if at least one $x_i \leq 0$

3 average $\bar{x}$

5 $1/\bar{x}$

7 Put

$$f(x,\mu,\sigma^2) = \frac{1}{\sqrt{2\pi\sigma^2}} e^{-(1/2\sigma^2)(x-\mu)^2};$$

then $\partial \log f/\partial\mu = (x-\mu)/\sigma$, $\partial \log f/\partial\sigma^2 = -1/2\sigma^2 + (x-\mu)^2/2\sigma^4$. The likelihood equations are

$$(x_1-\mu)/\sigma + \cdots + (x_n-\mu)/\sigma = 0,$$
$$-n/2\sigma^2 + ((x_1-\mu)^2 + \cdots + (x_n-\mu)^2)/2\sigma^4 = 0.$$

9 Note relation to Bernoulli trials with probability $p = F(x)$ of success.

11 Distribution $F(x) = 0$ for $x \le 0$; $= x^n$ for $0 < x < 1$; $= 1$ for $x \ge 1$. Density nx^{n-1} for $0 < x < 1$, and 0 elsewhere.

13 Distribution: $1 - (1 - F(x))^n$; density $n(1 - F(x))^{n-1} f(x)$

15 $e^4 - 1$ (approx. 54)

17 $3e^{-4} - 2e^{-6}$ (approx. 0.04999)

19 Apply the definitions of conditional probability and independence.

SECTION 9.4

1 (a)

−2	0	1	2	3	4	5	6
0.04	0.12	0.16	0.13	0.24	0.22	0.08	0.01

(b)

0	1	2	3	4	5	6	7	8
0.01	0.04	0.10	0.18	0.23	0.22	0.15	0.06	0.01

3 (a) 4.0, 1.2 (b) 3.8, 9.78

5 (a)

−3	−1	1	3
$\frac{1}{8}$	$\frac{3}{8}$	$\frac{3}{8}$	$\frac{1}{8}$

(b)

−3	−2	−1	0	1	2	3
0.008	0.060	0.186	0.305	0.279	0.135	0.027

(c)

0	1	2	3	4	5	6
1/27	3/27	6/27	7/27	6/27	3/27	1/27

7 (a) $h(t) = t$ for $0 < t \le 1$; $= 2 - t$ for $1 < t \le 2$; $= 0$ elsewhere
(b) density $= \frac{1}{2}t^2$ for $0 < t \le 1$; $= \frac{3}{4} - (t - \frac{3}{2})^2$ for $1 < t \le 2$; $= \frac{1}{2}(3-t)^2$ for $2 < t \le 3$; $= 0$ elsewhere

9 (a) $h(t) = \frac{2}{3}t^3$ for $0 < t < 1$; $= 4t - \frac{2}{3}t^3 - \frac{8}{3}$ for $1 \le t < 2$; $= 0$ elsewhere
(b) $h(t) = 0$ for $t \le 0$; $= \frac{1}{6}t^3 e^{-t}$ for $t > 0$
(c) normal density with mean 2, standard deviation $3\sqrt{2}$

11 (a) $P(Y \le x) = P(X \le (x-d)/c)$
(b) Apply this result with $c = 1/n$, $d = 0$.
(c) This involves the distribution function, not just the moments in (8) and (9) of Section 9.2.

13 $E(X_1 + \cdots + X_n) = E(X_1) + \cdots + E(X_n) = p + \cdots + p = np$. Similarly for the variance.

15 (a)

k	0	1	2	3
$P(X_1 + X_2 = k)$	$e^{-2\lambda}$	$2\lambda e^{-2\lambda}$	$2\lambda^2 e^{-2\lambda}$	$\frac{4}{3}\lambda^3 e^{-2\lambda}$

(b) Poisson with 2λ in place of λ.

17 Chebyshev bound: 0.1481; exact probability 0.0094

19 $E(h(X_1,X_2)) = \int\int g_1(x)g_2(y)f(x)f(y)\,dx\,dy$
$= \int g_1(x)f(x)\,dx \cdot \int g_2(x)f(x)\,dx$

21 (a) $F(x)$, $F(x)(1-F(x))/n$
(b) Law of Large Numbers

SECTION 9.5

1 (a) 1.2 ± 0.620 (b) 0.2 ± 0.98 (c) -1.3 ± 1.04 (d) 8.3 ± 2.06

3 (a) $\alpha = 0.0571$; $\beta = 0.0571$ (b) $\alpha = 0.0548$; $\beta = 0.0082$
(c) $\alpha = 0.1314$; $\beta = 0.0838$ (d) $\alpha = 0.0000$; $\beta = 0.0228$

5 (a) 68 (b) 166

7 20 ± 1.16

9 rejected

11 Yes, $c = 1.03$

13 No, $c = 48.0$

$$\Phi\left(\frac{c-\mu_0}{\sigma/\sqrt{n}}\right) = \Phi(-z_{2\alpha}) = \alpha$$

$$\Phi\left(\frac{-c}{\sigma/\sqrt{n}}\right) + 1 - \Phi\left(\frac{c}{\sigma/\sqrt{n}}\right) = \Phi(-z_\alpha) + 1 - \Phi(z_\alpha) = \alpha$$

Appendix

Table I Powers, roots, reciprocals*

N	N^2	$\sqrt{N}$	$\sqrt{10N}$	N^3	$\sqrt[3]{N}$	$\sqrt[3]{10N}$	$\sqrt[3]{100N}$	$1000/N$
1	1	1.00 000	3.16 228	1	1.00 000	2.15 443	4.64 159	1000.00
2	4	1.41 421	4.47 214	8	1.25 992	2.71 442	5.84 804	500.00 0
3	9	1.73 205	5.47 723	27	1.44 225	3.10 723	6.69 433	333.33 3
4	16	2.00 000	6.32 456	64	1.58 740	3.41 995	7.36 806	250.00 0
5	25	2.23 607	7.07 107	125	1.70 998	3.68 403	7.93 701	200.00 0
6	36	2.44 949	7.74 597	216	1.81 712	3.91 487	8.43 433	166.66 7
7	49	2.64 575	8.36 660	343	1.91 293	4.12 129	8.87 904	142.85 7
8	64	2.82 843	8.94 427	512	2.00 000	4.30 887	9.28 318	125.00 0
9	81	3.00 000	9.48 683	729	2.08 008	4.48 140	9.65 489	111.11 1
10	100	3.16 228	10.00 00	1 000	2.15 443	4.64 159	10.00 00	100.00 0
11	121	3.31 662	10.48 81	1 331	2.22 398	4.79 142	10.32 28	90.90 91
12	144	3.46 410	10.95 45	1 728	2.28 943	4.93 242	10.62 66	83.33 33
13	169	3.60 555	11.40 18	2 197	2.35 133	5.06 580	10.91 39	76.92 31
14	196	3.74 166	11.83 22	2 744	2.41 014	5.19 249	11.18 69	71.42 86
15	225	3.87 298	12.24 74	3 375	2.46 621	5.31 329	11.44 71	66.66 67
16	256	4.00 000	12.64 91	4 096	2.51 984	5.42 884	11.69 61	62.50 00
17	289	4.12 311	13.03 84	4 913	2.57 128	5.53 966	11.93 48	58.82 35
18	324	4.24 264	13.41 64	5 832	2.62 074	5.64 622	12.16 44	55.55 56
19	361	4.35 890	13.78 40	6 859	2.66 840	5.74 890	12.38 56	52.63 16
20	400	4.47 214	14.14 21	8 000	2.71 442	5.84 804	12.59 92	50.00 00
21	441	4.58 258	14.49 14	9 261	2.75 892	5.94 392	12.80 58	47.61 90
22	484	4.69 042	14.83 24	10 648	2.80 204	6.03 681	13.00 59	45.45 45
23	529	4.79 583	15.16 58	12 167	2.84 387	6.12 693	13.20 01	43.47 83
24	576	4.89 898	15.49 19	13 824	2.88 450	6.21 446	13.38 87	41.66 67
25	625	5.00 000	15.81 14	15 625	2.92 402	6.29 961	13.57 21	40.00 00
26	676	5.09 902	16.12 45	17 576	2.96 250	6.38 250	13.75 07	38.46 15
27	729	5.19 615	16.43 17	19 683	3.00 000	6.46 330	13.92 48	37.03 70
28	784	5.29 150	16.73 32	21 952	3.03 659	6.54 213	14.09 46	35.71 43
29	841	5.38 516	17.02 94	24 389	3.07 232	6.61 911	14.26 04	34.48 28
30	900	5.47 723	17.32 05	27 000	3.10 723	6.69 433	14.42 25	33.33 33
31	961	5.56 776	17.60 68	29 791	3.14 138	6.76 790	14.58 10	32.25 81
32	1 024	5.65 685	17.88 85	32 768	3.17 480	6.83 990	14.73 61	31.25 00
33	1 089	5.74 456	18.16 59	35 937	3.20 753	6.91 042	14.88 81	30.30 30
34	1 156	5.83 095	18.43 91	39 304	3.23 961	6.97 953	15.03 69	29.41 18
35	1 225	5.91 608	18.70 83	42 875	3.27 107	7.04 730	15.18 29	28.57 14
36	1 296	6.00 000	18.97 37	46 656	3.30 193	7.11 379	15.32 62	27.77 78
37	1 369	6.08 276	19.23 54	50 653	3.33 222	7.17 905	15.46 68	27.02 70
38	1 444	6.16 441	19.49 36	54 872	3.36 198	7.24 316	15.60 49	26.31 58
39	1 521	6.24 500	19.74 84	59 319	3.39 121	7.30 614	15.74 06	25.64 10
40	1 600	6.32 456	20.00 00	64 000	3.41 995	7.36 806	15.87 40	25.00 00
41	1 681	6.40 312	20.24 85	68 921	3.44 822	7.42 896	16.00 52	24.39 02
42	1 764	6.48 074	20.49 39	74 088	3.47 603	7.48 887	16.13 43	23.80 95
43	1 849	6.55 744	20.73 64	79 507	3.50 340	7.54 784	16.26 13	23.25 58
44	1 936	6.63 325	20.97 62	85 184	3.53 035	7.60 590	16.38 64	22.72 73
45	2 025	6.70 820	21.21 32	91 125	3.55 689	7.66 309	16.50 96	22.22 22
46	2 116	6.78 233	21.44 76	97 336	3.58 305	7.71 944	16.63 10	21.73 91
47	2 209	6.85 565	21.67 95	103 823	3.60 883	7.77 498	16.75 07	21.27 66
48	2 304	6.92 820	21.90 89	110 592	3.63 424	7.82 974	16.86 87	20.83 33
49	2 401	7.00 000	22.13 59	117 649	3.65 931	7.88 374	16.98 50	20.40 82
50	2 500	7.07 107	22.36 07	125 000	3.68 403	7.93 701	17.09 98	20.00 00

* From *Rinehart Mathematical Tables, Formulas and Curves,* compiled by Harold D. Larsen.

Table I (cont'd.) Powers, roots, reciprocals

N	N^2	$\sqrt{N}$	$\sqrt{10N}$	N^3	$\sqrt[3]{N}$	$\sqrt[3]{10N}$	$\sqrt[3]{100N}$	$1000/N$
51	2 601	7.14 143	22.58 32	132 651	3.70 843	7.98 957	17.21 30	19.60 78
52	2 704	7.21 110	22.80 35	140 608	3.73 251	8.04 145	17.32 48	19.23 08
53	2 809	7.28 011	23.02 17	148 877	3.75 629	8.09 267	17.43 51	18.86 79
54	2 916	7.34 847	23.23 79	157 464	3.77 976	8.14 325	17.54 41	18.51 85
55	3 025	7.41 620	23.45 21	166 375	3.80 295	8.19 321	17.65 17	18.18 18
56	3 136	7.48 331	23.66 43	175 616	3.82 586	8.24 257	17.75 81	17.85 71
57	3 249	7.54 983	23.87 47	185 193	3.84 850	8.29 134	17.86 32	17.54 39
58	3 364	7.61 577	24.08 32	195 112	3.87 088	8.33 955	17.96 70	17.24 14
59	3 481	7.68 115	24.28 99	205 379	3.89 300	8.38 721	18.06 97	16.94 92
60	3 600	7.74 597	24.49 49	216 000	3.91 487	8.43 433	18.17 12	16.66 67
61	3 721	7.81 025	24.69 82	226 981	3.93 650	8.48 093	18.27 16	16.39 34
62	3 844	7.87 401	24.89 98	238 328	3.95 789	8.52 702	18.37 09	16.12 90
63	3 969	7.93 725	25.09 98	250 047	3.97 906	8.57 262	18.46 91	15.87 30
64	4 096	8.00 000	25.29 82	262 144	4.00 000	8.61 774	18.56 64	15.62 50
65	4 225	8.06 226	25.49 51	274 625	4.02 073	8.66 239	18.66 26	15.38 46
66	4 356	8.12 404	25.69 05	287 496	4.04 124	8.70 659	18.75 78	15.15 15
67	4 489	8.18 535	25.88 44	300 763	4.06 155	8.75 034	18.85 20	14.92 54
68	4 624	8.24 621	26.07 68	314 432	4.08 166	8.79 366	18.94 54	14.70 59
69	4 761	8.30 662	26.26 79	328 509	4.10 157	8.83 656	19.03 78	14.49 28
70	4 900	8.36 660	26.45 75	343 000	4.12 129	8.87 904	19.12 93	14.28 57
71	5 041	8.42 615	26.64 58	357 911	4.14 082	8.92 112	19.22 00	14.08 45
72	5 184	8.48 528	26.83 28	373 248	4.16 017	8.96 281	19.30 98	13.88 89
73	5 329	8.54 400	27.01 85	389 017	4.17 934	9.00 411	19.39 88	13.69 86
74	5 476	8.60 233	27.20 29	405 224	4.19 834	9.04 504	19.48 70	13.51 35
75	5 625	8.66 025	27.38 61	421 875	4.21 716	9.08 560	19.57 43	13.33 33
76	5 776	8.71 780	27.56 81	438 976	4.23 582	9.12 581	19.66 10	13.15 79
77	5 929	8.77 496	27.74 89	456 533	4.25 432	9.16 566	19.74 68	12.98 70
78	6 084	8.83 176	27.92 85	474 552	4.27 266	9.20 516	19.83 19	12.82 05
79	6 241	8.88 819	28.10 69	493 039	4.29 084	9.24 434	19.91 63	12.65 82
80	6 400	8.94 427	28.28 43	512 000	4.30 887	9.28 318	20.00 00	12.50 00
81	6 561	9.00 000	28.46 05	531 441	4.32 675	9.32 170	20.08 30	12.34 57
82	6 724	9.05 539	28.63 56	551 368	4.34 448	9.35 990	20.16 53	12.19 51
83	6 889	9.11 043	28.80 97	571 787	4.36 207	9.39 780	20.24 69	12.04 82
84	7 056	9.16 515	28.98 28	592 704	4.37 952	9.43 539	20.32 79	11.90 48
85	7 225	9.21 954	29.15 48	614 125	4.39 683	9.47 268	20.40 83	11.76 47
86	7 396	9.27 362	29.32 58	636 056	4.41 400	9.50 969	20.48 80	11.62 79
87	7 569	9.32 738	29.49 58	658 503	4.43 105	9.54 640	20.56 71	11.49 43
88	7 744	9.38 083	29.66 48	681 472	4.44 796	9.58 284	20.64 56	11.36 36
89	7 921	9.43 398	29.83 29	704 969	4.46 475	9.61 900	20.72 35	11.23 60
90	8 100	9.48 683	30.00 00	729 000	4.48 140	9.65 489	20.80 08	11.11 11
91	8 281	9.53 939	30.16 62	753 571	4.49 794	9.69 052	20.87 76	10.98 90
92	8 464	9.59 166	30.33 15	778 688	4.51 436	9.72 589	20.95 38	10.86 96
93	8 649	9.64 365	30.49 59	804 357	4.53 065	9.76 100	21.02 94	10.75 27
94	8.836	9.69 536	30.65 94	830 684	4.54 684	9.79 586	21.10 45	10.63 83
95	9 025	9.74 679	30.82 21	857 375	4.56 290	9.83 048	21.17 91	10.52 63
96	9 216	9.79 796	30.98 39	884 736	4.57 886	9.86 485	21.25 32	10.41 67
97	9 409	9.84 886	31.14 48	912 673	4.59 470	9.89 898	21.32 67	10.30 93
98	9 604	9.89 949	31.30 50	941 192	4.61 044	9.93 288	21.39 97	10.20 41
99	9 801	9.94 987	31.46 43	970 299	4.62 607	9.96 655	21.47 23	10.10 10
100	10 000	10.00 000	31.62 28	1 000 000	4.64 159	10.00 000	21.54 43	10.00 00

Table II Values of exponential functions*

x	e^x	e^{-x}	x	e^x	e^{-x}	x	e^x	e^{-x}
0.00	1.0000	1.00 000	**0.50**	1.6487	.60 653	**1.00**	2.7183	.36 788
0.01	1.0101	0.99 005	0.51	1.6653	.60 050	1.01	2.7456	.36 422
0.02	1.0202	.98 020	0.52	1.6820	.59 452	1.02	2.7732	.36 059
0.03	1.0305	.97 045	0.53	1.6989	.58 860	1.03	2.8011	.35 701
0.04	1.0408	.96 079	0.54	1.7160	.58 275	1.04	2.8292	.35 345
0.05	1.0513	.95 123	**0.55**	1.7333	.57 695	**1.05**	2.8577	.34 994
0.06	1.0618	.94 176	0.56	1.7507	.57 121	1.06	2.8864	.34 646
0.07	1.0725	.93 239	0.57	1.7683	.56 553	1.07	2.9154	.34 301
0.08	1.0833	.92 312	0.58	1.7860	.55 990	1.08	2.9447	.33 960
0.09	1.0942	.91 393	0.59	1.8040	.55 433	1.09	2.9743	.33 622
0.10	1.1052	.90 484	**0.60**	1.8221	.54 881	**1.10**	3.0042	.33 287
0.11	1.1163	.89 583	0.61	1.8404	.54 335	1.11	3.0344	.32 956
0.12	1.1275	.88 692	0.62	1.8589	.53 794	1.12	3.0649	.32 628
0.13	1.1388	.87 810	0.63	1.8776	.53 259	1.13	3.0957	.32 303
0.14	1.1503	.86 936	0.64	1.8965	.52 729	1.14	3.1268	.31 982
0.15	1.1618	.86 071	**0.65**	1.9155	.52 205	**1.15**	3.1582	.31 664
0.16	1.1735	.85 214	0.66	1.9348	.51 685	1.16	3.1899	.31 349
0.17	1.1853	.84 366	0.67	1.9542	.51 171	1.17	3.2220	.31 037
0.18	1.1972	.83 527	0.68	1.9739	.50 662	1.18	3.2544	.30 728
0.19	1.2092	.82 696	0.69	1.9937	.50 158	1.19	3.2871	.30 422
0.20	1.2214	.81 873	**0.70**	2.0138	.49 659	**1.20**	3.3201	.30 119
0.21	1.2337	.81 058	0.71	2.0340	.49 164	1.21	3.3535	.29 820
0.22	1.2461	.80 252	0.72	2.0544	.48 675	1.22	3.3872	.29 523
0.23	1.2586	.79 453	0.73	2.0751	.48 191	1.23	3.4212	.29 229
0.24	1.2712	.78 663	0.74	2.0959	.47 711	1.24	3.4556	.28 938
0.25	1.2840	.77 880	**0.75**	2.1170	.47 237	**1.25**	3.4903	.28 650
0.26	1.2969	.77 105	0.76	2.1383	.46 767	1.26	3.5254	.28 365
0.27	1.3100	.76 338	0.77	2.1598	.46 301	1.27	3.5609	.28 083
0.28	1.3231	.75 578	0.78	2.1815	.45 841	1.28	3.5966	.27 804
0.29	1.3364	.74 826	0.79	2.2034	.45 384	1.29	3.6328	.27 527
0.30	1.3499	.74 082	**0.80**	2.2255	.44 933	**1.30**	3.6693	.27 253
0.31	1.3634	.73 345	0.81	2.2479	.44 486	1.31	3.7062	.26 982
0.32	1.3771	.72 615	0.82	2.2705	.44 043	1.32	3.7434	.26 714
0.33	1.3910	.71 892	0.83	2.2933	.43 605	1.33	3.7810	.26 448
0.34	1.4049	.71 177	0.84	2.3164	.43 171	1.34	3.8190	.26 185
0.35	1.4191	.70 469	**0.85**	2.3396	.42 741	**1.35**	3.8574	.25 924
0.36	1.4333	.69 768	0.86	2.3632	.42 316	1.36	3.8962	.25 666
0.37	1.4477	.69 073	0.87	2.3869	.41 895	1.37	3.9354	.25 411
0.38	1.4623	.68 386	0.88	2.4109	.41 478	1.38	3.9749	.25 158
0.39	1.4770	.67 706	0.89	2.4351	.41 066	1.39	4.0149	.24 908
0.40	1.4918	.67 032	**0.90**	2.4596	.40 657	**1.40**	4.0552	.24 660
0.41	1.5068	.66 365	0.91	2.4843	.40 252	1.41	4.0960	.24 414
0.42	1.5220	.65 705	0.92	2.5093	.39 852	1.42	4.1371	.24 171
0.43	1.5373	.65 051	0.93	2.5345	.39 455	1.43	4.1787	.23 931
0.44	1.5527	.64 404	0.94	2.5600	.39 063	1.44	4.2207	.23 693
0.45	1.5683	.63 763	**0.95**	2.5857	.38 674	**1.45**	4.2631	.23 457
0.46	1.5841	.63 128	0.96	2.6117	.38 289	1.46	4.3060	.23 224
0.47	1.6000	.62 500	0.97	2.6379	.37 908	1.47	4.3492	.22 993
0.48	1.6161	.61 878	0.98	2.6645	.37 531	1.48	4.3929	.22 764
0.49	1.6323	.61 263	0.99	2.6912	.37 158	1.49	4.4371	.22 537
0.50	1.6487	.60 653	**1.00**	2.7183	.36 788	**1.50**	4.4817	.22 313

Table II (cont'd.) Values of exponential functions

x	e^x	e^{-x}	x	e^x	e^{-x}	x	e^x	e^{-x}
1.50	4.4817	.22 313	**2.00**	7.3891	.13 534	**2.50**	12.182	.082 085
1.51	4.5267	.22 091	2.01	7.4633	.13 399	2.51	12.305	.081 268
1.52	4.5722	.21 871	2.02	7.5383	.13 266	2.52	12.429	.080 460
1.53	4.6182	.21 654	2.03	7.6141	.13 134	2.53	12.554	.079 659
1.54	4.6646	.21 438	2.04	7.6906	.13 003	2.54	12.680	.078 866
1.55	4.7115	.21 225	**2.05**	7.7679	.12 873	**2.55**	12.807	.078 082
1.56	4.7588	.21 014	2.06	7.8460	.12 745	2.56	12.936	.077 305
1.57	4.8066	.20 805	2.07	7.9248	.12 619	2.57	13.066	.076 536
1.58	4.8550	.20 598	2.08	8.0045	.12 493	2.58	13.197	.075 774
1.59	4.9037	.20 393	2.09	8.0849	.12 369	2.59	13.330	.075 020
1.60	4.9530	.20 190	**2.10**	8.1662	.12 246	**2.60**	13.464	.074 274
1.61	5.0028	.19 989	2.11	8.2482	.12 124	2.61	13.599	.073 535
1.62	5.0531	.19 790	2.12	8.3311	.12 003	2.62	13.736	.072 803
1.63	5.1039	.19 593	2.13	8.4149	.11 884	2.63	13.874	.072 078
1.64	5.1552	.19 398	2.14	8.4994	.11 765	2.64	14.013	.071 361
1.65	5.2070	.19 205	**2.15**	8.5849	.11 648	**2.65**	14.154	.070 651
1.66	5.2593	.19 014	2.16	8.6711	.11 533	2.66	14.296	.069 948
1.67	5.3122	.18 825	2.17	8.7583	.11 418	2.67	14.440	.069 252
1.68	5.3656	.18 637	2.18	8.8463	.11 304	2.68	14.585	.068 563
1.69	5.4195	.18 452	2.19	8.9352	.11 192	2.69	14.732	.067 881
1.70	5.4739	.18 268	**2.20**	9.0250	.11 080	**2.70**	14.880	.067 206
1.71	5.5290	.18 087	2.21	9.1157	.10 970	2.71	15.029	.066 537
1.72	5.5845	.17 907	2.22	9.2073	.10 861	2.72	15.180	.065 875
1.73	5.6407	.17 728	2.23	9.2999	.10 753	2.73	15.333	.065 219
1.74	5.6973	.17 552	2.24	9.3933	.10 646	2.74	15.487	.064 570
1.75	5.7546	.17 377	**2.25**	9.4877	.10 540	**2.75**	15.643	.063 928
1.76	5.8124	.17 204	2.26	9.5831	.10 435	2.76	15.800	.063 292
1.77	5.8709	.17 033	2.27	9.6794	.10 331	2.77	15.959	.062 662
1.78	5.9299	.16 864	2.28	9.7767	.10 228	2.78	16.119	.062 039
1.79	5.9895	.16 696	2.29	9.8749	.10 127	2.79	16.281	.061 421
1.80	6.0496	.16 530	**2.30**	9.9742	.10 026	**2.80**	16.445	.060 810
1.81	6.1104	.16 365	2.31	10.074	.09 9261	2.81	16.610	.060 205
1.82	6.1719	.16 203	2.32	10.176	.09 8274	2.82	16.777	.059 606
1.83	6.2339	.16 041	2.33	10.278	.09 7296	2.83	16.945	.059 013
1.84	6.2965	.15 882	2.34	10.381	.09 6328	2.84	17.116	.058 426
1.85	6.3598	.15 724	**2.35**	10.486	.09 5369	**2.85**	17.288	.057 844
1.86	6.4237	.15 567	2.36	10.591	.09 4420	2.86	17.462	.057 269
1.87	6.4883	.15 412	2.37	10.697	.09 3481	2.87	17.637	.056 699
1.88	6.5535	.15 259	2.38	10.805	.09 2551	2.88	17.814	.056 135
1.89	6.6194	.15 107	2.39	10.913	.09 1630	2.89	17.993	.055 576
1.90	6.6859	.14 957	**2.40**	11.023	.09 0718	**2.90**	18.174	.055 023
1.91	6.7531	.14 808	2.41	11.134	.08 9815	2.91	18.357	.054 476
1.92	6.8210	.14 661	2.42	11.246	.08 8922	2.92	18.541	.053 934
1.93	6.8895	.14 515	2.43	11.359	.08 8037	2.93	18.728	.053 397
1.94	6.9588	.14 370	2.44	11.473	.08 7161	2.94	18.916	.052 866
1.95	7.0287	.14 227	**2.45**	11.588	.08 6294	**2.95**	19.106	.052 340
1.96	7.0993	.14 086	2.46	11.705	.08 5435	2.96	19.298	.051 819
1.97	7.1707	.13 946	2.47	11.822	.08 4585	2.97	19.492	.051 303
1.98	7.2427	.13 807	2.48	11.941	.08 3743	2.98	19.688	.050 793
1.99	7.3155	.13 670	2.49	12.061	.08 2910	2.99	19.886	.050 287
2.00	7.3891	.13 534	**2.50**	12.182	.08 2085	**3.00**	20.086	.049 787

Table II (cont'd.) Values of exponential functions

x	e^x	e^{-x}	x	e^x	e^{-x}	x	e^x	e^{-x}
3.00	20.086	.04 9787	**3.50**	33.115	.030 197	**4.00**	54.598	.01 8316
3.01	20.287	.04 9292	3.51	33.448	.029 897	4.01	55.147	.01 8133
3.02	20.491	.04 8801	3.52	33.784	.029 599	4.02	55.701	.01 7953
3.03	20.697	.04 8316	3.53	34.124	.029 305	4.03	56.261	.01 7774
3.04	20.905	.04 7835	3.54	34.467	.029 013	4.04	56.826	.01 7597
3.05	21.115	.04 7359	**3.55**	34.813	.028 725	**4.05**	57.397	.01 7422
3.06	21.328	.04 6888	3.56	35.163	.028 439	4.06	57.974	.01 7249
3.07	21.542	.04 6421	3.57	35.517	.028 156	4.07	58.557	.01 7077
3.08	21.758	.04 5959	3.58	35.874	.027 876	4.08	59.145	.01 6907
3.09	21.977	.04 5502	3.59	36.234	.027 598	4.09	59.740	.01 6739
3.10	22.198	.04 5049	**3.60**	36.598	.027 324	**4.10**	60.340	.01 6573
3.11	22.421	.04 4601	3.61	36.966	.027 052	4.11	60.947	.01 6408
3.12	22.646	.04 4157	3.62	37.338	.026 783	4.12	61.559	.01 6245
3.13	22.874	.04 3718	3.63	37.713	.026 516	4.13	62.178	.01 6083
3.14	23.104	.04 3283	3.64	38.092	.026 252	4.14	62.803	.01 5923
3.15	23.336	.04 2852	**3.65**	38.475	.025 991	**4.15**	63.434	.01 5764
3.16	23.571	.04 2426	3.66	38.861	.025 733	4.16	64.072	.01 5608
3.17	23.807	.04 2004	3.67	39.252	.025 476	4.17	64.715	.01 5452
3.18	24.047	.04 1586	3.68	39.646	.025 223	4.18	65.366	.01 5299
3.19	24.288	.04 1172	3.69	40.045	.024 972	4.19	66.023	.01 5146
3.20	24.533	.04 0762	**3.70**	40.447	.024 724	**4.20**	66.686	.01 4996
3.21	24.779	.04 0357	3.71	40.854	.024 478	4.21	67.357	.01 4846
3.22	25.028	.03 9955	3.72	41.264	.024 234	4.22	68.033	.01 4699
3.23	25.280	.03 9557	3.73	41.679	.023 993	4.23	68.717	.01 4552
3.24	25.534	.03 9164	3.74	42.098	.023 754	4.24	69.408	.01 4408
3.25	25.790	.03 8774	**3.75**	42.521	.023 518	**4.25**	70.105	.01 4264
3.26	26.050	.03 8388	3.76	42.948	.023 284	4.26	70.810	.01 4122
3.27	26.311	.03 8006	3.77	43.380	.023 052	4.27	71.522	.01 3982
3.28	26.576	.03 7628	3.78	43.816	.022 823	4.28	72.240	.01 3843
3.29	26.843	.03 7254	3.79	44.256	.022 596	4.29	72.966	.01 3705
3.30	27.113	.03 6883	**3.80**	44.701	.022 371	**4.30**	73.700	.01 3569
3.31	27.385	.03 6516	3.81	45.150	.022 148	4.31	74.440	.01 3434
3.32	27.660	.03 6153	3.82	45.604	.021 928	4.32	75.189	.01 3300
3.33	27.938	.03 5793	3.83	46.063	.021 710	4.33	75.944	.01 3168
3.34	28.219	.03 5437	3.84	46.525	.021 494	4.34	76.708	.01 3037
3.35	28.503	.03 5084	**3.85**	46.993	.021 280	**4.35**	77.478	.01 2907
3.36	28.789	.03 4735	3.86	47.465	.021 068	4.36	78.257	.01 2778
3.37	29.079	.03 4390	3.87	47.942	.020 858	4.37	79.044	.01 2651
3.38	29.371	.03 4047	3.88	48.424	.020 651	4.38	79.838	.01 2525
3.39	29.666	.03 3709	3.89	48.911	.020 445	4.39	80.640	.01 2401
3.40	29.964	.03 3373	**3.90**	49.402	.020 242	**4.40**	81.451	.01 2277
3.41	30.265	.03 3041	3.91	49.899	.020 041	4.41	82.269	.01 2155
3.42	30.569	.03 2712	3.92	50.400	.019 841	4.42	83.096	.01 2034
3.43	30.877	.03 2387	3.93	50.907	.019 644	4.43	83.931	.01 1914
3.44	31.187	.03 2065	3.94	51.419	.019 448	4.44	84.775	.01 1796
3.45	31.500	.03 1746	**3.95**	51.935	.019 255	**4.45**	85.627	.01 1679
3.46	31.817	.03 1430	3.96	52.457	.019 063	4.46	86.488	.01 1562
3.47	32.137	.03 1117	3.97	52.985	.018 873	4.47	87.357	.01 1447
3.48	32.460	.03 0807	3.98	53.517	.018 686	4.48	88.235	.01 1333
3.49	32.786	.03 0501	3.99	54.055	.018 500	4.49	89.121	.01 1221
3.50	33.115	.03 0197	**4.00**	54.598	.018 316	**4.50**	90.017	.01 1109

Table II (cont'd.) Values of exponential functions

x	e^x	e^{-x}	x	e^x	e^{-x}	x	e^x	e^{-x}
4.50	90.017	.011 109	**5.00**	148.41	.00 67379	**7.50**	1 808.0	.000 5531
4.51	90.922	.010 998	5.05	156.02	.00 64093	7.55	1 900.7	.000 5261
4.52	91.836	.010 889	5.10	164.02	.00 60967	7.60	1 998.2	.000 5005
4.53	92.759	.010 781	5.15	172.43	.00 57994	7.65	2 100.6	.000 4760
4.54	93.691	.010 673	5.20	181.27	.00 55166	7.70	2 208.3	.000 4528
4.55	94.632	.010 567	**5.25**	190.57	.00 52475	**7.75**	2 321.6	.000 4307
4.56	95.583	.010 462	5.30	200.34	.00 49916	7.80	2 440.6	.000 4097
4.57	96.544	.010 358	5.35	210.61	.00 47482	7.85	2 565.7	.000 3898
4.58	97.514	.010 255	5.40	221.41	.00 45166	7.90	2 697.3	.000 3707
4.59	98.494	.010 153	5.45	232.76	.00 42963	7.95	2 835.6	.000 3527
4.60	99.484	.010 052	**5.50**	244.69	.00 40868	**8.00**	2 981.0	.000 3355
4.61	100.48	.009 9518	5.55	257.24	.00 38875	8.05	3 133.8	.000 3191
4.62	101.49	.009 8528	5.60	270.43	.00 36979	8.10	3 294.5	.000 3035
4.63	102.51	.009 7548	5.65	284.29	.00 35175	8.15	3 463.4	.000 2887
4.64	103.54	.009 6577	5.70	298.87	.00 33460	8.20	3 641.0	.000 2747
4.65	104.58	.009 5616	**5.75**	314.19	.00 31828	**8.25**	3 827.6	.000 2613
4.66	105.64	.009 4665	5.80	330.30	.00 30276	8.30	4 023.9	.000 2485
4.67	106.70	.009 3723	5.85	347.23	.00 28799	8.35	4 230.2	.000 2364
4.68	107.77	.009 2790	5.90	365.04	.00 27394	8.40	4 447.1	.000 2249
4.69	108.85	.009 1867	5.95	383.75	.00 26058	8.45	4 675.1	.000 2139
4.70	109.95	.009 0953	**6.00**	403.43	.00 24788	**8.50**	4 914.8	.000 2035
4.71	111.05	.009 0048	6.05	424.11	.00 23579	8.55	5 166.8	.000 1935
4.72	112.17	.008 9152	6.10	445.86	.00 22429	8.60	5 431.7	.000 1841
4.73	113.30	.008 8265	6.15	468.72	.00 21335	8.65	5 710.1	.000 1751
4.74	114.43	.008 7386	6.20	492.75	.00 20294	8.70	6 002.9	.000 1666
4.75	115.58	.008 6517	**6.25**	518.01	.00 19305	**8.75**	6 310.7	.000 1585
4.76	116.75	.008 5656	6.30	544.57	.00 18363	8.80	6 634.2	.000 1507
4.77	117.92	.008 4804	6.35	572.49	.00 17467	8.85	6 974.4	.000 1434
4.78	119.10	.008 3960	6.40	601.85	.00 16616	8.90	7 332.0	.000 1364
4.79	120.30	.008 3125	6.45	632.70	.00 15805	8.95	7 707.9	.000 1297
4.80	121.51	.008 2297	**6.50**	665.14	.00 15034	**9.00**	8 103.1	.000 1234
4.81	122.73	.008 1479	6.55	699.24	.00 14301	9.05	8 518.5	.000 1174
4.82	123.97	.008 0668	6.60	735.10	.00 13604	9.10	8 955.3	.000 1117
4.83	125.21	.007 9865	6.65	772.78	.00 12940	9.15	9 414.4	.000 1062
4.84	126.47	.007 9071	6.70	812.41	.00 12309	9.20	9 897.1	.000 1010
4.85	127.74	.007 8284	**6.75**	854.06	.00 11709	**9.25**	10 405	.000 0961
4.86	129.02	.007 7505	6.80	897.85	.00 11138	9.30	10 938	.000 0914
4.87	130.32	.007 6734	6.85	943.88	.00 10595	9.35	11 499	.000 0870
4.88	131.63	.007 5970	6.90	992.27	.00 10078	9.40	12 088	.000 0827
4.89	132.95	.007 5214	6.95	1 043.1	.00 09586	9.45	12 708	.000 0787
4.90	134.29	.007 4466	**7.00**	1 096.6	.00 09119	**9.50**	13 360	.000 0749
4.91	135.64	.007 3725	7.05	1 152.9	.00 08674	9.55	14 045	.000 0712
4.92	137.00	.007 2991	7.10	1 212.0	.00 08251	9.60	14 765	.000 0677
4.93	138.38	.007 2265	7.15	1 274.1	.00 07849	9.65	15 522	.000 0644
4.94	139.77	.007 1546	7.20	1 339.4	.00 07466	9.70	16 318	.000 0613
4.95	141.17	.007 0834	**7.25**	1 408.1	.00 07102	**9.75**	17 154	.000 0583
4.96	142.59	.007 0129	7.30	1 480.3	.00 06755	9.80	18 034	.000 0555
4.97	144.03	.006 9431	7.35	1 556.2	.00 06426	9.85	18 958	.000 0527
4.98	145.47	.006 8741	7.40	1 636.0	.00 06113	9.90	19 930	.000 0502
4.99	146.94	.006 8057	7.45	1 719.9	.00 05814	9.95	20 952	.000 0477
5.00	148.41	.006 7379	**7.50**	1 808.0	.00 05531	**10.00**	22 026	.000 0454

Table III Four-place natural logarithms*

N	.00	.01	.02	.03	.04	.05	.06	.07	.08	.09
1.0	0.0000	0.0100	0.0198	0.0296	0.0392	0.0488	0.0583	0.0677	0.0770	0.0862
1.1	0.0953	0.1044	0.1133	0.1222	0.1310	0.1398	0.1484	0.1570	0.1655	0.1740
1.2	0.1823	0.1906	0.1989	0.2070	0.2151	0.2231	0.2311	0.2390	0.2469	0.2546
1.3	0.2624	0.2700	0.2776	0.2852	0.2927	0.3001	0.3075	0.3148	0.3221	0.3293
1.4	0.3365	0.3436	0.3507	0.3577	0.3646	0.3716	0.3784	0.3853	0.3920	0.3988
1.5	0.4055	0.4121	0.4187	0.4253	0.4318	0.4383	0.4447	0.4511	0.4574	0.4637
1.6	0.4700	0.4762	0.4824	0.4886	0.4947	0.5008	0.5068	0.5128	0.5188	0.5247
1.7	0.5306	0.5365	0.5423	0.5481	0.5539	0.5596	0.5653	0.5710	0.5766	0.5822
1.8	0.5878	0.5933	0.5988	0.6043	0.6098	0.6152	0.6206	0.6259	0.6313	0.6366
1.9	0.6419	0.6471	0.6523	0.6575	0.6627	0.6678	0.6729	0.6780	0.6831	0.6881
2.0	0.6931	0.6981	0.7031	0.7080	0.7129	0.7178	0.7227	0.7275	0.7324	0.7372
2.1	0.7419	0.7467	0.7514	0.7561	0.7608	0.7655	0.7701	0.7747	0.7793	0.7839
2.2	0.7885	0.7930	0.7975	0.8020	0.8065	0.8109	0.8154	0.8198	0.8242	0.8286
2.3	0.8329	0.8372	0.8416	0.8459	0.8502	0.8544	0.8587	0.8629	0.8671	0.8713
2.4	0.8755	0.8796	0.8838	0.8879	0.8920	0.8961	0.9002	0.9042	0.9083	0.9123
2.5	0.9163	0.9203	0.9243	0.9282	0.9322	0.9361	0.9400	0.9439	0.9478	0.9517
2.6	0.9555	0.9594	0.9632	0.9670	0.9708	0.9746	0.9783	0.9821	0.9858	0.9895
2.7	0.9933	0.9969	1.0006	1.0043	1.0080	1.0116	1.0152	1.0188	1.0225	1.0260
2.8	1.0296	1.0332	1.0367	1.0403	1.0438	1.0473	1.0508	1.0543	1.0578	1.0613
2.9	1.0647	1.0682	1.0716	1.0750	1.0784	1.0818	1.0852	1.0886	1.0919	1.0953
3.0	1.0986	1.1019	1.1053	1.1086	1.1119	1.1151	1.1184	1.1217	1.1249	1.1282
3.1	1.1314	1.1346	1.1378	1.1410	1.1442	1.1474	1.1506	1.1537	1.1569	1.1600
3.2	1.1632	1.1663	1.1694	1.1725	1.1756	1.1787	1.1817	1.1848	1.1878	1.1909
3.3	1.1939	1.1969	1.2000	1.2030	1.2060	1.2090	1.2119	1.2149	1.2179	1.2208
3.4	1.2238	1.2267	1.2296	1.2326	1.2355	1.2384	1.2413	1.2442	1.2470	1.2499
3.5	1.2528	1.2556	1.2585	1.2613	1.2641	1.2669	1.2698	1.2726	1.2754	1.2782
3.6	1.2809	1.2837	1.2865	1.2892	1.2920	1.2947	1.2975	1.3002	1.3029	1.3056
3.7	1.3083	1.3110	1.3137	1.3164	1.3191	1.3218	1.3244	1.3271	1.3297	1.3324
3.8	1.3350	1.3376	1.3403	1.3429	1.3455	1.3481	1.3507	1.3533	1.3558	1.3584
3.9	1.3610	1.3635	1.3661	1.3686	1.3712	1.3737	1.3762	1.3788	1.3813	1.3838
4.0	1.3863	1.3888	1.3913	1.3938	1.3962	1.3987	1.4012	1.4036	1.4061	1.4085
4.1	1.4110	1.4134	1.4159	1.4183	1.4207	1.4231	1.4255	1.4279	1.4303	1.4327
4.2	1.4351	1.4375	1.4398	1.4422	1.4446	1.4469	1.4493	1.4516	1.4540	1.4563
4.3	1.4586	1.4609	1.4633	1.4656	1.4679	1.4702	1.4725	1.4748	1.4770	1.4793
4.4	1.4816	1.4839	1.4861	1.4884	1.4907	1.4929	1.4951	1.4974	1.4996	1.5019
4.5	1.5041	1.5063	1.5085	1.5107	1.5129	1.5151	1.5173	1.5195	1.5217	1.5239
4.6	1.5261	1.5282	1.5304	1.5326	1.5347	1.5369	1.5390	1.5412	1.5433	1.5454
4.7	1.5476	1.5497	1.5518	1.5539	1.5560	1.5581	1.5602	1.5623	1.5644	1.5665
4.8	1.5686	1.5707	1.5728	1.5748	1.5769	1.5790	1.5810	1.5831	1.5851	1.5872
4.9	1.5892	1.5913	1.5933	1.5953	1.5974	1.5994	1.6014	1.6034	1.6054	1.6074
5.0	1.6094	1.6114	1.6134	1.6154	1.6174	1.6194	1.6214	1.6233	1.6253	1.6273
5.1	1.6292	1.6312	1.6332	1.6351	1.6371	1.6390	1.6409	1.6429	1.6448	1.6467
5.2	1.6487	1.6506	1.6525	1.6544	1.6563	1.6582	1.6601	1.6620	1.6639	1.6658
5.3	1.6677	1.6696	1.6715	1.6734	1.6752	1.6771	1.6790	1.6808	1.6827	1.6845
5.4	1.6864	1.6882	1.6901	1.6919	1.6938	1.6956	1.6974	1.6993	1.7011	1.7029

$\ln .1 = .6974–3$ $\ln .01 = .3948–5$ $\ln .001 = .0922–7$

Table III (cont'd.) Four-place natural logarithms

N	.00	.01	.02	.03	.04	.05	.06	.07	.08	.09
5.5	1.7047	1.7066	1.7084	1.7102	1.7120	1.7138	1.7156	1.7174	1.7192	1.7210
5.6	1.7228	1.7246	1.7263	1.7281	1.7299	1.7317	1.7334	1.7352	1.7370	1.7387
5.7	1.7405	1.7422	1.7440	1.7457	1.7475	1.7492	1.7509	1.7527	1.7544	1.7561
5.8	1.7579	1.7596	1.7613	1.7630	1.7647	1.7664	1.7681	1.7699	1.7716	1.7733
5.9	1.7750	1.7766	1.7783	1.7800	1.7817	1.7834	1.7851	1.7867	1.7884	1.7901
6.0	1.7918	1.7934	1.7951	1.7967	1.7984	1.8001	1.8017	1.8034	1.8050	1.8066
6.1	1.8083	1.8099	1.8116	1.8132	1.8148	1.8165	1.8181	1.8197	1.8213	1.8229
6.2	1.8245	1.8262	1.8278	1.8294	1.8310	1.8326	1.8342	1.8358	1.8374	1.8390
6.3	1.8405	1.8421	1.8437	1.8453	1.8469	1.8485	1.8500	1.8516	1.8532	1.8547
6.4	1.8563	1.8579	1.8594	1.8610	1.8625	1.8641	1.8656	1.8672	1.8687	1.8703
6.5	1.8718	1.8733	1.8749	1.8764	1.8779	1.8795	1.8810	1.8825	1.8840	1.8856
6.6	1.8871	1.8886	1.8901	1.8916	1.8931	1.8946	1.8961	1.8976	1.8991	1.9006
6.7	1.9021	1.9036	1.9051	1.9066	1.9081	1.9095	1.9110	1.9125	1.9140	1.9155
6.8	1.9169	1.9184	1.9199	1.9213	1.9228	1.9242	1.9257	1.9272	1.9286	1.9301
6.9	1.9315	1.9330	1.9344	1.9359	1.9373	1.9387	1.9402	1.9416	1.9430	1.9445
7.0	1.9459	1.9473	1.9488	1.9502	1.9516	1.9530	1.9544	1.9559	1.9573	1.9587
7.1	1.9601	1.9615	1.9629	1.9643	1.9657	1.9671	1.9685	1.9699	1.9713	1.9727
7.2	1.9741	1.9755	1.9769	1.9782	1.9796	1.9810	1.9824	1.9838	1.9851	1.9865
7.3	1.9879	1.9892	1.9906	1.9920	1.9933	1.9947	1.9961	1.9974	1.9988	2.0001
7.4	2.0015	2.0028	2.0042	2.0055	2.0069	2.0082	2.0096	2.0109	2.0122	2.0136
7.5	2.0149	2.0162	2.0176	2.0189	2.0202	2.0215	2.0229	2.0242	2.0255	2.0268
7.6	2.0281	2.0295	2.0308	2.0321	2.0334	2.0347	2.0360	2.0373	2.0386	2.0399
7.7	2.0412	2.0425	2.0438	2.0451	2.0464	2.0477	2.0490	2.0503	2.0516	2.0528
7.8	2.0541	2.0554	2.0567	2.0580	2.0592	2.0605	2.0618	2.0631	2.0643	2.0656
7.9	2.0669	2.0681	2.0694	2.0707	2.0719	2.0732	2.0744	2.0757	2.0769	2.0782
8.0	2.0794	2.0807	2.0819	2.0832	2.0844	2.0857	2.0869	2.0882	2.0894	2.0906
8.1	2.0919	2.0931	2.0943	2.0956	2.0968	2.0980	2.0992	2.1005	2.1017	2.1029
8.2	2.1041	2.1054	2.1066	2.1078	2.1090	2.1102	2.1114	2.1126	2.1138	2.1150
8.3	2.1163	2.1175	2.1187	2.1199	2.1211	2.1223	2.1235	2.1247	2.1258	2.1270
8.4	2.1282	2.1294	2.1306	2.1318	2.1330	2.1342	2.1353	2.1365	2.1377	2.1389
8.5	2.1401	2.1412	2.1424	2.1436	2.1448	2.1459	2.1471	2.1483	2.1494	2.1506
8.6	2.1518	2.1529	2.1541	2.1552	2.1564	2.1576	2.1587	2.1599	2.1610	2.1622
8.7	2.1633	2.1645	2.1656	2.1668	2.1679	2.1691	2.1702	2.1713	2.1725	2.1736
8.8	2.1748	2.1759	2.1770	2.1782	2.1793	2.1804	2.1815	2.1827	2.1838	2.1849
8.9	2.1861	2.1872	2.1883	2.1894	2.1905	2.1917	2.1928	2.1939	2.1950	2.1961
9.0	2.1972	2.1983	2.1994	2.2006	2.2017	2.2028	2.2039	2.2050	2.2061	2.2072
9.1	2.2083	2.2094	2.2105	2.2116	2.2127	2.2138	2.2148	2.2159	2.2170	2.2181
9.2	2.2192	2.2203	2.2214	2.2225	2.2235	2.2246	2.2257	2.2268	2.2279	2.2289
9.3	2.2300	2.2311	2.2322	2.2332	2.2343	2.2354	2.2364	2.2375	2.2386	2.2396
9.4	2.2407	2.2418	2.2428	2.2439	2.2450	2.2460	2.2471	2.2481	2.2492	2.2502
9.5	2.2513	2.2523	2.2534	2.2544	2.2555	2.2565	2.2576	2.2586	2.2597	2.2607
9.6	2.2618	2.2628	2.2638	2.2649	2.2659	2.2670	2.2680	2.2690	2.2701	2.2711
9.7	2.2721	2.2732	2.2742	2.2752	2.2762	2.2773	2.2783	2.2793	2.2803	2.2814
9.8	2.2824	2.2834	2.2844	2.2854	2.2865	2.2875	2.2885	2.2895	2.2905	2.2915
9.9	2.2925	2.2935	2.2946	2.2956	2.2966	2.2976	2.2986	2.2996	2.3006	2.3016

ln .0001 = .7897–10 ln .00001 = .4871–12 ln .000 001 = .1845–14

Table IV Trigonometric Functions for Angles in Radians*

Rad.	Sin	Tan	Ctn	Cos	Rad.	Sin	Tan	Ctn	Cos
0.00	.00000	.00000	–	1.00000	**0.50**	.47943	.54630	1.8305	.87758
.01	.01000	.01000	99.997	0.99995	.51	.48818	.55936	1.7878	.87274
.02	.02000	.02000	49.993	.99980	.52	.49688	.57256	1.7465	.86782
.03	.03000	.03001	33.323	.99955	.53	.50553	.58592	1.7067	.86281
.04	.03999	.04002	24.987	.99920	.54	.51414	.59943	1.6683	.85771
.05	.04998	.05004	19.983	.99875	.55	.52269	.61311	1.6310	.85252
.06	.05996	.06007	16.647	.99820	.56	.53119	.62695	1.5950	.84726
.07	.06994	.07011	14.262	.99755	.57	.53963	.64097	1.5601	.84190
.08	.07991	.08017	12.473	.99680	.58	.54802	.65517	1.5263	.83646
.09	.08988	.09024	11.081	.99595	.59	.55636	.66956	1.4935	.83094
0.10	.09983	.10033	9.9666	.99500	**0.60**	.56464	.68414	1.4617	.82534
.11	.10978	.11045	9.0542	.99396	.61	.57287	.69892	1.4308	.81965
.12	.11971	.12058	8.2933	.99281	.62	.58104	.71391	1.4007	.81388
.13	.12963	.13074	7.6489	.99156	.63	.58914	.72911	1.3715	.80803
.14	.13954	.14092	7.0961	.99022	.64	.59720	.74454	1.3431	.80210
.15	.14944	.15114	6.6166	.98877	.65	.60519	.76020	1.3154	.79608
.16	.15932	.16138	6.1966	.98723	.66	.61312	.77610	1.2885	.78999
.17	.16918	.17166	5.8256	.98558	.67	.62099	.79225	1.2622	.78382
.18	.17903	.18197	5.4954	.98384	.68	.62879	.80866	1.2366	.77757
.19	.18886	.19232	5.1997	.98200	.69	.63654	.82534	1.2116	.77125
0.20	.19867	.20271	4.9332	.98007	**0.70**	.64422	.84229	1.1872	.76484
.21	.20846	.21314	4.6917	.97803	.71	.65183	.85953	1.1634	.75836
.22	.21823	.22362	4.4719	.97590	.72	.65938	.87707	1.1402	.75181
.23	.22798	.23414	4.2709	.97367	.73	.66687	.89492	1.1174	.74517
.24	.23770	.24472	4.0864	.97134	.74	.67429	.91309	1.0952	.73847
.25	.24740	.25534	3.9163	.96891	.75	.68164	.93160	1.0734	.73169
.26	.25708	.26602	3.7591	.96639	.76	.68892	.95045	1.0521	.72484
.27	.26673	.27676	3.6133	.96377	.77	.69614	.96967	1.0313	.71791
.28	.27636	.28755	3.4776	.96106	.78	.70328	.98926	1.0109	.71091
.29	.28595	.29841	3.3511	.95824	.79	.71035	1.0092	.99084	.70385
0.30	.29552	.30934	3.2327	.95534	**0.80**	.71736	1.0296	.97121	.69671
.31	.30506	.32033	3.1218	.95233	.81	.72429	1.0505	.95197	.68950
.32	.31457	.33139	3.0176	.94924	.82	.73115	1.0717	.93309	.68222
.33	.32404	.34252	2.9195	.94604	.83	.73793	1.0934	.91455	.67488
.34	.33349	.35374	2.8270	.94275	.84	.74464	1.1156	.89635	.66746
.35	.34290	.36503	2.7395	.93937	.85	.75128	1.1383	.87848	.65998
.36	.35227	.37640	2.6567	.93590	.86	.75784	1.1616	.86091	.65244
.37	.36162	.38786	2.5782	.93233	.87	.76433	1.1853	.84365	.64483
.38	.37092	.39941	2.5037	.92866	.88	.77074	1.2097	.82668	.63715
.39	.38019	.41105	2.4328	.92491	.89	.77707	1.2346	.80998	.62941
0.40	.38942	.42279	2.3652	.92106	**0.90**	.78333	1.2602	.79355	.62161
.41	.39861	.43463	2.3008	.91712	.91	.78950	1.2864	.77738	.61375
.42	.40776	.44657	2.2393	.91309	.92	.79560	1.3133	.76146	.60582
.43	.41687	.45862	2.1804	.90897	.93	.80162	1.3409	.74578	.59783
.44	.42594	.47078	2.1241	.90475	.94	.80756	1.3692	.73034	.58979
.45	.43497	.48306	2.0702	.90045	.95	.81342	1.3984	.71511	.58168
.46	.44395	.49545	2.0184	.89605	.96	.81919	1.4284	.70010	.57352
.47	.45289	.50797	1.9686	.89157	.97	.82489	1.4592	.68531	.56530
.48	.46178	.52061	1.9208	.88699	.98	.83050	1.4910	.67071	.55702
.49	.47063	.53339	1.8748	.88233	.99	.83603	1.5237	.65631	.54869
0.50	.47943	.54630	1.8305	.87758	**1.00**	.84147	1.5574	.64209	.54030

Table IV (cont'd.) Trigonometric Functions for Angles in Radians

Rad.	Sin	Tan	Ctn	Cos
1.00	.84147	1.5574	.64209	.54030
1.01	.84683	1.5922	.62806	.53186
1.02	.85211	1.6281	.61420	.52337
1.03	.85730	1.6652	.60051	.51482
1.04	.86240	1.7036	.58699	.50622
1.05	.86742	1.7433	.57362	.49757
1.06	.87236	1.7844	.56040	.48887
1.07	.87720	1.8270	.54734	.48012
1.08	.88196	1.8712	.53441	.47133
1.09	.88663	1.9171	.52162	.46249
1.10	.89121	1.9648	.50897	.45360
1.11	.89570	2.0143	.49644	.44466
1.12	.90010	2.0660	.48404	.43568
1.13	.90441	2.1198	.47175	.42666
1.14	.90863	2.1759	.45959	.41759
1.15	.91276	2.2345	.44753	.40849
1.16	.91680	2.2958	.43558	.39934
1.17	.92075	2.3600	.42373	.39015
1.18	.92461	2.4273	.41199	.38092
1.19	.92837	2.4979	.40034	.37166
1.20	.93204	2.5722	.38878	.36236
1.21	.93562	2.6503	.37731	.35302
1.22	.93910	2.7328	.36593	.34365
1.23	.94249	2.8198	.35463	.33424
1.24	.94578	2.9119	.34341	.32480
1.25	.94898	3.0096	.33227	.31532
1.26	.95209	3.1133	.32121	.30582
1.27	.95510	3.2236	.31021	.29628
1.28	.95802	3.3413	.29928	.28672
1.29	.96084	3.4672	.28842	.27712
1.30	.96356	3.6021	.27762	.26750

Rad.	Sin	Tan	Ctn	Cos
1.30	.96356	3.6021	.27762	.26750
1.31	.96618	3.7471	.26687	.25785
1.32	.96872	3.9033	.25619	.24818
1.33	.97115	4.0723	.24556	.23848
1.34	.97438	4.2556	.23498	.22875
1.35	.97572	4.4552	.22446	.21901
1.36	.97786	4.6734	.21398	.20924
1.37	.97991	4.9131	.20354	.19945
1.38	.98185	5.1774	.19315	.18964
1.39	.98370	5.4707	.18279	.17981
1.40	.98545	5.7979	.17248	.16997
1.41	.98710	6.1654	.16220	.16010
1.42	.98865	6.5811	.15195	.15023
1.43	.99010	7.0555	.14173	.14033
1.44	.99146	7.6018	.13155	.13042
1.45	.99271	8.2381	.12139	.12050
1.46	.99387	8.9886	.11125	.11057
1.47	.99492	9.8874	.10114	.10063
1.48	.99588	10.983	.09105	.09067
1.49	.99674	12.350	.08097	.08071
1.50	.99749	14.101	.07091	.07074
1.51	.99815	16.428	.06087	.06076
1.52	.99871	19.670	.05084	.05077
1.53	.99917	24.498	.04082	.04079
1.54	.99953	32.461	.03081	.03079
1.55	.99978	48.078	.02080	.02079
1.56	.99994	92.621	.01080	.01080
1.57	1.00000	1255.8	.00080	.00080
1.58	.99996	−108.65	−.00920	−.00920
1.59	.99982	−52.067	−.01921	−.01920
1.60	.99957	−34.233	−.02921	−.02920

Table V Trigonometric Functions for Angles in Degrees*

Degrees	Sine	Tangent	Cotangent	Cosine	
0	0	0	——	1.0000	**90**
1	.0175	.0175	57.290	.9998	89
2	.0349	.0349	28.636	.9994	88
3	.0523	.0524	19.081	.9986	87
4	.0698	.0699	14.301	.9976	86
5	.0872	.0875	11.430	.9962	**85**
6	.1045	.1051	9.5144	.9945	84
7	.1219	.1228	8.1443	.9925	83
8	.1392	.1405	7.1154	.9903	82
9	.1564	.1584	6.3138	.9877	81
10	.1736	.1763	5.6713	.9848	**80**
11	.1908	.1944	5.1446	.9816	79
12	.2079	.2126	4.7046	.9781	78
13	.2250	.2309	4.3315	.9744	77
14	.2419	.2493	4.0108	.9703	76
15	.2588	.2679	3.7321	.9659	**75**
16	.2756	.2867	3.4874	.9613	74
17	.2924	.3057	3.2709	.9563	73
18	.3090	.3249	3.0777	.9511	72
19	.3256	.3443	2.9042	.9455	71
20	.3420	.3640	2.7475	.9397	**70**
21	.3584	.3839	2.6051	.9336	69
22	.3746	.4040	2.4751	.9272	68
23	.3907	.4245	2.3559	.9205	67
24	.4067	.4452	2.2460	.9135	66
25	.4226	.4663	2.1445	.9063	**65**
26	.4384	.4877	2.0503	.8988	64
27	.4540	.5095	1.9626	.8910	63
28	.4695	.5317	1.8807	.8829	62
29	.4848	.5543	1.8040	.8746	61
30	.5000	.5774	1.7321	.8660	**60**
31	.5150	.6009	1.6643	.8572	59
32	.5299	.6249	1.6003	.8480	58
33	.5446	.6494	1.5399	.8387	57
34	.5592	.6745	1.4826	.8290	56
35	.5736	.7002	1.4281	.8192	**55**
36	.5878	.7265	1.3764	.8090	54
37	.6018	.7536	1.3270	.7986	53
38	.6157	.7813	1.2799	.7880	52
39	.6293	.8098	1.2349	.7771	51
40	.6428	.8391	1.1918	.7660	**50**
41	.6561	.8693	1.1504	.7547	49
42	.6691	.9004	1.1106	.7431	48
43	.6820	.9325	1.0724	.7314	47
44	.6947	.9657	1.0355	.7193	46
45	.7071	1.0000	1.0000	.7071	**45**
	Cosine	**Cotangent**	**Tangent**	**Sine**	**Degrees**

Table VI Binomial Probabilities for k Successes, n Trials*

						p					
n	k	.05	.10	.15	.20	.25	.30	.35	.40	.45	.50
1	0	.9500	.9000	.8500	.8000	.7500	.7000	.6500	.6000	.5500	.5000
	1	.0500	.1000	.1500	.2000	.2500	.3000	.3500	.4000	.4500	.5000
2	0	.9025	.8100	.7225	.6400	.5625	.4900	.4225	.3600	.3025	.2500
	1	.0950	.1800	.2550	.3200	.3750	.4200	.4550	.4800	.4950	.5000
	2	.0025	.0100	.0225	.0400	.0625	.0900	.1225	.1600	.2025	.2500
3	0	.8574	.7290	.6141	.5120	.4219	.3430	.2746	.2160	.1664	.1250
	1	.1354	.2430	.3251	.3840	.4219	.4410	.4436	.4320	.4084	.3750
	2	.0071	.0270	.0574	.0960	.1406	.1890	.2389	.2880	.3341	.3750
	3	.0001	.0010	.0034	.0080	.0156	.0270	.0429	.0640	.0911	.1250
4	0	.8145	.6561	.5220	.4096	.3164	.2401	.1785	.1296	.0915	.0625
	1	.1715	.2916	.3685	.4096	.4219	.4116	.3845	.3456	.2995	.2500
	2	.0135	.0486	.0975	.1536	.2109	.2646	.3105	.3456	.3675	.3750
	3	.0005	.0036	.0115	.0256	.0469	.0756	.1115	.1536	.2005	.2500
	4	.0000	.0001	.0005	.0016	.0039	.0081	.0150	.0256	.0410	.0625
5	0	.7738	.5905	.4437	.3277	.2373	.1681	.1160	.0778	.0503	.0312
	1	.2036	.3280	.3915	.4096	.3955	.3602	.3124	.2592	.2059	.1562
	2	.0214	.0729	.1382	.2048	.2637	.3087	.3364	.3456	.3369	.3125
	3	.0011	.0081	.0244	.0512	.0879	.1323	.1811	.2304	.2757	.3125
	4	.0000	.0004	.0022	.0064	.0146	.0284	.0488	.0768	.1128	.1562
	5	.0000	.0000	.0001	.0003	.0010	.0024	.0053	.0102	.0185	.0312
6	0	.7351	.5314	.3771	.2621	.1780	.1176	.0754	.0467	.0277	.0156
	1	.2321	.3543	.3993	.3932	.3560	.3025	.2437	.1866	.1359	.0938
	2	.0305	.0984	.1762	.2458	.2966	.3241	.3280	.3110	.2780	.2344
	3	.0021	.0146	.0415	.0819	.1318	.1852	.2355	.2765	.3032	.3125
	4	.0001	.0012	.0055	.0154	.0330	.0595	.0951	.1382	.1861	.2344
	5	.0000	.0001	.0004	.0015	.0044	.0102	.0205	.0369	.0609	.0938
	6	.0000	.0000	.0000	.0001	.0002	.0007	.0018	.0041	.0083	.0156
7	0	.6983	.4783	.3206	.2097	.1335	.0824	.0490	.0280	.0152	.0078
	1	.2573	.3720	.3960	.3670	.3115	.2471	.1848	.1306	.0872	.0547
	2	.0406	.1240	.2097	.2753	.3115	.3177	.2985	.2613	.2140	.1641
	3	.0036	.0230	.0617	.1147	.1730	.2269	.2679	.2903	.2918	.2734
	4	.0002	.0026	.0109	.0287	.0577	.0972	.1442	.1935	.2388	.2734
	5	.0000	.0002	.0012	.0043	.0115	.0250	.0466	.0774	.1172	.1641
	6	.0000	.0000	.0001	.0004	.0013	.0036	.0084	.0172	.0320	.0547
	7	.0000	.0000	.0000	.0000	.0001	.0002	.0006	.0016	.0037	.0078
8	0	.6634	.4305	.2725	.1678	.1001	.0576	.0319	.0168	.0084	.0039
	1	.2793	.3826	.3847	.3355	.2670	.1977	.1373	.0896	.0548	.0312
	2	.0515	.1488	.2376	.2936	.3115	.2965	.2587	.2090	.1569	.1094
	3	.0054	.0331	.0839	.1468	.2076	.2541	.2786	.2787	.2568	.2188
	4	.0004	.0046	.0185	.0459	.0865	.1361	.1875	.2322	.2627	.2734
	5	.0000	.0004	.0026	.0092	.0231	.0467	.0808	.1239	.1719	.2188
	6	.0000	.0000	.0002	.0011	.0038	.0100	.0217	.0413	.0703	.1094
	7	.0000	.0000	.0000	.0001	.0004	.0012	.0033	.0079	.0164	.0312
	8	.0000	.0000	.0000	.0000	.0000	.0001	.0002	.0007	.0017	.0039

Linear interpolations with respect to θ will in general be accurate at most to two decimal places.

* From *Handbook of Tables for Probability and Statistics,* second edition, William H. Beyer, editor. Copyright © 1968, by the Chemical Rubber Co. Used by permission of the Chemical Rubber Co.

Table VI (cont'd.) Binomial Probabilities for k Successes, n Trials

n	k	p = .05	.10	.15	.20	.25	.30	.35	.40	.45	.50
9	0	.6302	.3874	.2316	.1342	.0751	.0404	.0207	.0101	.0046	.0020
	1	.2985	.3874	.3679	.3020	.2253	.1556	.1004	.0605	.0339	.0176
	2	.0629	.1722	.2597	.3020	.3003	.2668	.2162	.1612	.1110	.0703
	3	.0077	.0446	.1069	.1762	.2336	.2668	.2716	.2508	.2119	.1641
	4	.0006	.0074	.0283	.0661	.1168	.1715	.2194	.2508	.2600	.2461
	5	.0000	.0008	.0050	.0165	.0389	.0735	.1181	.1672	.2128	.2461
	6	.0000	.0001	.0006	.0028	.0087	.0210	.0424	.0743	.1160	.1641
	7	.0000	.0000	.0000	.0003	.0012	.0039	.0098	.0212	.0407	.0703
	8	.0000	.0000	.0000	.0000	.0001	.0004	.0013	.0035	.0083	.0176
	9	.0000	.0000	.0000	.0000	.0000	.0000	.0001	.0003	.0008	.0020
10	0	.5987	.3487	.1969	.1074	.0563	.0282	.0135	.0060	.0025	.0010
	1	.3151	.3874	.3474	.2684	.1877	.1211	.0725	.0403	.0207	.0098
	2	.0746	.1937	.2759	.3020	.2816	.2335	.1757	.1209	.0763	.0439
	3	.0105	.0574	.1298	.2013	.2503	.2668	.2522	.2150	.1665	.1172
	4	.0010	.0112	.0401	.0881	.1460	.2001	.2377	.2508	.2384	.2051
	5	.0001	.0015	.0085	.0264	.0584	.1029	.1536	.2007	.2340	.2461
	6	.0000	.0001	.0012	.0055	.0162	.0368	.0689	.1115	.1596	.2051
	7	.0000	.0000	.0001	.0008	.0031	.0090	.0212	.0425	.0746	.1172
	8	.0000	.0000	.0000	.0001	.0004	.0014	.0043	.0106	.0229	.0439
	9	.0000	.0000	.0000	.0000	.0000	.0001	.0005	.0016	.0042	.0098
	10	.0000	.0000	.0000	.0000	.0000	.0000	.0000	.0001	.0003	.0010
11	0	.5688	.3138	.1673	.0859	.0422	.0198	.0088	.0036	.0014	.0004
	1	.3293	.3835	.3248	.2362	.1549	.0932	.0518	.0266	.0125	.0055
	2	.0867	.2131	.2866	.2953	.2581	.1998	.1395	.0887	.0513	.0269
	3	.0137	.0710	.1517	.2215	.2581	.2568	.2254	.1774	.1259	.0806
	4	.0014	.0158	.0536	.1107	.1721	.2201	.2428	.2365	.2060	.1611
	5	.0001	.0025	.0132	.0388	.0803	.1321	.1830	.2207	.2360	.2256
	6	.0000	.0003	.0023	.0097	.0268	.0566	.0985	.1471	.1931	.2256
	7	.0000	.0000	.0003	.0017	.0064	.0173	.0379	.0701	.1128	.1611
	8	.0000	.0000	.0000	.0002	.0011	.0037	.0102	.0234	.0462	.0806
	9	.0000	.0000	.0000	.0000	.0001	.0005	.0018	.0052	.0126	.0269
	10	.0000	.0000	.0000	.0000	.0000	.0000	.0002	.0007	.0021	.0054
	11	.0000	.0000	.0000	.0000	.0000	.0000	.0000	.0000	.0002	.0005
12	0	.5404	.2824	.1422	.0687	.0317	.0138	.0057	.0022	.0008	.0002
	1	.3413	.3766	.3012	.2062	.1267	.0712	.0368	.0174	.0075	.0029
	2	.0988	.2301	.2924	.2835	.2323	.1678	.1088	.0639	.0339	.0161
	3	.0173	.0852	.1720	.2362	.2581	.2397	.1954	.1419	.0923	.0537
	4	.0021	.0213	.0683	.1329	.1936	.2311	.2367	.2128	.1700	.1208
	5	.0002	.0038	.0193	.0532	.1032	.1585	.2039	.2270	.2225	.1934
	6	.0000	.0005	.0040	.0155	.0401	.0792	.1281	.1766	.2124	.2256
	7	.0000	.0000	.0006	.0033	.0115	.0291	.0591	.1009	.1489	.1934
	8	.0000	.0000	.0001	.0005	.0024	.0078	.0199	.0420	.0762	.1208
	9	.0000	.0000	.0000	.0001	.0004	.0015	.0048	.0125	.0277	.0537
	10	.0000	.0000	.0000	.0000	.0000	.0002	.0008	.0025	.0068	.0161
	11	.0000	.0000	.0000	.0000	.0000	.0000	.0001	.0003	.0010	.0029
	12	.0000	.0000	.0000	.0000	.0000	.0000	.0000	.0000	.0001	.0002

Table VI (cont'd.) Binomial Probabilities for *k* Successes, *n* Trials

						p					
n	*k*	.05	.10	.15	.20	.25	.30	.35	.40	.45	.50
13	0	.5133	.2542	.1209	.0550	.0238	.0097	.0037	.0013	.0004	.0001
	1	.3512	.3672	.2774	.1787	.1029	.0540	.0259	.0113	.0045	.0016
	2	.1109	.2448	.2937	.2680	.2059	.1388	.0836	.0453	.0220	.0095
	3	.0214	.0997	.1900	.2457	.2517	.2181	.1651	.1107	.0660	.0349
	4	.0028	.0277	.0838	.1535	.2097	.2337	.2222	.1845	.1350	.0873
	5	.0003	.0055	.0266	.0691	.1258	.1803	.2154	.2214	.1989	.1571
	6	.0000	.0008	.0063	.0230	.0559	.1030	.1546	.1968	.2169	.2095
	7	.0000	.0001	.0011	.0058	.0186	.0442	.0833	.1312	.1775	.2095
	8	.0000	.0000	.0001	.0011	.0047	.0142	.0336	.0656	.1089	.1571
	9	.0000	.0000	.0000	.0001	.0009	.0034	.0101	.0243	.0495	.0873
	10	.0000	.0000	.0000	.0000	.0001	.0006	.0022	.0065	.0162	.0349
	11	.0000	.0000	.0000	.0000	.0000	.0001	.0003	.0012	.0036	.0095
	12	.0000	.0000	.0000	.0000	.0000	.0000	.0000	.0001	.0005	.0016
	13	.0000	.0000	.0000	.0000	.0000	.0000	.0000	.0000	.0000	.0001
14	0	.4877	.2288	.1028	.0440	.0178	.0068	.0024	.0008	.0002	.0001
	1	.3593	.3559	.2539	.1539	.0832	.0407	.0181	.0073	.0027	.0009
	2	.1229	.2570	.2912	.2501	.1802	.1134	.0634	.0317	.0141	.0056
	3	.0259	.1142	.2056	.2501	.2402	.1943	.1366	.0845	.0462	.0222
	4	.0037	.0349	.0998	.1720	.2202	.2290	.2022	.1549	.1040	.0611
	5	.0004	.0078	.0352	.0860	.1468	.1963	.2178	.2066	.1701	.1222
	6	.0000	.0013	.0093	.0322	.0734	.1262	.1759	.2066	.2088	.1833
	7	.0000	.0002	.0019	.0092	.0280	.0618	.1082	.1574	.1952	.2095
	8	.0000	.0000	.0003	.0020	.0082	.0232	.0510	.0918	.1398	.1833
	9	.0000	.0000	.0000	.0003	.0018	.0066	.0183	.0408	.0762	.1222
	10	.0000	.0000	.0000	.0000	.0003	.0014	.0049	.0136	.0312	.0611
	11	.0000	.0000	.0000	.0000	.0000	.0002	.0010	.0033	.0093	.0222
	12	.0000	.0000	.0000	.0000	.0000	.0000	.0001	.0005	.0019	.0056
	13	.0000	.0000	.0000	.0000	.0000	.0000	.0000	.0001	.0002	.0009
	14	.0000	.0000	.0000	.0000	.0000	.0000	.0000	.0000	.0000	.0001
15	0	.4633	.2059	.0874	.0352	.0134	.0047	.0016	.0005	.0001	.0000
	1	.3658	.3432	.2312	.1319	.0668	.0305	.0126	.0047	.0016	.0005
	2	.1348	.2669	.2856	.2309	.1559	.0916	.0476	.0219	.0090	.0032
	3	.0307	.1285	.2184	.2501	.2252	.1700	.1110	.0634	.0318	.0139
	4	.0049	.0428	.1156	.1876	.2252	.2186	.1792	.1268	.0780	.0417
	5	.0006	.0105	.0449	.1032	.1651	.2061	.2123	.1859	.1404	.0916
	6	.0000	.0019	.0132	.0430	.0917	.1472	.1906	.2066	.1914	.1527
	7	.0000	.0003	.0030	.0138	.0393	.0811	.1319	.1771	.2013	.1964
	8	.0000	.0000	.0005	.0035	.0131	.0348	.0710	.1181	.1647	.1964
	9	.0000	.0000	.0001	.0007	.0034	.0116	.0298	.0612	.1048	.1527
	10	.0000	.0000	.0000	.0001	.0007	.0030	.0096	.0245	.0515	.0916
	11	.0000	.0000	.0000	.0000	.0001	.0006	.0024	.0074	.0191	.0417
	12	.0000	.0000	.0000	.0000	.0000	.0001	.0004	.0016	.0052	.0139
	13	.0000	.0000	.0000	.0000	.0000	.0000	.0001	.0003	.0010	.0032
	14	.0000	.0000	.0000	.0000	.0000	.0000	.0000	.0000	.0001	.0005
	15	.0000	.0000	.0000	.0000	.0000	.0000	.0000	.0000	.0000	.0000

Table VII Standard Normal Density Function*

x	$\phi(x)$	x	$\phi(x)$	x	$\phi(x)$	x	$\phi(x)$
.00	.3989	.50	.3521	1.00	.2420	1.50	.1295
.01	.3989	.51	.3503	1.01	.2396	1.51	.1276
.02	.3989	.52	.3485	1.02	.2371	1.52	.1257
.03	.3988	.53	.3467	1.03	.2347	1.53	.1238
.04	.3986	.54	.3448	1.04	.2323	1.54	.1219
.05	.3984	.55	.3429	1.05	.2299	1.55	.1200
.06	.3982	.56	.3410	1.06	.2275	1.56	.1182
.07	.3980	.57	.3391	1.07	.2251	1.57	.1163
.08	.3977	.58	.3372	1.08	.2227	1.58	.1145
.09	.3973	.59	.3352	1.09	.2203	1.59	.1127
.10	.3970	.60	.3332	1.10	.2179	1.60	.1109
.11	.3965	.61	.3312	1.11	.2155	1.61	.1092
.12	.3961	.62	.3292	1.12	.2131	1.62	.1074
.13	.3956	.63	.3271	1.13	.2107	1.63	.1057
.14	.3951	.64	.3251	1.14	.2083	1.64	.1040
.15	.3945	.65	.3230	1.15	.2059	1.65	.1023
.16	.3939	.66	.3209	1.16	.2036	1.66	.1006
.17	.3932	.67	.3187	1.17	.2012	1.67	.0989
.18	.3925	.68	.3166	1.18	.1989	1.68	.0973
.19	.3918	.69	.3144	1.19	.1965	1.69	.0957
.20	.3910	.70	.3123	1.20	.1942	1.70	.0940
.21	.3902	.71	.3101	1.21	.1919	1.71	.0925
.22	.3894	.72	.3079	1.22	.1895	1.72	.0909
.23	.3885	.73	.3056	1.23	.1872	1.73	.0893
.24	.3876	.74	.3034	1.24	.1849	1.74	.0878
.25	.3867	.75	.3011	1.25	.1826	1.75	.0863
.26	.3857	.76	.2989	1.26	.1804	1.76	.0848
.27	.3847	.77	.2966	1.27	.1781	1.77	.0833
.28	.3836	.78	.2943	1.28	.1758	1.78	.0818
.29	.3825	.79	.2920	1.29	.1736	1.79	.0804
.30	.3814	.80	.2897	1.30	.1714	1.80	.0790
.31	.3802	.81	.2874	1.31	.1691	1.81	.0775
.32	.3790	.82	.2850	1.32	.1669	1.82	.0761
.33	.3778	.83	.2827	1.33	.1647	1.83	.0748
.34	.3765	.84	.2803	1.34	.1626	1.84	.0734
.35	.3752	.85	.2780	1.35	.1604	1.85	.0721
.36	.3739	.86	.2756	1.36	.1582	1.86	.0707
.37	.3725	.87	.2732	1.37	.1561	1.87	.0694
.38	.3712	.88	.2709	1.38	.1539	1.88	.0681
.39	.3697	.89	.2685	1.39	.1518	1.89	.0669
.40	.3683	.90	.2661	1.40	.1497	1.90	.0656
.41	.3668	.91	.2637	1.41	.1476	1.91	.0644
.42	.3653	.92	.2613	1.42	.1456	1.92	.0632
.43	.3637	.93	.2589	1.43	.1435	1.93	.0620
.44	.3621	.94	.2565	1.44	.1415	1.94	.0608
.45	.3605	.95	.2541	1.45	.1394	1.95	.0596
.46	.3589	.96	.2516	1.46	.1374	1.96	.0584
.47	.3572	.97	.2492	1.47	.1354	1.97	.0573
.48	.3555	.98	.2468	1.48	.1334	1.98	.0562
.49	.3538	.99	.2444	1.49	.1315	1.99	.0551
.50	.3521	1.00	.2420	1.50	.1295	2.00	.0540

* From *Handbook of Tables for Probability and Statistics,* second edition, William H. Beyer, editor.

Table VII (cont'd.) Standard Normal Density Function

x	$\phi(x)$	x	$\phi(x)$	x	$\phi(x)$	x	$\phi(x)$
2.00	.0540	2.50	.0175	3.00	.0044	3.50	.0009
2.01	.0529	2.51	.0171	3.01	.0043	3.51	.0008
2.02	.0519	2.52	.0167	3.02	.0042	3.52	.0008
2.03	.0508	2.53	.0163	3.03	.0040	3.53	.0008
2.04	.0498	2.54	.0158	3.04	.0039	3.54	.0008
2.05	.0488	2.55	.0155	3.05	.0038	3.55	.0007
2.06	.0478	2.56	.0151	3.06	.0037	3.56	.0007
2.07	.0468	2.57	.0147	3.07	.0036	3.57	.0007
2.08	.0459	2.58	.0143	3.08	.0035	3.58	.0007
2.09	.0449	2.59	.0139	3.09	.0034	3.59	.0006
2.10	.0440	2.60	.0136	3.10	.0033	3.60	.0006
2.11	.0431	2.61	.0132	3.11	.0032	3.61	.0006
2.12	.0422	2.62	.0129	3.12	.0031	3.62	.0006
2.13	.0413	2.63	.0126	3.13	.0030	3.63	.0005
2.14	.0404	2.64	.0122	3.14	.0029	3.64	.0005
2.15	.0396	2.65	.0119	3.15	.0028	3.65	.0005
2.16	.0387	2.66	.0116	3.16	.0027	3.66	.0005
2.17	.0379	2.67	.0113	3.17	.0026	3.67	.0005
2.18	.0371	2.68	.0110	3.18	.0025	3.68	.0005
2.19	.0363	2.69	.0107	3.19	.0025	3.69	.0004
2.20	.0355	2.70	.0104	3.20	.0024	3.70	.0004
2.21	.0347	2.71	.0101	3.21	.0023	3.71	.0004
2.22	.0339	2.72	.0099	3.22	.0022	3.72	.0004
2.23	.0332	2.73	.0096	3.23	.0022	3.73	.0004
2.24	.0325	2.74	.0093	3.24	.0021	3.74	.0004
2.25	.0317	2.75	.0091	3.25	.0020	3.75	.0004
2.26	.0310	2.76	.0088	3.26	.0020	3.76	.0003
2.27	.0303	2.77	.0086	3.27	.0019	3.77	.0003
2.28	.0297	2.78	.0084	3.28	.0018	3.78	.0003
2.29	.0290	2.79	.0081	3.29	.0018	3.79	.0003
2.30	.0283	2.80	.0079	3.30	.0017	3.80	.0003
2.31	.0277	2.81	.0077	3.31	.0017	3.81	.0003
2.32	.0270	2.82	.0075	3.32	.0016	3.82	.0003
2.33	.0264	2.83	.0073	3.33	.0016	3.83	.0003
2.34	.0258	2.84	.0071	3.34	.0015	3.84	.0003
2.35	.0252	2.85	.0069	3.35	.0015	3.85	.0002
2.36	.0246	2.86	.0067	3.36	.0014	3.86	.0002
2.37	.0241	2.87	.0065	3.37	.0014	3.87	.0002
2.38	.0235	2.88	.0063	3.38	.0013	3.88	.0002
2.39	.0229	2.89	.0061	3.39	.0013	3.89	.0002
2.40	.0224	2.90	.0060	3.40	.0012	3.90	.0002
2.41	.0219	2.91	.0058	3.41	.0012	3.91	.0002
2.42	.0213	2.92	.0056	3.42	.0012	3.92	.0002
2.43	.0208	2.93	.0055	3.43	.0011	3.93	.0002
2.44	.0203	2.94	.0053	3.44	.0011	3.94	.0002
2.45	.0198	2.95	.0051	3.45	.0010	3.95	.0002
2.46	.0194	2.96	.0050	3.46	.0010	3.96	.0002
2.47	.0189	2.97	.0048	3.47	.0010	3.97	.0002
2.48	.0184	2.98	.0047	3.48	.0009	3.98	.0001
2.49	.0180	2.99	.0046	3.49	.0009	3.99	.0001
2.50	.0175	3.00	.0044	3.50	.0009	4.00	.0001

Table VIII Areas under the Standard Normal Curve*

z	*Second decimal place in z*									
	.00	.01	.02	.03	.04	.05	.06	.07	.08	.09
.0	.0000	.0040	.0080	.0120	.0160	.0199	.0239	.0279	.0319	.0359
.1	.0398	.0438	.0478	.0517	.0557	.0596	.0636	.0675	.0714	.0753
.2	.0793	.0832	.0871	.0910	.0948	.0987	.1026	.1064	.1103	.1141
.3	.1179	.1217	.1255	.1293	.1331	.1368	.1406	.1443	.1480	.1517
.4	.1554	.1591	.1628	.1664	.1700	.1736	.1772	.1808	.1844	.1879
.5	.1915	.1950	.1985	.2019	.2054	.2088	.2123	.2157	.2190	.2224
.6	.2257	.2291	.2324	.2357	.2389	.2422	.2454	.2486	.2517	.2549
.7	.2580	.2611	.2642	.2673	.2704	.2734	.2764	.2794	.2823	.2852
.8	.2881	.2910	.2939	.2967	.2995	.3023	.3051	.3078	.3106	.3133
.9	.3159	.3186	.3212	.3238	.3264	.3289	.3315	.3340	.3365	.3389
1.0	.3413	.3438	.3461	.3485	.3508	.3531	.3554	.3577	.3599	.3621
1.1	.3643	.3665	.3686	.3708	.3729	.3749	.3770	.3790	.3810	.3830
1.2	.3849	.3869	.3888	.3907	.3925	.3944	.3962	.3980	.3997	.4015
1.3	.4032	.4049	.4066	.4082	.4099	.4115	.4131	.4147	.4162	.4177
1.4	.4192	.4207	.4222	.4236	.4251	.4265	.4279	.4292	.4306	.4319
1.5	.4332	.4345	.4357	.4370	.4382	.4394	.4406	.4418	.4429	.4441
1.6	.4452	.4463	.4474	.4484	.4495	.4505	.4515	.4525	.4535	.4545
1.7	.4554	.4564	.4573	.4582	.4591	.4599	.4608	.4616	.4625	.4633
1.8	.4641	.4649	.4656	.4664	.4671	.4678	.4686	.4693	.4699	.4706
1.9	.4713	.4719	.4726	.4732	.4738	.4744	.4750	.4756	.4761	.4767
2.0	.4772	.4778	.4783	.4788	.4793	.4798	.4803	.4808	.4812	.4817
2.1	.4821	.4826	.4830	.4834	.4838	.4842	.4846	.4850	.4854	.4857
2.2	.4861	.4864	.4868	.4871	.4875	.4878	.4881	.4884	.4887	.4890
2.3	.4893	.4896	.4898	.4901	.4904	.4906	.4909	.4911	.4913	.4916
2.4	.4918	.4920	.4922	.4925	.4927	.4929	.4931	.4932	.4934	.4936
2.5	.4938	.4940	.4941	.4943	.4945	.4946	.4948	.4949	.4951	.4952
2.6	.4953	.4955	.4956	.4957	.4959	.4960	.4961	.4962	.4963	.4964
2.7	.4965	.4966	.4967	.4968	.4969	.4970	.4971	.4972	.4973	.4974
2.8	.4974	.4975	.4976	.4977	.4977	.4978	.4979	.4979	.4980	.4981
2.9	.4981	.4982	.4982	.4983	.4984	.4984	.4985	.4985	.4986	.4986
3.0	.4987	.4987	.4987	.4988	.4988	.4989	.4989	.4989	.4990	.4990
3.1	.4990	.4991	.4991	.4991	.4992	.4992	.4992	.4992	.4993	.4993
3.2	.4993	.4993	.4994	.4994	.4994	.4994	.4994	.4995	.4995	.4995
3.3	.4995	.4995	.4995	.4996	.4996	.4996	.4996	.4996	.4996	.4997
3.4	.4997	.4997	.4997	.4997	.4997	.4997	.4997	.4997	.4997	.4998
3.5	.4998									
4.0	.49997									
4.5	.499997									
5.0	.4999997									

Table IX Poisson Distribution*

k	λ = 0.1	0.2	0.3	0.4	0.5	0.6	0.7	0.8	0.9	1.0
0	.9048	.8187	.7408	.6703	.6065	.5488	.4966	.4493	.4066	.3679
1	.0905	.1637	.2222	.2681	.3033	.3293	.3476	.3595	.3659	.3679
2	.0045	.0164	.0333	.0536	.0758	.0988	.1217	.1438	.1647	.1839
3	.0002	.0011	.0033	.0072	.0126	.0198	.0284	.0383	.0494	.0613
4	.0000	.0001	.0003	.0007	.0016	.0030	.0050	.0077	.0111	.0153
5	.0000	.0000	.0000	.0001	.0002	.0004	.0007	.0012	.0020	.0031
6	.0000	.0000	.0000	.0000	.0000	.0000	.0001	.0002	.0003	.0005
7	.0000	.0000	.0000	.0000	.0000	.0000	.0000	.0000	.0000	.0001

k	λ = 1.1	1.2	1.3	1.4	1.5	1.6	1.7	1.8	1.9	2.0
0	.3329	.3012	.2725	.2466	.2231	.2019	.1827	.1653	.1496	.1353
1	.3662	.3614	.3543	.3452	.3347	.3230	.3106	.2975	.2842	.2707
2	.2014	.2169	.2303	.2417	.2510	.2584	.2640	.2678	.2700	.2707
3	.0738	.0867	.0998	.1128	.1255	.1378	.1496	.1607	.1710	.1804
4	.0203	.0260	.0324	.0395	.0471	.0551	.0636	.0723	.0812	.0902
5	.0045	.0062	.0084	.0111	.0141	.0176	.0216	.0260	.0309	.0361
6	.0008	.0012	.0018	.0026	.0035	.0047	.0061	.0078	.0098	.0120
7	.0001	.0002	.0003	.0005	.0008	.0011	.0015	.0020	.0027	.0034
8	.0000	.0000	.0001	.0001	.0001	.0002	.0003	.0005	.0006	.0009
9	.0000	.0000	.0000	.0000	.0000	.0000	.0001	.0001	.0001	.0002

k	λ = 2.1	2.2	2.3	2.4	2.5	2.6	2.7	2.8	2.9	3.0
0	.1225	.1108	.1003	.0907	.0821	.0743	.0672	.0608	.0550	.0498
1	.2572	.2438	.2306	.2177	.2052	.1931	.1815	.1703	.1596	.1494
2	.2700	.2681	.2652	.2613	.2565	.2510	.2450	.2384	.2314	.2240
3	.1890	.1966	.2033	.2090	.2138	.2176	.2205	.2225	.2237	.2240
4	.0992	.1082	.1169	.1254	.1336	.1414	.1488	.1557	.1622	.1680
5	.0417	.0476	.0538	.0602	.0668	.0735	.0804	.0872	.0940	.1008
6	.0146	.0174	.0206	.0241	.0278	.0319	.0362	.0407	.0455	.0504
7	.0044	.0055	.0068	.0083	.0099	.0118	.0139	.0163	.0188	.0216
8	.0011	.0015	.0019	.0025	.0031	.0038	.0047	.0057	.0068	.0081
9	.0003	.0004	.0005	.0007	.0009	.0011	.0014	.0018	.0022	.0027
10	.0001	.0001	.0001	.0002	.0002	.0003	.0004	.0005	.0006	.0008
11	.0000	.0000	.0000	.0000	.0000	.0001	.0001	.0001	.0002	.0002
12	.0000	.0000	.0000	.0000	.0000	.0000	.0000	.0000	.0000	.0001

k	λ = 3.1	3.2	3.3	3.4	3.5	3.6	3.7	3.8	3.9	4.0
0	.0450	.0408	.0369	.0334	.0302	.0273	.0247	.0224	.0202	.0183
1	.1397	.1304	.1217	.1135	.1057	.0984	.0915	.0850	.0789	.0733
2	.2165	.2087	.2008	.1929	.1850	.1771	.1692	.1615	.1539	.1465
3	.2237	.2226	.2209	.2186	.2158	.2125	.2087	.2046	.2001	.1954
4	.1734	.1781	.1823	.1858	.1888	.1912	.1931	.1944	.1951	.1954
5	.1075	.1140	.1203	.1264	.1322	.1377	.1429	.1477	.1522	.1563
6	.0555	.0608	.0662	.0716	.0771	.0826	.0881	.0936	.0989	.1042
7	.0246	.0278	.0312	.0348	.0385	.0425	.0466	.0508	.0551	.0595
8	.0095	.0111	.0129	.0148	.0169	.0191	.0215	.0241	.0269	.0298
9	.0033	.0040	.0047	.0056	.0066	.0076	.0089	.0102	.0116	.0132

* From *Handbook of Tables for Probability and Statistics,* second edition, William H. Beyer, editor.

Table IX (cont'd.) Poisson Distribution

k	λ 3.1	3.2	3.3	3.4	3.5	3.6	3.7	3.8	3.9	4.0
10	.0010	.0013	.0016	.0019	.0023	.0028	.0033	.0039	.0045	.0053
11	.0003	.0004	.0005	.0006	.0007	.0009	.0011	.0013	.0016	.0019
12	.0001	.0001	.0001	.0002	.0002	.0003	.0003	.0004	.0005	.0006
13	.0000	.0000	.0000	.0000	.0001	.0001	.0001	.0001	.0002	.0002
14	.0000	.0000	.0000	.0000	.0000	.0000	.0000	.0000	.0000	.0001

k	λ 4.1	4.2	4.3	4.4	4.5	4.6	4.7	4.8	4.9	5.0
0	.0166	.0150	.0136	.0123	.0111	.0101	.0091	.0082	.0074	.0067
1	.0679	.0630	.0583	.0540	.0500	.0462	.0427	.0395	.0365	.0337
2	.1393	.1323	.1254	.1188	.1125	.1063	.1005	.0948	.0894	.0842
3	.1904	.1852	.1798	.1743	.1687	.1631	.1574	.1517	.1460	.1404
4	.1951	.1944	.1933	.1917	.1898	.1875	.1849	.1820	.1789	.1755
5	.1600	.1633	.1662	.1687	.1708	.1725	.1738	.1747	.1753	.1755
6	.1093	.1143	.1191	.1237	.1281	.1323	.1362	.1398	.1432	.1462
7	.0640	.0686	.0732	.0778	.0824	.0869	.0914	.0959	.1002	.1044
8	.0328	.0360	.0393	.0428	.0463	.0500	.0537	.0575	.0614	.0653
9	.0150	.0168	.0188	.0209	.0232	.0255	.0280	.0307	.0334	.0363
10	.0061	.0071	.0081	.0092	.0104	.0118	.0132	.0147	.0164	.0181
11	.0023	.0027	.0032	.0037	.0043	.0049	.0056	.0064	.0073	.0082
12	.0008	.0009	.0011	.0014	.0016	.0019	.0022	.0026	.0030	.0034
13	.0002	.0003	.0004	.0005	.0006	.0007	.0008	.0009	.0011	.0013
14	.0001	.0001	.0001	.0001	.0002	.0002	.0003	.0003	.0004	.0005
15	.0000	.0000	.0000	.0000	.0001	.0001	.0001	.0001	.0001	.0002

k	λ 5.1	5.2	5.3	5.4	5.5	5.6	5.7	5.8	5.9	6.0
0	.0061	.0055	.0050	.0045	.0041	.0037	.0033	.0030	.0027	.0025
1	.0311	.0287	.0265	.0244	.0225	.0207	.0191	.0176	.0162	.0149
2	.0793	.0746	.0701	.0659	.0618	.0580	.0544	.0509	.0477	.0446
3	.1348	.1293	.1239	.1185	.1133	.1082	.1033	.0985	.0938	.0892
4	.1719	.1681	.1641	.1600	.1558	.1515	.1472	.1428	.1383	.1339
5	.1753	.1748	.1740	.1728	.1714	.1697	.1678	.1656	.1632	.1606
6	.1490	.1515	.1537	.1555	.1571	.1584	.1594	.1601	.1605	.1606
7	.1086	.1125	.1163	.1200	.1234	.1267	.1298	.1326	.1353	.1377
8	.0692	.0731	.0771	.0810	.0849	.0887	.0925	.0962	.0998	.1033
9	.0392	.0423	.0454	.0486	.0519	.0552	.0586	.0620	.0654	.0688
10	.0200	.0220	.0241	.0262	.0285	.0309	.0334	.0359	.0386	.0413
11	.0093	.0104	.0116	.0129	.0143	.0157	.0173	.0190	.0207	.0225
12	.0039	.0045	.0051	.0058	.0065	.0073	.0082	.0092	.0102	.0113
13	.0015	.0018	.0021	.0024	.0028	.0032	.0036	.0041	.0046	.0052
14	.0006	.0007	.0008	.0009	.0011	.0013	.0015	.0017	.0019	.0022
15	.0002	.0002	.0003	.0003	.0004	.0005	.0006	.0007	.0008	.0009
16	.0001	.0001	.0001	.0001	.0001	.0002	.0002	.0002	.0003	.0003
17	.0000	.0000	.0000	.0000	.0000	.0000	.0001	.0001	.0001	.0001

Table IX (cont'd.) Poisson Distribution

k	λ 6.1	6.2	6.3	6.4	6.5	6.6	6.7	6.8	6.9	7.0
0	.0022	.0020	.0018	.0017	.0015	.0014	.0012	.0011	.0010	.0009
1	.0137	.0126	.0116	.0106	.0098	.0090	.0082	.0076	.0070	.0064
2	.0417	.0390	.0364	.0340	.0318	.0296	.0276	.0258	.0240	.0223
3	.0848	.0806	.0765	.0726	.0688	.0652	.0617	.0584	.0552	.0521
4	.1294	.1249	.1205	.1162	.1118	.1076	.1034	.0992	.0952	.0912
5	.1579	.1549	.1519	.1487	.1454	.1420	.1385	.1349	.1314	.1277
6	.1605	.1601	.1595	.1586	.1575	.1562	.1546	.1529	.1511	.1490
7	.1399	.1418	.1435	.1450	.1462	.1472	.1480	.1486	.1489	.1490
8	.1066	.1099	.1130	.1160	.1188	.1215	.1240	.1263	.1284	.1304
9	.0723	.0757	.0791	.0825	.0858	.0891	.0923	.0954	.0985	.1014
10	.0441	.0469	.0498	.0528	.0558	.0588	.0618	.0649	.0679	.0710
11	.0245	.0265	.0285	.0307	.0330	.0353	.0377	.0401	.0426	.0452
12	.0124	.0137	.0150	.0164	.0179	.0194	.0210	.0227	.0245	.0264
13	.0058	.0065	.0073	.0081	.0089	.0098	.0108	.0119	.0130	.0142
14	.0025	.0029	.0033	.0037	.0041	.0046	.0052	.0058	.0064	.0071
15	.0010	.0012	.0014	.0016	.0018	.0020	.0023	.0026	.0029	.0033
16	.0004	.0005	.0005	.0006	.0007	.0008	.0010	.0011	.0013	.0014
17	.0001	.0002	.0002	.0002	.0003	.0003	.0004	.0004	.0005	.0006
18	.0000	.0001	.0001	.0001	.0001	.0001	.0001	.0002	.0002	.0002
19	.0000	.0000	.0000	.0000	.0000	.0000	.0000	.0001	.0001	.0001

k	λ 7.1	7.2	7.3	7.4	7.5	7.6	7.7	7.8	7.9	8.0
0	.0008	.0007	.0007	.0006	.0006	.0005	.0005	.0004	.0004	.0003
1	.0059	.0054	.0049	.0045	.0041	.0038	.0035	.0032	.0029	.0027
2	.0208	.0194	.0180	.0167	.0156	.0145	.0134	.0125	.0116	.0107
3	.0492	.0464	.0438	.0413	.0389	.0366	.0345	.0324	.0305	.0286
4	.0874	.0836	.0799	.0764	.0729	.0696	.0663	.0632	.0602	.0573
5	.1241	.1204	.1167	.1130	.1094	.1057	.1021	.0986	.0951	.0916
6	.1468	.1445	.1420	.1394	.1367	.1339	.1311	.1282	.1252	.1221
7	.1489	.1486	.1481	.1474	.1465	.1454	.1442	.1428	.1413	.1396
8	.1321	.1337	.1351	.1363	.1373	.1382	.1388	.1392	.1395	.1396
9	.1042	.1070	.1096	.1121	.1144	.1167	.1187	.1207	.1224	.1241
10	.0740	.0770	.0800	.0829	.0858	.0887	.0914	.0941	.0967	.0993
11	.0478	.0504	.0531	.0558	.0585	.0613	.0640	.0667	.0695	.0722
12	.0283	.0303	.0323	.0344	.0366	.0388	.0411	.0434	.0457	.0481
13	.0154	.0168	.0181	.0196	.0211	.0227	.0243	.0260	.0278	.0296
14	.0078	.0086	.0095	.0104	.0113	.0123	.0134	.0145	.0157	.0169
15	.0037	.0041	.0046	.0051	.0057	0062	.0069	.0075	.0083	.0090
16	.0016	.0019	.0021	.0024	.0026	.0030	.0033	.0037	.0041	.0045
17	.0007	.0008	.0009	.0010	.0012	.0013	.0015	.0017	.0019	.0021
18	.0003	.0003	.0004	.0004	.0005	.0006	.0006	.0007	.0008	.0009
19	.0001	.0001	.0001	.0002	.0002	.0002	.0003	.0003	.0003	.0004
20	.0000	.0000	.0001	.0001	.0001	.0001	.0001	.0001	.0001	.0002
21	.0000	.0000	.0000	.0000	.0000	.0000	.0000	.0000	.0001	.0001

Table IX (cont'd.) Poisson Distribution

k	λ 8.1	8.2	8.3	8.4	8.5	8.6	8.7	8.8	8.9	9.0
0	.0003	.0003	.0002	.0002	.0002	.0002	.0002	.0002	.0001	.0001
1	.0025	.0023	.0021	.0019	.0017	.0016	.0014	.0013	.0012	.0011
2	.0100	.0092	.0086	.0079	.0074	.0068	.0063	.0058	.0054	.0050
3	.0269	.0252	.0237	.0222	.0208	.0195	.0183	.0171	.0160	.0150
4	.0544	.0517	.0491	.0466	.0443	.0420	.0398	.0377	.0357	.0337
5	.0882	.0849	.0816	.0784	.0752	.0722	.0692	.0663	.0635	.0607
6	.1191	.1160	.1128	.1097	.1066	.1034	.1003	.0972	.0941	.0911
7	.1378	.1358	.1338	.1317	.1294	.1271	.1247	.1222	.1197	.1171
8	.1395	.1392	.1388	.1382	.1375	.1366	.1356	.1344	.1332	.1318
9	.1256	.1269	.1280	.1290	.1299	.1306	.1311	.1315	.1317	.1318
10	.1017	.1040	.1063	.1084	.1104	.1123	.1140	.1157	.1172	.1186
11	.0749	.0776	.0802	.0828	.0853	.0878	.0902	.0925	.0948	.0970
12	.0505	.0530	.0555	.0579	.0604	.0629	.0654	.0679	.0703	.0728
13	.0315	.0334	.0354	.0374	.0395	.0416	.0438	.0459	.0481	.0504
14	.0182	.0196	.0210	.0225	.0240	.0256	.0272	.0289	.0306	.0324
15	.0098	.0107	.0116	.0126	.0136	.0147	.0158	.0169	.0182	.0194
16	.0050	.0055	.0060	.0066	.0072	.0079	.0086	.0093	.0101	.0109
17	.0024	.0026	.0029	.0033	.0036	.0040	.0044	.0048	.0053	.0058
18	.0011	.0012	.0014	.0015	.0017	.0019	.0021	.0024	.0026	.0029
19	.0005	.0005	.0006	.0007	.0008	.0009	.0010	.0011	.0012	.0014
20	.0002	.0002	.0002	.0003	.0003	.0004	.0004	.0005	.0005	.0006
21	.0001	.0001	.0001	.0001	.0001	.0002	.0002	.0002	.0002	.0003
22	.0000	.0000	.0000	.0000	.0001	.0001	.0001	.0001	.0001	.0001

k	λ 9.1	9.2	9.3	9.4	9.5	9.6	9.7	9.8	9.9	10
0	.0001	.0001	.0001	.0001	.0001	.0001	.0001	.0001	.0001	.0000
1	.0010	.0009	.0009	.0008	.0007	.0007	.0006	.0005	.0005	.0005
2	.0046	.0043	.0040	.0037	.0034	.0031	.0029	.0027	.0025	.0023
3	.0140	.0131	.0123	.0115	.0107	.0100	.0093	.0087	.0081	.0076
4	.0319	.0302	.0285	.0269	.0254	.0240	.0226	.0213	.0201	.0189
5	.0581	.0555	.0530	.0506	.0483	.0460	.0439	.0418	.0398	.0378
6	.0881	.0851	.0822	.0793	.0764	.0736	.0709	.0682	.0656	.0631
7	.1145	.1118	.1091	.1064	.1037	.1010	.0982	.0955	.0928	.0901
8	.1302	.1286	.1269	.1251	.1232	.1212	.1191	.1170	.1148	.1126
9	.1317	.1315	.1311	.1306	.1300	.1293	.1284	.1274	.1263	.1251
10	.1198	.1210	.1219	.1228	.1235	.1241	.1245	.1249	.1250	.1251
11	.0991	.1012	.1031	.1049	.1067	.1083	.1098	.1112	.1125	.1137
12	.0752	.0776	.0799	.0822	.0844	.0866	.0888	.0908	.0928	.0948
13	.0526	.0549	.0572	.0594	.0617	.0640	.0662	.0685	.0707	.0729
14	.0342	.0361	.0380	.0399	.0419	.0439	.0459	.0479	.0500	.0521
15	.0208	.0221	.0235	.0250	.0265	.0281	.0297	.0313	.0330	.0347
16	.0118	.0127	.0137	.0147	.0157	.0168	.0180	.0192	.0204	.0217
17	.0063	.0069	.0075	.0081	.0088	.0095	.0103	.0111	.0119	.0128
18	.0032	.0035	.0039	.0042	.0046	.0051	.0055	.0060	.0065	.0071
19	.0015	.0017	.0019	.0021	.0023	.0026	.0028	.0031	.0034	.0037

Table IX (cont'd.) Poisson Distribution

	λ									
k	9.1	9.2	9.3	9.4	9.5	9.6	9.7	9.8	9.9	10
20	.0007	.0008	.0009	.0010	.0011	.0012	.0014	.0015	.0017	.0019
21	.0003	.0003	.0004	0004	.0005	.0006	.0006	.0007	.0008	.0009
22	.0001	.0001	.0002	.0002	.0002	.0002	.0003	.0003	.0004	.0004
23	.0000	.0001	.0001	.0001	.0001	.0001	.0001	.0001	.0002	.0002
24	.0000	.0000	.0000	.0000	.0000	.0000	.0000	.0001	.0001	.0001

Index